Soil Organic Matter and Feeding the Future

Advances in Soil Science

Series Editors:
Rattan Lal
BA Stewart

Global Climate Change and Cold Regions Ecosystems
R. Lal, J. M. Kimble, and B. A. Stewart

Assessment Methods for Soil Carbon
R. Lal, J. M. Kimble, R. F. Follett, and B. A. Stewart

Soil Erosion and Carbon Dynamics
E. J. Roose, R. Lal, C. Feller, B. Barthès, and B. A. Stewart

Soil Quality and Biofuel Production
R. Lal and B. A. Stewart

World Soil Resources and Food Security
R. Lal and B. A. Stewart

Soil Water and Agronomic Productivity
R. Lal and B. A. Stewart

Principles of Sustainable Soil Management in Agroecosystems
R. Lal and B. A. Stewart

Soil Management of Smallholder Agriculture
R. Lal and B. A. Stewart

Soil-Specific Farming: Precision Agriculture
R. Lal and B. A. Stewart

Soil Phosphorus
R. Lal and B. A. Stewart

Urban Soils
R. Lal and B. A. Stewart

Soil Nitrogen Uses and Environmental Impacts
R. Lal and B. A. Stewart

Soil and Climate
Rattan Lal and B. A. Stewart

Food Security and Soil Quality
R. Lal and B. A. Stewart

Soil Degradation and Restoration in Africa
Rattan Lal and B. A. Stewart

Soil and Fertilizers: Managing the Environmental Footprint
Rattan Lal

The Soil-Human Health-Nexus
Rattan Lal

Soil Organic Matter and Feeding the Future: Environmental and Agronomic Impacts
Rattan Lal

Soil Organic Carbon and Feeding the Future: Basic Soil Processes
Rattan Lal

For more information about this series, please visit: https://www.crcpress.com/Advances-in-Soil-Science/book-series/CRCADVSOILSCI

Soil Organic Matter and Feeding the Future

Environmental and Agronomic Impacts

Edited by

Rattan Lal

Distinguished University Professor of Soil Science, School of Environment and Natural Resources (SENR), College of Food, Agricultural and Environmental Sciences (CFAES), The Ohio State University

Director, CFAES Dr. Rattan Lal Carbon Management and Sequestration Center (C-MASC)

CRC Press
Taylor & Francis Group
Boca Raton London New York

CRC Press is an imprint of the
Taylor & Francis Group, an informa business

First edition published 2022
by CRC Press
6000 Broken Sound Parkway NW, Suite 300, Boca Raton, FL 33487-2742

and by CRC Press
2 Park Square, Milton Park, Abingdon, Oxon, OX14 4RN

CRC Press is an imprint of Taylor & Francis Group, LLC

Library of Congress Cataloging-in-Publication Data

Names: Lal, R., editor.
Title: Soil organic matter and feeding the future : environmental and agronomic impacts / edited by Rattan Lal.
Other titles: Advances in soil science (Boca Raton, Fla.).
Description: First edition. | Boca Raton, FL : CRC Press, 2022. | Series: Advances in soil sciences | Includes bibliographical references and index. | Summary: "Soil organic matter (SOM) is the primary determinant of soil functionality and is a source of numerous ecosystem services critical to human well-being and nature conservancy. This volume of Advances in Soil Sciences presents impacts of land use and soil management on soil organic carbon (SOC) dynamics and discusses how conservation through restorative land use and adoption of recommended management practices to create a positive soil-ecosystem carbon budget can increase carbon stock and soil health"-- Provided by publisher.
Identifiers: LCCN 2021028143 | ISBN 9780367609702 (hardback) | ISBN 9781032133706 (paperback) | ISBN 9781003102762 (ebook)
Subjects: LCSH: Soil science. | Agronomy--Effect of environment on.
Classification: LCC S591 .S6957 2022 | DDC 631.4--dc23
LC record available at https://lccn.loc.gov/2021028143

ISBN: 9780367609702 (hbk)
ISBN: 9781032133706 (pbk)
ISBN: 9781003102762 (ebk)

DOI: 10.1201/9781003102762

Typeset in Times
by Deanta Global Publishing Services, Chennai, India

Contents

Foreword

Growth in agricultural production since the 1960s is a global success story. The rate of increase of production of food grains, milk, fish, poultry, meat, and other products has exceeded that of the population growth. However, there has been a strong disconnect between agriculture and nature. Consequently, there are severe problems such as (1) soil degradation by accelerated erosion, (2) depletion of soil/terrestrial carbon stocks and emission of greenhouse gases from soil and farm operations, (3) sedimentation of water ways and eutrophication of surface waters, (4) algal bloom and anoxia of coastal waters, (5) mass extinction and the decline of biodiversity, (6) decrease in profitability of agriculture because of increase in off-farm inputs, and (7) Sustainable Development Goals (SDGs) of the Agenda 2030 of the United Nations are not on track. Meanwhile, the widespread problem of hunger and malnutrition is aggravated because of disruption in food production and supply chains by the COVID 19 pandemic. To the 2019 estimate of ~700 million (M) undernourished population of the world (one in 11 persons), another 80–130 M were added by the end of 2020. The prevalence of malnutrition, already impacting more than 2 billion (B) globally (~25% of the world population) in developed and developing countries, has also been aggravated by the increase in the intensity and severity of soil degradation by a range of physical, chemical, biological, and ecological processes. Therefore, sustainable management of soil under agroecosystems and restoration of soil health and functionality/quality are among high priorities needed to reverse soil degradation and set in motion the restorative trends. While scientific understanding of the cause–effect relationship (soil health vs. planetary health) must be increased, it is also important to translate the proven scientific knowledge into action. In this context, the importance of managing and enhancing soil organic matter (SOM) content cannot be over-emphasized.

Quantity, quality, and turnover time of SOM content are critical to soil health and functionality. The fact that SOM is "the elixir of life" has been recognized by ancient civilizations, and its sustainable management through appropriate farm management (e.g., green manuring, composting, complex rotations, and integration of crops with trees and livestock) has been practiced for millennia. The management of SOM has received less attention as a result of easier access and economic availability of chemical fertilizers, pesticides, and fossil fuel for the production and application of these options since the 1960s. With widespread problems of soil degradation and environmental pollution, however, it is now realized that (1) soil fertility cannot be managed just by indiscriminate use of chemical fertilizers or lime, (2) weed and pests cannot be managed just by adding herbicides and pesticides, (3) drought cannot be mitigated by flood-based irrigation, and (4) soil crusting and compaction cannot be alleviated just by plowing and sub-soiling. This is symptomatic management of alleviating soil-related constraints to agronomic production without consideration of the environmental impacts. Thus, restoration and sustainable management of soil health and functionality necessitate re-carbonization of SOM content and improvement of its quality and turnover time.

Therefore, this book is timely and pertinent because it addresses the management of SOM content in soils of agroecosystems. In addition to basic principles, the present volume also provides research information about case studies of countries around the world and deliberates policy issues to implement action plans for effective SOM management at the global, national, regional, and farm levels. The information collated in this book is based on the joint effort of many scientists, professors, experts, and researchers involved in the science and practice of SOM management and its impact on agronomic productivity and eco-efficiency of inputs, such as savings in fertilizer, her bicides, pesticides, and energy for tillage and other farm operations with a progressive increase of SOM contents in the root zone. Indeed, the volume is a reflection of many long years of professional experience of the authors in reviewing, analyzing, and synthesizing vast and dynamic field of expertise and innovations in SOM and its management. The authors have focused on many case studies that review a variety of modern tools and techniques for SOM management and sequestration and interpretation for informed decision making. The authors have explored and examined numerous advances, such as those related to precision farming, monitoring of SOM, impacts on soil quality, moisture regime, and management of inputs.

The authors have elucidated that system-based conservation agriculture can enhance SOM content and improve the overall prospects of agriculture. Hence, the book is a major contribution to the field of SOM and its management by highlighting the importance of applied agronomic research. This book is a pertinent reference resource for researchers, students, practitioners, and policymakers in soil science, agronomy, ecology, and management of natural resources with a specific focus on some global issues, such as food and nutritional security, adaptation and mitigation of climate change, soil quality, biodiversity, and the action plan for advancing the "Sustainable Development Goals" of the United Nations.

I thank the authors for sharing their knowledge and wisdom and for timely submission and revisions. I thank Ms. Maggie Willis and Ms. Janelle Watts for managing the flow of the manuscript and for checking format, references, and style, and for submitting it to the publisher. I also thank Ms. Randy Brehm and Ms. Julia Tanner of Taylor and Francis for their help and support in the timely publication of the book.

Sincerely,

Rattan Lal
Distinguished University Professor of Soil Science, SENR.
Director, Carbon Management and Sequestration Center
The Ohio State University, Columbus, OH, USA

Date: 10 April 2021

Editor

Rattan Lal, PhD, is Distinguished University Professor and Director of the CFAES Rattan Lal Center Carbon Management and Sequestration at The Ohio State University, and adjunct professor at the University of Iceland and IARI, New Delhi. He was president of the WASWAC (1987–1990), ISTRO (1988–1991), SSSA (2006–2008), and the IUSS (2017–2018). He researches soil C sequestration, conservation agriculture, soil health, soil erosion and C dynamics, soil structure, eco-intensification, soil restoration, and soils of the tropics. He has authored more than 1,000 journal articles, authored/edited more than 100 books, mentored 370 researchers, has an h-index of 167, and has total citations of 123,959. In a Stanford study (Ioannidis et al. 2019, 2020), he is ranked #110 globally among the world's top 2% of scientists, and #1 among scientists in Agronomy & Agriculture. Reuter Thompson ranked him #391 among the top 1000 climatologists. At the Inter-American Institute for Cooperation on Agriculture (IICA), he holds the positions of Chair in Soil Science and Goodwill Ambassador in Sustainable Development. He is a member of the 2021 UN Food System Summit Science Committee and Action Tracks 1 and 3. He has received the GCHERA World Agriculture Prize (2018), Glinka World Soil Prize (2018), Japan Prize (2019), IFFCO Award (2019), Arrell Global Food Innovation Award (2020), World Food Prize (2020), and Padma Shri Award (2021).

Contributors

Kokou Adambounou Amouzou
African Plant Nutrition Institute
Yamoussoukro, Côte d'Ivoire

Naeem Ahmad
College of Agronomy
Northwest A&F University
Yangling, Shaanxi, China

M. A. Ansari
ICAR Research Complex for NEH Region
Manipur Centre
Lamphelpat, Manipur, India

R. S. Antil
Amity Food & Agriculture Foundation
Amity University Uttar Pradesh
Noida, India

S. Babu
Indian Agricultural Research Institute
ICAR
New Delhi, India

Rajan Bhatt
Regional Research Station, Kapurthala
Punjab Agricultural University
Ludhiana, Punjab, India

Rosa M. Carbonell-Bojollo
IFAPA Centro Alameda del Obispo
Córdoba, Spain

Amitava Chatterjee
Soil Science (Dept. 7680)
North Dakota State University
Fargo, North Dakota, USA

Regis Chikowo
Plant Production Sciences and Technologies Department
University of Zimbabwe
Harare, Zimbabwe

Anup Das
ICAR Research Complex for NEH Region
Tripura Centre
Lembucherra, Tripura, India

Rolf Derpsch
Consultant
Paraguay

S. Devi
ICAR Research Complex for NEH Region
Umiam, Meghalaya, India

Mina Devkota
International Centre for Agricultural Research in the Dry Areas (ICARDA)
Rabat, Morocco

Sudarshan Dutta
African Plant Nutrition Institute
Mohammed VI Polytechnic University
Benguérir, Morocco

Theodor Friedrich
Food and Agriculture Organization of the United Nations (FAO)
Rome, Italy

Rajan Ghimire
Agricultural Science Center
New Mexico State University
Clovis, New Mexico, USA

P.K. Ghosh
National Institute of Biotic Stress Management
ICAR
Raipur, Chhattisgarh, India

Emilio Gonzalez-Sanchez
Universidad de Córdoba, Spain, European Conservation Agriculture Federation (ECAF)
Belgium and Asociación Española Agricultura de Conservación, Suelos Vivos (AEAC.SV)
Córdoba, Spain

Kodigal A. Gopinath
ICAR-Central Research Institute for Dryland Agriculture
Hyderabad, Telangana, India

David R. Huggins
USDA Agricultural Research Service
Northwest Sustainable Agroecosystems Research Unit
Pullman, WA, USA

Stephen Ichami
Department of Land Resource Management and Agricultural Technology, College of Agriculture and Veterinary Medicine, Faculty of Agriculture
University of Nairobi
Kabete, Nairobi, Kenya

Abhishek Jangir
ICAR-National Bureau of Soil Survey and Land Use Planning
Nagpur, India

Arun Jyoti Nath
Assam University
Silchar, India

Zheng-Rong Kan
College of Agronomy and Biotechnology, China Agricultural University
Key Laboratory of Farming System, Ministry of Agriculture of China
Beijing, China

B. K. Kandpal
ICAR Research Complex for NEH Region
Umiam, Meghalaya, India

Amir Kassam
University of Reading
Reading, UK

Xiangbin Kong
College of Land Science and Technology, China Agricultural University
Key Laboratory of Agricultural Land Quality, Monitoring of Natural Resources
Beijing, China

Sandeep Kumar
ICAR-Indian Institute of Pulses Research Regional Station
Phanda, Bhopal, India

Manoj Kumar Jhariya
Department of Farm Forestry, University Teaching Department
Sant Gahira Guru Vishwavidyalaya
Sarguja, Ambikapur, Chhattisgarh, India

Rattan Lal
CFAES Dr. Rattan Lal Carbon Management and Sequestration Center
The Ohio State University
Columbus, Ohio, USA

Ahmad Latif Virk
College of Agronomy and Biotechnology, China Agricultural University
Key Laboratory of Farming System, Ministry of Agriculture of China
Beijing, China

J. Layek
ICAR Research Complex for NEH Region
Umiam, Meghalaya, India

Ming Lei
College of Land Science and Technology, China Agricultural University
Key Laboratory of Agricultural Land Quality, Monitoring of Natural Resources
Beijing, China

Kaushik Majumdar
African Plant Nutrition Institute
Mohammed VI Polytechnic University
Benguérir, Morocco

Sunil Mandi
ICAR-Indian Agricultural Research Institute
New Delhi, India

Gaurav Mishra
Rainforest Research Institute
Jorhat, India

K. P. Mohapatra
National Bureau of Plant Genetic Resources
ICAR
New Delhi, India

Rachid Moussadek
International Centre for Agricultural Research in the Dry Areas (ICARDA)
Institut National de la Recherche Agronomique (INRA)
Rabat, Morocco

Rachid Mrabet
Institut National de la Recherche Agronomique (INRA)
Rabat, Morocco

James Mutegi
African Plant Nutrition Institute, ICIPE Compound
Nairobi, Kenya

R. P. Narwal
Chaudhary Charan Singh Haryana Agricultural University
Hisar, Haryana, India

Somanath Nayak
ICAR-Indian Agricultural Research Institute
New Delhi, India

Samuel Njoroge
African Plant Nutrition Institute, ICIPE Compound
Nairobi, Kenya

Qiuping Peng
Washington State University
Department of Crop and Soil Sciences
Pullman, WA, USA

Jian-Ying Qi
College of Agronomy and Biotechnology, China Agricultural University
Key Laboratory of Farming System, Ministry of Agriculture of China
Beijing, China

Bal Ram Singh
Faculty of Environmental Sciences and Natural Resource Management
Norwegian University of Life Sciences
Ås, Norway

Don C. Reicosky
Soil Scientist Emeritus
USDA-ARS
USA

Amit Roy
Executive Director
The Global Phosphorus Institute

S. Saha
ICAR Research Complex for NEH Region
Mizoram Centre
Kolasib, Mizoram, India

Upendra M. Sainju
Northern Plains Agricultural Research Laboratory
USDA, Agricultural Research Service
Sidney, Montana, USA

Seema Sheoran
ICAR-Indian Institute of Maize Research
Ludhiana, Punjab, India

G. W. Sileshi
Addis Ababa University
Addis Ababa, Ethiopia

R. Singh
ICAR Research Complex for NEH Region
Sikkim Centre
Tadong, Sikkim, India

Yashbir Singh Shivay
ICAR-Indian Agricultural Research Institute
New Delhi, India

Gulab Singh Yadav (deceased)
ICAR Research Complex for NEH Region
Tripura Centre, ICAR
Lembucherra, Tripura, India

Cherukumalli Srinivasa Rao
ICAR-National Academy of Agricultural Research Management
Hyderabad, Telangana, India

Xiaobing Sun
College of Land Science and Technology, China Agricultural University
Key Laboratory of Agricultural Land Quality, Monitoring of Natural Resources
Beijing, China

Ram Swaroop Meena
Department of Agronomy, Institute of Agricultural Sciences
Banaras Hindu University
Varanasi, India

Jun Wang
Shaanxi Key Laboratory of Earth Surface System and Environmental Carrying Capacity
College of Urban and Environmental Science, Northwest University
Xi'an 710127, China

Hai-Lin Zhang
College of Agronomy and Biotechnology, China Agricultural University
Key Laboratory of Farming System, Ministry of Agriculture of China
Beijing, China

Xin Zhao
College of Agronomy and Biotechnology, China Agricultural University
Key Laboratory of Farming System, Ministry of Agriculture of China
Beijing, China

Shamie Zingore
African Plant Nutrition Institute
Mohammed VI Polytechnic University
Benguérir, Morocco

1 Enhancing Fertilizer Use Efficiency by Managing Soil Health
Emerging Trends

Amit Roy

CONTENTS

LIST OF ABBREVIATIONS

B: Boron
CA: Conservation agriculture
Cl: Chlorine
Cu: Copper
DAP: Diammonium phosphate
FAO: Food and Agriculture Organization of the United Nations
Fe: Iron
FUE: Fertilizer use efficiency
GHG: Greenhouse gases
IFA: International Fertilizer Association
IFAD: International Funds for Agricultural Development
IFDC: International Fertilizer Development Center

DOI: 10.1201/9781003102762-1

IPNI:	International Plant Nutrition Institute
ISFM:	Integrated soil fertility management
K:	Potassium
MAP:	Monoammonium phosphate
Mn:	Manganese
Mo:	Molybdenum
N:	Nitrogen
NPK:	Nitrogen-phosphorus-potassium
NUE:	Nitrogen use efficiency
P:	Phosphorus
PFP:	Partial factor productivity
SOM:	Soil organic matter
SSP:	Single super phosphate
SSNM:	Site-specific nutrient management
TSP:	Triple super phosphate
UNIDO:	United Nations Industrial Development Organization
USDA:	US Department of Agriculture
WFP:	World Food Programme
WHO:	World Health Organization
Zn:	Zinc

1.1 INTRODUCTION

The quality of man's life is directly dependent upon that thin layer of the earth's crust we refer to as "soil." Soil supports our food and fiber production and, thus, the welfare of the world's inhabitants. Our management of the soil determines its productivity over the long term. Management has many facets including enhancing soil fertility, ensuring an adequate supply of plant nutrients for use by growing plants, etc. What we know as plant nutrients comprises 16 chemical elements that are generally found in the soil, but their supply can be depleted with year-after-year of growing crops. Also, some nutrients are lost by natural processes. Farmers, ranchers, and growers use chemical fertilizers to supplement depleted nutrients for optimum crop production.

Plant nutrients usually are grouped into three categories: primary, secondary, and micronutrients. The primary nutrients are nitrogen (N), phosphorus (P), and potassium (K). Secondary nutrients are calcium (Ca), magnesium (Mg), and sulfur (S). Micronutrients are boron (B), chlorine (Cl), copper (Cu), iron (Fe), manganese (Mn), molybdenum (Mo), and zinc (Zn); these are needed in much smaller amounts than other nutrients. Carbon (C), hydrogen (H), and oxygen (O) round out the 16.

Over the coming decades, global population growth, coupled with dietary changes, will drive a substantial increase in food demand. By 2050, we need to produce enough food to feed about 9.8 billion people—2.2 billion more than today. By 2100, Africa and South Asia will be home to nearly 9 billion of the 11 billion on the planet (FAO 2018).

Agricultural growth has been a key factor in the rapid decline of global poverty since the 1950s (Roser and Ortiz-Ospina 2013). The development and adoption of improved, high-yielding seed varieties and fertilizers, combined with agronomic management, were the drivers for dramatic increases in agricultural productivity in the US and Europe during the 1950s and in Asia and Latin America during the 1960s and 1970s. Between 1960 and 2000, Green Revolution seed and fertilizer technologies were responsible for yield increases of 208% for wheat, 109% for rice, 157% for maize, 78% for potatoes, and 36% for cassava across Asia and Latin America. In many regions, Green Revolution-driven intensification also saved new land from conversion to agriculture and allowed marginal lands to be spared from agricultural use (Pingali 2012).

As demand for food grows so does the application of fertilizers. Global fertilizer consumption has expanded rapidly to the current level of about 190 million metric tons of fertilizer nutrients that were used for crop production, with China, India, Brazil, and the US accounting for about 50% of the total amount of nutrients applied through fertilizers in the world (FAO 2019). Fertilizer consumption and, consequently, food production has increased in most countries but not in sub-Saharan Africa where per capita food production has decreased since the 1970s. Compared with 160 kg nutrients per hectare (kg/ha) in South Asia, 158 kg/ha in the European Union, 127 kg/ha in North America, and 140 kg/ha in Latin America and the Caribbean, sub-Saharan Africa uses only 18 kg of nutrients per hectare (FAO 2019). Sub-Saharan Africa (SSA) uses merely 2.2% of the world's fertilizers (IFASTAT 2019). This low rate of application of fertilizers, coupled with poor soil fertility in SSA, is a major contributor to low productivity and poverty, where nutrient balances for many farming systems are negative, i.e., the crops grown are extracting more nutrients than are returned to the soil through inorganic and organic fertilizers. Increases in food production still come primarily from the expansion of agriculture onto new, often marginal lands, and yield gaps remain high (Klucas 2015) (Figure 1.1).

After decades of progress, the number of malnourished people globally is beginning to increase; in 2019 it was 688 million, and it is projected to reach 841 million by 2030 (Table 1.1) (FAO et al. 2020). More than 50% of the world's malnourished live in Asia. Another third of the hungry are in sub-Saharan Africa, where undernourishment is increasing rapidly particularly in regions affected by drought and conflict. The outlook for 2030 for Africa is alarming, and the continent is significantly off track to achieve the Zero Hunger target of the Sustainable Development Goals, even without considering the impact of COVID-19.

The global agricultural systems are strained due to rising food demands and climate change coupled with limited land, water, and declining soil fertility. The COVID-19 pandemic's projected impact on the food system will further strain the agricultural systems. The pandemic has exposed risks to food and nutritional security for a large segment of the world population with possible disruption to the food supply chain. Compounding these disruptions are climate change, soil degradation, flood, and drought. Fundamental changes in agricultural systems will be required to nourish a growing population while reducing agriculture's environmental impact.

The global crop production driven by developing countries continues to increase in direct proportion to fertilizer application (Figure 1.2). This trend is not sustainable and results in nutrient losses to air and water, economic loss for the farmers, and loss of vital raw materials needed to produce fertilizers.

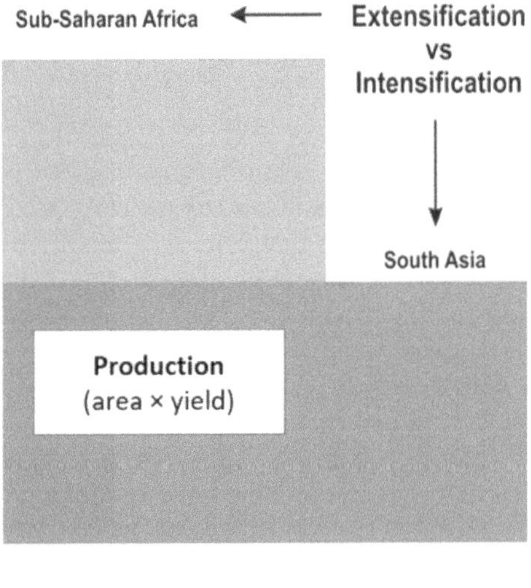

FIGURE 1.1 Comparison of factors causing increases in cereal production in Southeast Asia and Sub-Saharan Africa.

TABLE 1.1
Prevalence of Malnutrition in the World with Projections to 2030

	Number of Malnourished People (Millions)			
	2005	**2016**	**2019**	**2030***
WORLD	825.6	657.6	687.8	**841.4**
AFRICA	192.6	224.9	250.3	**433.2**
Northern Africa	18.3	14.4	15.6	21.4
Sub-Saharan Africa	174.3	210.5	234.7	**411.8**
ASIA	574.7	381.7	381.1	329.2
Central Asia	6.5	2.1	2	N.R.
Eastern Asia	118.6	N.R.	N.R.	N.R.
Southeast Asia	97.4	63.9	64.7	63
Southern Asia	328	256.2	257.3	203.6
Western Asia	24.3	29.2	30.8	42.1

N.R., Not Reported.
Do Not Reflect the Potential Impact of COVID-19 Pandemic (Roser and Ortiz-Ospina 2013).

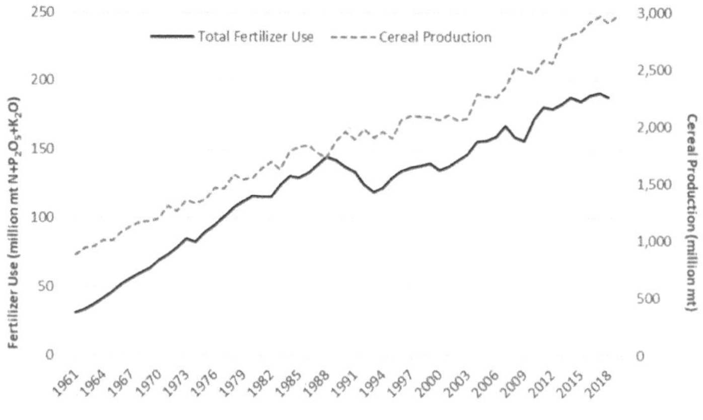

FIGURE 1.2 World total cereal production and fertilizer use (FAO 2019).

Global food production must increase to meet the needs of the growing population which is projected to be around 9.8 billion by 2025. The mid-century target for production increase of 70% has been recalibrated in the context of sustainable intensification and environmental imperatives. The current analysis shows that an increase of approximately 25–70% above current production levels may be sufficient to meet 2050 crop demand (Mitchell, Smith, Schipanski, Atwood, Mortensen 2017) to overcome undernutrition that continues to disproportionately affect children particularly in SSA and Southeast Asia. Simultaneously, nutrient losses and GHG emissions from agriculture should be on a declining path to ensure ecosystem services.

This chapter reviews the effects of soil health on fertilizer use efficiency, beginning with fertilizer products and their management, government policies on food production, and environmental issues from the use of fertilizers. Recent developments related to new fertilizers, the new thrust to improve soil fertility and rejuvenate degraded soils through better management practices, and meeting the demand for crops in the face of climate change while producing more nutritious food are all discussed in this chapter.

1.2 FERTILIZERS

Fertilizers provide plants with nutrients they need for growth and development. Plants live, grow, and reproduce by taking up water and nutrients, carbon dioxide from the air, and energy from the sun. Apart from carbon, hydrogen, and oxygen, which collectively make up 90–95% of the dry matter of all plants, other nutrients needed by plants come from the media in which they grow—essentially from the soil. To maintain soil fertility and productivity and prevent land degradation, nutrients taken up by crops must be replenished through the application of fertilizers, which are both manufactured and naturally available, including manure.

The beginning of our dependence on inorganic fertilizer can be traced back to the nineteenth century when Justus von Liebig articulated the theoretical foundations of crop production and when John Bennett Lawes began producing fertilizers containing phosphorus. However, only since the 1960s, when global starvation became a real possibility, have fertilizers assumed a predominant role in increasing agricultural productivity. Fertilizer was an integral part of the technological trinity—seed, water, and fertilizer—responsible for bringing about the Green Revolution that helped many densely populated countries, including India, China, and Indonesia, achieve food self-sufficiency in a short span of 20–25 years.

1.2.1 What Is a Fertilizer?

A commercial fertilizer is a material that contains at least one of the plant nutrients in a chemical form that, when applied to the soil, is soluble in the soil solution phase and assimilable by or "available" to plant roots. Most often, this implies chemical forms that are water soluble. In the case of phosphorus, solubility in special reagent solutions (citric acid, neutral ammonium citrate, or alkaline ammonium citrate) is often used as a guide for availability to plants. In the case of nitrogen, slow solubility in water may be more desirable from an environmental and efficiency standpoint than easy solubility. Fertilizer products are customarily designated by a series of numbers separated by dashes. This set of numbers is called the "grade" of the fertilizer product. Each of the numbers indicates the amount of a nutrient that the manufacturers guarantee is contained in the fertilizer. Usually, three numbers are used when giving the grade of the fertilizer product, and these three numbers always refer, in order, to the content of the primary nutrients: nitrogen, phosphorus, and potassium. If other nutrients are present, their content can also be indicated in the grade of the fertilizer product; each extra number is followed by the chemical symbol of the nutrient it represents.

1.2.1.1 Nitrogen

The ammonia produced through the Haber–Bosch process accounts for more than 90% of the world's nitrogen fertilizers, without which we would produce 48% less food (Erisman et al. 2008). Besides direct application (as a fertilizer), ammonia is also used as raw material in the production of urea, ammonium nitrate, and other N fertilizers, as well as in the production of monoammonium phosphate (MAP), diammonium phosphate (DAP), and other multi-nutrient fertilizers (Roy 2020). Fertilizers produced from ammonia provide farmers with a wide range of N source options for managing N to best meet their crop needs and logistical requirements.

1.2.1.2 Phosphates

Phosphorous-containing fertilizers are produced from mined phosphate rock found in several locations around the world (UNIDO, IFDC 1998). Phosphate rock is a finite, non-substitutable, nonrenewable resource. Phosphate deposits are found in every continent of the world except Antarctica. Recent estimates indicate a resource base between 290 billion tons to 460 billion tons with reserve of about 70 billion tons (economically extractable using current technologies) (Van Kauwenbergh 2010). Both reserves and resources are a "snapshot at a point in time" and not an indication of "all there is." This reserve/resources base can be extended by exploiting known, incompletely explored, and yet-unknown phosphate deposits at higher costs. Conservation and efficient use of this valuable natural resource and recycling of P is receiving increased attention through sustainable P management initiatives (Scholz et al. 2014). About 80% of mined phosphates is used for agriculture and the rest in other sectors e.g., electronics, food, detergents, etc.

Unprocessed phosphate rock may be applied as a source of P nutrition on acidic soils (soils below pH 5.5), but most (more than 90%) phosphate rock is processed for production of phosphate fertilizers by reacting it with either sulfuric or nitric acid. Phosphate products include single superphosphate (SSP), triple superphosphate (TSP), MAP, DAP, and nitric phosphates (Roy 2020).

1.2.1.3 Potassium

Potassium is of the least concern among the three primary nutrients. Not only is this element abundantly present in the earth's crust, but it can also be found in conveniently concentrated deposits in both deeply buried and near-surface sediments. Even the most conservative reserve base estimates indicate more than 400 years of reserve at the projected rate of use.

Potassium-containing fertilizers include potassium chloride (commercially known as potash), potassium sulfate, potassium magnesium sulfate (commercially known as sulfomag), and potassium nitrate.

N, P, and K fertilizers are the most widely used fertilizers, primarily because virtually all crops require significant amounts of these nutrients to maximize production. Crops also need sulfur, calcium, and magnesium (secondary nutrients) for development, but their uptake is usually lower than for the primary nutrients (N, P, and K). The micronutrients Zn, Fe, Mn, Cu, B, Mo, and Cl are needed in small amounts but are essential for plant growth. A deficiency of secondary and micronutrients usually becomes the limiting factor in crop production.

1.2.2 Use Efficiency of Fertilizers

Most of today's fertilizers were developed several decades ago and for temperate agricultures, with plant-available nutrient contents (mainly NPK) as high as possible (high analyses) with the intent of delivering them at the lowest cost to farmers. These fertilizers dramatically increased the agricultural productivity of temperate-zone countries. These same fertilizers were a part of the package that triggered the Green Revolution in Asia in the 1970s. Farmers broadcasted fertilizers according to blanket fertilizer application recommendations.

This blanket application has resulted in a progressive decrease in partial factor productivity (PFP) which indicates kilogram of grain produced per kilogram of fertilizer used (Figure 1.3). This progressive decrease in PFP is partly due to the fertilizers that do not match the need of the soil and plant, particularly the absence of secondary and micronutrients. Application timing and place are also contributing factors along with, in many cases, seriously degraded soils that limit agronomic efficiency and results in low fertilizer use efficiency (FUE) (especially nitrogen) as well. This results in air and water pollution that is adversely affecting human wellbeing.

Globally, more than 50% of the nitrogen fertilizers applied to cropland are lost to the environment (Zhang et al. 2015; Lassaletta et al. 2014) with adverse impacts on the quality of water, soil, air, biodiversity, ecosystems, and human health. More recent data show that the nitrogen use efficiency (NUE) for cereal in 2015 was 35, 41, 30, and 21% for the world, US, China, and India, respectively (Omara, Aula, Oyebiyi, Raun 2019).

The low NUE in China and India is due to a combination of factors including high N use, low use of other fertilizers (P, K, and micronutrients), government policies, and the prevalence of degraded and depleted soils. In the US, higher NUE could be due to better soil health and precision fertilizer application (Omara, Aula, Oyebiyi, Raun 2019). In contrast, the NUE in SSA is about 70% (Zhang, Davidson, Mauzerall, Searchinger, Dumas, Shen 2015). Despite a fertilizer application rate that is nearly an order of magnitude lower than the world average of 140 kg nutrients per hectare of arable land, the NUE remains high due to inherently low soil nutrient content and unmet nutrient needs of

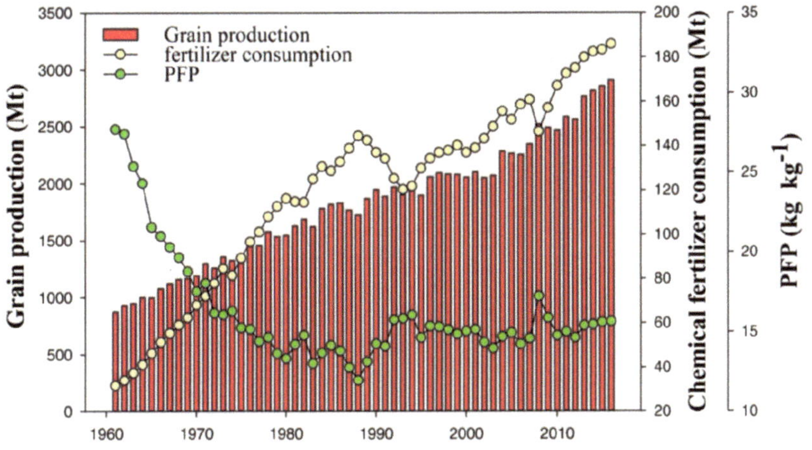

FIGURE 1.3 Global cereal production, fertilizer consumption, and partial factor productivity (PFP) (derived from FAO and IFA data).

the plant. Even though SSA's NUE is comparably high, SSA contributes to greenhouse gas (GHG) emissions through land clearing for agriculture (Our World in Data 2021).

The other primary nutrient (P) is usually lost from soils mainly through erosion. A recent study by Alewell et al. estimates that more than 50% of global P loss in agriculture is attributable to soil erosion (Alewell et al. 2020). Other losses include leaching in the drainage and soluble P runoff (Syers, Johnston, Curtin 2008, https://www.palmbeachpost.com/story/weather/hurricane/2020/07/16/crops-warns-of-lake-okeechobee-algae-bloom/41705629/). These losses combined with inefficiencies in the conversion of phosphate rock to current fertilizers results in less than 10% of mined phosphates in food and feed (Scholz, Roy, Brand, Hellums, Ulrich 2014). Another component that is more difficult to account for is what is commonly referred to as fixation reactions, which are P precipitation and other sorption processes with calcium in calcareous soils and iron and aluminum mostly in acidic soils.

1.2.3 Options to Improve Fertilizer Use Efficiency

In most countries, fertilizer application is based on optimum economic return anchored in crop response data averaged over large areas. But studies have shown that farmers' fields show large variability in terms of nutrient-supplying capacity and crop response to nutrients. Thus, blanket fertilizer application recommendations may lead farmers to overfertilize in some areas and under fertilize in others or to apply an improper balance of nutrients for their soil or crop. Thus, fertilizer applications based on blanket recommendations result in economic loss for the farmers, loss of energy and raw material resources needed to produce fertilizers, and air and water pollution. In general, the options to improving the economics of fertilizer use and minimizing air and water pollution are embodied in the 3P principles (1) **p**roduct, (2) **p**ractice, and (3) **p**olicy. These principles are interconnected and must be implemented simultaneously to attain the desired benefits.

1.2.3.1 Product

Product is new fertilizers with characteristics that release their nutrient content over an extended period to match the nutrient requirements of the crop. Two important special types of fertilizers are (1) slow- and controlled-release fertilizers with the release of nutrients over several months, and (2) stabilized fertilizers (fertilizers associated with nitrification or urease inhibitors), delaying either the nitrification of ammonia or the ammonification of urea (Roy 2015). Due to economic and environmental considerations, the slow- and controlled-release and stabilized N fertilizers have been much more important than potash, because unutilized potash remains available for subsequent crops. As mentioned above, phosphate losses are generally associated with soil erosion and runoff of soluble phosphates into the waterways. Recent incidences of excessive algal growth in inland waterways (Figure 1.4) highlight the need to manage nutrients (fertilizers and other sources) applied to agricultural lands. The development of slow-release phosphates is becoming important in the context of improving uptake by plants and reducing fixation by soils high in iron and aluminum (Weeks, Hettiarachchi 2019).

1.2.3.2 Practice

Practice is an approach to maximize economic, social, and environmental benefits. This includes soil and crop management to fit the different conditions encountered in cropping areas. Sometimes this is also referred to as site-specific nutrient management (SSNM) which, among others, includes a balanced use of all essential crop nutrients (Richards et al. 2015). The SSNM approach has proven to be an efficient way to reduce NOx emissions to the atmosphere and leaching losses to surface and sub-surface waters (Roy 2015).

Certainly, application of the right type of fertilizer, in the right amount, at the right time, and in the right place, commonly known as the 4Rs, ensures increased crop production, better utilization of nutrients, and reduced losses of nitrogen to the environment (Bruulselma, Fixen, Sulewski 2012). An advancement of the 4R principles is precision farming (also known as precision agriculture) that employs detailed, site-specific information to precisely manage production inputs to optimize production allowing for better use of resources to maintain the quality of the environment while improving sustainability of the food supply.

1.2.3.3 Policy

Policy is essential to improving productivity and FUE, increase farmers' income, and reduce environmental degradation. But without conducive policies, neither technologies nor management practices would be effective.

China and India, which collectively account for more than 30% of global nitrogen consumption, both have a relatively low NUE (ranging between 25 and 30%). In China, an input subsidy reached $17 billion in 2014, and the farmers' use of fertilizer-N exceeded the recommended amount by 20–60% in cereals and even more in vegetables whose NUE is around 15%. Farmers in China made

FIGURE 1.4 Algal growth in Lake Okeechobee, Florida, March 2020 (Source: *The Palm Beach Post*).

a conscious decision to shift to fruit and vegetable production partly because the crop-to-fertilizer price ratio is more attractive than that of cereal production (Zhang et al. 2015).

In India, fertilizers are highly subsidized, nitrogen more so than other nutrients, which results in the overuse of nitrogen-based fertilizers that distorts nutrient (nitrogen-phosphorus-potassium) ratios, all of which results in a low FUE, particularly NUE. To correct nutrient-ratio distortion, the Indian government introduced a nutrient-based subsidy in 2010. But the impact on improving FUE has been minimal since urea was excluded from the scheme (Roy 2020).

In Europe, NUE has progressively increased since the 1990s. In 2016, 30% more crops were produced with 20% less fertilizer-N application (Roy 2020). The European Union Common Agricultural Policy partly triggered this phenomenon by reducing crop subsidies and through the European Union Nitrates Directive, which limited manure application rates on agricultural land (Zhang et al. 2015).

While the 3Ps have shown improvements in FUE, higher efficiency can be achieved by improving soil health, which is inseparable from plant and animal productivity; soil also maintains or enhances water and air quality. Soil fertility is integral to soil health, and a study of its organic realm requires an ecosystem to gene approach. So, to enhance FUE, several interconnected steps are needed, including soil quality improvement through soil organic carbon (SOC) sequestration, alleviating drought stress by soil-water management, controlling soil erosion, adoption of conservation agriculture, using fertilizers tailored to soils, and adoption of the 4R principles, conservation agriculture (CA), and integrated soil fertility management. Healthy soils hold water and nutrients, and water stored in soils (green water) supports 90% of global and livestock production, particularly in Africa. Soil's ability to hold green water and absorb rainwater is directly related to its overall health. Healthy soils retain rainwater and more nutrients. In contrast, degraded soils hold less water and nutrients and are thus less effective in supporting agricultural activity and mitigating flooding and erosion. Soils also act as a carbon sink. The world's soils contain three times as much carbon as the world's vegetation and almost twice as much as the atmosphere. The healthier the soil, the more vegetation it can support and the more atmospheric carbon it can sequester. Even small declines in soil health can have measurable impacts on carbon dioxide levels in the atmosphere.

1.3 SOILS

There are several methods of improving soil health, but the two most prevalent practices are conservation agriculture and integrated soil fertility management.

1.3.1 Conservation Agriculture

Conservation agriculture (CA) is promoted to increase crop production and environmental sustainability—particularly soil health. The term encompasses many techniques for capturing and storing water and improving soil quality, thereby optimizing yields and profits through improvement of soil structure and rationalizing agricultural inputs. CA was developed in response to continuously declining land productivity under conventional soil tillage systems and has been defined differently by different authors. But a working definition of CA enunciated by the Food and Agriculture Organization of the United Nations (FAO) follows.

CA is an approach to managing agro-ecosystems for improved and sustained productivity, increased profits, and food security while preserving and enhancing the resource base and the environment. CA is characterized by the practical application of three linked principles, along with other complementary good agricultural practices of crop and production management, namely (Lal 2015):

1. Principle 1: Continuous absence of or minimal mechanical soil disturbance (implemented by the practice of no-till seeding or broadcasting of crop seeds, and direct placing of plants into untilled soil; and causing minimum soil disturbance from any cultural operation, harvest operation, or farm traffic);

2. Principle 2: Maintenance of a permanent biomass soil mulch cover on the ground surface (implemented by retaining crop biomass, root stocks and stubbles, and cover crops and other sources of ex situ biomass); and
3. Principle 3: Diversification of crop species (implemented by adopting a cropping system in which crops are rotated, and/or sequenced and/or associations involving annuals and perennial crops, including a balanced mix of legume and non-legume crops).

Established CA systems achieve yield levels comparable with and even higher than those under conventional intensive tillage systems. CA does not necessarily lead to yield penalties. Furthermore, the overall crop and biomass production within a season increases over time under CA management compared with tillage-based management. CA systems stimulate biological processes in the soil and above the soil surface, and, combined with reduced erosion and leaching, can lead to increased water and nutrients use efficiencies (FAO 2014). In addition, CA has been proved to sequester organic carbon in the soil at a rate ranging from about 0.1–0.5 mg/ha/year or more depending on the amount of biomass being returned, prevailing SOC content, thermal and moisture climate, length of growing season, soil type and fertility, cropping systems, and management practices (Kassam, Friedrich, Derpsch 2018; Corsi et al. 2014).

Tillage as a soil management concept was questioned for the first time in the 1930s, when dust bowls devastated wide areas of the midwestern US (Derpsch 1998). Ideas for reducing tillage and keeping soil covered with crop biomass followed, and the term conservation tillage was introduced for practices aimed at erosion control. In the 1940s, seeding machinery allowed direct seeding without any soil tillage. In the early 1970s, due to high erosion, "no-tillage" reached Brazil, where farmers and scientists together transformed the technology into the system which today is called CA. Yet it took another 20 years before CA reached significant adoption levels. From the early 1990s, the uptake of CA started growing exponentially, leading to a revolution in the agriculture of southern Brazil, Argentina, Paraguay, and Uruguay.

CA systems are now operational on all continents in all land-based agriculture; this confirms that CA principles are universally applicable to all agricultural landscapes and land uses with locally formulated and adapted practices. After a slow start, CA adoption continues to increase and in 2015/16, CA was practiced globally on about 180 M ha of cropland, corresponding to about 12.5% of the total global cropland compared with about 106 M ha in 2008/09 (Kassam, Friedrich, Derpsch 2018). This is equivalent to an increase of some 69% globally since 2008/09. In 2015/16, 78 countries had adopted CA systems, an increase of 42 additional countries since 2008/09. The average annual rate of global expansion of CA cropland area since 2008/2009 has been some 10.5 M ha. The largest extents of adoption are in South and North America, followed by Australia and New Zealand, Asia, Russia and Ukraine, Europe, and Africa; although CA has not been mainstreamed in the agricultural systems of these countries (Kassam, Friedrich, Derpsch 2018).

Thus, FUE can be improved through a CA system-based approach. Certainly, adoption of the 4Rs is extremely important as is the 4S concept proposed by Lal (2020). The 4S comprises soil management efforts to (1) sequester carbon, (2) restore soil health, (3) conserve soil water, and (4) manage soil temperature. In the US, corn productivity has progressively increased since 1990 through precision farming and better soil fertility management as evidenced by the increase in the partial factor productivity (expressed as kg grain/kg nutrients) since 1990 (Figure 1.5).

1.3.2 Integrated Soil Fertility Management

Integrated soil fertility management (ISFM) is a set of soil fertility management practices that encompass the use of fertilizer, organic inputs, and improved germplasm, combined with the knowledge of local practices and conditions, with the aim of improving FUE and crop productivity. All inputs need to be managed following sound agronomic and economic principles (Fairhurst 2012;

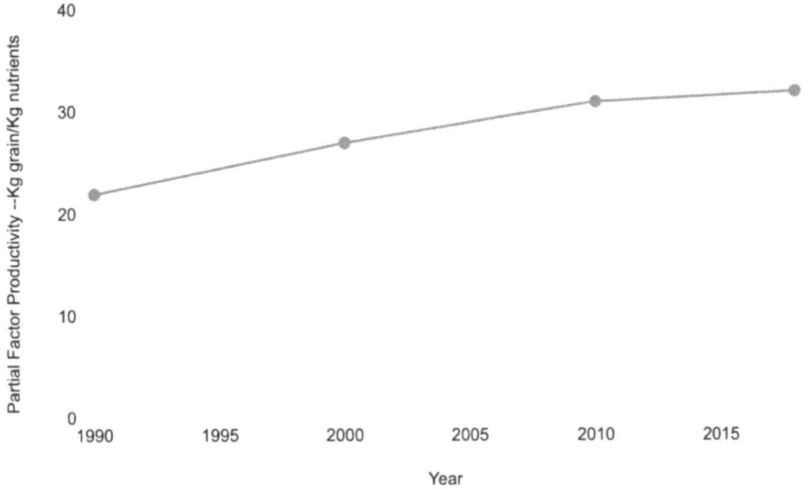

FIGURE 1.5 Increasing trend in the productivity of corn in the US (constructed from data of the US Department of Agriculture USDA).

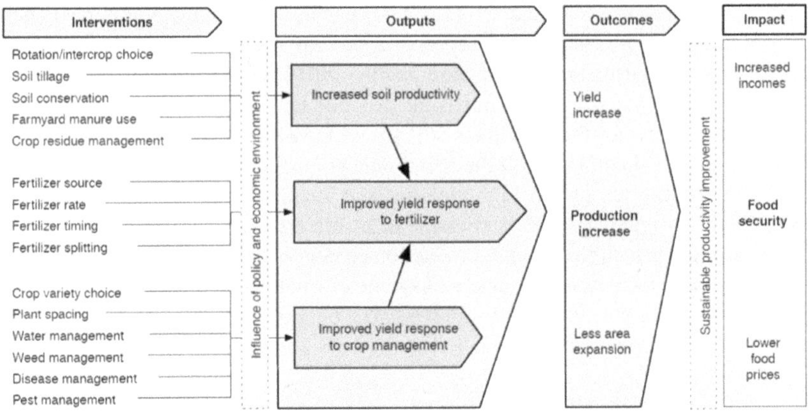

FIGURE 1.6 Integrated soil fertility management (ISFM), interventions, outputs, outcomes, and impacts (Fairhurst 2012).

Vanlauwe, Bationo, Chianu et al. 2010) The process is described in terms of interventions, outputs, outcomes, and impact in Figure 1.6.

When introduced successfully, productivity increases and less land is required to achieve a given level of production. The impact is the sustainable improvement of food security, increased farm incomes, and lower food prices, which benefit the urban population.

Thus, ISFM is an approach aimed at improving soil fertility by judiciously combining fertilizers with organic and mineral amendments together with soil and water conservation measures. The judicious combination of water, soil, organic and mineral amendments, and soil and fertilizer nutrients is expected to increase soil organic matter (SOM) content, and thus improve nutrient replenishment, soil fertility maintenance and, hence, soil water and nutrient holding capacity that can support more drought-tolerant cropping systems. The approach is underpinned by some key principles, including but not limited to (1) maximizing the use of available farm organic sources, (2)

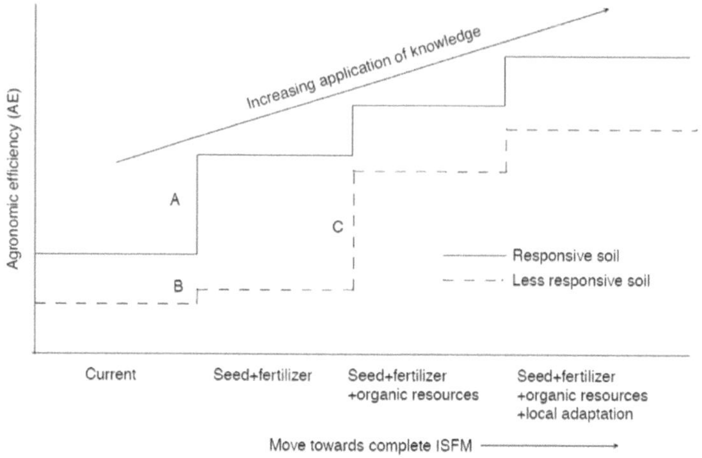

FIGURE 1.7 Relationship between the agronomic efficiency (AE) of fertilizers and organic resources and the implementation of various components of ISFM (Sanginga, Woomer 2009).

harnessing the synergetic effect of combined use of organic and mineral fertilizers, and (3) using available farm natural resources that are economically viable. As such, the types of ISFM can differ depending on climate, agro-ecological zone, cropping system, soil type, quantity and type of locally available resources, and farms typology. Consequently, ISFM has the greatest potential for impact in areas where there is a need for crop intensification due to an increasing population. Certainly, farmers' access to markets for their products will accelerate adoption of ISFM.

The definition of ISFM combines all the agronomic components necessary to make crops grow and yield well, including the use of high-yielding and healthy planting material, plant nutrients (whether supplied as organic materials or mineral fertilizers), and other soil amendments. The approach embraces the principles of plant production ecology where yield is a function of the interaction between genotype, environment, and management, where genotype is the seed or plants used in the farming system (local or improved varieties). The environment refers to the soils and climate in the location, and the management refers to the farmer's ability and skill in managing crops and the farming system.

A model is used to illustrate the impact of moving toward a more complete implementation of ISFM (Figure 1.7):

- The more complete the implementation of ISFM, the greater the value for agronomic efficiency.
- It is important to make a distinction between responsive soils and less responsive soils:
  - The response to seed and fertilizer inputs is large in responsive soils (point A).
  - The response to seed and fertilizer inputs is small in non-responsive "degraded" soils (point B).
  - Organic resources are required to make efficient use of fertilizer and improved seeds (point C).

1.4 AFRICA AND INTEGRATED SOIL FERTILITY MANAGEMENT

In the 1960s, the focus was the use of improved germplasm and fertilizer. This combination led to a rapid increase in food production, especially in Asia and Latin America where the responses to fertilizers were dramatic. However, when this strategy was introduced in Africa in the early 1980s, it had very little impact on the productivity in SSA because the soils were generally poor, particularly in SOM, and the high rate of fertilizer applications did not result in the increases in productivity seen in Asia and Latin America (Fairhurst 2012; Sanginga, Woomer 2009). The situation was further complicated by the removal of input subsidy, and fertilizers became unaffordable for the smallholder farmers (Sanginga, Woomer 2009). Hence, available organic matter became the only source of plant nutrients. But the lack of sufficient organic resources and intense labor requirements became constraining factors (Vanlauwe et al. 2010; Sanginga and Woomer 2009). Neither fertilizers nor organic resources alone are solutions to increasing agricultural productivity in Africa, but a combination of both, adapted to local conditions and practices along with locally adopted soil fertility management practices, would be needed to optimize the agronomic efficiency in crop production (Figure 1.8).

In Africa, access to affordable fertilizers remains a big constraint for smallholder farmers, particularly in remote areas. Judicious use of fertilizers is essential for increasing agricultural productivity in Africa. However, the fertilizer application rate is about 10–12% of the world average. So, crops use native nutrients in the soils to grow, resulting in the decline of soil fertility. It is estimated that, over a 30-year period, 132 million tons of N, 15 million tons of P, and 90 million tons of K have been lost from cultivated land in 37 African countries (Smaling 1993). To maintain crop production, the farmers resort to slash and burn agriculture, and marginal lands and wildlife habitats are converted to agricultural lands. The annual average forest clearance in Africa between 2010 and 2014 was 902,219 ha for croplands, pasture, and logging (Vanlauwe 2004).

Besides low application rate, a major issue preventing effective utilization of fertilizers by crop has been the "pan-territorial/blanket" application recommendations that fail to take into account differences in farmers' resource endowments (soil type, labor capacity, climate risk, etc.). Past fertilizer recommendations have been based on single major cash crops, such as maize, tea, and cotton and did not take into account complex farming systems, e.g., crop rotations, inter-cropping, etc. But this blanket recommendation is giving way to site- and crop-specific recommendations based on soil mapping that provides a more accurate nutrient status of the soils allowing the application of balanced, tailored fertilizers (Roy 2020).

Although the fertilizer application rate (kg nutrients/ha) in Africa is low, the NUE is ~70% (Zhang et al. 2015) compared with the world average of 35%. In Africa, the plants utilize most of the applied fertilizers and get the rest of their nutrients from soils that are already low in nutrients.

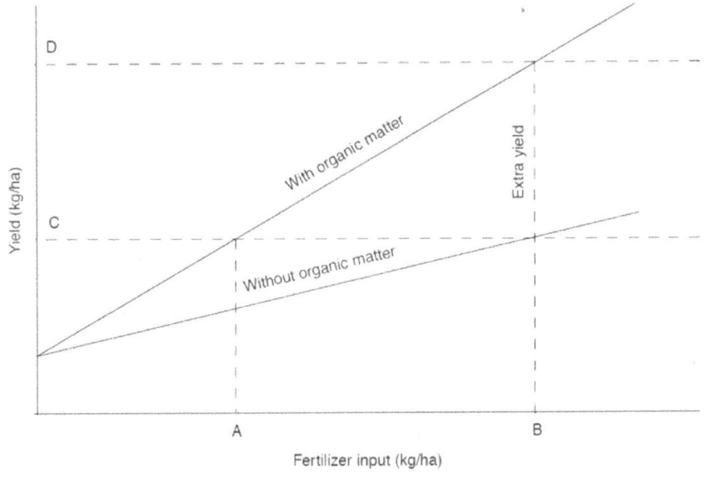

FIGURE 1.8 Positive interaction between fertilizer and organic inputs resulting in extra yield due to ISFM practices (Fairhurst 2012).

Besides, most African soils are inherently low in SOC content (<20–30 mg/kg). This is due to the low root growth of crops and natural vegetation but also the rapid turnover rates of organic materials with high soil temperature, persistent bush burning, and intense faunal activity, particularly termites (Bationo 2008). Africa needs to intensify its agricultural production to feed the growing population. This cannot be achieved simply by applying more fertilizers as has been done in other countries (mostly in Asia) without simultaneously improving soil health, particularly increasing SOM and soil carbon that are vital to retaining nutrients and moisture for growing plants more efficiently. The current African agricultural system is characterized by low yield and high NUE. The longer-term strategy for Africa is to shift crop production directly from a low-yield, high-NUE status to a high-yield, high-NUE status. This shift will require leapfrogging over the historical evolution of agricultural development by employing technologies and management practices that promote high NUE. One obvious starting point is the soil; improving its fertility is paramount to realizing the benefits of improved seeds and/or fertilizers. Thus, the approach in SSA should be to adopt ISFM tailored to the local conditions.

In Benin, West Africa, there is a discernable difference in the impact of ISFM practice (Figure 1.9). After three years of adopting ISFM, the experimental plots showed dramatic increases in yield of maize, sorghum, and irrigated rice over farmers' managed fields. For example, yield increases for maize, sorghum, and irrigated rice were 280%, 80%, and 83%, respectively (Breman

FIGURE 1.9 Impact of ISFM in Benin, West Africa (Breman 2021).

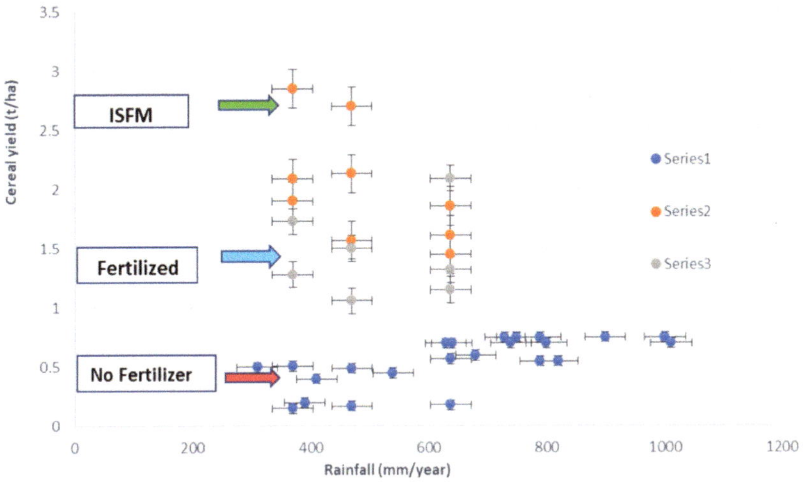

FIGURE 1.10 Improvement in cereal yields, through ISFM, of fertilized and non-fertilized plots in the Sahelian region of West Africa (Breman 2021).

2021). Analyses of several years of trials in the Sahel confirm the effectiveness of ISFM even in areas of low rainfall (Breman 2021). Cereal yields of ISFM plots were higher than fertilized and control plots (Figure 1.10). This again confirms that fertilizer alone is not the solution for soils with low fertility.

1.5 FERTILIZER POLICIES AND NUTRIENT USE EFFICIENCY

All countries institute policies to support their farmers, but these supports differ from country to country. Some policies are directed to improve productivity and NUE, while others are to increase production to feed its expanding populations. For example, in Europe more crop is produced with less fertilizer application by reducing crop subsidies and limited manure application. In most developed economies, farmers are provided direct and/or indirect support. Among policies, input subsidies (particularly for fertilizer) have a large impact on NUE. Developed economies do not subsidize inputs because of crop insurance schemes that pay in cases of crop failure, particularly whether-related. In contrast, developing economies subsidize inputs, thereby assuming the risk of crop failure. In Asian countries, subsidies for fertilizers are universal, and every farmer is entitled to subsidized fertilizers. As a result, in many cases, there is over application of fertilizers resulting in higher losses and lower NUE. For example, China and India are the two largest fertilizer consumers in the world, and their NUE is less than 25%.

In Africa, more than 70% of the farmers are smallholders, and the subsidy models are termed as “smart subsidies” because they are targeted to either crops or certain groups of farmers. In assessing the effectiveness of subsidies, whether universal or targeted, they have helped to increase production but not in proportion to the fertilizer application because of either low soil fertility or high nutrient losses from the inappropriate fertilizers or over application of fertilizers. Besides, there are no incentive(s) for farmers to maximize return from the use of fertilizers. Going forward, it is imperative for the fertilizer subsidies to be incentive-based with clear targets. Given the vital role of soil carbon in increasing nutrient efficiency, the increase in soil organic carbon in farmers’ fields is a logical choice. With recent advances in handheld, portable soil organic carbon measurement equipment that can be used in the field, the idea of farmers increasing soil organic carbon tied to fertilizer subsidy is possible and should be pursued (GreenBiz.com 2021).

1.6 CONCLUSIONS

Over the coming decades, global population growth, coupled with dietary changes, will drive a substantial increase in food demand. By 2050, we need to produce enough food to feed about 9.8 billion people—2.2 billion more than exist today. Compounding this challenge is the number of malnourished people that in 2019 was 688 million and is projected to reach 841 million by 2030 even without considering the impact of the COVID-19 pandemic. Agricultural growth has been a key factor in the rapid decline of global poverty since the 1950s, driven by the adoption of improved, high-yielding seed varieties, fertilizers, agronomic management, and conducive policies. These combinations of factors triggered the Green Revolution where fertilizer was a key ingredient.

Since the 1960s, global fertilizer consumption has increased dramatically, driven by the developing countries, and total cereal production has increased in proportion to fertilizer use without much consideration of the soils and plant needs. This has resulted in progressive decline of the PFP and NUE, mainly in the developing economies of Asia. Progressive decline in PFP and NUE results in air and water pollution. Sub-Saharan Africa is an exception, where the fertilizer application rate is only about 10% of the world average, and production increases are due to soil nutrient mining resulting in the rapid decline of soil health.

While management of efficient fertilizers has shown some improvements in the NUE, higher efficiencies can be achieved by improving soil quality through SOC sequestration, alleviating drought stress by soil-water management, controlling soil erosion, adoption of conservation agriculture, using fertilizers tailored to soils and crops, and adoption of the 4R principles, CA, and ISFM. Many governments in Africa have factored ISFM into their agricultural development programs because of promising results in improving soil fertility and increasing crop yields.

Fertilizers will be key to producing more food and keeping pace with population growth, particularly in developing countries. So, increasing the efficiency of their use is of paramount importance not only from the standpoint of profitability but also for reducing GHG and conserving raw material resources. Increasing SOC content is essential in improving nutrient use efficiency.

Most countries support their farmers directly or indirectly. In developing countries, however, input subsidy—particularly fertilizers—is most common. In Asia, fertilizer subsidies are universal; all farmers are eligible to receive them. While in Africa, subsidies are targeted to selected farmers and/or crops. In both subsidy schemes, farmers are not required to meet any targets from the use of subsidized fertilizers. Considering the importance of SOC in improving NUE, soil health, and reducing GHG, it is important that the quantum of fertilizer subsidies be contingent on farmers increasing the SOC content in their fields. With recent advances in handheld, portable SOC measurement equipment, the idea of making fertilizer subsidies incentive-based is within the realm of possibility.

REFERENCES

Alewell, C., Ringeval, B., Ballabio, C., et al. 2020. Global phosphorus shortage will be aggravated by soil erosion. *Nature Communications* 11, 4546. https://doi.org/10.1038/s41467-020-18326-7

Bationo, A. 2008. *Integrated Soil Fertility Management Options for Agricultural Intensification in the Sudano-Sahelian Zone of West Africa*. Academy Science Publishers, Nairobi.

Breman, Henk. 2021. Personal communications. www.henkbreman.com

Bruulselma, T. W., Fixen, P. E. and Sulewski, G. D. (eds). 2012. *4R Plant Nutrition Manual: A Manual for Improving the Management of Plant Nutrition*. International Plant Nutrition Institute (IPNI), Norcross, GA.

Corsi, S., Friedrich, T., Kassam, A. and Pisante, M. 2014. A review of carbon sequestration through conservation agriculture. In L. K. Heng, K. Sakadevan, G. Dercon and M. L. Nguyen (eds), *Proceedings of the International Symposium on Managing Soils for Food Security and Climate Change Adaptation and Mitigation*, FAO, Rome, 205–207.

Derpsch, R. 1998. Historical review of no-tillage cultivation of crops. In *Proceedings of the 1st JIRCAS Seminar on Soybean Research on No-Tillage Culture & Future Research Needs*, 1–18 March 5–6, 1998, Iguassu Falls, Brazil. Working Report No. 13, JIRCAS.

Erisman, J. W., Sutton, M. A., Galloway, J., Klimont, Z. and Winiwarter, W. 2008. How a century of ammonia synthesis changed the world. *Nature Geoscience* 1, 636–639. Macmillan Publishers Ltd., London, UK.

Fairhurst, T. (ed.). 2012. *Handbook for Integrated Soil Fertility Management.* Africa Soil Health Consortium, Nairobi, Kenya .

FAO. 2014. *What Is Conservation Agriculture?* FAO, Rome. http://www.fao.org/ag/ca/1a.html

FAO. 2018. *The Future of Food and Agriculture – Alternative Pathways to 2050.* FAO, Rome. http://www.fao.org/3/I8429EN/i8429en.pdf

FAO. 2019. *Fertilizer Consumption.* https://data.worldbank.org/indicator/Ag.CON.FERT.ZS?view=chart

FAO, IFAD, UNICEF, WFP and WHO. 2020. *The State of Food Security and Nutrition in the World 2020. Transforming Food Systems for Affordable Healthy Diets.* FAO, Rome. https://doi.org/10.4060/ca9692en

GreenBiz.com. 2021, February 11. *Measuring Soil Carbon Levels with Hand-Held Probe.* GreenBiz Forum Phoenix, USA.

https://www.palmbeachpost.com/story/weather/hurricane/2020/07/16/corps-warns-of-lake-okeechobee-algae-bloom/41705629/

IFASTAT Consumption Database. 2019. https://www.ifastat.org/databases/plant-nutrition

Kassam, A., Friedrich, T. and Derpsch, R. 2018. Global spread of conservation agriculture. *International Journal of Environmental Studies.* https://doi.org/10.1080/00207233.2018.1494927

Klucas, G. 2015, September 30. Yield gap study highlights potential for higher crop yields in Africa. *Nebraska Today*, University of Nebraska-Lincoln. https://news.unl.edu/newsrooms/article/yield-gap-study-highlights-potential-for-higher-crop-yields-in-africa/

Lal, R. 2015. A system approach to conservation agriculture. *Journal of Soil and Water Conservation* 70(4), 82A–88A.

Lal, R. 2020. Carbon-centric integrated nutrient management: A solution for enhancing farm productivity and carbon sequestration in India. *Indian Journal of Fertilizers* 16(4), 300–312.

Lassaletta, L., Billen, G., Grizzetti, B., Anglade, J. and Garnier, J. 2014. 50-year trends in nitrogen use efficiency of world cropping systems: The relationship between yield and nitrogen input to cropland. *Environmental Research Letters* 9(10), 105011.

Mitchell, H., Smith, R., Schipanski, M., Atwood, L. and Mortensen, D. 2017. Agriculture 2050: Recalibrating targets for sustainable intensification. *Bio Science* 67(4), 386–391.

Omara, P., Aula, L., Oyebiyi, F. and Raun, W.R. 2019. World cereal nitrogen use efficiency trends: review and current knowledge. *Agrosystems, Geosciences and Environment* 2, 180045. https://doi.org/10.2134/age2018.10.0045

Our World in Data. 2021. www.ourworldin data.org

Pingali, P. 2012. Green revolution: Impacts, limits, and the path ahead. *Proceedings of the National Academy of Sciences* 109(31), 12302–12308. https://doi.org/10.1072/pnas.0912953109

Richards, M. B., Butterbach-Bahl, K., Jat, M. L., Lipinski, B., Ortiz-Monasterio, I. and Sapkota, T. 2015. *Site-Specific Nutrient Management: Implementation Guidance for Policymakers and Investors.* Global Alliance for Climate-Smart Agriculture, Rome, Italy.

Roser, M. and Ortiz-Ospina, E. 2013, revised 2019. Global extreme poverty. https://ourworldindata.org/grapher/declining-global-poverty-share-1820-2015?time

Roy, A. H. 2015. Global fertilizer industry: Transitioning from volume to value. The 29th Francis new memorial Lecture. *Proceedings* 769. International Fertilizer Society, London, UK.

Roy, A. H. 2020. Tailored fertilizers: The need of the times. In *Soil and Fertilizers: Managing the Environmental Footprint*, ed. R. Lal. CRC Press. 2020. Boca Raton, Florida, USA, pp. 217–338.

Sanginga, N. and Woomer, P. (eds). 2009. *Integrated Soil Fertility Management in Africa: Principles, Practices and Developmental Process.* Tropical Soil Biology and Fertility Institute of the International Centre for Tropical Agriculture, Nairobi, 263 pp.

Scholz, R. W., Roy, A. H., Brand, F. S., Hellums, D. T. and Ulrich, A. E. (eds). 2014. *Sustainable Phosphorus Management—A Global Transdisciplinary Roadmap.* Springer, Dordrecht, the Netherlands.

Smaling, E. M. A. 1993. *An Agro-Ecological Framework of Integrated Nutrient Management with Special Reference to Kenya.* Wageningen Agricultural University, Wageningen, the Netherlands.

Syers, J. K., Johnston, A. E. and Curtin, D. 2008. *Efficiency of Soil and Fertilizer Phosphorus Use.* FAO fertilizer and plant nutrition bulletin 18. Food and Agriculture Organization of the United Nations, Rome, p. 108.

United Nations Industrial Development Organization (UNIDO) and International Fertilizer Development Center (IFDC). 1998. *Fertilizer Manual.* Kluwer Academic Publishers, Dordrecht, the Netherlands.

United States Department of Agriculture, National Agricultural Statistics Service. Yearly Chemical use Surveys. 1980–2018. *The USDA Chemical Use Surveys, 1980–2018.*

Van Kauwenbergh, S. J. 2010. World phosphate rock reserves and resources. *International Fertilizer Development Center Technical Bulletin T-75*. Muscle Shoals, Alabama, USA.

Vanlauwe, B. 2004. Integrated soil fertility management research at TSBF: The framework, the principles, and their application. In *Managing Nutrient Cycles to Sustain Soil Fertility in Sub-Saharan Africa* (pp. 25–42), ed. Bationo, A. Academy Science Publishers, A Division of African Academy of Sciences, Nairobi, Kenya.

Vanlauwe, B., Bationo, A., Chianu, J., et al. 2010. Integrated soil fertility management. *Outlook on Agriculture* 39, 17–24.

Weeks, Jr., J. J. and Hettiarachchi, M. 2019. A review of the latest in phosphorus fertilizer technology: Possibilities and pragmatism. *Journal of Environmental Quality* 48, 1300–1313.

Zhang, X., Davidson, E., Mauzerall, D., Searchinger, T., Dumas, P. and Shen, Y. 2015. Managing nitrogen for sustainable development. *Nature*. https://doi.org/10.1038/nature15743

2 Conservation Agriculture
Carbon and Conservation Centered Foundation for Sustainable Production

Don C. Reicosky and Amir Kassam

CONTENTS

DOI: 10.1201/9781003102762-2

LIST OF ACRONYMS AND ABBREVIATIONS

AEI—: agroecological intensification
BDA—: British Dietetic Association
C—: carbon
C:N—: carbon to nitrogen ratio
CA—: Conservation Agriculture
CA-CoP—: Conservation Agriculture Community of Practice
CAS—: Conservation Agriculture Systems
CC—: cover crop
CEC—: cation exchange capacity
CT—: conservation tillage
CTIC—: Conservation Tillage Information Center
ECAF—: European Conservation Agriculture Federation
FAO—: Food and Agriculture Organization
GHGs—: greenhouse gases
INM—: integrated nutrient management
NRCS—: Natural Resources Conservation Service
NT—: no-till
NUE—: nitrogen use efficiency
OM—: organic matter
RA—: regenerative agriculture
RELACO—: Red Latinoamericana de Labranza Conservacionista (Latin-American Network for Conservation Tillage)
RT—: reduced tillage
SDGs—: sustainable development goals
SH—: soil health
SOC—: soil organic carbon
SOM—: soil organic matter
TA—: tillage agriculture
US—: United States
USDA—: United States Department of Agriculture
WUE—: water use efficiency

2.1 INTRODUCTION

2.1.1 Global Population

Meeting the triple challenges of growing food for healthy diets, organizing its equitable access, and reducing environmental damage will require careful attention to the world's soils. Soils are a critical component of terrestrial ecosystems and a fundamental constituent for sustaining life on earth. Food security and agriculture rests on our living soils. We need climate-smart agriculture, making the transition to farming better suited to cope with impacts of climate variability and extremes that jeopardize global food security and natural resources (Branca et al. 2011; Gattinger et al. 2011; Lal 2014a).

2.1.2 Climate Mitigation

Ongoing loss and degradation of agricultural soil and water assets due to increasing extremes in precipitation will continue to challenge both rain-fed and irrigated agriculture unless innovative conservation methods are implemented (Larson et al. 1997; Smith 2011; Lal 2013b). Conservation

Agriculture (CA) has increasingly been endorsed as Climate Smart Agriculture, contributing to both climate change adaptation and mitigation (Harvey et al. 2013; Pretty and Bharucha 2014; Kassam et al. 2017; Lal 2013b; Gonzalez-Sanchez et al. 2020). This will require a global orientation and perspective (Hobbs and Govaerts 2010; Kassam et al. 2009, 2018; Friedrich et al. 2012; Kirkegaard et al. 2014; Hunt et al. 2020), a concerted, strategic action to protect our soil, water, and air quality through new more efficient and ecological conservation practices.

2.1.3 Environmental Degradation

Farmers around the world have relied on intensive tillage for the last 10,000 years (Lal et al. 2007; Montgomery, 2007a; Reusser et al. 2015). Although this approach has had many benefits, it also has serious problems, notably the resulting susceptibility of soil to erosion and degradation (Delgado et al. 2011). According to Pimentel et al. (1995), about 430 million ha—almost one-third of the global arable land area—has been lost to soil erosion. Efforts to control human-induced land degradation and soil erosion have been building on the ruins of the past tillage and monoculture concepts (Lal et al. 2007; Montgomery 2007a, b). Both Diamond (2005) and Montgomery (2007a) agree: agricultural sustainability—particularly the conservation of soils—is critical to our long-term survival. Innovative application of comprehensive soil conservation practices is required to respond to the multiple problems of soil degradation if the planet is to feed the population and stay healthy.

2.2 OBJECTIVE

The objective of this chapter is to review the impacts of soil organic matter (SOM) on crop yields and to demonstrate that the Conservation Agriculture Systems (CAS) umbrella of ecological conservation practices encompasses aspects of carbon (C) management and soil fertility for food production and security while clarifying alternate types of agriculture.

2.3 SOIL ORGANIC MATTER AND CARBON MANAGEMENT

2.3.1 Photosynthesis Converts Solar to Biochemical Energy

Solar radiation is the main driving force of the ecosystem, and it is converted by green plants into organic compounds through the process of photosynthesis. Briefly, C diffuses as CO_2 from the atmosphere into chloroplasts of leaves, where it is synthesized into simple sugars by photosynthesis. These sugars are translocated throughout the plant and used as building blocks for various other organic forms: proteins, complex carbohydrates, and lipids. The formation of energy-rich compounds with C-C and C-H bonds transforms the incident solar energy in a cascade of biochemical energy flowing to other organisms in the ecosystems (Volk 2003; Pelletier et al. 2011; Blaustein 2012). Photosynthesis and the subsequent use of its byproducts by other organisms' cycles C between the atmosphere and vegetation, soils, and oceans, while human energy consumption cycles C from fossil fuel pools to the atmosphere. As C flows between them, each of these different pools has the capacity to be either a temporary source or a sink. Plant C captured in photosynthesis is transient, cycling in the soil-plant-atmosphere system (Kane 2015). This continuous movement of C through the soil-food web means that C is constantly changing forms in the soil as it is incorporated into new organisms or converted into different organic compounds used by all living organisms.

Farming, in essence, is about managing C—using photosynthesis to trap solar energy in C-rich plant materials that can be used as food, feed, fuel, or fiber in human societies. A portion of this C—sometimes as much as half or more—is removed as harvested grain, forage, or other commodities, some of which may eventually be returned to the system as organic amendments in the form of compost, manure, biosolids, and biochar. The unharvested C (above ground and root biomass or "residues"), including root exudates, along with any amendments, remains on or in the soil and becomes part of the SOM (Banwart et al. 2019). Through the microbial process of decay, the C in residues, amendments, and SOM is oxidized back to CO_2 and returned to the atmosphere, thereby completing the cycle. However, within this cycle, rates of oxidation and decomposition as well as rates of accumulation of SOM depend on climatic conditions and how the soil is managed and the resulting biophysical conditions (Chenu et al. 2019; Banwart et al. 2019).

2.3.2 Carbon is the "C" that Starts "Conservation"

Soil is alive and is as vital to human survival as air, water, and the sun; its protection and enrichment with energy and organic carbon (OC) are needed to sustain soil life for the future sustainability of our planet. The C content of soil is vital not only because of its sheer mass as a reservoir, but also because of its function (Murphy 2014; Lal 2013a; Lal 2015d; Milne et al. 2015; Lehman et al. 2015; Oldfield et al. 2015; Powlson et al. 2015). This importance arises not from C itself, but from the broader assemblage—the SOM—in which C is embedded that provides numerous ecosystem services (Loveland and Webb 2003; Brussaard 2012; Milne et al. 2015; Schjønning et al. 2018; Chenu et al. 2019).

An important attribute of C is its ability to "connect" or chemically bond with many other elements. This ability of C to bond to itself and with other elements enables many combinations and molecules to form numerous organic compounds that serve our food energy needs. This attribute builds on biodiversity from cover crop mixes and diverse crop rotations and associations through a combination of "nutrient density" and "nutrient diversity" to provide a wide range of diverse, quality foods for us and our "biological partners" on earth to maintain our quality of life and food security (Green et al. 2020). While this ability to "connect" is critical for the major nutrient elements of N, P, and K, it is even more critical for the >20 micronutrients to support other major biological processes (Powlson et al. 2015; Barker and Pilbeam 2015; Crouse 2018).

C cycling in agricultural systems, specifically CO_2 transformations, is more active than other greenhouse gases (GHGs) due to its consumption during photosynthesis and use in plant production. Despite the fact that plant biomass residue will be decomposed, and a portion of residue C will be converted to CO_2, the net emission of CO_2 from agriculture is minor compared to the amount of CO_2 cycled within the system (Snyder et al. 2009). The relationship between C cycling and CO_2 emissions goes deeper than the decomposition of plant biomass residue and its subsequent role in CO_2 emissions. Plant primary production has been correlated with soil respiration across all the major biomes on Earth (Raich and Schlesinger 1992). In cotton production, soil respiration increases over the growing season following a pattern similar to primary productivity of the cotton plant (Li et al. 2012). One factor related to plant primary productivity is the release of exudates from plant roots into the rhizosphere. The release of carbonaceous root exudates has been reported to stimulate soil respiration rates (Landi et al. 2006) and increase microbial density around plant roots in the field (Baudoin et al. 2003). When considering agricultural CO_2 emissions, Koizumi et al. (1991) determined that the major source of CO_2 from soil was microbial and root respiration with 45% of total soil respiration reported to be due to rhizosphere respiration in corn (*Zea mays* L.) (Rochette et al. 1991), suggesting root exudation potentially plays a major role in soil respiration by adding an additional C source and increasing microbial density.

A fundamental facet of soil conservation is maintaining adequate reserves of soil C. Observations over thousands of years, as well as recent scientific studies, have shown that the productivity and functioning of a soil is directly related to its content of plant-derived organic matter (OM) (Allison 1973; Blanco-Canqui and Lal 2009a, b). The plant depends on the soil and the soil depends on the plant. C provides a critical linkage in this symbiotic relationship (Nielsen et al. 2011; Tardy et al. 2015; Jackson et al. 2017; Banwart et al. 2019; Chenu et al. 2019). Moreover, soil C has also been promoted recently as a repository of excess CO_2, by various "sequestration" strategies (Ingram and Fernandes 2001; Lal 2008). For these reasons, conserving soil is closely linked to conserving C required for numerous soil and ecosystem functions (Blanco-Canqui and Lal 2009a; Banwart et al. 2019). C is the "C" that starts "Conservation."

2.3.3 Soil Organic Matter and Nutrient Cycling

SOM provides many ecological benefits that include soil aggregates and structure, porosity, dark color, friability, "biologic glues," energy for the microbes, habitat for soil biology, along with nutrient cycling and storage. All of these individual benefits to the plants make it difficult to quantify specific mineral nutrient effects on subsequent plant growth and productivity. As much of SOM is derived mainly from plant biomass residues, it contains all of the essential plant nutrients and serves as a slow-release form and storehouse of plant nutrients to be recycled to the next plant generation (Schjønning et al. 2018). We now know that microbial byproducts, and dead cells comprise large amounts of stable SOM because they can become attached to mineral surfaces or physically protected within soil aggregates (Banwart et al. 2019; Veloso et al., 2020). Seiter and Horwath (2004), reviewing strategies for managing SOM to supply plant nutrients, present the complexities and challenges of managing SOM with emphasis on supplying plant nutrients. They conclude that increased SOM can lead to environmentally sound agricultural systems for sustained nutrient availability. SOM can provide a vast reservoir of nutrients for plants (Bruce et al. 1990; Power 1994; Brady and Weil 1999). The mineralization of SOM is the primary source of available N, P, and S in many natural ecosystems; however, quantitative details are challenging for determining cause and effect. The SOM also contributes to the soil cation exchange capacity (CEC), to the water holding capacity (Hudson 1994), and to structure development (Bronick and Lal 2005; Banwart et al. 2019), that all lead to increased water use efficiency (WUE) (Basche et al. 2016), which makes it difficult to identify a specific nutrient and amount contributing to increased productivity.

OM, and hence C, is closely associated with many critical functions of soil in ecosystem services (Reicosky 2020). Firstly, the SOM contains a vast storehouse of nutrients (Ellert and Bettany 1995; Powlson et al. 2015). For example, a soil with 175 Mg SOM/ha (100 Mg C/ha), might contain about 9,100 kg N/ha, 800 kg organic P/ha and 1,300 kg S/ha in organic form (Powlson et al. 2015). As the SOM decays, the nutrients are slowly released in mineral forms available for plant uptake. Because these nutrients are derived from plant materials, they occur in SOM roughly proportional to plant demands. And because rates of decay are favored by moisture and temperature conditions similar to those for plant growth, the slow rate of nutrient release is often synchronized with temporal patterns of plant demands. Aside from its being a repository of nutrients, SOM can also help retain nutrient ions by enhancement of soil CEC. Consequently, the SOM content of soils is often considered the preeminent indicator of SH, and the impetus for its enhancement has recently been further intensified (Lehman et al. 2015; Milne et al. 2015; Lal 2015c). Innovative and effective management of plant biomass for nutrients, water, and pests is therefore an essential aspect of CA. Further, practices like, appropriate crop rotations and/or intercropping emphasizing biodiversity and using perennial crops or diverse cover crop mixes, may also be critical elements of CA to extend the photosynthesizing period and capture more solar energy for nutrient cycling.

The C content of soil is vital not only because of its sheer mass as a reservoir, but also because of its function (Murphy 2014; Lal 2015c; Milne et al. 2015; Lehman et al. 2015; Oldfield et al. 2015). This importance arises not from C itself but from the broader assemblage of the SOM in which it is embedded. The C content of SOM, which also includes hydrogen, oxygen, and numerous nutrients, is remarkably constant (typically about 0.58 g C g^{-1}SOM), so C can be used as a measure and indicator of SOM (Allison 1973). Although sometimes referred to as one pool, the SOM contains a wide range of forms and is usually defined as all materials, living and dead, of biotic origin (Semenov et al. 2013; Franzluebbers 2010; Stockmann et al. 2013). Thus, it includes living microbes, senescing vegetative materials, including roots and exudates, as well as carbonaceous compounds derived from plant biomass applied centuries or millennia ago. To reflect this diversity, SOM is often subdivided into multiple fractions (Wander 2004), using a wide range of chemical and physical techniques, including acid hydrolysis, flotation, sieving, bioassay, and numerous combinations thereof. The intent is to subdivide the SOM into various categories or pools, based on susceptibility to decay, ranging from highly decomposable to the very stable or recalcitrant. This stability in SOM was once believed to be related to the chemical composition of the SOM molecules themselves, but growing evidence suggests that most OM are inherently decomposable, and recalcitrance is conferred by association with mineral nutrient components of the soil through adsorptive or protective mechanisms (Dungait et al. 2012). For example, association of SOM with adsorption sites on clay minerals may slow its decay, suggesting that soils have a finite "capacity" for affording protection to added C. When that capacity is exceeded, the soils are said to be "saturated" with C (Castellano et al. 2015; Lützow et al. 2006; Six et al. 2002; Stewart et al. 2007). Regardless of form and mechanisms involved, all SOM is subject to decay, although at very different rates, with turnover times ranging from days (for soluble exudates), to millennia (for highly stabilized C forms) (Janzen 2005). Although dissecting SOM into various fractions has provided invaluable insights, OM is best seen as a continuum of forms, decomposing at various rates recycled back to its simple mineral nutrient constituents that fertilize the next generation of plants (Lehmann and Kleber 2015).

Dhaliwal et al. (2019), concluded that the role of OM in availability and transformation of micronutrients (Zn, Cu, Fe, Mn, B, and Mo) in the soil is vital. Crop residues are reported to be important sources of many micronutrients. The chemical composition of plants can be used to assess nutritional status and the value of soil tests for assessing nutrition status. Build-up of OM in soil influences its physico-chemical properties which in turn governs the soil micronutrient dynamics affecting the distribution of micronutrients between soil colloids and solution by controlling sorption, specific surface area, CEC, and soluble complexes. The SOM is critical in maintaining soil quality/health because of its role in cycling micronutrients at low concentrations that act as a cofactor for various enzymes associated with the metabolism of organic molecules, such as carbohydrates, nucleic acids, proteins, and lipids (Barker and Pilbeam 2015). Dhaliwal et al. (2019) concluded that the inclusion of SOM management and organic amendments in integrated nutrient management (INM) should be highly encouraged to mitigate the worldwide phenomenon of micronutrient deficiencies in food grains. Major efforts are warranted to increase our understanding of the agroecological complexities with micronutrients and returning crop biomass directly or indirectly through other amendments, thus, harnessing the full benefits of micronutrients to crops to enhance the human diet.

For agri-food systems, optimizing nutrient density will require the adoption of emerging technologies (e.g., those related to soil health principles, CA, sustainable intensification, etc.) using both traditional crops and diverse cover crop mixes within this SOM management and nutrient cycling. Assessing nutritional, health, and environmental sustainability dimensions of agri-food production requires a better understanding of the nutrient recycle through plant management and the decomposition of plant biomass. Nutrient diversity metrics measure the diversity of nutrients within a food supply or production system. Currently, diversity metrics are infrequently used in food quality and sustainability assessments. Global nutrient diversity is changing, and diversity metrics can support targeted production-side interventions that account for nutritional impacts in addition to yield, mass, or economic impacts. While diversity is an important aspect of sustainable agri-food production, many metrics calculate diversity based on differences in food quantity instead of nutrients (Bogard et al. 2018). This could lead to unknown micronutrient deficiency risks. Although outside the scope of this chapter, food processing is also known to modify nutritional content of food, sometimes negatively, for example through grain polishing processing or producing refined flour from cereals and other grains from which much of its natural protein, oil, and fiber has been removed. Thus, modern diets that rely on highly processed foods have been shown to increase the risks of non-communicable diseases (i.e., diabetes, arteriosclerosis, inflammation), and certain types of cancer, through their impact on gut biomes and body metabolism. The current poor relationship between sustainable agriculture and the human diet must change to ensure that healthy food from sustainable agriculture based on CA systems is supporting healthy diets and nutrition that are based dominantly on whole foods and plant-based diets that are healthy for both humans and the planet (Millen et al. 2016; BDA 2018; Health Canada 2019; Willett et al. 2019; Kassam et al. 2020). Currently, the mismatch between land use and sustainable diets is significant given that some 80% of the agricultural land is used for livestock production which is not only a major cause of extensive land and environmental degradation (McMichael 2020; Mason and Kassam 2020) but is also known to be a significant contributor to human illnesses and diseases (Kassam et al. 2020) and must be addressed (Nobari 2020; Norberg-Hodge 2020).

Organic C, by nature, is dynamic; a C atom captured by photosynthesis and entering the soil as plant biomass is likely to reappear in the atmosphere again as CO_2 within a few months or years or even possibly centuries and may travel long distances. Only about one-third of the fresh crop biomass residue C remains in the soil after one year (Angers and Chenu 1997; Jenkinson and Rayner 1977; Voroney et al. 1989); and only 5–20% remains after 2 years in the soil (Buyanovsky and Wagner 1977; Broder and Wagner 1988). Reicosky and Janzen (2018) suggest that only a small fraction of the initial C input is "stabilized," however, this C is not permanently locked away but remains vulnerable to slow, gradual release over periods of decades to centuries, leaving a very small portion of C in the form of recalcitrant humus, a loose term for a group of C compounds that are extremely resistant to decomposition. As a result, SOM is susceptible to continuous decomposition and transformation through microbial activity and cannot be seen as a permanent "storehouse" for C and may be better visualized as C "flowing" through the soil-plant-atmosphere system ending up back in the atmosphere as CO_2 (Janzen 2006, 2015; Reicosky and Janzen 2018).

The rapidly growing scientific literature on various aspects of C storage in soils has given rise to the introduction of several terms when discussing the amounts of C that are, or could be, stored in soils. The term "carbon sequestration potential," in particular, is used with different meanings, sometimes referring to what might be possible given a certain set of management conditions with little regard to soil factors, which fundamentally determine C storage (Ingram and Fernandes 2001). C sequestration is in many ways a nebulous term, particularly when applied to the soil. The term has been widely used but seldom specifically defined (Olson et al. 2014b) to avoid ambiguity and potential confusion. Reicosky and Janzen (2018) briefly addressed C storage and C sequestration differences suggesting the wide differential in C dynamics may require that we consider C energy flow through the soil–plant-atmosphere system (Janzen 2006, 2015) making it sometimes difficult to clearly interpret the results in statements in the literature. Chenu et al. (2019) also stressed that soil organic carbon (SOC) storage should be clearly differentiated from SOC sequestration, as sequestration assumes net removal of atmospheric CO_2. They further suggested that sequestration be quantified for a given duration, chosen as 20 years by the IPCC (2006). Chenu et al. (2019), implicate the role of soil microorganisms in stabilization of SOM, microbial physiology, and soil biodiversity induced by agricultural practices. Stewart (2018) suggests that C is not sequestered in soil but that carbonaceous materials present in the soil are biologically decomposed organic compounds (SOM) that with time become increasingly resistant to further decomposition. Those organic compounds contain a wide range of macro and micronutrients released during the decomposition process to be recycled and used by the next generation of plants.

We need a more concerted effort to improve the production-side of agri-food systems. Much work remains to be done to understand locally specific controls and causative interactions of soil management, soil type, and climate on the SOM–yield relationship (Lal 2020b; Oldfield et al. 2019; Bai et al. 2019). The main C management strategy should, therefore, be directed toward minimizing losses (i.e., promoting SOC generation and protective measures) because this is fundamental to much of the science of soil conservation in the three core principles of CA. The loss of soil fertility around the world suggests a C failure that will be discussed in the context of CA, which is C-centered and C-driven. The role of SOM and its decomposition-releasing C and nutrients for the next generation of plants is critically important in maintaining sustainable production systems.

SOM anchors global terrestrial productivity and food, feed, and fiber supply. SOM retains water and soil nutrients and stores more global C than do plants and the atmosphere combined. SOM is also decomposed by microbes, returning CO_2, a GHG, to the atmosphere. Unfortunately, soil C stocks have been widely lost or degraded through land use changes and unsustainable forest and agricultural management practices. Therefore, restoration of SOM content in soils of agroecosystems is needed to reverse the degradation trends and enhance ecosystem functions and societal services (Banwart et al. 2015). To understand SOM structure and function and to maintain and restore SOM, we need a better appreciation of the saturation capacity and the retention of above- and belowground inputs of OM. Banwart et al. (2019) suggest root inputs are approximately five times more likely than an equivalent mass of aboveground biomass litter to be stabilized as SOM. Microbes, particularly fungi and bacteria, and soil faunal food webs strongly influence SOM decomposition at shallower depths, whereas mineral associations drive stabilization at depths >30 cm.

Soil biodiversity and SOM are critically important in providing ecosystem services (Laban et al. 2018). Both are required and vital to the way ecosystems function, and they largely determine the role of land in producing food, storing water, and mitigating climate change. Soil biodiversity reflects the mix of living organisms in the soil, including bacteria, fungi, protozoa, insects, worms, and other invertebrates and vertebrates that require SOM as their primary energy, nutrients, and habitat sources. Implementation of the CA principles in a system centered on C management enables farming in nature's way (Reicosky and Janzen 2018). Biodiversity and SOM are the "ecological" keys to unlocking the multiple economic and environmental benefits required for sustainable management, conservation, and restoration of ecosystems (Laban et al. 2018).

In CA systems SOM increases over time and so does soil's ability to retain nutrients and water. In fact, as SOM builds up, crop response to applied N changes such that less N is needed for the same yield. Carvalho et al. (2012) showed that with a soil under conventional intensive tillage agriculture (TA) containing 1% SOM, it took 160 kg N ha^{-1} to produce 3 Mg ha^{-1} of wheat grain. However, when the same soil built up to 2% SOM through continuous CA, 3 Mg ha^{-1} yield was obtained with only 37 kg N ha^{-1}. A crop-modeling exercise showed that at 3% SOM content, a yield of 3.5 Mg ha^{-1} would be achieved without any application of N that presumably was supplied by decomposing of the built-up SOM.

There is a general perception that SOM is essential for sustained food production (Lal 2004, 2009, 2010). Well-documented effects have been reported of SOM on soil structure (e.g., Bronick and Lal 2005), plant nutrition and soil microbial activity (Diacono and Montemurro 2010), and a multitude of other soil processes and functions. However, difficulties arise when searching for threshold SOM levels securing crop production across soil types and even within specific soil types. Target or threshold SOM levels for sustaining soil functions are valuable for farmers as well as other stakeholders involved in securing soil quality.

Wilhelm et al. (2004) had difficulties identifying the required return of plant biomass to soil in order to secure SOM contents sustaining corn production across different regions of the US. Van-Camp et al. (2004) performed a comprehensive review on how best to express target SOM values across European soils. They had to reject the option of identifying universal thresholds across soils and geographical regions. Instead, they suggested a procedure defining target values for a number of well-defined regional soil units identified through climate type, soil type (texture), and soil drainage. They acknowledged that the suggestion would imply hundreds of target SOM levels across Europe. That would certainly not be operational, and calls went out for a new conceptual understanding of the role of SOM in relation to crop production.

Recent studies indicate that the ratio of soil clay to SOC is a better index than SOC per se for defining critical thresholds in clay dispersibility and also other soil structural fingerprints, like the bulk density (Czyz et al. 2017; Dexter et al. 2008; Getahun et al. 2016; Schjønning et al. 2012). Thus, these studies seem to point out threshold ratios above which soil structural conditions are compromised. Further studies have indicated that the ratio between mineral fines <20μm (i.e., the clay + fine silt fraction) and SOC at high levels may be a similarly promising proxy of soil conditions jeopardizing structural stability (Jensen et al. 2017a; Schjønning et al. 2012). This is supported by other studies including the clay and 2–50 μm silt fractions in expressions with SOM to predict soil structural conditions (Pieri 1995, 2012; Pulido-Moncada et al. 2014, 2015; Reynolds et al. 2009). The benefit of including the silt fraction in the index above probably relates to the fact that silt particles have also been shown to interact significantly with SOC (Christensen 1992; Hassink 1997).

Soils with high ratios of mineral fines to SOC may therefore create problematic situations during tillage operations, where dispersed clay increases the probability of poor seedbeds and nonoptimal germination of seedlings. This, in turn, may negatively affect crop yields.

Lal's (2020b) synthesis of the available literature shows that an increase in SOM content of depleted/degraded soils can have a positive effect on crop yield. The specific cause and effect of the increased SOM is challenging because of the multiple benefits attributed to SOM. This makes the contribution of specific nutrients effect on yield difficult to separate from an increase in nitrogen availability (Carvalho et al. 2012), increase in selected micronutrients (Dhaliwal et al. 2019), increase in plant available water (Hudson 1994), increase in soil structure and subsequent rooting depths (Banwart et al. 2019), Lal (2020b) suggests the nominal threshold level of SOC required for optimum yield may be ~2% in temperate soil and ~1% in tropical soils and that additional research is needed to establish critical limits/ranges of SOM content for the diverse global soils, climates, and eco-regions (Oldfield et al. 2019).

2.4 INTEGRATING CONSERVATION AND AGRICULTURE

Conservation is a word to be understood holistically and respected, revered, and used to describe sustainable agriculture. Conservation and agriculture must blend/come together when describing our sustainable agricultural production systems. All aspects of conservation practices are integrated into agriculture when practicing CA (Kassam 2018). The Latin root of conserve means "to keep together." CA abides by these principles to keep the soil together as a living ecosystem that enables successful agriculture and food production and helps address climate change. As a noun, it also means "prevention of wasteful use of a resource." It has been defined as "planned management of a natural resource to prevent exploitation, destruction, or neglect." In biology, it actively focuses on protecting, restoring, regenerating, and enhancing and also responding to crises. Because CA encompasses ecological and biological aspects of a "living soil," soil conservation is appropriate because it preserves, protects, and regenerates the soil biological processes and their essential functions in agricultural ecosystems through prudent management (Harvey et al. 2017). Ecological conservation was originally focused on keeping things the way they were.

Agroecology is an approach to production that aspires to promote scientific principles and practices of ecology rather than rules or recipes to develop an agroecological production system out of a conventional farm in a stepwise transition process (Gliessman 2014). Agroecological management is claimed to lead to optimal recycling of nutrients and SOM turnover, closed energy flows, water and soil conservation, and balanced pest–natural enemy populations—all key processes in maintaining the agroecosystem's productivity and its self-sustaining capacity (Altieri 2002; Altieri and Nicholls 2012). The challenge to aligning modern agricultural systems with ecological principles is immense, especially in the current context of agricultural development where specialization, short-term productivity, and economic efficiency of non-biological processes are the main driving forces (Phelan 2009).

There is a deficiency in the clarity between current conservation concepts and knowledge of ecology and the prescription of conservation measures and development. Protecting ecosystem integrity and species diversity is at the core of all conservation or restoration management actions, especially in agricultural systems. Harvey et al. (2017) reviewed the linkage of conservation science with modern ecology. They suggest protection of biotic interactive networks within regional species pools described by Morales-Castilla et al. (2015) to lead to more appropriate conservation targets to maintain ecosystem integrity. This protection requires a direct link between conservation and ecology that serves as a science foundation for CA. Ecosystem processes and ultimately landscape-scale delivery of ecosystem societal services in agricultural systems are necessary to achieve the pressing conservation management and restoration of ecology goals of conserving biodiversity and ecosystem processes (Harvey et al. 2017).

Conservation clearly projects a sense of protection and preservation while encouraging efficiency of use of our natural resources that provide resilience and renewal or regeneration for sustainability, restoration, and enhancement. Optimizing nature conservation and ecology requires understanding the "biology" behind soil and water conservation systems. Agroecology as a science provides the scientific basis to address the production by a biodiverse agroecosystem able to sponsor its own functioning (Altieri 2002). The core principles of agroecology include recycling nutrients and C energy on the farm, rather than introducing external inputs; enhancing SOM and soil biological activity; diversifying plant species and genetic resources in agroecosystems over time and space; integrating crops and livestock, and optimizing interactions and productivity of the total farming system rather than the yields of individual species (Altieri 1995; Gliessman 2014). By breaking the monoculture nature of farming systems, agroecological diversification aims at mimicking ecological processes leading to optimal nutrient cycling and OM turnover, soil biological activation, closed energy flows, water and soil conservation and balanced pest–natural enemy populations. However, as a practice, agroecology is based on intensive tillage and soil management which acts counter to the objectives of ecological sustainability and biological optimization because tillage is used to mineralize OM to release nutrients for crop growth, a process which not only impoverishes SOM but also soil quality and function.

Conservation as a modifier of agriculture in CA systems appeals to a wide range of farmers because it preserves and protects the natural resources in a regenerative manner which is more self-protecting, with greater yields and yield stability as well as higher factor productivity—thus better and more reliable income (Kassam 2018). Conservation incorporates a biologically dynamic ecological foundation of production systems such that the natural resource base and its agricultural potential and ecosystem functions are conserved, enhanced, and maintained at the optimum level. This means that all three interlinked CA principles work together to provide an ecological and biological foundation (Harvey et al. 2017; Kassam 2018); and along with complementary good agricultural practices dealing with integrated soil, crop, nutrient, water, pest and energy management, to offer an optimal performance of maximum efficiency, adaptability and resilience (and profitability and stability), and maximum economic output with minimum input (FAO 2011, 2016; Kassam et al. 2013; Kassam 2018). We need to connect conservation practices with conservation principles to get a better and deeper understanding and acceptance of the terminology and description of Conservation Agriculture (Kassam 2018).

The notion of conservation of living resources—using them in such ways that vital stocks of plants and animals are maintained, and their benefits enjoyed by succeeding generations—is not new, but many conservation battles remain to be fought and won. Conservation progress has been lamentably slow, largely because it has been seen as peripheral to society's continuing quest for social and economic welfare based on a neoliberal economic system that seeks infinite growth and profit in a finite natural world with planetary boundaries but is inherently divisive and unable to meet the basic needs of all human beings (Kassam and Kassam 2020a).

2.4.1 Conservation in a Broader Context

Conservation is a unique term, sometimes ambiguous to some, that can be interpreted differently and may reflect personal biases that distort the intended meaning and scope. Conservation is about conserving land and life and is the act of protecting Earth's natural resources for current and future generations (Dumanski and Peiretti 2013). Merriam-Webster (2020) defines conservation as the protection and preservation of something, with minimal external inputs to be used efficiently for future reuse. Conservation is about protecting nature and our natural resources along with their attendant biological processes. Conservation is the act of protecting Earth's natural resources for current and future generations. Since agriculture depends on our natural resources (sun, soil, water, air, biodiversity, human intelligence), this discussion will emphasize conservation of our natural resources required for food and livelihood security. The combined concepts of "used efficiently" and permanence of our resources from the definition of conservation imply a resource to be repeatedly used by future generations. So, looking at the broad context of resource conservation, it has built in nature's "recycling, renewing, refreshing, and regenerating" these vital resources and just about everything biological. So how can we protect and preserve the resources while we "use" the resource? It seems reasonable that we can use a resource and then return it to its previous state through regeneration and enhanced conservation management.

Conservation generally refers to the act of consciously and efficiently using land and/or its natural resources. Conservation planning implies "long term" which points to sustainability, robust planning and strategies, science-based practices, observation, and data-driven, highly integrated with diverse conservation practices, diversity in conservation practices contributes to enhanced biodiversity, C and nutrient cycling, recycling and regeneration and a sense of stewardship (Dumanski and Peiretti 2013). Conservation is the care and protection of these resources so that they can persist for future generations. It includes maintaining diversity of species, genes, and ecosystems, as well as functions of the environment, such as water and nutrient cycling and storage.

Conservation embraces all aspects of the science of agroecology that provide a foundation for complex biological interactions (Harvey et al. 2017). Agroecology emphasizes the interrelatedness of all agroecosystem components and the complex dynamics of ecological processes that promote principles rather than rules or recipes to develop an agroecological production system out of a conventional farm in a stepwise transition process. Farmers are increasingly challenged to make use of their intellectual and communication skills throughout this period of transition because they have to optimize input-use efficiency, substitute synthetic with biological or organic inputs, and redesign the production system and its management. The emphasis is on the design of complex agroecosystems in which synergisms between numerous biological components replace or reduce production inputs by enhancing processes with proper management allows farmers to "naturally" provide the soil fertility, productivity, and crop protection of their farming systems (Altieri 1995).

The protection, preservation, management, or restoration of natural environments and the ecological communities that inhabit them must be considered in ecosystem management. Conservation is generally held to include the management of human use of natural resources for current public benefit and sustainable social and economic utilization. Conservation seeks the sustainable use of nature by humans for activities such as spiritual, cultural, transportation, logging, or mining, while preservation means protecting nature from human use.

Farmers' definition and views or interpretation of "conservation" may differ substantially from place to place and from that of the local experts most likely to be asked to contribute adoption estimates to global monitoring efforts, creating the potential for miscommunication (Findlater et al. 2019). There is therefore substantial interest in how CA is currently promoted, monitored, and evaluated. Inaccurate adoption estimates jeopardize CA's potential as a climate change adaptation strategy, creating illusory progress that may disincentivize further substantive efforts toward agricultural sustainability and climate resilience.

"Conservation is a state of harmony between men and land."

—Aldo Leopold

The need for conservation science stems from estimates suggesting species loss (Ceballos et al. 2015) and loss in biodiversity and habitat that will lead to the breakdown of ecosystem functioning, stability, and services (Isbell et al. 2019) thereby contributing to poverty and starvation, and will reset the course of evolution on this planet (Tilman et al. 2017; Millennium Ecosystem Assessment 2005). Palm et al. (2014) reviewed CA and ecosystem services in smallholder farming systems and concluded that other uses of crop biomass and biomass limitations would require careful consideration of CA systems based on different agroecological and social–economic conditions. Recognition of site-specific challenges requires conservation as a holistic concept that justifies a broad umbrella coverage of many conservation practices impacting crop biomass management. Conservation has an intrinsic appeal for resilience and regeneration. A conservation-based agriculture production approach serves as an umbrella for all individual conservation practices to improve soil health by working together synergistically for the common good as illustrated in Figure 2.1. The broad

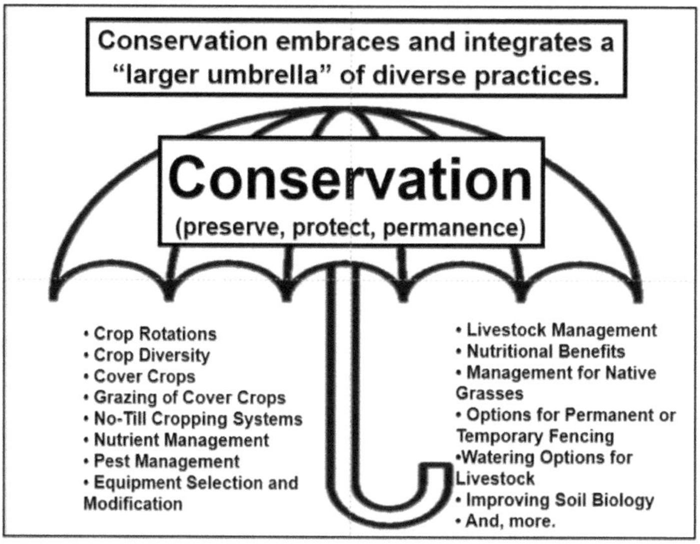

FIGURE 2.1 An umbrella of diverse conservation practices that embrace agroecology and contribute to a science foundation, diversity, and resiliency in integrated CAS.

umbrella of "conservation practices" enables the detailed description and substitution for many of the tenets of CA, based on site-specific characteristics or farmer preference, while maintaining the integration and simplicity and the combined biological power of the three interlinked core CA principles as the foundation of CA systems. Conservation enables agricultural biodiversity that is conserved and enhanced in CA systems along with wild and natural biodiversity.

To meet the goal of maintaining biodiversity, conservation practice requires on-the-ground actions. Decisions of what conservation management actions to implement were originally based on the experiential knowledge supplemented with innovative "gut" feel of conservation practitioners. Scientific evidence has begun to take precedence, however. To this effect, practitioners have developed a suite of optimal decision-making tools that minimize subjective human influences, such as project prioritization protocols (Joseph et al. 2009). The over-arching aim of these frameworks is to ensure that best practice is evidence-based, where evidence pertains to the effectiveness with which a conservation action is in line with the stated goals of conservation biology and agroecology.

2.5 CONSERVATION AGRICULTURE SYSTEMS

The historical evolution of CA systems was initiated with the recognition of severe soil erosion and degradation associated with conventional intensive tillage agriculture (Delgado et al. 2011; Kassam et al. 2020a). Documented evidence (Kassam et al. 2020a) shows that the term "conservation agriculture" was first coined in Spanish at the IV RELACO meeting in Morelia, Michoacán, Mexico, in 1997 (RELACO 1997). In English, RELACO stands for Latin-American Network for Conservation Tillage. Only one paper, i.e., Friedrich (1997) used the term "Conservation Agriculture" in the title. Conservation tillage reflected the old paradigm and mainstream thinking until the 1990s. During the RELACO meeting, Rolf Derpsch and Theodor Friedirich initiated a discussion to change the extended name of the organization from "Red Latinoamericana de Labranza Conservacionista" (Latin-American Network for Conservation Tillage) to "Red Latinoamericana de Agricultura Conservacionista" (Latin-American Network for Conservation Agriculture) arguing that one cannot conserve the soil while at the same time tilling it. Subsequently, Reicosky (2015) expressed concerns with the phrase "conservation tillage" because there is no "soil conservation" with any type of tillage that has led to modern day CA.

From there on, the two FAO representatives attending that meeting, Theodor Friedrich and José Benites, decided that CA was the term FAO should use for sustainable agriculture systems in the future. An interdisciplinary informal workgroup on CA was created in FAO with members from all divisions of the agricultural department. In 1998, FAO organized its first regional workshop on CA in Harare, where a code of practice for CA was drafted describing the three interlinked CA principles as we know them today and their practical application (Kassam et al. 2020a). In 2001, FAO, in partnership with the European Conservation Agriculture Federation (ECAF), a newly established regional network of national farmer associations and involved stakeholders, held the first World Congress on Conservation Agriculture in 2001 in Madrid, Spain WCCA 2001).

In July 2008, the *International Technical Workshop on Investing in Sustainable Crop Intensification: The Case for Improving Soil Health* was held at FAO in the Plant Production and Protection Division (FAO 2008) led by Amir Kassam, Theodor Friedrich, Francis Shaxson, Eric Kueneman, and Andrew McMillan. Most of the CA champions around the world, some 100 individuals, including those in FAO, attended to discuss and review the basis for successful adoption and spread of CA in different countries and regions, and how CA could be globalized.

The workshop prepared a global action plan to mainstream CA and recommended that a multi-stakeholder Conservation Agriculture Community of Practice (CA-CoP) communication platform, hosted by FAO, should be launched to facilitate the adoption and mainstreaming of CA globally (FAO 2008). Justification for CA as the best choice for sustainable production intensification was published based on the outcome of the international workshop (Kassam et al. 2009). The CA-CoP platform, hosted by FAO, was launched in January 2009 with Amir Kassam serving as the moderator. This platform has been running continuously since then and now reaches several thousands of subscribers and stakeholders around the world. Through the CA-CoP platform, all kinds of information about CA, including journal articles, official documents, and news, is distributed practically every day. When CA-CoP was established, the global CA cropland area was 106 M ha. In 2015/16 it was 180 M ha, an increase of some 70% (Kassam et al. 2018). More details of the global development of CA are given in Kassam et al (2018; 2020a).

Since then, CA has been promoted by institutions and organizations with the expectation that it can contribute to sustainable intensification (Hobbs 2007; Hobbs et al. 2008; FAO 2008; 2011) and has spread into all the regions of the world to become a global agricultural movement (Friedrich et al. 2012; Kassam et al. 2014a, b, 2018, 2020a). CA has been highlighted as a key path toward food security and a range of other economic, environmental, and social benefits (FAO 2008, 2011; Hobbs et al. 2008; Kassam et al. 2009, 2014a, 2014b; Kassam and Friedrich 2012; Corsi et al. 2012; Pretty and Bharucha 2014; Farooq and Siddique 2015; Jat et al. 2014; Lal 2015d; Reicosky and Janzen 2018; Kassam, 2019; 2020; Kassam and Kassam 2020b,c; Mitchell et al. 2019; Reicosky 2020).

Extensive literature and literally hundreds of long-term field experiments around the globe document the capability of conservation practices, e.g., no-till (Friedrich 2020), cover crops, tillage reduction (Ogle et al. 2005; Franzluebbers 2010; Kravchenko and Robertson 2011), mulch cover and cover crops, (Calegari et al. 2020; Abdalla et al. 2019; Poeplau and Don 2015), annual and perennial cropping systems (Hobbs et al. 2020; Conant et al. 2016; Ogle et al. 2005; Guo and Gifford 2002), and many others, to increase soil C contents. Hence the field experimental evidence that regenerative CA systems and practices can significantly increase soil C stocks is unequivocal. Of course, results vary for different combinations of climate and soil types and management systems (Corsi et al. 2012), but, in general, we understand the variability in responses from region to region and we can design regionally appropriate climate-smart regenerative agroecosystems based on CA.

CA is a broad inclusive term to describe systems that are based on the application of the three interlinked principles, namely: (i) continuous no or minimum mechanical soil disturbance (no-till seeding and weeding), (ii) permanent soil vegetative mulch cover (crop biomass including stubble and cover crops), and (iii) diverse cropping (agronomic and cover) based on rotations, sequences,

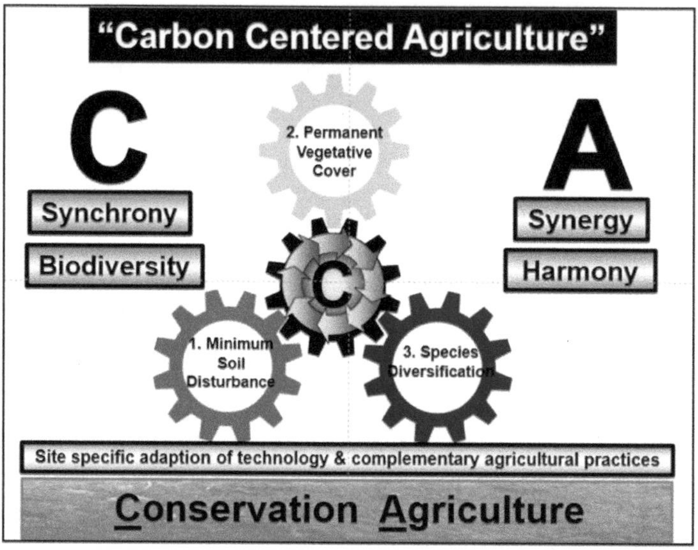

FIGURE 2.2 Conceptual diagram of the integration and synchrony of the three primary principles of CA in a C-centric system.

and associations that incorporate soil health concepts and merging good practices of modern farming and natural resource conservation profitably. While each of the principles may be considered and applied as a separate entity, it is the fluid integration of all three principles and their C-centric supportive soil and crop health practices in space and time across landscapes that are keys to an effective CA land use system. The natural interactions and diversity within the cropping system contribute to numerous economic, environmental and social benefits (Lal 2015b; Kassam et al. 2013, 2020b; Kassam and Kassam 2020b). The core principles of CA are universal, but the solutions are local and revolve around C cycling using a system's approach illustrated in Figure 2.2. CA represents a new way of thinking and taking care of agricultural lands, emulating nature, and integrating conservation principles and practices into an agricultural system that is regenerative, conserving, and much more resilient than tillage agriculture can ever be. CA is based on an ecosystem approach to sustainable agriculture production and land management and, by definition, integrates conservation into agriculture which results in greater output, factor productivity and profit while maintaining environmental quality and sustainability (Kassam 2019; Kassam and Kassam 2020b).

CA evolved in part as a response to the severe soil erosion that devastated the US Midwest in the 1930s. Currently, CA systems are practiced successfully across a range of agroecological conditions, soil types, and farm sizes (Derpsch et al. 2010), and are responding to a range of inter-related global crises, such as soil erosion and land degradation, food insecurity, rising costs of inputs and energy, climate change, the lack of ecosystem services, pervasive poverty, the need for a pro-poor sustainable agriculture alternative, and demands for food sovereignty and localization. CA has spread widely. In 2015/16, there were some 180 million hectares of global annual cropland (12.5% of global cropland) systems grown without tillage in CA systems, corresponding to an annual rate of increase globally of 10.5 million hectares since 2008/09 (Kassam et al. 2018). This represents an increase of 70% in CA cropland area since 2008/09. Current global area of CA croplands appears to be around 200 million hectares. Adoption varies greatly by region. CA covers some 45.5% of annual cropland in Australia and New Zealand, 63.2% of annual cropland across South America and 22.7% in North America. In contrast, adoption across Europe, Asia, and Africa is lower (covering only 5%, 4.1%, and 1.1%, respectively of annual cropland in each of the three continents in 2015/16) (Kassam et al.

2018); although it seems that CA cropland area has increased recently, and, in Africa, there seem to be several million hectares of CA cropping with trees in the semi-arid areas (Kassam et al., 2020c).

Wherever CA has been adopted it appears to have had both agricultural and environmental benefits. Yet CA represents a fundamental change in production system thinking. It has counterintuitive and often unrecognized elements that promote soil health, productive capacity, and ecosystem services. The practice of CA thus requires a deeper understanding of its ecological underpinnings in order to manage its various elements for sustainable intensification, where the aim is to optimize resource use and protect or enhance ecosystem processes in space and time over the long term. For these reasons CA is science-driven and knowledge- and management-intensive. CA constitutes principles and practices that can make a major contribution to sustainable production intensification and help to restore degraded agricultural land.

The farming practices required to implement these objectives will differ according to local conditions and needs but must integrate the CA core principles with conservation practices to produce food and other biological products in nature's image.

2.5.1 Minimum Soil Disturbance (Practices—No Tillage, Direct Seeding, Planting Green, No-till Weeding)

Soil C is at the center of this new agriculture. A permanently covered, non-tilled soil, is a quality soil that is full of life. The avoidance of tillage preserves the C and habitat of insects, earthworms, and millions of fungi, bacteria, and other microorganisms essential to the proper functioning of our ecosystems. Minimizing soil disturbance by mechanical tillage (once brought to good porous condition), and, whenever possible, seeding or planting directly into untilled soil maintains SOM, soil structure, and overall soil health and biological activity.

However, in order to maintain gains in soil C, it is important to continuously manage soils with no-till (NT) and maintain a positive C balance. Grandy and Robertson (2006) found that tilling a previously untilled soil quickly reversed nearly all the previously recorded gains by disrupting aggregates and exposing C molecules to microbial attack. In the last few decades, the adoption of NT continues to rise, however, many of the attributes and functional benefits of NT management may take several years to consistently and fully affect yields, profits, soil health and function and environmental outcomes. A global meta-analysis of NT productivity across a range of climates, soils, and crop types found that NT management consistently out-performed conventional tillage for rain-fed crops in dry climates, but, under moist climates or irrigated conditions, the differences were more variable (Pittelkow et al. 2015). Some of the variations in this meta-analysis may likely be due to the short duration of many of the studies, with few of the studies lasting 10 years or longer that had large seasonal variation impacting the production, and not all data came from CA systems. Cusser et al. (2020) tested the time required to detect significant differences following conversion of conventional tillage agriculture (TA) to continuous NT using power analysis and a moving window analytical approach to evaluate the development and stability of several attributes of continuous NT as compared to TA over 29 years. They found over a decade is needed to detect the consistent effects of NT in a rain-fed corn-soybean-wheat system. Both crop yield and soil water availability required 15 years or longer to generate patterns consistent with 29-year trends. These results underscore the importance of decadal and longer studies for revealing consistent dynamics and outcomes of NT agriculture, such as CA, shown to become more and more beneficial in the long term, suggesting that management recommendations based on short-term studies can be misleading.

2.5.2 Permanent Vegetative Mulch Cover

Crop biomass retained as vegetative mulch (practice-plant biomass, stubble, cover crops) on the soil has effects on many soil-related structural components and processes of the agroecosystem. The effects of permanent vegetative cover as mulch vary according to climate conditions, and factors such as biomass amount and type of biomass can influence crop responses to mulching. Enhancing and maintaining vegetative cover on the soil surface, using crops, cover crops, or crop biomass and stubble (roots stock with straw or stalk) protects the soil surface, conserves water and nutrients, promotes soil biological activity and habitat, and contributes to integrated weed and pest management. Ranaivoson et al. (2017) reviewed agroecological functions of crop biomass under CA and found that ~8 Mg ha^{-1} of biomass were needed to decrease soil water evaporation by about 30% compared to NT bare soil, and a minimum of at least 2 Mg ha^{-1} of biomass was required to achieve the maximum effect on soil water infiltration, water runoff, and soil erosion control. They found a weak response with increased vegetative cover on soil meso- and macrofauna abundance and soil macro-nutrient supply (N, P, and K). Noteworthy was the average increase in annual SOC gain with increased amounts of biomass mulch, with a mean of 0.38 Mg C ha^{-1} $year^{-1}$ and 4 to 5 Mg ha^{-1} of biomass supplied. Ranaivoson et al. (2017) suggest that optimal amounts of vegetative cover in CA will largely depend on existing constraints to crop production which can be addressed with mulching.

Carvalho et al. (2012), found that under Mediterranean climate NT reduced SOM mineralization but that maximum increase of SOC can only be achieved if crop biomass is retained on the field. Nitrogen use efficiency (NUE) was also significantly improved with greater SOC contents. Under TA (moldboard plow + disc) and reduced tillage (RT), SOC remained almost unchanged over the experimental period. NT on its own improved SOC by 18% in relation to CT and NT+ cereal straw retained increased SOC by 62%. There were no significant differences between CT and NT in the amount of biomass left in the field over the experimental period, but NT+ cereal straw retained significantly more biomass in relation to the other treatments.

Crop biomass returns or removal, biological oxidation rates, and soil erosion control the SOC losses from agricultural systems. Reicosky et al. (1995) reviewed tillage and biomass production impacts on SOC in agricultural systems. They concluded that SOC was controlled by crop biomass input that was largely determined by crop choice, fertilization, and climate. Reicosky et al. (2002) evaluated the impact of 30 years of continuous corn with moldboard plow tillage with and without biomass (silage removal) and at low and high nitrogen rates. The experimental land area was farmed with conventional tillage and rotation methods prior to the initiation of the experiment. Measurements of SOC changes over the 30 years showed the total C, total N, and C:N ratio of the soil remained unchanged over the 30-year study in the fertilized treatments. All four treatments produced essentially the same SOC content of 2.2% C in the 0–20 cm depth. The 30-year cumulative input of 241 Mg ha^{-1} of the added aboveground biomass yielded no differences in SOC when compared to those treatments with biomass (silage) removed. The results demonstrated the impact of intensive tillage with the moldboard plow that resulted in soil C losses irrespective of biomass removal as corn silage or addition of aboveground crop biomass where only the grain was harvested. These results suggest that no or minimum soil disturbance is required before any addition or removal of aboveground biomass can impact SOC accumulation in CA systems.

Permanent vegetative cover enables a variety of measures to mitigate soil erosion, improve water holding capacity, and increase SOM to improve soil health and boost crop yields. A key feature is the revision or reduction of soil disturbance (through tilling). In CA systems, direct seeding involves no tillage prior to sowing. CA consists of a group of management strategies to avoid or minimize soil disturbance, maintain soil cover, and rotate crops or grow them in mixtures that seek to maintain an optimum environment in the root zone in terms of soil temperature and water availability, soil structure, and biotic diversity and activity (Kassam et al. 2009).

2.5.3 Species Diversification

Conservation of soil biodiversity, SOC through sustainable farming practices (diverse rotations and associations including cover crops) is critically essential to improving agricultural soil health and productivity. CA is more about diversity in plant management because of the importance of C and

C cycling. Crop rotations and associations usually increase SOM content when compared with monocultures suggesting benefits (Odell et al. 1984; Dick et al. 1986a, b; Hargrove and Frye 1987; Havlin et al. 1990; Van Eerd et al. 2014). Crop rotation also influences soil C dynamics by affecting the proportions of the decomposable organic compounds returned to the soil, leading to potential gains in soil C (Hutchinson et al. 2007; Ogle et al. 2005). The guiding CA principles are that nature manages soil better with biodiversity and when it is left alone. The economic benefits are that we can dramatically reduce fossil fuels and synthetic inputs (fertilizers and pesticides) and water needs while increasing SOM and improving soil biology.

Enhancing soil biodiversity can be done by growing diverse crop mixtures and by extending crop rotations (Poeplau and Don 2015; Chatterjee et al. 2016). Introducing cool season and warm season species, legumes and grasses, shallow rooted and deep-rooted crops can promote diversity (Anderson 2005; Mokany et al. 2006; Burney et al. 2010; Kemper et al. 2011; Kätterer et al. 2011; Kell 2011; Paustian et al. 2016). The number and depth of earthworm pores, the depth, distribution, the type of root structure, and the synergistic relationship that is maintained without tillage disrupting the naturally developed bio-pore network provide multiple benefits minimizing compaction with respect to water storage and management (Williams and Weil 2004; Chen and Weil 2010; Kemper et al. 2011).

Diverse cover crop mixes are helpful with drainage management and enhancing evapotranspiration improving soils and crop production and as scavengers for residual N and P (Qi et al. 2011; Stavi et al. 2012; Zhang et al. 2015; Ferchaud and Mary 2016). Overwinter cover crops are beneficial in reducing surface water flow and extracting residual soil N left over in the crop root zone after harvest which leads to improvements in water quality (Drury et al. 2014; Dabney et al. 2001; Strock et al. 2004). Cereal rye is one of the most widely used cover crops, especially in the Corn Belt of the US, because of its winter-hardy characteristics and versatility (Singer et al. 2007). Sievers and Cook (2018), evaluated aboveground and belowground decomposition and nutrient release of two single winter cover crops, hairy vetch (*Vicia villosa* Roth) and cereal rye (*Secale cereale* L.), after termination with herbicides through a 16-week period during the cash crop growing season. Hairy vetch, with its lower C:N ratio, decomposed rapidly after termination in the spring, suggesting growers consider delaying termination until close to cash crop planting to decrease the risk of N leaching loss. Cereal rye residues decompose much more slowly and may also immobilize N because of its high C:N ratio. Constantin et al. (2010) reported that cover crops increased soil storage of organic N and decreased residual soil NO_3^- concentrations in the root zone by 34–52%, and reduced NO_3^- leaching below the root zone by 36–62%. Camarotto et al. (2020) looked at SOC stock over a 6-year period within a 0–50 cm soil profile. Their results suggested that CA changed the SOC distribution rather than the total amount of SOC and no benefit in SOC sequestration was observed with CA and cover crops (CCs). They hypothesize that a longer study period would be required to better understand the potential benefits of CA and CC on SOC storage as demonstrated by Cusser et al. (2020). Additionally, use of cover crops may mitigate climate change through the transformation of CO_2 into SOM (Reicosky and Forcella 1998; Delgado et al. 2011; Olson et al. 2014b).

Cover crops also help with integrated weed management in CA systems, reducing or eliminating the need for herbicides (Basch et al. 2020). For example, CCs preceding the main crop can be terminated by using a roller crimper before direct seeding the next crop. This technique, known as "planting green" is popular in South America but, more recently, research has increased and its application encouraged in North America too (Pieri et al. 2002; Duiker 2017; Gullickson, 2018). In addition to weed control, such CCs produce extra biomass, and, in the case of legume CCs, nitrogen is added to the soil and helps with nutrient cycling. In the case of smallholders, multi-purpose CCs in intercropped CA systems contribute to the total economic output but also help to control weeds and insect pests and pathogens. It also produces additional biomass and, in the case of legume cover crops, biologically fixed N is added to the soil. For example, Lablab bean (*Lablab purpureus* (L.) Sweet) intercropped with maize smothers weeds efficiently while providing protective ground cover, additional biomass and biological nitrogen to the soil (Owenya et al. 2011). In the case of several species of tick-trefoil or tick clover (silverleaf, *Desmodium uncinatum* and greenleaf, *Desmodium intortum*) as a CC with maize, it eliminates witchweed or witches weed (*Striga*), a pernicious semi-parasitic weed, and pushes away maize stem borer moths while adding nitrogen to the soil. This forms part of the push-pull system of integrated biological pest control in CA and non-CA systems in several African countries by smallholder farmers (Khan et al. 2017, 2020).

The "living soil" includes a jungle of creatures, seen and unseen, working underfoot performing countless critical functions (Brussaard 1999; Wall and Nielsen 2012; Bardgett and van der Putten 2014). Biodiversity is an element of sustainability and is necessary for harmony and stability in nature (Nielsen et al. 2011; Balvanera et al. 2014); understanding and nurturing this biodiversity is needed to ensure that management practices sustain ecosystem services (Lal 1991; Brussaard et al. 1997, 2012; Pereira et al. 2012; Costello et al. 2013; Pascual et al. 2015). The success of CA systems relies on and benefits from the interactive synergies between the biological, physical, hydrological, and chemical properties and processes in the soil that enhance C management (Jackson et al. 2017; Banwart et al. 2019; Chenu et al. 2019).

Root-derived C and microbial biomass make a significant contribution to SOM (Jackson et al. 2017). Enhancing biodiversity in the soil involves promoting microorganisms and mesofauna including bacteria, fungi, protozoa, nematodes, earthworms, termites, beetles, and spiders. Soil biodiversity facilitates the formation of stable micro and macro aggregates that create soil structure and porosity and add significant amount of microbial biomass and root biomass to SOM stock (Goss et al. 2017; Day et al. 2020). In the soil system, biodiversity is encouraged through NT, soil-mulch cover, as well as by diverse sources of root biomass (Jackson et al. 2017) and exudates that serve as substrate for microorganisms and mesofauna which enhance and support a number of soil-mediated ecosystem functions (Kumar et al. 2006; Eisenhauer et al. 2017, 2018; Baumert et al. 2018; Kassam et al., 2020b). Root biomass and root exudates, being made up of easy-to-ingest OM, is readily incorporated into the SOM pool whereas biomass mulch cover on the soil surface, also made up of harder-to-decompose material such as cellulose and lignin, remains for a relatively longer period of time, protecting the soil surface from climate and exposure (Jackson et al. 2017). Earthworms and termites are able to ingest this type of biomass and produce worm casts and excreta material which are rich in OM and nitrogen (Bhadauria and Saxena 2010; Jouquet et al., 2014; Bignell 2019).

In diverse cropping, root systems of different shapes and sizes allow the whole soil profile to be worked upon by roots in terms of nutrient and water acquisition and retention. With deeper and denser root systems, nutrients from lower layers are mobilized or stopped from leaching away and are cycled to be utilized by crops. Diverse root systems can contribute significant amounts of C from the root biomass and root exudates to the soil C pools at deeper layers. Also, along with root biomass contribution, deeper root systems allow microorganisms including earthworms to operate at deeper soil layers where roots provide substrate in the form of liquid C exudates that microbiomes are able to utilize and grow on, contributing the microbial biomass to the C pools when they die. Furthermore, a particular contribution to soil aggregate formation and stability is made by all microorganisms, particularly bacteria and *Glomeromycota* fungi (Hibbett et al. 2007), that use SOM and C compounds in root exudates to produce an array of organic compounds. A group of these compounds are glycoproteins, including glomalin, which serve as a biological glue to bind and hold soil particles and intermediate forms of OM into stable micro and macroaggregates (Veloso et al., 2020) that help to build soil structure and porosity and impart spongy behavioral characteristics to healthy soils.

2.5.4 Conservation Agriculture for Soil Carbon Storage

Human civilizations and the continuing sustainability are inextricably linked to soils and the wide array of the ecosystem services soils impacted by the loss of SOM. Natural climate solutions are conservation, restoration, and improved land management actions that increase C storage and minimize GHG emissions in landscapes across the globe (Bradford et al. 2019). Soils store C from the atmosphere to mediate atmospheric greenhouse gas levels; they filter contaminants (from infiltrating recharge) to deliver clean drinking water to aquifers; they provide habitat and maintain a microbial community and gene pool that decomposes and recycles dead OM and transforms nutrients into available forms for plants; they enable release of mineral nutrients from parent rock; and they enable soil structure to store and transmit water in ways that help prevent floods (Lal 2010). Credits for SOC storage and accumulation can become a structural part of the solution to mitigate climate change, where the societal value of soil C stored and of less GHG emissions should be based on all ecosystem services. Both short-term and long-term increases in SOC pools would need to be commoditized and traded based on both on-site and off-site societal benefits (Lal 2008). Crop biomass management (at ~45% C) should also be valued and subsidies considered.

C is the key chemical element of SOM that improves soil physical, chemical, and biological properties and processes (Corsi et al. 2012; Lal 2015d; Kane 2015; Sanderman et al. 2017; Reicosky 2020). Understanding the ecological and physical controls over SOM is critical for maintaining a balanced C cycle, increasing food and fiber production, and safeguarding species diversity and ecosystem services. Incentive payments will require some type of verification of C in the soil, encouraging management practices that increase soil C to help mitigate climate extremes. A policy platform that supports a voluntary and private market for ecosystem services provided by CA (that pays farmers and ranchers) is needed to improve soil health, store C, decrease erosion, and reduce soil and nutrient pollution. These policies will require substantial scientific evidence that the soil C is providing all of the pertinent ecosystem services being incentivized. It may turn out to be that "carbon" will be the crop that pays farmers to manage C for his/her farm and ecosystem services (Bradford et al. 2019).

There remains an ongoing debate on the relative merit of CA and NT for storing C and its relative merit affecting crop yield, environmental quality, and climate change mitigation (Sumberg et al. 2012; Giller et al. 2009; Powlson et al. 2014; Powlson et al. 2016; Pittelkow et al. 2015a; Pittelkow et al. 2015b). Powlson et al. (2014) argue that NT is beneficial for soil quality and adaptation of agriculture to climate change, but that its role in climate change mitigation is widely overstated. Powlson et al. (2016), using meta-analysis, found substantial variation in annual increases in SOC stock compared to conventional TA practice and a significant number of cases with no measurable increase. They reported that most SOC stock increases under CA are overestimates because of errors introduced by inappropriate soil sampling methodology and other benefits, as from crop diversification, are frequently overlooked when considering CA and warrant greater attention. Govaerts et al. (2009) reviewed 78 papers to analyze the potential impact of CA on C storage: in 40 cases the soil C stock was higher in NT as compared to TA, in 31 cases there was no significant difference, and in seven it was lower. Similar conclusions comparing CA and NT are provided by Pittelkow et al. (2015a, b) from a global meta-analysis. It is slowly becoming evident that NT alone does not work as well as the three principles operating continuously and simultaneously in CA systems.

Page et al. (2020) concluded the effects of CA on SOC worldwide are variable, with both increases in regions where soil and climatic conditions are favorable for biomass production and where the system does not negatively impact yield. Under these conditions, CA can lead to higher amounts of SOC relative to TA systems, particularly in the surface layer. Where greater SOC occurs, these are also often accompanied by improvements in soil structure, water infiltration, and soil water storage, as well as plant nutrient availability, microbial biomass and diversity, and yield. However, when CA is used in certain less favorable environments (e.g., cold, wet environments with poorly drained soils) or where the CA system may not yet have been well-adapted to local conditions, taking into account the specific agronomic, social, and environmental challenges present, then CA may take longer to become a successful system of management in all aspects, both practice and benefits. On the other hand, CA systems provide opportunities for innovation and biodiversity in coping with some of these adverse conditions with more flexibility than TA (Reicosky 2020).

A fundamental facet of soil conservation is maintaining adequate SOM. CA is a major player in our survival through C cycling and soil C storage. Biodiversity with C cycling and flow is an element of sustainability that is necessary for harmony and stability in nature (Reicosky and Janzen 2018). This biodiversity provided by diverse crop rotations and CC mixes must be understood and nurtured to enhance soil biodiversity and resilient soil ecosystem services. The difficulties associated with accurately quantifying changes in soil C stocks do not necessarily limit opportunities to increase soil C reserves. They do, however, emphasize the importance of following well-conceived sampling and analytical strategies, and the need for carefully evaluating earlier measurements to ensure they are not inadvertently biased by methodological oversights (Dick et al. 1986a, b; Dick et al.1996; Robertson et al.1999; Milne et al. 2010; Philibert et al. 2012; Derpsch et al. 2014; Brouder and Gomez-Macpherson 2014; Kladivko et al. 2014; Olson et al. 2014a; White et al. 2013).

CA, through enhanced soil C management, can contribute to climate change mitigation. Soil C concentrations and stocks have been reduced in agricultural soils following long-term use of TA (Lal et al. 2007; Sanderman et al. 2017; Sá et al. 2020). Adoption of practices such as diverse rotations, associations, and cover crops can protect and rebuild soil C. As a result of lack of clarity in some research methods and materials, and the terminology confusion associated with CT and NT loosely used as synonyms for CA (Corsi et al. 2012; Derpsch et al. 2014; Reicosky 2015; Reicosky and Janzen 2018; Reicosky 2020), early research on "Conservation Agriculture" was focused on comparing TA and NT and associated biomass soil cover without complete integration of the three primary principles. Only in the last 10–15 years within the SH movement have the concepts of crop biodiversity and cover crops been incorporated into the research analysis, providing the complement of three integrated principles in a comprehensive system engaged simultaneously. Recent reports support theories that C storage in cropland is mainly influenced by the amount of C input (Kell 2011; Lynch and Wojciechowski 2015; Paustian et al. 2016). Poeplau and Don (2015), using meta-analysis, found the time since introduction of CCs in rotations was linearly correlated with SOC stock change (R2 = 0.19) with an annual change rate of 0.32+/–0.08 Mg ha^{-1} yr^{-1} in a mean soil depth of 22 cm during the observed period of up to 54 years. Thus, the dynamic cover crop C input and storage proved to be efficient and an important C input in CA systems (Reicosky and Janzen 2018). Corsi et al. (2012), summarized many complex and interacting variables in CA and concluded the full benefits and advantages of CA after conversion from TA to CA with respect to SOC storage can usually be seen only in the medium- to longer-term after CA practices are well established and/or improved statistical techniques are employed (Cusser et al. 2020). The impact of this small increase in soil C is multiplied through the many enhanced crop adaptability relationships and ecosystem services enabled by the CA production system.

Corsi et al. (2012), Reicosky and Janzen (2018), Sá et al. (2020) and others have shown that CA enables higher rates of C storage in the soil compared with TA, where, in most cases, there is a continuous decrease in C content. The lack of C storage or loss reported in agricultural systems is often associated with any one or a combination of the following reasons: uncertain soil disturbance, monocropping, specific crop rotations, poor crop biomass management, and different soil depth sampling increments and extended sampling >30 cm. Another cause of variability includes insufficient soil disturbance details in the methods and materials section of published manuscripts along with insufficient details on agronomic management differences. In addition, different definitions and interpretations of CT improperly used in CA are common on a global scale which sometimes leads to misunderstandings in evaluation and use of scientific data and practical results, especially in the categorization of data input to comprehensive meta-analyses (Philibert et al. 2012). Use of terms indicating reduced tillage, minimum tillage, mulch tillage, and strip tillage each with their own unique "soil disturbance" are often grouped into the NT category improperly reflecting the absolute "minimum soil disturbance" defined in CA (Kassan et al. 2020a; Reicosky 2015; Reicosky 2020). The characteristics of CA make it one of the systems best able to contribute to climate change mitigation by reducing atmospheric GHG concentrations for small and large farmers. Gonzalez-Sanchez et al. (2019), using meta-analyses, estimated the potential annual C storage in African agricultural soils through CA amounts to 143 Tg of C per year. This figure represents about 93 times the current C storage rate suggesting further improvement with more CA in Africa.

2.6 TERMINOLOGY AND TYPES OF AGRICULTURE

Conservation, sustainability, soil health, and regenerative agriculture (RA) are terms typically tied to the soil resource. CA is already a very potent global force and source of knowledge for resilient agricultural production, environmental protection and rehabilitation, and climate change mitigation and adaptability (Farooq and Siddique 2015; Jat et al. 2014; Kassam et al. 2017; Kassam 2020). Other agricultural types can build on this success, and evolution and improvement in the CA systems and practices within forms of sustainable agriculture, sustainable intensification, soil health farming and regenerative agriculture to keep CA for C management and as a critical component of agricultural mitigation strategy into the future. Conservation, sustainability, soil health, organic, agroecology, and regenerative agriculture are terms typically tied to the soil resource. Building on CA with its expanded conservation practices and concepts seems a logical progression, however, there is need for clarity in conservation science, agroecology, communication, and understanding if we, as a society, are to continue our positive steps toward simultaneous food security, environmental protection, and sustainable diets.

Using a variety of names for new types of agriculture, a combination of conservation and agroecology has emerged, regarding CA essentially as a set of additional tools to fix the problems of conventional food production systems as illustrated in Figure 2.3. The schematic illustration represents how CA system principles also serve as the core foundation to other types of sustainable agriculture systems. Many researchers now see that conservation and agroecology must be connected and integrated as a way to transform conventional agriculture into more sustainable farming, without radically challenging underlying relations of the broader agriculture industry (Kassam 2019). With the current innovations in agricultural terminology, there is a lack of scientific details in the ambiguous "health metaphor" that leads to confusing and misunderstanding in management and policy discussions. Alignment on common definitions and a systemic approach that prevents CA from becoming just another "green communication marketing tool" that provides justification for a classical model of natural resource conservation and enhanced management measures. Attempts to apply and extend CA principles need improvement by (1) better collaboration between scientists and representatives from CA industries; (2) better communication of the risks of not implementing

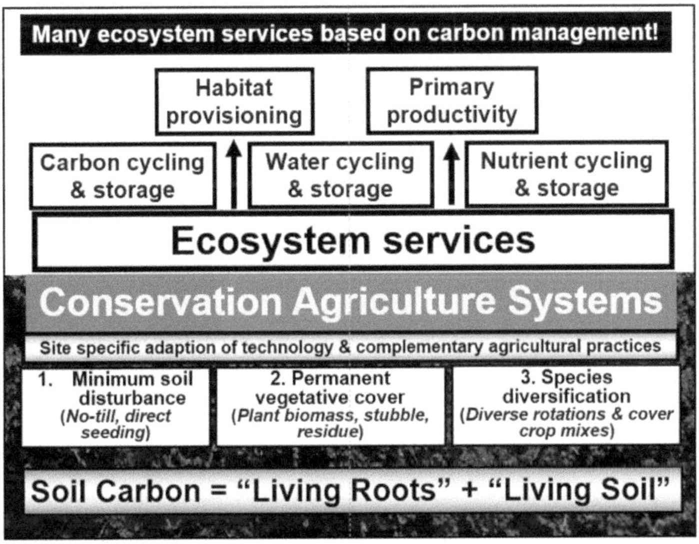

FIGURE 2.3 Schematic representation of the foundation of CAS that serves as core principles in other types of sustainable agriculture systems.

CA; and (3) ensuring CA projects and programs continue ecosystem management for food security and sustainable agricultural production.

Over the last 40 years CA has maintained a science-based foundation as evidenced by the large number of scientific publications in this field. As a result, Conservation Agriculture has been highlighted as a key path toward food security (FAO 2008, 2011; Hobbs et al. 2008; Kassam et al. 2009, 2014a, 2014b, 2017; Kassam and Friedrich 2012; Corsi et al. 2012; Pretty and Bharucha 2014; Farooq and Siddique, 2015; Jat et al. 2014; Lal 2015d; Reicosky and Janzen 2018; Mitchell, et al. 2019; Kassam 2019, 2020; Reicosky 2020). Much of the literature on different types of agriculture is in what may be called the "grey literature" as reports that are not generally peer-reviewed and are typically published in non-scientific magazines. At the same time, the three core principles of CA are generally agreed to and included in other types of agriculture, particularly mulch cover and crop diversification. All three principles of CA need to be strategically integrated into all different types of agriculture approaches. Some promoters of different types of agriculture do not have a science background that occasionally makes it challenging to link the value systems to explicit conservation practices that align with the CA principles.

There have been misunderstandings about what CA is: some groups have assumed that CA is only suitable for large-scale farmers (not smallholders); others think it does not embrace or is not compatible with organic farming. While it is true that globally some two-thirds of CA annual cropland area comprise larger-scale holdings and are managed by less than half a million farmers, the remaining CA area is farmed by millions of small- and medium-scale farmers on all continents in countries, such as Brazil, Paraguay, Italy, Spain, Mozambique, Zambia, Malawi, Kenya, Tanzania, Ghana, Morocco, Tunisia, Iran, India, and China. Many smallholder farmers practice organic CA, certified and uncertified. For example, in northern Mozambique and southern coastal Tanzania, thousands of farmers practice uncertified organic CA as agrochemicals are simply not available (Lalani et al. 2017). Also, biological practices, such as roller crimping and "planting green" for direct seeding (Duiker 2017; Gullickson, 2018), intercropping (Owenya et al. 2011), cover cropping, and push-pull practices for biological pest management (Khan et al. 2017, 2020) are part of CA systems that are used by smallholders in different continents. Legume crops in cropping systems, including as CCs (for biological nitrogen fixation and nutrient cycling) are commonly used in CA systems in industrialized countries and low-income countries. There is now the possibility of altering the fungi:bacteria ratio in the soil for biological weed management. Perennial CA systems (orchards, plantations, agroforestry) in Africa and Asia are also being managed biologically (Kassam et al. 2020c). However, much more research attention is needed on a regular basis worldwide to generate and spread knowledge that can (1) be applied in an adapted manner in CA systems to minimize the need for agrochemicals and energy and (2) eliminate as far as possible the need for synthetic agrochemical inputs. In general, in non-organic CA systems, a significant reduction in agrochemical inputs (some 50% or more) has become possible over time as soil and landscape health and SOM content is improved and effective integrated pest (weeds, insects, and pathogens), nutrient, and water management is established.

2.6.1 Conventional Intensive Tillage Agriculture

Reducing the amount of tillage or avoiding tillage in cropping systems on the prairies has long been recognized as an important strategy to reduce soil erosion and degradation and improve soil health and productivity. Soils with minimal levels of crop biomass on the surface are more susceptible to soil erosion, and the surface soil structure is less stable (Chang and Lindwall 1989). Tillage is the "mechanical manipulation of the soil for any purpose" (Soil Science Glossary Terms Committee 2008; Reicosky and Allmaras 2003). It has been perceived as important to seedbed preparation, weed control, and incorporation of agricultural chemicals or amendments. However, tillage destroys SOM through two inter-related processes. First, OM at depth in the soil is slower to decompose as soil temperature and moisture levels vary more slowly at depth and oxygen partial pressure is often lower. Inversion tillage with moldboard plowing brings this OM to the surface and decomposition through oxidation is speeded up by an order of magnitude. Tillage has upset nature's "power balance" emitting C as CO_2 resulting in soil C loss and increasing GHG emissions (from the combustion of fossil fuel)—a double negative for our C footprint.

Second, conventional intensive tillage agriculture (TA) competes with nature, disturbs the soil and soil biology; uses reduced crop diversification or monoculture techniques; uses a reductionist approach in management requiring higher inputs of fossil fuels, equipment expense, and maintenance, as well as synthetic fertilizers, herbicides, pesticides, and insecticides (Reicosky and Allmaras 2003). It would seem that modern TA has done little to integrate conservation into agriculture, as the main focus has been on increasing crop and livestock production, in terms of both yields and total output in rain-fed and irrigated agriculture based on the narrow Green Revolution approach. Farmers in different parts of the world have also been reacting to address and overcome the inherent degradation and loss in productivity. They, along with some extension agronomists and few machine companies, began to replace TA with NT agriculture and introduced soil and water conservation practices that eventually led to an alternate paradigm now generally referred to as CA.

Tillage practices receive attention for their role in changing SOC stocks over seasonal and decadal time scales (Moreira et al. 2016); moldboard and chisel plowing are implicated in large losses of SOC at regional scales, whereas CT or NT practices are suggested as means for preserving SOM (Sá, J.C. de Moraes et al. 2015; Lal et al. 2007; Reicosky and Lindstrom 1993; Reicosky 1997; Reicosky and Archer 2007). The act of tillage creates an instantaneous change in the soil environment by opening physical spaces between soil aggregates, breaking aggregates, allowing for greater exchange of GHGs between the soil and atmosphere, and inducing immediate abiotic oxidation of freshly exposed organic C-containing surfaces (Homyak et al. 2016). Recent work by Alcántara et al. (2016) documented deep plowing (55–90 cm) techniques can increase SOC on decadal time scales, and other reviews conclude that reports of increased SOC from lower intensity plowing are overstated due to historical bias toward shallow soil sampling (Baker et al. 2007; Powlson et al. 2014).

The ecological implications of inversion tillage with the moldboard plow in some instances appeared to last for a long time after the plowing is discontinued. For example, Isbell et al. (2019) found nearly 100 years following agricultural abandonment, local plant diversity recovers only incompletely, and plant productivity does not significantly recover. By 91 years after agricultural abandonment, despite gaining many local species, formerly plowed fields still had only three-quarters of the plant diversity and half of the plant productivity observed in a nearby remnant ecosystem that had never been plowed.

2.6.2 Conservation Tillage

Conservation tillage (CT) includes a broad set of practices with the goal of leaving some crop biomass on the soil surface to reduce erosion and increase water infiltration. Mannering and Fenster (1983) define CT as a broad term used with "any" tillage system with the primary objective of "reducing soil and water loss" and agree that this has caused confusion. The various practices described as "conservation tillage" have led to terminology confusion. Conservation tillage is often confused with NT or variants of CT described in vague terms, such as minimum tillage, mulch tillage, ridge tillage, strip tillage, contour plowing, and reduced tillage, where planting is achieved on specially prepared surfaces with various amounts of crop biomass (Hobbs 2007; Dumanski and Peiretti 2013; Derpsch et al. 2014). These definitions of CT and the use of jargon are sending a mixed and confusing message (CTIC 1999).

Conservation tillage (CT) or "any tillage sequence, the object of which is to minimize or reduce loss of soil and water; operationally, a tillage or tillage and planting combination which leaves a 30% or greater crop biomass cover on the surface," (Soil Science Glossary Terms Committee 2008) is ambiguously intermediate and covers a broad range of soil disturbance and crop residue incorporation between these two extremes (Derpsch et al. 2014). Reicosky (2015) concluded that CT, although well-intentioned, does not always provide the soil erosion protection and water conservation required. Present day implementation of CA principles and concepts will be a way to cope with food production during climate variations and extremes. Both the public and policy groups must understand the importance of CA with continuous no or minimum mechanical soil disturbance for sustainable food security for modern civilizations.

2.6.3 Sustainable Development

The concept of sustainable development has evolved over decades as a result of expanding human civilizations, widespread natural resource degradation, and the need for continuing sustainability of natural resources and habitats linked to soils as well as the large number of ecosystem services soils provide to society and nature. Du Pisani (2006) presents an in-depth history on the evolution of sustainable development with emphasis on notions of progress, growth, development, and sustainability in different social, economic, and philosophical contexts related to human use and degradation of soil resources. The concept of sustainable development was a compromise between agricultural expansion and conservation, with emphasis on expanding the conservation of the natural resources that support our existence. It is clear that the current global development and environmental goals of "sustainable development," the United Nation's SDGs (Keesstra et al. 2016), require principles of science to protect and maintain the environment and the associated ecosystem services dependent on our soils. The term sustainable development from an agricultural perspective translates into "sustainable agriculture" that implies a multidimensional approach to managing agroecosystems encompassing productivity, environmental, economic, and social components of sustainability; and CA has been shown to make a major contribution to sustainable agriculture development globally.

The consequences of intensive tillage agricultural production on our soil resources have been extensive with accelerated soil erosion, desertification, salinization, acidification, compaction, biodiversity loss, nutrient depletion, loss of SOM and degradation of soil quality. The loss of SOM has received more attention due to its critical role in the global C cycle and as a key component in sustainable food production and climate mitigation. Modern and traditional tillage farming has led to the degradation and loss of the natural resource base, including biodiversity in the soil, in the cropping system and above the ground, and over the landscape (Pretty 2002). In general, it has led to high levels of soil erosion, water pollution, and loss of ecosystem societal services such as clean water, C storage, C, nutrient, and water cycling; regulatory processes, habitats for wildlife and natural enemies of pests, and pollination services, etc. (Kassam et al. 2013).

Sustainable development concepts represented a new way of thinking about integrating resource conservation into agriculture that is conserving and much more resilient than conventional intensive tillage agriculture. Soil erosion and degradation is a common consequence of any type of tillage agriculture, the visible signs that led to NT concepts and operations. Sustainable development in agriculture has transitioned from utilizing conservation tillage, reduced tillage, forms of strip tillage, to the ultimate NT or direct seeding and agroecology (Altieri 1995) that opened up other aspects of conservation integrated into practicing sustainable intensification and CA, with its core of universally applicable principles. New attention is focused on the beneficial economic aspects of NT-enabled investments in agriculture for economic growth, with the possibility that low-income countries can avoid some of the mistakes that have been made in industrialized nations (Garnett et al. 2013).

2.6.4 Sustainable Intensification

The new paradigm of "sustainable production intensification," as elaborated on in FAO (2011), recognizes the need for a productive and remunerative agriculture which at the same time conserves and enhances the natural resource base and environment, and positively contributes to harnessing the environmental services. The balanced cycling of C between the atmosphere, plants, and soil is the basis of sustainability (Garnett et al. 2013). The profitability and sustainability of production systems—managed by both large- and small-scale farmers—derive from efficiency in both the use and conservation of available resources, applied in appropriately combined crop-soil-water-nutrient-pest-ecosystem management practices. These practices are locally devised and adapted to capture a range of productivity, socioeconomic and environmental co-benefits of agriculture and ecosystem services at the farm, landscape, and provincial, or national scale and beyond (Pretty 1997; Pretty 2002; Kassam et al. 2009, 2013; Godfray et al. 2010; Godfray and Garnett 2014; FAO 2010, 2011; Pretty and Bharucha, 2014; Foresight 2011).

Sustainable intensification is a process to increase agriculture yields without adverse impacts on the environment, taking the whole ecosystem into consideration. It aims for the same goals as CA. While CA offers much potential for sustainable intensification, scientific debate highlights a number of difficulties and contentions. Each component in the CA system or "package" requires interpretation (Stevenson et al. 2014), and the applicability and scalability of CA to smallholder systems has been questioned by some academic researchers, especially for smallholders in sub-Saharan developing country contexts (Giller et al. 2009, 2011; Stevenson et al. 2014). Sustainable solutions for food security must be holistic and address issues such as food accessibility—with wider consideration of issues related to equitable distribution of food and individual empowerment in the intensification decision process—and distributive and procedural justice (Loos et al. 2014). While the applicability of CA to sub-Saharan contexts has been called into question by a tiny number of individuals, some case studies nevertheless show beneficial and attractive socio-ecological outcomes, as CA cropland area managed by small farmers in 2015/16 increased by 211% since 2008/09 (Kassam et al. 2018) and has increased further since then. Comparisons between CA and conventional plots demonstrate yield increases in the former, as well as reduced soil loss, increased soil C content, and improved soil structure and water productivity (Marongwe et al. 2011; Owenya et al. 2011; Thierfelder et al. 2015). There is also evidence that there are now several millions of hectares under CA annual cropping with trees, particularly in the vast tracts of Sahel and semi-arid regions north and south of the equator in Africa (Garrity 2017; Kassam et al. 2020c).

Despite general agreement that meeting food demand without further loss of natural ecosystems requires sustainable intensification, there is little dialogue about the research agenda needed to achieve it (Cassman and Grassini 2020). Although progress toward sustainable intensification is behind schedule, we are optimistic that current trends can get back on course assuming a well-prioritized and adequately funded research portfolio and appropriate policies and institutions to support it (FAO 2011, 2016; Kassam et al. 2018, 2020a).

This agroecological intensification (AEI) system is based on the CA approach and includes three core practices: avoiding or minimizing soil disturbance, maintaining permanent soil cover, and crop diversification integrating crop rotations and associations. This system has been applied on a wide range of row crops, estimated to cover more than 180 million ha worldwide (Kassam et al. 2018) or about 12.5% of all land under annual crops. In addition, in Africa, CA systems in the semi-arid areas include leguminous trees, which can enhance biological nitrogen fixation and help to maintain vegetative soil cover (Garrity et al. 2010, Garrity 2017). Comparisons between CA and contrasting AEI farming systems have focused primarily on outcomes associated with minimal soil disturbance and mulching treatments versus TA, rather than with established CA systems.

2.6.5 Soil Health Farming

The core concept of "soil health" dates back over 100 years and has evolved ever since (Brevik 2018). Doran et al. (1996) defined soil health (SH) as "the capacity of soil to function as a vital living system, within an ecosystem and land use boundaries, to sustain plant and animal productivity, maintain or enhance water and air quality, and promote plant and animal health." In brief, SH, sometimes referred to as soil quality, is the continued capacity of soil to function as a vital living ecosystem that sustains plants, animals, and humans (NRCS 2017). Lehmann et al. (2020) provide a comprehensive review of the evolution of SH. They compared the definition and history of soil health with other soil concepts and indicated that quantifying soil health is still dominated by chemical indicators; however, the importance of soil biodiversity and SOM is receiving increased attention. Fundamental to the SH concept is the idea that soil is a living ecosystem, and the well-being of soil is essential for achieving ecosystem services, including high quality air and water, promoting a diverse biotic and microbial community structure, supporting a high level of crop productivity, and promoting human health.

The importance of the soil–human health nexus has also been recognized ever since the dawn of civilization (Howard 1947; Magdoff 2001; Sherwood and Uphoff 2000; Brevik 2014; Brevik and Sauer 2015; Pepper 2013; Oliver and Gregory 2015; Wall et al. 2015; Kemper and Lal 2017; Lehmann et al. 2020). Viewing soil as a living ecosystem transforms the way we think about soil and how to improve and manage it to advance our agricultural production without degrading environmental quality. Science-based SH is slowly evolving in response to information needs for addressing multiple complex problems including climate change, GHGs, and the C cycle; soil erosion and land degradation; crop disease and pests; depletion and pollution of water resources, and loss of biodiversity (FAO 2008; Eagle et al. 2010; Laishram et al. 2012; Tsiafouli et al. 2015; Stirling et al. 2016). A major goal of the soil health movement is to promote the science foundation for consumers, farmers, and policy makers to help educate people to understand the need to maintain more C in the soil (FAO 2008; Lal 2016; Reicosky and Janzen 2018; Lehmann et al. 2020).

The current "soil health movement" is shifting the emphasis toward ecological agriculture in the US. CA systems operating with complementary integrated production and management practices and SH systems are providing the same ecosystem services with emphasis on C management. The principles of CA (FAO 2008; Hobbs et al. 2008; Kibblewhite et al. 2008; Kassam et al. 2009; Kassam et al. 2014a,b; Lehman et al. 2015) and SH (NRCS 2017; USDA 2019; Moebius-Clune et al. 2016; Lal 2016; Al-Kaisi and Lal 2017) are essentially the same, except that SH has a fourth principle that includes a living root system as long as biologically possible. This principle is implicit in CA systems as explicitly highlighted in its principle of diverse cropping through rotations, sequences, and associations, and including cover crops in annual and perennial systems. While both CA and SH terminology are still evolving, CA terminology is used more globally while SH terminology is primarily used in the US and Canada (NRCS 2017; USDA 2018; Lal 2015b; Lal 2016; Moebius-Clune et al. 2016). SH advocates say interest is growing in nurturing the health of the vital natural resource, however, there's no standard way to measure SH or predict its potential for improvement.

2.6.6 Regenerative Agriculture

The terminology and concept of RA are still evolving. The term "regenerative agriculture" is, at present, a less formally articulated concept or approach to farming systems yet one that is also gaining attention in recent years as a design platform for future production systems. Newton et al. (2020) reviewed numerous journal articles and practitioner websites and found there were many definitions and descriptions of regenerative agriculture in usage today. The confusion or vagaries around RA are not likely to be cleared up in the near future. RA originated as a marketing term from the organic agriculture movement that would allow them to depart from organic principles and use some fertilizers or pesticides appropriate to the circumstances (Newton et al. 2020). The origin of RA in the US dates to the 1980s with Robert Rodale (Rodale 1983), and to its more recent revival to the Rodale Institute (2014), who defined RA by "its tendencies toward closed nutrient loops, greater diversity in the biological community, fewer annuals and more perennials, and greater reliance on internal rather than external resources." Early examples of scientific articles using the term regenerative were presented by Pretty (1995) and LaCanne and Lundgren (2018). Pretty (1995) draws together new empirical evidence from a diverse range of agroecological and community to show agricultural impacts of more sustainable practices and identifies the common elements of success in implementing sustainable agricultural practices. Promoters of the new Regenerative Agriculture Organic Certification policy that is contained in the 2018 US Farm Bill point to comparable yields for numerous crops including corn, wheat, rice, soybean, and sunflower with organic regenerative systems compared to conventional systems (De Ponti et al. 2012). More recently, some RA proponents seem to have added farm animals as a necessary element which has led to further confusion and controversy.

Lal (2020a) describes RA as a system for sustainable food production and climate mitigation that is soil-centric, rather than seed-centric, and extends the concept that the health of soil, plants, animals, people, and environment is one and indivisible. RA comprises system-based CA, which includes integrated NT farming, residue mulching, cover cropping, nutrient and pest management, complex rotations, and integration of crops with trees and livestock (Lal 2015d). Lal (2020a) states the goal of RA is to apply the concept of "more from less" (McAfee 2019) to agriculture and produce more from less (Lal 2013a; 2020a): less land area, less input of chemicals, less use of water, less emission of GHGs, less risk of soil degradation, and less use of fossil energy-based inputs.

Although RA has no universal definition so far, the term is often used to describe conservation practices aimed at promoting SH by restoring SOC. The definition of RA is still evolving following the science of agroecology now being recognized in all types of sustainable agriculture promoting C management (Newton et al. 2020). The world's soils store several times the amount of C in the atmosphere, acting as a natural "carbon sink." But globally, soil C stocks have been declining as a result of factors, such as the conversion of natural landscapes to croplands, tillage agriculture, overgrazing, and accelerated soil erosion by water and wind. One goal of regenerative practices is to use some of the C that plants have absorbed from the atmosphere to help restore soil C. Ranganathan et al. (2020) discuss practices people are calling RA, examine their climate change mitigation potential, and evaluate their place among other agriculture mitigation options. Conservation practices grouped under RA include NT and the use of cover crops. Other practices include diverse crop rotations, planting three or more crops in rotation, and rotating crops with livestock grazing. RA is built on a belief that says that we are depleting our soils and we need to do a lot better if we are going to mitigate climate extremes (Ranganathan et al. 2020).

We need to lift the value of clearly communicating mainstream conservation practices integrated within science-based CA principles as a comprehensive and universally applicable system. Just as CA and RA are occasionally used interchangeably, the words sustainable and regenerative are also often used interchangeably, which leads to confusion (Newton et al. 2020). The term sustainable might call for techniques that maintain SH that is still evolving in definition and metrics. The integrated core principles of CA systems are well established empirically and scientifically and are broadly used in most if not all agricultural ecosystems. It can serve as a foundation for RA because all the elements that have been included under RA by Lal (2020a) are within the scope of integrated CA as a comprehensive system. RA addresses the critical need to restore or regenerate soil health, re-balance the water and C cycles, create new topsoil, and grow food and other materials in a regenerative way that enables nature to sustain these systems. There is a need to understand the definition and core principles of RA and the same basic explanation and direction for an agriculture that works with protecting and utilizing natural resources efficiently, especially sun, soil, biodiversity, and air containing C, N, and H_2O. Apparently, this is because RA is still an evolving concept, and it is expected that it will continue to evolve.

2.7 DISCUSSION

Key responses in CA systems aim at reducing GHG emission-intensity, and maximizing co-benefits from adaptation and mitigation efforts, through better management of C and N cycles, increased resource-use efficiency, the conservation of C-rich landscapes, measures to strengthen resilience and—on the demand side—a reduction in food losses and improvements in diets. We as a society must work together to foster both the science and the practical art of natural resource conservation to maintain food security and planetary health. An enormous amount of information exists to explain the benefits of increased CA systems and practices—but, which practices work and which don't, and what are the necessary and sufficient conditions for their adoption and scaling? The renewed emphasis on sustainable and earlier evolution of CA with emphasis on conservation principles reawakened the interest of farmers in SH and RA.

Transitioning from conventional tillage agriculture systems to CA systems was motivated by excessive soil erosion associated with tillage that led pioneer farmers to NT or direct seeding as the first step to minimize the erosion problem. This coincidentally led to other environmental benefits: avoiding or minimizing soil disturbance, leaving the crop biomass on the surface as mulch to protect the soil from raindrop impact, erosion, enhance infiltration and water storage; and also providing a "slow-release" energy source for the soil organisms. The next step in CA evolution trended more into understanding soil biology, integrated pest management, and the use of cover crops with emphasis on C management and understanding a more complex soil system (Chenu et al. 2019). And during the transition to CA, the loss of biodiversity and resilience was also evident with the successful use of crop diversity including CCs, primarily for erosion control and scavenging N, also showed increases with more C input that reflected the net effect of enhancing all three primary principles of CA. CA concepts came together relatively rapidly as a result of the synergistic benefits of implementing the three principles simultaneously. Out of this success of the three primary principles, the SH movement evolved mainly because of the understanding of the soil as a living system. Noteworthy during this transition was the decrease in input costs (fuel, equipment, labor, repair and maintenance, synthetic fertilizers, herbicides, pesticides, etc.) due to the increase in resilience to both abiotic and biotic stresses, such as weeds, insects and pathogens; the increase in productivity; and the social, economic, and environmental benefits (increased profits, decreased erosion, increased water quality, enhance ecosystem services, decreased GHG emissions, etc.).

Overall soil biology is the foundation of SH, function, and resilience (Banwart et al. 2019). They provide a large array of physical, biological, chemical, hydrological benefits and resilience to the cropping systems and farmers in the fields and across landscapes and watersheds during crop production in space and time (Kassam et al. 2013, 2020b; Kassam and Kassam 2020c). These benefits are affected by historical management of the land, agroecological environment, terrain and gravity, type of cropping and farming systems, and quality of production and land use management. It is only when these crop and field-level benefits are being harnessed is it possible to also benefit from the linked watershed-level ecosystem processes that deliver societal services, such as clean water supply, regulated stream flows and rivers; minimization of soil erosion, flooding, and environmental pollution; C, nutrient, and water cycling; pollination; and overall system resilience (Basche et al. 2016).

CA is systems thinking for sustainable farming that can serve as a design platform for other types of agriculture (Lal 2015d). A system is a suite of practices and management methods that, when used together, result in additional improvements to SH that help address primary resource concerns. The elements of a CAS integration go beyond the minimum standards and specifications of a conservation practice. So when we talk about CA on annual cropland, we emphasize some of the descriptive adjectives associated with each conservation practice element, such as NT, direct seeding, crop rotations and associations, cover crop integration, and integrated pest (weeds, insects, and pathogens), nutrient, water, and energy management. CA conserves natural resources, biodiversity, and labor. It increases available soil water, reduces heat and drought stress, and builds up SH in the longer term (Kassam and Kassam 2020b, c). An array of conservation technology platforms enables flexible integration into CA systems that have a sustainable ecological foundation that utilizes and promotes time- and research-tested regenerative farming principles and conservation practices driving profitable on-farm conservation practice adoption, and culminating in positive outcomes for farm businesses, communities, and our shared natural resources. The scope to integrate additional overlays and components on to this ecological foundation to harness further gains is almost limitless.

The evolution of CA systems is a continuing emergent journey of learning and understanding the natural systems and our collaboration with Mother Nature to develop the synergistic benefits of conserving and protecting our resources for our food security and sustainable agricultural land use (Jackson et al. 2017). As part of the journey, we must continue our daily individual "re-search programs" of trial and error, understanding, and explaining how nature operates. We should view each growing season as a new experiment with notes and observations for management changes that may enhance our understanding of the complexities and marvels of CA for future food production and sustainable agroecosystem management.

The main conservation benefit from CA is greater natural self-protection and resilience as a result of the synergistic benefits from several conservation practices incorporated into a unified and synchronized CA system. These practices are focused on soil restoration and conservation of natural resources (i.e., sun, soil, water, air, intellect, biodiversity) that also restores and conserves as many ecosystem processes and ecological functions and life below and above the land surface as possible. The goal of CA is to enhance health and functionality of soil by restoring SOM content, strengthening processes which recycle C and mineral nutrients, and enhancing water storage and plant available water in the root zone. The latter is critical to enhancing resilience of soil and of agroecosystems against climate-induced droughts (e.g., pedological and agronomical damage). The initial process for incorporating conservation practices into a CA system can require a bit of trial and error, curiosity, and imagination. Applying proven methodologies will need to account for individual farming operations and make the necessary adaptations where appropriate. While the setup process can be challenging, the benefits can be substantial and enduring.

Our review and analysis suggest conservation as a modifier of agriculture is more comprehensive and acceptable because it includes many diverse science-based conservation practices that enable protection, preservation, and resilience from numerous synergistic interactions between the core principles. The popular perception that CA and RA are different and what specific traits make them different may still be evolving. We suggest CA terminology should be preferred over the list of new "marketing buzzwords" describing the types of agriculture. It is difficult to understand the psychology of why RA seems preferred over CA in the US. To some, RA seems to be CA under a different name.

> ***"The conservation of our natural resources and their proper use constitute the fundamental problem which underlies almost every other problem of our national life."***
>
> ***—Theodore Roosevelt***

Rational use and conservation of living resources must be the active concern of all humanity, if the world is to continue to develop in any reasonable degree of harmony with Nature (Harvey et al. 2017). Enhancing belowground and aboveground ecological interactions through integration of soil and plant management practices in CA constitutes a robust and sustainable path for optimizing agroecosystem functions and productivity. CA has many beneficial roles which in turn increases the survivability and sustainability as well as biodiversity of soil mesofauna. Conservation of soil biota is the key issue to make the soil fertile. There is a strong relationship between CA and soil mesofauna and food webs below and above the ground.

The broad umbrella of many conservation practices integrated into comprehensive Conservation Agricultural Systems enables resilience, sustainability, SH, regenerative agriculture, etc., utilizing Nature's resources that need protection, preservation, and efficient use if we are going to have truly sustainable food production. CA is, at its heart, an approach to farming that puts more back into the environment and society than it takes out. Farmers are increasingly challenged to make use of their intellectual and communication skills throughout this period of transition because they have to optimize conventional input-use efficiency and redesign the production system. Researchers are also under a constant challenge to open up more and more of the "bazaar" of opportunities that CA systems present, including those offered by soil biodiversity and aboveground biodiversity for transforming CA into biological and organic systems. The next set of breakthroughs with CA are needed in improving the mobilization of natural sources of nutrients in the soil based on novel ways of managing soil biodiversity. Also needed are biological approaches to crop protection based on novel ways of managing pest–predator relationships, allelopathy, and gene functions related to crop protection and resilience, including relationships of root systems with soil microbiology, and effects of these on crop responses to favorable and unfavorable growth and environmental development.

This chapter would not be considered complete if nothing was mentioned about the variable adoption and research performance of CA in research experiments. This is despite the generally accepted scientific and empirical evidence that CA principles are universally applicable, and that CA systems do offer improved productivity, economic, social, and environmental benefits to farmers and society at large (Kassam 2020). Also, it is generally the case that both the uptake of CA and the benefits are enhanced by farmer-led organizations and networks that generate civil society and social cohesion to generate the enabling environment conditions for progress (Kassam et al. 2014c). These include: existence of pioneers and champions, effective momentum for innovation, training, and learning; mutual support; input and output value chain services; sharing experiences; and organizing participatory and technical extension and strategic events. This type of networking is a partnership effort of several entities to help connect agricultural producers who can help each other save time and money by sharing their experiences. Mentor–mentee roles as a relationship where farmers share knowledge, advice, and resources and provide a stable supporting relationship minimizing the risk of trying something new. Dumanski et al. (2014) discuss the role of dedicated pioneers and champions in CA: farmers, researchers, extensionists, and service providers, who questioned tradition and implemented new theories and new technologies in soil conservation. These pioneer efforts were local and individual, but as the interest spread and more farmers came on board, they formed farmer-led local and national farmer associations, cooperatives, and clubs to promote the philosophies and concepts based on working together and learning by doing (McKell and Peiretti 2004). Much of the early success with CA is attributed to this farmer-to-farmer communication and direct sharing of information and experimental knowledge but also to the varied roles played by farmers and their communities, government policies and institutions that have supported CA innovation, the emergence of famer-led organizations (such as NT farmer associations), and, importantly, the contributions of individual farmers (Pape and Prokopy 2017). These aspects of CA adoption are further reviewed in Kassam et al. (2014c).

Understandably, there have been calls for improvements to the evidence base on CA, given that in most places in Europe, Africa, and Asia, CA adoption by farmers as a basis for production and livelihood improvement and research on CA systems and on different aspects of CA has been of recent origin (Brouder and Gomez-MacPherson 2014). Also, in most countries official statistics on CA are not collected as they do on conventional tillage agriculture. This notwithstanding, CA has continued to spread globally with more than half the countries of the world now engaged to promote CA research, training, and adoption (Kassam et al. 2018). Some meta-analyses and reviews also show that the evidence on yield impacts and C storage potential is mixed (Stevenson et al. 2014; Pittelkow et al. 2015; Corbeels et al. 2020). This may, in part, reflect the context sensitivity of CA, where outcomes depend on the precise combination of practices used, and may differ by crop type. It may also reflect the fact that biological systems that operate across the globe in different soils, climates, and agroecologies would be expected to show substantial variability in performance, which, in the case of CA, is also influenced by the length of time the CA change process has been operating. It would seem that most meta-analyses use data that are not sufficiently differentiated and therefore are based on mixed quality data, often not filtered sufficiently for aspects such as adherence to CA principles, site history, severity of soil degradation at the start of the process of transformation, agroecological potential of the location, land improvement such as removing the hard pan prior to setting up the CA experiment, length of time involved, farming system type, management quality of experiments, and expertise level of the experimenter. There is some evidence that NT may result in yield penalties in the short-term (e.g., Brouder and Gomez-MacPherson 2014), but generally the researchers concerned, educated, and trained in conventional tillage agriculture, do not have the expertise or the experience to explain the underlying reasons. This may hinder adoption among smallholders, who may "attribute" more value to immediate costs and benefits than those incurred in the future, as they must navigate precarious and pressing concerns over food and livelihood security (Giller et al. 2011). Other studies find yield increases to be consistent and stable over time. For example, Sorrenson (1997) and Derpsch (2008) compared conventional and CA systems over a decadal scale and found yield decreases over the period in conventional, and yield increases over the same period under CA, in addition to fewer inputs. This appears to be the case generally where conditions for developing and sustaining good quality CA systems are present (Jat et al. 2014; Thierfelder et al. 2015; Li et al. 2016; Kassam et al. 2017; Kassam 2020). There is also a need for more evidence on the precise implications of improved land management across agricultural sectors and agroecosystems (e.g., Morgan et al. 2010), using appropriate soil sampling strategies (Baker et al. 2007) and standardized research methodologies (Derpsch et al. 2014).

2.8 CONCLUSIONS AND SUMMARY

Soil needs to be considered as the foundation of an ecological system that is biologically alive and functional. It is a finite resource. Conservation of soil biodiversity and SOC through sustainable farming practices must be deemed essential to improve upon agricultural SH and productivity. Improved soil biodiversity is closely correlated with an increase in C stored in the soil, which is a natural C sink. It is imperative in reducing GHG emissions and mitigating climate extremes. The global C cycle includes a significant C stock in terrestrial soils, critically important in agricultural production. Global land and soil degradation are strongly linked to the depletion of this C stock. Conservation of the ecological properties in our agricultural ecosystems requires a C-centric form of CA.

CA principles (the interlinked application of continuous no or minimal soil disturbance, permanent vegetative mulch cover, and crop species diversity) are universally applicable in all land-based agricultural landscapes and cropping systems that implement agroecological principles. The application of the three core CA principles, which embrace all of the agroecological principles, will favor the transition to a C-neutral agriculture and the long-term achievement of sustainability within agricultural systems. CA can be considered a super-integrated system that can concretely contribute to tackling the challenges of providing continued enhanced ecosystem services in the 21st century.

The CA global movement, also known as the CA Community of Practice (CA-CoP), has built a legacy by caring for the soil, the planet, and your food and other biological products—you can join that legacy and be a part of the future we're building and protecting our resources together. CA System partners collaborate with Nature to obtain synergistic benefits, with minimum soil disturbance, protecting the soil surface with living and dead organic biomass, encouraging maximum biodiversity, and including a more holistic management approach that enables economic profits while saving the social and economic cost of environmental degradation by minimizing soil erosion, pollution, and environmental quality issues that are not presently accounted for because of the difficulty in measuring the impacts. Maintaining a science perspective in our food security is important. The scientific method emphasizes the sharing of data and research so that they can be duplicated, tested, and challenged. At the same time, the social forces within the CA-CoP emphasize the value of farmer-based innovations and empirical knowledge generated by farmers and participatory approaches to learning, sharing, and communication. Ensuring a continuous two-way flow of scientific and empirical information and knowledge will always remain key to maintaining effectiveness and efficacy in the growing CA global movement.

The importance of conservation in the cropping system referred to as CAS needs to get better acceptance of that terminology around the world. Conservation centered agricultural practices are C-centered agriculture practices. CA is taking all conservation practices to a higher level of integration and is a no-regret option that offers multiple benefits and deserves high-level visibility and immediate global implementation for future generations. Embracing the CA approach as a basis for a new paradigm of agriculture through a transparent, consistent, and science-based process is long overdue and will secure benefits to global agriculture well into the future. Although the conviction toward healthy soils, healthy landscapes, and healthy farms must come primarily from the farming community, this must be underpinned by the scientific, rural, and urban sectors and supported by the broader society and their needs and concerns for healthy diets and a healthy planet. At present, society and the planet are burdened with a large set of interconnected chronic crises. CA is capable of addressing most of these crises, but there must be a strong partnership among all the sectors in the food and agriculture systems to promote adoption and success of CA-related enabling approaches (Kassam and Kassam 2020a). We owe it to future generations.

ACKNOWLEDGMENTS

The authors acknowledge the discussions, interactions, and ideas of friends and colleagues, especially Rattan Lal for his review comments and suggestions, who have impacted our thinking for many years. We thank them for their indirect contribution to this manuscript.

BIBLIOGRAPHY

Abdalla, M., A. Hastings, K. Cheng, et al. 2019. A critical review of the impacts of cover crops on nitrogen leaching, net greenhouse gas balance and crop productivity. *Global Change Biology*, 25(8): 2530–2543.

Alcántara, V., A. Don, R. Well and R. Nieder. 2016. Deep ploughing increases agricultural soil organic matter stocks. *Global Change Biology*, 22: 2939–2956. doi: 10.1111/gcb.13289.

Al-Kaisi, M. M. and R. Lal. 2017. Conservation agriculture systems to mitigate climate variability effects on soil health. In M. M. Al-Kaisi and B. Lowery (eds.), *Soil Health and Intensification of Agroecosystems*. Elsevier Academic Press, London, pp. 79–103.

Allison, F. E. 1973. *Soil Organic Matter and Its Role in Crop Production*. Elsevier Scientific Publishing Company, Amsterdam.

Altieri, M. A. 1995. *Agroecology: The Science of Sustainable Agri-Culture*. Westview Press, Boulder, CO, 433 pp.

Altieri, M. A. 2002. Agroecology: The science of natural resource management for poor farmers in marginal environments. *Agriculture, Ecosystems and Environment*, 93: 1–24.

Altieri, M. A. and C. I. Nicholls. 2012. Agroecology: Scaling up for food sovereignty and resiliency. *Sustainable Agriculture Reviews*, 11: 1–29.

Anderson, R. L. 2005. Improving sustainability of cropping systems in the Central Great Plains. *Journal of Sustainable Agriculture*, 26(1): 97–114. doi: 10.1300/J064v26n01_08.

Angers, D. A. and C. Chenu. 1997. Dynamics of soil aggregation and C sequestration. In R. Lal (ed.), *Soil Processes and the Carbon Cycle*. CRC Press, Boca Raton, FL, pp. 199–206.

Bai, X., Y. Huang, W. Ren, et al. 2019. Responses of soil carbon sequestration to climate-smart agriculture practices: A meta-analysis. *Global Change Biology*, 25: 2591–2606. doi: 10.1111/gcb.14658.

Baker, J. M., T. E. Ochsner, R. T. Venterea and T. J. Griffis. 2007. Tillage and soil carbon sequestration—What do we really know? *Agriculture, Ecosystems & Environment*, 118: 1–5. doi: 10.1016/j.agee.2006.05.014.

Balvanera, P., I. Siddique, L. Dee, et al. 2014. Linking biodiversity and ecosystem services: Current uncertainties and the necessary next steps. *BioScience*, 64(1): 49–57. doi: 10.1093/biosci/bit003.

Banwart, S. A., N. P. Nikolaidis, Y. Zhu, C. L. Peacock and D. L. Sparks. 2019. Soil functions: Connecting earth's critical zone. *Annual Review of Earth and Planetary Sciences*, 47: 333–359. doi: 10.1146/annurev-earth-063016-020544.

Banwart, S. A., E. Noellemeyer, E. Milne, et al. 2015. The global challenge for soil carbon. In Banwart, S. A., Noellemeyer, E., and Milne, E. (eds.), *Soil Carbon: Science, Management and Policy for Multiple Benefits*. CABI, Wallingford, UK, pp. 1–9.

Bardgett, R. D. and W. H. Van der Putten. 2014. Belowground biodiversity and ecosystem functioning. *Nature*, 515(7528): 505–11.

Barker, A. V. and D. J. Pilbeam (eds.). 2015. *Handbook of Plant Nutrition.* 2nd ed. CRC Press/Taylor and Francis Group, Boca Raton, FL, pp. 773.

Basch, G., F. Teixera and S. Duiker. 2020. Weed management practices and benefits in conservation agriculture systems. In Kassam, A. (ed.), *Advances in Conservation Agriculture. Volume 2 – Practice and Benefits.* Burleigh Dodds, Cambridge, UK, pp. 105–142.

Basche, A. D., T. K. Kaspar, S. A. Archontoulis, et al. 2016. Soil water improvements with the long-term use of a cover crop. *Agricultural Water Management*, 172: 40–50. doi: 10.1016/j.agwat.2016.04.006.

Baudoin, E., E. Benizri and A. Guckert. 2003. Impact of artificial root exudates on the bacterial community structure in bulk soil and maize rhizosphere. *Soil Biology and Biochemistry*, 35: 1183–1192.

Baumert, V. L., N. A. Vasilyeva, A. A. Vladimirov, I. C. Meier, I. Kögel-Knabner and C. W. Mueller. 2018. Root exudates induce soil macroaggregation facilitated by fungi in subsoil. *Frontiers in Environmental Science*, 6(40): 1–17.

BDA. 2018. One blue dot: Environmentally sustainable diet toolkit. Retrieved from https://www.bda.uk.com/professional/resources/environmentally_sustainable_diet_toolkit_-_one_blue_dot.

Bhadauria, T. and K. G. Saxena. 2010. Role of earthworms in soil fertility maintenance through the production of biogenic structures. *Appl. Environ. Soil Sci.*, Volume 2010, Article ID 816073, 7pages. doi: 10.1155/2010/816073.

Bignell, D. E. 2019. Termite ecology in the first two decades of the 21st century: A review of reviews. *Insects*, 10(3): 60. doi: 10.3390/insects10030060.

Blanco-Canqui, H. and R. Lal. 2009a. Soil conservation and carbon dynamics. In Blanco-Canqui, H. and Lal, R. (eds.), *Principles of Soil Conservation and Management*, Springer, Dordrecht, pp. 449–476.

Blanco-Canqui, H. and R. Lal (eds). 2009b. *Principles of Soil Conservation and Management.* Springer, Dordrecht, pp. 617. doi: 10.1007/978-1-4020-8709-7.

Blaustein, R. 2012. Can biology transform our energy future? *Bioscience*, 62(2): 115–119.

Bogard, J. R., G. C. Marks, S. Wood and S. H. Thilsted. 2018. Measuring nutritional quality of agricultural production systems: Application to fish production. *Global Food Security*, 16: 54–64. doi: 10.1016/j.gfs.2017.09.004.

Bradford, M. A., C. J. Carey, L. Atwood, et al. 2019. Soil carbon science for policy and practice. *Nature Sustainability*, 2: 1070–1072. doi: 10.1038/s41893-019-0431-y.

Brady, N. C. and R. R. Weil. 1999. *The Nature and Properties of Soils.* 12th ed. Prentice-Hall, Upper Saddle River, NJ. 881 pp.

Branca, G., N. McCarthy, L. Lipper and M. C. Jolejole. 2011. *Climate-Smart Agriculture: A Synthesis of Empirical Evidence of Food Security and Mitigation Benefits from Improved Cropland Management.* Mitigation of Climate Change in Agriculture Series 3. Food and Agriculture Organization of the United Nations (FAO) Rome, p. 43.

Brevik, E. C. 2014. Soils and human health – An overview. In Brevik, E. C. and Burgess, L. C. (eds.), *Soils and Human Health.* 1st ed. CRC Press, Boca Raton, pp. 29–56.

Brevik, E. C. and T. J. Sauer. 2015. The past, present, and future of soils and human health studies. *Soil*, 1: 35–46.

Brevik, E.C. 2018. A brief history of the soil health concepts. *Soil Science Society of America Journal Field/Historical Notes.* SSSA Profile 82(6). https://profile.soils.org/files/soil-communication/documents/95_document1_d50a8524-dbf4-4856-8c62-4a03ede6c3d4.pdf

Broder, M. D. and G. H. Wagner. 1988. Microbial colonization and decomposition of corn, wheat and soybean residue. *Soil Science Society of America* 52: 112–117.

Bronick, C. J. and R. Lal. 2005. Soil structure and management: A review. *Geoderma*, 124: 3 –22.

Brouder, S. M. and H. Gomez-Macpherson. 2014. The impact of conservation agriculture on smallholder agricultural yields: A scoping review of the evidence. *Agriculture, Ecosystems and Environment*, 187: 11–32.

Bruce, R. R., G. W. Langdale and L. T. West. 1990. Modification of soil characteristics of degraded soil surfaces by biomass input and tillage affecting soil water regime. *Transactions 14th International Congress of Soil Science*, 6: 4–9.

Brussaard, L. 1999. On the mechanisms of interactions between earthworms and plants. *Pedobiologia*, 43: 880–885.

Brussaard, L. 2012. Ecosystem services provided by the soil biota. In: Wall, D. H., Bardgett, R. D., Behan-Pelletier, V., Herrick, J. E., Jones, T. H., Ritz, K., Six, J., Strong, D. R. and van der Putten, W. H. (eds.), *Soil Ecology and Ecosystem Services.* Oxford University Press, Oxford, UK, pp. 45–58.

Brussaard, L., V. M. Behan-Pelletier, D. E. Bignell, et al. 1997. Biodiversity and ecosystem functioning in soil. *Ambio – A Journal of the Human Environment*, 26: 563–570.

Burney, J. A., S. J. Davis and D. B. Lobell. 2010. Greenhouse gas mitigation by agricultural intensification. *Proceedings of the National Academy of Sciences of the United States of America*, 107(26): 12052–12057. doi: 10.1073/pnas.0914216107.

Buyanovsky, G. A. and G. H. Wagner. 1977. Crop residue input to soil organic matter on Sanborn field. In E. A. Paul, K. H. Paustian, E. T. Elliot and C. V. Cole (eds.), *Soil Organic Matter in Temperate Agroecosystems: Long-Term Experiments in North America*. CRC Press, Boca Raton, FL, pp. 73–83.

Calegari, A., T. Tiecher, E. B. Wutke, L. B. dos S. Canalli, R. Bunch and D. dos S. Rheinhemer. 2020. The role and management of soil mulch and cover crops in conservation agriculture systems. In Kassam, A. (ed.), *Advances in Conservation Agriculture Volume 1 Systems and Science*. Burleigh Dodds, Cambridge, UK, pp. 179–248.

Camarotto, C., I. Piccoli, N. Dal Ferro, et al. 2020. Have we reached the turning point? Looking for evidence of SOC increase under conservation agriculture and cover crop practices. *European Journal of Soil Science*. 71(6): 1050–1063. doi: 10.1111/ejss.12953.

Carvalho, M., G. Basch, J. M. G. Calado, et al. 2012. Long term effect of tillage system and crop residue management on soil carbon content of a Luvisol under rain-fed Mediterranean conditions. *Agrociencia, (Special Issue)*, 16(3): 183-187.

Cassman, K. G. 1999. Ecological intensification of cereal production systems: Yield potential, soil quality, and precision agriculture. *Proceedings of the National Academy of Sciences of the United States of America*, 96: 5952–5959.

Cassman, K. G. and P. Grassini. 2020. A global perspective on sustainable intensification research. *Nature Sustainability*, 3: 262–268.

Castellano, M. J., K. E. Mueller, D. C. Olk, J. E. Sawyer and J. Six. 2015. Integrating plant litter quality, soil organic matter stabilization, and the carbon saturation concept. *Global Change Biology*, 21(9): 3200–3209.

Ceballos, G., P. R. Ehrlich, A. D. Barnosky, A. García, R. M. Pringle and T. M. Palmer. 2015. Accelerated modern human-induced species losses: Entering the sixth mass extinction. *Science Advances*, 1: e1400253.

Chang, C. and C. W. Lindwall. 1989. Effect of long-term minimum tillage practices on some physical properties of a Chernozemic clay loam. *Canadian Journal of Soil Science*, 69: 443–449.

Chatterjee, A., K. Cooper, A. Klaustermeier, R. Awale and L. J. Cihacek. 2016. Does crop species diversity influence soil carbon and nitrogen pools? *Agronomy Journal*, 108(1): 427–32. doi: 10.2134/agronj2015.0316.

Chen, G. and R. R. Weil. 2010. Penetration of cover crop roots through compacted soils. *Plant and Soil*, 331(1–2): 31–43. doi: 10.1007/s11104-009-0223-7.

Chenu, C., D. A. Angers, P. Barré, D. Derrien, D. Arrouays and J. Balesdent. 2019. Increasing organic stocks in agricultural soils: Knowledge gaps and potential innovations. *Soil & Tillage Research*, 188: 41–52. doi: 10.1016/j.still.2018.04.011.

Christensen, B. T. 1992. Physical fractionation of soil and organic matter in primary particle size and density separates. In Stewart, B. A. (ed.), *Advances in Soil Science Series*. Springer-Verlag, New York, pp. 1–90.

Conant, R. T., C. E. P. Cerri, B. B. Osborne and K. Paustian. 2016. Grassland management impacts on soil carbon stocks: A new synthesis. *Ecological Applications*, 27: 662–668.

Conservation. 2020. Merriam-Webster.com Dictionary, Merriam-Webster. https://www.merriam-webster.com/dictionary/conservation (accessed 26 August 2020).

Constantin, J., B. Mary, F. Laurent, et al. 2010. Effects of catch crops, no till and reduced nitrogen fertilization on nitrogen leaching and balance in three long-term experiments. *Agriculture, Ecosystems and Environment*, 135(4): 268–278. doi: 10.1016/j.agee.2009.10.005.

Corbeels, M., K. Naudin, A. M. Whitbread, R. Kuhne and P. Letourmy. 2020. Limits of conservation agriculture to overcome low crop yields in sub-Saharan Africa. *Nature Food*, 11: 447–454.

Corsi, S., T. Friedrich, A. Kassam, et al. 2012. Soil organic carbon accumulation and greenhouse gas emission reductions from conservation agriculture: A literature review. *Integrated Crop Management*, vol. 16. Plant Production and Protection Division, Food and Agriculture Organization of the United Nations, Rome, 89 pp.

Costello, M. J., R. M. May and N. E. Stork. 2013. Can we name earth's species before they go extinct? *Science*, 339(6118): 413–416.

Crouse, D. A. 2018. Soils and plant nutrients. In: K. A. Moore and L. K. Bradley (eds.), *North Carolina Extension Gardener Handbook*. NC State Extension, Raleigh, NC. https://content.ces.ncsu.edu/extension-gardener-handbook/1-soils-and-plant-nutrients.

CTIC (Conservation Technology Information Center). 1999. *What's Conservation Tillage?* CTIC, Core 4 Program, Lafayette, IN.

Cusser, S., C. Bahlai, S. M. Swinton, G. P. Robertson and N. M. Haddad. 2020. Long-term research avoids spurious and misleading trends in sustainability attributes of no-till. *Global Change Biology*, 26(6): 3715–3725.

Czyz, E. A., J. Rejman, A. R. Dexter, et al. 2017. Clay-organic complexes in a Polish loess soil. *International Agrophysics*, 31: 447–452.

Dabney, S. M., J. A. Delgado and D. W. Reeves. 2001. Using winter cover crops to improve soil and water quality. *Communications in Soil Science and Plant Analysis*, 32(7–8): 1221–50. doi: 10.1081/CSS-100104110.

Day, S., A. Calegari, A. Santos, M. Cremonnes, L. Maia, W. Demetrio and M. L. C. Bartz. 2020. Biodiversity management practices and benefits in Conservation Agriculture systems. In Kassam (ed.), *Advances in Conservation Agriculture Volume 2 – Practice and Benefits*. Burleigh Dodds, Cambridge, UK, pp. 267–302.

De Ponti, T., B. Rijk and M. K. van Ittersum. 2012. The crop yield gap between organic and conventional agriculture. *Agriculture Systems*, 108: 1–9. doi: 10.1016/j.agsy.2011.12.004.

Delgado, J. A., P. M. Groffman, M. A. Nearing, et al. 2011. Conservation practices to mitigate and adapt to climate change. *Journal of Soil and Water Conservation*, 66(4): 118A–129A.

Derpsch, R. 2008. Critical steps to no-till adoption. In Goddard, T., Zoebisch, M. A., Gan, Y., Ellis, W., Watson, A., and Sombatpanit, S. (eds.), *No-Till Farming Systems*. World Association of Soil Water Conservation, WASWC, Bangkok, pp. 479–495.

Derpsch, R., A. J. Franzluebbers, S. W. Duiker, et al. 2014. Why do we need to standardize no-tillage research? *Soil & Tillage Research*, 137: 16–22.

Derpsch, R., T. Friedrich, A. Kassam and H. W. Li. 2010. Current status of adoption of no-till farming in the world and some of its main benefits. *International Journal of Agricultural and Biological Engineering*, 3: 1–25.

Dexter, A. R., G. Richard, D. Arrouays, et al. 2008. Complexed organic matter controls soil physical properties. *Geoderma*, 144: 620–627.

Dhaliwal, S. S., R. K. Naresh, A. Mandal, R. Singh and M. K. Dhaliwal. 2019. Dynamics and transformations of micronutrients in agricultural soils as influenced by organic matter build-up: A review. *Environmental and Sustainability Indicators*, 1–2: 100007. doi: 10.1016/j.indic.2019.100007.

Diacono, M. and F. Montemurro. 2010. Long-term effects of organic amendments on soil fertility. A review. *Agronomy for Sustainable Development*, 30: 401–422.

Diamond, J. 2005. *Collapse: How Societies Choose to Fail or Succeed*. Revised ed. Penguin Books, New York.

Dick, W. A., D. M. Van Doren, Jr., G. B. Triplett, Jr. and J. E. Henry. 1986a. Influence of long-term tillage and rotation combinations on crop yields and selected soil parameters: I. Mollic Ochraqualf. *Research Bulletin*, 1180. Ohio Agricultural Research & Development Center Library, Ohio State University, Wooster, OH.

Dick, W. A., D. M. Van Doren, Jr., G. B. Triplett, Jr. and J. E. Henry. 1986b. Influence of long-term tillage and rotation combinations on crop yields and selected soil parameters: II. Typic Fragiudalf. *Research Bulletun*, 1181. Ohio Agricultural Research & Development Center Library, Ohio State University, Wooster, OH.

Dick, R. P., D. R. Thomas and J. J. Halvorson. 1996. Standardized methods, sampling, and sample retreatment. In J. W. Doran and A. J. Jones (eds.), *Methods for Assessing Soil Quality*. Soil Science Society of America, Madison, WI, 107–121.

Doran, J. W., M. Sarrantonio and M. A. Liebig. 1996. Soil health and sustainability. *Advances in Agronomy*, 56:1–54. doi: 10.1016/S0065-2113(08)60178-9.

Drury, C. F., C. S. Tan, T. W. Welacky, et al. 2014. Reducing nitrate loss in tile drainage water with cover crops and water-table management systems. *Journal of Environmental Quality*, 43(2): 587–598. doi: 10.2134/jeq2012.0495.

Du Pisani, J. A. 2006. Sustainable development – Historical roots of the concept. *Environmental Sciences*, 3(2): 83–96. doi: 10.1080/15693430600688831.

Duiker, S. W. 2017. Planting green: A new cover crop management technique. *Field Crop News*. https://extension.psu.edu/planting-green-a-new-cover-crop-management-technique.

Dumanski, J. and R. Peiretti. 2013. Modern concepts of soil conservation. *International Soil and Water Conservation Research*, 1(1): 19–23.

Dumanski, J., D. C. Reicosky and R. A. Peiretti. 2014. Pioneers in soil conservation and conservation agriculture. *International Soil and Water Conservation Research (ISWCR)*, 2(1), March 2014: 107.

Dungait, J. A. J., D. W. Hopkins, A. S. Gregory and A. P. Whitmore. 2012. Soil organic matter turnover is governed by accessibility not recalcitrance. *Global Change Biology*, 18(6): 1781–1796.

Eagle, A. J., L. R. Henry, L. P. Olander, K. Haugen-Kozyra, N. Millar and G. P. Robertson. 2010. *Greenhouse Gas Mitigation Potential of Agricultural Land Management in the United States: A Synthesis of the Literature. Technical Working Group on Agricultural Greenhouse Gases (T-AGG) Report.* Nicholas Institute for Environmental Policy Solutions, Duke University, Durham, NC.

Eisenhauer, N., J. Hines, F. Isbell, et al. 2018. Plant diversity maintains multiple functions in future environments. *eLife*, 7: e41228. doi: 10.7554/eLife.41228.

Eisenhauer, N., A. Lanoue, T. Strecker, et al. 2017. Root biomass and exudates link plant diversity with soil bacterial and fungal biomass: Plant diversity maintains multiple soil functions in future environments. *Scientific Reports*, 7: 44641. doi: 10.1038/srep44641.

Ellert, B. H. and J. R. Bettany. 1995. Calculation of organic matter and nutrients stored in soils under contrasting management regimes. *Canadian Journal of Soil Science*, 75(4): 529–538.

FAO. 2008. *Investing in Sustainable Crop Intensification: The Case for Soil Health.* Report of the International Technical Workshop, FAO, Rome, July. Integrated Crop Management (vol. 6). FAO, Rome.

FAO. 2011. *Save and Grow, A Policymaker's Guide to Sustainable Intensification of Smallholder Crop Production.* Food and Agriculture Organization of the United Nations, Rome, 116 pp.

FAO. 2016. *Save and Grow in Practice: Maize, Rice and Wheat.* FAO, Rome.

Farooq, M. and K. H. M. Siddique. 2015. *Conservation Agriculture.* Springer International, Basel, Switzerland.

Ferchaud, F. and B. Mary. 2016. Drainage and nitrate leaching assessed during 7 years under perennial and annual bioenergy crops. *BioEnergy Research*, 9(2): 656–70. doi: 10.1007/s12155-015-9710-2.

Findlater, K. M., M. Kandlikar and T. Satterfield. 2019. Misunderstanding conservation agriculture: Challenges in promoting, monitoring and evaluating sustainable farming. *Environmental Science & Policy*, 100: 47–54. doi: 10.1016/j.envsci.2019.05.027.

Foresight, U. K. 2011. *The Future of Food and Farming.* Final Project Report. The Government Office for Science, London.

Franzluebbers, A. J. 2010. Achieving soil organic carbon sequestration with conservation agricultural systems in the southeastern United States. *Soil Science Society of America Journal*, 74: 347–357.

Friedrich, T. 1997. Potencial y aspectos de la mecanización agrícola en la Agricultura Conservacionista. In Claverán, A. R. and Rulfo, F. O. (eds.), *Memorias de la IV Reunión Bienal de la Red Latinoamericana de Agricultura Conservacionista*, RELACO, 17–20 November 1997, Michoacán, México. CENAPROS, INIFAP, SAGAR, FAO, pp. 155-1.

Friedrich, T. 2020. The role of no or minimum mechanical disturbance in conservation agriculture systems. In Kassam, A. (ed.), *Advances in Conservation Agriculture*, Volume 1, *Systems and Science.* Burleigh Dodds, Cambridge, UK, pp. 1–30.

Friedrich, T., R. Derpsch and A. H. Kassam. 2012. Overview of the global spread of conservation agriculture. *Field Actions Science Reports*, 6. http://factsreports.revues.org/1941#text.

Garbach, K., J. C. Milder, F. A. J. DeClerck, et al. 2017. Examining multi-functionality for crop yield and ecosystem services in five systems of agroecological intensification. *International Journal of Agricultural Sustainability*, 15(1): 11–28.

Garnett, T., M. C. Appleby, A. Balmford, et al. 2013. Sustainable intensification in agriculture: Premises and policies. *Science*, 341: 33–34.

Garrity, D. 2017. How to make conservation agriculture EverGreen. In Kassam A., Mkomwa, S., and Friedrich, T. (eds.), *Conservation Agriculture for Africa: Building Resilient Farming Systems in a Changing Climate.* CABI, Wallingford, UK, pp. 167–182.

Garrity, D., O. Akinnifesi, O. Ajayi, G. W. Sileshi, J. G. Mowo, A. Kalinganire, M. Larwanou and J. Bayala. 2010. Evergreen agriculture: A robust approach to sustainable food security in Africa. *Food Security*, 2(3): 197–214.

Gattinger, A., J. Jawtusch, A. Muller and P. Mäder. 2011. *No-Till Agriculture–A Climate Smart Solution?* Climate Change and Agriculture Report No. 2, MISEREOR e.V., Aachen, Germany.

Getahun, G. T., L. J. Munkholm and P. Schjønning. 2016. The influence of clay-to-carbon ratio on soil physical properties in a humid sandy loam soil with contrasting tillage and residue management. *Geoderma*, 264: 94–102.

Giller, K. E., M. Corbeels, J. Nyamangara, B. Triomphe, F. Affholder, E. Scopel and P. Tittonell. 2011. A research agenda to explore the role of conservation agriculture in African smallholder farming systems. *Field Crops Research*, 124: 468–472.

Giller, K. E., E. Witter, M. Corbeels and P. Tittonell. 2009. Conservation agriculture and smallholder farming in Africa: The heretics' view. *Field Crops Research*, 114(1): 23–34.

Gliessman, S. R. 2014. *Agroecology: The Ecology of Sustainable Food Systems.* 3rd ed. CRC Press, Boca Raton, FL, pp. 405.

Godfray, H. C. J., J. R. Beddington, I. R. Crute, et al. 2010. Food security: The challenge of feeding 9 billion people. *Science*, 327: 812–818.

Godfray, H. C. J. and T. Garnett. 2014. Food security and sustainable intensification. *Philosophical Transactions of the Royal Society of London Series B*, 369: 20120273.

Gonzalez-Sanchez, E. J., O. Veroz-Gonzalez, G. Conway, et al. 2019. Meta-analysis on carbon sequestration through conservation agriculture in Africa. *Soil and Tillage Research*, 190: 22–30. doi: 10.1016/j.still.2019.02.020.

Gonzalez-Sanchez, E. J., O. Veroz-Gonzalez, M. Morena-Garcia, J. A. Gil-Ribes, et al. 2020. Conservation agriculture: Climate change mitigation and adaptability benefits. In Kassam, A. (ed.), *Advances in Conservation Agriculture*, Volume 2 *Practice and Benefits*. Burleigh Dodds, Cambridge, UK, pp. 303–334.

Goss, M. J., M. Carvalho and I. Brito. 2017. *Functional Diversity of Mycorrhiza and Sustainable Agriculture: Management to Overcome Biotic and Abiotic Stress*. Academic Press, London.

Govaerts, B., N. Verhulst, A. Castellanos-Navarrete, K. D. Sayre, J. Dixon and L. Dendooven. 2009. Conservation agriculture and soil carbon sequestration: Between myth and farmer reality. *Critical Reviews in Plant Sciences*, 28(3): 97–122.

Grandy, A. S. and G. P. Robertson. 2006. Aggregation and organic matter protection following tillage of a previously uncultivated soil. *Soil Science Society of America Journal*, 70(4): 1398–1406.

Green, A., T. Nemecek, A. Chaudhary and A. Mathys. 2020. Assessing nutritional, health, and environmental sustainability dimensions of agri-food production. *Global Food Security*, 26: 100406. doi: 10.1016/j.gfs.2020.100406.

Guo, L. B. and R. M. Gifford. 2002. Soil carbon stocks and land use change: A meta analysis. *Global Change Biology*, 8(4): 345–360.

Gullickson, G. 2018. Planting into green cover crops. *Successful Farming*. https://www.agriculture.com/crops/soybeans/planting-into-green-cover-crops.

Hargrove, W. L. and W. W. Frye. 1987. The need for legume cover crops in conservation tillage production. In Power, D. F. (ed.), *The Role of Legumes in Conservation Tillage System*. Soil Conserv. Soc. AM, Ankeny, IA, pp. 1–4.

Harvey, C. A., M. Chacón, C. I. Donatti, et al. 2013. Climate-smart landscapes: Opportunities and challenges for integrating adaptation and mitigation in tropical agriculture. *Conservation Letters*, 7(2): 77–90. doi: 10.1111/conl.12066.

Harvey, E., I. Gounand, C. L. Ward and F. Altermatt. 2017. Bridging ecology and conservation: From ecological networks to ecosystem function. *Journal of Applied Ecology*, 54: 371–379. doi: 10.1111/1365-2664.12769.

Hassink, J. 1997. The capacity of soils to preserve organic C and N by their association with clay and silt particles. *Plant Soil*, 191: 77–87.

Havlin, J. L., D. E. Kissel, L. E. Maddux, M. M. Classen and J. H. Long. 1990. Crop rotation and tillage effects on soil organic carbon and nitrogen. *Soil Science Society of America Journal*, 54: 448–452.

Health Canada. 2019. Canada's dietary guidelines. Retrieved from https://food-guide.canada.ca/static/assets/pdf/CDG-EN-2018.pdf.

Hibbett, D. S. et al. 2007. A higher-level phylogenetic classification of the fungi. *Mycological Research*, 111: 509–547.

Hobbs, P. and B. Govaerts. 2010. How conservation agriculture can contribute to buffering climate change. In M. Reynolds (ed.), *Climate Change and Crop Production*. CABI, Wallingford, UK, pp. 177–199.

Hobbs, P., C. Thierfelder, P. Wall., R. Gupta and A. Calegari. 2020. The role of crop and cropping system management in conservation agriculture systems. In Kassam, A. (ed.), *Advances in Conservation Agriculture Volume 1 – Systems and Science*. Burleigh Dodds Science Publishing Limited, Cambridge, UK, pp. 249–278.

Hobbs, P. R. 2007. Conservation agriculture: What is it and why is it important for future sustainable food production? *The Journal of Agricultural Science*, 145(2): 127–137. doi: 10.1017/S0021859607006892.

Hobbs, P. R., K. Sayre and R. Gupta. 2008. The role of conservation agriculture in sustainable agriculture. *Philosophical Transactions of the Royal Society of London. Series B, Biological Sciences*, 363(1491): 543–555. doi: 10.1098/rstb.2007.2169.

Homyak, P. M., J. C. Blankinship, K. Marchus, D. M. Lucero, J. O. Sickman and J. P. Schimel. 2016. Aridity and plant uptake interact to make dryland soils hotspots for nitric oxide (NO) emissions. *Proceedings of the National Academy of Sciences of the United States of America*, 113: E2608–E2616.

Howard, A. 1947. *The Soil and Health: A Study of Organic Agriculture*. Devin-Adair Company, New York, pp. 307.

Hudson, B. D. 1994. Soil organic matter and available water capacity. *Journal of Soil and Water Conservation*, 49(2): 189–194.

Hunt, J. R., C. Celestina and J. A. Kirkegaard. 2020. The realities of climate change, conservation agriculture and soil carbon sequestration. *Global Change Biology*, 26: 3188–3189. doi: 10.1111/gcb.15082.

Hutchinson, J. J., C. A. Campbell and R. L. Desjardins. 2007. Some perspectives on carbon sequestration in agriculture. *Agricultural and Forest Meteorology*, 142(2): 288–302. doi: 10.1016/j.agrformet.2006.03.030.

Ingram, J. S. I. and E. C. M. Fernandes. 2001. Managing carbon sequestration in soils: Concepts and terminology. *Agriculture, Ecosystems and Environment*, 87: 111–117. doi: 10.1016/S0167-8809(01)00145-1.

IPCC. 2006. *IPCC Guidelines for National Greenhouse Gas Inventories*. Eds. Eggleston, H. S., Buendia, L., Miwa, K., Ngara, T., Tanabe, K. Prepared by the National Greenhouse Gas Inventories Programme, Japan.

Isbell, F., D. Tilman, P. B. Reich and A. T. Clark. 2019. Deficits of biodiversity and productivity linger a century after agricultural abandonment. *Nature Ecology & Evolution*, 3: 1533–1538. www.nature.com/natecolevol.

Jackson, R. B., K. Lajtha, S. E. Crow, G. Hugelius, M. G. Kramer and G. Piñeiro. 2017. The ecology of soil carbon: Pools, vulnerabilities, and biotic and abiotic controls. *Annual Review of Ecology, Evolution, and Systematics*, 48: 419–445.

Janzen, H. H. 2005. Soil carbon: A measure of ecosystem response in a changing world? *Canadian Journal of Soil Science*, 85, Special Issue: 467–480. doi: 10.4141/S04-081.

Janzen, H. H. 2006. The soil carbon dilemma: Shall we hoard it or use it? *Soil Biology and Biochemistry*, 38(3): 419–424.

Janzen, H. H. 2015. Beyond carbon sequestration: Soil as conduit of solar energy. *European Journal of Soil Science*, 66(1): 19–32. doi: 10.1111/ejss.12194/full.

Jat, R. A., K. L. Sahrawat, A. H. Kassam and T. Friedrich. 2014. Conservation agriculture for sustainable and resilient agriculture: Global status, prospects and challenges. In Jat, R. A., Sahrawat, K. L. and Kassam, A. H. (eds.), *Conservation Agriculture: Global Status, Prospects and Challenges*. CABI, Boston, MA, pp. 1–25.

Jenkinson, D. S. and J. H. Rayner. 1977. The turnover of soil organic matter in some of the Rothamsted classical experiments. *Soil Science*, 123(5): 298–305.

Joseph, L. N., R. F. Maloney and H. P. Possingham. 2009. Optimal allocation of resources among threatened species: A project prioritization protocol. *Conservation Biology*, 23(2): 328–338. doi: 10.1111/j.1523-739.2008.01124.x.

Jouquet, P., E. Blanchart and Y. Capowiez. 2014. Utilization of earthworms and termites for the restoration of ecosystem functioning. *Applied Soil Ecology*, 73: 34–40.

Kane, D. 2015. Carbon sequestration potential on agricultural lands: A review of current science and available practices. Special Report, *National Sustainable Agriculture Coalition, Breakthrough Strategies and Solutions, LLC*. http://sustainableagriculture.net/publications.

Kassam, A. 2019. Integrating conservation into agriculture. In Farooq, M. and Pisante, M. (eds.), *Innovations in Sustainable Agriculture*. Springer Nature, Switzerland AG, pp. 27–42. doi: 10.1007/978-3-030-23169-9_2.

Kassam, A. (ed.). 2020. *Advances and Conservation Agriculture Volume 1 – Systems and Science and Volume 2 – Practice and Benefits*. Burleigh Dodds, Cambridge, UK.

Kassam, A., G. Basch, T. Friedrich, et al. 2013. Sustainable soil management is more than what and how crops are grown. In: R. Lal and Stewart, R. A. (eds.), *Principles of Soil Management in Agro-Ecosystems. Advances in Soil Science*. CRC Press, Boca Raton, pp. 337–399.

Kassam, A., R. Derpsch and T. Friedrich. 2014a. Global achievements in soil and water conservation: The case of conservation agriculture. *International Soil and Water Conservation Research*, 2(1): 5–13.

Kassam, A., R. Derpsch and T. Friedrich. 2020a. Development of conservation agriculture systems globally. In Kassam, A. (ed.), *Advances in Conservation Agriculture*, Volume 1, *Systems and Science*. Burleigh Dodds, Cambridge, UK, pp. 31–86.

Kassam, A. and T. Friedrich. 2012. An ecologically sustainable approach to agricultural production intensification: Global perspectives and developments. *Field Actions Science Reports*, 6. http://factsreports.revues.org/1382.

Kassam, A., T. Friedrich, F. Shaxson, H. Bartz, I. Mello, J. Kienzle and J. Pretty. 2014c. The spread of conservation agriculture: Policy and institutional support for adoption and uptake. *Field Actions Science Reports*, 7: 1–12.

Kassam, A., T. Friedrich, F. Shaxson and J. Pretty. 2009. The spread of conservation agriculture: Justification, sustainability and uptake. *International Journal of Agricultural Sustainability*, 7(4): 292–320.

Kassam, A., E. J. Gonzalez-Sanchez, S. C. Cheak, et al. 2020c. Managing conservation agriculture systems: Orchards, plantations and agroforestry. In Kassam, A. (ed.), *Advances in Conservation Agriculture*, vol. 1. Burleigh Dodds, Cambridge, UK, pp. 327–358.

Kassam, A., E. J. Gonzalez-Sanchez, T. Goddard, et al. 2020b. Harnessing ecosystem services with conservation agriculture. In Kassam, A. (ed.), *Advances in Conservation Agriculture*, vol. 2. Burleigh Dodds, Cambridge, UK, pp. 391–418.
Kassam, A. and L. Kassam (eds.). 2020a. *Rethinking Food and Agriculture: New Ways Forward*. Elsevier, Woodhead Publishing, Duxford, UK, p. 444.
Kassam, A. and L. Kassam. 2020b. The need for conservation agriculture. In Kassam, A. (85i]8ed.), *Advances in Conservation Agriculture, Volume 1, Systems and Science*. Burleigh Dodds, Cambridge, UK, pp. 1–30.
Kassam, A. and L. Kassam. 2020c. Practice and benefits of conservation agriculture systems. In Kassam, A. (ed.), *Advances in Conservation Agriculture, Volume 2, Practice and Benefits*. Burleigh Dodds, Cambridge, UK, pp. 1–36.
Kassam, A., S. Mkomwa and T. Friedrich. 2017. *Conservation Agriculture in Africa: Building Resilient Farming Systsems in a Changing Climate*. CABI, Wallingford, UK.
Kassam, A., T. Friedrich and R. Derpsch. 2018. Global spread of conservation agriculture. *International Journal of Environmental Studies*. 76: 1–23. doi: 10.1080/00207233.2018.1494927.
Kassam, A., T. Friedrich, R. Derpsch and J. Kienzle. 2014b. Worldwide adoption of conservation agriculture. *6th World Congress on Conservation Agriculture*, 22–27 June 2014, Winnipeg, Canada.
Kassam, S., D. Jenkins, D. Bristor and Z. Kassam. 2020. Healthy diets as a guide to responsible food systems. In A. Kassam and L. Kassam (eds.), *Rethinking Food and Agriculture: New Ways Forward*. Elsevier, Woodhead Publishing, Duxford, UK, pp. 323–352.
Kätterer, T., M. A. Bolinder, O. Andrén, H. Kirchmann and L. Menichetti. 2011. Roots contribute more to refractory soil organic matter than aboveground crop residues, as revealed by a long-term field experiment. *Agriculture, Ecosystems and Environment*, 141(1–2): 184–192. doi: 10.1016/j.agee.2011.02.029.
Keesstra, S. D., J. Bouma, J. Wall, et al. 2016. The significance of soils and soil science towards realization of the United Nations Sustainable Development Goals. *SOIL*, 2: 111–128. doi: 10.5194/soil-2-111-2016.
Kell, D. B. 2011. Breeding crop plants with deep roots: Their role in sustainable carbon, nutrient and water sequestration. *Annals of Botany*, 108(3): 407–418. doi: 10.1093/aob/mcr175.
Kemper, K. J. and R. Lal. 2017. Pay dirt! human health depends on soil health. *Complementary Therapies in Medicine*, 32: A1–2. doi: 10.1016/j.ctim.2017.04.005.
Kemper, W. D., N. N. Schneider and T. R. Sinclair. 2011. No-till can increase earthworm populations and rooting depths. *Journal of Soil and Water Conservation*, 66(1): 13A–7A. doi: 10.2489/jswc.66.1.13A.
Khan, Z. R., C. A. Midega, J. O. Pittchar, A. Murage and J. Picket. 2017. Climate-smart push-pull: A conservation agriculture technology for food security and environmental sustainability in Africa. In Kassam, A., Mkomwa, S. and Friedrich, T. (eds.), *Conservation Agriculture for Africa: Building Resilient Farming Systems in a Changing Climate*. CABI, Wallingford, UK, pp. 151–166.
Khan, Z. R., A. W. Murage, J. O. Pittchar and C. A. Midega. 2020. Insect pest and disease management practices in conservation agriculture systems: A case of push-pull practice. In Kassam, A. (ed.), *Advances in Conservation Agriculture, vol. 2 – Practice and Benefits*. Burleigh Dodds, Cambridge, UK, pp. 143–168.
Kibblewhite, M. G., K. Ritz and M. J. Swift. 2008. Soil health in agricultural systems. *Philosophical Transactions of the Royal Society of London Series B*, 363(1492): 685–701. doi: 10.1098/rstb.2007.2178.
Kirkegaard, J. A., M. K. Conyers, J. R. Hunt, C. A. Kirkby, M. Watt and G. J. Rebetzke. 2014. Sense and nonsense in conservation agriculture: Principles, pragmatism and productivity in Australian mixed farming systems. *Agriculture, Ecosystems & Environment*, 187: 133–145.
Kladivko, E. J., M. J. Helmers, L. J. Abendroth, et al. 2014. Standardized research protocols enable transdisciplinary research of climate variation impacts in corn production systems. *Journal of Soil and Water Conservation*, 69(6): 532–542. doi: 10.2489/jswc.69.6.532.
Koizumi, H., T. Nakadai, Y. Usami, et al. 1991. Effect of carbon dioxide concentration on microbial respiration. *Soil Ecological Research*, 6: 227–232.
Kravchenko, A. N. and G. P. Robertson. 2011. Whole-profile soil carbon stocks: The danger of assuming too much from analyses of too little. *Soil Science Society of America Journal*, 75: 235–240.
Kumar, R., S. Pandey and A. Pandey. 2006. Plant roots and carbon sequestration. *Current Science*, 91: 885–890.
Laban, P., G. Metternicht and J. Davies. 2018. *Soil Biodiversity and Soil Organic Carbon: Keeping Drylands Alive*. IUCN, Gland, Switzerland. viii + 24p. www.iucn.org/resources/publications.
LaCanne, C. E. and J. G. Lundgren. 2018. Regenerative agriculture: Merging farming and natural resource conservation profitably. *PeerJ*, 6:e4428. doi: 10.7717/peerj.4428.

Laishram, J., K. G. Saxena, R. K. Maikhuri and K. S. Rao. 2012. Soil quality and soil health: A review. *International Journal of Ecology and Environmental Sciences*, 38(1): 19–37.

Lal, R. 2008. Carbon sequestration. *Philosophical Transactions of the Royal Society of London Series B*, 363: 815–830.

Lal, R. 2010. Managing soils and ecosystems for mitigating anthropogenic carbon emissions and advancing global food security. *BioScience*, 60(9): 708–721.

Lal, R. 2013a. Food security in a changing climate. *Ecohydrology & Hydrobiology*, 13(1): 8–21.

Lal, R. 2013b. Intensive agriculture and the soil carbon pool. In M. S. Kang and S. S. Banga (eds.), *Combating Climate Change: An Agricultural Perspective*. CRC Press, Taylor and Francis Group, Boca Raton, FL, pp. 59–72.

Lal, R. 2014a. Climate strategic soil management. *Challenges*, 5(1): 43–74. doi: 10.3390/challe5010043.

Lal, R. 2014b. Soil conservation and ecosystem services. *International Soil and Water Conservation Research*, 2(3): 36–47. doi: 10.1016/S2095-6339(15)30021-6.

Lal, R. 2015a. Restoring soil quality to mitigate soil degradation. *Sustainability*, 7(5): 5875–5895.

Lal, R. 2015b. A system approach to conservation agriculture. *Journal of Soil and Water Conservation*, 70(4): 82A–88A. doi: 10.2489/jswc.70.4.82A.

Lal, R. 2015c. Sequestering carbon and increasing productivity by conservation agriculture. *Journal of Soil and Water Conservation*, 70(3): 55A–62A. doi: 10.2489/jswc.70.3.55A.

Lal, R. 2016. Soil health and carbon management. *Food and Energy Security*, 5: 212–222.

Lal, R. 2020a. Regenerative agriculture for food and climate. *Journal of Soil and Water Conservation* 75: 123A. doi: 10.2489/jswc.2020.0620A.

Lal, R. 2020b. Soil organic matter content and crop yield. *Journal of Soil and Water Conservation*, 75(2): 27A–32A. doi: 10.2489/jswc.75.2.27A.

Lal, R., D. C. Reicosky and J. D. Hanson. 2007. Evolution of the plow over 10,000 years and the rationale for no-till farming. *Soil and Tillage Research*, 93: 1–12. doi: 10.1016/j.still.2006.11.004.

Lalani, B., P. Dorward and G. Holloway. 2017. Farm-level economic analysis: Is conservation agriculture helping the poor? *Ecological Economics*, 141: 144–153.

Landi, L., F. Valori, J. Ascher, et al. 2006. Root exudate effects on the bacterial communities, CO2 evolution, nitrogen transformations and ATP content of rhizosphere and bulk soils. *Soil Biology and Biochemistry*, 38: 509–516.

Larson, W. E., M. J. Lindstrom and T. E. Schumacher. 1997. The role of severe storms in soil erosion: A problem needing consideration. *Journal of Soil Water Conservation*, 52(2): 90–5.

Lehman, R. M., C. A. Cambardella, D. E. Stott, et al. 2015. Understanding and enhancing soil biological health: The solution for reversing soil degradation. *Sustainability*, 7(1): 988–1027. doi: 10.3390/su7010988.

Lehmann, J., D. A. Bossio, I. Kögel-Knabner, et al. 2020. The concept and future prospects of soil health. *Nature Reviews Earth and Environment*. 1:544–553. doi: 10.1038/s43017-020-0080-8.

Lehmann, J. and M. Kleber. 2015. The contentious nature of soil organic matter. *Nature*, 528(7580): 60–68. doi: 10.1038/nature16069.

Li, H. W., J. He, Z. P. Bharuchab, R. Lal and J. Pretty. 2016. Improving China's food and environmental security with conservation agriculture. *International Journal of Agricultural Sustainability*. Informa UK Limited, trading as Taylor & Francis Group, pp. 1–15. doi: 10.1080/14735903.2016.1170330.

Li, Z., R. Zhang, X. Wang, et al. 2012. Growing season carbon dioxide exchange in flooded non-mulching and non-flooded mulching cotton. *PLOS ONE*, 7: e50760.

Loos, J., D. J. Abson, M. J. Chappell, et al. 2014. Putting meaning back into "sustainable intensification". *Frontiers in Ecology and the Environment*, 12(6): 356–361.

Loveland, P. and J. Webb. 2003. Is there a critical level of organic matter in the agricultural soils of temperate regions: A review. *Soil & Tillage Research*, 70: 1–18.

Lützow, M. V., I. Kogel-Knabner, K. Ekschmitt, E. Matzner, G. Guggenberger, B. Marschner and H. Flessa. 2006. Stabilization of organic matter in temperate soils: Mechanisms and their relevance under different soil conditions—A review. *European Journal of Soil Science*, 57(4): 426–445.

Lynch, J. P. and T. Wojciechowski. 2015. Opportunities and challenges in the subsoil: Pathways to deeper rooted crops. *Journal of Experimental Botany*, 66(8): 2199–2210.

Magdoff, F. 2001. Concept, components, and strategies of soil health in agroecosystems. *Journal of Nematology*, 33: 169–172.

Maharjana, B., S. Das and B. S. Achary. 2020. Soil health gap: A concept to establish a benchmark for soil health management. *Global Ecology and Conservation*, 23: e01116. doi: 10.1016/j.gecco.2020.e01116.

Mannering, J. V. and C. R. Fenster. 1983. What is conservation tillage? *Journal of Soil and Water Conservation*, 38(3): 140–143.

Marongwe, L. S., K. Kwazira, M. Jenrich, C. Thierfelder, A. Kassam and T. Friedrich. 2011. An African success: The case of conservation agriculture in Zimbabwe. *International Journal of Agricultural Sustainability*, 9: 153–161.

Mason, J. and L. Kassam. 2020. Agriculture planted the seeds of alienation from nature. In A. Kassam and L. Kassam (eds.), *Rethinking Food and Agriculture: New Ways Forward*. Elsevier, Woodhead Publishing, Duxford, UK, pp. 31–52.

McAfee, A. 2019. *More from Less: The Surprising Story of How We Learned to Prosper Using Fewer Resources—And What Happens Next*. Simon and Schuster, New York, pp. 352.

McDonald, M. D., K. Lewis, G. L. Ritchie, et al. 2019. Carbon dioxide mitigation potential of conservation agriculture in a semi-arid agricultural region. *AIMS Agriculture and Food*, 4(1): 206–222. doi: 10.3934/agrfood.2019.1.206.

McKell, D. and R. A. Peiretti. 2004. *Promoting Soil Conservation and Conservation Agriculture through Farmer Associations*. ISCO 2004 - 13th International Soil Conservation Organisation Conference – Brisbane, July 2004.

McMichael, P. 2020. Political-economy of the global food and agriculture system. In A. Kassam and L. Kassam (eds.), *Rethinking Food and Agriculture: New Ways Forward*. Elsevier, Woodhead Publishing, Duxford, UK. pp. 53–76.

MEA - Millennium Ecosystem Assessment. 2005. *Ecosystems and Human Well-Being: Synthesis*. Island Press, Washington, DC.

Merriam-Webster. 2020. Merriam-Webster.com Dictionary, Merriam-Webster. https://www.merriam-webster.com/dictionary/conservation (accessed 26 August 2020).

Millen, B. E., S. Abrams, L. Adams-Campbell, C. A. Anderso, J. T. Brenna, W. W. Campbell, … A. H. Lichtenstein. 2016. The 2015 dietary guidelines advisory committee scientific report: Development and major conclusions. *Advances in Nutrition: An International Review Journal*. 7(3):438–444. doi: 10.3945/an.116.012120.

Milne, E., S. A. Banwart, E. Noellemeyer, et al. 2015. Soil carbon, multiple benefits. *Environmental Development*, 13: 33–38.

Milne, E., M. Sessay, K. Paustian, et al. 2010. Towards a standardized system for the reporting of carbon benefits in sustainable land management projects. Grassland carbon sequestration: Management, policy and economics (Proceedings of the Workshop on the role of grassland carbon sequestration in the mitigation of climate change, Rome, April 2009). *Integrated Crop Management*, 11: 105–117.

Miner, G. L., J. A. Delgado, C. E. Stewart and J. Ippolito. 2020. Soil health management practices and crop productivity. *Agricultural & Environmental Letters*, 5: e20023, 1–8.

Mitchell, J. P., D. C. Reicosky, E. A. Kueneman, J. Fisher and D. Beck. 2019. Conservation agriculture systems. *CAB Reviews*, 14(001): 25. https://www.cabi.org/cabreviews/review/20193184383.

Moebius-Clune, B. N., D. J. Moebius-Clune, B. K. Gugino, O. J. Idowu, et al. 2016. *Comprehensive Assessment of Soil Health – The Cornell Framework Manual*, edition 3.1. Cornell University, Geneva, NY. http://soilhealth.cals.cornell.edu/ (accessed 1 February 2019).

Mokany, K., R. J. Raison and A. S. Prokushkin. 2006. Critical analysis of root: Shoot ratios in terrestrial biomes. *Global Change Biology*, 12: 84–96.

Montgomery, D. R. 2007a. *Dirt: The Erosion of Civilizations*. University of California Press, Berkeley, CA.

Montgomery, D. R. 2007b. Soil erosion and agricultural sustainability. *Proceedings of the National Academy of Sciences of the United States of America*, 104(33), 268–272.

Morales-Castilla, I., M. G. Matias, D. Gravel and M. B. Araujo. 2015. Inferring biotic interactions from proxies. *Trends in Ecology & Evolution*, 30: 347–356.

Moreira, W. H., C. A. Tormena, D. L. Karlen, Á. P. DaSilva, T. Keller and E. Betioli Jr. 2016. Seasonal changes in soil physical properties under long-term no-tillage. *Soil and Tillage Research*, 160: 53–64. doi: 10.1016/j.still.2016.02.007.

Morgan, J. A., R. F. Follet, L. H. Allen, et al. 2010. Carbon sequestration in agricultural lands of the United States. *Journal of Soil and Water Conservation*, 65: 7A–13A.

Murphy, B. W. 2014. *Soil Organic Matter and Soil Function – Review of the Literature and Underlying Data*. Australian Government, Department of the Environment, Canberra, Australia.

Newton, P., N. Civita, L. Frankel-Goldwater, K. Bartel and C. Johns. 2020. What is regenerative agriculture? A review of scholar and practitioner definitions based on processes and outcomes. *Frontiers in Sustainable Food Systems*. 4: Article 577723. pp. 1–11. doi: 10.3389/fsufs.2020.577723.

Nichols, V., N. Verhulst, R. Cox and B. Govaerts. 2015. Weed dynamics and conservation agriculture principles: A review. *Field Crops Research*, 183: 56–68.

Nielsen, U. N., E. Ayres, D. H. Wall and R. D. Bardgett. 2011. Soil biodiversity and carbon cycling: A review and synthesis of studies examining diversity-function relationships. *Global Change Biology*, 62(1): 105–116.

Nobari, N. 2020. Social movements in the transformation of food and agriculture systems. In A. Kassam and L. Kassam (eds.), *Rethinking Food and Agriculture: New Ways Forward*. Elsevier, Woodhead Publishing, Duxford, UK, pp. 371–398.

Norberg-Hodge, H. 2020. Alternatives to the global food regime: Steps toward system transformation. In A. Kassam and L. Kassam (eds.), *Rethinking Food and Agriculture: New Ways Forward*, Elsevier, Woodhead Publishing, Duxford, UK, pp. 399–412.

NRCS. 2017. Soil health [WWW Document]. *NRCS*. https://www.nrcs.usda.gov/wps/portal/nrcs/main/soils/health/ (accessed 23 April 2020).

O'Dell, D., N. S. Eash, B. B. Hicks, et al. 2020. Conservation agriculture as a climate change mitigation strategy in Zimbabwe. *International Journal of Agricultural Sustainability*. 18(3): 250–265. doi: 10.1080/14735903.2020.1750254.

Odell, R. T., S. W. Melsted and W. M. Walker. 1984. Changes in organic carbon and nitrogen of Morrow Plot soils under different treatments. *Soil Science*, 137(3): 160–171.

Ogle, S. M., F. J. Breidt and K. Paustian. 2005. Agricultural management impacts on soil organic carbon storage under moist and dry climatic conditions of temperate and tropical regions. *Biogeochemistry*, 72: 87–121.

Oldfield, E. E., M. A. Bradford and S. A. Wood. 2019. Global meta-analysis of the relationship between soil organic matter and crop yields. *SOIL*, 5(1): 15–32. doi:doi: 10.5194/soil-5-15-2019.

Oldfield, E. E., S. A. Wood, C. A. Palm and M. A. Bradford. 2015. How much SOM is needed for sustainable agriculture? *Frontiers in Ecology and the Environment*, 13(10): 527–544.

Oliver, M. A. and P. J. Gregory. 2015. Soil, food security and human health: A review. *European Journal of Soil Science*, 66: 257–276.

Olson, K., S. A. Ebelhar and J. M. Lang. 2014b. Long-term effects of cover crops on crop yields, soil organic carbon stocks and sequestration. *Open Journal of Soil Science*, 4: 284–92.

Olson, K. R., M. M. Al-Kaisi, R. Lal and B. Lowery. 2014a. Experimental consideration, treatments, and methods in determining soil organic carbon sequestration rates. *Soil Science Society of America Journal*, 78(2): 348–360. doi: 10.2136/sssaj2013.09.0412.

Owenya, M. Z., L. W. Mariki, J. Kienzle, T. Friedrich and A. Kassam. 2011. Conservation agriculture (CA) in Tanzania: The case of the Mwangaza BCA farmer field school, Rhotia Village, Karatu District, Arusha. *International Journal of Agricultural Sustainability*, 9: 145–152.

Page, K. L., Y. P. Dang and R. C. Dalal. 2020. The ability of conservation agriculture to conserve soil organic carbon and the subsequent impact on soil physical, chemical, and biological properties and yield. *Frontiers in Sustainable Food Systems*, 4: 31. doi: 10.3389/fsufs.2020.00031.

Palm, C., H. Blanco-Canqui, F. De Clerck, L. Gatere and P. Grace. 2014. Conservation agriculture and ecosystem services: An overview. *Agriculture, Ecosystems & Environment*, 187: 87–105.

Pape, A. and L. S. Prokopy. 2017. Delivering on the potential of formal farmer networks: Insights from Indiana. *Journal of Soil and Water Conservation*, 72(5): 463–470. www.swcs.org.

Pascual, U., M. Termansen, K. Hedlund, et al. 2015. On the value of soil biodiversity and ecosystem services. *Ecosystem Services*, 15: 11–8.

Paustian, K., J. Lehmann, S. Ogle, D. Reay, G. P. Robertson and P. Smith. 2016. Climate-smart soils. *Nature*, 532: 49–57. doi: 10.1038/nature17174.

Pelletier, N., E. Audsley, S. Brodt, et al. 2011. Energy intensity of agriculture and food systems. *Annual Review of Environment and Resources*, 36 (1): 223–46.

Pepper, I. L. 2013. The soil health-human health nexus. *Critical Reviews in Environmental Science and Technology*, 43(24): 2617–2652.

Pereira, H. M., L. M. Navarro and I. S. Martins. 2012. Global biodiversity change: The bad, the good, and the unknown. *Annual Review of Environment and Resources*, 37(1): 25–50.

Phelan, P. L. 2009. Ecology-based agriculture and the next green revolution: Is modern agriculture exempt from the laws of ecology? In Bohlen, P. J. and House, G. (eds.), *Sustainable Agroecosystem Management: Integrating Ecology, Economics, and Society*. CRC Press: Boca Raton, FL, pp. 97–135.

Philibert, A., C. Loyce and D. Makowski. 2012. Assessment of the quality of meta-analysis in agronomy. *Agriculture, Ecosystems and Environment*, 148: 72–82.

Pieri, C. J. M. 1995. Long term soil management experiments in semiarid Francophone Africa. In Lal, R. and Stewart, B. A. (eds.), *Soil Management: Experimental Basis for Sustainability and Environmental Quality*. Papers Presented at a Workshop Held June 16–18, 1993 at Ohio State University. Advances in Soil Science Series. Boca Raton, pp. 225–266.

Pieri, C. J. M. 2012. *Fertility of Soils: A Future for Farming in the West African Savannah*. Springer-Verlag, Berlin, Germany, 348 pp.

Pieri, C. J. M., G. Evers, J. Landers, P. O'Connell and E. Terry. 2002. *No-Till Farming for Sustainable Rural Development*. Agriculture & Rural Development Working Paper. World Bank, Washington, DC.

Pimentel, D., C. Harvey, P. Resosudarmo, et al. 1995. Environmental and economic cost of soil erosion and conservation benefits. *Science*, 267(5201): 1117–1123.

Pittelkow, C. M., B. A. Linquist, M. E. Lundy, et al. 2015a. When does no-till yield more? A global meta-analysis. *Field Crops Research*, 183: 156–168. doi: 10.1016/j.fcr.2015.07.020.

Pittelkow, C. M., X. Liang, B. A. Linquist, et al. 2015b. Productivity limits and potentials of the principles of conservation agriculture. *Nature*, 517(7534): 365–368. doi: 10.1038/nature13809.

Poeplau, C. and A. Don. 2015. Carbon sequestration in agricultural soils via cultivation of cover crops – A meta-analysis. *Agriculture, Ecosystems & Environment*, 200: 33–41.

Power, J. F. 1994. Understanding the nutrient cycling process. *Journal of Soil and Water Conservation*, 49: 16–23.

Powlson, D. S., Z. Cai and P. Lemanceau. 2015. Soil carbon dynamics and nutrient cycling. In S. A. Banwart, E. Noellemeyer, and E. Milne (eds.), *Soil Carbon: Science, Management and Policy for Multiple Benefits*. CAB International, Wallingford, UK, pp. 98–107.

Powlson, D. S., C. M. Stirling, M. L. Jat, et al. 2014. Limited potential of no-till agriculture for climate change mitigation. *Nature Climate Change*, 4: 678. doi: 10.1038/nclimate2292.

Powlson, D. S., C. M. Stirling, C. Thierfelder, R. P. White and M. L. Jat. 2016. Does conservation agriculture deliver climate change mitigation through soil carbon sequestration in tropical agro-ecosystems? *Agriculture, Ecosystems and Environment* 220: 164–174.

Pretty, J. 1995. *Regenerating Agriculture: Policies and Practice for Sustainability and Self-Reliance*. Earthscan Publications, London, UK, pp. 329.

Pretty, J. 1997. The sustainable intensification of agriculture. *Natural Resources Forum*, 21(4): 247–256.

Pretty, J. 2002. *Agri-Culture: Reconnecting People, Land and Nature*. Earthscan, London, p. 261.

Pretty, J. and Z. P. Bharucha. 2014. Sustainable intensification in agricultural systems. *Annals of Botany*, 114(8): 1571–1596. doi: 10.1093/aob/mcu205.

Pulido-Moncada, M., D. Gabriels, W. Cornelis and D. Lobo. 2015. Comparing aggregate stability tests for soil physical quality indicators. *Land Degradation and Development*, 26: 843–852.

Pulido-Moncada, M., D. Gabriels, D. Lobo, et al. 2014. A comparison of methods to assess susceptibility to soil sealing. *Geoderma*, 226: 397–404.

Qi, Z., M. J. Helmers, R. D. Christianson and C. H. Pederson. 2011. Nitrate-nitrogen losses through subsurface drainage under various agricultural land covers. *Journal of Environmental Quality*, 40(5): 1578–1585. doi: 10.2134/jeq2011.0151.

Raich, J. W. and W. H. Schlesinger. 1992. The global carbon dioxide flux in soil respiration and its relationship to vegetation and climate. *Tellus*, 44: 81–99.

Ranaivoson, L., K. Naudin, A. Ripoche, et al. 2017. Agro-ecological functions of crop residues under conservation agriculture. A review. *Agronomy for Sustainable Development*, 37(26): 1–17. doi: 10.1007/s13593-017-0432-z.

Ranganathan, J., R. Waite, T. Searchinger, et al. 2020. Regenerative agriculture: Good for soil health, but limited potential to mitigate climate change. https://www.wri.org/blog/2020/05/regenerative-agriculture-climate-change.

Reicosky, D. C. 1997a. Tillage methods and carbon dioxide loss: Fall versus spring tillage. In R. Lal, J. M. Kimble, R. F. Follett, and B. A. Stewart (eds.), *Management of Carbon Sequestration in Soil*. CRC Press, Boca Raton, FL, pp. 99–111.

Reicosky, D. C. 1997b. Tillage-induced CO_2 emission from soil. *Nutrient Cycling in Agroecosystems*, 49: 273–285.

Reicosky, D. C. 2003. Conservation agriculture: Global environmental benefits of soil carbon management. In García Torres, L., Benites, J., Martínez-Vilela, A., and Holgado-Cabrera, A. (eds.), *Conservation Agriculture*. Springer, Dordrecht, pp. 3–12.

Reicosky, D. C. 2015. Conservation tillage is not conservation agriculture. *Journal of Soil and Water Conservation*, 70(4): 103A–108A.

Reicosky, D. C. 2020. Conservation agriculture systems: Soil health and landscape management. In Kassam, A. (ed.), *Advances in Conservation Agriculture, Volume 1, Systems and Science*. Burleigh Dodds, Cambridge, UK, pp. 87–154.

Reicosky, D. C. and R. R. Allmaras. 2003. Advances in tillage research in North American cropping systems. In A. Shrestha (ed.), *Cropping Systems: Trends and Advances*. Haworth Press, Inc., New York, pp. 75–125.

Reicosky, D. C. and D. W. Archer. 2007. Moldboard plow tillage depth and short-term carbon dioxide release. *Soil Tillage Research*, 94: 109–121.

Reicosky, D. C., S. D. Evans, C. A. Cambardella, et al. 2002. Continuous corn with moldboard tillage: Residue and fertility effects on soil carbon. *Journal of Soil and Water Conservation*, 57(5): 277–284.

Reicosky, D. C. and F. Forcella. 1998. Cover crop and soil quality interactions in agroecosystems. *Journal of Soil and Water Conservation*, 53(3): 224–229.

Reicosky, D. C. and H. H. Janzen. 2018. Conservation agriculture: Maintaining land productivity and health by managing carbon flows. In R. Lal and B. A. Stewart (eds.), *Soil and Climate. Advances in Soil Science*. Taylor and Francis Group, LLC, CRC Press, Boca Raton, pp. 131–161.

Reicosky, D. C. and W. D. Kemper. 1995. Effects of crop residue on infiltration, evaporation, and water-use efficiency. In W. C. Moldenhauer and L. N. Mielke (eds.), *Crop Residue Management to Reduce Erosion and Improve Soil Quality*. USDA-ARS, pp. 65–72, Conservation Research Report Number 42, North Central Region.

Reicosky, D. C., W. D. Kemper, G. W. Langdale, C. L. Douglas, Jr. and P. E. Rasmussen. 1995. Soil organic matter changes resulting from tillage and biomass production. *Journal of Soil and Water Conservation*, 50(3): 253–261.

Reicosky, D. C. and M. J. Lindstrom. 1993. Fall tillage method: Effect on short-term carbon dioxide flux from soil. *Agronomy Journal*, 85: 1237–1243.

Reicosky, D. C. and M. J. Lindstrom. 1995. The impact of fall tillage on short-term carbon dioxide flux. In R. Lal et al. (eds.), *Soils and Global Change*. Lewis Publishers, Chelsea, MI, pp. 177–187.

Reicosky, D. C., T. J. Sauer and J. L. Hatfield. 2011. Challenging balance between productivity and environmental quality: Tillage impacts. In J. L. Hatfield and T. J. Sauer (eds.), *Editors Soil Management: Building a Stable Base for Agriculture*. American Society of Agronomy and Soil Science Society of America, Madison, WI, pp. 13–38.

RELACO. 1997. Proceedings of the bi-annual meeting of the Latin-American network for conservation tillage. In Claverán, A. R. and Rulfo, F. O. (eds.), *Memorias de la IV Reunión Bienal de la Red Latinoamericana de Agricultura Conservacionista*, RELACO, 17–20 November, Morelia, Michoacán, México. CENAPROS, INIFAP, SAGAR, FAO, 361p.

Reusser, P., P. Bierman and D. Rood. 2015. Quantifying human impacts on rates of erosion and sediment transport at a landscape scale. *Geology*, 43(2): 171–4. doi: 10.1130/G36272.1.

Reynolds, W. D., C. F. Drury, C. S. Tan, C. A. Fox and X. M. Yang. 2009. Use of indicators and pore volume–function characteristics to quantify soil physical quality. *Geoderma*, 152: 252–263.

Robertson, G. P., D. C. Coleman, C. S. Bledsoe, et al. (eds.). 1999. *Standard Soil Methods for Long-Term Ecological Research*. Oxford University Press, New York.

Rochette, P., R. Desjardins and E. J. Pattey. 1991. Spatial and temporal variability of soil respiration in agricultural fields. *Canadian Journal of Soil Science*, 71: 189–196.

Rockström, J., J. Williams, G. Daily, et al. 2017. Sustainable intensification of agriculture for human prosperity and global sustainability. *Ambio*, 46: 4–17.

Rodale Institute. 2014. *Regenerative Organic Agriculture and Climate Change: A Down-to-Earth Solution to Global Warming*. Rodale Institute, Inc., Kutztown, PA.

Rodale, R. 1983. Breaking new ground: The search for a sustainable agriculture. *The Futurist*, 17: 15–20.

Sá, J. C. de M., L. Séguy, F. Tivet, R. Lal, et al. 2015. Carbon depletion by plowing and its restoration by no-till cropping systems in oxisols of subtropical and tropical agro-ecoregions in Brazil. *Land Degradation and Development*, 26: 531–543. doi: 10.1002/ldr.2218.

Sá, J. C. de M., F. Tivet, R. Lal, A. deO. Ferreira, C. Breidid, T. M. Inagaki and D. P. Goncalves. 2020. Carbon management practices and benefits in conservation agriculture systems: Carbon sequestration rates. In Kassam, A. (ed.), *Advances in Conservation Agriculture: Volume 2 – Practice and Benefits*. Burleigh Dodds, Cambridge, UK, pp. 199–228.

Sanderman, J., T. Hengl and G. J. Fiske. 2017. Soil carbon debt of 12,000 years of human land use. *Proceedings of the National Academy of Sciences of the United States of America*, 114(36): 9575–9580.

Schjønning, P., L. W. de Jonge, L. J. Munkholm, et al. 2012. Clay dispersibility and soil friability—testing the soil clay-to-carbon saturation concept. *Vadose Zone Journal*, 11(1): vzj2011.0067. https://doi.org/10.2136/vzj2011.0067

Schjønning, P., J. L. Jensen, S. Bruun, et al. 2018. The role of soil organic matter for maintaining crop yields: Evidence for a renewed conceptual basis. *Advances in Agronomy*, 150: 35–79. Academic Press. doi: 10.1016/bs.agron.2018.03.001.

Seiter, S. and W. R. Horwath. 2004. Strategies for managing soil organic matter to supply plant nutrients. In F. Magdoff and R. R. Weil (eds.), *Soil Organic Matter in Sustainable Agriculture*. CRC Press, Boca Raton, FL, pp. 269–293.

Semenov, V. M., A. S. Tulina, N. A. Semenova and L. A. Ivannikova. 2013. Humification and nonhumification pathways of the organic matter stabilization in soil: A review. *Eurasian Soil Science*, 46(4): 355–68.

Sherwood, S. and N. Uphoff. 2000. Soil health: Research, practice and policy for a more regenerative agriculture. *Applied Soil Ecology*, 15: 85–97.

Sievers, T. and R. L. Cook. 2018. Aboveground and root decomposition of cereal rye and hairy vetch cover crops. *Soil Science Society of America Journal*, 82(1): 147–155. doi: 10.2136/sssaj2017.05.0139.

Singer, J. W., S. M. Nusser and C. J. Alf. 2007. Are cover crops being used in the US corn belt? *Journal of Soil and Water Conservation*, 62(5): 353–358.

Six, J., R. T. Conant, E. A. Paul and K. Paustian. 2002. Stabilization mechanisms of soil organic matter: Implications for C-saturation of soils. *Plant and Soil*, 241(2): 155–176.

Smith, M. D. 2011. The ecological role of climate extremes: Current understanding and future prospects. *Journal of Ecology*, 99(3): 651–655.

Snyder, C. S., T. W. Bruulsema, T. L. Jensen, et al. 2009. Review of greenhouse gas emissions from crop production systems and fertilizer management effects. *Agriculture, Ecosystems and Environment*, 133: 247–266.

Soil Science Glossary Terms Committee. 2008. *Glossary of Soil Science Terms*. Soil Science Society of America, Madison, WI.

Sommer, R., B. K. Paul, J. Mukalama and J. Kihara. 2018. Reducing losses but failing to sequester carbon in soils–the case of conservation agriculture and integrated soil fertility management in the humid tropical agro-ecosystem of Western Kenya. *Agriculture, Ecosystems and Environment*, 254: 82–91.

Sorrenson, W. J. 1997. Financial and economic implications of no-tillage and crop rotations compared to conventional cropping systems. *TCI Occasional Paper Series No.9*, FAO, Rome.

Stavi, I., R. Lal, S. Jones and R. C. Reeder. 2012. Implications of cover crops for soil quality and geodiversity in a humid-temperate region in the Mid-western USA. *Land Degradation and Development*, 23(4): 322–330. doi: 10.1002/ldr.2148.

Stevenson, J. R., R. Serraj and K. G. Cassman. 2014. Evaluating conservation agriculture for small-scale farmers in Sub-Saharan Africa and South Asia. *Agriculture, Ecosystems and Environment*, 187: 1–10.

Stewart, B. A. 2018. Nutrient requirements for soil carbon sequestration. In R. Lal and B. A. Stewart (eds.), *Soil and Climate. Advances in Soil Science*. Taylor and Francis Group, LLC, CRC Press, Boca Raton, pp. 163–170.

Stewart, C. E., K. Paustian, R. T. Conant, A. F. Plante and J. Six. 2007. Soil carbon saturation: Concept, evidence and evaluation. *Biogeochemistry*, 86(1): 19–31.

Stirling, G., H. Hayden, T. Pattison and M. Stirling. 2016. *Soil Health, Soil Biology, Soilborne Diseases and Sustainable Agriculture. A Guide*. CSIRO Publishing, Clayton South, VIC 3169 Australia, p. 280.

Stockmann, U., M. A. Adams, J. W. Crawford, et al. 2013. The knowns, known unknowns and unknowns of sequestration of soil organic carbon. *Agriculture, Ecosystems and Environment*, 164: 80–99.

Strock, J. S., P. M. Porter and M. P. Russelle. 2004. Cover cropping to reduce nitrate loss through subsurface drainage in the northern UC Corn Belt. *Journal of Environmental Quality*, 33(3): 1010–6. doi: 10.2134/jeq2004.1010.

Sumberg, J., J. Thompson and P. Woodhouse. 2012. Contested agronomy: Agricultural research in a changing world. In J. Sumberg and J. Thompson (eds.), *Contested Agronomy: Agricultural Research in a Changing World*. Routledge, London, pp. 1–21.

Tardy, V., A. Spor, O. Mathieu, et al. 2015. Shifts in microbial diversity through land use intensity as drivers of carbon mineralization in soil. *Soil Biology and Biochemistry*, 90: 204–13.

Thierfelder, C., R. Matemba-Mutasa and L. Rusinamhodzi. 2015. Yield response of maize (Zea mays L.) to conservation agriculture in Southern Africa. *Soil & Tillage Research*, 146: 232–240.

Tiessen, H., E. Cuevas and P. Chacon. 1994. The role of soil organic matter in sustaining soil fertility. *Nature*, 371: 783–785.

Tilman, D., M. Clark, D. R. Williams, K. Kimmel, S. Polasky and C. Packer. 2017. Future threats to biodiversity and pathways to their prevention. *Nature*, 546: 73–81.

Tsiafouli, M. A., E. Thébault, S. P. Sgardelis, et al. 2015. Intensive agriculture reduces soil biodiversity across Europe. *Global Change Biology*, 21: 973–985.

United States Department of Agriculture (USDA). 2018. *Soil Health Nuggets.* USDA, Washington, DC. https://www.nrcs.usda.gov/Internet/FSE_DOCUMENTS/stelprdb1101660.pdf.

USDA-NRCS. 2019. Soil health. https://www.nrcs.usda.gov/wps/portal/nrcs/main/soils/health/

Van Eerd, L. L., K. A. Congreves, A. Hayes, A. Verhallen and D. C. Hooker. 2014. Long-term tillage and crop rotation effects on soil quality, organic carbon, and total nitrogen. *Canadian Journal of Soil Science*, 94(3): 303–15. doi: 10.4141/cjss2013-093.

Van-Camp, L., B. Bujarrabal, A. R. Gentile, et al. 2004. *Reports of the Technical Working Groups Established under the Thematic Strategy for Soil Protection.* EUR 21319 EN/5. Office for Official Publications of the European Communities, Luxembourg, p. 872.

Veloso, M. G., D. A. Angers, M. H. Chantigny and C. Bayer. 2020. Carbon accumulation and aggregation are mediated by fungi in tropical soil under conservation agriculture. *Geoderma*, 363: 114159. doi: 10.1016/j.geoderma.2019.114159.

Volk, T. 2003. *Gaia's Body – Toward a Physiology of Earth.* Cambridge, MA: MIT Press.

Voroney, R. P., E. A. Paul and D. W. Anderson. 1989. Decomposition of wheat straw and stabilization of microbial producers. *Canadian Journal of Soil Science*, 69(1): 63–77.

Wall, D. H., R. D. Bardgett and E. F. Kelly. 2010. Biodiversity in the dark. *Nature Geoscience*, 3(5): 297–298.

Wall, D. H. and U. N. Nielsen. 2012. Biodiversity and ecosystem services: Is it the same below ground? *Nature Education Knowledge*, 3(12): 8. http://www.nature.com/scitable/knowledge/library/biodiversity-and-ecosystem-services-is-it-the-96677163.

Wall, D. H., U. N. Nielsen and J. Six. 2015. Soil biodiversity and human health. *Nature*, 528: 69–76.

Wander, M. 2004. Soil organic matter fractions and their relevance to soil function. In F. Magdoff and R. R. Weil (eds.), *Soil Organic Matter in Sustainable Agriculture.* CRC Press, Boca Raton, FL, pp. 67–102.

WCCA. 2001. Proceedings of the Ist World Congress on Conservation Agriculture: Environment, Farmers Experiences, Innovations, Socio-economy, Policy. Garcia-Torres, L., Benites, J., Martinez-Vilela, A. and Holgado-Cabrera, A. (eds). Madrid, Spain. October 1–5, 2001. ECAF and FAO.

Wei, W., Y. Yan, J. Cao, P. Christie, F. Zhang and M. Fan. 2016. Effects of combined application of organic amendments and fertilizers on crop yield and soil organic matter: An integrated analysis of long-term experiments. *Agriculture, Ecosystems and Environment*, 225: 86–92.

White, J. W., L. A. Hunt, K. J. Boote, et al. 2013. Integrated description of agricultural field experiments and production: The ICASA Version 2.0 data standards. *Computers and Electronics in Agriculture*, 96: 1–12.

Wilhelm, W. W., J. M. F. Johnson, J. L. Hatfield, et al. 2004. Crop and soil productivity response to corn residue removal: A literature review. *Agronomy Journal*, 96(1): 1–17.

Willett, W., J. Rockström, B. Loken, M. Springmann, T. Lang, S. Vermeulen, … C. Murray. 2019. Food in the anthropocene: The EAT–Lancet Commission on healthy diets from sustainable food systems. *Lancet*, 6736: 3–49. doi: 10.1016/ S0140-6736(18)31788-4.

Williams, S. M. and R. Weil. 2004. Cover crop root channels may alleviate soil compaction effects on soybean crop. *Soil Science Society of America Journal*, 68(4): 1403–9.

Zhang, T. Q., C. S. Tan, Z. M. Zheng and C. F. Drury. 2015. Tile drainage phosphorus loss with 4852long-term consistent cropping systems and fertilization. *Journal of Environmental Quality*, 44(2): 503–11. doi: 10.2134/jeq2014.04.0188.

Zhang, Y., X. Li, E. Gregorich, et al. 2020. Tillage and cropping effects on soil organic carbon: Biodegradation and storage in density and size fractions. *European Journal of Soil Science*, First published: 29 February 2020. doi: 10.1111/ejss.12949.

3 Relating Soil Organic Carbon Fractions to Crop Yield and Quality with Cover Crops

Upendra M. Sainju, Rajan Ghimire, and Jun Wang

CONTENTS

3.1 INTRODUCTION

Cover cropping is primarily practiced to control soil erosion by providing vegetative cover over bare soil and enhancing soil aggregation (Blanco-Canqui et al. 2013). The additional above- and belowground crop residues provided by cover crops increase soil organic matter (SOM) and C sequestration compared to no cover crop (Kuo et al. 1997; Sainju et al. 2003; Poeplau and Don 2015). Furthermore, cover crops grown after harvest of cash crops use residual soil N, thereby reducing N leaching compared to no cover crop (McCracken et al. 1994; Sainju and Singh 1997). Cover crops also enhance soil microbial biomass and community structure, thereby increasing nutrient cycling and C sequestration compared to no cover crop (Muhammed et al. 2020; Thapa et al. 2021). Other benefits of cover crops include reduced weed and pest infections, increased soil aggregation and structure, and enhanced water holding and infiltration capacity compared to no cover crop (Sainju et al. 2003; Blanco-Canqui et al. 2013; Brozovic et al. 2018).

Cover crop species differ on agroecosystem benefits. Legume cover crops fix N from the atmosphere, supply N from their residues, reduce N fertilization rates, and enhance succeeding crop yields and quality compared to nonlegume or no cover crop (Kuo et al. 1995; Hill et al. 2016; Sainju et al. 2018a). On the other hand, nonlegume cover crops increase soil organic C (SOC) due to greater above- and belowground biomass production (Kuo et al. 1997), reduce N leaching by extracting more soil residual N (McCracken et al. 1994; Sainju and Singh 1997), and improve soil structure by enhancing soil aggregation than legume or no cover crop (Sainju et al. 2000). As single cover crop species does not provide all of these benefits, a mixture of legume and nonlegume cover crops may do so. Some researchers (Kuo and Sainju 1998; Vaughn and Evanylo 1998; Sainju et al. 2005a, 2018b) have reported that a mixture of legume and nonlegume cover crops enhanced cover crop biomass yield and SOC compared to either cover crop species alone and sustained succeeding crop yields compared to nonlegume cover crops. Others (Poeplau and Don 2015; Brozovic et al. 2018)

DOI: 10.1201/9781003102762-3

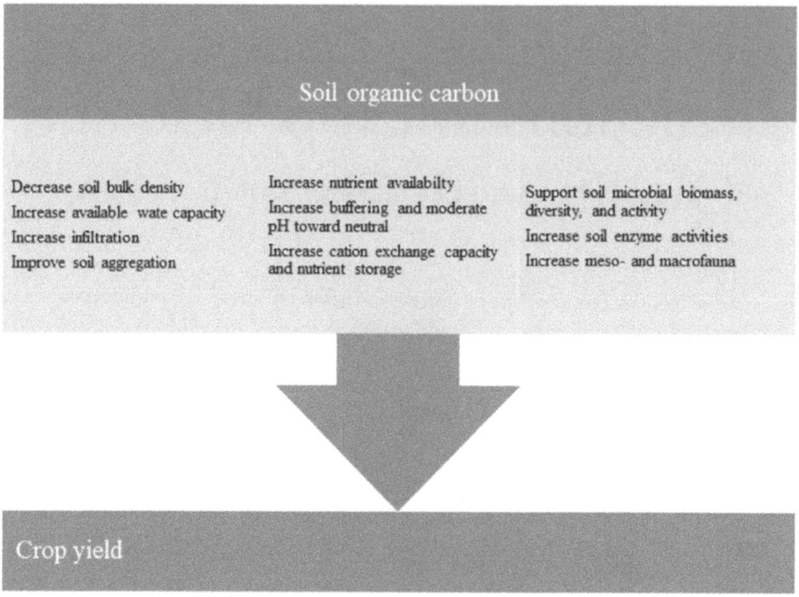

FIGURE 3.1 Multifunctionality of soil organic carbon (SOC) affecting crop yields.

found that a mixture of legume, nonlegume, and oilseed cover crops increased soil biodiversity compared to single cover crop species or no cover crop.

It often takes a longer time to observe changes in SOC due to management practices, such as cover cropping, because of its large pool size and inherent spatial variability (Franzluebbers et al. 1995; Salinas-Garcia et al. 1997). Labile C fractions, such as particulate organic C (POC), potential C mineralization (PCM), microbial biomass C (MBC), $KMnO_4$-extractable C, and carbohydrate, change rapidly due to management practices and have been identified as early indicators of changes in SOM that relate to crop yield (Franzluebbers et al. 1995; Salinas-Garcia et al. 1997; Franzluebbers 2016). The quality of SOM as observed by changes in labile C fractions is often a good indicator of soil health and crop production as opposed to organic matter quantity represented by SOC content (Franzluebbers 2016; van Es and Karlen 2019).

Although cover crops have a positive impact on soil labile and non-labile C fractions compared to no cover crops, there is a paucity of information about how these C fractions affect crop yields and quality due to cover cropping (Robertson et al. 2014; Oldfield et al. 2019). As stated above, nonlegume cover crops can increase SOC better than legume cover crops, but reduce crop yields due to N immobilization because of their higher residue C/N ratio. As a result, an increased N fertilization rate is required to sustain crop yields and quality following nonlegume cover crops. The SOM as indicated by SOC increases crop yields by enhancing many soil properties (Figure 3.1). The objective of this article is to deliberate biomass production potentials and N accumulation of cover crop species; the effect of cover crop on succeeding crop yields, nutritional quality, and soil labile and non-labile C fractions; and the relationship between soil C fractions and crop yields, protein concentration, and N uptake from data collated and synthesized from around the globe.

3.2 COVER CROP BIOMASS AND NITROGEN ACCUMULATION

Cover crop biomass varied among cover crop species, but biomass was usually greater with nonlegume than legume cover crops (Tables 3.1A and 3.1B). Although cover crop growth is affected by soil and climatic conditions in a region, yielding favorably with medium soil texture, mild winter

TABLE 3.1A
Cover Crop Biomass Yield and N Accumulation as Affected by Cover Crop Species in Various Regions

Location	Cover Crop	Biomass Yield ($Mg\ ha^{-1}$)	N Accumulation ($kg\ N\ ha^{-1}$)	Reference
Maryland, USA	Hairy vetch	2.96	100	Clark et al. (1994)
	Rye	5.00	74	
	Hairy vetch + rye	5.79	145	
Maryland, USA	Hairy vetch	4.00	158	Decker et al. (1994)
	Austrian winter pea	3.30	127	
	Crimson clover	4.30	115	
	Wheat	3.05	39	
Kentucky, USA	Hairy vetch	5.10	209	Ebelhar et al. (1994)
	Big flower vetch	1.90	60	
	Crimson clover	2.40	56	
	Rye	3.40	56	
Georgia, USA	Rye	4.03	38	Hargrove (1986)
	Crimson clover	7.17	170	
	Subterranean clover	4.00	114	
	Hairy vetch	4.25	153	
	Common vetch	4.30	134	
Kentucky, USA	Rye	1.50	14	Huntington et al. (1985)
	Hairy vetch	3.30	125	
North Carolina, USA	Austrian winter pea	1.96	51	Kamprath et al. (1968)
	Oat + hairy vetch	2.92	81	
	Hairy vetch	3.01	120	
Washington, USA	Rye	5.30	60	Kuo et al. (1997)
	Ryegrass	7.10	56	
	Hairy vetch	3.20	120	
	Austrian winter pea	3.90	100	
	Canola	3.30	44	
	Weeds	2.10	30	
Georgia, USA	Hairy vetch	3.40	128	McVay et al. (1989)
	Crimson clover	3.50	108	
	Wheat	1.80	32	
Alabama, USA	Weeds	2.70	31	Touchton et al. (1984)
	Crimson clover	4.50	95	
	Common vetch	4.90	118	
Argentina	Rye	4.19		Frasier et al. (2017)
	Vetch	3.66		
	Rye + vetch	5.37		

air temperature, and adequate precipitation evenly distributed throughout the growing season, biomass of nonlegume cover crops is enhanced with increased soil residual N (Mesinger et al. 1991; McCracken et al. 1994; Sainju et al. 2003). In contrast, legume cover crops do not respond to soil residual N level, as they fix N from the atmosphere. In fact, legume cover crops fix more N when soil residual N is low (Mesinger et al. 1991; McCracken et al. 1994; Sainju et al. 2003). Numerous researchers (Mesinger et al. 1991; Clark et al. 1994; Sainju et al. 2005a) reported that rye (nonlegume) biomass and N accumulation are influenced by soil residual N, but hairy vetch (legume)

TABLE 3.1B
Cover Crop Biomass Yield and N Accumulation as Affected by Cover Crop Species in Various Regions

Location	Cover Crop	Biomass Yield (Mg ha^{-1})	N Accumulation (kg N ha^{-1})	Reference
California, USA	Purple vetch	5.30	178	Shennan (1992)
	Lana wollywood vetch	6.30	220	
	Oat + purple vetch	6.30	130	
	Faba bean + vetch	5.10	174	
	Weeds	2.90	51	
	Austrian winter pea	3.20	110	
	Oat	5.40	52	
China	Huai bean	2.34		Zhang et al. (2016)
	Soybean	3.07		
	Mung bean	2.19		
Spain	Barley	5.16		Gabriel et al. (2014)
	Vetch	2.32		
	Barley + vetch	6.20		
California, USA	Barley	10.14		Mitchell et al. (1999)
	Vetch	5.59		
	Barley + vetch	9.10		
Georgia, USA	Weeds	1.22	21	Sainju et al. (2005a)
	Rye	4.05	42	
	Hairy vetch	4.23	136	
	Rye + hairy vetch	6.62	193	
Georgia, USA	Rye	3.03		Sainju et al. (2007)
	Blend[a]	1.85		
	Rye + blend	4.00		
	Winter weed	0.83		
	Crimson clover	2.15		
Georgia, USA	Control	0.89	11	Sainju et al. (2018a)
	Rye	1.72	16	
	Hairy vetch	2.13	42	
	Hairy vetch + rye	2.83	48	
Georgia, USA	Control	0.99	11	Sainju et al. (2018b)
	Rye	1.99	25	
	Hairy vetch	2.49	49	
	Hairy vetch + rye	4.18	59	

[a] Blend is a mixture of balansa clover, crimson clover, and hairy vetch.

biomass and N accumulation become more competitive during low soil residual N. An analysis of cover crop biomass across the globe showed that nonlegume cover crops produced 18% greater biomass than legume cover crops (Table 3.2). The control treatment consisted of weeds that had lower biomass than other cover crops. Several researchers (Kuo et al. 1995; Sainju et al. 1998, 2005b) also found that cover crop root biomass followed trends similar to shoot biomass, with greater root biomass in nonlegume than legume cover crops.

Although few studies have reported biomass and N accumulation for cover crop mixtures, studies with the mixtures showed that mixed legume and nonlegume cover crops had even greater biomass than either cover crop species alone (Tables 3.1A and 3.1B). Legume and nonlegume cover

TABLE 3.2
Mean[a] Cover Crop Biomass and N Accumulation, Succeeding Crop Yield and N Uptake, and Soil Organic C Concentration and Stock as Affected by Cover Crop Species

	Cover Crop		Succeeding Crop			Soil Organic C	
Cover Crop Species	Biomass Yield ($Mg\ ha^{-1}$)	N Accumulation ($kg\ N\ ha^{-1}$)	Yield ($Mg\ ha^{-1}$)	N Uptake ($kg\ N\ ha^{-1}$)	Protein Concentration ($g\ kg^{-1}$)	Concentration ($g\ C\ kg^{-1}$)	Stock ($Mg\ C\ ha^{-1}$)
Control	1.98 (±0.85)c[b]	28 (±15)b	11.3 (±3.8)b	36 (12)b	20 (±2)c	8.8 (±3.1)c	20.9 (±8.3)b
Legume	3.49 (±1.46)bc	117 (±50)a	14.0 (±4.2)a	115 (±35)a	51 (±5)a	10.5 (±3.2)b	22.4 (±6.7)ab
Nonlegume	4.12 (±2.11)ab	42 (±18)b	8.2 (±2.8)c	55 (±20)b	42 (±5)ab	12.9 (±4.8)a	22.3 (±7.2)ab
Mixture[c]	5.33 (±1.91)a	109 (±56)a	12.3 (±4.3)b	57 (±22)b	29 (±3)bc	13.9 (± 2.8)a	23.7 (±7.9)a

[a] Mean (± standard deviation).
[b] Numbers followed by different letters within a column are significantly different at $P \leq 0.05$.
[c] Legume + nonlegume mixture.

crop mixture had 23 and 35% greater biomass than nonlegume and legume cover crops, respectively (Table 3.2). Some researchers (Ta and Farris 1987; Russell and Hargrove 1989; Ranells and Wagger 1996; Sainju et al. 2005a) have proposed several mechanisms for the increased biomass in cover crop mixture: (1) transfer of N from legume to nonlegume cover crops during growth, enhancing the growth of the nonlegumes, (2) upright growth habit of nonlegumes providing an excellent scaffold for the viny growth habit of legumes, thereby intercepting more light and reducing the growth competition between the two species, and (3) higher seeding rates of cover crops in the mixture compared to monocultures, resulting in increased biomass. As greater biomass increases C input to the soil, soil C fractions are also expected to be greater with cover crop mixture than individual species or the control with no cover crop.

Several researchers (Kuo et al. 1997; Sainju et al. 2005b) have also reported greater belowground (root) biomass with nonlegume than legume cover crops, results that are proportional to aboveground biomass. The mixture of legume and nonlegume cover crops also showed greater belowground biomass than either species alone (Sainju et al. 2005b). Belowground biomass and associated root exudates are important sources of C inputs for enhancing SOC and labile C fractions, as producers often remove aboveground biomass for animal feedstock, litter, bioenergy production, and other uses. Roots can contribute as much as 12% of SOC, 31% of water-soluble C, and 52% of MBC in a growing season (Liang et al. 2002). Roots play a greater role in the C cycle, and have relatively greater influence on SOC than aboveground biomass (Boone 1994; Norby and Cotrufo 1998).

Nitrogen accumulation in cover crop biomass is greater with legumes than nonlegumes or no cover crop due to higher tissue N concentration (Tables 3.1A and 3.1B). The overall analysis indicated that legume cover crops accumulated 64 and 76% more N in biomass than nonlegumes and weeds in the control treatment, respectively (Table 3.2). Cover crop biomass N accumulation, however, was only 7% greater with legumes than mixed legume and nonlegume cover crops, and the difference was not statistically significant. Increased N transfer from legumes to nonlegumes during their growth can increase N concentration of nonlegumes in the mixture compared to individual species, resulting in increased N accumulation in the mixture (Ranells and Wagger 1996). As with cover crop biomass, increased N contribution from cover crop residue is also expected to enhance N uptake by the succeeding crops.

3.3 SUCCEEDING CASH CROP YIELDS, PROTEIN CONCENTRATION, AND NITROGEN UPTAKE

As with the cover crop biomass and N uptake, succeeding cash crop yields, protein concentration, and N uptake also varied among regions due to differences in soil and climatic conditions (Tables 3.3A, 3.3B, 3.3C, 3.3D, and 3.3E). Succeeding crop yields, protein concentration, and N uptake responded well to cover crops in humid and subhumid regions where air temperature and precipitation were favorable for crop growth. In these regions, legume cover crops increased succeeding crop yields, protein concentration, and N uptake better than nonlegumes or no cover crop due to increased N supply from legume residues. There was an increase in succeeding crop yields, protein concentration, and N uptake with legumes compared to no cover crop, ranging from –0.21 to 4.70 Mg ha^{-1}, –5 to 28 g kg^{-1}, and –2 to 106 kg N ha^{-1}, respectively (Table 3.4). Corresponding values for nonlegumes compared to no cover crop were –2.40 to 0.70 Mg ha^{-1}, –9 to 35 g kg^{-1}, and –11 to 39 kg N ha^{-1}, respectively. An overall analysis of the effect of cover crop on succeeding crop yields demonstrated that crop yield was 41 and 19% greater with legume than nonlegume or no cover crop, respectively (Table 3.2). Protein concentration was 18 and 61% greater with legume than nonlegume or no cover crop, respectively. Similarly, N uptake by succeeding crops was 52 and 69% greater with legume than nonlegume or no cover crop, respectively. Nonlegume cover crops depressed succeeding crop yields compared to legumes or no cover crop, possibly due to N immobilization as a result of the greater C/N ratio of nonlegume residues. Residues with a higher C/N ratio immobilize N that becomes unavailable to succeeding crops during their growth, resulting in reduced crop yields (Kuo and Sainju 1998; Sainju et al. 2005b, 2006).

The legume and nonlegume cover crop mixture produced succeeding crop yields and N uptake that were in between those produced by individual cover crop species—but not much different from those produced by legume cover crops (Tables 3.3A, 3.3B, 3.3C, and 3.3D). This is due to the nonsignificant difference in N supplied by legume and mixed cover crop residues (Table 3.2), as succeeding crop yields and N uptake depend on N supplied by cover crops (Kuo et al. 1995; Gabriel et al. 2019; Sainju et al. 2019). The mixed legume and nonlegume cover crops increased succeeding crop yields by –0.18 to 3.35 Mg ha^{-1}, protein concentration by –5 to 19 g kg^{-1}, and N uptake by –1 to 63 kg N ha^{-1} compared to no cover crop (Table 3.4). Our analysis indicated that the legume and nonlegume cover crop mixture increased succeeding crop yield by 33% compared to the nonlegume cover crop, but they were reduced by 14% compared to legume cover crop alone (Table 3.2). Similarly, N uptake by succeeding crops was 4% greater in the cover crop mixture than nonlegume cover crop, but it was 50% lower than legume cover crop. In contrast, protein concentration decreased by 45 to 76% in the cover crop mixture compared to legume and nonlegume cover crops. The large difference in protein concentration and N uptake by succeeding crops among legume, nonlegume, and mixed cover crops could be due to the lack of inadequate data available for analysis, as most studies did not measure protein concentration and N uptake by succeeding crops following cover crops.

An exception, however, occurred with cotton, where lint yield and N uptake were lower with legume (hairy vetch) than nonlegume (rye) cover crop (Table 3.3A; Sainju et al. 2006). Sainju et al. (2006) found that cotton lint yield responded negatively to N availability from legume cover crop residue and N fertilizer and that increased N availability from legume cover crop increased cotton biomass at the expense of lint yield. Similarly, Boquet et al. (2004) reported that cotton lint yield increased with wheat cover crop, but decreased with hairy vetch cover crop. However, cotton lint yield and N uptake with mixed legume and nonlegume cover crops were in between those produced by either species alone. Similar results occurred when the response crop following cover crop was a legume, such as soybean, which yielded lower following legume than following nonlegume cover crop (Restovich et al. 2012). Legumes following legumes do not respond well compared

TABLE 3.3A
Effect of Cover Crop on Succeeding Cash Crop Yield, Protein Concentration, and N Uptake in Various Regions

Location	Cover Crop	Cash Crop	Cash Crop Yield ($Mg\ ha^{-1}$)	Cash Crop N Uptake ($kg\ N\ ha^{-1}$)	Cash Crop Protein Conc. ($g\ kg^{-1}$)	Reference
Maryland, USA	Hairy vetch	Corn	8.05	172	134	Decker et al. (1994)
	Austrian winter pea		8.20	173	132	
	Crimson clover		8.05	164	127	
	Winter wheat		6.70	147	137	
	Control		7.0	148	132	
Kentucky, USA	Rye	Corn	0.90	17	119	Huntington et al. (1985)
	Hairy vetch		4.10	110	168	
	Control		0.40	14	219	
Georgia, USA	Hairy vetch	Corn	8.5	90	66	McVay et al. (1989)
	Wheat		3.50	30	54	
Washington, USA	Rye	Silage corn	7.40	112	95	Kuo et al. (1995)
	Ryegrass		7.20	60	52	
	Hairy vetch		12.30	179	91	
	Austrian winter pea		9.60	118	77	
	Canola		7.80	78	63	
	Control		7.60	73	60	
Georgia, USA	Control	Cotton	0.89	14	98	Sainju et al. (2006)
	Rye		0.91	15	103	
	Hairy vetch		0.68	12	110	
	Hairy vetch + rye		0.71	13	114	
	Control	Sorghum	2.80	43	96	
	Rye		2.30	32	87	
	Hairy vetch		3.50	60	107	
	Hairy vetch + rye		4.00	58	91	
Georgia, USA	Rye	Tomato	16.30	32	82	Sainju et al. (2000)
	Hairy vetch		35.90	77	89	
	Crimson clover		35.50	77	89	
	Control		18.70	40	89	
Slovenia	Control		9.63	125	81	Kramberger et al. (2009)
	Italian ryegrass		9.47	121	80	
	Winter rape		9.69	134	86	
	Subterranean clover		10.55	153	91	
	Crimson clover		10.62	154	91	

to following nonlegumes because legumes do not fix much N when soil N availability is high and chances of being infested by weeds and pests are greater (Miller et al. 2011).

In arid and semiarid regions with limited precipitation, cover crops had negative effect on succeeding crop yields (Tables 3.3B, 3.3C, 3.3D, and 3.3E). Several researchers (Lenssen et al. 2007; Nielsen et al. 2016) reported that in the semiarid central and northern Great Plains (Montana and Colorado, USA), succeeding crop yields were lower following cover crops than following fallow due to increased water use by cover crops, leaving reduced soil water available for succeeding

TABLE 3.3B
Effect of Cover Crop on Succeeding Cash Crop Yield, Protein Concentration, and N Uptake in Various Regions

Location	Cover crop	Cash crop	Cash crop yield ($Mg\ ha^{-1}$)	Cash crop N uptake ($kg\ N\ ha^{-1}$)	Cash crop protein conc. ($g\ kg^{-1}$)	Reference
Spain	Vetch	Corn	12.6	162	80	Gabriel and Quemada (2011)
	Barley		12.1	153	79	
	Control		11.4	144	79	
Argentina	Rye	Sorghum	11.36	80	44	Frasier et al. (2017)
	Vetch		13.05	147	70	
	Rye + vetch		13.65	133	61	
	Control		10.50	70	42	

crops. In these regions, soil water availability greatly affects dryland crop yields. Alternate-year fallowing is mostly practiced to conserve soil water, mineralize nutrients, and control weeds in such regions (Lenssen et al. 2007; Nielsen et al. 2016). In years with above-average precipitation, however, cover crops did not significantly depress succeeding crop yields compared to no cover crop in these regions (Nielsen et al. 2016). Similarly, with irrigation, Mesbah et al. (2019) did not observe a significant difference in succeeding crop yields between cover cropping and no cover cropping treatments in semiarid regions.

3.4 NITROGEN FERTILIZER REDUCTION DUE TO COVER CROPS

Because of N supplied by cover crop residues, N fertilization rates to succeeding crops can be lower following cover crops compared to following no cover crop. This is especially true for legume cover crops which fix N from the atmosphere and have higher tissue N concentration than nonlegume cover crops. The reduction in N fertilization rates to succeeding crops for legume cover crops ranged from 15 to 200 kg N ha^{-1} (Table 3.5). This indicates that succeeding crop yields following legume cover crops can be similar to those produced by N fertilization rates of 15 to 200 kg N ha^{-1}, depending on the performance of legumes, due to variations in soil and climatic conditions. Similarly, a mixture of legume and nonlegume cover crops can reduce N fertilization rates to succeeding crops by 93 to 179 kg N ha^{-1} (Table 3.5). Following nonlegume cover crops, however, N fertilization rates to succeeding crops can be similar or greater than following no cover crop, primarily due to N immobilization by nonlegume cover crop residues (Kuo and Sainju 1998; Sainju et al. 2005b, 2008).

3.5 SOIL ORGANIC CARBON

The SOC concentration also varied with cover crops in various regions (Tables 3.6A, 3.6B, and 3.6C), which is usually reflected by turnover rates of cover crop residues that are controlled by C/N ratios (Kuo et al. 1997). Residues with lower C/N ratios, such as legume cover crops, decompose more rapidly than residues with higher ratios, such as nonlegume cover crops, resulting in lower SOC at the 0–15 cm depth with legume than with nonlegume cover crops (Kuo et al. 1997). However, factors, such as soil temperature, water content, pH, and soil texture, also affect the rate of decomposition of residue in the soil (Kuo et al. 1997; Sainju et al. 2018a, Zhang et al. 2016). In

TABLE 3.3C
Effect of Cover Crop on Succeeding Cash Crop Yield, Protein Concentration, and N Uptake in Various Regions

Location	Cover Crop	Cash Crop	Cash Crop Yield ($Mg\ ha^{-1}$)	Reference
Maryland, USA	Hairy vetch	Tomato	104.00	Kelly et al. (1995)
	Control		80.00	
California. USA	Purple vetch	Corn	12.30	Shennan (1992)
	Oat + purple vetch		10.60	
	Austrian winter pea		11.10	
	Oat		6.80	
	Bell bean		10.20	
	Purple vetch	Tomato	103.00	
	Oat + purple vetch		90.00	
	Austrian winter pea		77.00	
	Oat		62.00	
	Bell bean		84.00	
Alabama, USA	Control	Cotton	0.60	Touchton et al. (1984)
	Crimson clover		0.90	
	Common vetch		0.80	
Spain	Barley	Corn	12.14	Gabriel et al. (2019)
	Hairy vetch		12.65	
China	Huai bean	Wheat	4.58	Zhang et al. (2016)
	Soybean		4.43	
	Mung bean		3.96	
China	Mung bean	Wheat	3.14	Zhang et al. (2013)
	Soybean		3.17	
	Huai bean		3.24	
Maryland, USA	Radish	Soybean	1.75	Williams and Weil (2004)
	Rye		1.82	
New York, USA	Barley	Soybean	1.60	Williams et al. (2000)
	Rye		1.54	
	Triticale		1.31	
	Wheat		1.92	
China	Rye	Wheat	3.15	Zhang and Wang (2018)
	Huai bean		2.70	
	Rye + huai bean		2.69	
China	Canola	Wheat	6.54	Xue et al. (2018)
	Soybean		4.96	

contrast, SOC stock at 0–30 cm was greater with nonlegume than legume cover crops (Tables 3.6A and 3.6C). As SOC stock is calculated by multiplying SOC concentration by the bulk density and the thickness of the soil layer, variations in soil bulk density among cover crops can affect SOC stock differently from SOC concentration. Increases or decreases in SOC concentration and stock, due to cover crop compared to no cover crop, varied with cover crop species, tillage practice, succeeding crop type, and soil depth (Tables 3.7A and 3.7B). Legume cover crops increased SOC concentration at 0–30 cm by 0.1 to 3.8 $g\ C\ kg^{-1}$ and SOC stock by −0.3 to 1.6 $Mg\ C\ ha^{-1}$ compared to no cover crop. Nonlegume cover crops increased SOC concentration at 0–30 cm by 0.7 to 1.64

TABLE 3.3D
Effect of Cover Crop on Succeeding Crop Yield, Protein Concentration, and N Uptake in Various Regions

Location	Cover Crop	Cash Crop	Cash Crop Yield ($Mg\ ha^{-1}$)	Reference
Argentina	Control	Corn	7.50	Restovich et al. (2012)
	Oat		7.10	
	Barley		7.60	
	Ryegrass		6.60	
	Oat + vetch		8.70	
	Vetch		8.60	
	Canola		8.10	
	Radish		8.40	
	Control	Soybean	5.20	
	Oat		5.10	
	Ryegrass		4.70	
	Barley		5.10	
	Oat + vetch		4.80	
	Vetch		5.00	
	Canola		4.90	
	Radish		5.30	
New Mexico, USA	Pea	Wheat	6.17	Ghimire et al. (2019)
	Oat		6.27	
	Canola		5.68	
	Pea + oat		6.19	
	Pea + canola		6.36	
	Pea + oat + canola		5.96	
	Pea + oat + canola + hairy vetch + radish + barley		6.61	
Argentina	Rye	Sorghum	11.36	Frasier et al. (2017)
	Vetch		13.05	
	Rye + vetch		13.65	
Colorado, USA	Cover crop		3.88	Nielson et al. (2016)
	Control		4.32	
Maryland, USA	Rye	Corn	10.32	Clark et al. (1997)
	Rye + hairy vetch		9.90	
	Hairy vetch		10.40	
Georgia, USA	Crimson clover	Corn	13.26	Sanders et al. (2018)
	Rye		13.04	
South Dakota. USA	Turnip + lentil + pea + millet	Corn	6.62	Reese et al. (2014)
	Radish + canola		13.97	

g C kg^{-1} and SOC stock by −0.7 to 1.9 Mg C ha^{-1}. Increases in SOC stock at 0–30 cm with legume and nonlegume cover crop mixtures compared to no cover crop ranged from −1.8 to 2.0 Mg C ha^{-1}.

The overall analysis of the effect of cover crop on SOC concentration revealed that SOC concentration at 0–30 cm was 19 and 32% greater with nonlegume than legume and no cover crop, respectively (Table 3.2). In contrast, soil C stock at 0–30 cm was similar between legume and nonlegume cover crops, but was 5–6% greater with cover crop than without. The SOC concentration at 0–30 cm

TABLE 3.3E
Effect of Cover Crop on Succeeding Crop Yield, Protein Concentration, and N Uptake in Various Regions

Location	Cover crop	Succeeding crop	Yield (Mg ha^{-1})	Reference
Georgia, USA	Rye	Eggplant	21.0	Sainju et. al. (2003)
	Hairy vetch		52.1	
	Crimson clover		45.5	
	Control		23.7	
Montana, USA	Lentil	Wheat	0.58	Lenssen et al. (2007)
	Mustard		0.53	
	Chickpea		0.70	
	Pea		0.63	
Georgia, USA	Control	Forage sorghum	15.50	Sainju et al. (2018a)
	Rye		14.00	
	Hairy vetch		14.50	
	Hairy vetch + rye		15.60	

was 7 and 24% greater with mixed legume and nonlegume cover crops than nonlegume and legume cover crops, respectively. Similarly, SOC stock at 0–30 cm was 6 and 5% greater with mixed cover crops than nonlegume and legume cover crops, respectively. Jian et al. (2020), in a meta-analysis, reported that cover cropping increased SOC stock at 0–15 cm by 16% compared to no cover cropping and that mixed legume and nonlegume cover cropping further enhanced the stock. Increased biomass residue contributed by cover crops (Table 3.2) likely increased SOC concentration and stock with nonlegume and mixed cover crops compared to no cover crop. Several researchers (Kuo et al. 1997, Sainju et al. 2000, 2003) reported that a higher C/N ratio of nonlegume cover crop residue also resulted in greater SOC at 0–30 cm concentration and stock with nonlegume and the mixture than with legume cover crop only due to the reduced decomposition of the residue. The results demonstrate that a mixture of legume and nonlegume cover crops may enhance soil C sequestration compared to either species alone or no cover crop.

3.6 SOIL LABILE CARBON FRACTIONS

Soil labile C fractions are those fractions of SOC that readily change within a crop growing season due to management practices when changes in SOC take a long time to occur due to spatial heterogeneity (Franzluebbers et al. 1995). These fractions are regarded as a measurement of SOM quality rather than quantity and are important indicators of soil health (Franzluebbers 2016; van Es and Karlen 2019). These fractions include PCM, MBC, POC, $KMnO_4$-extractable C, and carbohydrate. Limited data, however, are available about the effect of cover crop on soil labile C fractions.

Kuo et al. (1997) observed that nonlegume cover crops, such as rye and ryegrass, increased soil carbohydrate compared to legume cover crops, such as Austrian winter pea and hairy vetch, and the control with no cover crop at 0–15 cm, but were not different among cover crop species at 15–30 cm in Washington, USA (Table 3.8A). Ghimire et al. (2019) reported greater PCM at 0–15 cm with pea + oat and pea + canola mixtures than pea + canola + oat mixture and the control with no cover crop in New Mexico, USA. Similarly, Sainju et al. (2003) found greater PCM at 0–20 cm with rye and crimson clover than hairy vetch and the control in central Georgia, USA (Table 3.8B). In another study, this one in southern Georgia USA, however, Sainju et al. (2007) found greater PCM at 0–15

TABLE 3.4
Increase or Decrease in Crop Yield, Protein Concentration, and N Uptake of Succeeding Crops with Cover Crops Compared to No Cover Crop

Location	Cover Crop	Succeeding Crop	Succeeding Crop Yield (Mg ha^{-1})	Succeeding Crop N Uptake (kg N ha^{-1})	Protein Conc. (g kg^{-1})	Reference
Maryland, USA	Hairy vetch	Corn	1.05	24	2	Decker et al. (1994)
	Austrian winter pea		1.20	25	0	
	Crimson clover		1.05	16	−5	
	Winter wheat		−0.30	−1	5	
Washington, USA	Rye	Silage corn	−0.20	39	35	Kuo et al. (1995)
	Ryegrass		−0.40	−7	−8	
	Hairy vetch		4.70	106	31	
	Austrian winter pea		2.00	45	17	
	Canola		0.20	5	3	
Argentina	Rye	Sorghum	−0.86	10	−2	Frasier et al. (2017)
	Vetch		2.55	77	28	
	Rye + vetch		3.35	63	19	
Georgia, USA	Rye	Cotton	0.02	1	5	Sainju et al. (2006)
	Hairy vetch		−0.21	−2	12	
	Hairy vetch + rye		−0.18	−1	16	
	Rye	Sorghum	−0.50	−11	−9	
	Hairy vetch		0.70	17	11	
	Hairy vetch + rye		1.20	15	−5	
Georgia, USA	Rye	Tomato	−2.40	−8	−7	Sainju et al. (2000)
	Hairy vetch		17.2	37	0	
	Crimson clover		16.8	37	0	
Spain	Vetch	Corn	1.20	18	−1	Gabriel and Quemada (2011)
	Barley		0.70	9	−1	
Slovenia	Italian ryegrass	Corn	−0.16	−4	−1	Kramberger et al. (2009)
	Winter rape		0.09	9	5	
	Subclover		0.92	28	10	
	Crimson clover		0.99	29	10	

cm with crimson clover, blend (balansa clover + crimson clover + hairy vetch), and rye + blend than rye cover crop.

Soil MBC at 0–20 cm was greater with rye than hairy vetch and the control in central Georgia, USA (Sainju et al. 2003; Table 3.8B). Similarly, MBC at 0–15 cm was greater with rye than blend and crimson clover in southern Georgia, USA (Sainju et al. 2007). The POC at 0–20 cm was greater with cover crops than the control, but was not different among legume and nonlegume cover crops (Sainju et al. 2003). In New Mexico, USA, Ghimire et al. (2019) reported that $KMnO_4$-extractable C at 0–15 cm was greater with cover crop mixture (pea + oat + canola + hairy vetch + radish + barley) than pea and canola. The results demonstrate that nonlegume and a mixture of legume and

TABLE 3.5
Reduction in N Fertilization Rates to Succeeding Crops due to Legume Cover Crops for Sustainable Crop Production

Cover Crop	Succeeding Crop	Reduction in N Fertilization Rate (kg N ha^{-1})	Reference
Crimson clover	Corn	97	Wagger (1989)
Hairy vetch		105	
Pea	Corn	53	McVay et al. (1989)
Hairy vetch	Cotton	68	Touchton et al. (1984)
Crimson clover		68	
Hairy vetch	Sorghum	89	Breman and Wright (1984)
Hairy vetch	Sorghum	97	Hargrove (1986)
Common vetch		61	
Crimson clover		92	
Subterranean clover		61	
Hairy vetch	Corn	78	Varco et al. (1984)
Hairy vetch	Corn	87	Ebelhar et al. (1994)
Big flower vetch		15	
Crimson clover		22	
Hairy vetch	Corn	200	Worsham (1986)
Crimson clover		100	
Hairy vetch	Silage corn	200	Flannery (1981)
Hairy vetch	Cotton	84	Brown et al. (1985)
Crimson clover		51	
Crimson clover	Corn	56	Buntley (1986)
Hairy vetch	Corn	66	Herbeck et al. (1987)
Big flower vetch		64	
Hairy vetch	Barley	112	Meyer (1987)
Hairy vetch	Cotton	149	Sainju et al. (2006)
Hairy vetch + rye		93	
Hairy vetch	Sorghum	107	
Hairy vetch + rye		179	

nonlegume cover crops enhanced soil labile C fractions probably due to increased biomass residue inputs compared to legume and the control with no cover crop, similar to those observed for SOC.

3.7 RELATIONSHIP BETWEEN SOIL CARBON FRACTIONS AND CROP YIELD DUE TO COVER CROPS

The current analysis showed that SOC concentration at 0-30 cm was nonlinearly related to crop yield (Figure 3.2A). Crop yield decreased as SOC concentration increased from 4 to 12 g C kg^{-1}. Crop yield increased only when SOC concentration was $\geq$ 12.0 g C kg^{-1}. About 38% of variability in crop yield is explained by SOC concentration. In contrast, crop N uptake linearly increased with increased SOC concentration at 0–30 cm (Figure 3.2B). An increase in SOC concentration by 1.0 g C kg^{-1} increased N uptake by 4.4 kg N ha^{-1}. The SOC concentration explained 28% of variability in crop N uptake. Protein concentration in the crop, however, declined nonlinearly with SOC concentration at 0–30 cm from 4.5 to 12.0 g C kg^{-1}, after which it gradually increased (Figure 3.2C).

TABLE 3.6A
Cover Crop Effect on Soil Organic C (SOC) Under Different Crops in Various Regions

Location	Cover Crop	Succeeding Crop	Tillage[a]	Soil Depth	SOC Concentration (g C kg^{-1})	SOC Stock (Mg C ha^{-1})	Reference
Georgia, USA	Rye	Tomato	CT	0–20 cm	6.19	18.4	Sainju et al. (2000, 2002)
	Hairy vetch				5.92	17.9	
	Crimson clover				5.77	17.8	
	Control				4.55	16.5	
Georgia, USA	Rye	Cotton	ST	0–15 cm	11.50	17.6	Sainju et al. (2007)
	Blend[b]				12.46	17.5	
	Rye + Blend				11.93	18.4	
	Crimson clover				10.90	16.4	
Georgia, USA	Weeds	Cotton/ sorghum	NT	0–30 cm		25.4	Sainju et al. (2005a)
	Rye					25.5	
	Hairy vetch					26.1	
	Hairy vetch + rye					24.8	
	Weeds		ST			24.2	
	Rye					25.5	
	Hairy vetch					24.6	
	Hairy vetch + rye					25.5	
	Weeds		CT			21.6	
	Rye					23.5	
	Hairy vetch					22.6	
	Hairy vetch + rye					23.6	
Kentucky, USA	Control	Corn	NT	0–7.5 cm	10.6		Frye et al. (1986)
	Hairy vetch				13.5		
	Big flower vetch				12.7		
	Rye				11.5		

[a] Tillage are CT, conventional tillage; ST, strip tillage; and NT, no tillage.
[b] Blend is a mixture of balansa clover, crimson clover, and hairy vetch.

Protein concentration explained about 23% of variability in SOC concentration. However, SOC stock at 0–30 cm was not related to crop yield, protein concentration, and N uptake.

Some researchers (Touchton et al. 1984; Hargrove 1986; McVay et al. 1989) have reported that legume cover crops enhanced both SOC and succeeding crop yields compared to nonlegume cover crops. Others (Kuo et al. 1995, 1997; Sainju et al. 2003), however, reported that nonlegume cover crops increased SOC, but decreased succeeding crop yields compared to legume cover crops. Differences in biomass yields and N-supplying capacity of cover crops resulted in variation in SOC and succeeding crop yields among cover crops. They, however, also observed greater SOC with mixed legume and nonlegume cover crops than legume cover crops alone, but similar crop yields between these treatments. As a result, they suggested that a mixture of legume and nonlegume cover

TABLE 3.6B
Cover Crop Effect on Soil Organic C (SOC) Under Different Crops in Various Regions

					SOC		
Location	Cover Crop	Succeeding Crop	Tillage[a]	Soil Depth	Concentration ($g\ C\ kg^{-1}$)	Stock ($Mg\ C\ ha^{-1}$)	Reference
Georgia, USA	Control	Sorghum	NT	0–15 cm	6.35		Hargrove (1986)
	Rye				7.05		
	Crimson clover				6.65		
	Hairy vetch				7.60		
	Common vetch				7.65		
Washington, USA	Control	Corn	CT	0–15 cm	15.7		Kuo et al. (1997)
	Austrian winter pea				16.6		
	Hairy vetch				15.8		
	Canola				15.4		
	Rye				16.6		
	Ryegrass				16.6		
Georgia, USA	Control	Corn	CT	0–10 cm	8.63		McVay et al. (1989)
	Wheat				9.38		
	Crimson clover				10.05		
	Hairy vetch				9.68		
Alabama, USA	Control	Cotton	CT	0–10 cm	7.0		Touchton et al. (1984)
	Crimson clover				8.7		
	Hairy vetch				10.8		
China	Rye	Wheat	CT	0–20 cm	12.5		Zhang and Wang (2018)
	Huai bean				11.9		
	Rye + Huai bean				12.7		
China	Huai bean	Wheat	CT	0–20 cm	8.79		Zhang et al. (2016)
	Soybean				8.62		
	Mung bean				8.43		

[a] Tillage are CT, conventional tillage; ST, strip tillage; and NT, no tillage.

crops might enhance SOC while sustaining crop yield. Our overall analysis of cover crops also indicated that mixed legume and nonlegume cover crops increased SOC concentration and stock compared to legume or nonlegume cover crop alone, while succeeding crop yield for the cover crop mixture was in between the individual cover crops (Table 3.2).

Crop yield depends on soil water availability, which is affected by SOC. It is likely that the adequate amount of soil water available for enhancing crop yield occurs when SOC concentration is $\geq 12.0\ g\ C\ kg^{-1}$. The decline in crop yield when SOC increased from 4 to 12 $g\ C\ kg^{-1}$ (Figure 3.2A) could be due to N immobilization when nonlegume cover crop residues with higher C/N ratios were

TABLE 3.6C
Cover Crop Effect on Soil Organic C (SOC) Under Different Crops in Various Regions

					SOC		
Location	Cover Crop	Succeeding Crop	Tillage[a]	Soil Depth	Concentration ($g\ C\ kg^{-1}$)	Stock ($Mg\ C\ ha^{-1}$)	Reference
Argentina	Wheat	Corn	CT	0–20 cm	16.73		Duval et al. (2016)
	Oat				16.87		
	Oat + vetch				17.20		
	Vetch				16.63		
New Mexico, USA	Control	Wheat	NT	0–15 cm		13.8	Ghimire et al. (2019)
	Pea					12.8	
	Oat					14.0	
	Canola					12.4	
	Pea + Oat					13.0	
	Pea + canola					12.0	
	Pea + oat + canola					13.0	
	Pea + oat + canola + hairy vetch + radish + barley					12.8	
Georgia, USA	Control	Forage sorghum	NT	0–30 cm		31.1	Sainju et al. (2018a)
	Rye					30.4	
	Hairy vetch					30.8	
	Hairy vetch + rye					32.6	
Georgia, USA	Control	Sweet sorghum	NT	0–30 cm		33.9	Sainju et al. (2018b)
	Rye					33.6	
	Hairy vetch					35.5	
	Hairy vetch + rye					33.5	

[a] Tillage are CT, conventional tillage; ST, strip tillage; and NT, no tillage.

returned to the soil. As shown in Table 3.2, nonlegume and mixed legume and nonlegume cover crops had greater biomass than legumes. Because crop residue with a higher C/N ratio can immobilize soil N, which becomes unavailable to plants (Kuo and Sainju 1998), it is likely that decreased crop yield when SOM increased from 4 to 12 g C kg^{-1} was due to N immobilization. As the turnover rate of nonlegume residue increased, N could have mineralized from the residue, resulting in increased succeeding crop yield when SOC was ≥12 g C kg^{-1} (Kuo et al. 1997). Bauer and Black (1994) found that an increase in SOC stock by 580 kg C ha^{-1} to a depth of 3 cm increased spring wheat yield by 15.6 kg ha^{-1} due to improved soil structure and increased soil water-nutrient-crop productivity relationships. A global meta-analysis by Oldfield et al. (2019) on the relative effect size of SOC showed that the largest gains in the yield of major cereal crops occurred with SOC concentration between 1 and 20 g C kg^{-1}. Sainju (2014) showed that an increase in SOC stock to a depth of 5 cm by 1 Mg C ha^{-1} increased crop yield by 0.47 Mg ha^{-1}. To a depth of 120 cm, an increase in SOC stock by 1 Mg C ha^{-1} increased crop yield by 0.02 Mg ha^{-1}. Several others (Ghimire et al.

TABLE 3.7A
Increase or Decrease in Soil Organic C (SOC) with Cover Crop Compared to No Cover Crop Under Different Crops in Various Regions

Location	Cover Crop	Succeeding Crop	Tillage[a]	Soil Depth	SOC Conc. (g C kg^{-1})	SOC Stock (Mg C ha^{-1})	Reference
Georgia, USA	Rye	Tomato	CT	0–20 cm	1.64	1.9	Sainju et al. (2000)
	Hairy vetch				1.37	1.4	
	Crimson clover				1.22	1.3	
Georgia, USA	Rye	Cotton/ sorghum				0.1	
	Hairy vetch					0.7	
	Hairy vetch + rye					−0.6	
	Rye					1.3	
	Hairy vetch					0.4	
	Hairy vetch + rye					1.3	
	Rye					1.9	
	Hairy vetch					1.0	
	Hairy vetch + rye					2.0	
Kentucky, USA	Hairy vetch	Corn	NT	0–7.5 cm	2.9		Frye et al. (1986)
	Big flower vetch				2.1		
	Rye				0.9		
Georgia, USA	Rye	Sorghum	NT	0–15 cm	0.70		Hargrove (1986)
	Crimson clover				0.30		
	Hairy vetch				1.25		
	Common vetch				1.30		
Washington, USA	Austrian winter pea	Corn	CT	0–15 cm	0.90		Kuo et al. (1997)
	Hairy vetch				0.10		
	Canola				−0.30		
	Rye				0.90		
	Ryegrass				0.90		

[a] Tillage are CT, conventional tillage; ST, strip tillage; and NT, no tillage.
[b] Blend is a mixture of balansa clover, crimson clover, and hairy vetch.

2018; Sainju et al. 2009, 2016) found that increased SOC due to enhanced cover cropping intensity increased crop yield. Long-term experiments conducted in Illinois and Missouri, USA also revealed positive relationships among crop biomass C input, SOC concentration, and crop yields (Miles and Brown 2011, Nafziger and Dunker 2011).

The SOC also may help to recycle nutrients. The linear relationship between SOC concentration and crop N uptake (Figure 3.2B) suggests that enhanced N availability from organic matter mineralization can increase crop N uptake. Availability of N from SOM, however, depends on microbial activity, which is affected by soil temperature and water content. An exception, however, can occur in soils with high organic matter content, such as peat and muck soils, where high water content may

TABLE 3.7B
Increase or Decrease in Soil Organic C (SOC) with Cover Crop Compared to No Cover Crop Under Different Crops in Various Regions

					SOC		
Location	Cover Crop	Succeeding Crop	Tillage[a]	Soil Depth	Concentration ($g\ C\ kg^{-1}$)	Stock ($Mg\ C\ ha^{-1}$)	Reference
New Mexico, USA	Pea	Wheat	NT	0–15 cm		0.1	Ghimire et al. (2019)
	Oat					0.2	
	Canola					−1.4	
	Pea + Oat					−0.8	
	Pea + canola					−1.8	
	Pea + oat + canola					−0.8	
	Pea + oat + canola + hairy vetch + radish + barley					−1.0	
Georgia, USA	Rye	Forage sorghum	NT	0–30 cm		−0.7	Sainju et al. (2018a)
	Hairy vetch					−0.3	
	Hairy vetch + rye					1.5	
Georgia, USA	Wheat	Corn	CT	0–10 cm	0.75		McVay et al. (1989)
	Crimson clover				1.42		
	Hairy vetch				1.05		
Alabama, USA	Crimson clover	Cotton	CT	0–10 cm	1.7		Touchton et al. (1984)
	Hairy vetch				3.8		
Georgia, USA	Rye	Sweet sorghum	NT	0–30 cm		−0.3	Sainju et al. (2018b)
	Hairy vetch					1.6	
	Hairy vetch + rye					−0.4	

[a] Tillage are CT, conventional tillage; ST, strip tillage; and NT, no tillage.

reduce N availability due to anaerobic conditions. Similar to crop yield, the decline in crop protein concentration due to increased SOC concentration from 4 to 12 g C kg^{-1} (Figure 3.2C) was probably due to reduced N availability from N immobilization when cover crop residue with a higher C/N ratio was added to the soil. At SOC >12 g kg^{-1}, enhanced mineralization of residue may have increased soil N availability, resulting in increased protein concentration in crops.

Unlike SOC, limited availability of data resulted in reduced information about the relationship between soil labile C fractions and crop yield. Labile C fractions, however, related better than SOC to crop yield (Figure 3.3). Soil POC, PCM, and MBC were nonlinearly, but $KMnO_4$-extractable C linearly related to crop yield (Figure 3.3). Crop yield increased as POC increased to 3.35 Mg C ha^{-1}, after which it declined (Figure 3.3A). The POC explained 56% of the variability in crop yield. Similarly, crop yield increased as PCM increased to 420 kg C ha^{-1}, followed by a decline (Figure 3.3B). Crop yield also increased as MBC increased to 1430 kg C ha^{-1}, followed by a decline (Figure 3.3C). Both PCM and MBC explained 82 and 83% variability in crop yield, respectively.

TABLE 3.8A
Soil Labile C Fractions as Affected by Cover Crops in Various Regions

Location	Cover Crop	Soil Depth	Soil Labile C	Reference
			Carbohydrate (g kg^{-1})	
Washington, USA	Control	0–15 cm	3.45	Kuo et al. (1997)
	Rye		3.95	
	Austrian winter pea		3.55	
	Ryegrass		3.75	
	Hairy vetch		3.45	
	Canola		3.30	
	Control	15–30 cm	3.45	
	Rye		3.65	
	Austrian winter pea		3.45	
	Ryegrass		3.50	
	Hairy vetch		3.25	
	Canola		3.20	
			Potential C mineralization (kg C ha^{-1})	
New Mexico, USA	Control	0–15 cm	133	Ghimire et al. (2019)
	Pea		200	
	Oat		210	
	Canola		213	
	Pea + Oat		200	
	Pea + canola		213	
	Pea + oat + canola		170	
	Mixture[a]		205	
			$KMnO_4$-oxidizable C (kg C ha^{-1})	
New Mexico, USA	Control	0–15 cm	800	Ghimire et al. (2019)
	Pea		770	
	Oat		838	
	Canola		758	
	Pea + Oat		840	
	Pea + canola		780	
	Pea + oat + canola		810	
	Mixture[a]		850	

[a] Pea + oat + canola + hairy vetch + radish + barley.

Similarly, $KMnO_4$-extractable C explained 38% of variability in crop yield (Figure 3.3D). An increase in $KMnO_4$-extractable C by 1 kg C ha^{-1} increased crop yield by 0.005 Mg ha^{-1}.

The opposite nonlinear trends in the relationship between crop yield vs. SOC and crop yield vs. soil labile C fractions, except $KMnO_4$-extractable C, suggest that labile and non-labile C fractions behave differently to crop yield (Figures. 3.2 and 3.3). As crop yield increased, SOC reduced, but labile C fractions increased. The reverse was true when crop yield decreased. While SOC can influence crop yield by altering soil water content and nutrient levels, labile C fractions, such as PCM and MBC, can affect microbial biomass and activity and nutrient cycling. Increased SOC does not necessarily mean enhanced crop yield, but organic C has to be mineralized by microorganisms to recycle nutrients, especially N, to be available to plants. Crop yield, protein concentration, and N uptake are influenced both by soil water and nutrient availability. Usually, increased crop yield is associated with decreased protein concentration because they are negatively related (Lenssen et

TABLE 3.8B
Soil Labile C Fractions as Affected by Cover Crops in Various Regions

Location	Cover Crop	Soil Depth	Soil Labile C	Reference
			Potential C mineralization (kg C ha^{-1})	
Georgia, USA	Rye	0–20 cm	575	Sainju et al. (2003)
	Hairy vetch		420	
	Crimson clover		575	
	Control		283	
			Microbial biomass C (kg ha^{-1})	
	Rye		1550	Sainju et al. (2003)
	Hairy vetch		1469	
	Crimson clover		1530	
	Control		1298	
			Particulate organic C (Mg C ha^{-1})	
	Rye		3.6	Sainju et al. (2003)
	Hairy vetch		3.5	
	Crimson clover		3.6	
	Control		3.1	
			Potential C mineralization (kg C ha^{-1})	
Georgia, USA	Rye	0–15 cm	118	Sainju et al. (2007)
	Blend[a]		138	
	Rye + blend		132	
	Crimson clover		144	
			Microbial biomass C (kg ha^{-1})	
	Rye		454	Sainju et al. (2007)
	Blend[a]		409	
	Rye + blend		430	
	Crimson clover		402	

[a] Blend is a mixture of balansa clover, crimson clover, and hairy vetch.

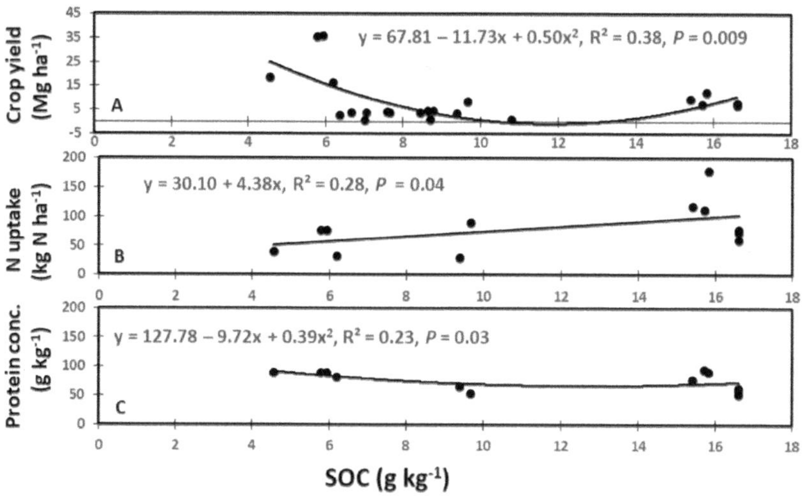

FIGURE 3.2 Relationship between (A) crop yield, (B) N uptake, and (C) protein concentration and soil organic C (SOC).

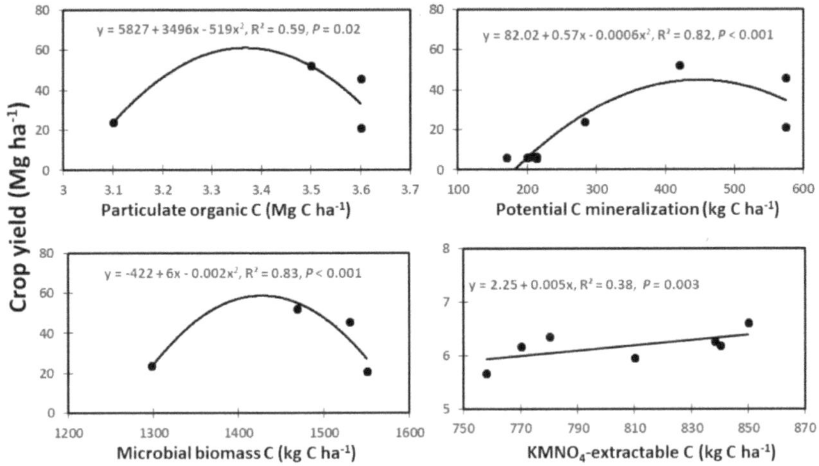

FIGURE 3.3 Relationship between crop yield and (A) soil particulate organic C (POC), (B) potential C mineralization (PCM), (C) microbial biomass C (MBC), and (D) $KMnO_4$-extractable C.

al. 2007; Miller et al. 2011). Adequate soil water and nutrients are required to achieve economical crop yield, but their excess levels can reduce yield. While anaerobic conditions reduce root growth of most crops (except rice), thereby reducing yields, crops are able to extract only 50% of nutrients applied from fertilizers. Unused N from mineralization of crop residue and SOM and N fertilizers remain as soil residual N that can degrade soil and environmental quality through acidification, N leaching, and greenhouse gas (N_2O) emissions. Excessive levels of some nutrients, such as Al, can also be toxic to crop growth. Several researchers (Franzluebbers and Haney 2006; Franzluebbers 2016) have reported that PCM measured from CO_2 flush after 1–3 days incubation is related to crop yield. These results suggest that appropriate combination of cover crops, such as a legume and non-legume mixture, may be required to obtain a balance between soil labile and non-labile C fractions to achieve desirable crop yields and quality.

3.8 CHALLENGES AND OPPORTUNITIES FOR ENHANCING SOIL CARBON AND CROP YIELD WITH COVER CROPS

Although cover crops provide numerous ecosystem services, particularly increasing soil C sequestration and crop yield compared to no cover crop, some challenges remain for cover crops to be adopted by producers in order to be profitable. Producers will not adopt improved management technology unless they are convinced that the technology will enhance farm income and their livelihood. Some of the benefits from adopting cover crops will be short-term, such as increasing crop yield, while others will be long-term, such as enhancing soil health and quality. Some of the challenges and benefits of growing cover crops are:

1. It is expensive to grow cover crops, as producers need to buy cover crop seeds and plant them. The additional expenses in growing cover crops should be justified by increased income from enhanced succeeding crop yields and quality and improvement in soil health.
2. Cover crops do not grow well in all regions. Winter cover crops grow best in humid and subhumid regions with mild winter and adequate precipitation distributed uniformly throughout the growing season. In arid and semiarid regions where precipitation is limited and growing season is shorter, cover crops do not perform well. Even when cover crops

produce significant biomass in these regions during years with above-average precipitation, cover crops depress succeeding crop yields by extracting soil water and reducing the water available for succeeding crops. Applying irrigation to grow cover crops may not justify the cost associated with the practice unless dramatic increases in crop yield and soil health occur in these regions.

3. There should be incentives to grow cover crops from the government, society, and cooperatives in regions where cover crops can provide ecosystem services. These incentives could be additional income from increasing C sequestration due to cover crops from C credit markets, reducing N leaching and soil erosion, or improving soil health.
4. The benefits of using cover crops should be clearly explained to the producers by conducting economic analysis. Economic analysis of ecosystem services, such as C sequestration, soil erosion, N leaching, greenhouse gas emissions, and biodiversity, remains a challenge.
5. The type of cover crop should perform well in a region and be socially acceptable to producers. Some producers might want to harvest aboveground biomass of cover crops for animal feedstock and short-term economic benefit. Such practices, however, reduce SOC by reducing crop residue C input. This should be clearly explained to the producers.
6. Some producers may want to let livestock graze over the cover crop biomass in the field. Although such a system has been shown to increase SOC (Franzluebbers 2007), more research is needed to evaluate the effect of animal grazing of cover crops on SOC, crop yields, and quality.

3.9 CONCLUSIONS

Cover crops varied in biomass production and N accumulation that affected soil labile and non-labile C fractions and succeeding crop yields and quality. Legume cover crops increased N supply, reduced N fertilizer rates, and enhanced succeeding crop yields and quality compared to nonlegumes or the control with no cover crop, but were not effective in enhancing soil labile and non-labile C fractions compared to nonlegume cover crops. In contrast, nonlegume cover crops produced greater biomass and enhanced SOC and labile C fractions than legumes or no cover crop. Cover crops reduced N fertilization rates by as much as 15–200 kg N ha^{-1} for sustaining succeeding crop yields. A mixture of legume and nonlegume cover crops was more effective in enhancing SOC and sustaining crop yields and quality than either species alone. Crop yield and protein concentration increased when SOC decreased due to C mineralization, but labile C fractions increased due to cover crops. Labile C fractions related better than SOC to crop yield. Although cover crops can provide many agronomical and environmental benefits, producers should be fully acquainted with the challenges and opportunities of growing cover crops that affect farm income and their livelihood.

REFERENCES

Bauer, A., and A.L. Black. 1994. Quantification of the effect of soil organic matter content on soil productivity. *Soil Science Society of America Journal* 58(1):185–193.

Blanco-Canqui, H., J.D. Holman, A.J. Schlegel, J. Tatarko, and T.M. Shaver. 2013. Replacing fallow with cover crops in a semiarid soil: Effects on soil properties. *Soil Science Society of America Journal* 77(3):1026–1034.

Boone, R.D. 1994. Light-fraction soil organic matter: Origin and contribution to net nitrogen mineralization. *Soil Biology and Biochemistry* 26(11):1459–1468.

Boquet, D.J., R.L. Hutchinson, and G.A. Breitenbeck. 2004. Long-term tillage, cover crop, and nitrogen rate effects on cotton: Yield and fiber properties. *Agronomy Journal* 96(5):1436–1442.

Breman, J.W., and D.L. Wright. 1984. Using winter legume mulches as a nitrogen source for no-tillage corn and sorghum production. In: J.T. Touchton (ed.), Proceedings of the 7th Annual Southeastern No-tillage Systems Conference, Auburn, AL, pp. 6–17.

Brown, S.M., T. Whitewell, J.T. Touchton, and C.H. Burmester. 1985. Conservation tillage system for cotton production. *Soil Science Society of America Journal* 49(5):1256–1260.

Brozovic, B., D. Jug, I. Jug, B. Durdevic, V. Vukadinovic, V. Tadic, and B. Stipsevic. 2018. Influence of winter cover crop incorporation on weed infestation on popcorn maize organic production. *Agriculturae Conspectus Scientificus* 83:77–81.

Buntley, G.J. 1986. Tennessee no-tillage updates. In: R.E. Phillips (ed). Southern Region No-Till Conference. Southern Region Series Bulletin 30, Lexinton, KY, pp. 100–102.

Clark, A.J., A.M. Decker, and J.J. Meisinger. 1994. Seeding rate and kill date effects on hairy vetch-cereal rye cover crop mixtures for corn production. *Agronomy Journal* 86(6):1065–1070.

Clark, A.J., A.M. Decker, J.J. Meisinger, and M.S. Mcintosh. 1997. Kill date of vetch, rye, and a vetch-rye mixture: II. Soil moisture and corn yield. *Agronomy Journal* 89(3):434–441.

Decker, A.M., A.J. Clark, J.J. Meisinger, F.R. Mullford, and M.S. McIntosh. 1994. Legume cover crop contributions to no-tillage corn production. *Agronomy Journal* 86(1):126–135.

Duval, M.S., J.A. Galantini, J.E. Capuro, and J.M. Martinez. 2016. Winter cover crops in soybean monoculture: Effects on soil organic carbon and its fractions. *Soil and Tillage Research* 161:95–105.

Ebelhar, S.A., W.W. Frye, and R.L. Blevins. 1994. Nitrogen from legume cover crops for no-tillage corn. *Agronomy Journal* 76:51–55.

Flannery, R.L. 1981. Conventional vs. no-tillage corn silage production. *Better Crops* 66:3–6.

Franzluebbers, A.J. 2007. Integrated crop-livestock systems in the southeastern USA. *Agronomy Journal* 99:361–372.

Franzluebbers, A.J. 2016. Should soil testing services measure soil biological activity? *Agricultural and Environmental Letters* 1(1):150009, https://doi.org/10.2134/ael2015.11.0009.

Franzluebbers, A.J., and R.I. Haney. 2006. *Assessing Soil Quality in Organic Agriculture.* Critical Issues Report 2006. 2. Eugene, OR: The Organic Center Enterprise.

Franzluebbers, A.J., F.M. Hons, and D.A. Zuberer. 1995. Soil organic carbon, microbial biomass, and mineralizable carbon and nitrogen in sorghum. *Soil Science Society of America Journal* 59(2):460–466.

Frasier, I., E. Noellenmeyer, N. Amirotti, and A. Quiroga. 2017. Vetch-rye biculture is a sustainable alternative for enhanced nitrogen availability and low leaching losses in a no-till cover crop system. *Field Crops Research* 214:101–112.

Frye, W.W., J.H. Herbeck, and R.L. Blevins. 1986. Nitrogen from legume cover crops for no-tillage corn. In: *1986 Grain Research.* Lexington, KY: University of Kentucky.

Gabriel, J.L., and M. Quemada. 2011. Replacing bare fallow with cover crops in a maize cropping system: Yield, N uptake, and fertilizer rate. *European Journal of Agronomy* 34(3):133–433.

Gabriel, J.L., M. Vanclooster, and M. Quemada. 2014. Integrating water, nitrogen, and salinity in sustainable irrigated systems: Cover crops versus fallow. *Journal of Irrigation and Drainage Engineering* 140(9):175–187.

Gabriel, J.L., M. Quemada, D. Martin-Lammerding, and M. Vanclooster. 2019. Assessing the cover crop effect on soil hydraulic properties by inverse modelling in a 10-year field trial. *Agricultural Water Management* 222:62–71.

Ghimire, R., S. Machado, and P. Bista. 2018. Decline in soil organic carbon and nitrogen limits yield in wheat-fallow systems. *Plant and Soil* 422(1–2):423–435.

Ghimire, R., B. Ghimire, A.O. Mesbah, U.M. Sainju, and O.J. Idowu. 2019. Soil health response of cover crops in winter wheat-fallow system. *Agronomy Journal* 111(4):2108–2115.

Hargrove, W.L. 1986. Winter legumes as a nitrogen source for no-till grain sorghum 1. *Agronomy Journal* 78(1):70–74.

Herbeck, J.H., W.W. Frye, and R.L. Blevins. 1987. Nitrogen from legume cover crops for no-till corn and grain sorghum. In: J.F. Power (ed.), *The Role of Legumes in Conservation Tillage Systems.* Ankeny, IA: Soil and Water Conservation Society, pp. 51–52.

Hill, E.C., K.A. Renner, and C.L. Sprague. 2016. Cover crop impact on nitrogen availability and dry bean in an organic system. *Agronomy Journal* 108(1):329–341.

Huntington, T.G., J.H. Grove, and W.W. Frye. 1985. Release and recovery of nitrogen from winter annual cover crops in no-till corn production. *Communications in Soil Science and Plant Analysis* 16(2):193–211.

Jian, J., X. Du, M.S. Reiter, and R.D. Stewart. 2020. A meta-analysis of global cropland soil carbon changes due to cover cropping. *Soil Biology and Biochemistry* 143: 107735, https://doi.org/10.1016/j.soilbio.2020.107735.

Kamprath, E.J., W.V. Chandler, and B.A. Krantz. 1968. Winter cover crops: Their effects on corn yield and soil properties. In: *North Carolina Agriculture Experiment Station Technical Bulletin 19.* Raleigh, NC: North Carolina State University.

Kelly, T.C., Y.C. Lu, A.A. Abdul-Baki, and J.R. Teasdale. 1995. Economics of hairy vetch mulch system for producing fresh-market tomato in the Mid-Atlantic region. *Journal of American Society of Horticultural Science* 120:850–860.

Kramberger, B., A. Gselman, M. Janekovic, and B. Braeko. 2009. Effect of cover crop on soil mineral nitrogen and on the yield and nitrogen content of maize. *European Journal of Agronomy* 31:103–109.

Kuo, S., and U.M. Sainju. 1998. Nitrogen mineralization and availability of mixed leguminous and non-leguminous cover crop residues in soil. *Biology and Fertility of Soils* 26(4):346–353.

Kuo, S., E.J. Jellum, and U.M. Sainju. 1995. The effect of winter cover cropping on soil and water quality. In: Proceeding Western Nutrient Management Conference. Salt Lake City, UT, pp. 54–64.

Kuo, S., U.M. Sainju, and E.J. Jellum. 1997. Winter cover crop effects on organic carbon and carbohydrate in soil. *Soil Science Society of America Journal* 61(1):145–152.

Lenssen, A.W., G.D. Johnson, and G.R. Carlson. 2007. Cropping sequence and tillage system influences annual crop production and water use in semiarid Montana, USA. *Field Crops Research* 100(1):32–43.

Liang, B.C., X.L.Wang, and B.L.Ma. 2002. Maize root-induced change in soil organic carbon pools. *Soil Science Society of America Journal* 66(3):845–847.

McCracken, D.V., M.S. Smith, J.H. Grove, C.T. McKown, and R.L. Blevins. 1994. Nitrate leaching influenced by cover cropping and nitrogen source. *Soil Science Society of America Journal* 58(5):1476–1483.

McVay, K.A., D.E. Ratcliffe, and W.L. Hargrove. 1989. Winter legume effects on soil properties and nitrogen fertilizer requirements. *Soil Science Society of America Journal* 53(6):1856–1862.

Mesbah, A.O., A. Nilahyane, B. Ghimire, L. Beck, and R. Ghimire. 2019. Efficacy of cover crops on weed suppression, wheat yield, and water conservation in winter wheat-sorghum-fallow. *Crop Science* 59(4):1745–1752.

Mesinger, J.J., W.L. Hargrove, R.L. Mikkelsen, Jr., J.R. Williams, and V.W. Benson. 1991. Effects of cover crops on ground water quality. In: W.L. Hargrove (ed.), *Cover Crops for Clean Water.* Ankeny, IA: Soil and Water Conservation Society, pp. 57–68.

Meyer, D.W. 1987. Influence of green-manured, hayed, or grain legumes on grain yield and quality of barley crop in the northern Great Plains. In: J.D. Power (ed.), *The Role of Legumes in Conservation Tillage Systems.* Ankeny, IA: Soil and Water Conservation Society, pp. 94–95.

Miles, R.J., and J.R. Brown. 2011. The Sanborn field experiment: Implications for long-term soil organic carbon levels. *Agronomy Journal* 103(1):268–278.

Miller, P.R., E.J. Lighthiser, C.A. Jones, J.A. Holmes, T.L. Rick, and J.M. Wraith. 2011. Pea green manure management affects organic winter wheat yield and quality in semiarid Montana. *Canadian Journal of Plant Science* 91(3):497–508.

Mitchell, J.P., D.W. Peters, and C. Shennan. 1999. Changes in soil water storage in winter fallowed and cover cropped soils. *Journal of Sustainable Agriculture* 15(2–3):19–31.

Muhammad, I., J. Wang, U.M. Sainju, S. Zhang, F. Zhao, and A. Khan. 2020. Cover cropping enhances soil microbial biomass and affects microbial community structure: A meta-analysis. *Geoderma* 381:114696, https://doi.org/10.1016/j.geoderma.2020.114696.

Nafziger, E.D., and R.E. Dunker. 2011. Soil organic carbon trends over 100 years in the Morrow plots. *Agronomy Journal* 103(1):261–267.

Nielsen, D.C., D.J. Lyon, R.K. Higgins, G.W. Hergert, J.D. Holman, and M.F. Vigil. 2016. Cover crop effect on subsequent wheat yield in the central Great Plains. *Agronomy Journal* 108(1):243–256.

Norby, R.J., and M.F. Cotrufo. 1998. A question of litter quality. *Nature* 396(6706):17–18.

Oldfield, E.E., M.A. Bradford, and S.A. Wood. 2019. Global meta-analysis of the relationship between soil organic matter and crop yields. *Soil* 5(1):15–32.

Poeplau, C., and A.J.A. Don. 2015. Carbon sequestration in agricultural soils via cultivation of cover crops. A meta-analysis. *Agriculture, Ecosystems and Environment* 200:33–41.

Ranells, N.N., and M.G. Wagger. 1996. Nitrogen release from grass and legume cover crop monocultures and bicultures. *Agronomy Journal* 88(5):777–782.

Reese, C.L., D.E. Clay, S.A. Clay, A.D. Bich, A.C. Kennedy, S.A. Hansen, and J. Moriles. 2014. Winter cover crops impact on corn production in semiarid regions. *Agronomy Journal* 106(4):1479–1488.

Restovich, S., A. Andriulo, and S. Portela. 2012. Introduction of cover crops in a maize–soybean rotation of the Humid Pampas: Effect on nitrogen and water dynamics. *Field Crops Research* 128:62–70.

Robertson, P.G., K.L. Gross, S.K. Hamilton, D.A. Landis, T.M. Schmidt, S.S. Snapp, and S.M. Swinton. 2014. Farming for ecosystem services: An ecological approach to production agriculture. *BioScience* 64(5):404–415.

Russell, M.P., and W.L. Hargrove. 1989. Cropping systems: Ecology and management. In: R.F. Follett (ed.), *Nitrogen Management and Groundwater Protection*. New York: Elsevier Science Publication Company, pp. 277–317.

Sainju, U.M. 2014. Cropping sequence and nitrogen fertilization impact on surface residue, soil carbon sequestration, and crop yields. *Agronomy Journal* 106(4):1231–1242.

Sainju, U.M., and B.P. Singh. 1997. Winter cover crops for sustainable agricultural systems: Influence on soil properties, water quality, and crop yields. *Hortscience* 32(1):21–28.

Sainju, U.M., B.P. Singh, and W.F. Whitehead. 1998. Cover crop root distribution and its effects on soil nitrogen cycling. *Agronomy Journal* 90(4):511–518.

Sainju, U.M., B.P. Singh, and W.F. Whitehead. 2000. Cover crops and nitrogen fertilization effects on soil carbon and nitrogen and tomato yield. *Canadian Journal of Soil Science* 80(3):523–532.

Sainju, U.M., B.P. Singh, and W.F. Whitehead. 2002. Long-term effects of tillage, cover crops, and nitrogen fertilization on soil organic carbon and nitrogen concentrations in sandy loam soils in Georgia, USA. *Soil and Tillage Research* 63(3–4):167–179.

Sainju, U.M., W.F. Whitehead, and B.P. Singh. 2003. Cover crops and nitrogen fertilization effects on soil aggregation and carbon and nitrogen pools. *Canadian Journal of Soil Science* 83(2):155–165.

Sainju, U.M., W.F. Whitehead, and B.P. Singh. 2005a. Biculture legume-cereal cover crops for enhanced biomass yield and carbon and nitrogen. *Agronomy Journal* 97(5):1403–1412.

Sainju, U.M., W.F. Whitehead, and B.P. Singh. 2005b. Tillage, cover crops, and nitrogen fertilization effects on cotton and sorghum root biomass, carbon, and nitrogen. *Agronomy Journal* 97(5):1279–1290.

Sainju, U.M., B.P. Singh, W.F. Whitehead, and S. Wang. 2006. Tillage, cover crops, and nitrogen fertilization effects on soil nitrogen and cotton and sorghum yields. *European Journal of Agronomy* 25(4):372–382.

Sainju, U.M., H.H. Schomberg, B.P. Singh, W.F. Whitehead, P.G. Tillman, and S.L. Lachnicht-Weyers. 2007. Cover crops effects on soil carbon pools under conservation tillage cotton. *Soil and Tillage Research* 96(1–2):205–218.

Sainju, U.M., T. Caesar-TonThat, A.W. Lenssen, and R.G. Evans. 2009. Dryland crop yields and soil organic matter as influenced by long-term tillage and cropping sequence. *Agronomy Journal* 101(2):243–251.

Sainju, U.M., A.W. Lenssen, B.L. Allen, W.B. Stevens, and J.D. Jabro. 2016. Soil total carbon and crop yield affected by crop rotation and cultural practice. *Agronomy Journal* 109(1):1–9.

Sainju, U.M., H.P. Singh, B.P. Singh, W.F. Whitehead, A. Chiluwal, and R. Paudel. 2018a. Cover crop and nitrogen fertilization influence soil carbon and nitrogen under bioenergy sweet sorghum. *Agronomy Journal* 110(2):463–471.

Sainju, U.M., H.P. Singh, B.P. Singh, W.F. Whitehead, W. Chiluwal, and R. Paudel. 2018b. Soil carbon and nitrogen under bioenergy forage sorghum influenced by cover crop and nitrogen fertilization. *Agrosystems, Geosciences and Environment* 1(1):180004, https://doi.org/10.2134/age2018.03.0004.

Sainju, U.M., A.W. Lenssen, B.L. Allen, J.D. Jabro, and W.B. Stevens. 2019. Pea growth, yield, and quality in different crop rotations and cultural practices. *Agrosystems, Geosciences and Environment* 2(1), 180041.

Salinas-Garcia, J.R., F.M. Hons, and J.E. Matocha. 1997. Long-term effects of tillage and fertilization on soil organic matter dynamics. *Soil Science Society of America Journal* 61(1):152–159.

Sanders, Z.P., J.S. Andrews, and N.S. Hill. 2018. Water use efficiency in living mulch and annual cover crop corn production systems. *Agronomy Journal* 110(3):1128–1135.

Shennan, C. 1992. Cover crops, nitrogen cycling, and soil properties in semi-irrigated vegetable production systems. *Hortscience* 27(7):749–754.

Ta, T.C., and M.A. Farris. 1987. Species variation in the fixation and transfer of N from legumes to associated grasses. *Plant and Soil* 98(2):265–274.

Thapa, V.R., R. Ghimire, V. Acosta-Martinez, M. Marsalis, and M. Schipanski. 2021. Cover crop effects on soil microbial community structure and enzyme activities. *Applied Soil Ecology* 157:103735, https://doi.org/10.1016/j.apsoil.2020.103735.

Touchton, J.T., D.H. Rickerl, R.H. Walker, and C.E. Snipes. 1984. Winter legumes as a nitrogen source for no-tillage corn. *Soil and Tillage Research* 4(4):391–401.

van Es, H.M., and D.L. Karlen. 2019. Reanalysis validates soil health indicator sensitivity and correlation with long-term crop yields. *Soil Science Society of America Journal* 83:721–733.

Varco, J.J., W.W. Frye, J.H. Herbeck, and R.L. Blevins. 1984. Effects of legume cover crops on yield of no-till corn. In: J.J. Touchton (ed.), Proceedings of the 7th Annual Southeast No-Tillage System Conference, Auburn, AL, pp. 40–56.

Vaughan, J.D., and G.K. Evanylo. 1998. Corn response to cover crop species, spring desiccation time, and residue management. *Agronomy Journal*:536–544.

Wagger, M.G. 1989. Time of desiccation effects on plant composition and subsequent nitrogen release from several winter annual cover crops. *Agronomy Journal* 81(2):236–241.

Williams, M.M., D.A. Mortensen, and J.W. Doran. 2000. No-tillage soybean performance in cover crops for weed management in the western Corn Belt. *Journal of Soil and Water Conservation* 55:79–84.

Williams, S.M., and R.R. Weil. 2004. Cover crop root channels may alleviate soil compaction effects on soybean crop. *Soil Science Society of America Journal* 68(4):1403–1409.

Worsham, A.D. 1986. No-tillage research update: North Carolina. In: R.E. Philipps (ed.), Southern Region No-Till Conference. Southern Region Series Bulletin 319, Lexington, KY.

Xue, N., Z. Yang, and Z. Gao. 2018. Effects of green manures during fallow on moisture and nutrients of soil and winter wheat yield on the Loss Plateau of China. *Emirates Journal of Food and Agriculture* 29:978–987.

Zhang, D.B., P.W. Yao, Z. Na, W.D. Cao, S.Q. Zhang, Y.Y. Li, and Y.J. Gao. 2016. Soil water balance and water-use efficiency of dryland wheat in different precipitation years in response to green manure approach. *Science Reports* 6, 26856. https://doi.org/10.1038/srep26856.

Zhang, Q., and J. Wang. 2018. Effects of cover crop and nitrogen fertilization on soil moisture and crop yield in a dryland winter wheat field. *Agricultural Research in Arid Areas* 36:120–124.

Zhang, S., V. Sadras, X. Chen, and F. Zhang. 2013. Water-use efficiency of dryland wheat in the Loess Plateau in response to soil and crop management. *Field Crops Research* 151:9–18.

4 Global Spread of Conservation Agriculture for Enhancing Soil Organic Matter, Soil Health, Productivity, and Ecosystem Services

Amir Kassam, Emilio Gonzalez-Sanchez, Rosa M. Carbonell-Bojollo, Theodor Friedrich, and Rolf Derpsch

CONTENTS

DOI: 10.1201/9781003102762-4

LIST OF ABBREVIATIONS

AAA: Adaptation of African Agriculture
CA: Conservation Agriculture
CA-CoP: Global Conservation Agriculture Community of Practice
CCP: Conservation cropping protocol
CEC: Cation exchange capacity
COP: Conference of the Parties
CSA: Climate smart agriculture
CT: Conventional tillage
ECOSOC: Economic and Social Council
ESs: Ecosystem services
FAO: Food and Agriculture Organization of the United Nations
GHG: Greenhouse gas
HLPF: High-level Political Forum
LDN: Land degradation neutrality
MEA: Millennium ecosystem assessment
NT: No-till or no tillage
OM: Organic matter
PRB: Permanent raised beds
RA: Regenerative agriculture
RELACO: Latin-American Network for Conservation Tillage
SARD: Sustainable Agriculture and Rural Development
SDGs: Sustainable development goals
SOC: Soil organic carbon
SOM: Soil organic matter
UNCBD: UN Convention on Biological Diversity
UNCCD: UN Convention to Combat Desertification
UNCSD: UN Commission for Sustainable Development
UNFCCC: UN Framework Convention on Climate Change

4.1 INTRODUCTION

4.1.1 Conventional Tillage Agriculture and Its Consequences

During the past three decades, it has become clear that agricultural productivity depends on maintaining soil health and functions which involve managing soil as a complex biological ecosystem. When a soil is used for tillage agriculture, it is not possible to sustain soil health because tillage destroys soil life; all the associated biological processes damage soil structure and pore volume, prevent the enhancement and maintenance of soil organic matter, and debilitate all soil functions at the field and landscape level. The symptoms of unsustainable tillage agriculture include soil erosion and degradation, loss of crop yields and system resilience, dysfunctional ecosystem functions, and environmental pollution (MEA, 2005; FAO, 2011b; Kassam and Kassam, 2020a).

Soil health and functions, and consequently crop productivity, depend largely on a continuous supply of soil organic matter (SOM) to the soil microorganisms, such as bacteria, fungi, earthworms, and termites, who use and transform it into more stable forms, enhancing SOM content and soil biological, physical, chemical, and hydrological properties, as well as increasing the soil's ability to retain and release nutrients and water for crop growth. The organic matter in the soil is used by mycorrhiza fungi to produce glomalin, which is a binding and cementing substance for creating stable soil aggregates and building soil structure and porosity, which is important for enhancing water infiltration. The SOM itself as it builds up in the soil also serves as a pool of biological forms of plant nutrients that are not lost through volatilization or leaching.

One of the most serious environmental consequences of the mechanical disturbance of soil by tillage agriculture has been the decomposition and mineralization of SOM and the accelerated emission of CO_2. This process leads to concomitant loss of SOM, soil health and functions, and crop productivity. At field and landscape scales, the soil loses its resilience and succumbs to erosion and land degradation, with water unable to infiltrate into the soil due to impact of rain drops causing detachment of small soil particles that clog the pores and form a surface seal and because of compaction and hard pans in the sub-surface soil layers including the destruction of the biopores or macropores that facilitate internal drainage for the soil. With this degradation, the hydrological behavior of the watershed deteriorates, with water not draining into the deeper groundwater, seriously affecting streamflow regulation and water quality. Under intensive conventional tillage agriculture, increased runoff and soil erosion carry with the sediments microorganisms and fertilizer that contaminates both ground and surface water resources.

In essence, soils and landscapes under conventional tillage agriculture are unable to function normally because of mechanical disturbance by tillage leading to worldwide soil erosion and land degradation (Oldeman et al. 1991; MEA, 2005; Montgomery, 2007; Vlek et al. 2010; Nkonya et al. 2016). Along with its other weaknesses, such as poor soil health and crop biomass management and limited or poor cropping diversity, conventional tillage agriculture can never deliver all the required ecosystem services (ESs) that are needed to sustain normal quality of life of the society and the agroecosystem itself, such as clean water, climate regulation, control of erosion, carbon sequestration, water, nutrient, and carbon cycling; biological nitrogen fixation; and resilience against insect pests and diseases through natural control mechanisms (Kassam and Kassam, 2020a). In fact, the loss in agroecological potential of land due to tillage and poor soil and cropping system management soon leads to sub-optimal yield ceilings despite the application of excessive amounts of fertilizer (Brisson et al. 2010). In situations where inputs are not available, loss in yields is so severe that it leads to extensive land abandonment after only a few years of cultivation.

What has become clear, based on research and empirical evidence, is that the ability of farmers to fully mobilize and sustain land and crop productivity potentials depends on managing and getting a complex system of bio-geo-pedo-hydro processes to operate optimally within the soil-cropland-atmosphere system in space and time across the farming system. It has also become clear that for farmers to achieve sustainable production, the production system must ecologically function as closely as possible to natural vegetation systems. This means that the production system must be capable of maintaining and enhancing soil health and functions which determine its productivity. Furthermore, soil health and functions depend on maintaining its SOM at the highest possible level. This allows soil biology to play its part in building and maintaining soil structure and in meeting plant needs for nutrients and water and protecting against pests (Lindwall and Sonntag, 2010; Kassam et al. 2013; Reicosky, 2020). These multiple processes and their functions and maintaining a high degree of biodiversity not only sustain productivity and output at a high level, but also impart resilience to the system against biotic and abiotic stresses. This is not possible with conventional tillage systems, even at a high level of agrochemical inputs.

4.1.2 What Needs to Change?

To address the weaknesses in tillage agriculture, farmers began, initially in North and South America and subsequently in all other continents, to develop and practice no-till (NT) agriculture involving the complete avoidance of tillage except on a narrow furrow where the seeds are deposited. These NT systems keep the soil covered with vegetative biomass, protecting the soil against erosion by serving as substrate for soil microorganisms, and managing diverse cropping systems with crop rotations and associations. These changes, to move away from tillage farming, began in the 1930s after the Dust Bowl period in the mid-western USA which lasted for about a decade. Initially, the soil and water conservation program of the USDA focused on a mechanical engineering approach to erosion control based on terracing, contour bunds, and contour plowing, some of which involved large amounts of earth moving with special equipment. However, a breakthrough was achieved when the sweep (field cultivator, which undercuts crop residues) and other machines for minimum and reduced tillage practices, which would bury smaller amounts of residues, were developed. The concept of stubble mulching and reduced tillage practices was born. All these practices were described in the 1940s and 1950s under the general term of conservation tillage. Subsequently, in the 1960s and 1970s, it became clear that combining NT practice with retaining as much crop biomass as possible on the soil surface and diversifying the cropping systems assisted in building soil health, reducing weed infestation and the occurrence of insect pests and pathogens. However, the term "conservation tillage" continued to be used to embrace all these new practices. It was at the IV RELACO (Latin-American Network for Conservation Tillage) in 1997 in Mexico when it was proposed to replace the term "conservation tillage" with Conservation Agriculture (CA). The argument was that it was not possible to conserve the soil while at the same time tilling it. The term CA was adopted by the Food and Agriculture Organization (FAO) of the United Nations as a basis for describing sustainable agriculture systems. In 1998, the FAO organized the first regional workshop on CA in Harare where a code of practice for CA was drafted describing the three interlinked principles and their practical applications as we know them today (Kassam et al. 2020a).

Since then, a strong CA movement has developed around the world led by all kinds of champions including farmers, scientists, development experts, policy makers, public sector officials, and private sector service providers. This has led to the establishment of many regional and national NT or CA associations during the past two decades to (1) promote the adoption and spread of CA, and (2) build institutional and policy capacity to sustain innovation, knowledge and technology generation, and research and private sector support. In 2001, the First World Congress on CA was held in Madrid, Spain; since then World Congresses have been held every three years with the Eighth World Congress scheduled to be held in June 2021 in Switzerland.

In 2009, a Global Conservation Agriculture Community of Practice (Global CA-CoP) was established with a communication platform hosted by the FAO to facilitate the global mainstreaming of CA. During the past two decades, CA has spread across all continents and most agroecoregions. It is practiced by smallholders and large-scale farmers in the tropics, sub-tropics, and temperate regions covering annual and perennial cropping systems, including crops with trees, orchards, plantations, agroforestry, and mixed crop–animal systems. As CA practices and expertise have developed over the years based on practical experience and research, so has the scientific evidence of the large array of productivity, economic, environmental, social, and regenerative benefits offered by CA to farmers, society, and agroecosystems. CA principles, practices, and experiences worldwide have become the foundation for change and a source of inspiration to other approaches to agricultural production such as organic farming, agroecological, and regenerative agriculture (RA). These approaches have borrowed and keep borrowing from the CA experience including integrating CA principles and practices into their own approaches (Kassam and Kassam, 2020b). In the case of RA, this borrowing has been almost "total," thus presenting CA foundations and systems to their audiences under a different name.

4.2 CONSERVATION AGRICULTURE

4.2.1 Principles and Practices

CA is an alternate paradigm of agricultural production and land use that is applicable to all land-based agricultural production systems in rain-fed and irrigated farming, including annual, perennial and mixed systems, orchards, agroforestry and plantation systems, crop–livestock systems, and pasture and rangeland systems. According to the FAO (http://www.fao.org/conservation-agriculture), CA is described as an ecosystem approach to regenerative sustainable agriculture and land management based on the practical application of context-specific and locally-adapted three interlinked principles, namely (Kassam et al. 2019, 2020a):

(i) ***Continuous no or minimum mechanical soil disturbance.*** This principle is implemented by the practice of NT seeding or broadcasting of crop seeds, direct placement of planting material into untilled soil, and NT weeding, causing minimal soil disturbance from any cultural operation, harvest operation, or farm traffic. Broadcast seed sowing and planting crops directly into untilled soil reduces erosion; reduces the loss of SOM and disruptive mechanical cutting and smearing of pressure faces; promotes soil biodiversity and microbiological processes; protects soil structure and connected pores; avoids impairing movement of gases and water through the soil; promotes overall soil health and functions, including improved retention of soil moisture, plant nutrients, and soil carbon; reduces labor and energy requirements and greenhouse gas (GHG) emissions; and contributes to integrated weed, insect pest, pathogen and nutrient management as well as overall resilience and sustainability.

(ii) ***Permanent maintenance of a vegetative mulch cover on the soil surface.*** This principle is implemented by retaining crop biomass, root stocks and stubbles, and biomass from cover crops. Use of crop residues (including stubbles) and cover crops reduces soil erosion; protects the soil surface; conserves water and nutrients; supplies organic matter and carbon to the soil system; promotes soil biodiversity and microbiological activity to enhance and maintain soil health and functions including structure and aggregate stability (resulting from glomalin production by mycorrhiza), improved capture and retention of water, plant nutrients, and soil carbon; and contributes to integrated weed, insect pest, pathogen and nutrient management, as well as overall resilience and sustainability.

(iii) ***Diversification of species in cropping systems.*** This principle is implemented by adopting economically, environmentally, and socially adapted crops in rotations, and/or sequences and/or associations involving annual and perennial crops, including a balanced mix of legume and non-legume crops and cover crops. Use of diversified cropping systems contributes to diversity in rooting morphology and root composition; enhances soil biodiversity and microbiological activity; builds up soil organic matter; enhances crop nutrition and crop protection through the suppression of pathogens, diseases, insect pests and weeds; and contributes to integrated weed, insect pest, pathogen, and nutrient management, as well as overall resilience and sustainability. Crops can include annuals, short-term perennials, trees, shrubs, nitrogen-fixing legumes, and pastures, as appropriate. By including livestock and trees in the cropping system, nutrient recycling and environmental services are enhanced.

The above principles and associated reference practices are applied along with other complementary good agricultural production and land management practices of integrated crop, soil, nutrients, water, pest (weeds, insects, pathogens), and energy management to optimize the whole production system at the farm level. At the landscape and watershed levels, CA systems enable the harnessing of a range of ecosystem functions and societal services, including clean water, carbon sequestration, carbon, water, and nutrient cycling; surface and groundwater regulation; control of erosion; increased biodiversity and food webs; and pollination services (Kassam et al. 2020b). CA systems are present in all continents and across a wide range of agroecologies in temperate, sub-tropical, and tropical regions.

CA systems offer a large range of productivity, economic, environmental, and social benefits to farmers, their communities, and society in general, globally (Kassam, 2020a, b). CA systems are regenerative, resilient, and self-protecting. In functionally degraded agricultural soils under use, they build SOM and restore and sustain soil health and functions upon which soil productivity and ESs depend (Corsi et al. 2012; Gonzalez-Sanchez et al. 2017, 2019; Sá et al. 2020a, b). For agricultural lands that are abandoned to cropping, CA systems can help to rehabilitate and restore them (Amado et al. 2020).

4.2.2 Global Adoption and Spread of CA

Globally, CA is practiced across some 80 countries, and, in 2015/16, the CA area covered over 180 M ha (12.5% of global annual cropland) with global distribution as shown in Table 4.1 (Kassam et al. 2019). Some 69.9 M ha (38.7%) of the total global area under CA is in South America, corresponding to some 63.2% of the cropland in the region; and some 63.2 M ha (35.0%) is in North America, mainly in the USA and Canada, corresponding to 28.1% of the cropland of the region. Some 22.7 M ha (12.6%) is in Australia and New Zealand, corresponding to 45.4% of the cropland, and some 13.9 M ha (7.7%) is in Asia, corresponding to 4.1% of the cropland in the region. Some 10.8 M ha (6.0%) of the total global CA area is in the rest of the world, comprising 5.7 M ha in Russia and Ukraine, 3.6 M ha in Europe, and 1.5 M ha in Africa, corresponding to about 3.6%, 3.2%, and 1.1% of their total cropland area, respectively. In Europe, the 3.6 M ha of CA is spread across 29 countries and corresponds to 2% of the global CA cropland area. In Africa, the 1.51 M ha of CA is spread across 17 countries although more CA area does exist in Africa which has not been documented, such as in Senegal, Mali, Benin, Ghana, Burkina Faso, Niger, Ethiopia, Rwanda, and South Africa. In Asia, the 13.93 M ha is spread across 18 countries but again there is more CA area in Asia that has not been recorded, such as in Jordon, India, Nepal, Malaysia, Philippines, and Mongolia (Kassam et al. 2019). Overall, the CA area is about equally divided between the Global South and the Global North.

4.3 ENHANCING SOIL ORGANIC MATTER

Improving SOM in agricultural soils is essential for a sustainable crop production. SOM is composed of soil microbes including bacteria and fungi, decaying material from once-living organisms, such as plant and animal tissues, and products formed from their decomposition. SOM is a heterogeneous mixture of materials that range in stages of decomposition from fresh plant residues to highly decomposed material known as humus (Table 4.2).

Even if SOM is just 2–10% of most soil's mass, it has an important role in the physical, chemical, and biological functions of agricultural soils. In fact, SOM contributes to nutrient retention and turnover, soil structure, moisture retention and availability, degradation of pollutants, carbon sequestration, and soil resilience.

TABLE 4.1

Cropland Area Under CA (M ha) by Region in 2015/16; CA Area as a Percentage of Global Total Cropland, and CA Area as a Percentage of Cropland of Each Region (*Source:* Kassam et al. 2019)

Region	CA Cropland Area (M Ha)	Percent of Global CA Cropland Area	Percent of Cropland Area in the Region
South America	69.90	38.7	63.2
North America	63.18	35.0	28.1
Australia & NZ	22.67	12.6	45.5
Asia	13.93	7.7	4.1
Russia & Ukraine	5.70	3.2	3.6
Europe	3.56	2.0	5.0
Africa	1.51	0.8	1.1
Global total	**180.45**	**100**	**12.5**

TABLE 4.2
Size, Turnover Time, and Composition of the Soil Organic Matter Fractions

Fraction	Size Micrometers (μm) and Millimeters (mm)	Turnover Time	Composition
Dissolved organic matter	<45μm (in solution)	Minutes to days	Soluble root exudates, simple sugars, and decomposition by-products. It generally makes up less than 5% of total soil organic matter.
Particulate organic matter	53μm–2mm	2–50 years	Fresh or decomposing plant and animal matter with identifiable cell structure. Makes up 2–25% of total soil organic matter.
Humus	<53μm	Decadal (10s to 100s of years)	Older, decayed organic compounds that have resisted decomposition. Can make up more than 50% of total soil organic matter.
Resistant organic matter	<53μm–2mm	100s to 1000s of years	Relatively inert material, such as chemically resistant materials or organic remnants (e.g. charcoal). Can be up to 10% of soil organic matter.

4.3.1 The Connection between SOM and SOC

Soil organic matter (SOM) and soil organic carbon (SOC) are closely linked. SOM is made of organic compounds that are highly enriched in carbon (Ontl and Schulte, 2012; Griffin and Edwards 2020), and so SOC is about 58% of SOM (Minasny et al. 2020). As an indicator for soil health, SOC is important for its contributions to food production, mitigation and adaptation to climate change, and the achievement of the (UN) Sustainable Development Goals (SDGs) (FAO, 2017).

While SOM is difficult to measure directly, SOC is a measurable component of SOM. Therefore, laboratories tend to report SOC. SOC refers only to the carbon component of organic compounds. The concentration of SOC is generally considered an indicator of soil quality because of its agronomic and ecological functions. Factors that influence SOC content include land use, soil properties, rainfall, temperature, crop characteristics (such as the amount and fate of crop residues or root distribution), and soil management practices, like NT (Carbonell-Bojollo et al. 2011).

Thus, SOM stock can be estimated from SOC content. Considering that about 58% of the mass of organic matter can be counted as carbon, we can estimate the percentage of SOM from the percentage of SOC, by using the Van Bemmelen conversion factor 1.724 (obtained from the relation 100/58) (Tabatabai, 1996). This conversion factor can vary in different soils, but 1.724 provides a reasonable estimate of SOM for most agricultural purposes. An example follows.

$$\text{SOM}\ (\%) = \text{SOC}\ (\%) \times 1.724$$

If more than a percentage, a quantity is needed, we would need to convert percentage to weight for a given depth and area.

$$\text{SOC stock in Mg of carbon per hectare}\ \left(\text{Mg C ha}^{-1}\right)$$

$$= \left(\text{SOC\%}\right) \times \left(\text{mass of soil in a given volume}\right)$$

For example, a soil with a SOC of 1.5% (0.015) and a bulk density of 1.2 g/cm^3 (equivalent to 1.2 Mg/m^3), would have SOC to a depth of 20 cm (0.2 m) per hectare (10,000 m^2) of:

$$(0.015 \text{ C}) \times (1.2 \text{ Mg/m}^3 \times 0.2 \text{ m} \times 10\ 000 \text{ m}^2/\text{ha}) = 36 \text{ Mg C ha}^{-1}.$$

Using the conversion factor of 1.724, the amount of SOM would be: 36 × 1.724 = 62.06 Mg.

4.3.2 How to Improve SOM and SOC with CA and Why It Is Essential

SOC is an important component for the functioning of agroecosystems, and its presence is central to the concept of sustainable maintenance of soil (Ramesh et al. 2019). Agricultural systems that improve SOC, would also improve SOM. But the opposite is also true. Changes in land management systems directly affect SOC content (Ordóñez Fernández et al. 2007) and conventional agriculture, based on soil tillage has contributed to 30–50% of SOC depletion in the last two decades of the 20th century (Reicosky, 2011). So, the conventional tillage-based way of treating agricultural soils seems to be underperforming when it comes to SOC management. Conversely, CA has demonstrated that a more soil-friendly alternative system is able to change the trend of degrading agricultural soils.

Soil is the largest terrestrial carbon sink and contains 2–3 times more carbon than the carbon in the atmosphere and vegetation, respectively. Therefore, a meager change in soil carbon sequestration will have a drastic impact on the global carbon cycle (Ramesh et al. 2019).

There is general consensus about the positive effect of CA with regard to SOC increase in the topsoil layers, which in global warming terms means greater carbon sequestration in soils. (Corsi et al. 2012; Sá et al. 2020a, b). In Wooster, Ohio, USA, the development of SOM content in the longest-running soil experiment (1962–2003) showed that in soybean–maize rotations in fields under plow tillage, the SOM decreased from 14 g kg^{-1} in the top soil in 1962 to less than 10 g kg^{-1} and continued to decrease even after 44 years. On the other hand, in the fields under CA systems, SOM increased from 13 g kg^{-1} in 1962 to 27 g kg^{-1} in 2003 and was still continuing to increase even after 44 years of experimentation (Mestelan et al. 2006; Derpsch, 2008b). Similar results were also obtained from another long-term experiment at Hoytville, Ohio. The carbon pool was also greater for CA than plow tillage at depths from 31 to 148 cm. At both sites, CA systems consistently produced higher yields than in plow-tilled fields (Mestelan et al. 2006; Derpsch, 2008b). This aspect of yield differences between CA systems and tillage systems is elaborated further for all regions in section 4.2. However, in a meta-analysis, Haddaway et al. (2017) found that C stock increased under NT compared to the high-intensity tillage in the upper soil (0–30 cm) around 4.6 Mg ha^{-1} (0.78–8.43 Mg ha^{-1}, 95% CI) over $\geq$ 10 years, while no effect was detected in the full soil profile. Derpsch et al. (2014) highlighted the need to standardize studies, as, in some cases, studies neglecting the potential of CA might not include the three principles of CA. Authors concluded that higher SOC stocks or concentrations in the upper soil not only promote a more productive soil with higher biological activity but also provide resilience to extreme weather conditions (Haddaway et al. 2017).

Numerous environmental benefits may result from agricultural activities that sequester soil C and contribute to environmental security. As part of no-regret strategies, practices that sequester soil C help reduce soil erosion and improve water quality and are consistent with more sustainable and less chemically dependent agriculture. As we learn more about soil C storage and its central role in direct environmental benefits, we must understand the secondary environmental benefits and what they mean to production agriculture. Increasing soil C storage can increase water infiltration, increase fertility and nutrient cycling, decrease wind and water erosion, minimize compaction, and enhance water quality. Decreased C emissions impede pesticide movement and generally enhance environmental quality. The sum of each individual benefit adds to a total package with major significance on a global scale. Incorporating C storage in conservation planning demonstrates concern for our global resources and presents a positive role for soil C that will have a major impact on our future quality of life (Reicosky, 2003). Actually, inversion tillage increases the flow of CO_2 into the atmosphere, decreasing the SOC content of the soil (Carbonell-Bojollo et al. 2019; Schlesinger and Andrews, 2000; Lopez-Garrido et al. 2011). In tilled soils, SOM progressively decreases, as a large part is removed annually (harvest) and a good part is lost due to mineralization, especially if tillage is very aggressive, such as that caused by the inversion of the soil profile due to moldboard plow use (Wallace, 1994).

The loss of SOM from a soil, in addition to a negative effect on the balance between the different carbon reservoirs, also affects the quality and fertility of the soils themselves, and their fertility can be seriously compromised. This is due to the great influence of organic matter (OM) on soil properties: physical, chemical, and biological, necessary for the development of its functions (Bauer and Black, 1994; Magfoff and Weil, 2004).

SOM is essential for the physical fertility of a soil since it intervenes in the formation and stability of aggregates (Gajri et al. 2002). It is considered a source of energy for plants and soil organisms (Brady and Weil, 2002). The amount, diversity, and activity of the macro and meso fauna of the soil and microorganisms is directly related to the amount of OM.

SOM is an indicator of the state of soil health (Reicosky, 2020). High contents improve the cohesion between the different elements of the soil, which increases their bonding strength. In CA, part of the harvest remains including roots degrade slowly, which translates into an increase in the OM content of the soil. The increase in the first few centimeters of the soil profile increases the nutrient reserves (Rhoton, 2000), which can be released gradually at a different rate from that in conventional tillage soil (Fox and Bandel, 1987).

4.3.3 SOM and a Better Soil Structure

The global deterioration of the edaphic structure is due to the fact that the compaction of the soil limits the growth of the roots, storage capacity, fertility, biological activity, and stability. This is why soil compaction constitutes one of the main threats to the soil identified in the European Union's Soil Thematic Strategy for Soil Protection (COM, 2020). Compaction also increases runoff which can end up in floods. Therefore, managing soil compaction is essential. It occurs when soil is subjected to mechanical pressure resulting from the use of heavy machinery and excessive grazing, especially if the soil is wet (Huber et al. 2008). Another factor that affects the soil structure is the formation of surface crusts, which cause low structural stability (Micó et al. 2006).

The implementation of CA implies a series of benefits on the physical properties of the soil, which contributes to the improvement of the soil structure. Thus, by keeping the soil unaltered thanks to the discontinuation of tillage operations, generation of a network of channels is favored as a result of the degradation of the roots, which together with the higher quantity of worms in a soil under CA, form what is known as biopores. These are preferential channels through which the water infiltrates deeply. According to López-Garrido (2011), CA favors properties related to soil structure, such as aggregate size distribution and aggregation index, among others. SOM is capital in all the processes that occur in the soil and in its quality, being widely accepted as an indicator of soil quality (Podmanicky et al. 2011). An increase in the SOM content means a better structure and favors structural stability. CA, due to the increase in SOM, contributes to the structural improvement of the soil (Reicosky, 2020).

4.3.4 SOM and Soil's Natural Fertility

The degradation or loss of the quality and quantity of soil is conditioned by the occurrence of various processes, such as erosion, salinization, contamination, drainage, acidification, loss of structure, compaction or a combination of them. Inadequate management of land use can lead to the occurrence of one or more of these processes, which negatively affect its quality and, therefore, its productive capacity or natural fertility. One of the examples of inadequate management of land use is intensive tillage, since plowing contributes to the degradation phenomena mentioned above.

Therefore, poor agricultural practices negatively affect the OM content and are contrary to the return of nutrient extractions, thus leading to not optimizing the use of fertilizers. In general, the edaphic parameters that are traditionally used to evaluate the fertility level of SOM, availability of nitrogen-phosphorus-potassium evolve favorably when the soil stops being mechanically cultivated and crop biomass and stubble is retained on the soil surface in the productive management of crops. Therefore, CA is able to improve the fertility of the soil since the slow decomposition of the superficial plant remains and gives rise to a superficial layer rich in humus, which, through its mineralization, will make available to the crops the nutrients contained therein (Roldán et al. 2003; Riley et al. 2005; Diekow et al. 2005; Reicosky, 2020).

SOM dynamics play a very important role in the natural fertility of the soil. When SOM is mineralized, nutrients become available to the plants. Together with clay particles, this decomposing SOM constitutes the colloidal fraction of the soil, responsible on the one hand for its chemical fertility and on the other for the development of soil structure or soil aggregates that increase the resistance of the soil against erosion. (Ordoñez Fernandez et al. 2010).

Within the SOM of the soil is the particulate OM, closely related to the development of the soil structure and that is very easily destroyed by tillage (Mrabet et al. 2001). That is why CA, where plant remains are slowly degraded, induces an OM dynamics similar to that which occurs in natural ecosystems.

Long-term studies show an increase in nitrogen in soil under CA (Lacasta Dutoit et al. 2005; Sombrero et al. 2006; Ordóñez Fernandez et al. 2007). However, the intensity and extent depend on the climate, soil type and the crop rotation. When legumes are included in the rotation in CA, they notably enrich the soil with organic nitrogen that, in the long term and due to mineralization processes, is available to the crops. Lacasta Dutoit and Meco (2005) and Sombrero et al. (2006) found N increases above 25% after 10 years or more of CA.

Regarding phosphorus (P), the minimum soil disturbance in CA leads to a greater efficiency of the phosphate fertilizer, increasing the concentration and availability of P, due to the stratification of SOM in the surface horizon. Ordóñez Fernandez et al. (2007) and Saavedra et al. (2007) observed that, after more than 19 years of CA, the concentrations of P and potassium (K) available for cultivation increased on the surface horizon compared to conventional tillage.

4.4 OPTIMIZING SOIL HEALTH AND PRODUCTIVITY

4.4.1 Soil Health

Soil health is the capacity of soil to function as a vital living system, with ecosystem and land use boundaries, to sustain plant and animal productivity, maintain or enhance water and air quality, and promote plant and animal health (Doran and Zeiss, 2000). The state and dynamics of soil health and fertility are the result of both the diversity of live and dead vegetation and microorganisms in CA systems and the resulting changes in the health profile of the soil (changes in chemical, physical, hydrological, and biological properties and their derivative functional characteristics, such as moisture and nutrient retention and supply, porosity and aggregate stability, load-bearing capacity, aeration, drainage, temperature regulation, soil formation, etc.) (Kassam et al. 2013).

The concepts of soil health and quality have been expanded to include not only its productive capacity but also the environmental functions that soils can perform, such as acting as a filter (protecting both surface and groundwater from plant protection products or industrial and municipal waste) or a sink for carbon sequestration, thus mitigating climate change. CA is an example of climate smart agriculture (CSA) and a sustainable agricultural system that could reinforce a number of ecosystem services while contributing positively to household food security (Nyanga et al. 2020). Soil health is largely governed by SOM content, which is dynamic and responds effectively to changes in soil management. Apart from areas with a heavy surplus of animal manure, the SOM of many cultivated soils across is diminishing as a result of intensive agriculture.

SOC favors the formation of aggregates of soil particles and consequently intervenes in the distribution of the pore space of the same, affecting various physical properties, such as usable humidity, aeration capacity, and movement of water and gases in the soil. In addition, the organic carbon of the soil, formed by compounds of different chemical nature and state of decomposition, intervenes in the chemical properties of the soil, increases the cation exchange capacity (CEC) and the buffer capacity on the soil reaction (pH). As products of the mineralization of SOM, various nutrients are released for plants, many of which are provided in a deficient way by soil minerals. SOC through its effects on the physical, chemical, hydrological, and biological properties of the soil and its functions are determinant of its productivity (Schjønning et al. 2018).

The above soil health characteristics and economic, environmental, and social benefits are elaborated in detail from a global perspective in Kassam et al. (2013, 2020b). Healthy soils offer *in-situ* ecosystem services at the field and farm level in terms of crop production and farm output that are beneficial for crop health and performance, farmers, and society. Healthy soils at the field and farm level also are part of the landscape and watershed level ecosystem services, the *ex-situ* services, that benefit the environment and society, as these are linked to and derived from the field and farm level ecosystem functions offered by healthy soils (Kassam et al. 2020b). These aspects are further elaborated in Section 4.5.

4.4.2 Productivity

In the literature, it is generally considered that an increase in SOM improves soil health and functional properties and, as a consequence, increases crop productivity in terms of both yield (area productivity) and input factor productivity (a measure of utilization efficiency of inputs, such as nutrients, water, and labor). However, the literature that directly links SOC or SOM and productivity in quantitative terms is not extensive partly because soil organic matter builds up relatively slowly over time, requiring longer-term experiments. Bauer and Black (1994) directly quantified the effect of SOM content on productivity in a trial carried out in North Dakota, USA. In their study, wheat production was compared over four years in plots with different levels of soil organic matter. The authors conclude that the contribution of 1 Mg ha^{-1} of soil organic matter within the surface 30.5 cm soil layer is equivalent to the inherent productivity of the soil of 35.2 kg ha^{-1} and 15.6 kg ha^{-1} of aboveground crop dry matter and wheat grain respectively (Bauer and Black, 1994).

A study carried out in Serbia by Seremesic et al. (2011) provided an average value of SOC productivity, although the factors that were compared were not the different SOC contents, but different rotations, with and without fertilization. This was a long-term study (1970–2007) carried out in the Pannonian Plain of Serbia that related wheat production with SOC levels. SOC production and levels were compared for the periods 1970–1980 and 2000–2007 (Table 4.3).

TABLE 4.3
Yield, Soil Organic Carbon, and Soil Organic Carbon Productivity (*Source:* Seremesic et al. 2011)

Crop Rotation	Fertilization	Wheat Yield (Mg ha^{-1}) 1970–80	2000–07	Differ	SOC (Mg ha^{-1}) 1970–80	2000–07	Differ	SOC Productivity (Mg ha^{-1})
Wheat monocropping	Yes	4.370	2.940	−1.430	69.0	65.5	−3.50	0.410
Wheat-Maize-Soja	Yes	5.030	4.840	−0.190	68.4	65.1	−3.30	0.058
Wheat-Maize	Yes	4.630	4.200	−0.430	66.5	63.9	−2.58	0.160
Wheat-Maize-Soja	No	1.660	1.530	−0.130	60.6	55.0	−5.60	0.024
Wheat-Maize	No	1.160	0.980	−0.180	56.5	51.5	−5.04	0.036
Average		**3.370**	**2.900**	**−0.470**	**64.2**	**60.2**	**−4.00**	**0.138**

According to Seremesic et al. (2011), the loss of SOC observed in all cases was due to the intensification of tillage, the reduced amount of organic and mineral fertilizers used, and the burning and removal of crop residues. This decrease in SOC led to a decrease in the productivity potential of wheat production. According to the study, the average productivity of 1 Mg ha^{-1} of soil organic carbon corresponds to 138 kg ha^{-1} of wheat. The authors conclude that by using proper soil management techniques in the three-year rotation, the loss of SOC in the area could be halted and possibly reversed. In order not to depend on external inputs, farmers should use their own biomass resources produced from the cropping systems to increase SOC. In CA, the permanent cover with stubble and crop and root biomass is essential to provide soil life with OM. Even if information in the literature is not extensive, some studies show the increase in SOC and yield in CA in annual crops. According to a review, in CA an increase of 29.18% in SOC resulted in a better yield by 11.91% (Mapama, 2018).

Often it also has been shown that as SOM increases, there is an improvement in crop nutrient response and yield. A good example of improved nutrient productivity in CA is from Carvalho et al. (2005, 2012) in Portugal, who showed that, with a soil under conventional tillage containing 1% SOM, it took 160 kg N ha^{-1} to produce 3 Mg ha^{-1} of wheat grain (Figure 4.1). However, when the same soil had built up 2% SOM through continuous CA, the same yield of 3 Mg ha^{-1} was obtained with 37 kg N ha^{-1}. A crop modeling exercise showed that at 3% SOM content, a yield of 3.5 Mg ha^{-1} would be achieved without any application of N, implying that the additional nitrogen was available from SOM in CA soils due to accumulation of SOM.

In general, it has been difficult to show consistent yield advantage of CA annual cropping systems based on reductive scientific experiments, often conducted using questionable methods or claiming to be a CA experiment when in reality it may not be CA. In general, it takes time to transform from a tillage system to CA, and it is often not possible to show yield advantages based on short-term experiments. Also, with some exceptions, it has not been possible to show consistent yield advantage based on meta-analyses across all agroecologies (Pittelkow et al. 2014; Corbeels et al. 2020), partly because of the use of mixed quality data sets where the exact characteristics and context of each data pair is not adequately known or understood. It is also partly due to the particular data sets selected or not selected for the analyses.

For example, in the meta-analysis conducted by Pittelkow et al. (2014), 60% of the data sets were from North America whereas good quality data sets from experienced research workers in Southern Africa were not included. Some data sets were included from so-called "conservation

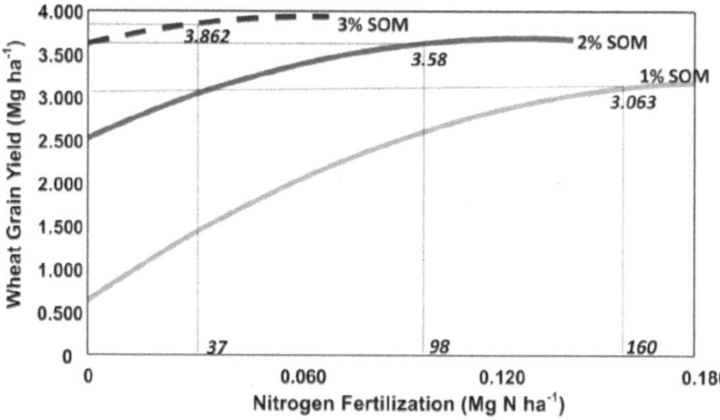

FIGURE 4.1 Yield response to nitrogen fertilization in wheat in Portugal. *Notes:* Values in italic represent the economically optimal N-fertilization rate and the respective yield for the different levels of soil organic matter (SOM) (0–30 cm depth). The dashed line is the modeled yield response for a SOM level of 3% Italic font highlights the most economic N-fertilization amounts and respective yields (*Source:* Carvalho et al. 2012).

tillage" experiments which would include a range of non-CA systems including forms of tillage systems whose data sets should not be included in such analyses. In the case of Corbeels et al. (2020), data sets from countries such as South Africa, which has 25% of its cropland area under good quality CA, were not included. Several data sets that were not from CA systems or from countries where there has not been adequate research and extension work conducted on CA, were included whereas good quality data sets from countries where there is significant adoption and spread of CA were left. Oddly enough, Corbeels et al. (2020) admit that where the three principles of CA were applied, there was significant yield advantage which casts a shadow on the purpose and reliability of Corbeels et al. (2020) meta-analysis as well as that of Pittelkow et al. (2014).

The above has led to concerns being expressed about the scientific value of such questionable processing of insufficiently differentiated and poor-quality database. There is also concern being expressed at the blunt manner in which meta-analysis and similar analyses are being conducted such that they are not capable of correctly identify reliable data sets and correctly handling the baseline and treatment variables in space and time that are expected to be present in CA systems operating on lands with different initial soil quality status and with a range of farm power and diversity of practical management expertise and experience. CA systems, whether in transition or at advanced stage, by definition are complex and management intensive systems that are biologically and ecologically dynamic in space and time. Their performance is not only heavily dependent on starting soil quality conditions and the experience of farmers but also on the dynamics of changes in soil health improvements that can continue for a long time before a new equilibrium status is established (Derpsch, 2008a; Kassam and Kassam, 2020c).

On the other hand, many good quality individual studies show that CA systems offer yield advantages compared with conventional tillage systems in all agroecological regions and in all continents. In the same vein, millions of farmers who practice CA in different continents and ecologies all are able to cite yield advantage as being one of the several reasons why they prefer to practice CA. For example, for Parana State in Brazil, Derpsch et al. (1986) and Calegari et al. (1995) reported 14% and 34% greater yields in CA wheat and soybean, respectively, compared to conventional system in a 7-year trial, from 1977–1984, comparing the performance of these crops in conventional systems and in CA system. Also, over the period 1976/77 to 2010/11, maize area in Brazil has remained around 40–50 M ha whereas yields increased by more than 50% because of the transformation of maize-based cropping systems from tillage production to CA systems (Calegari et al. 2014). Similar results have been reported from Argentina where between 1990 and 2010 with relatively small increases of the area planted with grain crops would result in a huge increase in grain output during this period of steady growth of the area under CA (Albertengo et al. 2014).

Sorrenson et al. (1997) reported differences in crop yields, as well as crop fertilizer and herbicide usage (the most significant items of farm costs) observed on the farms studied under both conventional tillage systems and CA systems. In general, depending on the crop, yields under conventional system were following a declining trend, while the reverse was occurring under CA when used in combination with green manure cover crops and applying all three principles of CA. Based on detailed analysis of the case study farms, as well as published research data from Parana, Brazil (analyzed by Sorrenson and Montoya, 1991), crop yields under conventional system decline over a period of about 10 years by between 5 and 15% (depending on the crop) while, over approximately the same time period under CA, they increase between 5 and 20% (again depending on the crop). These trends in crop yields were found to impact strongly on farm incomes. Savings in herbicide and fertilizer inputs, per crop, under CA compared to conventional systems which are partially dependent on the crop rotation being followed, range from 30 to 50%, respectively, over approximately the same period and significantly affect farm variable costs and profits.

The yield and financial impacts of good quality CA were analyzed in detail by Sorrenson et al. (1997) in Paraguay, where a CA-based land improvement program was implemented. The financial performance of a typical large-sized farm (135 ha), in the Itapúa region, was studied over a 10-year period from the late 1980s while the quality of the CA system was being improved. Farm model results for the overall farm, as well as separately for each crop and for each crop rotation, were produced. The farm model was based on thorough analysis of data from the case study farm (and others) spanning a number of years. Therefore, the results can be confidently considered indicative of what was actually being realized in practice by farmers in terms of improved yields and financial returns, as a result of practicing good quality CA systems.

The same authors also analyzed changes in farm income and costs which are also based on actual farmer experience in the region. Farm costs (both variable and fixed costs, the latter exclusive of the cost of NT equipment) increased under the CA system, compared to conventional tillage, but these increases were less than the corresponding increases in gross farm income. Net farm income increased considerably under CA, from US$ 9,770 in year 1 to US$ 33,700 in year 10. Under conventional tillage, it decreased from US$ 7,300 to US$ 1,100. In each of these years, net farm income was higher under CA than under conventional tillage. A similar study of a medium-sized farm (45 ha) showed that in year six, the conventional tillage farm was making an annual loss of US$ 158 compared to a profit of US$ 20,043 in the CA farm. By year 10 the conventional tillage farm was losing US$ 3,013 annually compared to an annual profit of US$ 31,142 in the CA farm.

Risks, defined as the probability of net farm income falling below zero in any given year, were also analyzed by Sorrenson et al. (1997), who concluded that farm risks decreased considerably following the adoption of a CA system, compared to conventional tillage. The main reasons for this were: (1) higher and more stable yields in CA, due to improved soil structure, higher water infiltration, and reduced soil erosion as well as increased soil moisture retention and reduced pests and diseases; (2) the impact of lowering farm income when soybean and wheat prices fall is smaller under CA, because it is possible to diversify into other cash crops; (3) reduced fuel use under CA, lowering the impact of any increase in the price of fuel; (4) over time, lower fertilizer and herbicide costs per crop under CA, as the impact of green manure crops and reduced fallow periods between crops take effect. Sorrenson et al. (1997) concluded that a situation, such as the one presented here, whereby highly attractive financial returns are accompanied by a lowering of risk, is rarely encountered in agricultural technology development. Generally, more profitable technologies carry with them concomitantly higher risks which necessitate farmers to weigh-up accepting these in exchange for higher profits, as opposed to operating at lower average profit but with lower risk. The value of the data presented above from the study by Sorrenson et al. (1997) is reinforced by the fact that in Paraguay the area of crops under CA system when they presented the report in 1997 was some 1 M ha while, ten years later in 2007, the area had increased to 2.6 M ha and now most cropping is performed by farmers under CA systems and the current area has expanded to more than 3.5 M ha.

Yield response of maize to CA cropping systems in Southern Africa covering Zimbabwe, Malawi, Mozambique, and Zambia was reviewed by Thierfelder et al. (2015). They analyzed crop performance in 26 sites which varied in average rainfall from 493 mm to 1471 mm and covering soil types ranging from sandy and sandy loam to loamy sand and sandy clay loam. Overall, results showed that there was a greater maize yield response (80%) on CA than on conventional tillage treatments across all sites and seasons. The relative difference between all tested CA systems showed that most systems in the different target communities were above the 1:1 line with some few exceptions. In some years, treatments in Malawi, Zambia, and Zimbabwe exceeded the 1:2 line indicating that there was more than double as much maize grain yield on the CA systems than on the corresponding conventional control treatment. It is remarkable that CA outperformed conventional tillage treatments in the very low, medium, and very high yielding environments. Also, in general, yield differences increased over time (7–10 years) across sites in all four countries. In another long-term study in Zambia, maize yields were on average 75% and 91% higher in CA systems than in conventional systems (Thierfelder et al. 2013). In the case of Zimbabwe, CA systems based on animal traction excelled under the circumstances of low fertility and erratic rainfall significantly increased maize productivity by up to 235% (1.761 Mg ha^{-1}) and legume productivity by 173% (0.265 Mg ha^{-1}) as compared to the conventional control treatment (Thierfelder et al. 2014). In Mozambique, long-term trends showed that direct seeded manual CA treatments out yielded conventional tillage treatments in up to 89% of cases on maize and in 90% of cases on legume in direct yield comparisons (Thierfelder et al. 2015).

In Tanzania, in 2004 when the CA for sustainable agriculture and rural development (CA SARD) project of FAO intervened in Rhotia village, Arusha District, Tanzania, farmers were taught about CA as a sustainable way of growing crops and managing soil health that conserves the soil and maintains or enhances soil fertility and productive capacity (Owenya et al. 2011). The CA cropping system involving maize, pigeon pea, and lablab was effective in the fight against hunger and poverty. The yields under CA were generally higher and farmers noted that intercropping of maize with cover crops (pigeon pea and lablab) provided three harvests per season instead of two. They also learned that the increase in crop production was brought about by improved soil health and water management under CA. Yields increased from 1.25 Mg ha^{-1} (2004) under conventional farmer practice to 7.0 Mg ha^{-1} (2009) under CA, a more than four-fold increase in five years. Yield advantages with CA have also been reported from other countries in Africa including South Africa, Mozambique, Kenya, Ghana, Burkina Faso, and North African countries (Jat et al. 2014; Kassam et al. 2017).

In South Asia, yield advantage in CA systems have been shown to be modest, ranging from no yield difference to some 50% greater yields in India (Farooq et al. 2020; Jat et al. 2019, 2020). However, even with modest yield advantage, economic advantage is significant due to the reduced cost of inputs, particularly energy, water, and labor. In Bangladesh, Haque et al. (2018) reported a yield increase of 28% for lentil, 19% for mustard and 6% for wheat by farmers who adopted CA planting, and profit increases by 47% for lentil, 55% for maize, 46% for mustard and 76% for wheat due to adoption of CA planting. In Pakistan, CA system resulted in a 12–15% increase in crop yield (Gill et al. 2002; Ahmad, 2009). The impact evaluation of bed and furrow irrigation technology also revealed better crop yield and water use efficiency compared to the traditional method (Gill and Awan, 2009).

In China, where there is now a substantial area under CA systems, both long-term studies and data from field research shows that cereal productivity can increase with CA. For example, in North China, 15-year mean winter wheat yield under CA was 18.8% higher at 3.61 Mg ha^{-1} compared with 3.04 Mg ha^{-1} for traditional tillage (Li et al. 2007), and yield increases of 9% for wheat (to 4.8 Mg ha^{-1}) and 9% for summer maize (to 10.1 Mg ha^{-1}) were observed (He et al. 2011). Li et al. (2016) conducted a meta-analysis of published research on CA in China (papers in both Chinese and English). The data set of 60 papers included 75 paired trials and 395 observations and allowed disaggregation of effects by crop (wheat, maize and rice), annual precipitation (<500 mm, 500–800 mm, >800 mm), and experimental duration (2–4, 5–9, ≥10 years). Their analysis shows that the greatest yield increases occurred in maize and in dryland zones, with yield increases the greatest in the longest duration studies. The mean increase in yields by >4.8% is observed for all zones and durations (4.5% or 263 kg ha^{-1} wheat, 8.3% or 424 kg ha^{-1} maize, 1.7% or 250 kg ha^{-1} rice). The mean increase in yields for all crops with CA practiced for ≥10 years is >7.0% (394 kg ha^{-1}), indicating that productivity continues to increase as the natural capital of the system is enhanced over time. Yield advantages from CA have been reported from other countries in Asia including Kazakhstan, Uzbekistan, Iran, Syria, Lebanon, Bangladesh, Laos, Cambodia, Vietnam, and DPR Korea (Jat et al. 2014; Kassam, 2020a).

Elsewhere, in North America, Europe, and Australia, yield and economic advantages from CA systems have been documented in several publications, such as by Lindwall and Sonntag (2010) and Lafond et al. (2014) for Canada, by Duiker and Thomason (2014) for the USA, by Friedrich et al. (2014) for Europe, and by Rochecouste and Crabtree (2014) for Australia. The time-related pattern of yield advantage in CA in these regions appear to be similar to those in South America, Africa, and Asia. There can be situations where significant yield advantages can take time to manifest themselves because of the already high yields being obtained under tillage agriculture or because of the degraded soil health conditions that can take time to improve. Overall, yield and other productivity advantages as well as economic, environmental, and social benefits in CA systems have been reported in scientific literature as well as from most if not all large-scale and small-scale farmers globally. This scientific and empirical evidence is strongest in areas where CA systems have been established for longer periods of time. This applies to both annual cropland systems as well as perennial systems, such as orchards, plantations, and cropland with trees (Kassam et al. 2017; Kassam, 2020a).

Tillage-based organic agriculture, as a system that is based on the application of organic fertilizer, such as animal manure, composts, and other forms of biofertilizers has been shown to increase SOC content levels but conventional organic systems rely heavily on tillage to mineralize the organic fertilizers to release plant nutrients. This does not allow carbon sequestration to occur at rates that are normally attained in CA systems. However, at the higher production levels, yields are generally lower compared to CA and also compared to conventional tillage agriculture. The economic compensation through subsidies or higher market prices can be of interest to some farmers (Pardo et al. 2014, Gragera-Facundo et al. 2010), but this does not solve the basic problem of soil degradation and poor quality of soil-mediated environmental services from tillage-based organic farming. CA is the system that combines the capacities to increase the levels of SOC while improving the productivity of crops and ecosystem services. Thus, it is important that CA-based organic or biological production is promoted where possible instead of tillage-based organic production (Chambert and Sarthou, 2020).

In CA systems, interlinked practices of NT, mulch cover, and diversified cropping systems already provide a foundation for integrated nutrient and pest (weed, insect, pathogen) management that reduces or minimizes the need for agrochemicals. Inclusion in the CA systems of legume crops and cover crop mixtures for biological nitrogen fixation, building biological stocks of nutrients, and practices—such as planting green, mixed cropping, push–pull techniques, and soil microorganism management—further improve biological pest control and nutrient and water availability. Such robust and self-reliant biological CA systems are being practiced in different parts of the world by large-scale and small-scale farmers including resource-poor farmers relying on manual and animal farm power. This is an area that deserves far more attention in the future so that all production systems begin to converge toward CA systems that are either organic or biological, using minimum amounts of agrochemicals (Kassam and Kassam, 2020b, c).

4.5 FACILITATING ECOSYSTEM SERVICES

The Millennium Ecosystem Assessment (MEA, 2005) identified several categories of public goods benefits that are provided "freely" by nature to society, including farmers, through the processes that sustain ecosystem functions. These environmental benefits are referred to as ESs because of their usefulness to society and life on earth in infinite ways. Originally they were referred to as "goods and services" but, since the MEA, they are all referred to as "services." They include the *provisioning* services of clean drinking water, edible and non-edible biological products, and processes that decompose and transform organic matter, carbon sequestration, biologically fixed nitrogen; *regulatory* services that control air quality, soil erosion, pest and diseases, natural hazards, and pollination; *supporting* services of soil formation, nutrient, carbon and water cycling, biodiversity, primary production, and atmospheric circulation; and *cultural services* related to recreation, sacred and spiritual relationships with nature, and particular sites or phenomenon (Figure 4.2, MEA, 2005). None of these services need any human intervention for their existence and performance. They are part of the natural world and would remain so even without us being around. With our interventions and activities, ecosystem processes and functions are drastically affected. In the case of tillage agriculture, most, if not all, soil-mediated ecosystem processes and functions are debilitated and some made dysfunctional, generating negative consequences. Hence the term "negative externalities" is often used to describe such occurrences.

Interventions from humankind have resulted in altered ecosystems and, in many cases, have led to ecosystem degradation as noted by the MEA. Some 89% of our ecosystems are degraded or severely degraded. Only about 11% of ecosystems are functioning adequately (MEA, 2005).

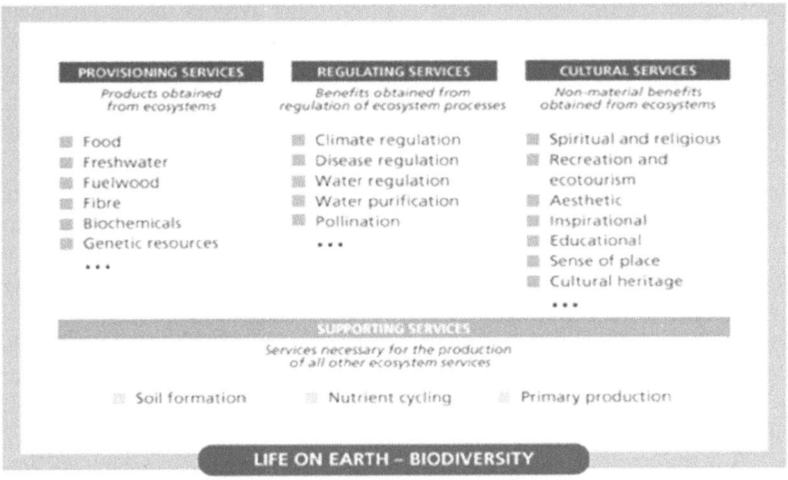

FIGURE 4.2 Ecosystem services from nature (*Source:* MEA, 2005).

Humankind's interventions have led to loss in the natural land resource base (including soil, water, biodiversity, atmosphere, etc.). Interventions dealing with agriculture production through tillage agriculture, both traditional and modern, have led to loss of organic matter, compaction and extensive erosion, land degradation, and recurrent droughts and floods.

In the case of agricultural land use, until the beginning of this century, relatively little direct attention was paid by farmers and service providers including research, education, and community development regarding the need to sustain ecosystem functions and processes. Little attention was paid to the necessity of harnessing ESs from agricultural lands, except for yields of biological products from plants and animals. Ecosystem functions and processes in agricultural landscapes operate at various nested and linked levels, from the level of the production field—affecting soil health and functions as well as crop and agricultural productivity and outputs—to the farm level, affecting overall farm productivity potential and profitability—to the landscape and watershed level, affecting large-scale services to society, such as clean water resources and control land degradation—to the regional level and beyond, affecting landscape and regional level ecosystem functioning and services to societies and the natural world.

Only in recent years, it has become clearly understood that in agricultural landscapes, provisioning ESs to meet the needs of food and agricultural products for society can only be delivered efficiently and without risks of harming ecosystem functions and processes when the related regulatory and supporting ESs are also allowed to operate effectively and efficiently. It is now clear that this is not possible in mechanically tilled agricultural landscapes. This is because mechanical tillage and tillage-based agriculture debilitates all the soil- and land-mediated supporting and regulatory functions and processes and hence the related services to meet the societal and environmental needs of provisioning services (Montgomery, 2007; Kassam et al. 2013). For example, to maintain the quality of water resources at a particular site or within a watershed, water cycling must function normally. This would include the maximum capture of rainfall by the land surface upon which the drops fall, maximum infiltration, absorption, and retention of the water in the soil profile, and the rest or excess being drained to lower layers and into the groundwater and deeper aquifers, from where it would flow into streams and rivers in a regulated manner. The space–time relationship of the hydrological processes would depend on the rainfall patterns, on the land use and ground cover, on the soil type and depth, on the topography and landform, and on the geomorphology.

However, in tillage-based agriculture, the regular mechanical disturbance from tillage and of bare and exposed soils, covered with poor crop diversity, soon leads to loss of soil health and functions, soil surface sealing, and soil losing its porosity and structure, making it unable to maintain adequate infiltration rate and retention and drainage characteristics. This is mainly the result of a reduction and elimination of deep burrowing soil mesofauna, such as the earthworm. With time, there is increasing runoff and erosion of topsoil, and runoff water loaded with sediment and agrochemicals enters the streams and rivers within the watershed at a very rapid rate, sometimes, causing floods, clogging stream and river channels, and causing water and landscape pollution. Sediment and chemical load in river water affect the quality of water in inland water bodies and reservoirs as well as seas and oceans into which rivers flow, eventually creating dead zones. Water percolating through the soil matrix of tilled soil, in the absence of sufficient OM, carries soluble soil nutrients to the aquifers, leading to groundwater pollution with nitrates, phosphates, and agrochemicals in areas of intensive agricultural production.

This section illustrates two types of ecosystem services that are harnessed by CA systems, namely, those that are mobilized and operate in the crop fields and those that are harnessed and delivered across watersheds and regional landscapes. The former services relate to soil health and functions and positively affect crop performance and biological products; and the latter relate to large-scale ecosystem services, such as water resource quality, carbon sequestration, and soil conservation which represent positive externalities that sustainable agriculture should provide to local communities and to the society at large. The examples of both types of ecosystem services are taken from Kassam et al. (2013, 2020b).

4.5.1 Field and Farm Scale Services

CA practices that related to three CA principles are now widely used in a range of farming systems in all continents on some 12.5% (180 M ha) of the global cropland (Kassam et al. 2019). They enhance SOM and soil health and add to sustainability of production and soil systems and generate a range of ESs in the field which relate to soil resilience and productivity. They also improve overall soil conditions and result in beneficial outcomes for production, ecosystem services, and socioeconomic conditions. However, to achieve the sustainable *output intensification* desired by producers, these CA practices need to be complemented by additional best practices related to integrated crop, soil, nutrient, water, pest, time, labor, and energy management. This whole production system enables the sustainable *intensification of output* to occur, rather than the *intensification of inputs,* and the optimization of the need and utilization of production inputs of seeds, agrochemicals, water, energy, time, and labor. Such sustainable production systems are biologically dynamic and ecologically alive, offering farmers many possible combinations of practices to choose from and adapt, according to their local production conditions and constraints (Pretty, 2002; Kassam et al. 2009, 2011a, 2014, 2017, 2020b, c; Pretty et al. 2011; Kassam, 2020a, b; Godfray et al. 2010; FAO, 2011a, 2016).

Thus, sustainable crop production intensification is the combination of all of these improved practices applied in a timely and efficient manner within the ecological underpinning created by the application of the three interlinked CA principles adapted to local conditions and constraints. For this, the ensuring of soil stability, and the favoring of self-recuperation of appropriate soil structural and functional conditions, are essential. Thus, sustainable soil management depends on more than how and what crops are grown. It depends on the dynamic ecological and biological foundation that is created by the production systems itself through its interaction with the self-protecting and self-regenerating capacity of all the interacting parts of the agroecology (FAO, 2008; Kassam et al. 2013, 2020b).

In agricultural crop fields, a number of ESs within the *provisioning* and *regulating* categories are mediated in the soil root zone locally that relate to soil life, health, and functions, as well as crop health and protection, and therefore to crop growth, productivity, and system resilience. The contributions of practices that implement the technical principles of CA—NT, mulch cover, and crop rotations/associations including legume crops—to important ESs have been discussed in Kassam et al. (2011b, 2014, 2020b) and FAO (2011a, 2012). They show which of the field-based ESs are facilitated by each of the core practices of CA systems. There is a long list of in-situ ecosystem services that can operate in healthy agricultural soils that are necessary for good crop performance and soil functions. However, all such conditions and services are debilitated and made sub-optimal or dysfunctional in agricultural soils that are managed under conventional tillage agriculture with bare soil and monocropping.

On the other hand, agricultural land that is managed using the CA approach facilitates ecosystem functions and services, particularly those services related to provisioning, regulating and supporting, derived as a result of improved conditions in the soil volume used by plant roots (Kassam et al. 2009, 2013). The improvement in the porosity of the soil is affected by the actions of the soil biota, which are present in greater abundance in the soil under CA. The mulch on the soil surface in CA systems, protects against the compacting and erosive effects of heavy rain, buffers temperature fluctuations, and provides energy and nutrients to the organisms below the soil surface.

Legumes in CA rotations provide increased availability of biologically fixed nitrogen in the soil, thus diminishing the need for large amounts of applied nitrogenous fertilizers. Society gains from CA regardless of farm size by increased water infiltration, diminished erosion and runoff, less downstream sedimentation and flood damage to infrastructure, better recharge of groundwater, and more regular stream flow throughout the year, resulting in a more assured community water supply and quality. Cleaner civic water supplies also reduce costs of treatment for urban/domestic use. CA also contributes to increased stability of food supplies due to greater resilience of crops in the face of drought. The rural community ultimately benefits from better nutrition, overall health, and less pressure on curative health services.

In CA systems, rotations and associations of crops also encourage agrobiodiversity as each crop will attract different overlapping spectra of microorganisms. The optimization of populations, range of species, and effects of the soil-inhabiting biota is encouraged by the recycling of crop biomass, stubbles, and other organic matter that provides the substrate for their metabolism. Many of the known wild pollinators are soil inhabiting, which find again a habitat under CA. Rotations and associations of crops inhibit the buildup of weeds, insect pests, and pathogens by interrupting their life cycles, making them more vulnerable to natural predator species and contributing development-inhibiting allelochemicals. The same crop mixtures, sequences, and rotations provide aboveground mixed habitats for insects, mammals, and birds. Under CA systems, it is possible to harness many of the above ESs from nature mainly because the ecosystem functions that generate these services are enhanced and protected, so that agriculture is not in competition with nature but works in harmony with it.

As has been described in some detail in Kassam and Kassam (2020c), CA systems that offer full productivity and economic, environmental, and social benefits cannot be established overnight. Depending on the severity of degradation of the soil, the conventional tillage systems must undergo a transformation into a CA system in space and time in order to harness the benefits and changes that improve productivity (and hence profit), efficiency, resilience, and ESs. Often, it is necessary to deal with the detrimental subsurface compaction and hard pans at the outset; otherwise, they have been demonstrated to have significant hindering effects on plant production and yield as well as several other soil-mediated ESs, due to increased physical impedance of roots and loss of pore space for retaining water at plant-available water potentials, or decreased water percolation to deeper soil layers and into groundwater or aquifers.

Equally, if soils are hindered from performing some of their basic functions because of conventional tillage, soil-mediated ecosystem services at the watershed and regional levels, such as water resource quality and quantity, water, nutrient, and carbon cycling, carbon sequestration, and erosion control, cannot be harnessed optimally. Hence, it is important to physically engender a more permeable *sub*surface before anything else can improve significantly. This has the dual effects noted in Kassam and Kassam. (2020c) of "*Maximize rainwater infiltration, minimize runoff.*" In practice, this means that the ground may need to be physically broken at the beginning of the process of transformation to CA with pickaxes, chisels or rippers, subsoilers, and/or para-plows. The process then has to continue biologically and ecologically to turn the soil into a healthy "carbon sponge" with the help of soil microorganisms including bacteria, mycorrhiza, and earthworms, as the practice of no or minimum soil disturbance, return of crop root and shoot biomass into the soil, and diversified cropping systems involving legumes and crops of different root systems, all continue to assist the soil to enhance its health and ecosystem functions. This is particularly the case for soil compactions, which with today's machinery traffic might be beyond the reach of mechanical subsoilers and can only be broken up in time by the action of roots and soil mesofauna (Friedrich, 2020).

4.5.2 Watershed and Landscape Scale Ecosystem Services

When the field level ecosystem services are reproduced across farms in a contiguous micro-catchment within a landscape or a region, the ESs provided at the landscape and watershed level—such as clean water, sequestration of carbon, avoidance of erosion, and runoff—become more evident (Lindwall and Sonntag, 2010; Kassam et al. 2011b, 2013, 2020b; Kassam and Kassam, 2020b). The co-benefits of more water infiltrating into the ground beyond the depth of plant roots are perceptible in terms of more regular stream flow from groundwater throughout the year, and/or more reliable yields of water from wells and boreholes. The benefits of carbon capture become apparent in terms of the darkening of soil color and crumbly feel of the soil, accompanied by improvements in crop growth, plus much less erosion and hence less deposition of sediment in adjacent waterways and into reservoirs.

Society and nature gain from CA regardless of farm size by diminished erosion and runoff, less downstream sedimentation and flood damage to infrastructure, better recharge of groundwater, and more regular stream flow throughout the year, resulting in a more assured community water supply and quality. In CA systems, the sequences and rotations of crops also encourage agrobiodiversity, as each crop will attract different overlapping spectra of microorganisms.

The interlinked CA practices of no or minimal soil disturbance, organic soil cover, and diversified crop associations and rotations, facilitate the harnessing of ESs on agricultural landscapes and watersheds. During the last couple of decades or so, harnessing CA-based ecosystem services is operating or being tested in different parts of the world (Kassam et al. 2013, 2020b). Several good examples include (1) the agricultural carbon sequestration scheme in Alberta, Canada (Swallow and Goddard, 2016); (2) the hydrological services from Paraná Basin III in Brazil located in the western part of Paraná state on the Paraguay border (Mello et al. 2020); (3) control of soil conservation and wind erosion in China (Li et al. 2011, 2016); and (4) control of erosion and restoration of soil and landscape health in Andalusia, Spain (Gonzalez-Sanchez et al. 2015). These examples were described in Kassam et al. (2013, 2020b) and are outlined in the following sections.

4.5.2.1 Canada: Carbon Sequestration Scheme in Alberta

Canada has the fifth largest area under CA systems with some 19.9 M ha (Kassam et al. 2019). In 2007 the Alberta government made amendments to their earlier climate change policy that required large, final emitters to reduce their annual GHG output. In order to reach their compliance targets, facilities had options, one of which was to purchase offset credits from Alberta sources using government approved protocols. One of the agriculture opportunities at the time was the development and application of an NT annual cropping protocol that recognized the increase in soil carbon sequestration with the adoption of NT cropping systems (Goddard et al. 2009). The updated version now named the Conservation Cropping Protocol (CPP) also includes a provision for reduction of summer fallow. In 2007, it was the first soil sequestration protocol in the world if not the first agriculture protocol for a GHG emissions compliance system. Other examples did not exist, and guidance literature was sparse. Swallow and Goddard (2016) have tracked the interconnectivity of international, national, and regional nested networks in formulating policy.

Unique features of the CPP include the use of pooled coefficients for soil carbon sequestration derived from the application of the national emissions inventory to soil map units combined with crops and years to average variance and risks across large areas thus avoiding the prohibitive transaction costs of measuring GHG state or flux changes at the site level. Another feature is the use of private aggregator companies to assemble fields and farms into larger projects that attract buyers. They also are responsible for third-party verification of projects and ensuring the project is correctly represented on the registry where tonnes are serialized and monitored by the Canadian Standards Association.

The government set a compliance price in 2007 of Cdn$15 per tonne CO_2e which increased to Cdn$20 in 2015 and Cdn$30 in 2016. Facilities that could not comply with the reduction target could purchase offsets or pay into a "tech fund" which invested in GHG mitigation research projects. In the decade following 2007, there were slightly more than 12 Tg from CPP projects, or that many tonnes of verified soil carbon sequestration. At a price of Cdn$15 that represents Cdn$180,000,000 or Cdn$180 million that flowed from large emitters to farmers and the agribusiness sector.

The role of aggregators was key in assembling projects and working with verifiers to improve tracking systems and lowering costs. Experienced aggregators have provided valuable insight into protocol revisions and "desk testing" new protocols in the final design stages. Farmers have become more carbon conscious and aware of the records needed to verify their change in practices. The comfort level with protocols, partial life cycle accounting, and offset projects has increased the knowledge and skills of specialists such that they have moved into more integrated life cycle analysis and sustainable certifications with confidence and understanding beyond just talking about the topics.

Baig and Gamache (2011) have reviewed some of the many other co-benefits of CA on the Canadian prairies. Soil carbon imparts many properties of soils and biological function and is often utilized as a key indicator for ESs value. Biologic carbon sequestration represents so much more than mitigation of GHGs, and public policy needs to incentivize behavior such as CA adoption for not only mitigation of emissions but to reward farmers for the private costs of delivering a public good.

4.5.2.2 Brazil: Watershed Services in the Paraná Basin III

Brazil has long experience in CA and currently has more than 32.0 M ha under various CA crop systems. As part of a strategy for improvement, conservation, and sustainable use of natural resources, the Itaipú Dam *Programa Cultivando Água Boa* (cultivating good water) established a partnership with farmers to achieve their goals in the Paraná Basin III located in the western part of Paraná state on the border with Paraguay (Mello and van Raij, 2006; ITAIPU, 2017; Laurent et al. 2011; Mello et al. 2020). Part of the goals related to the need to minimize runoff and erosion, chemical pollution, sediment load of water in streams, and eutrophication of water and silting in the Itaipú, the Itaipú Dam is one of the largest operational hydroelectric energy producers in the world. The dam works to integrate energy, environment, and biodiversity to achieve sustainable development. The Itaipú Dam is situated in one of the most productive agricultural regions in Brazil and Paraguay. Therefore, the water bodies that connect the territory with the reservoir are subject to environmental threats arising from agriculture, such as use of pesticides and fertilizer, waste from livestock production, and soil erosion (ITAIPU, 2017).

Thus, the dam's reservoir depends on the sustainable use and management of soil and water in the watershed/catchment for efficient electricity generation. Sediments, plant nutrients, and animal manure entering the reservoir are a result of inappropriate land use and pollute the water used by the turbines to generate electricity. This phenomenon shortens the useful lifespan of the reservoir and increases the maintenance costs of the power-generating turbines, thereby increasing electricity generation costs. Therefore, payments could be made through a program that would improve the conditions of electricity generation. The spatial unit covered by this program is the whole watershed. Functioning as a community, and joining many farmers in the watershed, they reached a scale where environmental impact can be monitored with suitable indicators to establish a system for payment for ESs.

One of the partnerships established in the *Cultivando Água Boa* program and developed through an agreement with the Brazilian NT farmers' federation (FEBRAPDP) was the Participatory

Methodology for Quality Assessment in Conservation Agriculture (Laurent et al. 2011; Roloff et al. 2011; ITAIPU, 2019; Mello et al. 2020). It was based on the former positive experiences with CA-based catchment development in Brazil. The program was the partners' plan to measure the impacts of farm management through a scoring system indicating how much each farm is contributing to the improvements of the water quality (Mello et al. 2020) (https://www.itaipu.gov.br/en/the-environment/cultivating-good-water).

What is most important is that the change from bad land use practice to good quality CA, and the resultant significant improvement in soil health and functions and in watershed hydrology, led to improved water resource quality. As a result, the operating life of the Itaipú Dam has increased by at least four-fold. Consequently, the *Cultivando Água Boa* program was awarded the "Water for Life" UN Water Best Practices Award 2015 as the world's best water management practice (Category 1) because the program is a permanent community participation movement in which Itaipú works together with the rural society to change its values in addition to mitigating and correcting environmental liabilities (ITAIPU, 2019).

A recent review (Mello et al. 2020) of the above experience concluded the following:

As a result of adopting CA production systems, farmers have been able to minimize soil erosion and runoff, increase water infiltration into the soil, and maximize water retention for production. At the same time, CA-based land use has led to more productive agriculture for farmers and has been a major driving force for economic and rural development.

In terms of new, transferable knowledge about efficient water management for rain-fed production and system sustainability, the Paraná Basin III development offers several policy and technical lessons:

(i) Tillage agriculture is not suitable for sustainable land management.
(ii) Sustainable agricultural production intensification and protection of ecosystem functions and services are not mutually exclusive.
(iii) Community-based soil and water conservation programs can be developed in a socially equitable and self-empowering manner for both smallholders and larger-scale farmers, as well as civil society.
(iv) The achievements of the Itaipú project have been dependent upon the goodwill, investments, and cooperation of several ministries (energy, water, agriculture, environment, social development, health, and education), all of whom worked constructively together for the benefit of all. This emphasizes the importance of intersectoral approaches.
(v) Land and water related developments on the Brazilian side of the Itaipú reservoir also took place in Paraguay, a country dominated by smallholder farmers at that time, many of them relying on animal farm power.

Effectively achieving a multi-sectoral approach to economic, social, and environmental development at all levels in the Paraná Basin III, given its size and 1 million inhabitants, is in essence equivalent to managing a small nation. A large proportion of the watershed population rely on agriculture and related sectors for their livelihoods. Productive and sustainable agriculture is not only a key driver for economic and social change—it is also a necessary foundation for both a good quality of life and for delivering soil- and land-mediated ecosystem services to society. The agriculture and rural development achievements in the Paraná Basin III, and those of the Itaipú hydropower complex, have been nationally and internationally recognized. Increasing the working lifespan of the complex by more than five-fold bodes well for the future and gives hope to its population that the Paraná III watershed has the potential to remain an area of opportunity for the foreseeable future.

4.5.2.3 China: Soil Conservation and Erosion Control

China is paying more attention to tilled soils being more susceptible to water and wind erosion. The best solution to control water and soil erosion is to eliminate tillage. Huge dust storms originating in West China and reaching Korea and Japan, creating serious air pollution problems in China, have been identified to originate from cropland during land preparation times. Practices which improve water use efficiency and natural resource management by reducing runoff and erosion are of great importance. Therefore, the adoption of CA practices, providing more residue cover, and minimum soil disturbance, has received considerable attention. As a consequence, agricultural land that is managed with CA systems in China has grown to some 9 M ha (Li et al. 2016).

Conventional tillage (CT) in dryland farming areas of northern China includes moldboard plowing to a depth of about 20 cm, followed by disc harrowing, hoeing, rolling, and leveling. Before plowing, all the crop residues in the fields are removed for animals and/or fuel. Long-term moldboard plowing and residue removal/burning have increased the risks of and actual wind and water erosion and the formation of hard pans in the deep soil layer. It has also resulted in poor soil physical and chemical properties and high inputs of energy and labor which leads to low farmer incomes. In some parts of northern China, particularly in the North China Plain, burning crop residue has increased during the last decades, which caused high levels of air pollution. Therefore, the government is pushing to accelerate the development and spread of various kinds of CA systems and related practices, such as controlled traffic and permanent raised beds (PRBs), to protect and improve soil, water, and air quality as ecosystem societal services (Bai et al. 2008; Gao et al. 2008; Chen et al. 2008; Kassam et al. 2009, 2013; He et al. 2012; Kautz et al. 2013; Ding et al. 2013; Lal, 2014).

Effect of Conservation Agriculture on Soil Fertility: The results from long-term (>5 years) field experiments showed that CA significantly improved SOM (>7%) and NPK near the soil surface as compared to CT in different regions of China. Moreover, accumulated SOM had a strong relationship with the implementation time of CA. For example, Zhang et al. (2019) after 26 years of experimentation found that SOM contents under CA practices were >20 and >60% greater than that under CT treatment, respectively. In addition, the low aggregate stability of CT soils resulting from frequent soil disturbance via intensive tillage affected NPK accumulation. Under CA, the physical protection provided by OM is increased leading to a significant NPK increase. The SOM, N, P, and K in soil are critical nutrients for crop production and, to a large extent, hold the land productivity and crop yields. Increased soil fertility is helpful for the improvement of crop yields in a long-term CA system. Li et al. (2016), based on meta-analysis, showed that the mean increase in crops yields with CA practiced for a period of ≥10 years was >7.0% relative to CT, indicating that productivity continues to increase as the natural capital of the system is enhanced over time.

Effect of Conservation Agriculture on Water Erosion: Li et al. (2011), Wang et al. (2014a), and Tan (2015) demonstrated that CA treatments significantly reduced runoff and soil erosion by >25 and >12%, respectively. Furthermore, a meta-analysis of CA in northern China (Li et al. 2016) showed that adoption of CA reduced water runoff and erosion by >40%, increased soil water-holding capacity by >3%, and water use efficiency by >10% for most crops and agroecologies. A rainfall simulation experiment was also carried out to study the effect of CA on water erosion, and the results showed that, under rainfall intensities from 10 to 80 mm h^{-1}, CA delayed runoff start time and reduced runoff by 17.9 to 38.7%.

Effect of Conservation Agriculture on Wind Erosion: CA reduced soil erosion by wind as was shown by Li et al. (2011), Wang et al. (2014a), Tan (2015), Chen et al. (2016), and Lin et al. (2017). Soil erosion under CA was significantly lower than under CT. Wind tunnel simulation experiments by Yang et al. (2019) and Lin et al. (2017) showed that CA reduced mean soil loss by 72.6%. Under natural wind erosion conditions, Wang et al. (2014b) indicated that CA reduced soil loss by >50%, compared to CT during an 11-hour dust storm event. The reduction of soil loss will vary in CA systems depending on the measurement duration, weather conditions, and erosive winds.

4.5.2.4 Spain: Soil and Water Conservation in Andalusia

Groundcovers in woody crops are considered an essential part of CA-based perennial systems in Spain (Gonzalez Sanchez et al. 2015). Among the woody perennial crops, olive orchards are an important agroecosystem in Spain, where 2.7 M ha or 16% of the surface area is covered with olive groves. In some regions, olive groves are essential for the economy and the development of the society, i.e. in Andalusia, where it accounts for 60% of the Spanish olive-growing area. The value of the production of the olive oil sector was around €1,886 million on average from 2007 to 2012 (MAPA, 2019).

Historically, olive cropping has been concentrated on hilly lands, where soil erosion happens to be a severe and widespread problem. Locally, historical soil loss rates have been reported to reach up to 184 Mg $ha^{-1}yr^{-1}$ (Vanwalleghem et al. 2010, 2011). Erratic but high-intensity rainfall (especially during winter), as well as the management of the orchards lies at the origin of soil erosion. Commercial olive orchards, mainly grown under rain-fed conditions, are characterized by extremely adverse management conditions as farmers tend to till intensively to avoid competition with weeds for tree water and nutrient uptake. Therefore, effective conservation strategies, such as the establishment of groundcovers, are not easily adopted by farmers. Indeed, conventional tillage has been the dominant management system in olive orchards over the last decades. The combination of this human-induced low vegetation cover with the steep slope gradients on which these orchards are located, together with the high-intensity rainfall events that characterize the Mediterranean climate, explains why high soil erosion rates have been associated with olive oil production (Beaufoy, 2001).

The spread of CA expansion in woody crops in Spain was facilitated by a first CAP agri-environmental scheme. It was introduced in Andalusia in the late 90s and early 2000s aiming to fight soil erosion in olive orchards mainly by groundcover between trees and natural vegetation on the land borders. This is considered a key milestone in the development of CA systems in woody crops, as it helped to double the CA area of olive crops over a period of a decade or so from 0.43M ha in 2006 to 0.83 M ha in 2018. This happened without any substantial economic support to producers. During the same period, the area of perennial CA systems (all woody crops), including olives, increased by 56% from 0.83 M ha to 1.30 M ha. The proportion of CA perennial area to total perennial area increased from 16.5% in 2006 to 25.1% in 2018. Confidence in the new system and economic feasibility were the main reasons for the shift from the tillage system to CA. It was proven that CA is a win–win option not only for annual crops but also for woody crops (Garcia-Torres, 2001). Other soil erosion control practices were also promoted including contour bunding and the maintenance of pruning residues on the ground surface in the interrow space (Franco and Calatrava, 2006).

In general, there is still more room for the extension of policy measures to mitigate and reverse soil erosion and degradation (Calatrava et al. 2011). In some cases, farmers do not till the soil between tree lines, using herbicides to control weeds instead. However, this leads to bare soil with no protection, which means no improvement in SOM content or soil functions. Findings of Gomez et al. (2011) showed that bare soil, even though untilled, is capable of more runoff and sediment load in olive groves under certain conditions which should be seriously taken into account when designing soil and water conservation strategies. The faster adoption of CA-based groundcover systems instead of NT with bare soil can therefore be considered an important step toward the mitigation of soil erosion and degradation. These are significant ecosystem services mobilized by producers on farms with CA systems which also benefit the society at large.

4.6 POLICY AND INSTITUTIONAL IMPLICATIONS

CA systems are expanding at an annual rate of more than 10 M ha globally. However, at national levels, rates of CA adoption are different depending on historical and prevailing conditions. Some 50% of the CA area is in low-income countries, and 50% is in the industrialized countries. Globally, some 70% of the CA area is under large-scale farming whereas the rest is under small-scale farming, operated by millions of farmers (Kassam et al. 2019). In recent years adoption of CA is increasing in Europe, Asia, and Africa as more attention is being paid to support and help farmers to move away from tillage farming. However, scaling requires policy and institutional support varying from short-term financial assistance to farmers and assisting service providers to diversify their services to CA farmers. It also requires promoting the establishment of national and sub-national CA associations and local CA clubs and services (Kassam et al. 2014). A major requirement for scaling is the need for training of farmers and field experts to demonstrate and champion the transformation of conventional tillage farming to CA. This must be backed by good quality research and education on all aspects of CA in the context of adoption by farmers but also in terms of benefits to local and national economy, improvement in food security, employment, livelihoods, and quality of life.

Many countries, such as Brazil, Argentina, Paraguay, Uruguay, Kazakhstan, China, Morocco, South Africa, Namibia, Kenya, Zambia, Mozambique, Switzerland, Italy, and Spain are providing policy support at the provincial or national levels to farmers to accelerate the adoption and development of CA. Some of them have also mobilized institutional support in terms of extension, research, education, and private sector service provisioning. However, much more effort is needed to enable nations to aim at transforming all their agriculture sectors and establish a process of mainstreaming CA. Mainstreaming CA would aim at realigning all relevant institutions and stakeholders toward the provisioning of a full range of the required support to CA farmers and to the agriculture and food sectors. It would also mean providing financial and non-financial incentives to farmers to change and improve their farming practices and to harness or generate ecosystem services for local communities and for society as a whole. Furthermore, it would also mean assisting farmers and the agricultural sector to meet national and international targets in terms of climate change adaptation and mitigation, environmental quality, land and ecosystem restoration, and food security. Global players, such as the United Nations with the Sustainable Development Goals (SDGs), and in particular the FAO, once a driver for the CA movement, must undertake much more effort to mainstream CA as the core of sustainable land management among their member countries' policies.

In Europe, the Farm to Fork Strategy (European Commission, 2020), as essential guidance to the Common Agricultural Policy under the European Union's Green Deal, aims to accelerate our transition to a sustainable food system that should have a neutral or positive environmental impact; help to mitigate climate change and adapt to its impacts; reverse the loss of biodiversity; ensure food security, nutrition and public health, making sure that everyone has access to sufficient, safe, nutritious, and sustainable food; and preserve affordability of food while generating fairer economic returns, fostering competitiveness of the EU supply sector, and promoting fair trade.

The use of chemical pesticides in agriculture contributes to soil, water, air pollution, and biodiversity loss and can harm non-target plants, insects, birds, mammals, and amphibians. The European Commission will take additional action to reduce the overall use and risk of chemical pesticides by 50% and the use of more hazardous pesticides by 50% by 2030. To pave the way to alternatives and maintain farmers' incomes, the Commission will take a number of steps. It will revise the Sustainable Use of Pesticides Directive, enhance provisions on integrated pest management (IPM) and promote greater use of safe alternative ways of protecting harvests from pests and diseases. IPM encourages the use of alternative control techniques, such as crop rotation and mechanical weeding, and will be one of the main tools in reducing the use of, and dependency on, chemical pesticides, in general, and the use of more hazardous pesticides in particular. Agricultural practices that reduce the use of pesticides through the CAP will be of paramount importance and the strategic plans should reflect this transition and promote access to advice. The Commission will also facilitate the placing on the market of pesticides containing biologically active substances and reinforce the environmental risk assessment of pesticides. CA deliverables, based on solid scientific results, can help achieve major objectives in the European Common Agricultural Policy (CAP).

In a study published by the government of Spain, coefficients of increase of SOC were linked to the potential areas that different systems could be transformed into in the country. Table 4.4 summarizes the results of the increase in organic carbon in the soil, its translation into CO_2, the area eligible for each system according to the suitability of crops in Spain, the storage of OC, and the increase that they would suppose on the carbon currently stored in the soil. These increases in SOC can be transformed into potentially avoided CO_2 emissions ($C \times 44/12 = CO_2$). In terms of public investment and efficiency through EU agri-environmental schemes, CA is the system of least cost per t CO_2.

TABLE 4.4

Summary of Agricultural Practices/Systems, Their Effect on the Increase of Organic Carbon in the Soil and the Cost According to Agri-environmental Schemes in Spain (*Adapted from:* Mapama, 2018)

Agricultural Measures	SOC Increase (Mg C ha^{-1} yr^{-1})	CO_2 Stored in Soil as C (Mg CO_2 ha^{-1} yr^{-1})	Current Content SOC (Mg C ha^{-1})	Potential Area Adopting the Agricultural Measures (ha)	SOC Potential Increase (Mg C yr^{-1})	CO_2 Emissions Avoided (Mg CO_2 yr^{-1})	Public Investment (€ ha^{-1})	Cost (€ Mg^{-1} CO_2)
Organic farming in annual crops[1]	0.60	2.20	41.78	11,852.508	7,111.505	26,075.518	220.35	100.16
Organic farming in woody crops[2]	0.60	2.20	36.06	4,961.981	2,977.189	10,916.358	306.55	139.34
Conservation Agriculture in annual crops[3]	0.86	3.15	41.78	11,852.508	10,193.157	37,374.909	76.76	24.34
Conservation Agriculture in woody crops[4]	1.95	7.15	36.06	4,961.981	9,675.863	35,478.164	181.91	25.44
Compost[5]	0.07	0.26	39.72	12,684.249	53.362	195.661	240.45	936.82
Urban residues[6]	0.14	0.52	39.72	16,814.489	48.069	176.253	124,69	239.04
Manure/livestock residues[7]	0.71	2.60	39.72	2,509.103	1,781.463	6,532.031	210,00	80.67

1. Potential achievable if the majority of the surface of annual crops were in organic farming.
2. Potential achievable if the majority of the woody crop area were in organic farming.
3. Potential achievable if the majority of the area under annual crops were under conservation agriculture.
4. Potential achievable if the majority of the woody crop area were under conservation agriculture.
5. The potential that could be reached assuming that there was no limitation to the access of this type of residue.
6. The potential that could be reached assuming that there was no limitation to the access of this type of residue.
7. The potential that could be reached assuming that there was no limitation to the access of this type of residue.

There is no doubt that SOM is a necessary driver of soil health and functions. It is also responsible for achieving optimum production performance, in terms of maximum output with minimum input, and for harnessing all the soil-mediated ecosystem services at the on-farm production level as well as at the watershed and regional landscape level. Overall, SOM is accountable for sustaining soil resilience and for regeneration and restoration of degraded lands.

In the context of climate change and climate smart agriculture, CA has a crucial role to play in terms of facilitating the sequestration of carbon in the soil. In other words, CA is essential if the objective is to convert agricultural soils into carbon sinks. Under conventional tillage agriculture, the soil is a source of carbon and other greenhouse gases, and life cycle has a large carbon footprint. Under CA, the situation is the opposite. Thus, one of the major future roles of CA is in the development of climate-smart agriculture nationally and globally. CA is already being used to serve in carbon offset trading schemes e.g. Alberta, Canada, and in low-carbon programs (e.g. Brazil).

In the past, education and training have played a major institutional role in modernizing agriculture and all the support sectors. This has involved producing graduates, professionals, and field experts to build public and private sector capacity for research and innovation, service provisioning, and development of input sectors and output value chains. Also, it has involved generating employment in land use, manufacturing, agro-processing, and service sectors. CA represents a new paradigm of agriculture and calls for a different way of thinking about land use, production systems, sustainability, resilience, climate-smart agriculture, agricultural development, and restoration of land degradation. In the medium to longer-term, agricultural transformation led by CA farmers is not sufficient condition for achieving sustainability. Education and research in CA must also accompany CA adoption by farmers in order to achieve total transformation. In this regard it is pertinent to mention here that the no-tillage technology, which is a precondition for applying CA, is at this time (2020) not in the curricula of many universities and is not a question that matters in research and technology development in many countries of the world (e.g. Germany). To mention this fact is useful to understand that, in most countries, the introduction of CA can be hard work and it could be that people involved may be called "crazy," as the pioneers were called in Brazil and other countries in South America. Still, today, the majority of farmers worldwide would not be amused if you took their plows away.

4.7 RELEVANCE OF CA TO GLOBAL GOVERNANCE

At the UN Earth Summit in Rio in 1992, three major global environmental threats to sustainable development were recognized. These were climate change, loss of biodiversity, and land degradation and desertification. To address these global challenges, three United Nations conventions were established to help understand the nature of the threats and formulate possible solutions; legally binding agreements, the Rio Conventions, were opened for signature. They are: The UN Framework Convention on Climate Change (UNFCCC); The UN Convention on Biological Diversity (UNCBD); and The UN Convention to Combat Desertification (UNCCD). In addition, the Earth Summit resulted in the Rio Declaration on Environment and Development, Agenda 21, and Forest Principles.

To ensure compliance with the Rio agreements, particularly the Rio Declaration on the Environment and Development and Agenda 21, delegates to the Earth Summit established the UN Commission for Sustainable Development (UNCSD). In 2013, the UNCSD was replaced by the High-level Political Forum (HLPF) on Sustainable Development that meets every year as part of the Economic and Social Council (ECOSOC) meetings, and every fourth year as part of the UN General Assembly meetings. To coordinate activities related to sustainable development, a UN Sustainable Development Group with a Development Coordination office has been established. The main contribution of the HLPF and the Sustainable Development Group has been the formulation of the 2030 Agenda for Sustainable Development with its 17 SDGs which were adopted in September 2015.

It is realized that all SDGs are interconnected. However, there are several SDGs that can benefit directly from agricultural land use that is based on CA systems. These are: SDG 1 (no poverty), 2 (zero hunger), 3 (human health), 4 (education), 5 (gender), 6 (water), 7 (energy), 8 (economic growth and employment), 12 (sustainable consumption and production), 13 (climate action), 14 (marine resources), and 15 (terrestrial ecosystems). The key benefits of CA systems are elaborated in Kassam and Kassam (2020b) showing that they can contribute to the achievement of the above SDGs.

CA systems contribute to SDGs through increased productivity and profit, improved soil health, crop diversification and healthier diets, eliminating drudgery, facilitating women farmers' work, and reducing child labor in agriculture, reduced or minimized use of pesticides and herbicides, reduced machinery, labor, and energy use and costs, decreased soil erosion and land degradation, increased biomass for livestock, greater adaptability to climate change, climate change mitigation, and rehabilitation of degraded lands and ESs, as well as enhanced ecosystem services to society. Every nation is committed to achieving the SDGs, thus agriculture based on CA can become part of the solution, provided national policies and global policies can integrate CA as a key component of sustainable agricultural land use development.

The three UN Conventions to address the challenges of climate change, loss of biodiversity, and land desertification and degradation are being implemented through their respective Conference of the Parties (COP) processes. These COP processes have yet to embrace CA-based agricultural land use as a significant part of the solution. For climate change, at COP 21, the 4 per 1000 proposal from the French government was accepted, and this initiative is expected to facilitate the process of increasing soil carbon content by 0.4% per year which is equivalent to the total amount of carbon being emitted annually. However, the 4 per 1000 initiative has yet to establish how this can be done, along with the supporting scientific evidence. Agricultural land use based on good quality CA systems has a major contribution to make to climate change mitigation by sequestering carbon in the soil. It also has a major contribution to make to climate change adaptability (Gonzalez-Sanchez et al. 2012, 2017, 2019). CA systems are acknowledged to be climate-smart. Given that there are already 180 M ha of CA cropland systems globally, it is important that the role of CA in climate change mitigation and adaptability be recognized in national and international policies and institutional land use strategies. This way they would be responding to international climate change agreements, such as COP 21 in Paris and COP 22 in Marrakesh. At COP22, the Moroccan government's Triple A initiative of Adaptation of African Agriculture (AAA) was accepted. Again, CA land uses have a major role to play in implementing the Triple A initiative. The Moroccan government itself has also proposed to support the transformation of 2 M ha of cropland currently under tillage agriculture to CA systems over a period of five years.

Similarly, the role of CA in reversing land degradation, enhancing soil health and productivity, as well as rehabilitating soil-mediated ecosystem services is elaborated in Kassam et al. (2013) and Kassam (2020b) with successful examples from Brazil, Canada, Spain, China, and Australia. Recently, under the UNCCD, a framework for Land Degradation Neutrality (LDN) has been proposed (Cowie et al. 2018). Given that land use based on CA systems is capable of reversing land degradation, as well as rehabilitating abandoned agricultural lands, it is important that UNCCD's Land Neutrality Framework incorporates the CA dimensions. In this approach, solutions to combating desertification and land degradation in agricultural lands are based on CA regenerative and restoration processes.

Equally, the role of CA in enhancing agrobiodiversity as well as biodiversity in general is being increasingly documented. CA systems are known to enhance soil health and biology. This is based on increasing soil biodiversity in terms of promoting soil microorganisms as well as meso fauna, including earthworms, termites, spiders, beetles, pollinators, frogs, and other soil-inhabiting organisms. Over time, CA systems create a soil habitat for all kinds of organisms that can play a useful role in nutrient, water, and carbon cycling, as well as integrated pest management. Similarly, CA systems are known to promote biodiversity above the ground through their diversified cropping systems and mulch cover, both of which contribute not only to enhancing soil biodiversity but also aboveground biodiversity including natural enemies of pests. Through this increase in soil and crop biodiversity, CA systems can offer greater resilience to biotic and abiotic stresses. CBD and FAO have recognized the role of CA in enhancing biodiversity as seen from their recent documents (e.g. FAO, 2019).

In light of the above, it is only logical to view CA as offering a common land use management framework to achieve the objective of the three UN Conventions and contribute to achieving the SDGs.

4.8 CONCLUSIONS

CA offers a global opportunity to all nations to transform their out-of-date degrading agriculture into regenerative and sustainable agriculture. Global scientific and empirical evidence shows that CA is applicable to all land-based crops and production systems—rain-fed or irrigated, organic or non-organic, annual or perennial or mixed, orchards, plantations, and agroforestry and pastures; all types of farmers, smallholder or large-scale, all types of farm power—manual, animal traction and motorized, across all climates—tropics, sub-tropics, and temperate, and across arid, semi-arid, sub-humid, and humid moisture regimes. CA systems operate from sea level to 3,000 m altitude, and with soils that are >90% sand to >90% clay.

Scientific studies and analyses have shown that CA has considerable potential to enhance SOM, and thereby improve soil health, productivity, and ESs. There need not be any negative trade-offs in established CA systems between productivity and ESs, which is the case with conventional tillage agriculture. CA is climate-smart but smarter in many other ways, offering a large range of productivity, economic, environmental, and social benefits to farmers and society. CA systems have been shown to regenerate soil biology and restore soil functions, reestablish favorable pest–predator dynamics, enhance biodiversity and ecosystem processes at the field and watershed level, minimize production inputs, and optimize production. CA systems are therefore resilient, versatile, and adaptable and can make a major contribution to achieving the SDGs globally.

CA is at an opportune moment. The public has begun to become environmentally conscious and in favor of production practices with a small environmental footprint and co-benefits for the public good. Increasingly, international conventions are looking to agriculture for support on a number of policies from economic diversity to climate change to the environment. CA has found adoption around the world in a wide range of farm and national circumstances. Scientific evidence has grown to validate CA systems and support policy or institutional developments which can align and strengthen other policies. The emergence of civil society and private sectors as a policy influencer, increasing demand for greater localization and equity, inclusive and pro-poor development strategies and climate change all invite opportunities for CA to be incorporated in national development plans and well aligned with multiple international interests.

CA has a bright future ahead. Often the development of practice, policy, and institutions has been from the bottom up. More recently, however, we have seen more and more top-down support being directed toward CA by public, private, and civil sector institutions. CA stakeholders and influencers need to link across in a collaborative fashion to enable adaptive and robust policies to be implemented at all levels. Stakeholders are farmers, government, academia and industry, and NGOs, and all should be involved to develop CA policy and implement action.

REFERENCES

Ahmad, S. 2009. Water availability and future water requirements. In: Project Management and Policy Implementation Unit of the Ministry of Water and Power (MOWP) Water conservation, present situation and future strategies. Proc. MOWP, Planning Commission and Asian Development Bank National Seminar, 21 May 2009, Project Management and Policy Implementation Unit MOWP, Islamabad, Pakistan. pp. 43–62.

Albertengo, J., Bellos, C., Giraudo, M.B., Peiretti, R., Permingeat, H. and Wall, L. 2014. Conservation agriculture in Argentina. In: Jat, R.A., Sahrawat, K.L. and Kassam, A.H. (eds.), *Conservation Agriculture: Global Prospects and Challenges*. CABI, Wallingford, UK, 352–374

Amado, T.J.C., Martins da Costa, C.A., dos Anjos Leal, O. and Pott, L.P. 2020. Chapter 14: Rehabilitating degraded and abandoned agricultural lands with conservation agriculture systems. In: Kassam, A. (ed.), *Advances in Conservation Agriculture*. Volume 2: Practice and Benefit. Burleigh Dodds, Cambridge, UK, 419–464.

Bai, Y.H., Chen, F., Li, H.W., Chen, H., He, J., Wang, Q.J., Tullberg, J.N. and Gong, Y.S. 2008. Traffic and tillage effects on wheat production on the Loess Plateau of China: 2. Soil physical properties. *Australian Journal of Soil Research* 46:652–658.

Baig, M.N. and Gamache, P.M. 2011. *The Economic, Agronomic and Environmental Impact of No-Till on the Canadian Prairies*. Volume 2. Alberta Reduced Tillage Linkages, Canada.

Bauer, A. and AL.L. Black. 1994. Quantification of the effect of soil organic matter content on soil productivity. *Soil Science Society of America Journal* 58(1):185–193.

Beaufoy, G. 2001. The Environmental Impact of Olive Oil Production in the European Union: Practical Options for Improving the Environmental Impact. In: European Forum on Nature Conservation and Pastoralism. http://ec.europa.eu/environment/agriculture/pdf/oliveoil.pdf

Brady, N.C. and R.R. Weil. 2002. Carbon balance in the soil-plant-atmosphere system. *The Nature and Properties of Soils* 524–542.

Brisson, N., Gate, P., Gouache, D., Charmet, G., Oury, F.X. and Huard, F. 2010. Why are wheat yields stagnating in Europe? A comprehensive data analysis for France. *Field Crops Research* 119:201–12.

Calatrava, J., Barbera, G.G. and Castillo, V.M. 2011. Farming practices and policy measures for agricultural soil conservation in semi-arid Mediterranean areas: the case of the guadalentin basin in southeast Spain. *Land Degradation & Development* 22 (1):58–69.

Calegari, A., Ferro, M., Grzesiuk, and Jacinto Jr, L. 1995. *Plantio directo e totacao de culturas. Experimencia em latossolo roxo, 1985–1992*. Cooperativa dos Cefeicultores e Agropecularistas de Maringá Ltd, Maringá, PR, Brazil, 64 pp.

Calegari, A., de Araújo, A.G., Costa, A., Fuents-Lanillo, Junior, R.C. and dos Santos, D.R. 2014. Conservation Agriculture in Brazil. In: Jat, R.A., Sahrawat, K.L. and Kassam, A.H. (eds.), *Conservation Agriculture: Global Prospects and Challenges*. CABI, Wallingford, UK, 54–88.

Carbonell-Bojollo, R., González-Sánchez, E.J., Veroz-González, O. and R. Ordóñez-Fernández. 2011. Soil management systems and short term CO_2 emissions in a clayey soil in southern Spain. *Science of the Total Environment* 409(15):2929–2935.

Carbonell-Bojollo, R., Veroz-Gonzalez, O., Ordoñez-Fernandez, R., Moreno-Garcia, M., Basch, G., Kassam, A., de Torres, M.A.R.-R. and E.J. Gonzalez-Sanchez. 2019. The effect of conservation agriculture and environmental factors on CO_2 emissions in a rain-fed crop rotation. *Sustainability* 11(14):3955.

Carvalho, M., Basch, G., Alpendre, P., Brand, O.M., Santos, F. and M. Figo. 2005. Nitrogen management model for wheat production. In: Online of the Conference on Technological Innovation on Mediterranean Agricultural Systems. University of Evora, Evora, Portugal.

Carvalho, M., Basch, G., Calado, J.M.G. and Barros, J.F.C. 2012. Long term effect of tillage system and crop residue management on soil carbon content of a Luvisol under rain-fed Mediterranean conditions. *Agrociencia*, (Special Issue), 183–187.

Chabert, A. and J.P. Sarthou. 2020. Conservation agriculture as a promising trade-off between conventional and organic agriculture in bundling ecosystem services, *Agriculture, Ecosystems & Environment* 292(2020):106815.

Chen, H., Bai, Y.H., Wang, Q.J., Chen, F., Li, H.W., Tullberg, J.N., Murray, J.R., Gao, H.W. and Gong, Y.S. 2008. Traffic and tillage effects on wheat production on the Loess Plateau of China: 1. Crop yield and SOM. *Australian Journal of Soil Research* 46:645–651.

Chen, M.F., Wang, H.X., Yu, G.Q., et al. 2016. Study on residual stubble erosion control effect of peanut-maize intercropping model in wind-sand semi-arid region. *Liaoning Agricultural Sciences* 1:73–75.

COM. 2020. Towards a thematic strategy for soil protection. *Communication from the Commission to the Council, the European Parliament, the Economic and Social Committee and the Committee of the Regions*. COM 2002, 179 final. Commission of the European Communities.

Corbeels, M., Naudin, K., Whitbread, A.M., Kuhne, R. and Letourmy, P. 2020. Limits of Conservation Agriculture to overcome low crop yields in sub-Saharan Africa. *Nature Food* 1:447–454.

Corsi, S., Freidrich, T., Kassam, A., Pisante, M. and Sa, J. de M. 2012. Soil organic carbon accumulation and greenhouse gas emission reductions from conservation agriculture: A literature review. *Integrated Crop Management* Volume 16. FAO, Rome.

Cowie, A.L., Orr, B.J., Castillo Sanchez, V.M., Chasek, P., Crossman, N.D., Erleweinf, A., Louwagie, G., Maron, M., Metternichti, G.I., Minelli, S., Tengberg, A.E., Walter, S. and Welton, S. 2018. Land in balance: The scientific conceptual framework for land degradation neutrality. *Environmental Science and Policy* 79:25–35.

Derpsch, R. 2008a. No-tillage and conservation agriculture: A progress report. In: Goddard, T., Zoebisch, M.A., Gan Y., Ellis, W., Warson, A. and Sombatpanit, S. (eds.), *No-Till Farming Systems*. Special publication no. 3, WASWAC, Bangkok, Thailand.

Derpsch, R. 2008b. No-tillage and mulch cover. *Leading Edge* 7(2):421–430.

Derpsch, R., Sidiras, N. and Roth, C.H. 1986. Results of studies made from 1977 to 1984 to control erosion by cover crops and tillage techniques in Parana, Brazil. *Soil and Tillage Research* 8:253–263.

Derpsch, R., Franzluebbers, A.J., Duiker, S.W., et al. 2014. Why do we need to standardize no-tillage research? *Soil and Tillage Research* 137:16–22.

Diekow, J., Mielniczuk, J., Knicher, H., Bayer, C., Dick, D.P. and Kogel-Knaber, I. 2005. Soil C and N stocks as affected by cropping systems and nitrogen fertilization in a southern Brazil Acrisol managed under no-tillage for 17 years. *Soil and Tillage Research* 81:87–95.

Ding, F., Hu, Y.L., Li, L.J., Li, A., Shi, S., Lian, P.Y., et al. 2013. Changes in soil organic carbon and total nitrogen stocks after conversion of meadow to cropland in Northeast China. *Plant Soil* 373:659–672.

Doran, J.W. and Zeiss, M.R. 2000. Soil health and sustainability: Managing the biotic component of soil quality. *Applied Soil Ecology* 15:3–11.

Duiker, S.W. and Thomason, W. 2014. Conservation agriculture in the USA. In: Jat, R.A., Sahrawat, R.K. and Kassam, A.H. (eds.), *Conservation Agriculture: Global Prospects and Challenges*. CABI, Wallingford, UK, 26–53.

European Commission. 2020. Farm to fork: For a fair, healthy and environmentally-friendly food system. https://ec.europa.eu/food/sites/food/files/safety/docs/f2f_action-plan_2020_strategy-info_en.pdf (accessed September 11, 2020).

FAO. 2008. Investing in sustainable crop intensification: The case for soil health. In: *Report of the International Technical Workshop, FAO, Rome, July* Volume 6: Integrated Crop Management. Rome, FAO. Available at http://www.fao.org/ag/ca/.

FAO. 2011a. *Save and Grow. A Policymakers' Guide to the Sustainable Intensification of Smallholder Crop Production*. FAO, Rome, 98 pp. www.fao.org/ag/save-and-grow/.

FAO. 2011b. *The State of the World's Land and Water Resources for Food and Agriculture (SOLAW): Managing Systems at Risks*. Summary Report. FAO, Rome. 47 pp.

FAO 2012. *Soil Organic Carbon Accumulation and Greenhouse Gas Emission Reductions from Conservation Agriculture: A Literature Review* Volume 16: Integrated Crop Management. FAO, Rome.

FAO. 2016. *Save and Grow in Practice*: *Maize*, *Rice* and *Wheat*. FAO, Rome.

FAO. 2017. Soil organic carbon: The hidden potential. Food and Agriculture Organization of the United Nations. http://www.fao.org/documents/card/es/c/ed16dbf7-b777-4d07-8790-798604fd490a/ (accessed September 15, 2020).

FAO. 2019. The State of World's Biodiversity for Food and Agriculture. FAO, Rome. http://www.fao.org/3/CA3129EN/ca3129en.pdf

Farooq, M., Nawaz, A., Saharawat, Y.S., Reeves, T. and Siddique, K. 2020. Crop and cropping systems management practices and benefits in conservation agriculture systems. In: Kassam, A. (ed.), *Advances in Conservation Agriculture* Volume 2: Practice and Benefits. Burleigh Dodds, Cambridge, UK, 37–74.

Fox, R.H. and V.A. Bandel. 1987. Nitrogen utilization with no-tillage. In: Spragne, M. A and Triplett, G.B. (eds.), *Mo Tillage and Surface Tillage Agriculture*. Wiley Interscience Publication, New York, 115–148.

Franco, J.A. and Calatrava, J. 2006. Adoption of soil erosion control practices in Southern Spanish olive groves. In: Online of the International Association of Agricultural Economists, Gold Coast, Australia, August 12–18. 16pp.

Friedrich, T. 2020. Chapter 4: The role of no or minimum mechanical soil disturbance in conservation agriculture systems. In: Kassam, A. (ed.), *Advances in Conservation Agriculture* Volume 1: Systems and Science. Burleigh Dodds, Cambridge, UK, 155–178.

Friedrich, T., Kassam, A. and Corsi, S. 2014. Conservation agriculture in Europe. conservation agriculture: In: Jat, R.A., Sahrawat, R.K. and Kassam, A.H. (eds.), *Global Prospects and Challenges*. CABI, Wallingford, UK, 127–179.

Gajri, P.R., Arora, V.K. and S.S. Prihar. 2002. *Tillage for Sustainable Cropping*. International Book Distributing Co. Lucknow, India.

Gao, H., Li, H.W. and Li, W.Y. 2008. Development of conservation tillage. *Transactions of the Chinese Society for Agricultural Machinery* 39(9):43–48.

García-Torres, L. 2001. *Agricultura de Conservación en el Olivar: Cubiertas Vegetales*. Asociación Española Agricultura de Conservación. Suelos Vivos (AEAC/SV), Córdoba, Spain (in Spanish).

Gill, M.A., Ashraf, C.M. and Ahmad, M. 2002. Pakistan's experience in resource conservation technologies for combating drought. In: M.A. Kahlown, N. Yasmin and M. Ashraf (eds), Drought and Water Management Strategies. Proc (Pub. # 122-2002 of PCRWR). Pakistan Council of Research in Water Resources (PCRWR), SAARC Workshop, September 16–18 2002, Pakistan Council of Research in Water Resources, Lahore, Pakistan. 61–75.

Gill, M.A., and Awan, N.A. 2009. Water conservation through resource conservation technologies. In: Project Management and Policy Implementation Unit (PMPIU) of the Ministry of Water and Power Water conservation, present situation and future strategies. Online of the Ministry of Water and Power, Planning Commission and Asian Development Bank National Seminar, 21 May 2009. Project Management and Policy Implementation Unit (PMPIU), Ministry of Water and Power, Islamabad, Pakistan. 76–88.

Goddard, T., Haugen-Kozyra, K. and Ridge, A. 2009. Conservation agriculture protocols for green house gas offsets in a working carbon market. Paper presented at the IV World Congress on Conservation Agriculture, 3-7 February 2009, New Delhi, India.

Godfray, C., Beddington, J.R. and Crute, I.R. 2010. Food security: the challenge of feeding 9 billion people. *Science* 327:812–818.

Gomez, J.A., Llewellyn, C., Basch, G., Sutton, B., Dyson, J.S. and Jones, C.A. 2011. The effects of cover crops and conventional tillage on soil and runoff loss in vineyards and olive groves in several Mediterranean countries. *Soil Use and Management* 27, 502–514.

González-Sánchez, E.J., Ordóñez-Fernández, R., Carbonell-Bojollo, R., Veroz-González, O. and Gil-Ribes, J. 2012. Meta-analysis on atmospheric carbon capture in Spain through the use of conservation agriculture. *Soil and Tillage Research* 122:52–60.

Gonzalez-Sanchez, E.J., Veroz-Gonzalez, O., Blanco-Roldan, G.L., Marquez-Garcia, F., Carbonell-Bojollo, R. 2015. A renewed view of conservation agriculture and its evolution over the last decade in Spain. *Soil and Tillage Research* 146(PB):204–212.

González-Sánchez, E.J., Moreno-García, M., Kassam, A., Holgado-Cabrera, A., TriviñoTarradas, P., Carbonell-Bojollo, R., Pisante, M., Veroz-González, O. and Basch, G. 2017. Conservation agriculture: Making climate change mitigation and adaptation real in Europe. European Conservation Agriculture Federation (ECAF). Available at: http://www.ecaf.org/downloads/books/23-conservation-agriculture-climate-cha nge-report/file.

González-Sánchez, E.J., Veroz-Gonzalez, O., Conway, G., Moreno-Garcia, M., Kassam, A., Mkomwa, S., Ordoñez-Fernandez, R., Triviño-Tarradas, P. and Carbonell-Bojollo, R. 2019. Meta-analysis on carbon sequestration through Conservation Agriculture in Africa. *Soil and Tillage Research* 190:22–30. doi:10.1016/j.still.2019.02.020.

Gragera-Facundo, J., Daza-Delgado, C., Gil-Torralvo, C.G., Gutiérrez-Perera, J.M., and A. Esteban-Perdigón. 2010. Comparing the yield of three pepper cultivars in two growing systems, organic and conventional, in Extremadura (Spain). In: International Symposium on XXVIII International Horticultural Congress on Science and Horticulture for People (IHC2010), 933, Lisbon, Portugal, pp. 131–135.

Griffin, E. and T. Edwards. 2020. What is soil organic carbon? https://www.agric.wa.gov.au/measuring-and-assessing-soils/what-soil-organic-carbon (accessed September 10, 2020).

Haddaway, N.R., Hedlund, K., Jackson, L.E., et al. 2017. How does tillage intensity affect soil organic carbon? A systematic review. *Environmental Evidence* 6:30.

Haque, M.E., Bell, R.W., Jahiruddin, M., Hossain, M.M.; Rahman, M.M., Begum, M., Hossen, M.A., Salahin, N., Zahan, T., and Hossain, M.M. 2018. *Manual for Smallholders' Conservation Agriculture in Rice-based Systems*. Murdoch University, Perth, Australia, p. 108.

He, J., Li, H.W., Rasaily, R.G., Wang, Q.J., Cai, G.H., Su, Y.B., . . . Liu, L. J. 2011. Soil properties and crop yields after 11 years of no tillage farming in wheat-maize cropping system in North China Plain. *Soil and Tillage Research* 113:48–54.

He, J., McHugh, A. D., Li, H. W., Wang, Q. J., Li, W. Y., Rasaily, R. G. and Li, H. 2012. Permanent raised beds improved soil structure and yield of spring wheat in arid north-western China. *Soil Use and Management* 28:536–543.

Huber, S., Prokop, G., Arrouays, D., et al. 2008. *Environmental Assessment of Soil Monitoring Volume I Indicators and Criteria*. EUR 23490 EN/1. Office for Official Publication of the European Communities, Luxembourg, 339 pp.

ITAIPU. 2017. Itaipu dam works to combine energy, environment and biodiversity. Itaipú Dam statement on commitment at COP 23, External Statement on 8 November 2017, Budapest, Hungary (https://unfccc.int/news/itaipu-dam-works-to-combine-energy-environment-and-biodiversity)

ITAIPU. 2019. Cultivando Agua Boa (Growing Good Water). https://www.itaipu.gov.br/en/the-environment/cultivating-good-water (accessed 4 July 2019)

Jat, H.S., Datta, A., Choudhary, M., Sharma, P.C., Yadav, A.K., Choudhary, V., Gathala, M.K., Jat, M.L. and McDonald, A. 2019. Climate Smart Agriculture practices improve soil organic carbon pools, biological properties and crop productivity in cereal-based systems of North-West India. *Catena* 181:104059.

Jat, H.S., Choudhary, K.M., Nandal, D.P., Yadav, A.K., Poonia, T., Yadvinder-Singh, Sharma, P.C., and Jat, M.L. 2020. Conservation agriculture-based sustainable intensification of cereal systems leads to energy conservation, higher productivity and farm profitability. *Environmental Management* 65(6):774–786. https://doi.org/10.1007/s00267-020-01273-w

Jat, R.A., Sahrawat, K.L. and Kassam, A.H. 2014. *Conservation Agriculture: Global Prospects and Challenges*. CABI, Wallingford, UK.

Kassam, A. (ed.) 2020a. *Advances in Conservation Agriculture*. Volume 1: Systems and Science. Burleigh Dodds, Cambridge, UK.

Kassam, A. (ed.) 2020b. *Advances in Conservation Agriculture*. Volume 2: Practice and Benefits. Burleigh Dodds, Cambridge, UK.

Kassam, A. and Kassam, L. 2020a. Chapter 1: The need for conservation agriculture. In: Kassam, A. (ed.), *Advances in Conservation Agriculture*. Volume 1: Systems and Science. Burleigh Dodds, Cambridge, UK, 1–30.

Kassam, A. and Kassam, L. 2020b. Paradigms of agriculture. In: Kassam, A. and Kassam, L. (eds.), *Rethinking Food and Agriculture*. New Ways Forward. Elsevier, Amsterdam.

Kassam, A. and Kassam, L. 2020c. Chapter 1: Practice and benefits of conservation agriculture. In: Kassam, A. (ed.), *Advances in Conservation Agriculture*. Volume 2: Systems and Science. Burleigh Dodds, Cambridge, UK, 1–36.

Kassam, A., Friedrich, T. Shaxson, T. F., Reeves, T., Pretty, J. and Sà, J.C. de M. 2011a. Production systems for sustainable intensification: integrated productivity with ecosystem services. *Technikfolgenabschatzung: Theorie und Praxis* 2:39–45.

Kassam, A., Friedrich, T., Shaxson, F., Bartz, H., Mello, I., Kienzle, J. and Pretty, J. 2014. The spread of conservation agriculture: policy and institutional support for adoption and uptake. *Field Actions Science Reports* 7. http:// factsreports.revues.org/372.

Kassam, A., Mkomwa, S. and Friedrich, T. 2017. *Conservation Agriculture for Africa: Building Resilient Farming Systems in a Changing Climate*. CABI, Wallingford, UK.

Kassam, A., Derpsch, R. and Friedrich, T. 2020a. Development of conservation agriculture systems globally. In: Kassam, A. (ed.), *Advances in Conservation Agriculture*. Burleigh Dodds, Cambridge, UK, 32–86.

Kassam, A., Gonzalez-Sanchez, E., Goddard, T., Hongwen, L., Mello, I., Mkomwa, S., Shaxson, F. and Freidrich, T. 2020b. Harnessing ecosystem services with conservation agriculture. In: Kassam, A. (ed.), Advances in Conservation Agriculture. Volume 2: Practice and Benefits. Burleigh Dodds, Cambridge, UK, 391–418.

Kassam A.H., Friedrich, T., Shaxson, T.F. and Pretty, J.N. 2009. The spread of conservation agriculture: Justification, sustainability and uptake. *International Journal of Agriculture Sustainability* 7(4):292–320.

Kassam, A.H., Mello, I., Goddard, T., Friedrich, T., Laurent, F., Reeves, T. and Hansmann, B. 2011b. Harnessing ecosystem services with conservation agriculture in Canada and Brazil. In: 5th World Congress of Conservation Agriculture incorporating 3rd Farming Systems Design Conference, September 2011, Brisbane, Australia.

Kassam, A.H., Basch, G., Friedrich, T., Shaxson, F., Goddard, T., Amado, T., Crabtree, B., Hongwen, L., Mello, I., Pisante, M. and Mkomwa, S. 2013. Sustainable soil management is more than what and how crops are grown, In: R. Lal and B.A. Stewart (eds.), *Principles of Soil Management in Agro-Ecosystems. Advances in Soil Science*. CRC Press, Taylor & Francis Group, Boca Raton, FL.

Kassam, A.H., Friedrich, T. and Derpsch, R. 2019. Global spread of conservation agriculture. *International Journal of Environmental Studies* 76 (1):29–51 https://doi.org/10.1080/00207233.2018.1494927.

Kautz, T., Amelung, W., Ewert, F., Gaiser, T., Horn, R., Jahn, R., et al. 2013. Nutrient acquisition from arable subsoils in temperate climates: A review. *Soil Biology & Biochemistry* 57:1003–1022.

Lacasta, Dutoit, C. and Meco, M. R. 2005. Interacción entre laboreo y rotaciones de cultivos en ambientes semiáridos en la producción del cereal y las propiedades químicas y bioquímicas del suelo. In: *Actas I Congreso Internacional sobre Agricultura de Conservación*. Asociación Española Agricultura de Conservación/Suelos Vivos, Cordoba, Spain, pp. 423–428.

Lacasta, Dutoit, C., Meco, M., R. and Maire, N. 2005. Evolución de las producciones y de los parámetros químicos y bioquímicos del suelo, en un agrosistema de cereales, sometidos a diferentes manejos de suelo durante 21 años. IN: *Actas I Congreso Internacional sobre Agricultura de Conservación*. Asociación Española Agricultura de Conservación / Suelos Vivos, Cordoba, Spain, pp. 429–436.

Lafond, G.P., Clayton, G.W. and Fowler, B.D. 2014. Conservation agriculture on the Canadian Prairies. In: Jat, R.A., Sahrawat, K.L. and Kassam, A.H. (eds.), *Conservation Agriculture: Global Prospects and Challenges*. CABI, Wallingford, UK, 89–107.

Lal, R. 2014. Societal value of soil carbon. *Journal of Soil and Water Conservation* 69:186A–192A.

Laurent, F., Leturcq, G., Mello, I., Corbonnois, J. and Verdum, R. 2011. La diffusion du semis direct au Brésil, diversité des pratiques et logiques territoriales: l'exemple de la regiόn d'Itaipu au Paraná. *Confins* 12. http://confins.revues.org/7143

Li, H.W., Gao, H.W., Wu, H. D., Li, W. Y., Wang, X. Y. and He, J. 2007. Effects of 15 years of conservation tillage on soil structure and productivity of wheat cultivation in northern China. *Australian Journal of Soil Research* 45:344–350.

Li, H. W., He, J., Bharuchab, Z. P., Lal, R. and Pretty, J. 2016. Improving China's food and environmental security with conservation agriculture. *International Journal of Agricultural Sustainability*. http://dx.doi.org/10.1080/14735903.2016.1170330

Li, L. L., Huang, G. B., Zhang, R. Z., Bill, B., Guangdi, L. and Kwong, Y. C. 2011. Benefits of conservation agriculture on soil and water conservation and its progress in China. *Agricultural Sciences in China* 10:850–859 (in Chinese with English abstract).

Lin, Y., Li, H. P. and Xiao, B. 2017. Influencing factors of wind erosion and their quantitative relationships with erosion rate in black soil area of Northeast China. *Journal of Soil and Water Conservation* 31(4):44–50. DOI: 10.13870/j.cnki.stbcxb.2017.04.008.

Lindwall, C.W. and B. Sonntag (eds.) 2010. *Landscape Transformed: The History of Conservation Tillage and Direct Seeding*. Knowledge Impact in Society, University of Saskatchewan, Saskatoon, Canada.

López-Garrido, S., Madejón, E., Murillo, J. M. and F. Moreno. 2011. Soil quality alteration by moldboard plowing in a commercial farm devoted to no-tillage under Mediterranean conditions. *Agriculture, Ecosystem and Environment* 140 (1–2):30, 182–190.

Magfoff, F. and Weil, R.R. 2004. Soil organic matter management strategies. In: Magfoff, F. and Weil, R.R. (Eds.) *Soil Organic Matter in Sustainable Agriculture* CRC Press, Boca Raton, Florida, USA, pp. 45–65.

MAPA. 2019. Encuesta sobre Superficies y Rendimientos de Cultivos 2018 (Ministry for Agriculture of Spain, Survey on surfaces and crop yields). https://www.mapa.gob.es/es/estadistica/temas/estadisticas-agrarias/boletin2018_tcm30-504212.pdf (in Spanish)

Mapama. 2018. Iniciativa 4 por mil: el carbono orgánico del suelo como herramienta de mitigación y adaptación al cambio climático en España. *Ministerio de Agricultura y Pesca, Alimentación y Medio Ambiente*. https://www.miteco.gob.es/images/es/4por1000_tcm30-438109.pdf (accessed September 10, 2020).

Millennium Ecosystem Assessment (MEA). 2005. *Ecosystems and Human Well-Being: Synthesis*. Millennium Ecosystem Assessment. Island Press, Washington, DC.

Mello, I. and van Raij, B. 2006. No-till for sustainable agriculture in Brazil. *Online of the World Association of Soil and Water Conservation* P1:49–57.

Mello, I., Roloff, G., Laurent, F., Gonzalez-Sanchez, E. and Kassam, A. 2020. Sustainable land management with conservation agriculture for rain-fed production: The case of Paraná III Watershed (Itaipu dam) in Brazil. *Contribution to the Background Report to the FAO ESA 2020 SOFA Report on Agricultural Water Management, under a Contract between the Swedish University of Agricultural Sciences (SLU) and the European Conservation Agriculture Federation (ECAF)*. June 2020.

Mestelan, S.A., Smeck, N.E., Durkalski, J.T. and Dick, W.A. 2006, Changes in soil profile properties as affected by 44 years of continuous no-tillage. In: Online of the 17th ISTRO Conference, Kiel, Germany, 28 August–3 September 2006.

Micó, C., Recatalá, L., Peris, M. and J. Sánchez. 2006. Assessing heavy metal sources in agricultural soils of a European Mediterranean area by multivariate analysis. *Chemosphere 65*:863–872.

Minasny, B., McBratney, A.B., Wadoux, A.M.J.-C., Nyak Akoeb, E. and T. Sabrina. 2020. Precocious 19th century soil carbon science. *Geoderma Regional* 22: e00306

Montgomery, D.R. 2007. *Dirt: The Erosion of Civilizations*. Berkeley and Los Angeles: University of California Press. 285 pp.

Mrabet, R., Saber, N., El-Brahli, A., Lahlou, S. and F. Bessam. 2001. Total, particulate organic matter, and structural stability of a Calcixeroll soil under different wheat rotations and tillage systems in a semiarid area of Morocco. *Soil and Tillage Research* 57:225–235.

Nkonya, E., Mirzabaev, A. and von Braun, J. (eds.) 2016. *Economics of Land Degradation and Improvement: A Global Assessment for Sustainable Development*. IFPRI and ZEF. Springer Open, New York, 695 pp. (Online). http://link.springer.com/book/10.1007/978-3- 319-19168-3/page/1

Nyanga, P.H., Umar, B.B., Chibamba, D., Mubanga, K.H., Kunda-Wamuwi, C.F. and B.M. Mushili. 2020. Chapter 6: Reinforcing ecosystem services through conservation agriculture in sustainable food systems.

In: L. Rusinamhodzi (ed.), *The Role of Ecosystem Services in Sustainable Food Systems*. Academic Press, San Diego, CA, 119–133.

Oldeman, L.R., Hakkeling, R.T.A. and Sombroek, W.G. 1991. *World Map of the Human-Induced Soil Degradation: An Explanatory Note*. International Soil Reference and Information Centre (ISRIC), Wageningen, the Netherlands.

Ontl, T. A. and Schulte, L.A. 2012. Soil carbon storage. *Nature Education Knowledge* 3(10):35.

Ordóñez Fernández R., González P., Giráldez J.V. and F. Perea. 2007. Soil properties and crop yields after 21 years of direct drilling trials in Southern Spain. *Soil and Tillage Research* 94, 47–54.

Ordóñez Fernández, R, González Fernández, P. and R. Carbonell Bojollo. 2010. Evolución de la materia orgánica y propiedades químicas del suelo. In: *Aspectos agronómicos y medioambientales de la Agricultura de Conservación*. Eumedia y Ministerio Medio Ambiente y Medio Rural y Marino, Madrid, Spain, pp. 67–76.

Owenya, M. Z., Mariki, W. L., Kienzle, J., Friedrich, T. and Kassam, A. 2011. Conservation Agriculture (CA) in Tanzania: The case of the Mwangaza CA farmer field school (FFS), Rhotia Village, Karatu District, Arusha. *International Journal of Agricultural Sustainability* 9:145–52.

Pardo, G., Perea, F., Martínez, Y. and J.M. Urbano. 2014. Economic profitability analysis of rain-fed organic farming in SW Spain. *Outlook on Agriculture* 43(2):115–122.

Pittelkow, C., Xinqiang, L., Lindquist, B., van Groeningen, K.J., Lee, J., Lundy, M.E., van Gestel, N., Six, J., Venterea, R.T. and van Kessel, C. 2014. Productivity limits and potentials of the principles of conservation agriculture. *Nature Letter* 517:365–368.

Podmanicky, L., Balázs, K., Belényesi, M., Centeri, Cs., Kristóf, D. and N. Kohlheb. 2011. Modelling soil quality changes in Europe. An impact assessment of land use change on soil quality in Europe. *Ecological Indicators* 11 (1):4–15.

Pretty, J.N. 2002. *Agri-Culture: Reconnecting People, Land and Nature*. Earthscan, London.

Pretty, J.N., Toulmin, C. and Williams, S. 2011. Sustainable intensification in African agriculture. *International Journal of Agricultural Sustainability* 9(1):5–24.

Ramesh, T., Bolan, N.S., Kirkham, M.B.E., et al. 2019. Chapter 1: Soil organic carbon dynamics: Impact of land use changes and management practices: A review. In: D. L. Sparks (ed.), *Advances in Agronomy* Volume 156. Academic Press, San Diego, CA, 1–107.

Reicosky, D.C. 2003. Conservation agriculture: Global environmental benefits of soil carbon management. In: L. García-Torres, J. Benites, A. Martínez-Vilela, A. Holgado-Cabrera (eds.), *Conservation Agriculture*. Springer, Dordrecht.

Reicosky, D.C. 2011. Conservation agriculture: Global environmental benefits of soil carbon management. In: Fifth World Congress on Conservation Agriculture, Brisbane, Australia. Vol. 1. ACIAR, Canberra, ACT, pp. 3–12.

Reicosky, D.C. 2020. Conservation agriculture systems: Soil health and landscape management. In: Kassam, A. (ed.), *Advances in Conservation Agriculture Volume: Science and Systems*. Burleigh Dodds, Cambridge, UK, 87–154.

Rhoton, F. E. 2000. Influence of time on soil response to no-till practices. Soil Science Society of America Journal 64:700–709.

Riley, H.C.F., Bleken, M.A., Abrahamsen, S., Bergjord, A.K., and A.K. Bakken. 2005. Effects of alternative tillage systems on soil quality and yield of spring cereals on silty clay loam and sandy loam soils in cool, wet climate of central Norway. *Soil and Tillage Research* 80:79–93.

Rochecouste, J. and Crabtree, W.L. 2014. Conservation agriculture in Australian dryland cropping. In: Jat, R.A., Sahrawat, R.K. and Kassam, A.H. (eds.), *Conservation Agriculture: Global Prospects and Challenges*. CABI, Wallingford, UK, 108–126.

Roldan, A., Caravaca, F., Hernández, M.T., et al. 2003. No-tillage, crop residue additions, legume cover cropping effects on soil quality characteristic under maize in Patzcuaro watershed (Mexico). *Soil and Tillage Research* 72:65–73.

Roloff, G., Lutz, R.A. T. and Mello, I. 2011. An index to rate the quality of no-till systems: A conceptual framework. In: 5th World Congress of Conservation Agriculture incorporating 3rd Farming Systems Design Conference, September 2011, Brisbane, Australia.

Sa, J.C. de M., Lal, R., Ferreira, A. de O., Briedis, C., Inagaki, T.M., Goncalves, D.P. and Romaniw, J. 2020a. Chapter 7: Carbon management practices and benefits in Conservation Agriculture systems: Carbon sequestration rates. In: Kassam, A. (ed.), *Advances in Conservation Agriculture*. Burleigh Dodds, Cambridge, UK, 199–228.

Sa, J.C. de M., Lal, R., Ferreira, A. de O., Briedis, C., Inagaki, T.M., Goncalves, D.P. and Romaniw, J. 2020b. Chapter 8: Carbon management practices and benefits in conservation agriculture systems: Soil organic carbon fractions. In: Kassam, A. (ed.), *Advances in Conservation Agriculture.* Burleigh Dodds, Cambridge, UK, 229–266.

Saavedra, C., Velasco, J., Pajuelo, P., Perea, F. and A. Delgado. 2007. Effects of tillage on phosphorus release potential in a Spanish vertisol. *Soil Science Society of America Journal* 71:56–63.

Schjønning, P., Jensen, J.L., Bruun, S., et al. 2018. Chapter 2: The role of soil organic matter for maintaining crop yields: Evidence for a renewed conceptual basis. In: D. L. Sparks (ed.), *Advances in Agronomy*, Vol 150. Academic Press, San Diego, CA, 35–79.

Schlesinger, W.H. and J.A. Andrews. 2000. Soil respiration and the global carbon cycle. *Biogeochemistry* 48(1), 7–20.

Seremesic, S., Milosev, D., Djalovic, I., Zeremski, T., and J. Ninkov. 2011. Management of soil organic carbon in maintaining soil productivity and yield stability of winter wheat. *Plant, Soil and Environment* 57(5):216–221.

Sombrero, A., De Benito, A., González, I., and M.A. Álvarez. 2006. Influencia del laboreo sobre las propiedades químicas del suelo en Agricultura de Conservación. *Agricultura de Conservación* 2:34–38.

Sorrenson, W.J. & Montoya, L.J. 1991. Chapter about the economics of tillage practices In: Derpsch, R., Sidiras, C. H. N. and Kopke, U. (eds.), *Controle Da Erosão No Paraná, Brasil: Sistemas De Cobertura Do Solo, Plantio Direto E Preparo Conservacionista Do Solo.* GTZ, Eschborn, 165–192.

Sorrenson, W.J., Portillo, J.L. & Núñez, M. 1997. *Economics of No-tillage and Crop Rotations. Policy and Investment Implications.* Final Report, 1 October 1997. FAO Report No.: 97/075 ISP-PAR. [Cited 28 May 2019]. http://www.rolf-derpsch.com/en/no-till/economics-mechanized/

Swallow, B.M. and Goddard, T.W. 2016. Development of Alberta's greenhouse gas offset system within Canadian and international climate policy contexts. *International Journal of Climate Change Strategies and Management* 8(3):318–337.

Tabatabai, M.A. 1996. Soil organic matter testing: An overview. In: F Magdoff, et al. (eds.), *Soil Organic Matter: Analysis and Interpretation*, 1–10. SSSA Spec. Pub. 46. American Society of Agronomy, Madison, WI.

Tan, C. J. 2015. Effects of conservation tillage on the soil nutrients maintenance and control of water erosion of sloping farmland. Master Dissertation. Northwest A&F University, Yangling, China (in Chinese with English abstract).

Thierfelder, C., Mwila, M., Rusinamhodzi, L., 2013. Conservation agriculture in eastern and southern provinces of Zambia: long-term effects on soil quality and maize productivity. *Soil and Tillage Research* 126, 246–258.

Thierfelder, C., Rusinamhodzi, L., Ngwira, A.R., Mupangwa, W., Nyagumbo, I., Kassie, G.T., Cairns, J. E. 2014. Conservation agriculture in Southern Africa: Advances in knowledge. *Renewable Agriculture and Food Systems* 30(4): 1–21. doi:10.1017/S1742170513000550

Thierfelder, C., Matemba-Mutasa, R. and Rusinamhodzi, L. 2015. Yield response of maize (Zea mays L.) to conservation agriculture in Southern Africa. *Soil and Tillage Research* 146, 232–240.

Vanwalleghem, T., Laguna, A., Giraldez, J. V. and Jimenez-Hornero, F. J. 2010. Applying a simple methodology to assess historical soil erosion in olive orchards. *Geomorphology* 114 (3):294–302.

Vanwalleghem, T., Amate, J. I., Gonzalez de Molina, M., Fernandez, D.S. and Alfonso Gomez, J. 2011. Quantifying the effect of historical soil management on soil erosion rates in Mediterranean olive orchards. *Agriculture Ecosystems & Environment* 142 (3–4):341–351.

Vlek, P.L.G., Le, Q.B. and Tamene, L. 2010. Assessment of land degradation, its possible causes and threat to food security in sub-Saharan Africa. In: Rattan Lal and B.A. Stewart (eds.), *Advances in Soil Science: Food Security and Soil Quality.* CRC Press, Boca Raton, FL, 57–86.

Wallace, A. 1994. Soil organic matter is essential to solving soil and environmental problems. *Communications in Soil Science and Plant Analysis* 25:15–28.

Wang, L. X., Jiang, X., Li, Y. and Ji, H. Y. 2014a. Effects of conservation tillage on soil water erosion and soil compaction degree. *Journal of Irrigation and Drainage* 33:83–85.

Wang, R.D. Xiao, D.P. Chang C.P., et al. 2014b. Estimation on farmland soil loss by single wind erosion using improved particle-size distribution comparison method. *Transactions of the CSAE* 30:278–285 (in Chinese with English abstract).

Yang, C.H., Feng, F.X., Cai, Q., et al. 2019. Soil wind erosion in wheat/maize intercropping with zero tillage and straw retention. *Journal of Desert Research* 39:9–15 (in Chinese with English abstract).

Zhang, G. Y., Lyu, B. B., Zhang, L. P., Liu, Z., Fan, Q. L, Wei, M. F., Yao, Z., Yuan, J. W. and Chai, Y. J. 2019. Effect of long-term no-tillage with stubble on soil fertility and diversity of prokaryotic microbiome in dryland wheat soils on the Loess Plateau, China. *Chinese Journal of Eco-Agriculture* 27:358–368 (in Chinese with English abstract).

5 The Effects of Soil Organic Matter and Organic Resource Management on Maize Productivity and Fertilizer Use Efficiencies in Africa

Shamie Zingore, Samuel Njoroge, Stephen Ichami, Kokou Adambounou Amouzou, James Mutegi, Regis Chikowo, Sudarshan Dutta, and Kaushik Majumdar

CONTENTS

5.1 INTRODUCTION

Sub-Saharan Africa (SSA) is experiencing rising food insecurity as a result of low crop productivity and a rapidly growing population (FAO and ECA 2018; Van Ittersum et al. 2016). Yields of cereal crops in SSA are currently less than 30% of attainable yields (Van Ittersum et al. 2013). The yields of maize, the most important cereal crop in SSA, have stagnated at about 2 Mg ha^{-1} compared to 3.5 Mg ha^{-1} in Latin America and South Asia, 5.2 Mg ha^{-1} in China, and more than 15 Mg ha^{-1}

DOI: 10.1201/9781003102762-5

in Europe (Pasuquin et al. 2014; Van Ittersum et al. 2013). The low levels of crop productivity are exacerbated by an inherently poor soil resource base that has been subjected to soil fertility depletion and land degradation for many decades due to poor management including low nutrient application (Tittonell and Giller 2013; Van der Velde et al. 2014; Muller et al. 2012). It is estimated that a large part of the arable land in SSA is degraded, with large areas characterized by critically low soil organic matter (SOM) (Bationo and Fening 2018). Low soil fertility presents one of the most pressing challenges for sustainably increasing crop productivity in SSA. Under these conditions, increased use and judicious management of fertilizer and organic nutrient resources are essential to optimizing crop productivity, using fertilizer efficiently, and recapitalizing SOM and nutrients.

Crop productivity and nutrient use efficiencies depend on several interacting crop, environmental, and management factors that include crop and varieties, soil and climate conditions, and nutrient management practices. Crop fertilizer response and returns on fertilizer investment in smallholder maize farming in most of SSA exhibit a high degree of variability and uncertainty due to poor soil fertility conditions and seasonal weather variability (Cairns et al. 2013; Rurinda et al. 2020). Improved understanding of the influence of soil fertility on crop productivity and fertilizer use efficiency is critical for targeting appropriate crop and nutrient management interventions for specific growing conditions. Soil fertility is intricately associated with SOM, the status of which is governed by biomass input and management, mineralization, leaching, and erosion (Bationo and Fenning 2018). Due to its strong influence on soil biological, chemical, and physical properties, as well as crucial soil functions, SOM is regarded as an important indicator of soil fertility with direct implications for crop productivity, food security, as well as human livelihoods (Lorenz and Lal 2016; Amundson et al. 2015; Koch et al. 2013; McBratney et al. 2014; Lal 2016). SOM is associated with increased microbial biomass and enzyme activity (Chander et al. 1998) and serves as a natural reservoir of nitrogen, phosphorus, sulfur, and other nutrients (Donahue et al. 1983). SOM also supports and regulates essential soil ecosystem services such as nutrient cycling, water infiltration and storage, improves soil structure, and protects the soil against erosion (Bayala et al. 2018; Kihara et al. 2020; Lal 2016).

Recent reviews by Schjønning et al. (2018) and Oldfield et al. (2019) provide evidence of the role of SOM in maintaining crop yields in Danish farming systems and at a global level. The role of SOM in enhancing the productive capacity of the soils in SSA has been highlighted in several studies (e.g. Bayala et al. 2018; Mtangadura et al. 2017; Nandwa 2001; Zingore et al. 2007a). Studies in SSA have also shown that SOM in smallholder maize production systems is highly variable (Castellanos-Navarrete et al. 2014; Corbeels et al. 2019; Okeyo et al. 2009; Zingore et al. 2008). Variability in SOM can be attributed to differences in management practices, soil types, and landscape (Godde et al. 2016). Despite the general trend of depletion of soil organic carbon (SOC) across smallholder farms (Bationo et al. 2007; Smaling et al. 1997), differences in management practices at the farm and field scales result in complex short-range differences in SOM that account for profound heterogeneity in crop productivity and fertilizer use efficiency within production systems (Diarisso et al. 2016; Tittonell et al. 2013; Zingore et al. 2007b; Chikowo et al. 2014). A major challenge for maize production intensification in SSA is the high occurrence of unproductive degraded soils that respond poorly to the application of fertilizer and organic resources, a phenomenon that has been attributed to depletion of SOM (Ichami et al. 2018; Zingore 2011).

Maize is one of the most important food crops in SSA occupying about 15% of all cropped land and accounting for about 40% of the total cereal crop production (FAO 2020). Maize also receives the largest proportion of the fertilizers used in SSA. From economic and environmental perspectives, mineral fertilizers and organic resources should be targeted optimally within the cropping systems to ensure high productivity and high nutrient use efficiencies. The relationship between SOM and crop productivity and fertilizer use efficiency in smallholder farming systems is not well established and is confounded by complex soil constraints (Kihara et al. 2016) that severely curtail the scope for improved nutrient management decisions. This paper reviews the status and determinants of the SOM status in maize-based farming systems and examines its critical role for maintaining the fertility and productivity of soils in SSA and the implications for nutrient management decision support. We also review previous studies and case study datasets of maize fertilizer trials to assess the extent to which SOM regulates maize productivity and fertilizer use efficiency and gain insights into the role of integrated soil fertility management practices in optimizing the agronomic performance of smallholder maize-based farming systems in SSA.

5.1.1 Review Approach

A review of key publications on the status and drivers of SOM in crop production systems in SSA was conducted in the first literature review step. This was followed by a review of the conceptual relationships between SOC, nutrient management practices and maize yields, fertilizer responses, and nutrient use efficiencies. Subsequently, a detailed meta-analysis explored the general trends and relationships of SOC with maize yields and yield response to fertilizer application. For this step, we searched online literature in Google, ScienceDirect, and Mendeley databases using strings related to SOM or SOC or maize yield change or smallholder farming systems influenced by agronomic nutrient management practices. We restricted the search to English language articles published for SSA between 1970 and 2020. In the second approach we purposely searched past meta-analysis articles and associated databases (e.g. Ichami et al. 2018; Kihara et al. 2017; Rusinamhodzi et al. 2011). The meta-analyses associated data were uploaded into repositories like the Harvard Dataverse (https://dataverse.harvard.edu/.). This data was further searched using the aforementioned criteria for additional datasets across SSA.

Although there are several studies that examined soil fertility degradation in smallholder farming landscapes in SSA, there is little standardization in the sampling design or conducted soil analysis. To avoid wide variation in the dataset, specific studies were selected based on agronomic management. Data from searched fertilization experiments across countries in SSA were collated using the above-mentioned approaches. Data from different fertilizer trials were extracted and key variables, such as maize yield of control plots, reported variances (standard deviation or coefficient of variation); maize yield of fertilized treatment and fertilizer rate were included in the database. We also included known agronomic management practices (factors): the cropping systems, conservational tillage management, methods of fertilizer application, and integrated soil fertility management (ISFM) options (germplasm, inorganic fertilizers, combined application) in the database. Each row in the database contained maize yield from paired treatments including a fertilized and an unfertilized control treatment, with their corresponding agronomic management parameters, only when recorded. Thus, this database facilitated the evaluation of the effect of regulating SOC management on maize yield response to fertilizer application for a wide range of production factors.

Statistical analyses were conducted in the R computing environment (R Core Team 2020). Fertilizer response (FR) was computed using the "metafor" package (Viechtbauer 2010). The FR was calculated as a ratio (natural log (ln)) of the mean maize yield of the fertilized plot ($\overline{x}_t$ in kg ha^{-1}) and the mean maize yield of the unfertilized control plot ($\overline{x}_c$ in kg ha^{-1}) following Hedges et al. (1999):

$$\ln FR = \ln\left(\frac{\overline{x}_t}{\overline{x}_c}\right). \tag{5.1}$$

Correlations between maize yield for the control, SOC concentration, and fertilizer response were analyzed using the "ggplot2" package (Wickham 2016). Weighted means for fertilizer response and SOC were calculated separately and used to evaluate the influence of agronomic management practices on their variability. A random effect meta-regression model was used to quantify variability in the dataset prior to the analysis. Multiple regression models were developed using FR as the response variable to deduce the relation and interaction between SOC and maize yield in the unfertilized control plot.

Analysis of data at the continental level was confounded by high variability and complexity in datasets. We therefore complemented the meta-analysis study by a review of specific case studies in southern, East, and West Africa to draw insights from relatively uniform geographic zones that are representative of major maize production systems across the continent. The case studies within regions also provided datasets from trials with consistent treatments and comparable methods for SOC analysis.

5.2 THE STATUS OF SOIL ORGANIC MATTER IN SUB-SAHARAN AFRICAN CROPPING SYSTEMS

Sub-Saharan Africa covers an expansive area of 24.6 million km^2, with a wide array of soil and land management types. The predominant soils are characterized by low inherent fertility with small natural nutrient reserves due to a bedrock that consists mostly of granites and gneiss (Tully et al. 2015; van Straaten 2011). The rocks are some of the oldest in the world and are highly weathered, giving rise to a large area of soils with clay fractions dominated by kaolinite and sesquioxide, low nutrient content, high acidity, and high P-fixing capacities (Deckers et al. 2000). The nutrient stocks in uncultivated soils under native vegetation are small due to very low (<10%) weatherable minerals. Small areas of more fertile soil are associated with recent volcanic activity that occurred mainly in the eastern Africa highland areas (Bationo et al. 1998; Brouwer and Powell 1998).

The low productivity of native forest and woodlands and rapid soil organic carbon (SOC) turnover stimulated by high soil temperature and microfauna activity, in the tropical and sub-tropical agroecological zones, account for low SOC contents even under the natural vegetation state (Bationo et al. 2003; Zingore et al. 2005). Texture is a major driver of SOC contents under natural vegetation due to the influence of clay and silt fractions on SOC stabilization (Korschens et al. 1998). In soils of the tropics, the clay and silt contents, particularly the kaolinitic types, have a major bearing on the level of SOC (Christensen 1992; Feller and Beare 1997; Zinn et al. 2007). The contents of SOC also depend on the agroecological conditions, which influence the quantity and quality of organic matter inputs available. For example, in West Africa, the SOC contents in the topsoil layer under native vegetation average 1.2% for the humid zone, decrease to 0.7% for the sub-humid zone, and to 0.4% in the semi-arid zone (Windmeijer and Andriesse 1993). The variable agroecological and soil conditions give rise to low but highly variable SOC stocks at the continental scale, with SOC contents ranging from very low levels in the deserts of the Sahara, the Kalahari, and the Horn of Africa, to relatively high in the wetlands and floodplain areas (Figure 5.1). Within arable areas, SOC contents in over 80% of the arable land are often below the general critical levels of 2%. In the agroecological zones that are important for maize production, the soils of the tropical zone contain slightly more SOC than the soils of the sub-humid savanna zones.

Driven by rapid population growth and the corresponding increase in anthropic pressure on arable land, forests and protected areas and woodlands are increasingly being converted to agricultural production to meet the growing food demand (Forkuor 2014; Liniger et al. 2011), relying largely on the beneficial effects of SOM and nutrient reserves (Bationo and Fening 2018). Land use change when native vegetation areas are converted to agriculture is commonly associated with loss of SOC. Davidson and Ackerman (1993) reported a 20–30% global mean decline of SOC stock with the introduction of agriculture. Greater decline in SOC is often observed in smallholder maize production systems of SSA. For example, studies using chronosequences of land clearance showed that more than 50% of the SOC originally in uncultivated soils was lost in the initial five years of cultivation under low-input continuous maize production. This results in critically low SOC equilibrium levels that are 50–80% below the threshold levels under native vegetation (Zingore et al. 2005; Nyberg et al. 2012). The rapid loss of SOC and degradation of soil quality in the low-input systems are due to several factors, including reduced recycling of organic matter, accelerated decomposition, exposure of aggregate protected organic matter and accelerated soil erosion (Feller et al.

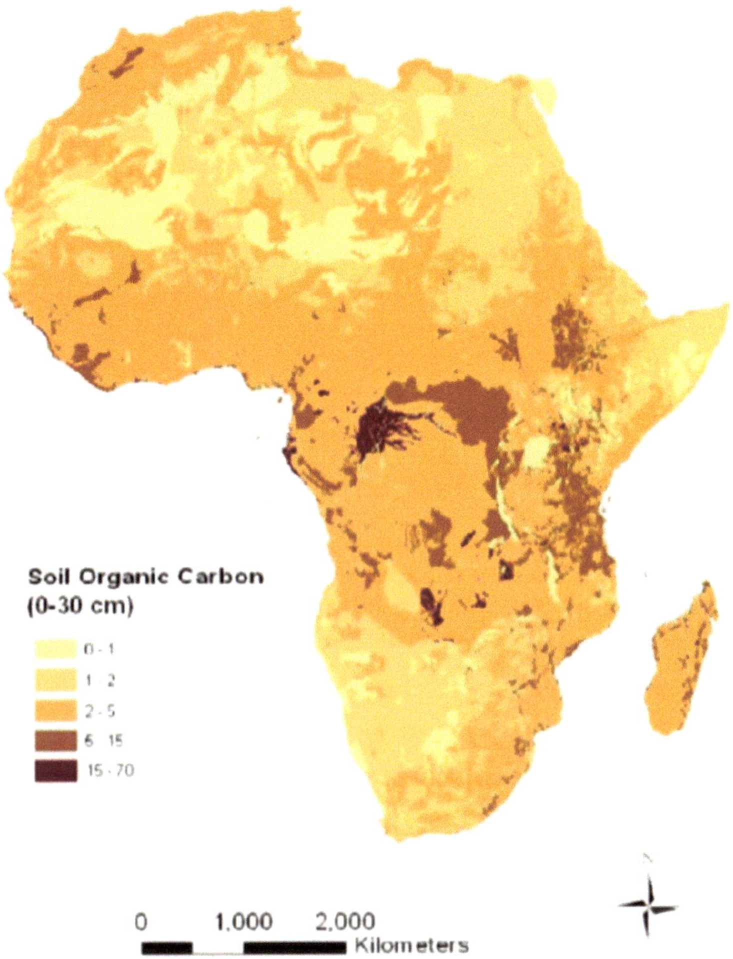

FIGURE 5.1 Spatial distribution of SOC stocks (g kg^{-1}) in Africa (*Source:* Henry et al. 2009).

1996). Similar to threshold SOC levels in uncultivated soils, equilibrium levels in cultivated soils are highly dependent on the clay and silt contents of the soils. Small concentrations of SOC occur in sandy soils that offer little physical protection (Hassink 1997). In a long-term trial conducted for 90 years, sandy soils with 3% clay content resulted in an equilibrium SOC of 0.7%, but equilibrium levels were threefold higher at 2% for soils with 21% clay (Körchens et al. 1998). Zingore et al. (2005) showed even lower equilibrium levels of 0.5% and 0.9% for soils with 3–5% and 32% clay contents, respectively, in low-input maize production systems in Zimbabwe.

Despite the overall low SOC contents that characterize low-input maize production systems in SSA, SOC levels at the farm- and field-scale are highly nuanced and variable. Smallholder farming systems in SSA are characterized by a wide diversity of farming households, and the availability and use intensity of organic and mineral fertilizer nutrient inputs varies widely between farm types, leading to variation in soil fertility and SOM at the farm level (Zingore et al. 2007a; Tittonell et al. 2005). Resource-endowed farmers use larger amounts of mineral fertilizers and manure, giving rise to fields with higher SOM levels than fields of resource-constrained farmers (Mtambanengwe and Mapfumo 2005; Tittonell et al. 2005). Mtambanengwe and Mapfumo (2005) showed that mean

SOC contents were >0.6% for fields receiving high organic matter and fertilizer inputs and were less than 0.46% for fields cultivated continuously with low nutrient application. SOC contents in fields belonging to resource-endowed farmers were 28% higher than those for fields of resource-constrained farmers, confirming differences in organic matter management practices linked to farm types. Differences in fertility status of fields due to management have been observed to increase with increasing time of cultivation, suggesting that the observed SOC gradients are a cumulative effect of years of differential resource management by different farm types. Differences in management intensity also occur within farms. This arises from preferential nutrient application to sections of fields on the farm, usually those close to homesteads (Zingore et al. 2007b). Differential resource management within and between farms is a key driver for SOC levels, with major implications on nutrient use efficiencies and crop productivity (Smaling and Braun 1996; Njoroge et al. 2018).

5.3 NITROGEN USE EFFICIENCY IN THE CONTEXT OF INTEGRATED SOIL FERTILITY MANAGEMENT

There is recognition that sustainable crop production intensification in smallholder farming systems in SSA will require effective integrated soil fertility management (ISFM) technologies that not only supply nutrients required by crops, but also capitalize on the complementarity between fertilizer and organic nutrient resources and rebuild soil nutrient stocks and soil organic matter in the long term (Vanlauwe et al. 2010; Kumwenda et al. 1997). While increased use of mineral fertilizers is indispensable, available mineral fertilizers on their own are inadequate to sustainably increase crop yields and optimize fertilizer use efficiency. The limitations of fertilizer can be attributed to the multiple and complex biophysical and socioeconomic constraints and their interactions on the crop productivity of smallholder farms (Dutta et al. 2020). Organic resources provide an essential complimentary nutrient resource due to their multiple benefits that include supply of nutrients, buffering of soil pH and increasing soil organic matter (Bayu et al. 2005). Symbiotic biological N_2 fixation (BNF) is an option for supplementing fertilizer N due to supply of organic resources with high N contents (Giller 2001). Several technologies based on BNF, including agroforestry, green manures, and rotation or intercropping of cereal crops with grain legumes can provide substantial N and act as a complementary source of N.

ISFM encompasses

> a set of soil fertility management practices that necessarily include the use of fertilizer, organic inputs, and improved germplasm combined with the knowledge on how to adapt these practices to local conditions, aiming at maximizing agronomic use efficiency of the applied nutrients and improving crop productivity.
>
> **(Vanlauwe et al. 2010)**

The ISFM concept focuses on increased crop productivity and agronomic fertilizer use efficiency as key performance indicators and considers differences in the productive capacity and response to fertilizer of different soils (Figure 5.2). Responsive soils, which support significant crop productivity response to fertilizer, as depicted by the inflection point A are distinguished from the non-responsive ones, inflection point B, which are characterized by low crop productivity and low fertilizer response, even when improved high-yielding crop varieties are used. In the case of non-responsive soils, management practices for correcting complex constraints and soil fertility restoration through, for example, application of organic resources, are a prerequisite for increasing crop productivity and fertilizer use efficiency (inflection point C) (Zingore et al. 2008; Zingore 2011). Mineral and organic nutrient resources have different effects on soil fertility and crop production, and their use in combination can have synergistic effects, which lead to greater nutrient use efficiencies than when used separately, both in the short and long term (Giller 2002; Vanlauwe 2004).

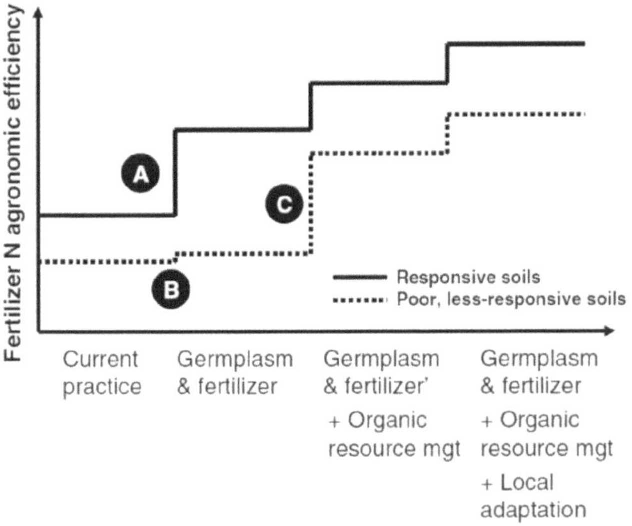

FIGURE 5.2 Conceptual framework of ISFM, with the conceptual relationship with agronomic use efficiency. Framework of ISFM showing the conceptual relationship between management practices and agronomic nitrogen use efficiency. (*Source:* Vanlauwe et al. 2010).

5.4 FERTILIZER RESPONSE AND AGRONOMIC N USE EFFICIENCY PATTERNS

Agronomic fertilizer use efficiency is intricately related to soil quality. A conceptual illustration of the relationship between SOC, used as a proxy for soil quality, and crop yields and agronomic nitrogen use efficiency (AEN) postulates three categories of response that can form the basis of optimizing fertilizer management; namely (i) non-responsive soils with very low SOC contents, (ii) non-responsive soils with very high SOC contents, and (iii) high responsive soils with moderate SOC contents (Tittonell et al. 2008; Vanlauwe et al. 2010; Musinguzi et al. 2013). Two zones of SOC conditions that respond poorly to fertilizer and display low N use efficiency are identified in Figure 5.3. At the lower end of the SOC spectrum, very low SOC levels are associated with complex chemical, physical, and biological constraints that severely limit fertilizer response. At the other end of the spectrum, very high SOC levels result in high N mineralization rates and sufficient soil N supply to meet the N required to achieve the attainable yields. Soils in the "non-responsive fertile" category are not common in smallholder farming systems in SSA and are only found in small hotspots of nutrient accumulation in fields that receive high amounts of organic resources and fertilizer (Tittonell et al. 2008). For resource-constrained systems, strategies for nutrient management optimization should focus on striking a balance between maintaining SOC above a critical value to ensure high agronomic fertilizer use efficiency, while avoiding concentration of nutrient resources to levels that occlude viable N use efficiency.

In practice, the establishment of critical SOC contents is confounded by highly variable maize yield responses to N application along the gradients of SOC as a consequence of the complexity of management, soil, and climatic factors. Analysis of AEN in fertilizer response trials in maize production in East, West, and southern Africa showed highly variable values, ranging from −4 kg grain kg^{-1} N to 146 kg grain kg^{-1} N, depending on agronomic management standards, type of maize varieties used, combination of fertilizer with organic and inorganic fertilizer, and targeting fertilizer to fields managed differently in the past (Vanlauwe et al. 2010).

AEN on the majority of smallholder farms in SSA is often less than 15 kg grain kg^{-1} N against the >40 kg grain kg^{-1} N that is frequently achieved under optimally managed on-station experiments (Mushayi et al. 1998). Despite the wide variability in soil fertility status, fertilizer recommendations

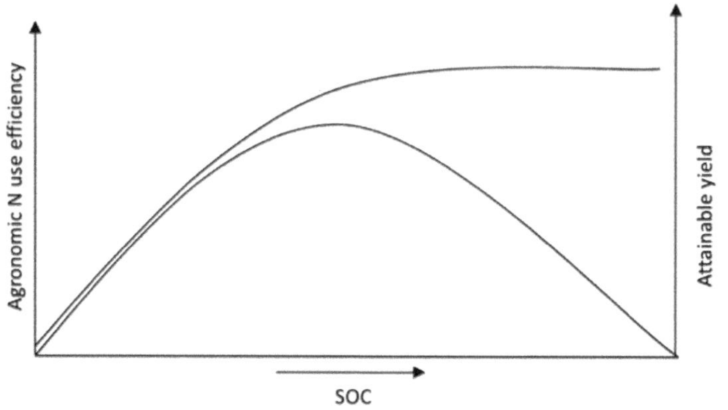

FIGURE 5.3 A conceptual relationship between SOC levels, yields and agronomic N use efficiency of crops (Adapted from: Musinguzi et al. 2013).

TABLE 5.1
Effects of Fertilizer on Maize Yields (Mg ha^{-1}) on Fields Differing in Fertility Status in Selected Countries in Southern Africa

	Unfertilized Maize		Fertilized Maize[1]		Agronomic Fertilizer Use Efficiency	
Country	Depleted Fields	Fertile Fields	Depleted Fields	Fertile Fields	Depleted Fields	Fertile Fields
Malawi	0.5	2.9	3.0	5.2	27	25
Zambia	0.6	1.0	2.9	5.6	23	46
Zimbabwe	0.2	1.8	1.8	7.5	14	51

(*Source:* Siame et al. 1997; Benson 1999; Weil and Mugogo 2000; Snapp and Silim 2002; Chikowo et al. 2004; Robertson et al. 2005; Zingore et al. 2007b)

[1] Data for experiments with high fertilizer application rates (>100 kg N ha^{-1}), to target non-limiting N for attainable yields.

for maize in SSA are mainly blanket, based on attainable yields, as defined by, for example, rainfall. This overlooks key factors that affect fertilizer use efficiency and different access to inputs by farmers. Soil fertility, as determined by management history, strongly affects fertilizer use efficiency and a wide range of studies in southern Africa have shown that fertilizer use efficiency is higher on fertile soils compared to degraded soils (Table 5.1). The exception here was in Malawi, where the difference in agronomic fertilizer use efficiency between fertile and degraded fields was small, mainly due to relatively lower fertility on fields considered to be fertile. Infertile soils cultivated continuously with low additions of organic matter exhibit problems of soil acidity and multiple deficiencies of nutrients, including Ca, Mg, and micronutrients, which reduce maize response to application of N and P fertilizer (Zingore et al. 2008).

The essential role of mineral fertilizers is not only limited to increasing crop yields, but also to provide pathways for sustainable soil fertility management through increased organic matter production. A case study in Zimbabwe showed that commercial farmers were able to sustain high productivity under maize monoculture through large fertilizer dressings of between 160 and 200 kg N ha^{-1} yr^{-1} and 30–35 kg P ha^{-1} yr^{-1} (Figure 5.4; Zingore et al. 2005). Under such systems, maize grain yields of 7–8 Mg ha^{-1} were achieved, with a concomitant 10–12 Mg ha^{-1} of stover. In a positive feedback process, large portions of the stover were returned to the soil and incorporated, resulting in soils with substantially larger SOC on commercial farms than under smallholder agriculture, where

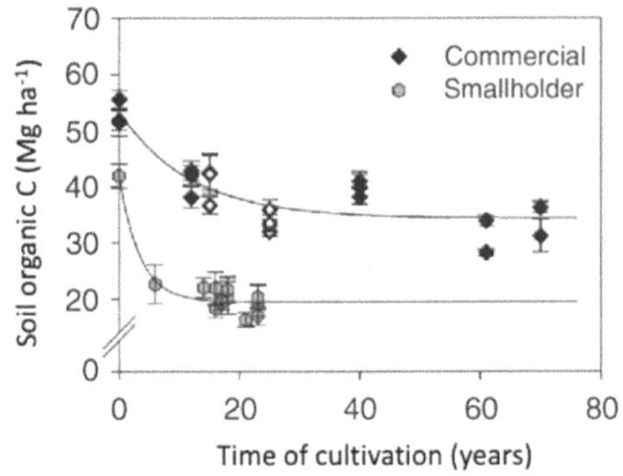

FIGURE 5.4 Comparison of long-term soil organic matter dynamics in smallholder and commercial maize production systems on luvisol in Zimbabwe(*Source:* Zingore et al. 2005).

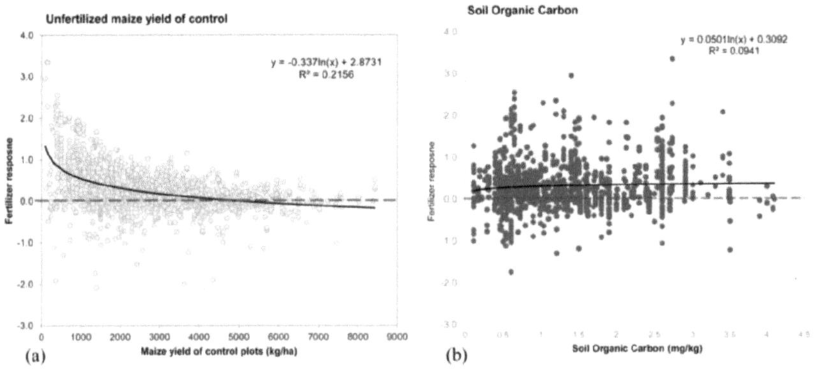

FIGURE 5.5 Relationship between fertilizer responses and (a) control maize yield (Mg ha^{-1}) as well as (b) soil organic carbon (mg kg^{-1}). The red line shows the zero line.

mineral fertilizers were not used (Figure 5.4). Improving crop yields on smallholdings by increased use of mineral fertilizers can therefore provide a pathway for enhancing food production as well as improving soil fertility and sequestering more carbon in the soil.

5.5 INFLUENCE OF SOIL ORGANIC MATTER ON FERTILIZER USE EFFICIENCY

5.5.1 Relationship between Fertilizer Response, Soil Organic Carbon and Maize Yields

Analysis of data from maize fertilizer response trials showed a nonlinear decline in fertilizer response with increasing yields in unfertilized control plots (Figure 5.5a; Table 5.2). Maize grain yield in the control plots varied substantially among experimental sites, ranging from 0.1 to 8.4 Mg ha^{-1}, highlighting the complex heterogeneity in management, soil, and environmental growing conditions. The relationship between unfertilized control maize yield and fertilizer response was significant ($P = 0.0016$), however, with a coefficient of determination of only 21% (Table 5.2). On the contrary, the relationship between unfertilized control maize yield and SOC contents (Figure 5.5b) was statistically insignificant ($P = 0.643$) and explained only 9% variation in the fertilizer response (Table 5.2). Multiple regression combining SOC with unfertilized control maize yields as explanatory variables, explained 25% of the variation in fertilizer response. Chikowo et al. (2011) similarly showed poor correlation between soil N supply and nutrient use efficiencies, but strong correlation

TABLE 5.2
Multiple Regression Model Results Indicating the Influence, and Interactions of Soil Organic Carbon (SOC), Unfertilized Maize Yields of Control Plots and Applied Nitrogen (N) from Inorganic Fertilizer Source on Fertilizer Response Across Africa

Predictor	Estimate	Standard Error	t Value	*P* Value	Statistical Significance
Intercept	0.899	0.111	8.094	0.000	***
SOC	−0.069	0.059	−1.167	0.024	**
Applied Nitrogen (kg/ha)	−0.001	0.001	−0.515	0.607	
Control (Mg/ha)	−0.180	0.038	−4.747	0.000	***
SOC*Control	0.002	0.021	0.092	0.927	
SOC* Applied Nitrogen	0.001	0.001	0.804	0.422	
Applied Nitrogen *Control	0.000	0.000	0.887	0.375	
SOC* Applied Nitrogen* Control	0.000	0.000	−0.091	0.0928	.
R^2 value	0.22				
Adjusted R^2	0.23				

SOC, soil organic carbon; N, nitrogen (kg/ha); Control, unfertilized maize yields (Mg ha^{-1}). Significance level codes: 0 "***" 0.001 "**" 0.01 "*" 0.05 "." 0.1 " " 1.

between soil N supply and control yields. The combination of SOC and unfertilized control maize yield improved the prediction of maize productivity of smallholder farming systems in SSA, suggesting that there are complex yield-limiting factors that are more reliably expressed through control yields but not strongly related to the SOC contents. High variability in fertilizer response was observed at the lower range of control yields and SOC contents, which is due to the increased complexity of soil constraints with decreasing soil fertility.

The poor correlation between SOC and fertilizer response derived from multiple continental-scale studies can also be the result of differences in field trial management and laboratory analysis methods and standards. Focusing on specific case studies and sites can assist in untangling the complex patterns and elucidating the role of SOC in maize productivity under similar climatic conditions, consistent experimental treatments, and SOC analysis methods.

5.5.2 Southern Africa Case Studies

5.5.2.1 Case Study 1A: Kurwakumire et al. 2014; Kafesu et al. 2018

The Zimbabwean agricultural landscape is dominated by infertile sandy soils derived from granitic parent material (Nyamapfene 1991). The moderating effect of topography or occasional non-granite parent material results in pockets of soils with appreciable clay content. While nutrient gradients in soils with similar textures generally develop due to differentials in soil fertility management, there are also inherent nutrient gradients that are linked to soil texture differences within landscapes. Some soils have degraded to a point where strategic targeting of scarce nutrient resources by smallholder farmers on different field types becomes crucial to ensuring viable fertilizer use efficiencies.

To assess fertilization strategies for optimizing crop productivity and NUE in maize production on heterogeneous light-textured soils under rain-fed conditions in Zimbabwe, Kurwakumire et al. (2014) established a nutrient omission study during two cropping seasons, across three on-farm sites with SOC ranging from 0.35 to 0.89%, and clay content between 4 and 19%. Fertilizer responses were influenced by both nutrient management and initial soil fertility (Table 5.3). Overall, this study established that balanced nutrient management had an overriding effect on maize grain and water productivity but only for soils that had SOC >0.44%. Water and soil nutrient interactions are much

TABLE 5.3
Nitrogen and Phosphorus Agronomic Efficiencies [A] and N and P Recovery Efficiencies [B] as Influenced by Nutrient Management and SOC Content, Wedza District, Zimbabwe

[A] N and P Agronomic Efficiencies

	SOC (g kg^{-1})		AE_N			AE_P	
		NK	NPS	NPKS	NPS	PKS	NPKS
Site		kg grain kg^{-1} N applied			kg grain kg^{-1} P applied		
Chingwa	3.5	7	16.0	17.0	31.5	2	35.5
Chinhengo	5.4	12.1	35.2	31.4	51.8	13.3	51.4
Muhwati	8.9	14.1	29.9	36.3	50.5	14.1	52.4

[B] N and P recovery efficiencies

Site	SOC (g kg^{-1})		RE_N			RE_P	
		NK	NPS	NPKS	NPS	PKS	NPKS
		Fraction N uptake (kg kg^{-1})			Fraction P uptake (kg kg^{-1})		
Chingwa	3.5	0.03	0.31	0.35	0.10	0.11	0.12
Chinhengo	5.4	0.40	0.71	0.72	0.15	0.01	0.16
Muhwati	8.9	0.44	0.79	0.83	0.15	0.02	0.15

more important on sandy soils due to the limited water-holding capacity. This study also recommended rainfall-responsive N fertilizer application in response to near real-time rainfall received as a strategy to reduce losses related to fertilizer investments under the risky rain-fed crop production systems, especially on coarse-textured soils inherently low in SOC.

In a related study under similar soil and agroecological conditions, Kafesu et al. (2018) showed identical trends of yield responses and fertilizer use efficiency. Baseline yields (<1 Mg ha^{-1}) and attainable yields (<2 Mg ha^{-1}) were low in fields with less than 0.4% SOC (Figures 5.6a, b), which translated into very low AEN (Figure 5.6c). Small increases in SOC with a range of 0.4–0.6% resulted in strong increases in baseline yields, attainable yields, and AEN by more than 100%, signifying the important connection between SOC, land degradation, and crop productivity on the granitic sandy soils, with critical SOC values estimated to be in the range of 0.4–0.5% (Zingore et al. 2008; Mtambanengwe and Mapfumo 2005).

5.5.2.2 Case Study 1B: Zingore et al. 2007a; 2008; 2011

These studies traced soil fertility indicators in different farmers' fields. As expected, SOC contents were larger on farms located on the clay soils compared with those on the granitic sands, irrespective of farm typology. This was linked to the high capacity for physico-chemical C stabilization in soils richer in clay and silt, and the resultant higher soil C thresholds that remain stable over time even under poor organic matter input (Feller and Beare 1997). The studies also established that, for farms with fields on the same soil type, SOC content in plots closest to the homesteads was significantly larger on farms in the higher resource endowment categories.

Livestock ownership has a huge influence on SOC cycling on farms. On average, cattle owners use between 5 and 10 Mg manure per farm annually, which provides 50–100 kg N and 15–30 kg P (Zingore et al. 2007a). To derive soil fertility niches meaningful for targeting fertilizer recommendations on the sandy soils, Zingore et al. (2011) classified the homefields, midfields, and outfields into three fertility zones (FZs 1–3) that captured the soil fertility variability within and across farms. FZ1 consisted of the most fertile fields closest to homesteads that received large additions of manure and fertilizer. FZ2 covered medium soil fertility fields that received moderate applications of fertilizer and manure in the past. More than 50% of the cultivated area fell in the FZ3 category,

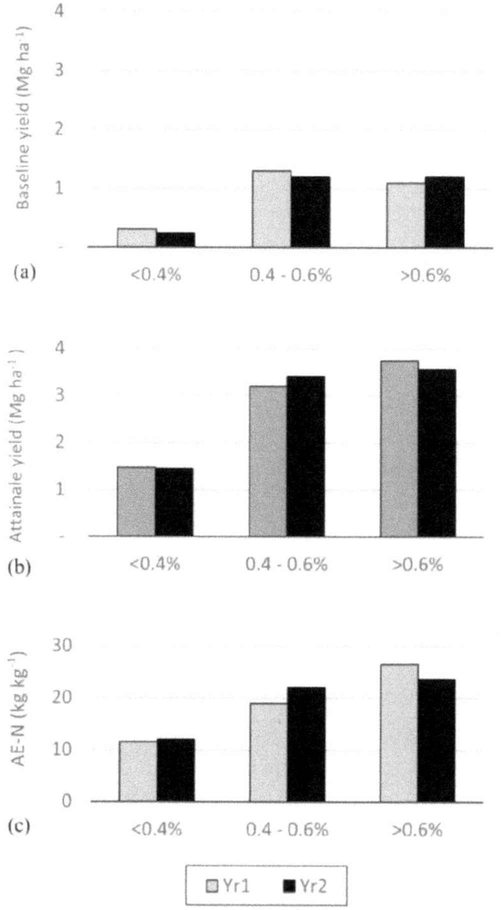

FIGURE 5.6 Baseline yield, attainable yields, and agronomic nitrogen use efficiencies for soil organic carbon ranges on a granitic sandy soil in Zimbabwe.

consisting of infertile fields that were cultivated for long periods with little addition of mineral fertilizers or organic nutrient resources. Maize yields attainable with good management practices decreased from about 6 Mg ha^{-1} in the FZ1 to 4.5 Mg ha^{-1} in the FZ2 and 1 Mg ha^{-1} in the FZ3. The low attainable yields in the FZ3 were associated with soil degradation and multiple constraints to maize productivity, including multiple nutrient deficiencies and soil acidity.

5.5.3 East Africa Case Studies

5.5.3.1 Case Study 2A: Vanlauwe et al. 2006

This study quantified maize grain yield responses to applied fertilizer N, P, and K using on-farm nutrient omission trials established in fields differing in initial soil fertility. Selected farms differed in farmer resource endowment, while, within farms, experimental fields differed in distance from the homestead. Farmer resource endowment and field location from the homestead have been identified as are key attributes influencing the intensity of mineral and organic resource applications in smallholder farming systems of SSA (Tittonell et al. 2005; Zingore et al. 2007a). Mean soil SOC levels were significantly influenced by field location from the homestead, with homefields showing significantly ($p<0.05$) larger SOC contents than all outfields and the low and high resource-endowed midfields (Figure 5.7). Within field types, differences in SOC contents were also observed, with medium resource-endowed homefields and midfields having significantly ($p<0.05$) larger SOC contents than low resource-endowed homefields and midfields, respectively (Figure 5.7). Observed differences in SOC contents between fields differing in proximity to the homestead are in line with findings by Zingore et al. (2007b) for smallholder farming systems in Zimbabwe. In the western Kenya region, such differences have been primarily linked to the preferential application of animal manure, the principal source of organic inputs in the western Kenya region (Castellanos-Navarrete et al. 2015), in fields closer to homesteads where livestock are housed (Tittonell et al. 2005). Between farms, differences in SOC contents based on farmer resource endowment reflected the strong influence of farmer access to mineral fertilizer and organic resources on SOC status.

Maize grain yield responses to fertilizer N applications differed between and within farms (Figure 5.8a). Within farms, grain yield response to N was generally largest in outfields and smallest in homefields (Figure 5.8a). These were characterized by the smallest and largest SOC contents, respectively (Figure 5.7), demonstrating strong effects of field SOC status on yield response to fertilizer N applications. Between fields, mean yield response to N differed based on farmer resource endowment, with contrasting patterns observed based on the field location (Figure 5.8a). For example, while outfields were generally characterized by the largest mean responses to N given their lower SOC status, N response in low resource-endowed outfields was much smaller compared to the medium and high resource-endowed outfields (Figure 5.8a). On the other hand, while homefields were generally characterized by smaller yield response to N, mean yield response in low resource-endowed homefields was close to 1.5 Mg ha^{-1}, but <0.5 Mg ha^{-1} in high resource-endowed homefields.

Patterns of mean agronomic efficiency of applied N (AEN) were in line with observed yield response patterns (Figure 5.8b). Low SOC status outfields had the highest AEN, with values >15 kg kg^{-1} observed in the medium and high resource-endowed outfields (Figure 5.8b). High resource-endowed homefields which were characterized by mean SOC content >1.5% accounted for the lowest AEN (<5 kg kg^{-1}), demonstrating the limited fertilizer use efficiency attained in high fertility soils. Observed patterns in yield response and in AEN are typical of expected yield and AEN across the SOC spectrum (Figure 5.3). At the lower end of the SOC spectrum, the smaller response to N and mean AEN of about <12 kg kg^{-1} in low resource outfields characterized by SOC contents <1.5 could be an indicator of additional biophysical constraints, typical of low fertility and less

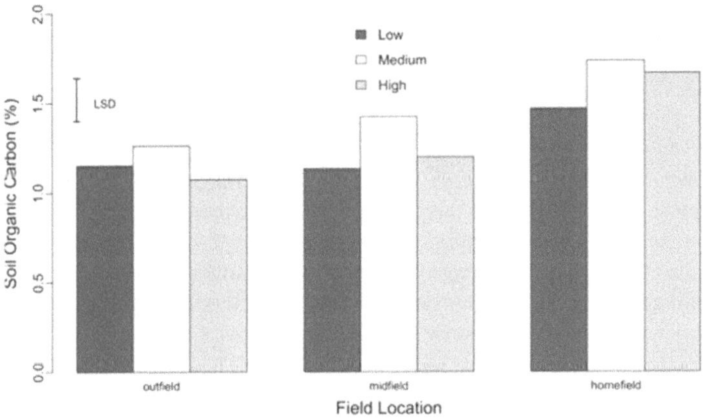

FIGURE 5.7 Differences in mean soil organic carbon contents (%) between fields (n=54) differing in distance from the homestead (outfields, midfields, and homefields), and in farmer resource endowment (low, medium, and high), in western Kenya.

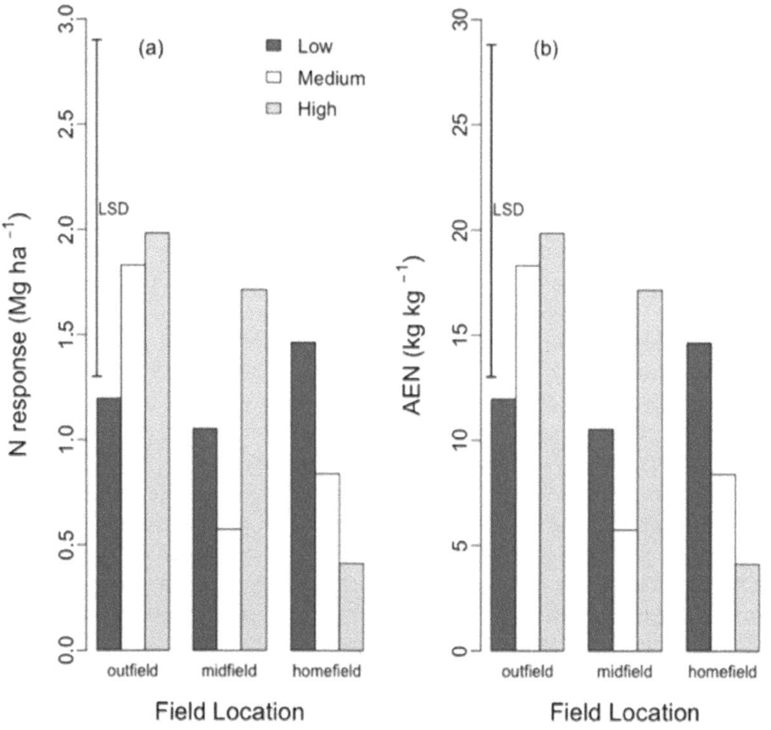

FIGURE 5.8 Differences in (a) mean maize grain yield response to fertilizer N (Mg ha^{-1}) and (b) mean agronomic efficiency of applied N (kg kg^{-1}), between fields (n=54) differing in distance from the homestead (outfields, midfields, and homefields), and in farmer resource endowment (low, medium, and high), in western Kenya (*Source:* Vanlauwe et al. 2006).

responsive fields (Vanlauwe et al. 2010). At the other end of the SOC spectrum, small yield response to fertilizer N and mean AEN <5 kg kg^{-1} in high resource homefields characterized by SOC contents >1.5 can be attributed to higher N mineralization rates and higher soil N supply characteristic of non-responsive fertile fields (Vanlauwe et al. 2010). While not very common in smallholder systems of SSA, the occurrence of such non-responsive fertile fields have been related to nutrient accumulation in fields that receive high applications of organic resources and fertilizer (Tittonell et al. 2008; Vanlauwe et al. 2010) which is expected of the high farmer resource-endowed homefields in this study.

While observed yield response and AEN patterns were generally in line with differences in field SOC status, substantial differences in yield response to N and AEN between fields with similar mean SOC contents, but differing in farmer resource endowment, were observed. This could be related to differences in the composition of SOC contents between fields and in soil conditions resulting from differences in field management that are not immediately captured by SOC measurements. This highlights the need for more detailed field characterization beyond SOC measurements for improved predictions of yield response to fertilizer N.

5.5.3.2 Case Study 2B: Njoroge et al. 2017

On-farm nutrient omission trials conducted over six consecutive cropping seasons in western Kenya allow for the assessment of initial field SOC status on spatial-temporal patterns in yield response to fertilizer N applications (Figure 5.9). Fields in this study differed in past manure application history and in initial SOC status. SOC contents were generally in the range of 1.5–2.5%, with the

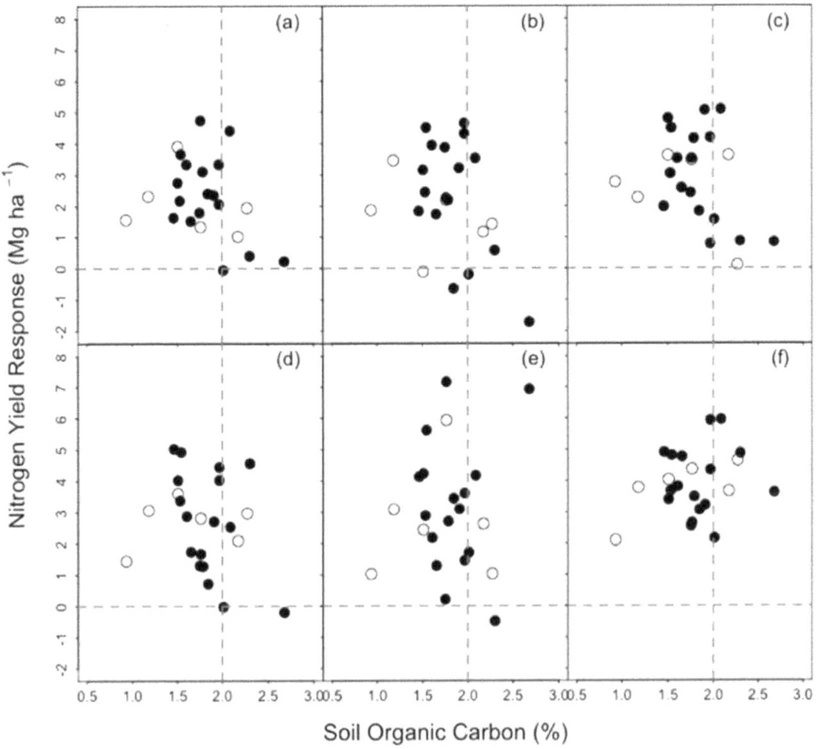

FIGURE 5.9 Temporal and spatial patterns in maize grain yield response to fertilizer N (Mg ha^{-1}) in on-farm trials (n=24) conducted across six consecutive seasons (long rainy season 2013 to short rainy season 2015: LR 2013 (a), SR 2013 (b), LR 2014 (c), SR 2014 (d), LR 2015 (e) and SR 2015 (f)), in fields differing in past manure applications in western Kenya. White circles represent fields without any farmer-applied manure in any of the three seasons preceding the experiment, while black circles represent fields with some farmer-applied manure within the three seasons preceding the experiment.

exception of one field without past manure application that had SOC contents <1%, while one field with manure applied in the past had SOC contents >2.5% (Figure 5.9). In the first three cropping seasons, yield responses to N were small in fields with SOC content >2%, while responses in fields with SOC contents ranging from 1.5 to 2% were larger but highly variable (Figures 5.9a, 5.9b, and 5.9c). However, in the subsequent three cropping seasons, yield response to N in fields with initial SOC contents >2% increased, while responses in fields with initial SOC contents ranging from 1.5 to 2% remained highly variable, though the variability appeared to decline in the last cropping season (Figures 5.9d, 5.9e, and 5.9f). In each season, the field with SOC contents <1% showed a moderate yield response to N that was consistently less than half of the largest yield response in a particular season variable (Figures 5.9a, 5.9b, 5.9c, 5.9d, 5.9e, and 5.9f).

The observed initial lower yield response to N fertilization in fields with initial SOC contents >2% can be attributed to larger soil N supply in these fields. Continued cropping without fertilizer N application in these fields, however, led to a decline in soil N supply, resulting in the observed larger response to N in later seasons. This indicates fragility of SOC and soil N supply in such fields and further demonstrates that such fertile non-responsive fields can rapidly lose their fertile non-responsive status if applications of organic resources and mineral N are stopped. Therefore, while such fields may initially only require moderate quantities of fertilizer N to enhance AEN, larger fertilizer N applications or the co-application of available organic resources with moderate amounts of fertilizer N may be required to maintain soil fertility and enhance AEN over time. The consistently lower yield response to N in the field with initial SOC contents <1% reaffirms the constrained response to fertilizer N under very low SOC contents, which has been attributed to the presence of additional biophysical constraints (Vanlauwe et al. 2010). Such fields may therefore require targeted measures, such as the application of large quantities of high-quality organic resources, to improve SOC contents and address biophysical constraints for enhanced yield responses to fertilizer N applications to be attained. Zingore et al. (2008) showed that regular applications of large amounts of animal manure were required to significantly increase SOC, pH, available P, and base saturation and to restore crop productivity in degraded fields. The large variability in yield response to N observed for fields with initial SOC contents ranging from 1.5 to 2% demonstrates the huge uncertainty in the expected yield response to N at moderate SOC contents as previously observed for Case Study 1. Even in fields that had a history of manure application, yield responses to N were highly variable over space and time. This can be related to the large variability in the nutrient content and in the frequency and timing of manure applications often observed in smallholder farming systems of SSA (Rufino et al. 2006), resulting in strong differences in potential soil N supply at moderate SOC levels. This reinforces the need to additionally characterize fields for improved predictions of expected yield response to fertilizer N.

Mean AEN between fields differing in past manure applications were not substantially different and appeared to increase over time for both past manure application categories (Table 5.4). However, observed differences in the range of AEN values, and in mean and range of PK and NPK yields, indicated contrasting AEN patterns based on past manure applications. For example, while near-similar mean AEN values >15 kg kg^{-1} were observed for both field categories in the short rainy season of 2014, mean PK and NPK yields differed by up to 0.9 Mg ha^{-1} (Table 5.4), reflecting substantial differences in crop productivity and soil N supply between these two field types. Increasing mean AEN values over time for both field types further reflect the expected changes in soil N supply following continued cropping with no N application. This is corroborated by the declining mean PK yields for both field categories over time, while mean yields with NPK are fairly stable (Table 5.4). These findings suggest the need to assess other parameters, such as actual yields attained in addition to AEN prior to making decisions on management practices required to enhance AEN. For example, while low AEN at low yields in low fertility fields may require the co-application of fertilizer N with available organic resources to enhance AEN, low AEN at high yields in high fertility fields may require adjustments in N fertilizer applications to enhance AEN by accounting for larger soil N supply.

Given the expected challenges of measuring SOC contents in smallholder farms of SSA due to limitations in access to soil analysis, findings from these case studies in western Kenya illustrate opportunities for estimating soil SOC status based on proxy indicators. The clear differences in mean SOC contents, based on resource endowment and field location from the homestead, demonstrate the potential for using such factors as reliable indicators of field-level SOC contents. By coupling such indicators with other relatively easy-to-assess factors, such as soil texture, which is a major driver of SOC contents (Feller and Beare 1997), reasonable estimates of the SOC status of different fields can be made for decision support in nutrient management. This can allow for improved prediction of expected yield response to fertilizer N applications, resulting in enhancements in fertilizer N use efficiency. Incorporation of such proxy indicators into decision support tools that additionally include information on locally attainable yields and expected ranges of yield responses to fertilizer N, can potentially result in further improvements in fertilizer use efficiency at the field level.

5.5.4 West Africa Case Study

Management of SOM to enhance crop productivity and cropping systems sustainability has received considerable research attention in West Africa due to the high prevalence of soils with

TABLE 5.4
Mean and Range of Agronomic N Efficiency, and Maize Grain Yield (Mg ha^{-1}) in Plots Fertilized with PK and NPK, Respectively, Across Six Consecutive Cropping Seasons in Fields Differing in Past Manure Application Practices in On-Farm Trials Conducted in Western Kenya. Seasons 1 – 6 Refer to LR 2013, SR 2013, LR 2014, SR 2014, LR 2015 and SR 2015 Respectively, While LR and SR Refer to the Long and Short Rainy Seasons Respectively

Parameter	Past Management	Season					
		LR 2013	SR 2013	LR 2014	SR 2014	LR 2015	SR 2015
AEN (kg kg^{-1})	(−) Manure	13.4	11.1	17.6	17.7	17.9	25.1
		(6.8-26)	(-0.7-23)	(0.7-24.1)	(9.6-24)	(6.8-39.5)	(14-31)
	(+) Manure	15.6	15.6	19.9	17.6	21.3	26.4
		(-0.4-31.6)	(-11.4-30.9)	(5.2-33.9)	(-1.3-33.5)	(-3.2-47.7)	(14.4-39.7)
PK yield (Mg ha^{-1})	(−) Manure	2.9	2.5	2.2	1.9	2	0.9
		(1.2-5.3)	(1.3-3.9)	(0.75-3.72)	(1.3-2.3)	(0.8-3.1)	(0.4-1.5)
	(+) Manure	3.4	2.8	2.9	2.8	2.9	1.5
		(1.4-4.8)	(0.9-5.5)	(0.9-5.3)	(1.6-6.5)	(1.1-5.2)	(1.5-3.8)
NPK yield (Mg ha^{-1})	(−) Manure	4.9	4.2	4.8	4.6	4.6	4.7
		(2.8-7.2)	(3.1-5.8)	(3.3-7.2)	(3.7-4.9)	(1.8-9)	(2.5-5.8)
	(+) Manure	5.7	5.2	5.9	5.4	6.1	5.5
		(2.2-8.2)	(3.3-7.3)	(2.9-7.7)	(2.9-8.2)	(2.8-9.4)	(3.3-6.8)

very low SOC contents across agroecological zones. Synergetic effects of crop residues retention and fertilizer application for enhanced maize yield and NUE were demonstrated in the region (Adetunji 1997; Amouzou et al. 2018; Buerkert et al. 1996; Naab et al. 2015). In Benin, Amouzou et al. (2018) recorded significantly higher AEN for maize with recommended inorganic fertilizer rates plus residue retention (37% in 2015 and 34% in 2014) than with fertilizer intensification alone, while the P agronomic efficiency (AEP) remained statistically similar between the two fertilization regimes (Table 5.5). Furthermore, no substantial differences in N and P apparent recovery were observed between the intensification regime and the integrated option for maize. Despite the differences in applied amounts of inorganic fertilizer under the intensification regime (80 kg N ha^{-1}, 26 kg P ha^{-1}, and 30 kg K ha^{-1}) and integrated option (44 kg N ha^{-1}, 15 kg P ha^{-1}, and 17.5 kg K ha^{-1} + residues retention), the similar N and P apparent recovery of maize underscored the consequent nutrient release potential of crop residue recycling and its impact on fertilizer use efficiencies. The nutrient release benefits of returned crop residues were demonstrated by Naab et al. (2015) under degraded soils in semi-arid conditions of Ghana. The authors showed a significant correlation between the amount of crop residues returned to the soil over four seasons and SOC ($r = 0.82$; $P = 0.007$), total soil N ($r = 0.75$; $P = 0.020$), and total soil P ($r = 0.69$; $P = 0.039$).

In Nigeria, maize yield increased by 254 kg ha^{-1} with every 1 Mg increase in SOC pool in a 0–15 cm layer (Lal, 2006), fitting in the range of 30–300 kg ha^{-1} yield increase reported for regions of Africa, Asia, and Latin America. Analysis of the effects of SOC on maize yields for data from the nutrient omission trials for Zimbabwe and Kenya showed values that at the lower end of the range determined by Lal et al. (2006) and are much lower than the rate of change observed for Nigeria (Table 5.6). The lower values for Kenya (20 kg maize Mg^{-1} SOC) than Zimbabwe (56 kg maize Mg^{-1} SOC) are likely due to the relatively high SOC range in the Kenya sites. The benefits of increasing SOC are less pronounced at high ranges of SOC due to more optimal soil chemical, physical and biological conditions above critical values (Zingore et al. 2008). The increase of N supply with change in SOC was also low for the Kenya and Zimbabwe sites (Table 5.6).

TABLE 5.5
Effect of Crop Residue Retention and Inorganic Fertilizer Use on Grain Yield, Nitrogen and Phosphorus Agronomic Efficiency (AE) and Apparent Recovery (RE) of Maize During the 2014 and 2015 Cropping Seasons at Ouri Yori, Benin

Growing Seasons	Fertilizer Regimes	Yield (Mg ha^{-1})	NUE Ratio		PUE Ratio	
			AE	RE	AE	RE
2014	Control (no fertilizer with residue removal)	1.4c				
	Recommended fertilizer plus residues	2.4b	24.7a	0.88a	62.1a	0.83a
	Fertilizer intensification	2.9a	19.3b	0.66a	59.2a	0.77a
	Low use of fertilizer	1.9c				
2015	Control (no fertilizer with residue removal)	1.7b				
	Recommended fertilizer plus residues	3.1a	33.8a	0.92a	74.6a	1.27a
	Fertilizer intensification	3.7a	25.8b	0.73a	79.4a	1.16a
	Low use of fertilizer	1.2b				
P-values SMS		<0.001	<0.001	0.130	0.935	0.699
P-values Seasons		0.024	0.003	0.482	0.182	<0.001
P-values SMS x Seasons		<0.001	0.636	0.811	0.751	0.702

(*Source:* Amouzou et al. 2018).

Means in a column within a season with similar letters are not significantly different at the 5% level according to the LSD test.

TABLE 5.6
The Relationship Between Soil Organic Carbon and Maize Yields for Case Study Datasets for the Sites in Kenya and Zimbabwe

Site	SOC Range	Maize Yield Range in Control Plots	Change of Maize Yield with SOC	Mean N Supply	Change of N Supply with SOC
	g kg^{-1}	Mg ha^{-1}	Kg Mg^{-1} SOC	Kg ha^{-1}	Kg Mg^{-1} SOC
Sandy soils with SOC (Zimbabwe)	2.3–21.6	0.1–4.1	56.0	10.5	3.8
Loamy and clayey soils with high SOC (Kenya)	5.9–100.8	0.7–4.6	20.0	20.0	1.7

5.6 BENEFITS OF SOIL ORGANIC MATTER FOR ENHANCING YIELD QUALITY AND SOIL WATER STORAGE

Beyond the function in supporting yield and N supply, there is evidence of the critical role of SOM in improved yield quality. Manzeke et al. (2019) showed that the bioavailability of Zn was strongly influenced by soil organic matter and organic resource management. The higher available Zn in the soil corresponded well with higher grain Zn contents (Manzeke et al. 2014). Similarly, Wood et al. (2018) showed strong effects of soil organic matter on the protein and Zn content of wheat in Ethiopia. An increase of SOM by 1% led to an increase in soil Zn that could meet Zn dietary requirements of an additional person for each 2 ha of cropland (Wood et al. 2018).

Soil surface retention of maize residues is a core component of conservation agriculture that has been extensively promoted by development agencies as one of the practices that result in increased productivity and fertilizer use efficiency. Despite mixed results in smallholder farming systems in SSA, there are indications that retention of residues offers good prospects for increasing resilience to moisture stress and improve fertilizer use efficiencies (Rusinamhodzi et al. 2011).

5.7 INCREASING FERTILIZER USE EFFICIENCY WITH COMBINED APPLICATION OF FERTILIZER AND MANURE AND LEGUME-BASED TECHNOLOGIES

A synthesis of results from field trials in Zimbabwe showed that sole application of manure increased maize productivity above control plots in fields of low and medium soil fertility status but not in fields with high soil fertility status (Figure 5.10). At a large scale, manure alone is insufficient to substantially increase maize productivity due to the limited amounts available. There is therefore a need to apply manure in combination with fertilizers, not only because of potential synergic benefits but also because of limited availability of both types of nutrient resources (Vanlauwe and Giller 2006). When combined with mineral fertilizers, manure led to increased productivity above fertilizer treatments alone, and the absolute increase in maize yields decreased with increasing soil fertility status. Combined application of manure with mineral N fertilizers has been recommended for maximization of short-term benefits of manure. This is particularly important on degraded fields where crops respond poorly to fertilizer application due to other limiting factors not addressed by fertilizers (Zingore et al. 2008). Under such conditions, application of large amounts of manure for several seasons may be necessary to restore SOM and associated soil physical fertility, and balance soil nutrients to facilitate good fertilizer response (Bayu et al. 2005).

Various legume-based technologies, such as rotation of cereal crops with grain legumes, improved fallows, alley cropping, and green manures have been advocated as viable options for providing

TABLE 5.7

Potential Targeting of ISFM Technologies Based on Analysis of Farming Systems in Sub-Humid Maize Production Agroecological Zones in Africa

Land Size	Labor Availability	Cattle Owner	General Characteristics	Target ISFM Technologies
<0.5 ha	High/low	No	Degraded soils. Land major constraints. Subsistence-oriented farms and farmers not self-food sufficient. Off-farm income an important source of livelihood.	Maize-grain legumes intercrops with small rates of fertilizer. Multi-nutrient fertilizer to correct nutrient deficiencies in degraded soil in combination with adapted grain legume germplasm. Incorporation of crop residues to improve soil fertility.
0.5–3 ha	High	Yes	Existence of soil fertility gradients. Farmers self-food sufficient and produce crops for market.	Redistribution of manure to reduce gradients of soil fertility. Manure applied in combination with moderate rates of N fertilizer. Rotation of maize with grain legumes. Inclusion of cash crops to improve income.
0.5–3 ha	Low	Yes	Existence of soil fertility gradients. Farmers self-food sufficient and produce crops for market.	Concentrate manure on small, well managed areas. Manure applied in combination with moderate rates of N fertilizer. Rotation of maize with grain legumes. Inclusion of cash crops to improve income.
0.5–3 ha	High	No	Low soil fertility. Farmers self-food sufficient.	Small rates of multi-nutrient fertilizer. Rotation of maize with grain legumes. Inclusion of cash crops to improve income. Incorporation of crop residues to improve soil fertility.
0.5–3 ha	Low	No	Low soil fertility. Farmers self-food sufficient.	Conservation agriculture in combination with small rates of multi-nutrient fertilizer to save labor and improve soil fertility. Rotation of maize with grain legumes. Inclusion of cash crops to improve income.
>3 ha	High/low	Yes	Land non-limiting. Highly variable characteristics.	Depending on labor availability and cattle ownership, strategies for farms 0.5–3 ha are applicable limited land area. Use of green manure and agroforestry technologies to enrich fallow land not allocated for crop production.

(*Synthesized from:* Shepherd and Soule 1998; Mtambanengwe and Mapfumo 2005; Waithaka et al. 2006; Zingore et al. 2007b; Zingore et al. 2009).

supplementary N to cereal crops through biological N fixation (Giller et al. 1997). Grain legumes have the advantage of providing income and protein for human nutrition, as well as, potentially supplying N to cereal crops grown in a rotation. Across SSA, grain legumes rotated with maize increased grain yield by 30–50% (Franke et al. 2017; Hulugalle and Lal 1986; Sogbedji et al. 2006; Uzoh et al. 2019; Snapp et al. 2002; Kerr et al. 2007). In particular, velvet bean (*Mucuna pruriens* var. *utilis*) rotation with maize increased grain yield of maize by about 40% in Togo (Fofana et al. 2004; Sogbedji et al. 2006) and over 100% in Nigeria (Uzoh et al. 2019) and Benin (Versteeg et al. 1998) compared to continuous sole maize cropping. Agroforestry technologies, such as alley cropping or hedgerow intercropping with N-fixing trees (*Senna siamea*, *Gliricidia sepium*, *Leucaena*

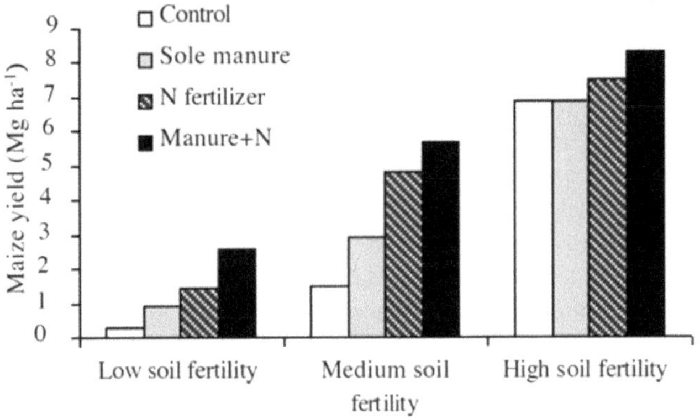

FIGURE 5.10 Effects of manure, N fertilizer, and their combination on productivity of maize in sub-humid northeast Zimbabwe. In the studies cited, manure is applied as an alternative to basal fertilizer. (*Source:* Grant, 1967; Mugwira and Murwira, 1997; Mtambanengwe and Mapfumo 2005; Zingore et al. 2007a).

leucocephala, *Albizia lebbeck*, etc.), have been widely tested to augment both soil fertility and maize productivity in SSA (Sileshi et al. 2008; Akinnifesi et al. 2007). Alley cropping supplemented with mineral fertilizer proved to regenerate and maintain soil fertility on degraded and non-degraded soils (Aihou et al. 1998). However, the positive effects through the green manuring of leguminous tree prunings were observed under timely pruning and correct management conditions. On the other hand, delayed pruning resulted in maize yield losses of up to 60% in farmers' fields (Versteeg et al. 1998). Investigation of the impacts of sole and combined application of tree pruning and urea fertilizer by Vanlauwe et al. (2001) showed higher N recovery from urea when combined with Leucaena leucocephala prunings in the derived savanna zone of Benin (44%) and Togo (32%) compared to the sole urea treatments (22% in Benin and 15% in Togo).

To be relevant and effective, various ISFM technologies must be aligned with the biophysical and socioeconomic opportunities and constraints associated with various farm types. Land and cattle ownership, availability of labor and access to input and output markets are some of the key determinants of use of ISFM technologies by farmers. Considering factors that are endogenous at the farm level, a synthesis of results from farm-level based experiments and analysis tools in maize-based cropping systems have shown that technological options can be targeted to households of different resource endowment (Table 5.7). Targeted participatory experiments, with farmers in different wealth categories and nested within value chain opportunities, are necessary to evaluate ISFM options including different rates and types of fertilizer, crop combinations, integrated use of manure and fertilizer and agroforestry to refine the general guidelines for specific sites and farms. While the tools to analyze farming systems have been mainly used as a platform for research to improve understanding of the complexity of smallholder farming systems, there is increasing scope for their use in guiding ISFM research to be more appropriate for smallholder farms. Such targeted ISFM research is beneficial not only to increase nutrient use efficiency at the farm level but also to enable innovation for strengthening value chains. Further they are useful to inform policy makers on matters, such as status of soil health, land degradation, amounts and formulation of fertilizers required in different regions, and for promotion of complementary ISFM technologies and targeting farmers for policy interventions to improve crop productivity, profitability, and fertilizer use efficiencies at scale.

5.8 CONCLUSIONS

Soil fertility depletion and land degradation associated with inherent and management factors, pose a major challenge for crop production intensification and efficient use of fertilizers in SSA.

Large areas of agricultural soils in the region are characterized by critically low SOC levels due to predominance of coarse-textured soils with kaolinitic clays and continuous cultivation with inadequate additions of inorganic and organic nutrient resources. Despite the general low status of SOC, socioeconomic diversity between farms influence management practices that drive complex differences in SOC contents with major implications on maize productivity and fertilizer use efficiency. The complex interactions between environmental and management factors obscure clear relations between SOM and maize yield response to fertilizer and fertilizer use efficiencies at the regional level. Maize yield response to fertilizer was more closely correlated with yields in unfertilized control plots that express more fully the constraints of other soil yield-limiting factors. However, close association of SOC with maize yields and agronomic N use efficiencies were evident in specific case studies under uniform agroecological conditions in East, West, and southern Africa. Sandy soils in Zimbabwe with very low SOC levels (<1%) were highly susceptible to degradation with a critical SOC level of 0.46%, below which the yields in control plots declined to less than 1 Mg ha^{-1}, and fertilizer use efficiencies were very poor. In Kenya, SOC contents ranged from 1 to 3%, with clear patterns of degraded non-responsive soils with <1.2% SOC, variable responses in mid-range SOC categories, and non-responsive fertile soils with high SOC soils (>2%). Various options of organic resource management showed synergic interactions with mineral fertilizer and are essential for enhancing maize productivity, fertilizer use efficiency, and soil organic matter under the low SOM conditions of the soils in SSA. Organic resources play an important role in restoration and maintenance of soil fertility due to multiple benefits that include, supply of nutrients, increase in soil pH and SOM. Use of animal manure and retention of maize crop residues are technologies with good potential for synergizing the effects of fertilizer use. Symbiotic biological N_2-fixation is also an important option for supplying high-quality organic resources for improving soil N budgets and supplementing N derived from fertilizers. When combined with mineral fertilizer, these technologies have the potential to improve productivity and fertilizer use efficiency by 10–60%, depending on soil fertility conditions and fertilizer application rates. However, the suitability, relevance and potential impact of various fertilizer management strategies and ISFM technologies on maize productivity and fertilizer use efficiency hinges on site- and farm-specific socioeconomic, soil, and climatic conditions.

REFERENCES

Adetunji, M.T., 1997. Organic residue management, soil nutrient changes and maize yield in humid ultisol. *Nutr. Cycl. Agroecosyst.* 47(3), 189–199

Aihou, K., Sanginga, N., Vanlauwe, B., Lyasse, O., Diels, J., Merckx, R., 1998. Alley cropping in the moist savanna of West-Africa: I. Restoration and maintenance of soil fertility on 'terre de barre' soils in Bénin Republic. *Agrofor. Syst.* 42(3), 213–227. https://doi.org/10.1023/A:1006114116095

Akinnifesi, F.K., Makumba, W., Sileshi, G., Ajayi, O.C., Mweta, D., 2007. Synergistic effect of inorganic N and P fertilizers and organic inputs from Gliricidia sepium on productivity of intercropped maize in Southern Malawi. *Plant Soil* 294(1–2), 203–217

Amouzou, K.A., Naab, J.B., Lamers, J.P.A., Becker, M., 2018. Productivity and nutrient use efficiency of maize, sorghum, and cotton in the West African Dry Savanna. *J. Plant Nutr. Soil Sci.* 2018, 1–14. https://doi.org/10.1002/jpln.201700139

Amundson, R., Berhe, A.A., Hopmans, J.W., Olson, C., Sztein, A.E., Sparks, D.L., 2015. Soil science. Soil and human security in the 21st century. *Science*. https://doi.org/10.1126/science.1261071

Bationo, A., Fening, J.O., 2018. Soil organic carbon and proper fertilizer recommendation. In: Bationo, A., Ngaradoum, D., Youl, S., Lompo, F., Fening, Joseph Opoku (eds.), *Improving the Profitability, Sustainability and Efficiency of Nutrients Through Site Specific Fertilizer Recommendations in West Africa Agro-Ecosystems*: Volume 1. Springer, Cham, pp. 1–10. https://doi.org/10.1007/978-3-319-58789-9_1

Bationo, A., Lompo, F., Koala, S., 1998. Research on nutrient flows and balances in West Africa: State-of-the-art. *Agric. Ecosyst. Environ.* 71(1–3), 19–36

Bationo, A., Mokwunye, U., Vlek, P.L.G., Koala, S., Shapiro, B.I., 2003. Soil fertility management for sustainable land use in the West African Sudano-Sahelian zone. In: M.P. Gichuru et al. (eds.), *Soil Fertility*

Management in Africa: A Regional Perspective. Academy of Science Publishers (ASP) and Tropical Soil Biology and Fertility of CIAT, Nairobi, Kenya

Bationo, A., Kihara, J., Vanlauwe, B., Waswa, B., Kimetu, J., 2007. Soil organic carbon dynamics, functions and management in West African agro-ecosystems. *Agric. Syst.* 94(1), 13–25. https://doi.org/10.1016/j.agsy.2005.08.011

Bayala, J., Kalinganire, A., Sileshi, G.W., Tondoh, J.E., 2018. Soil organic carbon and nitrogen in agroforestry systems in sub-Saharan Africa: A review. In: *Improving the Profitability, Sustainability and Efficiency of Nutrients Through Site Specific Fertilizer Recommendations in West Africa Agro-Ecosystems.* Springer, Cham. https://doi.org/10.1007/978-3-319-58789-9_4

Bayu, W., Rethman, N.F.G., Hammes, P.S., 2005. The role of manure in sustainable soil fertility management in sub-Saharan Africa. *J. Sustain. Agric.* 25(2), 113–136

Benson, T., 1999. *Validating and Strengthening the Area-Specific Fertilizer Recommendations for Hybrid Maize Grown by Malawian Smallholders. A Research Report of the Results of the nationwide1997/98 Maize Fertilizer Recommendations Demonstration.* Ministry of Agriculture, Malawi

Brouwer, J., Powell, J.M., 1998. Micro-topography and leaching: Possibilities for making more efficient use of nutrients in African agriculture. In: E.M.A. Smaling (ed.), *Nutrient Balances as Indicators of Productivity and Sustainability in sub-Saharan African Agriculture. Agric. Ecosyst. Environ.* 71(1–3), 229–239

Buerkert, A., Michels, K., Lamers, J.P.A., Marschner, H., Bationo, A., 1996. *Anti-Erosive, Soil Physical, and Nutritional Effects of Crop Residues, in: Wind Erosion in Niger: Implications and Control Measures in a Millet-Based Farming System.* Kluwers Academic Publishers, Dortrecht, The Netherlands, pp. 123–138

Cairns, J.E., Hellin, J., Sonder, K., Araus, J.L., MacRobert, J.F., Thierfelder, C., Prasanna, B.M., 2013. Adapting maize production to climate change in sub-Saharan Africa. *Food Sec.* 5(3), 345–360

Castellanos-Navarrete, A., Tittonell, P., Rufino, M.C., Giller, K.E., 2014. Feeding, crop residue and manure management for integrated soil fertility management: A case study from Kenya. *Agric. Syst.*, 124, 24–35. https://doi.org/10.1016/j.agsy.2014.03.001

Castellanos-Navarrete, A., Tittonell, P., Rufino, M.C., Giller, K.E., 2015. Feeding, crop residue and manure management for integrated soil fertility management: A case study from Kenya. *Agric. Syst.* 134, 24–35

Chander, K., Goyal, S., Nandal, D.P., Kapoor, K.K., 1998. Soil organic matter, microbial biomass and enzyme activities in a tropical agroforestry system. *Biol. Fertil. Soils* 27, 168–172. https://doi.org/10.1007/s003740050416

Chikowo, R., Zingore, S., Snapp, S., Johnston, A., 2014. Farm typologies, soil fertility variability and nutrient management in smallholder farming in Sub-Saharan Africa. *Nutr. Cycl. Agroecosyst.* 100, 1–18. https://doi.org/10.1007/s10705-014-9632-y

Chikowo, R., Corbeels, M., Mapfumo, P., Tittonell, P., Vanlauwe, B., Giller, K.E., 2011. Nutrient capture efficiencies and crop responses to soil fertility management in sub-Saharan Africa. *Nutr. Cycl. Agroecosyst.* 88, 59–77. https://doi.org/10.1007/978-90-481-2543-2_59

Christensen, B.T., 1992. Physical fractionation of soil organic matter in primary particle size and density separates. *Adv. Soil Sci.* 20, 1–89. http://doi.org/10.1007/978-1-4612-2930-8_1

Corbeels, M., Cardinael, R., Naudin, K., Guibert, H., Torquebiau, E., 2019. The 4 per 1000 goal and soil carbon storage under agroforestry and conservation agriculture systems in sub-Saharan Africa. *Soil Till. Res.* 188, 16–26. https://doi.org/10.1016/j.still.2018.02.015

Davidson, E.A., Ackerman, I.L., 1993. Changes in soil carbon inventories following cultivation of previously untilled soils. *Biogeochem* 20, 161–193. https://doi.org/10.1007/BF00000786

Deckers, J., Laker, M., Vanherreweghe, S., Vanclooster, M., Swennen, R., Cappuyns, V., 2000. 'State of the art on soil-related geo-medical issues in the world'. In: J. Låg (ed.), *Geomedical Problems in Developing Countries.* Norwegian Academy of Science and Letters, Oslo

Diarisso, T., Corbeels, M., Andrieu, N., Djamen, P., Douzet, J.-M., Tittonell, P., 2016. Soil variability and crop yield gaps in two village landscapes of Burkina Faso. *Nutr. Cycl. Agroecosyst.* 105(3), 199–216. https://doi.org/10.1007/s10705-015-9705-6

Donahue, R.L., Miller, R.W., Shickluna, J.C., et al. 1983. *Soils. An Introduction to Soils and Plant Growth.* Prentice-Hall, Inc., Hoboken, NJ.

Dutta, S., Chakraborty, S., Banerjee, H., Goswami, R., Majumdar, K., Li, B., Jat, M.L., 2020. Maize yield in smallholder agriculture system: An approach integrating socio-economic and crop management factors. *PLOS ONE* 15(2), e0229100

Feller, C., Beare, M.H., 1997. Physical control of soil organic matter dynamics in the tropics. *Geoderma* 79(1–4), 69–116

Feller, C., Albrecht, A., Tessier, D., 1996. Aggregation and organic matter storage in kaolinitic and smectitic soils of the tropics. In: M.R. Carter, Stewart, B.A. (eds.), *Structure and Organic Matter Storage in Agricultural Soils. Advances in Soil Science.* Lewis, New York, pp. 309–359

Fofana, B., Breman, H., Carsky, R.J., Van Reuler, H., Tamelokpo, A.F., Gnakpenou, K.D., 2004. Using mucuna and P fertilizer to increase maize grain yield and N fertilizer use efficiency in the coastal savanna of Togo. *Nutr. Cycl. Agroecosyst.* 68(3), 213–222. https://doi.org/10.1023/B:FRES.0000019457.43677.08

Food and Agriculture Organization of the United Nations, 2020. *FAOSTAT Database.* FAO, Rome, Italy

Food and Agriculture Organization of the United Nations (FAO) and European Commission on Agriculture (ECA), 2018. Regional Overview of Food Security and Nutrition. Addressing the Threat from Climate Variability and Extremes for Food Security and Nutrition. FAO, Accra, 116 pp

Forkuor, G., 2014. *Agricultural Land Use Mapping in West Africa Using Multi-Sensor Satellite Imagery.* University of Würzburg, Bavaria, Germany

Franke, A.C., van den Brand, G.J., Vanlauwe, B., Giller, K.E., 2017. Sustainable intensification through rotations with grain legumes in Sub-Saharan Africa: A review. *Agric. Ecosyst. Environ.* 172–185. https://doi.org/10.1016/j.agee.2017.09.029

Giller, K.E., 2001. *Nitrogen Fixation in Tropical Cropping Systems.* CAB International, Wallingford, UK

Giller, K.E., 2002. Targeting management of organic resources and mineral fertilizers: Can we match scientists' fantasies with farmers' realities. In: B. Vanlauwe, Sanginga, N., Diels, J., Merckx, R. (eds.), *Balanced Nutrient Management Systems for the Moist Savanna and Humid Forest Zones of Africa.* CAB International, Wallingford, UK, pp. 155–192

Giller, K.E., Cadish, G., Ehaliotis, C., Adams, E., Sakala, W.D., Mafongoya, P.L., 1997. Building soil nitrogen capital in Africa. In: R.J. Buresh, Sanchez, P.A., Calhoun, F. (eds.), *Replenishing Soil Fertility in Africa.* Soil Science Society of America Special Publication No. 51. Soil Science Society of America, Madison, WI, pp. 151–192

Godde, C.M., Thorburn, P.J., Biggs, J.S., Meier, E.A., 2016. Understanding the impacts of soil, climate, and farming practices on soil organic carbon sequestration: A simulation study in Australia. *Front. Plant Sci.* 7, 661. https://doi.org/10.3389/fpls.2016.00661

Hassink, J., 1997. The capacity of soils to preserve organic C and N by their association with clay and silt particles. *Plant Soil* 191(1), 77–87

Hedges, L.V., Gurevitch, J., Curtis, P.S., 1999. The meta-analysis of response ratios in experimental ecology. *Ecology* 80, 1150–1156. https://doi.org/10.1890/0012-9658(1999)080[1150:TMAORR]2.0.CO;2

Hulugalle, N.R., Lal, R., 1986. Root growth of maize in a compacted gravelly tropical alfisol as affected by rotation with a woody perennial. *Field Crops Res.* 13, 33–44. https://doi.org/10.1016/0378-4290(86)90005-5

Ichami, S.M., Shepherd, K.D., Sila, A.M., Stoorvogel, J.J., Hoffland, E., 2018. Fertilizer response and nitrogen use efficiency in African smallholder maize farms. *Nutr. Cycl. Agroecosyst.* 113, 1–19. https://doi.org/10.1007/s10705-018-9958-y

Kafesu, N., Chikowo, R., Mazarura, U., Gwenzi, W., Snapp, S., Zingore, S., 2018. Comparative fertilization effects on maize productivity under conservation and conventional tillage on sandy soils in a smallholder cropping system in Zimbabwe. *Field Crops Res.* 218, 106–114

Kerr, R.B., Snapp, S.S., Chirwa, M., Shumba, L., Msachi, R., 2007. Participatory research on legume diversification with Malawian smallholder farmers for improved human nutrition and soil fertility. *Exp. Agric.* 43(4), 437–453

Kihara, J., Nziguheba, G., Zingore, S., Coulibaly, A., Esilaba, A., Kabambe, V., Njoroge, S., Palm, C., Huising, J., 2016. Understanding variability in crop response to fertilizer and amendments in sub-Saharan Africa. *Agric. Ecosyst. Environ.* 229, 1–12

Kihara, J., Sileshi, G.W., Nziguheba, G., Kinyua, M., Zingore, S., Sommer, R., 2017. Application of secondary nutrients and micronutrients increases crop yields in sub-Saharan Africa. *Agron. Sustain. Dev.* 37(4), 25. https://doi.org/10.1007/s13593-017-0431-0

Kihara, J., Bolo, P., Kinyua, M., Nyawira, S.S., Sommer, R., 2020. Soil health and ecosystem services: Lessons from sub-Sahara Africa (SSA). *Geoderma* 370, 114342. https://doi.org/10.1016/j.geoderma.2020.114342

Koch, A., Mcbratney, A., Adams, M., Field, D., Hill, R., Crawford, J., Minasny, B., Lal, R., Abbott, L., O'Donnell, A., Angers, D., Baldock, J., Barbier, E., Binkley, D., Parton, W., Wall, D.H., Bird, M., Bouma, J., Chenu, C., Flora, C.B., Goulding, K., Grunwald, S., Hempel, J., Jastrow, J., Lehmann, J., Lorenz, K., Morgan, C.L., Rice, C.W., Whitehead, D., Young, I., Zimmermann, M., 2013. Soil security: Solving the global soil crisis. *Glob. Policy* 4, 434–441. https://doi.org/10.1111/1758-5899.12096

Körschens, M., Weigel, A., Schulz, E., 1998. Turnover of soil organic matter (SOM) and long-term balances: Tools for evaluating sustainable productivity of soils. *Z. Pflanzenernührung Bodenkd.* 161(4), 409–424
Kumwenda, J.D.T., Waddington, S.R., Snapp, S.S., Jones, R.B., Blackie, M.J., 1997. Soil fertility management in the smallholder maize-based cropping systems of Africa. In: *The Emerging Maize Revolution in Africa: The Role of Technology, Institution and Policy.* Michigan State University, East Lansing, MI
Kurwakumire, N., Chikowo, R., Mtambanengwe, F., Mapfumo, P., Snapp, S., Johnston, A., Zingore, S., 2014. Maize productivity and nutrient and water use efficiencies across soil fertility domains on smallholder farms in Zimbabwe. *Field Crops Res.* 164, 136–147
Lal, R., 2006. Enhancing crop yields in the developing countries through restoration of the soil organic carbon pool in agricultural lands. *Land Degrad. Dev.* 17(2), 197–209. https://doi.org/10.1002/ldr.696
Lal, R., 2016. Soil health and carbon management. *Food Energy Secur.* 5, 212–222 https://doi.org/10.1002/fes3.96
Liniger, H.P., Mekdaschi Studer, R., Hauert, C., Gurtner, M., 2011. *Sustainable Land Management in Practice: Guidelines and Best Practices for Sub-Saharan Africa.* TerrAfrica, World Overview of Conservation Approaches and Technologies (WOCAT) and Food and Agriculture Organization of the United Nations (FAO), Rome, Italy.
Lorenz, K., Lal, R., 2016. *Soil organic carbon: An appropriate indicator to monitor trends of land and soil degradation within the SDG framework?* Umweltbundesamt, Dessau-Roßlau. https://doi.org/10.3929/ethz-b-000238666
Manzeke, G.M., Mtambanengwe, F., Nezomba, H., Mapfumo, P., 2014. Zinc fertilization influence on maize productivity and grain nutritional quality under integrated soil fertility management in Zimbabwe. *Field Crops Res.* 166, 128–134
Manzeke, M.G., Mtambanengwe, F., Watts, M.J., Hamilton, E.M., Lark, R.M., Broadley, M.R., Mapfumo, P., 2019. Fertilizer manage-ment and soil type influence grain zinc and iron concentration under contrasting smallholder cropping systems in Zimbabwe. *Sci. Rep.* 9, 6445. https://10.1038/s41598-019-42828-0
McBratney, A., Field, D.J., Koch, A., 2014. The dimensions of soil security. *Geoderma* 213, 203–213. https://doi.org/10.1016/j.geoderma.2013.08.013
Mtambanengwe, F., Mapfumo, P., 2005. Organic matter management as an underlying cause for soil fertility gradients on smallholder farms in Zimbabwe. *Nutr. Cycl. Agroecosyst.* 73(2–3), 227–243
Mtangadura, T.J., Mtambanengwe, F., Nezomba, H., Rurinda, J., Mapfumo, P., 2017. Why organic resources and current fertilizer formulations in Southern Africa cannot sustain maize productivity: Evidence from a long-term experiment in Zimbabwe. *PLoS ONE* 12(8): e0182840. https://doi.org/10.1371/journal.pone.018284
Mugwira, L.M., Murwira, H.K., 1997. Use of cattle manure to improve soil fertility in Zimbabwe; past and current research and future research needs. Soil Fertility Network Research Results, Working Paper No 2. Soil Fertility Network for maize-based cropping systems in Malawi and Zimbabwe, 18 pp.
Muller, N.D., Gerber, J.S., Johnston, M., Ray, D.K., Ramankutty, N., Foley, J.A., 2012. Closing yield gaps through nutrient and water management. *Nature* 490(7419), 254–257
Mushayi, P., Waddington, S.R., Chiduza, C., 1998. Low efficiency of nitrogen use by maize on smallholder farms in sub-humid Zimbabwe, Maize Production Technology for the Future: Challenges and Opportunities. In: Proceedings of the Sixth Eastern and Southern Africa Regional Maize Conference, 21–25 September 1998. CIMMYT and EARO, Addis Ababa, pp. 278–281
Musinguzi, P., Tenywa, J.S., Ebanyat, P., Tenywa, M.M., Mubiru, N.D., Basamba, T.A., Leip, A., 2013. Soil organic carbon thresholds and nitrogen management in tropical agroecosystems: Concepts and prospects. *Sustain. Dev.* 6, 31–43
Naab, J.B., Mahama, G.Y., Koo, J., Jones, J.W., Boote, K.J., 2015. Nitrogen and phosphorus fertilization with crop residue retention enhances crop productivity, soil organic carbon, and total soil nitrogen concentrations in sandy-loam soils in Ghana. *Nutr. Cycl. Agroecosyst.* 102(1), 33–43. https://doi.org/10.1007/s10705-015-9675-8
Nandwa, S.M., 2001. Soil organic carbon (SOC) management for sustainable productivity of cropping and agro-forestry systems in Eastern and Southern Africa. *Nutr. Cycl. Agroecosyst.* 61, 143–158. https://doi.org/10.1023/A:1013386710419
Njoroge, R., Otinga, A.N., Okalebo, J.R., Pepela, M., Merckx, R., 2018. Maize (Zea mays L.) response to secondary and micronutrients for profitable N, P and K fertilizer use in poorly responsive soils. *Agronomy* 8(4): 49. https://doi.org/10.3390/agronomy8040049
Njoroge, S., Schut, A.G.T., Giller, K.E., Zingore, S., 2017. Strong spatial-temporal patterns in maize yield response to nutrient additions in African smallholder farms. *Field Crops Res.* 214, 321–330

Nyamapfene, K.W., 1991. *Soils of Zimbabwe*. Nehanda Publishers (Pvt) Ltd, Harare, Zimbabwe

Nyberg, G., Bargués Tobella, A., Kinyangi, J., Ilstedt, U., 2012. Soil property changes over a 120-yr chronosequence from forest to agriculture in western Kenya. *Hydrol. Earth Syst. Sci.* 16(7), 2085–2094

Okeyo, J.M., Shepherd, K.D., Wamicha, W., Shisanya, C., 2009. *Spatial Variation in Soil Organic Carbon within Smallholder Farms in Western Kenya: A Geospatial Approach. African Crop Science Journal* 14, 27–36.

Oldfield, E.E., Bradford, M.A., Wood, S.A., 2019. Global meta-analysis of the relationship between soil organic matter and crop yields. *Soil* 5, 15–32. https://doi.org/10.5194/soil-5-15-2019

Pasuquin, J.M., Pampolino, M.F., Witt, C., Dobermann, A., Oberthur, T., Fisher, M.J., Inubushi, K., 2014. Closing yield gaps in maize production in Southeast Asia through site-specific nutrient management. *Field Crops Res.* 156, 219–230

R Core Team, 2020. R: A language and environment for statistical computing. R Foundation for Statistical Computing, Vienna, Austria. https://www.R-project.org/

Robertson, M.J., Sakala, W., Benson, T., Shamudzarira, Z., 2005. Simulating response of maize to previous velvet bean (Mucuna pruriens) crop and nitrogen fertilizer in Malawi. *Field Crops Res.* 91, 91–105. https://doi.org/10.1016/j.fcr.2004.06.009

Rufino, M.C., Rowe, E.C., Delve, R.J., Giller, K.E., 2006. Nitrogen cycling efficiencies through resource-poor African crop-livestock systems. *Agric. Ecosyst. Environ.* 112(4), 261–282

Rurinda, J., Zingore, S., Jibrin, J.M., Balemi, T., Masuki, K., Andersson, J.A., Pampolino, M.F., Mohammed, I., Mutegi, J., Kamara, A.Y., Vanlauwe, B., Craufurd, P.Q., 2020. Science-based decision support for formulating crop fertilizer recommendations in sub-Saharan Africa. *Agric. Syst.* 180, 102790

Rusinamhodzi, L., Corbeels, M., Van Wijk, M.T., Rufino, M.C., Nyamangara, J., Giller, K.E., 2011. A meta-analysis of long-term effects of conservation agriculture on maize grain yield under rain-fed conditions. *Agron. Sustain. Dev.* 31, 657. https://doi.org/10.1007/s13593-011-0040-2

Schjønning, P., Jensen, J.L., Bruun, S., Jensen, L.S., Christensen, B.T., Munkholm, L.J., Oelofse, M., Baby, S., Knudsen, L., 2018. The role of soil organic matter for maintaining crop yields: Evidence for a renewed conceptual basis. *Adv. Agron.* 150, 35–79. https://doi.org/10.1016/bs.agron.2018.03.001

Shepherd, K.D., Soule, M.J., 1998. Soil fertility management in Western Kenya: Dynamic simulation of productivity, profitability and sustainability at different resource endowment levels. *Agric. Ecosyst. Environ.* 71, 131–145. https://doi.org/10.1016/S0167-8809(98)00136-4

Siame, J., Willey, R.W., Morse, S., 1997. A study of the partitioning of applied nitrogen between maize and beans in intercropping. *Exp. Agric.* 33(1), 35–41

Sileshi, G., Akinnifesi, F.K., Ajayi, O.C., Place, F., 2008. Meta-analysis of maize yield response to woody and herbaceous legumes in sub-Saharan Africa. *Plant Soil* 307(1–2), 1–19

Smaling, E.M.A., Braun, A.R., 1996. Soil fertility research in sub-Saharan Africa: New dimensions, new challenges. *Comm. Soil. Sci.* 7, 365–386

Smaling, E.A., Nandwa, S., Janssen, B., 1997. Soil fertility in Africa is at Stake. In: Buresh, R. J., Sanchez, P.A. (eds.), *Replenishing Soil Fertility in Africa.* SSSA special publication. Soil Science Society of America and America Society of Agronomy, Madison, WI

Snapp, S.S., Silim, S.N., 2002. Farmer preferences and legume intensification for low nutrient environments. *Plant Soil* 245(1), 181–192

Snapp, S.S., Rohrbach, D.D., Simtowe, F., Freeman, H.A., 2002. Sustainable soil management options for Malawi: Can smallholder farmers grow more legumes? *Agric. Ecosyst. Environ.* 91(1–3), 159–174

Sogbedji, J.M., van Es, H.M., Agbeko, K.L., 2006. Cover cropping and nutrient management strategies for maize production in western Africa. *Agron. J.* 98(4), 883. https://doi.org/10.2134/agronj2005.0025

Tittonell, P., Giller, K.E., 2013. When yield gaps are poverty traps: The paradigm of ecological intensification in African smallholder agriculture. F. Crop. *Res* 143, 76–90. https://doi.org/10.1016/j.fcr.2012.10.007

Tittonell, P., Vanlauwe, B., Leffelaar, P.A., Shepherd, K.D., Giller, K.E., 2005. Exploring diversity in soil fertility management of smallholder farms in western Kenya: II. Within-farm variability in resource allocation, nutrient flows and soil fertility status. *Agric. Ecosyst. Environ.* 110(3–4), 166–184

Tittonell, P., Vanlauwe, B., Corbeels, M., Giller, K.E., 2008. Yield gaps, nutrient use efficiencies and response to fertilisers by maize across heterogeneous smallholder farms of western Kenya. *Plant Soil* 313(1–2), 19–37

Tittonell, P., Muriuki, A., Klapwijk, C.J., Shepherd, K.D., Coe, R., Vanlauwe, B., 2013. Soil heterogeneity and soil fertility gradients in smallholder farms of the East African highlands. *Soil Sci. Soc. Am. J.* 77(2), 525–538. https://doi.org/10.2136/sssaj2012.0250

Tully, K., Sullivan, C., Weil, R., Sanchez, P., 2015. The State of soil degradation in sub-Saharan Africa: Baselines, trajectories, and solutions. *Sustainability* 7(6), 6523–6552. https://doi.org/10.3390/su7066523

Uzoh, I.M., Igwe, C.A., Okebalama, C.B., Babalola, O.O., 2019. Legume-maize rotation effect on maize productivity and soil fertility parameters under selected agronomic practices in a sandy loam soil. *Sci. Rep.* 9(1), 8539. https://doi.org/10.1038/s41598-019-43679-5

Van der Velde, M., Folberth, C., Balkovič, J., Ciais, P., Fritz, S., Janssens, I.A., Obersteiner, M., See, L., Skalský, R., Xiong, W., Peñuelas, J., 2014. African crop yield reductions due to increasingly unbalanced nitrogen and phosphorus consumption. *Glob. Chang. Biol.* 20(4), 1278–1288. https://doi.org/10.1111/gcb.12481

Van Ittersum, M.K., Cassman, K.G., Grassini, P., Wolf, J., Tittonell, P., Hochman, Z., 2013. Yield gap analysis with local to global relevance-a review. *Field Crops Res.* 143, 4–17

Van Ittersum, M.K., Van Bussel, L.G.J., Wolf, J., Grassini, P., Van Wart, J., Guilpart, N., Claessens, L., De Groot, H., Wiebe, K., Mason-D'Croz, D., Yang, H., Boogaard, H., Van Oort, P.A.J., Van Loon, M.P., Saito, K., Adimo, O., Adjei-Nsiah, S., Agali, A., Bala, A., Chikowo, R., Kaizzi, K., Kouressy, M., Makoi, J.H.J.R., Ouattara, K., Tesfaye, K., Cassman, K.G., 2016. Can sub-Saharan Africa feed itself? *Proc. Natl. Acad. Sci. U. S. A.* 113(52), 14964–14969

van Straaten, P., 2011. The geological basis of farming in Africa. In: Bationo, A., Waswa, B., Okeyo, J., Maina, F., Kihara, J. (eds.), *Innovations as Key to the Green Revolution in Africa*. Springer, Dordrecht. https://doi.org/10.1007/978-90-481-2543-2_3

Vanlauwe, B., 2004. Integrated soil fertility management research at TSBF: The framework, the principles and their application. In: A. Bationo (ed.), *Managing Nutrient Cycles to Sustain Soil Fertility in Sub-Saharan Africa*. Academy Science Publishers, Nairobi, Kenya, pp. 25–42

Vanlauwe, B., Aihou, K., Houngnandan, P., Diels, J., Sanginga, N., Merckx, R., 2001. Nitrogen management in 'adequate' input maize-based agriculture in the derived savanna benchmark zone of Benin Republic. *Plant Soil* 228(1), 61–71

Vanlauwe, B., Giller, K.E., 2006. Popular myths around soil fertility management in sub-Saharan Africa. *Agric. Ecosyst. Environ.* 116, 34–46. https://doi:10.1016/j.agee.2006.03.016

Vanlauwe, B., Tittonell, P., Mukalama, J., 2006. Within-farm soil fertility gradients affect response of maize to fertiliser application in western Kenya. *Nutr. Cycl. Agroecosyst.* 76(2–3), 171–182

Vanlauwe, B., Bationo, A., Chianu, J., Giller, K.E., Merckx, R., Mokwunye, U., Ohiokpehai, O., Pypers, P., Tabo, R., Shepherd, K.D., Smaling, E.M.A., Woomer, P.L., Sanginga, N., 2010. Integrated soil fertility management: Operational definition and consequences for implementation and dissemination. *Outlook Agric.* 39(1), 17–24. https://doi.org/10.5367/000000010791169998

Versteeg, M.N., Amadji, F., Eteka, A., Gogan, A., Koudokpon, V., 1998. Farmers' adoptability of Mucuna fallowing and agroforestry technologies in the coastal savanna of Benin. *Agric. Syst.* 56(3), 269–287. https://doi.org/10.1016/S0308-521X(97)00041-3

Viechtbauer, W., 2010. Conducting meta-analyses in R with the metafor package. *J. Stat. Software* 36, 1–48

Waithaka, M.M., Thornton, P.K., Herrero, M., Shepherd, K., 2006. Bio-economic evaluation of farmers perceptions of viable farms in western Kenya. *Agric. Syst.* 90(1–3), 243–271

Wickham, H., 2016. *ggplot2 Elegant Graphics for Data Analysis (Use R!)*. Springer, New York. https://doi.org/10.1007/978-0-387-98141-3

Wiel, R.R., Mughogho, S.K., 2000. Sulfur nutrition of maize in four regions of Malawi. *Agron. J.* 92(4), 649–656

Windmeijer, P.N., Andriesse, W., 1993. *Inland Valleys in West Africa: An Agroecological Characterization of Rice-Growing Environments*. Publication, 52. International Institute for Land Reclamation and Improvement (ILRI), Wageningen

Wood, S.A., Tirfessa, D., Baudron, F., 2018. Soil organic matter underlies crop nutritional quality and productivity in smallholder agriculture. *Agric. Ecosyst. Environ.* 266, 100–108

Zingore, S., 2011. Maize productivity and response to fertilizer use as affected by soil fertility variability, manure application, and cropping system. *Better Crop* 95, 4–6

Zingore, S., Manyame, C., Nyamugafata, P., Giller, K.E., 2005. Long-term changes in organic matter of woodland soils cleared for arable cropping in Zimbabwe. *Eur. J. Soil Sci.* 57, 727–736

Zingore, S., Murwira, H.K., Delve, R.J., Giller, K.E., 2007a. Influence of nutrient management strategies on variability of soil fertility, crop yields and nutrient balances on smallholder farms in Zimbabwe. *Agric. Ecosyst. Environ.* 119(1–2), 112–126. https://doi.org/10.1016/j.agee.2006.06.019

Zingore, S., Murwira, H.K., Delve, R.J., Giller, K.E., 2007b. Soil type, management history and current resource allocation: Three dimensions regulating variability in crop productivity on African smallholder farms. *Field Crops. Res.* 101(3), 296–305. https://doi.org/10.1016/j.fcr.2006.12.006

Zingore, S., Delve, R., Nyamangara, J., Giller, K., 2008. Multiple benefits of manure: The key to maintenance of soil fertility and restoration of depleted sandy soils on African smallholder farms. *Nutr. Cycl. Agroecosyst.* 80(3), 267–282

Zingore, S., Gonzalez-Estrada, E., Delve, R.J., Dimes, J., Herrero, M., Giller, K.E., 2009. An integrated evaluation of resource use options for smallholder farms in Zimbabwe. *Agric. Syst.* 101, 57–68

Zinn, Y.L., Lal, R., Bighama, J.M., Resck, D.V.S., 2007. Edaphic controls on soil organic carbon retention in the Brazilian cerrado: Texture and mineralogy. *Soil Sci. Soc. Am. J.* 71(4), 1204–1214. http://doi.org/10.2136/sssaj2006.0014

6 Cover Cropping for Managing Soil Organic Carbon Content

Amitava Chatterjee

CONTENTS

6.1 INTRODUCTION

Agricultural land use is responsible for the deterioration of soil organic carbon (SOC) over time through the accelerated rates of mineralization and erosion (Lal 2018). This decline in SOC is most evident under intensive soil management practices, like conventional tillage. In addition, the lack of ground cover during fallow and after harvest leads to loss of soil organic matter (SOM) through erosion. Subtracting current (2010) SOC stocks from historic SOC stocks, Sanderman et al. (2017) found that 37 Pg C, 75 Pg C, and 133 Pg C was lost due to land use change in the upper 0.3 m, 1 m, and 2 m of soil, respectively. It is a challenge to enhance productivity with minimum impact on SOC (Wittwer et al. 2017). Worldwide, croplands could sequester between 0.90 and 1.85 Pg C yr^{-1}, which is 26–53% of the 4 per 1000 initiative; about 94% of SOC within 15.9 million km^2 under agricultural land use could potentially be used to sequester C through improved sustainable management practices (Zomer et al. 2017). The depletion of SOC pools has resulted from three major processes: decomposition, leaching, and erosion, so the most feasible option to increase SOC storage should therefore be to increase C inputs (Poeplau and Don 2015; Chenu et al. 2019).

Incorporation of cover crop residues increases biomass C return and protects C losses from erosion (Lal 2015). Several researchers like, Wagger et al. (1998), Williams and Weil (2004), and Hubbard et al. (2013), have established that cover crops offer the opportunity to enhance soil quality by their positive influence on C dynamics via SOM maintenance and nutrient cycling. Hubbard et al. (2013) reported that biomass residue input was critical in adding C through retaining nitrogen (N) in SOM and improving soil physical properties in the very sandy soils of the southwestern Coastal Plain region of the United States. Kaye and Quemada (2017) identified three pathways of cover crop management influencing climate change: (1) cover crop could mitigate warming by 100–150 g CO_2 e m^{-2} yr^{-1}; (2) change in surface albedo might mitigate 12–46 g CO_{2e} m^{-2} yr^{-1} over a 100-yr time horizon and (3) cover crop could also enable climate change adaptation through reduced erosion, increased soil water management options during drought or flooding, and retention of N mineralized due to warming. In 2020, growers are predicted to plant approximately 8.1 million ha of the cover crop; assuming cover crop C sequestration potential of 2.02 Mg CO_2 e ha^{-1} yr^{-1}, it could be estimated to offset 60×10^6 Mg CO_2 e yr^{-1}, which is equivalent to the amount of greenhouse gases emitted by 12.8 million passenger vehicles, 14.9 coal powered plants, or 6.6 million homes annually (Tellatin and Myers 2018). While the concept of cover crop is not new, an enumeration of its role

DOI: 10.1201/9781003102762-6

in SOC sequestration is not nearly as extensive as that on nitrogen and other ecosystem services or ESs (Wagger et al. 1998, Tellatin and Myers 2018). Two recent meta-analysis studies by Poeplau and Don (2015) and Jian et al. (2020) have estimated that the mean rate of SOC sequestration from cover cropping was 0.12±0.03 (n= 61) and 0.16±0.06 (n=131) Pg C yr^{-1}, respectively. Both studies indicated that substantially fewer data were available for the changes in SOC for soil layers below 30 cm, and most studies were reported after less than 5 years of data collection; Poeplau and Don (2015) predicted that the steady state could be reached after 155 yrs of cover crop cultivation. The potential of cover crop to increase SOC depends on three processes: (1) quantity of above- and belowground biomass input, (2) rate of biomass decomposition, and (3) cover crop-derived C assimilation rate in the SOC pool. In this chapter, factors controlling these three processes are discussed at length to understand the current advances and identify the knowledge gaps to formulate management decisions and enhance the SOC sequestration potential of cover crop management practice.

6.2 COVER CROP BIOMASS C INPUT

Criteria for the significant amount of cover crop biomass input depends on the cover crops' ease of establishment, growth rate speed, and disease and stress tolerance (Table 6.1). However, these

TABLE 6.1
Characteristics of Cover Crop Species Contribute to Soil Organic Carbon Sequestration Potential

Cover Crop Species	Scientific Name	Characterisitics Linked to Improve Soil Carbon
Cool season and annual grass		
Sorguhm-Sudangrass hybrid	*Sorghum bicolor × S. bicolor var. Sudanese*	Unrivaled for adding organic matter. These tall, fast-growing, heat-loving annual grasses produce ample biomass. Roots fractured subsoil compaction.
Annual ryegrass	*Lolium multiflorum*	Ryegrass is a quick-growing, with an extensive dense yet shallow root system. It improves water infiltration and soil tilth.
Rye	*Secale cereale*	The hardiest of cereals, rye can be seeded in late fall and still provide considerable dry matter. It can grow on infertile, sandy, or acidic soil or on poorly prepared land and best in cool temperate zones. An excellent source of residue in no-till and minimum tillage systems and as a straw source, rye provides 11 Mg ha^{-1}.
Barley	*Hordeum vulgare*	Inexpensive and easy to grow, barley provides exceptional erosion control in semi-arid regions and in light soils. Deep fibrous root system can reach as deep as 2 m.
Brassicas		
Mustard/Canola	*Brassica spp.*	They can provide greater than 80% soil coverage. Depending on growing condition and soil type, they produce biomass up to 9 Mg $ha^{-}1$, and roots can grow up to 1.8 m.
Annual legume		
Cowpeas	*Vigna unguiculata*	Thrive in hot and moist zones, protect soil from erosion. It can fix about 145 kg N ha^{-1}. Leafy and prostrate cultivars are best for erosion prevention.
Crimson clover	*Trifolium incarnatum*	With its rapid and robust growth, it provides season-long erosion control.
Hairy Vetch	*Vicia villosa*	Hairy vetch can improve root zone water recharge over winter by reducing runoff and allowing more water to penetrate the soil profile through macropores created by residue. Hairy vetch improves topsoil tilth creating a loose and friable soil structure.

characteristics are further governed by plant phenology or traits, growing conditions, and management decisions (planting, seeding rate, termination method, etc.). Wendling et al. (2016) divided 20 cover crop species into five groups on the basis of their shoot and root phenology. First, the "biomass" group is characterized by high shoot and root biomass, and comprises sunflower (*Helianthus annuus*), faba bean (*Vicia faba*), and white mustard (*Sinapis alba*). Second, the "length" group was characterized by high root length density and comprises turnip (*Brassica rapa campestris*) and niger (*Guizotia abyssinica*). Third, the "intermediate" group is comprised of Indian mustard (*Brassica juncea*), oilseed radish (*Raphanus sativus oleiformis*), oat (*Avena strigosa*), and berseem clover (*Trifolium alexandrium*). Fourth, the "diameter" group is characterized by large root diameter and high leaf dry matter content, and comprises foxtail millet (*Setaria italica*), sorghum (*Sorghum sudanese*), buckwheat (*Fagopyrum esculentum*), and flax (*Linum usiatissium*). The fifth group, "specific leaf area," is characterized by high root tissue density and nitrogen and phosphorus concentration, and is comprised of lentil (*Lens nigricans*), common vetch (*Vicia sativa*), and hemp (*Cannabis sativa*). Based on 389 studies, Ruis et al. (2019) established the following sequence of cover crop species based on biomass production: sorghum (*Sorghum* sp.) (5.99 Mg ha^{-1}) >sunn hemp (*Crotalaria juncea* L.) (5.77 Mg ha^{-1}) >millet (*Pennisetum glaucum* L.) (4.95 Mg ha^{-1}) >rye (Secale cereale L.) (4.93 Mg ha^{-1})>crimson clover (3.53 Mg ha^{-1}) >barley =hairy vetch (3.22 Mg ha^{-1}) >annual ryegrass (3.02 Mg ha^{-1}) >oat (3.00 Mg ha^{-1}). However, the popularity of cover crop varies with region. Singer (2008) surveyed the cover crop adoption in the Corn Belt. They found that winter wheat and cereal rye are dominant in Indiana and Illinois; cereal rye and oat in Iowa; and oat in Minnesota. Overall, growers prefer cover crops that overwinter (68% of response) and fix nitrogen (64%).

Studies in the United States clearly indicate that cereal species produce significantly higher above and belowground biomass than oilseed or legume (Table 6.2). Appelgate et al. (2017) found that rye and rye mixtures produced the highest spring aboveground biomass (758 kg ha^{-1}). Also, rye accounted for more than 79% of spring aboveground biomass accumulation, which was 50% higher than produced by triticale and camelina monoculture. Kuo et al. (1997) also reported annual ryegrass and cereal rye had consistently higher aboveground biomass followed by hairy vetch and Austrian winter pea before incorporation (Table 6.2). Moreover, Wortman et al. (2012) found that monocultures of mustard family (2418 kg ha^{-1}) were twice as productive as legume monocultures (1216 kg ha^{-1}) and concluded that mustard species (Idagold mustard, Pacific Gold mustard [*Brassica juncea* L.], Czern and Coss, oilseed radish, and dwarf essex rape [*B. napus* L.] were fast-growing, competitive, and productive cover crops well suited for early spring growth in the western Corn Belt, and legume species (hairy vetch, field peas, crimson clover, and chickling vetch) were generally less impressive. Mixing species with complementary traits is a key strategy for increasing biomass production. Maximum biomass production benefits can be achieved from cover crop mixtures that include species exhibiting differences in architecture, physiology, and/or phenology (Finney et al. 2016, Blesh 2018). In Georgia, Sainju et al. (2005) reported that biomass C (2260–3512 kg ha^{-1}) contents in biculture (rye + vetch) were greater than that in monocultures under cotton-based production systems (Table 6.2). Ranells and Wagger (1996) reported that the mean cover crop dry matter production was in the order of rye + hairy vetch >hairy vetch >rye-crimson clover >rye >crimson clover. When grown in biculture with rye, hairy vetch accumulated more dry matter compared with crimson clover, both as a proportion of the biculture and as dry matter yield (Table 6.2). However, Appelgate et al. 2017, Finney et al. 2016, and Murrell et al. 2017 reported an opposite trend. Appelgate et al. (2017) concluded that cover crop mixtures had no advantage over monocultures except for increasing the fall stand density. Murrell et al. (2017) observed that cereal rye had the highest spring biomass in all mixtures, so the relative contribution of biomass from other species was comparatively low. From a study of 18 cover crop treatments preceding corn (*Zea mays* L.), Finney et al. (2016) found that cover crop mixtures did not produce more biomass than that by high-yielding monocultures; and concluded that opportunities to increase biomass-driven services with mixtures might be limited. Murrell et al. (2017) summarized that: (1) cover crop mixtures

TABLE 6.2
Biomass Carbon Input (Mg C ha^{-1}) by Different Cover Crop Species and Mixtures Across the United States

Location	Soil Texture	Cover Crop	Biomass C (Mg ha^{-1}) Above	Below	Total	Reference
Washington, USA	Silt loam	Rye	1.63	0.88	2.51	Kuo et al. (1997)
		Winter pea	1.02	0.68	1.70	
		Ryegrass	1.91	1.25	3.16	
		Vetch	1.21	0.43	1.64	
		Canola	1.14	0.33	1.47	
Georgia, USA	Sandy loam	Rye	1.80	0.15	1.94	Sainju et al. (2005)
		Vetch	1.69	0.12	1.80	
		Vetch + Rye	2.82	0.18	3.01	
North Carolina, USA	Loamy sand	Rye			1.68	Ranells & Wagger (1996)
		Crimson clover			1.43	
		Hairy vetch			1.68	
		Rye +Clover			1.93	
		Rye+ Vetch			2.28	
Iowa, USA	Loamy	Rye			0.345	Appelgate et al. (2017)
		Triticale			0.176	
		Camelina			0.122	
		Turnip			0.051	
		Hairy vetch			0.034	
		Rye + Canola			0.338	
		Rye + Camelina			0.329	
		Rye + Vetch			0.366	
		Rye + Camelina + Vetch			0.319	
		Triticale + Camelina			0.193	
		Triticale+ Vetch			0.137	
		Triticale+ Camelina + Vetch			0.223	

maintained higher diversity when allowed sufficient growth in fall; (2) cereal rye dominated mixtures in spring; (3) grasses over-performed in mixtures compared to their biomass production in monoculture; (4) brassicas underperformed in mixtures compared to monocultures; and (5) legume growth in the mixture was dependent on species and planting time. Tools for growers to predict biomass along with models predicting mixture composition in different environmental conditions could increase the biomass C input (Blesh 2018).

Besides species, cover crop biomass production also depends on other factors, like soil characteristics, cropping systems, and management. Beehler et al. (2017) found that rye biomass production was the lowest in depressions compared to that at summit and side slope positions. The cold and wet conditions in depressional sites delayed growth, in contrast to warm conditions at summit and side slope positions. Cover crop biomass production depends on crop rotation. Ruis et al. (2019) found that cover crop biomass production follows the order of vegetables >small grain-other systems >maize-other systems-small grain >maize >small grain. Lengthening the duration of cover

TABLE 6.3
Changes in Cover Crop Biomass (kg ha^{-1}) and C:N Ratio with Duration (Adapted from: Clark et al. 1997)

	Rye		Vetch		Vetch + Rye	
	Mass	C:N	Mass	C:N	Mass	C:N
Fall	510	13	490	10	700	11
Late March	1290	23	1890	10	2590	13
Early April	1990	30	2840	10	3480	12
Late April	3210	42	3350	11	5060	14
Early May			4500	11	6090	14

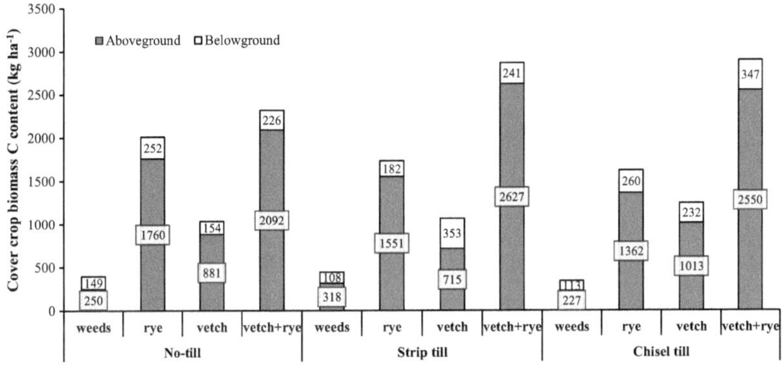

FIGURE 6.1 Effect of tillage practices and cover crop species on above- and belowground cover crop biomass carbon input (kg ha^{-1}) (Adapted from: Sainju et al. 2005).

crop growing season increases the cover crop biomass production (Table 6.3). Clark et al. (1997) found that cover crop dry matter yield and N content of all cover crops increased as spring kill was delayed. Early to mid-fall planting with late spring or summer termination produces the highest, whereas fall planting and termination yield the lowest biomass contribution (Ruis et al. 2019). Cover cropping effects increase with decrease in management intensity (Wittwer et al. 2017). Sainju et al. (2005) observed that aboveground biomass yield and C and N content in cover crops were greater in strip till with biculture than those in other treatments, except in chisel till with vetch and biculture. However, the belowground biomass yield was greater in chisel till with biculture than that in no-till, strip till, and chisel till with weeds (Figure 6.1). After 12 years, Olson et al. (2014) determined that C gain from cover crop was 10.2, 5.9, and 1.2 Mg C ha^{-1} (within 75 cm) under no-till, chisel, and moldboard plow, respectively.

6.3 DECOMPOSITION AND MINERALIZATION OF COVER CROP BIOMASS C

For maximizing the potential of cover crops to sequester C, it is essential to develop an understanding of the kinetics of C turnover during residue decomposition controlled primarily by C:N ratio and concentrations of lignin and other structural carbohydrates in various cropping systems (Wagger et al. 1998, Poeplau and Don 2015). The C:N ratios of above- and belowground portions of different cover crop species are presented in Table 6.4. Cereals have a higher C:N ratio than that of oilseed and legume. However, the C:N ratio, particularly for the same cereal species can have a wide range,

TABLE 6.4
Carbon to Nitrogen Ratio of Cover Crop Species

Location	Common Name	Latin Name	C:N Ratio		Reference
	Poaceae		Shoot	Root	
Santa Fe, Argentina	Oat	*Avena sativa* L.	39.0		Duval et al. (2016)
Pennsylvania, USA			20.5		Finney et al. (2016)
California, USA			33.6		Hu et al. (1997)
Georgia, USA			45.8		Quemada & Cabrera (1995)
Iowa, USA			9.6		Applegate et al. (2017)
Illinois, USA	Cereal rye	*Secale cereale* L.	34.7	40.3	Sievers & Cook (2018)
Georgia, USA			63.9		Quemada & Cabrera (1995)
Iowa, USA			12.3		Appelgate et al. (2017)
Georgia, USA			42.0	38.7	Sainju et al. (2005)
North Carolina, USA			40.0		Ranells & Wagger (1996)
Pennsylvania, USA			44.6		Finney et al. (2016)
Washington, USA			34.5	63	Kuo et al. (1997)
	Brassicaceae				
Pennsylvania, USA	Forage radish	*Raphanus sativus* L. *oleiformis*	15.4		Finney et al. (2016)
Pennsylvania, USA	Canola	*Brassica napus* L.	24.1		Finney et al. (2016)
Iowa, USA			8.65		Appelgate et al. (2017)
Washington, USA			28	76	Kuo et al. (1997)
Iowa, USA	Camelina	*Camelina sativa* L. *Crantz*	9.2		Applegate et al. (2017)
Iowa, USA	Turnip	*Brassica rapa L.*	7.7		Applegate et al. (2017)
	Fabaceae				
Pennsylvania, USA	Hairy Vetch	*Vicia villosa* Roth	9.5		Finney et al. (2016)
North Carolina, USA			11		Ranells & Wagger (1996)
Georgia, USA			11.3	17.7	Sainju et al. (2005)
Iowa, USA			9.6		Applegate et al. (2017)
Illinois, USA			9.5	17.3	Sievers & Cook (2018)
Washington, USA			10.0	26.5	Kuo et al. (1997)
California, USA			13.3		Hu et al. (1997)
Pennsylvania, USA	Sunhemp	*Crotalaria juncea L.*	10.5		Finney et al. (2016)
Pennsylvania, USA	Red clover	*Trifolium pratense L.*	10.1		Finney et al. (2016)
North Carolina, USA		*Trifolium incarnatum*	17		Ranells & Wagger (1996)

depending on total dry matter produced and time of termination. Leguminous cover crops have a narrower C:N ratio than cereal and oilseed because of the biological N fixation. Legume residues decompose faster; they also can absorb low available nutrients in the soil profile and can help in increasing the concentration of nutrients in surface layers of soil (Fageria et al. 2005). Above- and belowground portions have different C:N ratios; generally, belowground has a higher C:N ratio than the top portion (Table 6.4). These differences in the C:N ratio of cover crop influence decomposition. Hairy vetch aboveground (decomposition rate or k=0.4505) and root (k=0.6821) biomass decomposed at a faster rate than aboveground (k= 0.1368) and root (k=0.1866) biomass of cereal rye; growers should consider delaying termination of hairy vetch until close to cash crop planting (Sievers and Cook 2018).

Early termination of the cover crop results in a narrower C:N ratio and rapid decomposition, but the total aboveground biomass production is reduced. Clark et al. (1997) observed that with the delay in harvesting, the C:N ratio of rye and mixture with vetch became wider due to immobilization of N, but the C:N ratio of vetch residues hardly changes over time (Table 6.3). Small grain

TABLE 6.5
Concentrations of Carbon, Nitrogen, Cellulose, Hemicellulose and Lignin of Cover Crop Species

Cover Crop	Portion	C ($g\ kg^{-1}$)	N ($g\ kg^{-1}$)	Cellulose	Hemicellulose	Lignin	Reference
Oat	Aboveground	484	12.4	294	318	42	Duval et al. (2016)
Rye	Shoot	399	11.5	317	271	24.1	Sievers & Cook (2018)
Rye	Root	335	8.3	477	324	57.3	Sievers & Cook (2018)
Rye		452	11	254	259	35	Ranells & Wagger (1996)
Hairy vetch		443	40	202	77	88	Ranells & Wagger (1996)
Hairy vetch	Shoot	399	41.9	232	64	58.2	Sievers & Cook (2018)
Hairy vetch	Root	286	16.5	531	144	104.1	Sievers & Cook (2018)
Crimson clover		438	26	215	117	58	Ranells & Wagger (1996)

cover crops are usually killed at a stage when the C:N ratio is >30:1 (Reeves 1994). Five months after cover crop termination, belowground cover crop inputs were three times more likely to remain in the soil than the aboveground inputs; given six times greater abundance of belowground cover crop biomass-derived C than shoots in the soil represents three times greater relative contribution (Austin et al. 2017). Cover crops like spring barley are winter-killed, whereas some cover crops like fall rye and annual ryegrass remain vegetative over the winter months and continue growth in early spring. Liu et al. (2005) observed that annual ryegrass had greater shoot biomass and root mass and length density than fall rye. Wagger et al. (1998) found higher initial rates of C mineralization for fall compared to spring harvested vetch; nearly 35% of the added C in fall vetch had mineralized compared to 50% of that amount from spring vetch.

Besides the C:N ratio, the chemical composition of plant residues is considered an important determinant of their subsequent C turnover in soils. Cover crop residues represent a wide range of cellulose, hemicellulose, and lignin concentrations (Table 6.5). Concentrations of structural polymers, like lignin and cellulose, evidently segregate different cover crop species. Cereals, oat, and rye have comparatively higher cellulose and hemicellulose and lower lignin than legumes and vetch (Table 6.5). Concentrations of cellulose, hemicellulose, and lignin are higher for the root than shoot portion for both rye and hairy vetch. Lower C:N ratios and lignin: N ratios showed the different qualities in vetch residues; approximately 40–50% of vetch residue might be degraded 30 days after incorporation (Duval et al. 2016). Leaves from crimson clover, rye, wheat, and oat may have a lower C:N ratio and lignin concentration than stems, while the C:N ratio of crimson clover (leaves and stems) averages approximately 60% lower than the mean of grasses (21:1 vs. 53:1) (Wagger et al. 1998).

6.4 IMPACT ON SOC SEQUESTRATION

Soil OC may be divided into two parts: recalcitrant, which is mainly humic substances and is relatively stable; and labile organic C, which comes from cover crops (shoots, roots, root senescence, and exudates), a major microbial C source resulting in an increase of production of microbially derived binding agents (Liu et al. 2005). The time since introduction of cover crops in crop rotations is linearly correlated with SOC stock change (R^2=0.19) (Poeplau and Don 2015). Using ^{14}C isotope labeling, Mutegi et al. (2013) estimated that more than a 30-yr period of continuous autumn fodder radish would contribute 4.9 Mg C ha^{-1} with a residence time of more than 20 yrs of storage in soil. Annual return of approximately 5000 kg ha^{-1} of cover crop biomass or 2000 kg C ha^{-1} (assuming C concentration of 450 g C kg^{-1}) is considered to be the maintenance rate to avoid reduction of SOC concentration (Kuo et al. 1997). Recently, two meta-analyses on the consequences of cover cropping on SOC, by Poeplau and Don (2015) and Jian et al. (2020), have estimated SOC sequestration potential of 0.32±0.08 and 0.56 Mg C $ha^{-1}yr^{-1}$, respectively. Using RothC C turnover model, Poeplau and Don (2015) predicted that a new steady state of SOC stock would be reached after 155 yrs of cover crop cultivation with a mean SOC stock accumulation of 16.7±1.5 Mg ha^{-1} for a soil depth of 22 cm. They also estimated that cover crop has global SOC sequestration potential of 0.12 ±0.03 Pg C yr^{-1}, which would compensate for 8% of the direct annual greenhouse gas emission from agriculture. An increase in SOC after cover crop incorporation is most successful for rotation with bare fallow because cover crop can increase the C input and reduce C loss through erosion (Kaye and Quemada 2017). Plaza-Bonilla et al. (2016) determined that under bare fallow treatment, loss of SOC was −595 and −735 kg C $ha^{-1}yr^{-1}$ under one and two grain legumes (as cash crop), respectively; and the rate of SOC loss was reduced by 13 and 67% after cover crop incorporation, respectively. Cover crop biomass positively affected SOC changes while the C:N ratio of cover crop biomass was negatively correlated with SOC changes (Jian et al. 2020). Poeplau and Don (2015) found that only 13 out of 139 plots showed SOC depletion after introducing cover crops, which might be due to the priming effect, microbial degradation of stable C compounds due to increase in substrate supply through cover crop residues, and masking of small changes in SOC due to spatial heterogeneity.

Cover crop biomass production is significantly controlled by climate, cover crop species, soil type, and management practices (tillage, fertilizer, irrigation). Jian et al. (2020) found that medium-textured soils had the highest SOC stocks (39 and 37 Mg ha^{-1} with and without cover crops, respectively). Further, fine-textured soils showed the greatest increase in SOC after the inclusion of cover crop (mean change of 39.5%), whereas coarse textured soils (11.4%) and medium-textured soils (10.3%) had comparatively smaller changes in SOC. In comparison, soils in temperate climates had greater changes (18.7%) than those in tropical climates (7.2%) (Jian et al. 2020). Quantity of biomass input and quality (C:N ratio) driven by cover crop species interact with growing conditions and are major drivers in C dynamics. Kuo et al. (1997) reported that only rye and ryegrass produced a quantity of SOC from both above and belowground biomass that exceeded the maintenance rate (2 Mg C ha^{-1} yr^{-1}) and significantly increased SOC concentrations relative to the control. After six growing seasons of cover crops on soybean monoculture, more than 3.0 Mg C ha^{-1} yr^{-1} were supplied by gramineous cover crop species and less than 2 Mg C ha^{-1} yr^{-1} by the vetch treatment. Further, greater C input by gramineous species enhanced water storage and increased C input than vetch under no-till in Pampas Agriudolls, Argentina (Duval et al. 2016). On the contrary, Mazzoncini et al. (2011) found that non-legume (rye and brown mustard), low N supply legume (crimson and squarrosum clover), and high N supply legume (subterranean clover and hairy vetch) cover crops increased SOC content by 0.17, 0.41 and 0.43 Mg C ha^{-1} yr^{-1} in a long-term experiment (1993 to 2008) in central Italy (Figure 6.2). However, Zhou et al. (2016) found similar increases in soil C derived from legume (8.7–11.4%) and non-legume (9.0–11.3%).

The complementary effect of cover crops used in conjunction with conservation tillage had been well established (Reeves 1994). Based on a 12-yr (2000–2012) study in Illinois, Olson et al. (2014) reported that cover crop treatments had more SOC stock than that without cover crops for the same soil layer and tillage treatment (Figure 6.3). Using the pre-treatment baseline 2000 SOC stocks, the SOC gains were highest with the combination of no-till and cover crop. Wittwer et al. (2017) reported that cover crop effects on crop yield were the highest in the organic system with reduced tillage (+24%), intermediate in the organic system with tillage (13%), and in the conventional system with no tillage (8%); and the lowest in the conventional system with tillage (2%). Soil OC changes from cover cropping are associated with improvements in soil quality, specifically with decreased run off and erosion (Ladoni et a. 2016, Jian et al. 2020). Ladoni et al. (2016) observed that the highest level of SOC at depressional sites, followed by that at summit and side slopes, due to the higher aboveground inputs of C from both red clover and rye residues in depressions than on slopes. There were considerable differences in the magnitudes of topographical trends probably resulting from spatial distribution of soil C in the fields due to erosion. Thus, preferential placement of cover crops

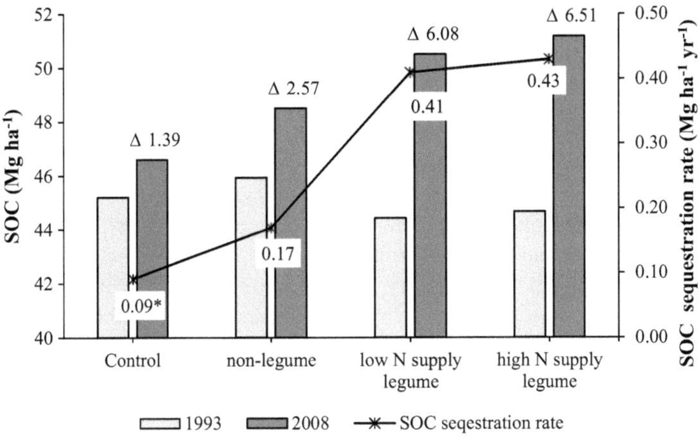

FIGURE 6.2 Mean effect of cover crop type on soil organic carbon (Mg ha^{-1}) and difference (Δ) in SOC (Mg ha^{-1}) between 1993 and 2008, and SOC sequestration rate (Mg C ha^{-1} yr^{-1}) (* indicates the values) in a loam soil in central Italy (Adapted from: Mazzoncini et al. 2011).

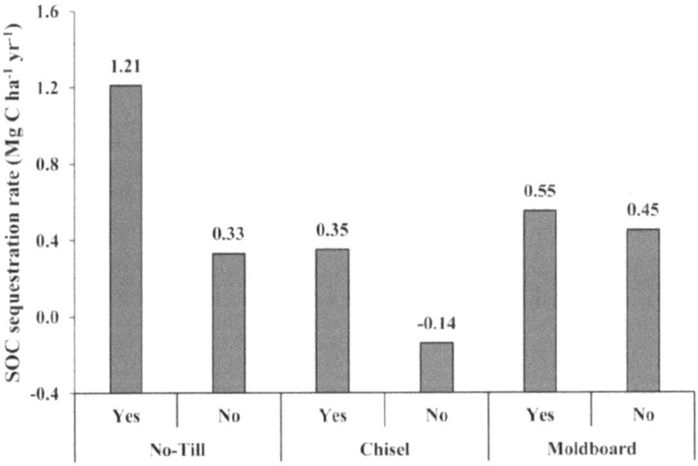

FIGURE 6.3 Effect of tillage practices (no-till, chisel, and moldboard) with (yes) and without (no) cover crops on SOC sequestration rate (Mg C ha^{-1} yr^{-1}) after 12 years in Illinois, United States (Adapted from Olson et al. 2014).

on eroded low-fertility elements of the relief can be a preferred strategy. Williams and Weil (2004) reported that rye left a thick mulch cover, resulting in conservation of soil water early in season and root channels left by forage radish may have provided soybean roots (growing under plowpan soil) with low resistance paths to subsoil water.

6.5 PATHWAYS OF INCREASING SOC

Cover crops' role in SOC contribution can be divided into three major pathways: (1) improving soil structure and aggregation (Fageria et al. 2005, Lal 2015), (2) increasing the diversity of C substrate (Kuo et al. 1997), and (3) encouraging the biological activity (Zhou et al. 2016). Soil structure is essential for water infiltration, aeration, and plant root development (Fageria et al. 2005), and soil aggregate stability is one of the most responsive parameters to cover crop management (Blanco-Canqui et al. 2015). Sometimes, the adoption of cover crops has no immediate effect on SOC, but it supports functions to improve aggregate stability. Large quantities of organic constituents released from cover crop roots result in the production of large amounts of polysaccharide binding agents in the surface soils, increasing the stability of soil aggregates (Liu et al. 2005). Glomalin produced by arbuscular mycorrhizal fungi (AMF) is a strong cementing agent that enhances aggregate stability (Lal 2015). Bowles et al. (2017) found that less intensive tillage and winter cover cropping similarly increased AMF colonization of summer annual cash crop roots by ~30%; growers can optimize the AMF role by adopting legume cover crops and no-till systems. Steele et al. (2012) reported 35% and 41% increase in aggregate stability for two consecutive years under the winter annual cereal cover crops but there was no increase in total organic matter and labile organic matter in the Coastal Plain soils of Maryland. Austin et al. (2017) observed that five months following cover crop termination, belowground cover crop inputs were three times more likely to remain in the soil than were aboveground inputs, and much of the root-derived C was in mineral associated soil fractions. In degraded silty clay loam soils in British Columbia, Liu et al. (2005) found that soils under the annual ryegrass and fall rye had a greater mean weight diameter, and percentage of water stable (2–6 mm) aggregates than control. Furthermore, annual ryegrass and fall rye increased the dilute acid-extractable polysaccharide fraction, which represents active binding agents (Figure 6.4). After four years of cereal rye, Rorick and Kladivko (2017) reported that wet soil aggregate mean weight diameter increased 66% and 29% over control within 0–10 and 10–20 cm depths, respectively, in

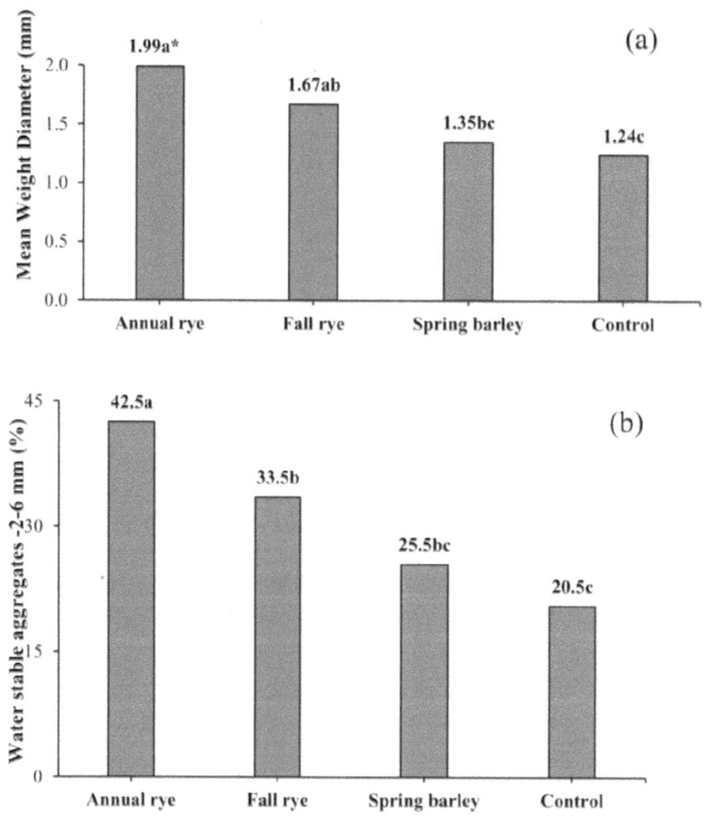

FIGURE 6.4 Changes in (a) mean weight diameter and (b) water stable aggregates (2–6 mm diameter) with cover crop options (Adapted from Liu et al. 2005).

a typical midwestern corn-soybean rotation. Po et al. (2009) found that bare fallow system did not show any increase in mean weight diameter after 3 years, and the only increase was observed for the potato–wheat–red clover system, in the sandy loam soils of Michigan. However, the extent of cover crop effect on soil aggregates depends on the cover crop species and probably on its root structure. Garcia-Gonzalez et al. (2018) found that water stable aggregates in the top 5 cm of soil were larger under barley cover crop than those in the vetch or the fallow; barley maintained a positive pattern for 10 years of study. After 34 years, vetch cover crop with no-tillage increased the wet aggregate stability by 13% at 0–15 cm and significantly improved the initial infiltration rate and field saturated hydraulic conductivity (Nouri et al. 2019).

Addition of cover crop residues can increase the light density fraction of organic matter (Ding et al. 2006). The heavy fraction of carbohydrate, composed mainly of microbial extracellular polysaccharides, increased in response to C inputs from cover crops; and an increase in carbohydrate concentration was well correlated with the soil microbial biomass C levels (Kuo et al. 1997). Wang et al. (2017) observed that permanganate oxidizable C (POXC) was significantly greater following radish (535.7 mg kg^{-1}) than following no cover crop (418.2 mg kg^{-1}) and forage radish had significant impact not only on the surface (0–30 cm) but also for deep horizons (90–105 cm). Positive effects of cover crop on microbial biomass C were reported across continents. Mendes et al. (1999) found that significantly greater soil microbial biomass C under cereal and legume cover cropping than the fallow treatment, in the Willamette silt loam soils of Oregon. Varela et al. (2014) found that cover crops (oat, rye) increased microbial biomass quantity and activity in both soil and residue samples of continuous no-till soybean in Argentina. Legume, non-legume, and mixed cover crop species increased microbial biomass C by 17.3–18.2, 15.6–18.9, and 20.9 kg ha^{-1}, respectively, relative to without any cover crop in temperate Australia (Zhou et al. 2016).

6.6 CONCLUSION

Cover cropping has grabbed the attention of farmers and conservation professionals over the past decade, and many federal agencies and farming and environmental stewardship entities have envisioned its wide adoption. According to the fifth annual cover crop survey (2016–2017) by the Sustainable Agriculture Research and Education (SARE) program and the Conservation Technology Information Center (CTIC), 88% (n =2102 farmers) of growers have adopted cover crop; 86% of respondents have identified soil health as the key benefit of cover crops; and 54% of the surveyed farmers believe that soil health benefits from cover crops began in the first year of use (CTIC, 2017). Soil OC is the main criteria for preserving soil health, but studies to determine the long-term (>10 yrs) effect of cover cropping on SOC are minimal; moreover, they lack depth. Major limitations of current literature on the cover cropping and SOC are the following: (1) very few studies have been conducted on a long-term basis, (2) most observations did not look beyond plow layer (>30 cm); and (3) critical ancillary data like bulk density were missing (Poeplau and Don 2015, Jian et al. 2020). Besides a well-designed experimental design, enumeration of the cover cropping benefit on SOC needs to supply the following information: (1) above- and belowground cover crop and cash crop biomass C input (Mg ha^{-1}); (2) initial and final profile (90 cm soil depth) SOC concentration (g C m^{-2}); (3) soil bulk density (Mg m^{-3}) with incremental depth similar to SOC concentration; (4) cover crop (seeding rate, planting, and killing dates) and cash crop management (tillage, fertilizer, and irrigation); and (5) weather data (daily rainfall and minimum and maximum soil temperature).

Selection of species for monoculture (cereal, legume, and oilseed) and proportion of mixing of these species play a major role in stand establishment and biomass production (Murrell et al. 2017) and hence C input (Figure 6.5). However, the SOC sequestration rate is controlled not only by the quantity but also by the quality (C:N ratio). Future studies should focus on tailoring the recipe of cover crop species (with the consideration of local climate and cropping rotation) and measuring the changes in SOC under different combinations over time. Decomposition rate and C mineralization from above- and belowground components of cover cropping can improve the understanding

FIGURE 6.5 Interseeded cover crops, (a) flax, (b) winter rye, (c) Austrian pea, in sugarbeet, and field after sugarbeet harvest (d) without cover crop, (e) with winter rye and (f) over-winter growth of rye in the same field located at Ada, Minnesota, United States.

and minimize the knowledge gaps regarding the cover crops' ultimate contribution to SOC pools. Tillage practices directly interact with the residue decomposition, and studies on the effect of conservation tillage practices on SOC turnover rate under different cover crops are minimal. A significant amount of SOC is lost through wind and water erosion, and there is a dire need for the design of long-term field studies to determine the integrated effect of conservation tillage practices and cover copping in preserving C loss from erosion (Figure 6.5). From the long-term tillage effect on SOC, we have learned that a small magnitude of changes in SOC has been overshadowed due to spatial heterogeneity; an intensive sampling protocol should be undertaken to minimize the variability. Most of the current crop modeling has not included the cover cropping in their structure under different scenarios due to lack of datasets; availability of more data from cover crop studies will make it possible for the future prediction of changes in SOC under cover cropping.

REFERENCES

Appelgate, S. R., A. W. Lenssen, M. H. Wiedenhoeft, and T. C. Kaspar. 2017. "Cover crop options and mixes for upper midwest corn-soybean systems." *Agronomy Journal* 109(3):968–984. doi: 10.2134/agronj2016.08.0453.

Austin, E. E., K. Wickings, M. D. McDaniel, G. P. Robertson, and A. S. Grandy. 2017. "Cover crop root contributions to soil carbon in a no-till corn bioenergy cropping system." *Global Change Biology Bioenergy* 9(7):1252–1263. doi: 10.1111/gcbb.12428.

Beehler, J., J. Fry, W. Negassa, and A. Kravchenko. 2017. "Impact of cover crop on soil carbon accrual in topographically diverse terrain." *Journal of Soil & Water Conservation* 72(3):272–279. doi: 10.2489/jswc.72.3.272.

Blanco-Canqui, H., T. M. Shaver, J. L. Lindquist, C. A. Shapiro, R. W. Elmore, C. A. Francis, and G. W. Hergert. 2015. "Cover crops and ecosystem services: Insights from studies in temperate soils." *Agronomy Journal* 107(6):2449–2474. doi: 10.2134/agronj15.0086.

Blesh, J. 2018. "Functional traits in cover crop mixtures: Biological nitrogen fixation and multifunctionality." *Journal of Applied Ecology* 55(1):38–48. doi: 10.1111/1365-2664.13011.

Bowles, Timothy M., Louise E. Jackson, Malina Loeher, and Timothy R. Cavagnaro. 2017. "Ecological intensification and arbuscular mycorrhizas: A meta-analysis of tillage and cover crop effects." *Journal of Applied Ecology* 54(6):1785–1793.

Chenu, C., D. A. Angers, P. Barre, D. Derrien, D. Arrouays, and J. Balesdent. 2019. "Increasing organic stocks in agricultural soils: Knowledge gaps and potential innovations." *Soil & Tillage Research* 188:41–52. doi: 10.1016/j.still.2018.04.011.

Clark, A. J., A. M. Decker, J. J. Meisinger, and M. S. McIntosh. 1997. "Kill date of vetch, rye, and a vetch-rye mixture. 1. Cover crop and corn nitrogen." *Agronomy Journal* 89(3):427–434.

Conservation Technology Information Center (CTIC). 2017. *Report of the 2016–17 National Cover Crop Survey*. Joint Publication of the Conservation Technology Information Center, the North Central Region Sustainable Agriculture Research and Education Program, and the American Seed Trade Association: West Lafayette, IN. Available online: https://www.sare.org/Learning-Center/Topic-Rooms/Cover-Crops/Cover-Crop-Surveys.

Ding, G. W., X. B. Liu, S. Herbert, J. Novak, D. Amarasiriwardena, and B. S. Xing. 2006. "Effect of cover crop management on soil organic matter." *Geoderma* 130(3–4):229–239. doi: 10.1016/j.geoderma.2005.01.019.

Duval, M. E., J. A. Galantini, J. E. Capurro, and J. M. Martinez. 2016. "Winter cover crops in soybean monoculture: Effects on soil organic carbon and its fractions." *Soil & Tillage Research* 161:95–105. doi: 10.1016/j.still.2016.04.006.

Fageria, N. K., V. C. Baligar, and B. A. Bailey. 2005. "Role of cover crops in improving soil and row crop productivity." *Communications in Soil Science & Plant Analysis* 36(19–20):2733–2757. doi: 10.1080/00103620500303939.

Finney, D. M., C. M. White, and J. P. Kaye. 2016. "Biomass production and carbon/nitrogen ratio influence ecosystem services from cover crop mixtures." *Agronomy Journal* 108(1):39–52. doi: 10.2134/agronj15.0182.

Garcia-Gonzalez, I., C. Hontoria, J. L. Gabriel, M. Alonso-Ayuso, and M. Quemada. 2018. "Cover crops to mitigate soil degradation and enhance soil functionality in irrigated land." *Geoderma* 322:81–88. doi: 10.1016/j.geoderma.2018.02.024.

Hu, S., N. J. Grunwald, A. H. C. Van Bruggen, G. R. Gamble, L. E. Drinkwater, C. Shennan, and M. W. Demment. 1997. Short-term effects of cover crop incorporation on soil carbon pools and nitrogen availability. *Soil Science Society of America Journal* 61(3):901–911.

Hubbard, R. K., T. C. Strickland, and S. Phatak. 2013. "Effects of cover crop systems on soil physical properties and carbon/nitrogen relationships in the coastal plain of southeastern USA." *Soil & Tillage Research* 126:276–283. doi: 10.1016/j.still.2012.07.009.

Jian, J. S., X. Du, M. S. Reiter, and R. D. Stewart. 2020. "A meta-analysis of global cropland soil carbon changes due to cover cropping." *Soil Biology & Biochemistry* 143: 107735. doi: 10.1016/j.soilbio.2020.107735.

Kaye, J. P., and M. Quemada. 2017. "Using cover crops to mitigate and adapt to climate change. A review." *Agronomy for Sustainable Development* 37(1):4. doi: 10.1007/s13593-016-0410-x.

Kuo, S., U. M. Sainju, and E. J. Jellum. 1997. "Winter cover crop effects on soil organic carbon and carbohydrate in soil." *Soil Science Society of America Journal* 61(1):145–152. doi: 10.2136/sssaj1997.03615995006100010022x.

Ladoni, M., A. Basir, P. G. Robertson, and A. N. Kravchenko. 2016. "Scaling-up: Cover crops differentially influence soil carbon in agricultural fields with diverse topography." *Agriculture, Ecosystems & Environment* 225:93–103. doi: 10.1016/j.agee.2016.03.021.

Lal, R. 2015. "Soil carbon sequestration and aggregation by cover cropping." *Journal of Soil & Water Conservation* 70(6):329–339. doi: 10.2489/jswc.70.6.329.

Lal, R. 2018. "Digging deeper: A holistic perspective of factors affecting soil organic carbon sequestration in agroecosystems." *Global Change Biology* 24(8):3285–3301. doi: 10.1111/gcb.14054.

Liu, A. G., B. L. Ma, and A. A. Bomke. 2005. "Effects of cover crops on soil aggregate stability, total organic carbon, and polysaccharides." *Soil Science Society of America Journal* 69(6):2041–2048. doi: 10.2136/sssaj2005.0032.

Mazzoncini, M., T. B. Sapkota, P. Barberi, D. Antichi, and R. Risaliti. 2011. "Long-term effect of tillage, nitrogen fertilization and cover crops on soil organic carbon and total nitrogen content." *Soil & Tillage Research* 114(2):165–174. doi: 10.1016/j.still.2011.05.001.

Mendes, I. C., A. K. Bandick, R. P. Dick, and P. J. Bottomley. 1999. "Microbial biomass and activities in soil aggregates affected by winter cover crops." *Soil Science Society of America Journal* 63(4):873–881. doi: 10.2136/sssaj1999.634873x.

Murrell, E. G., M. E. Schipanski, D. M. Finney, M. C. Hunter, M. Burgess, J. C. LaChance, B. Baraibar, C. M. White, D. A. Mortensen, and J. P. Kaye. 2017. "Achieving diverse cover crop mixtures: Effects of planting date and seeding rate." *Agronomy Journal* 109(1):259–271. doi: 10.2134/agronj2016.03.0174.

Mutegi, J. K., B. M. Petersen, and L. J. Munkholm. 2013. "Carbon turnover and sequestration potential of fodder radish cover crop." *Soil Use & Management* 29(2):191–198. doi: 10.1111/sum.12038.

Nouri, A., J. Lee, X. H. Yin, D. D. Tyler, and A. M. Saxton. 2019. "Thirty-four years of no-tillage and cover crops improve soil quality and increase cotton yield in alfisols, Southeastern USA." *Geoderma* 337:998–1008. doi: 10.1016/j.geoderma.2018.10.016.

Olson, K., S. A. Ebelhar, and J. M. Lang. 2014. Long-term effect of cover crops on crop yields, soil organic carbon stocks and sequestration. *Open Journal of Soil Science* 4:284–292. http://dx.doi.org/10.4236/ojss.2014.48030.

Plaza-Bonilla, D., J. M. Nolot, S. Passot, D. Raffaillac, and E. Justes. 2016. "Grain legume-based rotations managed under conventional tillage need cover crops to mitigate soil organic matter losses." *Soil & Tillage Research* 156:33–43. doi: 10.1016/j.still.2015.09.021.

Po, E. A., S. S. Snapp, and A. Kravchenko. 2009. "Rotational and cover crop determinants of soil structural stability and carbon in a potato system." *Agronomy Journal* 101(1):175–183. doi: 10.2134/agronj2008.0184.

Poeplau, C., and A. Don. 2015. "Carbon sequestration in agricultural soils via cultivation of cover crops: A meta-analysis." *Agriculture, Ecosystems & Environment* 200:33–41. doi: 10.1016/j.agee.2014.10.024.

Quemada, M., and M. L. Cabrera 1995. CERES-N model predictions of nitrogen mineralized from cover crop residues. *Soil Science Society of America Journal* 59(4):1059–1065.

Ranells, N. N., and M. G. Wagger. 1996. "Nitrogen release from grass and legume cover crop monocultures and bicultures." *Agronomy Journal* 88(5):777–782. doi: 10.2134/agronj1996.00021962008800050015x.

Reeves, D. W. 1994. "Cover crops and rotations." In: *Crops Residue Management, Advances in Soil Science*, ed. J. L. Hatfield and B. A. Stewart, 125–172. Lewis Publishers: Boca Raton, FL.

Rorick, J. D., and E. J. Kladivko. 2017. "Cereal rye cover crop effects on soil carbon and physical properties in southeastern Indiana." *Journal of Soil & Water Conservation* 72(3):260–265. doi: 10.2489/jswc.72.3.260.

Ruis, S. J., H. Blanco-Canqui, C. F. Creech, K. Koehler-Cole, R. W. Elmore, and C. A. Francis. 2019. "Cover crop biomass production in temperate Agroecozones." *Agronomy Journal* 111(4):1535–1551. doi: 10.2134/agronj2018.08.0535.

Sainju, U. M., W. F. Whitehead, and B. P. Singh. 2005. "Biculture legume-cereal cover crops for enhanced biomass yield and carbon and nitrogen." *Agronomy Journal* 97(5):1403–1412. doi: 10.2134/agronj2004.0274.

Sanderman, J., T. Hengl, and G. J. Fiske. 2017. "Soil carbon debt of 12,000 years of human land use." *Proceedings of the National Academy of Sciences of the United States of America* 114(36):9575–9580. doi: 10.1073/pnas.1706103114.

Sievers, T., and R. L. Cook. 2018. "Aboveground and root decomposition of cereal rye and hairy vetch cover crops." *Soil Science Society of America Journal* 82(1):147–155. doi: 10.2136/sssaj2017.05.0139.

Singer, J. W. 2008. "Corn belt assessment of cover crop management and preferences." *Agronomy Journal* 100(6):1670–1672. doi: 10.2134/agronj2008.0151.

Steele, M. K., F. J. Coale, and R. L. Hill. 2012. "Winter annual cover crop impacts on no-till soil physical properties and organic matter." *Soil Science Society of America Journal* 76(6):2164–2173. doi: 10.2136/sssaj2012.0008.

Tellatin, S., and R. L. Myers. 2018. "Cover crop impacts on US cropland carbon sequestration." *Journal of Soil & Water Conservation* 73(5):117a–121a. doi: 10.2489/jswc.73.5.117A.

Varela, M. F., C. M. Scianca, M. A. Taboada, and G. Rubio. 2014. "Cover crop effects on soybean residue decomposition and P release in no-tillage systems of Argentina." *Soil & Tillage Research* 143:59–66. doi: 10.1016/j.still.2014.05.005.

Wagger, M. G., M. L. Cabrera, and N. N. Ranells. 1998. "Nitrogen and carbon cycling in relation to cover crop residue quality." *Journal of Soil & Water Conservation* 53(3):214–218.

Wang, F., R. R. Weil, and X. X. Nan. 2017. "Total and permanganate-oxidizable organic carbon in the corn rooting zone of US Coastal Plain soils as affected by forage radish cover crops and N fertilizer." *Soil & Tillage Research* 165:247–257. doi: 10.1016/j.still.2016.08.022.

Wendling, M., L. Buchi, C. Amosse, S. Sinaj, A. Walter, and R. Charles. 2016. "Influence of root and leaf traits on the uptake of nutrients in cover crops." *Plant & Soil* 409(1–2):419–434. doi: 10.1007/s11104-016-2974-2.

Williams, S. M., and R. R. Weil. 2004. “Crop cover root channels may alleviate soil compaction effects on soybean crop.” *Soil Science Society of America Journal* 68(4):1403–1409. doi: 10.2136/sssaj2004.1403.

Wittwer, R. A., B. Dorn, W. Jossi, and M. G. A. van der Heijden. 2017. “Cover crops support ecological intensification of arable cropping systems.” *Scientific Reports* 7:41911. doi: 10.1038/srep41911.

Wortman, S. E., C. A. Francis, and J. L. Lindquist. 2012. “Cover crop mixtures for the western corn belt: Opportunities for increased productivity and stability.” *Agronomy Journal* 104(3):699–705. doi: 10.2134/agronj2011.0422.

Zhou, X. Q., H. W. Wu, G. D. Li, and C. R. Chen. 2016. “Short-term contributions of cover crop surface residue return to soil carbon and nitrogen contents in temperate Australia.” *Environmental Science & Pollution Research International* 23(22):23175–23183. doi: 10.1007/s11356-016-7549-5.

Zomer, R. J., D. A. Bossio, R. Sommer, and L. V. Verchot. 2017. “Global sequestration potential of increased organic carbon in cropland soils.” *Scientific Reports* 7(1):15554. doi: 10.1038/s41598-017-15794-8.

7 Fertilizer Use in the North China Plains for Improving Soil Organic Matter Content and Crop Yield

Xiangbin Kong, Xiaobing Sun, and Ming Lei

CONTENTS

ABBREVIATIONS

AP: Available phosphorus
AK: Available potassium
SOM: Soil organic matter
TN: Total nitrogen

DOI: 10.1201/9781003102762-7

7.1 INTRODUCTION

The global population will increase to nine billion by 2050, and global food demand will increase to 4 billion tons, and food demand will increase by 56% compared to 2010 (FAO, 2017; Godfray et al. 2010; Searchinger et al. 2019). Fertilizer has been widely used in agricultural systems to obtain higher crop yields, especially in China with a high population and less cultivated land area (Du et al. 2020; Lai et al. 2020). Global fertilizer use will continue to increase to meet the projected food demand, and it is widely believed that global N fertilizer for crop production may increase 1.4 times over the next 40 years (Faostat, 2014; Feng et al. 2016). However, excessive application of fertilizer can have adverse impacts on farmland biology and the ecological environment (Baker et al. 2013). To ensure national food security, China's fertilizer use increased from 8.84 million Mg in 1978 to 58.59 million Mg in 2017, at the average annual rate of growth of 5.1%, which is much higher than the global average growth rate (Department of Population. *Social, Science and Technology Statistics* of *National Bureau of Statistics of China, 1979-2018*). In recent years, the Chinese government has adopted scientific fertilization policies to reduce the amount of fertilizer use, and these policies have played a strong role in increasing food production and optimizing the ecological environment (Chen et al. 2014; Li et al. 2012; Hu et al. 2019). However, fertilizers are still an important factor in increasing grain production (Bedada et al. 2014; Zhang et al. 2018). Therefore, it is important to understand the impact of fertilizer use on the soil organic matter (SOM) content and crop yield.

Fertilizer application has become a conventional agronomic practice for increasing the SOM contentand crop yield in agroecosystems (Gai et al. 2018; Kong et al. 2014). Indeed, SOM is an important part of sustainable soil quality and productivity (Bogunovic et al. 2018; Fan et al. 2020). Human activities, such as fertilization, will have a profound impact on SOM even over the short time horizon of decadal scale (Benbi et al. 2012; Lal, 2004). China's agricultural activities have undergone major changes since the 1980s (Kong et al. 2014). Thus, SOM content has been affected by both agricultural activities (e.g. fertilizer, tillage, irrigation, rotation) and natural factors (e.g. climate, topography, soil, landscape) (Egodawatta et al. 2012; Cao et al. 2016; Zhu et al. 2019). The high input of fertilizers, pesticides, and labor has enhanced crop yield, along with a strong impact on SOM content (Egodawatta et al. 2012; Gong et al. 2009; Wei et al. 2016). With the ever-increasing intensification of cultivated land resources, changes in SOM stock and quality will also affect agricultural production and sustainability—especially in China's highly intensive agricultural regions. Therefore, changes in SOM content under a highly intensive agricultural system and its relationship with the amount of fertilizer use can play an important role in the rational use of fertilizer and the moderation of SOM content.

A review of available information from China shows that fertilizer use has a positive effect on crop yield (Du et al. 2020; Gong et al. 2009; Wei et al. 2016). However, it is precisely because of this those farmers use a large amount of fertilizer while they lack awareness of the harm that is caused by excessive use of fertilizer to the ecological environment (Baker et al. 2013; Chen et al. 2014; Kahrl et al. 2010; Zheng et al. 2020). Many studies have shown that the excessive use of fertilizers is a threat to soil, water, and the ecological environment (Baker et al. 2013; Gan, 2016). Yan et al. (2019) pointed out that the excessive application of chemical fertilizers in farmland and untreated livestock and poultry industry emissions have caused serious non-point source pollution in China. Kwanowaa et al. (2020) analyzed the factors affecting pollution by China's agriculture and found that the use of chemical fertilizers is one of the important factors that affect the quality of surface and groundwater. Baker et al. (2013) observed that fertilizers are an important stressor for amphibians in agriculturally dominated systems, and certain chemical classes are likely to harm amphibians more than others. Since the 2010s, the Chinese government has also realized that excessive use of fertilizers has harmed the ecological environment, and issued the "Action Plan for Zero Growth in Fertilizer Use by 2020."In addition, farmers in China use fertilizers based on soil tests [illegible] them in a targeted manner to improve the efficiency of fertilizer utilization and effectively [illegible] of application (Liu et al. 2019). However, understanding the science of the impact of fertilizer application on crop yield can lead to effective and rational use, which could eventually lead to reducing the harm caused to the ecological environment.

The North China Plain (NCP) is China's principal grain production area, contributing 60–80% of wheat and 35–40% of corn to the national annual production (Song et al. 2015; Kong et al. 2016). Fertilizer use in the NCP has increased crop yield, but the interaction between fertilizer and crop yield needs to be understood. The NCP is a highly intensive agricultural area, and the processes underlying its highintensity arable land utilization have strongly impacted SOM content and quality (Xu et al. 2006; Zhou et al. 2016). However, the relationship between the use of fertilizers and the content/quality of SOM is complex, and the effect of fertilizer on the changes in the SOM dynamic is still not clear. This chapter is based on a case study done in Quzhou County, a typical county in the North China Plain. A specific object of the study is to understand the impact of fertilizer application on the SOM content and crop yield.

7.2 MATERIALS AND METHODS

7.2.1 Study Area

Quzhou County is located in northeastern Handan City, southern Hebei Province, in the upper reaches of the Heilonggang watershed (36°34′45″–57′57″N, 114°50′30″–115°13′30″E), and covers a total area of 676.81 km^2 (Figure 7.1). The region has warm temperatures and a continental climate with a mean annual temperature and precipitation of 13.1°C and 566 mm, respectively (Wang et al. 2014). The soil parent material is alluvial and pluvial sediments of rivers, with the main soil types being fluvo-aquic, cinnamon, and solonetz, based on Chinese Soil Taxonomy (CRGCST, 2001). Wheat, corn, and cotton are the main crops in Quzhou County. Agriculture in Quzhou County was once restricted by alkalinity and salinity, but the cultivated land quality has improved over time as these problems were effectively addressed (Hu et al. 2005; Xu et al. 2019). As a typical county

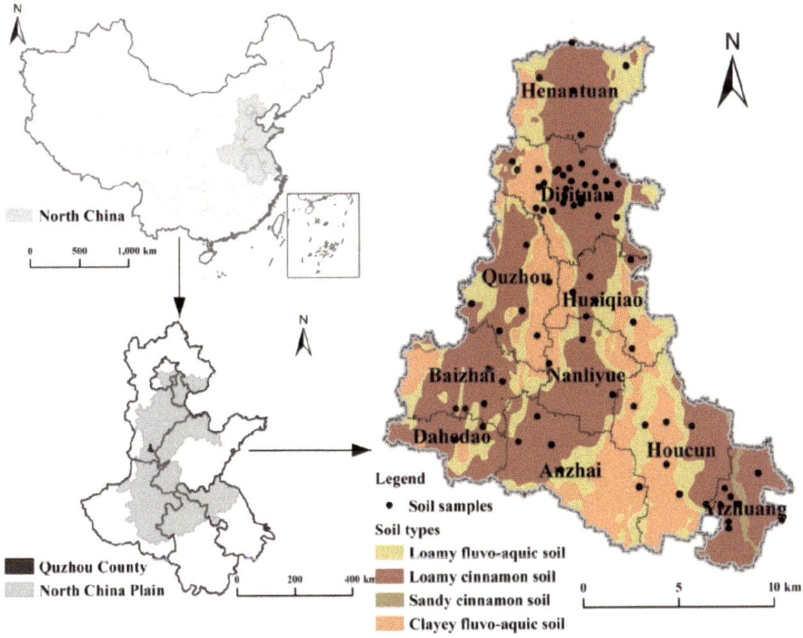

FIGURE 7.1 Map of the study area and soil sample collection sites.

in the intensive agricultural areas of the NCP, this county has a high multiple cropping index and requires a large capital investment. With intensive use of fertilizers and pesticides, grain yield has continued to increase, and the crop yield per unit area increased from 675 kg·ha^{-1} at the beginning of the People's Republic of China (PRC) to 7500 kg·ha^{-1} at present (Sun, et al. 2019), and the county's crop yield is 1.82 times the national average yield (DPSSTS of NBSC, 2018).

7.2.2 Soil Types and Distribution

The soil in Quzhou County includes fluvo-aquic soil, saline soil and cinnamon soil. Among them, the fluvo-aquic soil is the most predominant and accounts for 98.2% of the total land area. On the basis of soil texture, there are four soil types in Quzhou County: loamy fluvo-aquic soil (LFA), loamy cinnamon soil (LCI), sandy cinnamon soil (SCI), and clayey fluvo-aquic soil (CFA). However, spatial distribution of these textural classes varies widely. Loamy fluvo-aquic soil is distributed in all towns and villages of the county and, in general, follows a north–south strip distribution with the change in terrain. Loamy cinnamon soil is distributed from south to north, and is also distributed widely in the southwest and southeast. Sandy cinnamon soil is primarily found along ancient river channels and natural dikes around Houcun Town and Yizhuang Township. Clayey fluvo-aquic soil is mainly distributed in the midwest and southeast of the county (Figure 7.1). Therefore, this article is mainly focused on assessing the differences in soil nutrient SOM content and grain yield in three soil types (LFA, LCI, and CFA) because the area under sandy cinnamon soil accounts for only 0.72% of the total cultivated land.

7.2.3 Data Collection

This research involves data of soil analyses and those from a statistical yearbook containing data of soil type and land use. The soil sampling data include 79 soil sample points from three years: 1980, 2000, and 2018. The soil data were obtained for three years: May–June 1980, April–May 2000, and June–July 2018 through field sampling and analyses based on laboratory testing methods. Soil samples were collected from April to July, mainly to avoid the influence of the application of fertilizer on soil nutrient and SOM content. A set of 79 soil samples were collected each year, and a total of 237 soil samples were collected over 3 years. The basic process of soil collection involved selecting four points within a radius of 200m^2 from the same soil sample point to obtain soil surface samples (0–20 cm), mix them evenly, and use them as test samples. Soil samples were air-dried, ground, and passed through a 2-mm sieve for testing and analysis. Total soil nitrogen (TN), soil available phosphorus (AP), and soil available potassium (AK) were tested bythe Kjeldahl method (Lu, 1999), sodium bicarbonate extraction, colorimetric detection method (Olsen and Sommers, 1982); and ammonium acetate extraction, flame photometer detection method, respectively (Lu, 1999). SOM content was tested by the potassium dichromate oxidation method (Lu, 1999). Crop yield data were obtained from the statistical yearbook of Quzhou County (Quzhou of Bureau of Statistics, 1982–2018). To avoid deviations in the statistics of a single year in 1980, 2000, and 2018, this chapter uses the average grain output for three consecutive years before and after these years as the final crop yield. At the same time, this chapter selects existing typical sample points to obtain its fertilizer data. Among them, fertilizers mainly include basic types such as nitrogen fertilizer, phosphate fertilizer, potash fertilizer, and compound fertilizer.Data on soil type and land use were obtained from the Natural Resources and Planning Bureau of Quzhou County.

7.2.4 Data Analysis

The basic data of the experiment were processed using Excel 2016, and SPSS 21.0 was used to conduct a one-way analysis of variance (ANOVA) to assess the different characteristics of soil nutrient, SOM content, and crop yield for different soil types in 1980, 2000, and 2018. Pearson's correlation coefficient was used to test the significance of related parameters.

7.3 RESULTS

7.3.1 Characteristics of Soil Nutrient, Soil Organic Matter, and Crop Yield

7.3.1.1 Soil Nutrient Content Change

The soil nutrient content in Quzhou County showed an overall increasing trend from 1980 to 2018, but the magnitude of increase differed among soil nutrient types and had significant stages (Table 7.1). Soil nutrient content TN and AP increased from 0.52 g kg^{-1} and 4.84 mg kg^{-1} in 1980 to 1.29 g kg^{-1} and 37.99 mg kg^{-1} in 2018, respectively. However, AK decreased and then increased from 168.28 mg kg^{-1} in 1980 to 103.38 mg kg^{-1} in 2000—and finally increased to 230.20 mg kg^{-1} in 2018. The standard deviations of TN, AP, and AK in 1980, 2000, and 2018 all showed an increasing trend, but the coefficient of variation (CV) of different nutrients differed among years. The CV of TN in 1980, 2000, and 2018 ranged from 19.53% to 27.72%, showing moderate variation (Wilding, 1985). Except for AK in 1980, the degree of variation of each nutrient in other years exhibited high variations. With regard to changes in soil nutrient content in different periods (Figure 7.2), the

TABLE 7.1
Summary Statistics of Soil Nutrient Content in Quzhou County

Fertilizer	Year	Mean	Minimum	Maximum	Standard Deviation	CV(%)	Kurtosis	Skewness
TN	1980	0.52	0.28	0.80	0.11	21.52	−0.36	0.19
(g kg^{-1})	2000	1.05	0.66	1.75	0.21	19.53	1.37	0.94
	2018	1.29	0.29	2.51	0.36	27.72	1.59	0.47
AP	1980	4.84	1.00	13.50	2.15	44.38	3.81	1.67
(mg kg^{-1})	2000	21.86	1.80	80.80	16.73	76.57	3.14	1.76
	2018	37.99	4.50	94.90	20.03	52.71	−0.26	0.62
AK	1980	168.28	103.00	285.00	43.06	25.59	−0.24	0.86
(mg kg^{-1})	2000	103.38	48.00	235.00	49.95	48.31	0.24	1.18
	2018	230.20	91.90	514.00	103.45	44.94	0.78	1.09

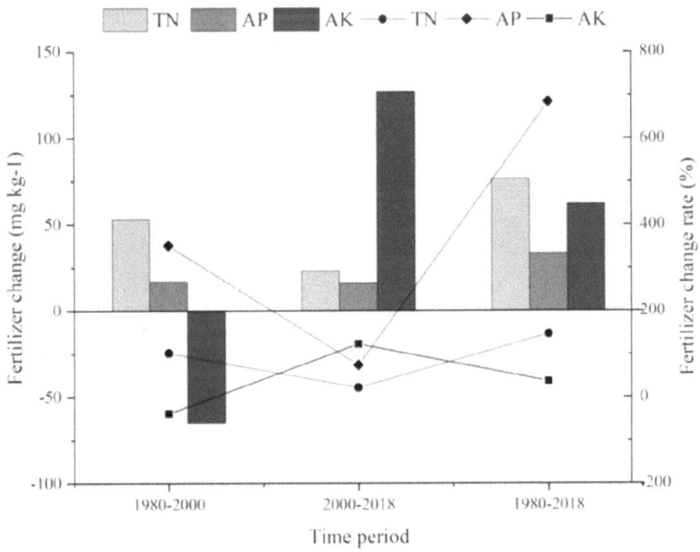

FIGURE 7.2 Temporal trends in soil nutrient content among different periods from 1980 to 2018. *Note:* In order to show the comparability of graphics, the value of TN was enlarged 100 times.

growth rates of AP, TN, and AK from 1980 to 2018 were 684.38%, 146.14%, and 36.80%, respectively. But the magnitude of change differed among different periods. Both TN and AP showed a growth trend in 1980–2000 and 2000–2018, and the growth rate in 1980 was higher than that in 2000–2018. However, the growth rate of AK from 1980 to 2000 was –38.57%, while that of AK from 2000 to 2018 was 122.67%, and the overall growth rate of AK was 36.80% from 1980 to 2018.

7.3.1.2 Soil Organic Matter

The average SOM content in Quzhou County showed a continuously increasing trend between 1980 and 2018 from 8.46 g kg^{-1} in 1980 to 17.97 g kg^{-1} in 2018 (Table 7.2). The standard deviation and CV of the SOM content in Quzhou County showed an increasing trend from 1980 to 2018, and SOM content had a moderate variation (Wilding, 1985). Based on the data of changes of SOM content in different periods (Figure 7.3), the growth and growth rate of SOM content in Quzhou County increased continuously from 1980 to 2000 and 2000 to 2018, and its growth rate increased from 40.75% to 50.87%. Overall, the average increase in the SOM content from 1980 to 2018 was 9.51 g kg^{-1}, and the growth rate was as high as 112.35%.

7.3.1.3 Crop Yield

From 1980 to 2018, the average crop yield in Quzhou County showed a trend of linear growth from 2,447kg ha^{-1} in 1980 to 7,486 kg ha^{-1} in 2018, with an increase of 5038 kg ha^{-1} and a growth rate of 205.9% (Table 7.3 and Figure 7.4). In 1980, 2000, and 2018, the standard deviation and CV of crop yield in Quzhou County decreased successively. The degree of variation in crop yield in 1980 was moderate, while the degree of variation in grain production in 2000 and 2018 was relatively weak (Wilding, 1985). However, a growth rate of grain yield in 2000–2018 was 51.8% higher than that in 1980–2000 (Figure 7.4).

TABLE 7.2
Summary Statistics of SOM Contents in Quzhou County (g kg^{-1}).

Year	Mean	Minimum	Maximum	Standard Deviation	CV(%)	Kurtosis	Skewness
1980	8.46	4.90	13.10	1.72	20.36	–0.12	0.38
2000	11.91	5.84	19.81	2.55	21.43	0.80	0.01
2018	17.97	5.76	31.00	5.11	28.42	–0.02	0.28

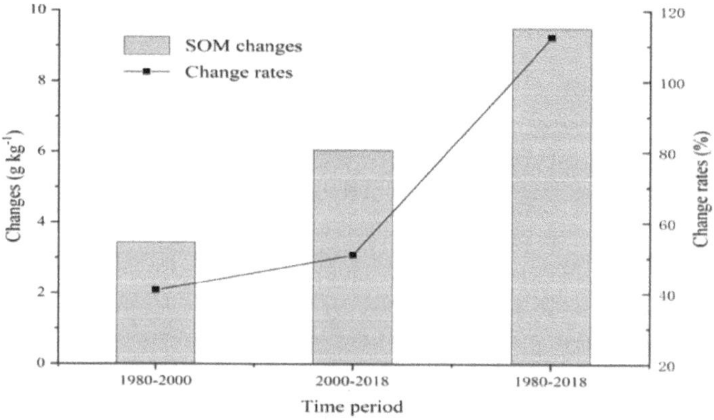

FIGURE 7.3 Changes in the soil organic matter content in different periods between 1980 and 2018.

TABLE 7.3
Summary Statistics of Crop Yield in Quzhou County (kg ha^{-1}).

Year	Mean	Minimum	Maximum	Standard Deviation	CV(%)	Kurtosis	Skewness
1980	2447	1163	4358	731.05	29.87	–0.38	0.55
2000	4911	2886	6494	673.05	13.71	2.39	–0.75
2018	7486	7245	7962	153.17	2.05	0.23	0.79

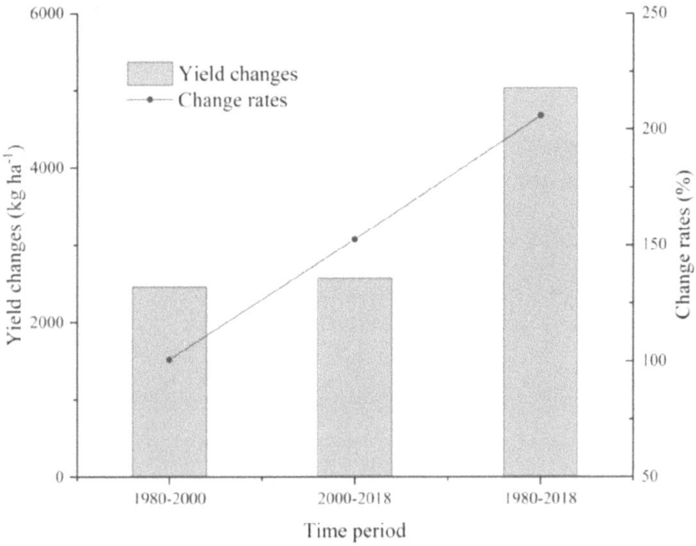

FIGURE 7.4 Temporal changes in crop yield over different periods from 1980 to 2018.

From the comparison between SOM and grain production (Figure 7.5), both SOM and grain production in Quzhou County showed an increasing trend from 1980 to 2018, and the increased rate of SOM was about half of that of grain production, but the increase characteristics are different at different stages. From 1980 to 2000, the increased rate of SOM in Quzhou County was much lower than the increased rate of crop yield. Among them, the increase rate of SOM was about 40%, while the increase rate of grain output was higher than 100%. From 2000 to 2018, the increase rate of organic matter and crop yield in Quzhou County was around 50%.

It can be seen from Figure 7.6 that both crop yield and fertilizer use in Quzhou County have shown a steady increase. Among them, the fertilizer use increased from 161 kg ha^{-1} in 1980 to 226 kg ha^{-1} in 2018, with a growth rate of 107.46%. It can be seen that there is a significant positive correlation between the use of fertilizers and crop yield, and the increase in fertilizers has a positive effect on the increase in crop yield.

It can be seen from Table 7.4 that the average increase in SOM content and the average increase in crop yield in Quzhou County from 1980 to 2000 and 2000 to 2018 both showed an increasing trend. Among them, the average increase in SOM and the average increase in crop yield from 2000 to 2018 were both greater than the corresponding increase in 1980–2000, but the increase in food production was relatively small. From the perspective of the contribution rate of SOM to crop yield, the contribution rate of SOM increase per unit of Quzhou County to crop yield from 1980 to 2018 was 35.37 kg g^{-1}. Besides, the increase in SOM contributed more to food production in 1980–2000 than that in 2000–2018.

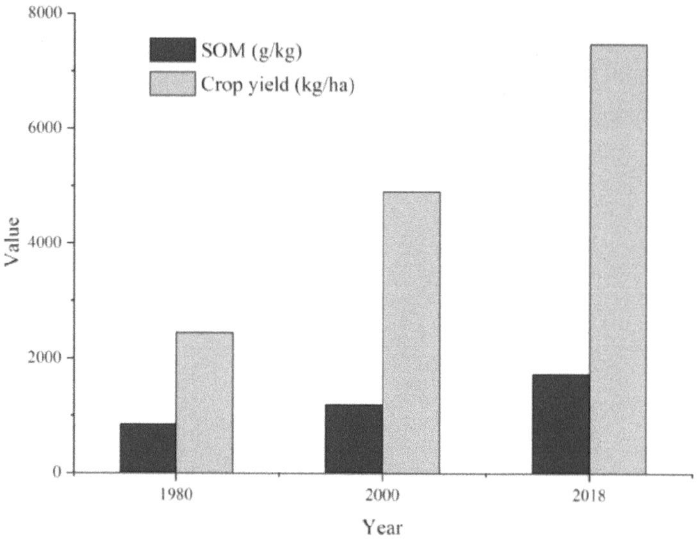

FIGURE 7.5 Average status of SOM and crop yield in Quzhou County from 1980 to 2017. *Note:* To show the comparability of graphics, the value of SOM was enlarged 100 times.

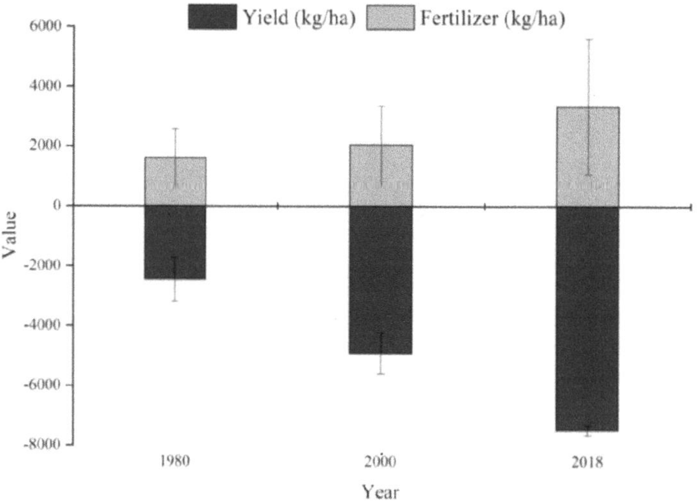

FIGURE 7.6 Characteristics of the relationship between fertilizer usage and crop yield in Quzhou County from 1980 to 2018. *Note:* To show the comparability of graphics, the value of fertilizer was enlarged 100 times.

7.3.2 Relationship between Soil Nutrient and SOM

7.3.2.1 Overall Relationship of Soil Nutrient on the SOM Content in Quzhou County

The regression results of the SOM content and TN, AP, and AK in 1980, 2000, and 2018 passed the significance test ($p<0.01$), indicating that soil nutrient had a significant effect on the SOM content (Table 7.5). However, the results showed that the influence of different soil nutrients on the SOM content was different as is indicated by different regression coefficients. In 1980, TN, AP, and AK simultaneously affected the SOM content, but TN had a relatively larger impact than that of AP

TABLE. 7.4
The Quantitative Relationship Between SOM and Grain Production

Year	SOM Average Increase(g/kg)	Average Increase in Crop Yield per Hectare(kg/ha)	SOM's Contribution to Grain Growth(kg)
1980–2000	3.49	2437.61	46.56
2000–2018	5.99	2592.03	28.85
1980–2018	9.48	5029.64	35.37

TABLE 7.5
Regression Results of Soil Organic Matter Content and Soil Nutrient

	1980		2000		2018	
Variables	B	Sig.	B	Sig.	B	Sig.
TN	8.14	0.000***	4.97	0.001***	12.22	0.000***
AP	0.13	0.047**	0.03	0.050**	0.02	0.249
AK	0.01	0.007***	−0.01	0.215	0.01	0.274
Constant	1.96	0.003***	6.62	0.000***	0.95	0.295
R^2	0.587		0.236		0.837	
Sig.	0.000***		0.000***		0.000***	

Note: *, **, *** significant at $p < 0.1$, $p < 0.05$, and $p < 0.01$, respectively.

TABLE 7.6
Regression Equations of the Changing Trend of Soil Organic Matter and Soil Nutrient for Different Periods

Time period	Line model	R^2	*P*
1980–2000	SOM=4.167TN+1.256	0.133	0.001
	SOM=0.052AP+2.608	0.131	0.001
	SOM=−0.012AK+2.640	0.055	0.023
2000–2018	SOM=10.035TN−1.682	0.521	0.000
	SOM=0.054AP+5.119	0.071	0.011
	SOM=0.012AK+4.389	0.062	0.017
1980–2018	SOM=11.476TN+0.704	0.763	0.000
	SOM=0.109AP+5.867	0.225	0.000
	SOM=0.019AK+8.359	0.157	0.000

and AK. In 2000 and 2018, TN had a significant impact on the SOM content, but AP and AK had a lesser impact. In general, from 1980 to 2018, TN had a significant impact on SOM, but the impact of AP and AK on SOM gradually decreased over time.

The regression results of the changes in the SOM content and the changes in soil nutrients passed the significance test ($p<0.05$), indicating that the changes in SOM follow a specific relationship with the changes in soil nutrients (Table 7.6). The change of TN has a greater impact on the change of SOM content; its regression coefficients were relatively high, and it gradually increased from 1980 to 2018. However, changes in AK and AP had a relatively low degree of influence on the SOM content. Except for AK changes in 1980–2000 that had a negative effect on changes in SOM content, AK and AP have positive effects on changes in SOM content in other years.

7.3.2.2 Effects of Soil Nutrients on SOM in Different Soil Types

Effects of TN on SOM content differed significantly among different soil types in 1980, and the average TN of LFA, LCI, and CFA were 0.57 g kg^{-1}, 0.49 g kg^{-1}, and 0.63 g kg^{-1}, respectively. However, differences in TN among different soil types in 2000 and 2018 were not obvious (Table 7.7 and Figure 7.4). The difference of AP among different soil types in 1980, 2000, and 2018 was also not obvious, and the average AP of LFA, LCI, and CFA in the three years were 4.97 mg kg^{-1}, 22.10 mg kg^{-1}, and 34.81 mg kg^{-1}, respectively. However, the differences in AK among different soil types in 1980, 2000, and 2018 were all significant, but the soil types with the highest AK in different years were different. The average AK of LFA was the highest in 1980, with a value of 202.7 mg kg^{-1}. The average AK of LFA was the highest in 1980, with a value of 202.7 mg kg^{-1}, but the soil type with the highest average AK in 2000 and 2018 was CFA, with values of 142.2 mg kg^{-1} and 311.0 mg kg^{-1}, respectively. However, the average AK of LCI in 1980, 2000, and 2018 was relatively low, with values of 151.9 mg kg^{-1}, 92.6 mg kg^{-1}, and 207.0 mg kg^{-1}, respectively (Figure 7.7).

There were significant differences in SOM content for different soil types in 1980 and 2018, but the differences in SOM of different soil types in 2000 were not significant (Table 7.8). The average SOM content of CFA was the highest in 1980 and 2018, with average values of 10.4 g kg^{-1} and 21.3 g kg^{-1}, respectively. However, the soil types with the lowest average SOM content in 1980 and 2018 were LFA and LCI, respectively, and their average SOM content was 8.0 g kg^{-1} and 17.1g kg^{-1}, respectively. The significance of coefficient of the ANOVA of SOM content of different soil types in 2000 was 0.067 ($p>0.05$), indicating that the difference in SOM content of different soil types was not obvious. The average SOM content of LFA, LCI, and CFA was 12.2 g kg^{-1}, 11.5 g kg^{-1}, and 13.2 g kg^{-1}, respectively.

Soil nutrients had different effects on the SOM content in different soil types (Table 7.9). In 1980, 2000, and 2018, the influence of soilnutrients on the SOM content in soil type LFA was not significant ($p>0.05$). In 1980, 2000, and 2018, the influence of soil nutrient on SOM content in soil type LCI was relatively significant, except that the influence of AK on SOM content in 2000 did not pass the significance test ($p=0.885$). Regarding the CFA soil type, TN in 1980, AP in 2000, and TN and AP in 2018 had significant effects on SOM content, while soil nutrient regimes had no significant effects on SOM content in other years.

7.3.3 Impact of Soil Nutrient on Crop Yield

7.3.3.1 Overall Impact of Soil Nutrients on Crop Yield in Quzhou County

Soil nutrient fertility had a positive impact on crop yield, but the impact varied among different years (Figure 7.8). The increase in AP in 1980 had a positive effect on increase of crop yield,

TABLE 7.7
ANOVA of the TN, AP, and AK for Different Soil Types Among Years

	TN		AP		AK	
Year	**F**	**Sig.**	**F**	**Sig.**	**F**	**Sig.**
1980	11.358	0.000	0.766	0.468	26.798	0.000
2000	1.027	0.363	0.025	0.976	6.810	0.002
2018	2.430	0.095	1.316	0.274	6.930	0.002

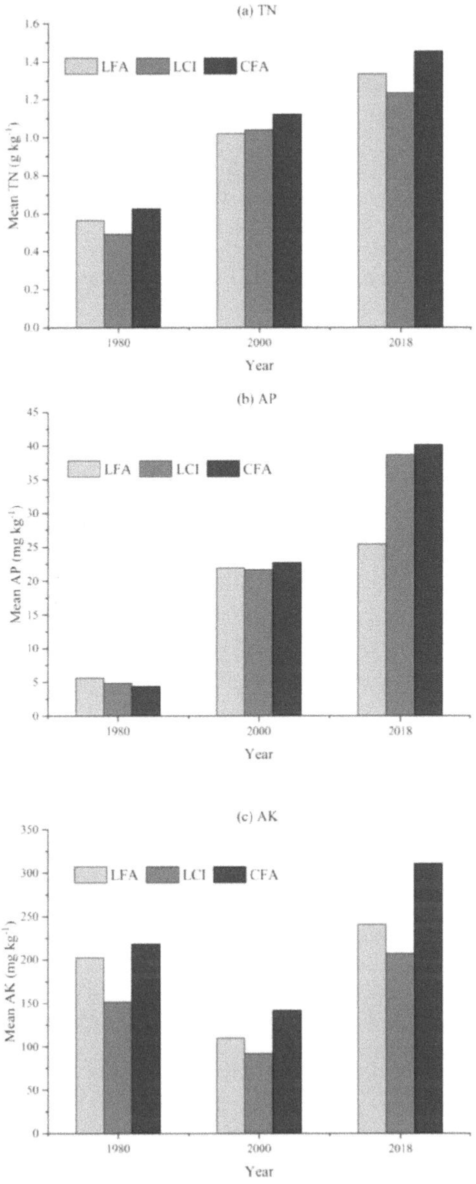

FIGURE 7.7 Mean soil nutrient content of different soil types: (a) TN, (b) AP, (c) AK.

while TN and AK had a negative effect on crop yield. This trend may be because the salinization of Quzhou County was relatively high at that time, which to some extent hindered the effect of nutrients on crop yield (Kong et al. 2003b). The increase in TN, AP, and AK in 2000 and 2018 had a positive effect on increasing crop yield. However, TN, AP, and AK have different effects on the increase of crop yield. The impact of TN and AK on crop yield first increased and then decreased, and the impact of AP on crop yield showed a steadily decreasing trend, which is closely related to the natural environment and human activities.

TABLE 7.8
ANOVA of Soil Organic Matter in Different Soil Types (g kg^{-1})

Soil Types		Number	Mean	Standard Deviation	F	Sig.
1980	LFA	6	8.57	1.65	16.241	0.000
	LCI	58	7.96	1.56		
	CFA	15	10.37	0.82		
2000	LFA	6	12.15	1.48	2.801	0.067
	LCI	58	11.54	2.51		
	CFA	15	13.24	2.71		
2018	LFA	6	18.02	2.10	4.496	0.014
	LCI	55	17.05	5.02		
	CFA	15	21.31	5.08		

TABLE 7.9
Regression Results of Soil Nutrient on SOM Content for Different Soil Types

Soil Type	Year	Line Model	R^2	P
LFA	1980	SOM=10.508TN+2.6298	0.3435	0.281
		SOM=0.1192AP+7.8938	0.0225	0.800
		SOM=–0.0114AK+10.88	0.0505	0.715
	2000	SOM=5.1946TN+6.8446	0.4944	0.293
		SOM=0.0618AP+10.794	0.5971	0.247
		SOM=–0.0094AK+13.182	0.0642	0.935
	2018	SOM=6.9822TN+8.6954	0.2674	0.322
		SOM=–0.0754AP+19.936	0.1700	0.171
		SOM=–0.013AK+21.133	0.2417	0.366
LCI	1980	SOM=12.004TN+2.0564	0.5806	0.000
		SOM=0.4548AP+5.7373	0.4791	0.000
		SOM=0.0235AK+4.3895	0.2313	0.000
	2000	SOM=6.8969TN+4.3666	0.3156	0.000
		SOM=0.047AP+10.524	0.1018	0.015
		SOM=0.0007AK+11.477	0.0002	0.885
	2018	SOM=12.245TN+1.9287	0.8455	0.000
		SOM=0.1011AP+13.13	0.1871	0.002
		SOM=0.0309AK+10.661	0.3133	0.000
CFA	1980	SOM=–0.1595TN+10.467	0.0004	0.001
		SOM=0.0574AP+10.115	0.0034	0.131
		SOM=0.004AK+9.4959	0.0362	0.633
	2000	SOM=–0.4739TN+13.772	0.0014	0.899
		SOM=0.1101AP+10.741	0.4166	0.011
		SOM=–0.1076AK+15.746	0.1071	0.225
	2018	SOM=16.544TN+2.7858	0.8796	0.000
		SOM=0.2322AP+11.979	0.5232	0.003
		SOM=0.0211AK+14.766	0.2404	0.150

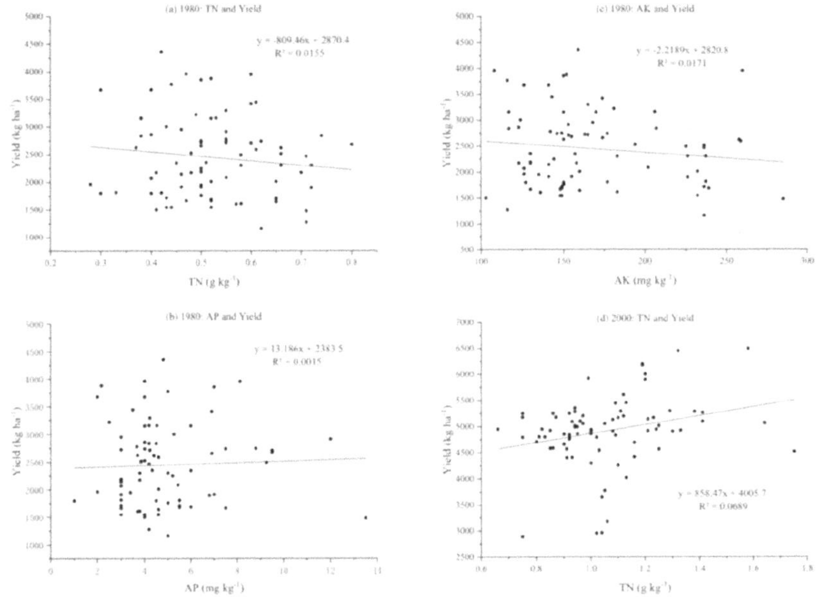

FIGURES 7.8 Regression curves of crop yield and soil nutrients in Quzhou County in 1980, 2000, and 2018 a, b, and c represent the regression relationship between TN, AP, and AK and crop yied in 1980; d, e, and f represent the regression relationship between TN, AP, and AK and crop yield in 2000; g, h, and i represent the regression relationship between TN, AP, and AK and crop yield in 2018.

7.3.3.2 Impact of Soil Nutrient on Crop Yield in Different Soil Types

Crop yields for different soil types in 1980 had obvious differences, while the differences among soil types in 2000 and 2018 were not obvious (Table 7.10). The crop yield of LCI was relatively high in 1980, with an average grain output of 2561kg ha^{-1}, while the crop yield of LFA was relatively low, with an average grain yield of 1863kg ha^{-1}. In 2000 and 2018, the CFA soil type had relatively high grain yields, with values of 5109 kg ha^{-1} and 7512 kg ha^{-1}, respectively. However, the soil types with relatively low crop yields in 2000 and 2018 were LCI and LFA.

Table 7.10 Soil nutrient had different effects on the crop yield in different soil types (Table 7.11). The AP in 1980 and 2018 and the AK in 2000 had significant effects on crop yield at $p<0.01$, $p<0.05$, and $p<0.05$, respectively for the LFA soil type. For the LCI soil type, AK in 2018 had a significant impact on grain yield at $p<0.1$, while for CFA, AP in 1980 and TN in 2000 had significant effects on grain yield at $p<0.05$. Soil nutrient levels had some effect on grain yield for soil types in other years, but these effects were not statistically significant.

7.4 DISCUSSION

7.4.1 Soil Nutrient in the Intensive Agricultural Region of the NCP

Soil nutrient is an important factor to promote crop yield, but the influence of soil nutrient on crop yield is different in different regions (Faostat, 2014; Feng et al. 2016). The NCP is an important grain production region in China, and the application of fertilizer affected soil nutrients, the SOM content, and crop yield. However, due to the influence of multiple factors (e.g. natural factors, socio-economic, agricultural policies, and technological level), changes in the soil nutrient and its impact on SOM and crop yield have temporal and spatial differences (Zhang et al. 2013, 2020a). This chapter focuses on Quzhou County, a typical county of the NCP, as an example (or a case study) to reveal

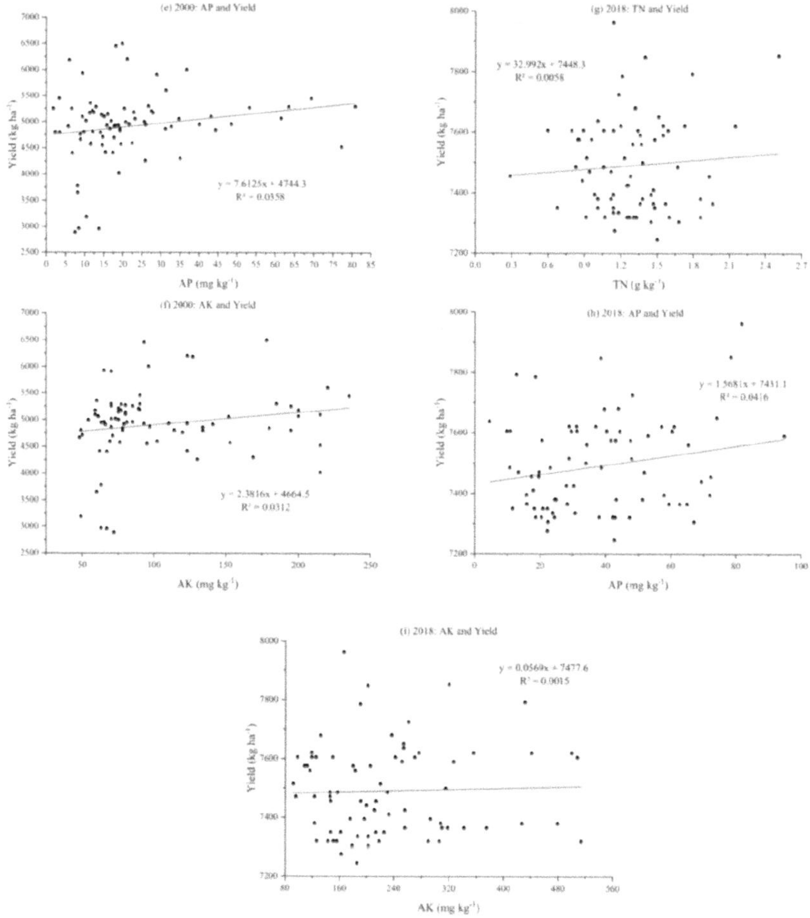

FIGURES 7.8 Continued

TABLE 7.10
ANOVA of Crop Yield for Different Soil Types

Soil types		Number	Mean	Standard deviation	F	Sig.
1980	LFA	6	1862.50	376.46	3.396	0.039
	LCI	58	2560.60	750.32		
	CFA	15	2243.50	624.99		
2000	LFA	6	4914.54	376.47	0.820	0.444
	LCI	58	4858.97	709.48		
	CFA	15	5109.25	608.70		
2018	LFA	6	7402.50	110.53	1.119	0.332
	LCI	58	7487.51	157.55		
	CFA	15	7512.30	146.80		

TABLE 7.11
Regression Results of Soil Nutrient and Crop Yield for Different Soil Types

Soil Type	Year	Line Model	R^2	P
LFA	1980	YIELD=1457.5TN+1039	0.1270	0.507
		YIELD=165.45AP+928.28	0.8335	0.006
		YIELD=5.236AK+801.34	0.2043	0.419
	2000	YIELD=697.59TN+4201.8	0.1375	0.714
		YIELD=0.0573AP+4913.3	0.0001	0.137
		YIELD=−8.5483AK+5856.3	0.8277	0.039
	2018	YIELD=−196.57TN+7664.9	0.0765	0.484
		YIELD=3.0233AP+7325.5	0.0987	0.017
		YIELD=1.0811AK+7142.5	0.6072	0.152
LCI	1980	YIELD=−181.12TN+26496	0.0006	0.815
		YIELD=17.592AP+2474.8	0.0031	0.718
		YIELD=−0.2967AK+2605.7	0.0002	0.876
	2000	YIELD=600.7TN+4234.2	0.0300	0.177
		YIELD=8.5934AP+4673.1	0.0427	0.123
		YIELD=1.6883AK+4702.6	0.0121	0.476
	2018	YIELD=72.436TN+7405	0.0295	0.173
		YIELD=1.6691AP+7429.8	0.0509	0.072
		YIELD=0.2349AK+7445.8	0.0181	0.299
CFA	1980	YIELD=−574.22TN+2603	0.0095	0.165
		YIELD=−344.83AP+3755	0.2089	0.017
		YIELD=2.3254AK+1736.6	0.0221	0.549
	2000	YIELD=1613.5TN+3298.9	0.3177	0.030
		YIELD=5.771AP+4978.2	0.0227	0.680
		YIELD=5.047AK+4391.6	0.1740	0.200
	2018	YIELD=−202.75TN+7807.6	0.1580	0.299
		YIELD=1.0967AP+7556.4	0.0140	0.786
		YIELD=−0.5598AK+7686.4	0.2032	0.235

the impact of soil nutrient on SOM content and crop yield in different fertilizer use stages from 1980 to 2018, which could be used to scientifically and rationally adjust fertilizer use to enhance SOM content, crop yield, and soil nutrients.

The data presented herein show that the growth rates of TN, AP, and AK in Quzhou County from 1980 to 2018 were 146.1%, 684.4%, and 36.8%, respectively, and changes in soil nutrients had significant differences. From 2000 to 2018, both TN and AP showed an increasing trend, while AK showed a trend of first decreasing and then increasing, which was closely related to the characteristics of soil background and human activities (Kong et al. 2006). During the 1960s and 1970s, Quzhou County was a typical region prone to drought, flood, and saline-alkali conditions, with extremely serious salinization and a relatively low soil nutrient content. However, with the comprehensive treatment, salinization is no longer an issue in this county (Figure 7.9). Coupled with the use of organic fertilizers and compound fertilizers, soil nutrient contents of TN and AP have increased (Kong et al. 2003a, b; Lu et al. 2020). However, the soil content of AK in Quzhou County decreased from 1980 to 2000, mainly due to the relatively high contents of AK in the soil and non-application of potash fertilizer for the longterm (Kong et al. 2003a, b; Sun et al. 2019). Since the 21st century, with the improvement of agricultural science and technology, Quzhou County has eliminated

FIGURE 7.9 Typical cultivated land before and after salinization treatment in Quzhou County: (a) saline-alkali cultivated land; (b) cultivated land in agriculturalization.

land salinization through comprehensive land consolidation and restoration and has increased crop yield by increasing the application of fertilizers. Therefore, fertilizers used in Quzhou County also showed a growth trend from 2000 to 2018. In recent years, however, with the support of China Agricultural University, Quzhou County has focused on improving the use efficiency of fertilizers, rather than simply increasing the rate of fertilizers input, which has played a positive role in promoting the sustainable development of food production and the ecological environment (Chen et al. 2020; Chang et al. 2020; Yin al. 2018).

7.4.2 Effect of Soil Nutrient on SOM in the Intensive Agricultural Region of the NCP

The SOM content is closely related to the natural environment, land use, rate and chemical fertilizer application, and the long-term high intensity of arable land use leading to changes of SOM content in the NCP (Fan et al. 2020; Wang et al. 2015; Zhang et al. 2020b). The average SOM content of Quzhou County increased from 8.46g kg^{-1} in 1980 to 17.97g kg^{-1} in 2018, with the average rate of increase of 146.1%. The FAO (2014) showed that average SOM content of the Quzhou County of57.1g kg^{-1} of SOM content is only 57% of the global average (32.5 g kg^{-1}). Furthermore, the average SOM content in Quzhou County is slightly lower than that of the national average SOM content. The SOM content of dry land in the NCP increased slightly, from 16.9 g kg^{-1} in 1980 to 17.7 g kg^{-1} in 2010. However, the SOM content of paddy fields decreased significantly from 21.0 g kg^{-1} in 1980 to 12.9 g kg^{-1} in 2010. Based on the grading standard of the second national soil survey (Zhou et al. 2017), the content of SOM in Quzhou County increased from the fifth grade in 1980 to the fourth grade in 2018, and the SOM content showed an increasing trend from 1980 to 2018. Due to the predominance of dry land in Quzhou County, the average SOM content is slightly higher than that in the NCP.

The data presented herein show that the change of the application of fertilizer from 1980 to 2018 had a significant impact on SOM content, but the effects of soil nutrient contents on SOM content were different. These results are similar to the related research results of Du et al.(2020).The high initial level of SOM content in Quzhou County also limited the rate of increase of the SOM content (Tian et al. 2020). However, as the degree of salinization decreased, along with the increase in fertilizer use from 1980 to 2018, the SOM content showed an increasing trend. Furthermore, the SOM content is affected by both natural and human factors, such as pH, management measures, and crop types (He et al. 2011; Zhang et al. 2018). Indeed, human activities play a key role in changing the SOM content. Since the beginning of the 21st century, with the continuous improvement of the level of agricultural science and technology, Quzhou County has effectively promoted the increase of food production through scientific technology and the input offertilization, while reducing the impact of excessive use of chemical fertilizers on the ecological environment (Li et al. 2020).

Therefore, the use of fertilizers at this stage has a positive effect on SOM content, but different soil nutrient content has different effects on SOM content and its dynamics.

7.4.3 Effect of Change of Fertilizer Use from 1980 to 2018 on Crop Yield

Fertilizer use can significantly increase grain production, and the increase in grain production in China is closely related to the high amount of fertilizer application (Cai et al. 2019; Jannoura et al. 2014). The NCP is an important grain production region in China, and there is a close relationship between the continuous increase in grain output and the increase in fertilizer input (Du et al. 2020). The crop yield in Quzhou County increased from 247.4 kg ha^{-1} in 1980 to 7485.8 kg ha^{-1} in 2018. The growth rate of grain yield is as high as 205.9%, which is much higher than the 105.2% growth rate of China for the same period (Department of Population. *Social, Science and Technology Statistics* of *National Bureau of Statistics of China, 1979-2018*).

The influence offertilizerson grain production is closely related to the natural environment, social economy, and technological level (Cai et al. 2019; Kong et al. 2014; Wei et al. 2016). Although the fertilizer use in Quzhou County had a phased effect on crop yield, fertilizers had a positive effect on the increase of crop yield. From the 1960s to the 1970s, the salinization of land in Quzhou County was very serious, severely restricting the utilization of cultivated land resources and the increase in grain production (Lu et al. 2020). The degree of salinization in Quzhou County affected the use of fertilizers for grain production to a certain extent. Previously, different fertilizers had different effects on grain production. However, with the comprehensive treatment of salinization, the effect offertilizer use on crop yield continues from 1980 to 2018, but to a certain extent, an incremental increase incrop yield with a unit increase in input of fertilizer is relatively stable. In addition, increasing food production requires the combined effect of multiple natural and human factors. Therefore, fertilizer use has been an important factor in increasing food production; and other factors, such as pesticides, irrigation, and farmland roads have also had an important impact on food production since the beginning of the 21st century. However, excessive use of fertilizers has adverse effects on water, soil, and air. In recent years, Quzhou County has promoted the effective absorption of fertilizers through the scientific use of fertilizer application and efficient agricultural management, reducing the impact on the ecological environment while progressively increasing food production (Bedada et al. 2014; Liu et al. 2013; Singh et al. 1996; Zhao et al. 2019). Therefore, with the scientific application of fertilizers and the improvement of agricultural science and technology, application of fertilizers has and continues to play a vital role in food production.

7.5 CONCLUSIONS

This study shows that the average increase of TN, AP, and AK in Quzhou County from 1980 to 2018 was 146%, 684%, and 37%, respectively; TN and AP showed an increasing trend, while AK showed a trend of first decreasing and then increasing. The SOM content increased from 8.46 g kg^{-1} in 1980 to 17.97 g kg^{-1} in 2018, with an increase of 112.35%. From 1980 to 2018, crop yield increased from 2447kg ha^{-1} to 7486 kg ha^{-1}, with an increase of 206%. Soil nutrienthad a significant impact on SOM, but the response followed a sequence of well-defined phases. TN has an important influence on SOM, while AP and AK had a relatively weak influence on the SOM from 2000 to 2018. Fertilizer use had a significant impact on crop yield, but due to the impact of salinization, TN and AK had no significant impact on crop yield in 1980. Since then, fertilization has had a significant impact on crop yield. From the perspective of different soil types, AK differed significantly. In 1980, the distribution of TN in different soil types was significantly different; in 1980, the distribution of SOM and crop yield in various soil types was different. In other years, the distribution of variables in different soil types was not significant. There was no obvious difference in the effects of soil nutrienton SOM and crop yield among different soil types.

ACKNOWLEDGMENT

This work was supported by the National Social Science Foundation of China (No. 41771561), Ministry of Education Research and Planning Foundation for Humanities and Social Sciences (No. 17YJA630040), and the Major Projects of China National Social Science Fund (No. 19ZDA096).

REFERENCES

Baker, N. J., Bancroft, B. A., Garcia, T. S., 2013. A meta-analysis of the effects of pesticides and fertilizers on survival and growth of amphibians. *Science of the Total Environment*449: 150–156.

Bedada, W., Karltun, E., Lemenih, M., Tolera, M., 2014. Long-term addition of compost and NP fertilizer increases crop yield and improves soil quality in experiments on smallholder farms. *Agriculture, Ecosystems and Environment*195: 193–201.

Benbi, D. K., Toor, A. S., Kumar, S., 2012. Management of organic amendments in rice wheat cropping system determines the pool where carbon is sequestered. *Plant and Soil*360(1–2): 145–162.

Bogunovic, I., Trevisani, S., Pereira, P., Vukadinovic, V., 2018. Mapping soil organic matter in the Baranja region (Croatia): Geological and anthropic forcing parameters. *Science of the Total Environment*643: 335–345.

Cai, A. D., Xu, M. G., Wang, B. R., Zhang, W. J., Liang, G. P., Hou, E. Q., Luo, Y. Q., 2019. Manure acts as a better fertilizer for increasing crop yields than synthetic fertilizer does by improving soil fertility. *Soil and Tillage Research*189: 168–175.

Cao, X. H., Long, H. Y., Lei, Q. L., Liu, J., Zhang, J. Z., Zhang, W. J., Wu, S. X., 2016. Spatio-temporal variations in organic carbon density and carbon sequestration potential in the topsoil of Hebei Province, China. *Journal of Integrative Agriculture*15(11): 2627–2638.

Chang, N. J., Zhai, Z., Li, H., Wang, L. G., Deng, J., 2020. Impacts of nitrogen management and organic matter application on nitrous oxide emissions and soil organic carbon from spring maize fields in the North China Plain. *Soil and Tillage Research*196:104441.

Chen, S. Y., Zhang, X. Y., Shao, L. W., Sun, H. Y., Niu, J. F., Liu, X. W., 2020. Effects of straw and manure management on soil and crop performance in North China Plain. *Catena*187:104359.

Chen, X. P., Cui, Z. L., Fan, M. S. , Vitousek, P., Zhao, M., Ma, W. Q., Wang, Z. L., Zhang, W. J., Yan, X. Y., Yang, J. C., Deng, X. P., Gao, Q., Zhang, Q., Guo, S. W., Ren, J., Li, S. Q., Ye, Y. L., Wang, Z. H., Huang, J. L., Tang, Q. Y., Sun, Y. X., Peng, X. L., Zhang, J. W. , He, M. R., Zhu, Y. J., Xue, J. Q., Wang, G. L., Wu, L., An, N., Wu, L. Q., Lin Ma, Zhang, W. F., Zhang, F. S, , 2014. Producing more grain with lower environmental costs. *Nature*514(7523): 486–489.

Cooperative Research Group on Chinese Soil Taxonomy (CRGCST), 2001. *Chinese Soil Taxonomy.* Science Press, Beijing.

Department of Population. *Social, Science and Technology Statistics (DPSSTS), National Bureau of Statistics of China (NBSC), China Population Statistics Yearbook, 1979–2018.* China Statistical Press, Beijing.

Du, Y. D., Cui, B. J., Zhang, Q., Wang, Z., Sun, J., Niu, W. Q., 2020. Effects of manure fertilizer on crop yield and soil properties in China: A meta-analysis. *Catena*193: 104617.

Egodawatta, W. C. P., Sangakkara, U. R., Stamp, P., 2012. Impact of green manure and mineral fertilizer inputs on soil organic matter and crop productivity in a sloping landscape of Sri Lanka. *Field Crops Research*129: 21–27.

Fan, M. M., Lal, R., Zhang, H., Margenot, A. J., Wu, J. T., Wu, P. B., Zhang, L. M., Yao, J. T., Cheng, F. R., Gao, C., 2020. Variability and determinants of soil organic matter under different land uses and soil types in eastern China. *Soil and Tillage Research*193: 104617.

FAO, 2017. The future of food and agriculture–Trends and challenges. Rome.

FAOSTAT, 2014. Online [Internet]. Retrieved August 6, 2020, from. http://faostat.fao.org/site/291/default.aspx. FAO, Rome.

Feng, J. F., Li, F. B., Deng, A. X., Feng, X. M., Fang, F. P., Zhang, W. J., 2016. Integrated assessment of the impact of enhanced-efficiency nitrogen fertilizer on N2O emission and crop yield. *Agriculture, Ecosystems and Environment*231: 218–228.

Gai, X. P., Liu, H. B., Liu, J., Zhai, L. M., Yang, B., Wu, S. X., Ren, T. Z., Lei, Q. L., Wang, H. Y., 2018. Long-term benefits of combining chemical fertilizer and manure applications on crop yields and soil carbon and nitrogen stocks in North China Plain. *Agricultural Water Management*208: 384–392.

Gan, L. L., 2016. Environmental risks of fertilizer use and the prevention and control measures in Chinese rural areas. *Chimica Oggi-Chemistry Today*34: 33–38.

Godfray, H. C. J., Beddington, J. R., Crute, I. R., Haddad, L., Lawrence, D., Muir, J. F., Pretty, J., Robinson, S., Thomas, S. M., Toulmin, C., 2010. Food security: The challenge of feeding 9 billion people. *Science*327(5967): 812–818.

Gong, W., Yan, X. Y., Wang, J. Y., Hu, T. X., Gong, Y. B., 2009. Long-term manure and fertilizer effects on soil organic matter fractions and microbes under a wheat–maize cropping system in northern China. *Geoderma*149(3–4): 318–324.

He, J., Li, H. W., Rabi, G. R., Wang, Q. J., Cai, G. H., Su, Y. B., Qiao, X. D., Liu, L. J., 2011. Soil properties and crop yields after 11 years of no tillage farming in wheat–maize cropping system in North China Plain. *Soil and Tillage Research*113(1): 48–54.

Hu, K. L., Huang, Y. F., Li, H., Li, B. G., Chen, D. L., White, R. E., 2005. Spatial variability of shallow groundwater level, electrical conductivity and nitrate concentration, and risk assessment of nitrate contamination in North China Plain. *Environment International*31(6): 896–903.

Hu, L. X., Zhang, X. H., Zhou, Y. H., 2019. Farm size and fertilizer sustainable use: An empirical study in Jiangsu, China. *Journal of Integrative Agriculture*18(12): 2898–2909.

Jannoura, R., Joergensen, R. G., Bruns, C., 2014. Organic fertilizer effects on growth, crop yield,and soil microbial biomass indices in sole and intercropping peas and oats under organic farming conditions. *European Journal of Agronomy*52: 259–270.

Kahrl, F., Li, Y. J., Su, Y. F., Tennigkeit, T., Wilkes, A., Xu, J. C., 2010. Greenhouse gas emissions from nitrogen fertilizer use in China. *Environmental Science and Policy*1(3): 688–694.

Kong, X. B., Zhang, F. R., Qi, W., Xu, Y., 2003a. The influence of land use change on soil fertility in intensive agricultural region: A case study of Quzhou County, Hebei. *Acta Geographica Sinica*58(3): 333–342.

Kong, X. B., Zhang, F. R., Xu, Y., Qi, W., 2003b. Changes in soil fertilities in saline soil region of Quzhou County, Hebei Province. *Rural Eco-Environment*3: 35–37.

Kong, X. B., Zhang, F. R., Qi, W., Xu, Y., Jiang, G. H., 2006. Influence of land use change on soil nutrients in an intensive agricultural region of North China. *Soil and Tillage Research*88(1–2): 85–94.

Kong, X. B., Lal, R., Li, B. G., Liu, H. B., Li, K. J., Feng, G. L., Zhang, Q. P., Zhang, B. B., 2014. Fertilizer intensification and its impacts in China's HHH plains. *Advances in Agronomy*125: 135–169.

Kong, X. B., Zhang, X. L., Lal, R., Zhang, F. R., Chen, X. H., Niu, Z. G., Han, L., Song, W.2016. *Groundwater Depletion by Agricultural Intensification in China's HHH Plains, Since 1980S*: Volume 135. Academic Press, Cambridge, MA, pp. 59–106.

Kwanowaa, Y. O., Furaha, R. K., Yan, L., Wei, D. M., 2020. Effects of planting field on groundwater and surface water pollution in China. *CLEAN: Soil Air Water*48:1900452.

Lai, Z. H., Chen, M. Q., Liu, T. J., 2020. Changes in and prospects for cultivated land use since the reform and opening up in China. *Land Use Policy*97:104781.

Lal, R., 2004. Soil carbon sequestration impacts on global climate change and food security. *Science*304(5677): 1623–1627.

Li, S., Wu, J. C., Wang, X. Q., Ma, L., 2020. Economic and environmental sustainability of maize-wheat rotation production when substituting mineral fertilizers with manure in the North China Plain. *Journal of Cleaner Production*271: 122683.

Li, Y., Fredrich, K., Pan, J. J., David, R. H., Su, Y. F., Andreas, W., Xu, J. C., 2012. Fertilizer use patterns in Yunnan Province, China: Implications for agricultural and environmental policy. *Agricultural Systems*110: 78–89.

Liu, C. A., Li, F. R., Zhou, L. M., Zhang, R. H., Jia, Y., Lin, S. L., Wang, L. J., Siddique, K. H. M., Li, F. M., 2013. Effect of organic manure and fertilizer on soil water and crop yields in newly-built terraces with loess soils in a semi-arid environment. *Agricultural Water Management*117: 123–132.

Liu, Y., Menjivar, J. R., Zhang, L., Zhang, J. B., Swisher, M. E., 2019. Technical training and rice farmers' adoption of low-carbon management practices: The case of soil testing and formulated fertilization technologies in Hubei, China. *Journal of Cleaner Production*226: 454–462.

Lu, R.1999. *Analytical Methods of Soil and Agricultural Chemistry*. China Agricultural Science and Technology Press, Beijing, pp. 107–240.

Lu, T. P., Wang, X. J., Xu, M. G., Yu, Z. T., Luo, Y. M., Smith, P., 2020. Dynamics of pedogenic carbonate in the cropland of the North China Plain: Influences of intensive cropping and salinization. *Agriculture, Ecosystems and Environment*292: 106820.

Olsen, S. R., Sommers, L. E., 1982. Phosphorus. In: Page, A.L., Miller, R.H., Keeney, D.R. (Eds.), *Methods of Soil Analysis*. Agronomy Series No. 9,Part 2. 2nd ed. Soil Science Society of America, Inc., Madison, WI, pp. 403–430.

Quzhou County Statistics Bureau, 1982–2018. *Quzhou County Statistical Yearbook, 1982–2018, Guangping: Guangping printing factory.*

Searchinger, T., Waite, R., Hanson, C., Ranganathan, J., Matthews, E., 2019. *World Resources Report: Creating a Sustainable Food Future* (Final Report).

Singh, S., Singh, S., Pannu, C. J. S., Mittalz, J. P., 1996. Fertilizer energy use and crop yield relationships for wheat in the Punjab ecosystems. *Energy Conversion and Management*37(10): 1547–1553.

Song, W., Deng, X. Z., Yuan, Y. W., Wang, Z., Li, Z. H., 2015. Impacts of land-use change on valued ecosystem service in rapidly urbanized North China Plain. *Ecological Modelling*318: 245–253.

Sun, X. B., Zhang, Q. P., Kong, X. B., Wen, L. Y., Zhao, J., Liu, F. J., 2019. Spatiotemporal characteristics of cultivated soil fertility in the intensive agricultural region of North China: A case study of Quzhou County in Hebei Province. *Chinese Journal of Eco-Agriculture*27(12): 1857–1869.

Tian, K., Zhang, B. E., Zhang, H. D., Huang, B., Darilek, J. L., Zhao, Y. C., Yang, J. S., 2020. Evaluation of soil quality in major grain-producing region of the North China Plain: Integrating minimum data set and established critical limits. *Ecological Indicators*117: 106613.

Wang, C., Li, X. L., Gong, T. T., Zhang, H. Y., 2014. Life cycle assessment of wheat-maize rotation system emphasizing high crop yield and high resource use efficiency in Quzhou County. *Journal of Cleaner Production*68: 56–63.

Wang, J. Y., Yan, X. Y., Gong, W., 2015. Effect of long-term fertilization on soil productivity on the North China plain. *Pedosphere*25(3): 450–458.

Wei, W. L., Yan, Y., Cao, J., Christie, P., Zhang, F. S., Fan, M. S., 2016. Effects of combined application of organic amendments and fertilizers on crop yield and soil organic matter: An integrated analysis of long-term experiments. *Agriculture, Ecosystems and Environment*225: 86–92.

Wilding, L. P., 1985. Spatial variability: Its documentation, accommodation, and implication to soil surveys. In: Nielsen, D.R., Bouma, J. (Eds.), *Soil Spatial Variability*. Pudoc, Wageningen, Netherlands.

Xu, E. Q., Wang, R., Zhang, H. Q., Yu, Z. X., 2019. Coupling index of water consumption and soil fertility correlated with winter wheat production in North China Region. *Ecological Indicators*102: 154–165.

Xu, Y., Zhang, F. R., Hao, X. Y., Wang, J. K., Wang, R., Kong, X. B., 2006. Influence of management practices on soil organic matter changes in the Northern China plain and Northeastern China. *Soil and Tillage Research*86(2): 230–236.

Yan, L., Wang, Y. Y., Tumbalam, P., Zhang, T. L., Gao, Q., Zhang, W. J., Wei, D. M., Yaa, O. K., 2019. Spatiotemporal distribution of chemical fertilizer application and manure application potential in China. *Environmental Engineering Science*36(10): 1337–1348.

Yin, H. J., Zhao, W. Q., Li, T., Cheng, X. Y., Liu, Q., 2018. Balancing straw returning and chemical fertilizers in China: Role of straw nutrient resources. *Renewable and Sustainable Energy Reviews*81: 2695–2702.

Zhang, H. Y., Niu, L. A., Hu, K. L. , Hao, J. M., Li, F., Gao, Z. Q., Wang, X., 2020a. Influence of tillage, straw-returning and mineral fertilization on the stability and associated organic content of soil aggregates in the North China plain. *Agronomy*10(7): 951.

Zhang, J., Balkovič, J., Azevedo, L. B., Skalský, R., Bouwman, A. F., Xu, G., Wang, J. Z., Xu, M. G., Yu, C. Q., 2018. Analyzing and modelling the effect of long-term fertilizer management on crop yield and soil organic carbon in China. *Science of the Total Environment*627: 361–372.

Zhang, W. X., Wang, X. J., Lu, T. P., Shi, H. J., Zhao, Y. J., 2020b. Influences of soil properties and hydrological processes on soil carbon dynamics in the cropland of North China Plain Influences of soil properties and hydrological processes on soil carbon dynamics in the cropland of North China Plain. *Agriculture, Ecosystems and Environment*295:106886.

Zhang, X. Y., Chen, L. D., Li, Q., Qi, X., Yang, S., 2013. Increase in soil nutrients in intensively managed cash-crop agricultural ecosystems in the Guanting Reservoir catchment, Beijing, China. *Geoderma*193–194: 102–108.

Zhao, H. L., Ning, P., Chen, Y. L., Liu, J. F., Ghaffar, Abdul, Shar, Tian X. H., Shi, J. L., 2019. Effect of straw amendment modes on soil organic carbon, nitrogen sequestration and crop yield on the North-Central Plain of China. *Soil Use and Management*35(3): 511–525.

Zheng, W. L., Luo, B. L., Hu, X. Y., 2020. The determinants of farmers' fertilizers and pesticides use behavior in China: An explanation based on label effect. *Journal of Cleaner Production*272: 123054.

Zhou, W., Wang, W. j., Zhang, B., Xiao, L., Lv, H. L., He, X. Y., 2017. Soil fertility evaluation for urban forests and green spaces in Changchun City. *Acta Ecologica Sinica*37(4): 1211–1220.

Zhou, Y., Biswas, A., Ma, Z. Q., Lu, Y. L., Chen, Q. X., Shi, Z., 2016. Revealing the scale-specific controls of soil organic matter at large scale in Northeast and North China Plain. *Geoderma*271: 71–79.

Zhu, E., Deng, J., Zhou, M., Gan, M., Jiang, R., Wang, K., Shahtahmassebi, A. R., 2019. Carbon emissions induced by land-use and land-cover change from 1970 to 2010 in Zhejiang, China. *Science of the Total Environment*646: 930–939.

8 Managing Rain-fed Rice Farms for Improving Soil Health and Advancing Food Security

A Meta-Analysis

Abhishek Jangir, Gaurav Mishra, G.W. Sileshi, Gulab Singh Yadav, and Arun Jyoti Nath

CONTENTS

LIST OF ABBREVIATIONS

BD: Bulk density
C: Carbon
EA: East Asia
FYM: Farmyard manure
GR: Green Revolution
Ha: Hectare
INM: Integrated nutrient management
K: Potassium
Kg: Kilogram
LA: Latin America
LMM: Linear mixed model
M: Million
MENA: Middle East and North Africa
Mg: Mega gram
MWD: Mean weight diameter
N: Nitrogen
P: Phosphorus
REML: Restricted maximum likelihood

DOI: 10.1201/9781003102762-8

RR: Response ratio
SA: South Asia
SE: Standard error
SOC: Soil organic carbon
SOM: Soil organic matter
SSA: Sub-Saharan Africa
T: Treatment

8.1 INTRODUCTION

About 80% of global agriculture is rain-fed and, as such, it is considered very important for global food security. Despite its immense potential, rain-fed agriculture is vulnerable to droughts, water scarcity, increasing population, and climate change (Srinivasarao et al. 2015). The management practices under rain-fed agriculture vary regionally, but they are essential, as they can meet most of the food demands of poor communities in developing countries. The proportion of rain-fed agriculture to entire agriculture is 95% in Sub-Saharan Africa (SSA), 90% in Latin America (LA), 60% in South Asia (SA), 65% in East Asia (EA), and 75% in the Middle East and North Africa (MENA) (FAO 2003). In India, agriculture was the primary source of employment before independence. Nearly 72.4% of the population in 1952 depended on agriculture for income generation, which decreased to 52.1% by 2014 due to the growth of the industrial and service sectors. However, there is no alternative to agriculture to feed India's fast-growing population, which will be 1.35 billion (B) by 2025 and 1.7 B by 2050 (UN 2020). Although the Green Revolution (GR) achieved great success in Indian agriculture during the 1960s, there are still some challenges caused by the ever-decreasing per capita availability of land and the growing population. Hence, the intensification of rain-fed agriculture is a prerequisite to feed the burgeoning population.

Rice (*Oryza sativa* L.) is an important cereal crop and is cultivated in different ecosystems, viz., uplands (including shifting cultivation areas), lowlands, and deep-water conditions (Devi et al. 2008). It is a staple food grain crop in the Asia-Pacific region, including India (Ladha et al. 2009), and is the staple food for half of the world population (Fuhrmann et al. 2019). India occupies an important position, in both area and production of rice. It has the largest area (42.9 M ha), accounting for about 27.1% of the world's total rice-growing area. The information regarding the area, production, and productivity under different farming conditions is summarized in Table 8.1.

In India, rain-fed rice is cultivated in Assam, Bihar, West Bengal, Odisha, Eastern parts of Madhya Pradesh and Uttar Pradesh, and the northeastern hill region (Rao et al. 2008). Rain-fed rice is cultivated on over 20 M ha of the land or 55% of the total rice area. Of this, upland rice cultivation

TABLE 8.1
Rice Ecologies, Area, Production, and Productivity in India

	Area	Yield*
Rice ecology	**M ha**	**Mg ha^{-1}**
1. Irrigated	26.0–28.0	3.5–4.5
2. Uplands	4.8–6.0	0.8–1.1
3. Rain-fed lowlands	12.0–14.5	2.5–3.0
4. Flood prone	1.3–2.0	1.3–1.6

Source: NRRI (2020)
*Estimated values are based on previous year's production data.

accounts for 13.5% while lowland rice cultivation accounts for 32.4% of the total rice area. About 5–6 M ha of the land is under upland rain-fed rice, which accounts for 13.5% of the total area under rice crop in India, while rain-fed lowland rice is grown on around 14.4 M ha, mostly in Eastern India, and the area accounts for 32.4% of the total rice area of the country (Mahajan et al. 2017). The productivity of rice in India is low (2.93 Mg ha^{-1}) due to the very low productivity of the rain-fed rice system, which accounts for 55% of the total rice area and has very low productivity (1 Mg ha^{-1}) (Chandrasekaran 2008). Despite its significance, a detailed comparative evaluation of nutrient management practices in rain-fed rice is still lacking.

The productivity of rain-fed rice depends on the indigenous soil nutrients available and the adopted nutrient management practices. Several attempts have been made to develop certain nutrient management practices, which could significantly increase yields and have the potential to maintain soil health and sustain food security. Rice ecosystems are also crucial to soil carbon (C) sequestration (Nath et al. 2016). In rice fields, decomposition of soil organic matter (SOM) slows down due to prolonged submergence and higher production, which also contributes to the buildup of soil organic carbon (SOC) (Rajkishore et al. 2015; Meetei et al. 2020).

A detailed comparative evaluation of these nutrient management practices is needed to provide insights into soil health improvement and soil C sequestration in rain-fed rice agroecosystems. Therefore, a goal of this research was to provide a comparative synthesis of integrated nutrient management (INM), that is, inorganic and organic nutriment management in terms of SOC storage, nutrient stocks, and rice productivity in the rain-fed cultivation systems. The specific objective was to document the potential of different nutrient management practices for enhancing rice productivity; provide a comparative synthesis of INM, inorganic and organic nutriment management, on SOC storage, nutrient stocks, and rice productivity in the rain-fed cultivation systems; and upscale it to other similar regions of the world to sustain soil health and advance food security of rain-fed rice farms.

8.2 METHODS OF DATA COLLECTION

8.2.1 About the Nutrient Management Practices

This study used a review and meta-analysis of published data to compare three nutrient management practices. Of these, INM involves the combined use of mineral fertilizers (NPK) with organic resources, such as cattle manure, crop residues, urban/rural wastes, composts, green manures, and bio-fertilizers (Antil 2012). The basic concept of INM is sustaining soil and crop productivity by optimizing all possible plant nutrient sources in an integrated manner. INM is an approach that seeks to increase the quality of production and protect the environment for posterity. Those studies reported the application of recommended dose of chemical fertilizers (NPK) (at 75% ratio) along with farmyard manure (FYM) or green manure (at 25% ratio) were screened for this study.

Synthetic fertilizer is a fundamental component of inorganic nutrient management. It is often either underused or overused in the absence of soil test-based fertilization assessments. There is a multiplicity of methods and techniques currently available for determining nutrient requirements, emphasizing the importance attributed to an awareness of fertilizer requirements. The fertilizer requirement of annual or perennial crops depends upon the crop grown and whether the purpose is to grow food or feed crop. Several studies were conducted to establish recommended doses of chemical fertilizers (NPK) and arrived at the general recommendation of N: 90, P: 30, K: 40 for the comparative study of INM and organic treatments.

Organic sources play a critical role in short-term nutrient availability and long-term SOM maintenance (Pang and Letey 2000). Organic farming's primary aim is to sustain soil health using biological wastes, organic wastes, and bio-fertilizers. Using organic manures promotes sustainable crop production and a pollution-free environment. Studies that reported 100% organic as solely

FYM/sesbania (*Sesbania grandiflora*)/rice straw/wheat straw/cow dung were screened for the comparative study of INM and inorganic treatments.

For data to be included in this analysis, the following criteria were used: (1) published in a refereed journal, book chapter, or peer-reviewed proceedings during the period 1980–2020 in India, (2) based on rain-fed rice systems, (3) reported data on soil physical and chemical properties under different management practices, and (4) reported the mean values as numerical or graphical data. Further screening was also performed on the data of the selected publications for analysis to meet the quality standards of meta-analysis.

8.2.2 Data Collected and Effect Sizes Used

The key data collected from the publications that qualified for meta-analysis included rice grain yield, rice straw yield, soil bulk density (BD), SOC, nitrogen (N), phosphorus (P), and potassium (K) contents. Then the N, P, and K stocks were calculated using the BD, soil depth, and nutrient concentration values.

Meta-analysis requires the use of an effect size index that provides a meaningful summary of the studies included in the analysis. In this analysis, change (δ) in measured variables between the treatment (T) and control was expressed as a percentage of the control, i.e. δ = 100*(T-*Control*)/*Control* because of its ease of interpretation. In addition, response ratio (RR = *T/Control*) was used because of its application in ecology (Hedges et al. 1999) and agriculture, where yields from treatment and control are compared. The RR quantifies the proportionate change that results from an experimental manipulation (Hedges et al. 1999).

8.2.3 Data Analysis

Before analysis, RR was log-transformed to ensure normality (Hedges et al. 1999). Considering the unequal sample sizes among treatments, all data analysis was implemented within a linear mixed model (LMM) framework using the MIXED procedure in the Statistical Analysis System (SAS). Nutrient management practices were held as the fixed effects and the study as the random effects. The covariance structure of the data was modeled using the method of restricted (residual) maximum likelihood (REML). Given the large imbalance in the sample sizes, the Kenward–Roger method was used for approximating the degrees of freedom. The method estimates the standard errors of fixed effects by the Prasad–Rao–Jeske–Kackar–Harville method. Statistically significant differences between nutrient management practices were based on 95% confidence and *P* value of 0.05.

The probability of obtaining a given value of yield and response ratios was estimated by using the different nutrient management practices. Then the cumulative probability distribution of yields and RR were plotted. A cumulative distribution function X was deemed dominating Z in the first order if the distribution of X lies everywhere to the right of the cumulative distribution of Z.

8.3 RESULTS AND DISCUSSION

8.3.1 Variation in Soil Properties with Nutrients Management Practices

Table 8.2 shows a comparison of soil physico-chemical properties between the three different nutrient management practices in rain-fed rice systems. The nutrient management practices significantly influenced soil BD, with INM maintaining an optimum BD level (1.25 gm cm^{-3}), whereas the inorganic nutrient management practices increased soil compaction (1.45 gm cm^{-3}). The analysis also revealed that different nutrient management practices significantly influenced SOC content. Organic inputs resulted in higher SOC content (1.54%) than the inorganic (1.38%) and INM (1.46%). Similar to SOC, available N stocks were also significantly affected by nutrient management practices. INM

TABLE 8.2
Influence of Integrated Nutrient Management (INM), Inorganic, and Organic Nutriment Management on Soil Properties Under Rain-fed Rice Cultivation in India

Variable	Type	Estimate (SE)	*P* Value
BD (gm cm^{-3})	INM	1.25 (0.05)	<0.0001
	Inorganic	1.45 (0.06)	
	Organic	1.32 (0.05)	
SOC (%)	INM	1.46 (0.8)	<0.0001
	Inorganic	1.38 (0.6)	
	Organic	1.54 (0.7)	
N stocks (kg ha^{-1})	INM	244.2 (61.8)	<0.0001
	Inorganic	132.73 (59.2)	
	Organic	164.71 (60.1)	
P stocks (kg ha^{-1})	INM	16.9 (12.0)	0.6763
	Inorganic	26.8 (6.9)	
	Organic	19.4 (8.9)	
K stocks (kg ha^{-1})	INM	565.4 (67.9)	<0.0001
	Inorganic	131.4 (62.6)	
	Organic	303.5 (64.4)	

maintained higher levels of N stock (244 kg ha^{-1}) compared with organic (165 kg ha^{-1}) and inorganic (133 kg ha^{-1}) nutrient management. Furthermore, K stocks were significantly highest under INM (565 kg ha^{-1}) followed by that under organic (304 kg ha^{-1}) and inorganic (131 kg ha^{-1}) nutrient management. On the other hand, available P stocks did not vary significantly among nutrient management practices.

In rice production systems, soils are managed by puddling (continuous submergence for the entire growth cycle) and sometimes by alternate wetting and drying conditions. Figure 8.1 shows the physiographic differences in the upland and lowland rice farms in India. Therefore, management of paddy soil fertility through inorganic, organic, and INM may influence soil properties and SOC content. This is why soils of rain-fed rice production managed under diverse nutrient management have different soil properties. The present study demonstrated that INM and organic nutrient management systems reduced BD of paddy soils compared with that of inorganic nutrient management attributed to the addition of more organic inputs. Application of high levels of organic amendment improves SOM content (Yimer et al. 2006), which in turn enhances granulation, mean weight diameter (MWD) of aggregates, and lower BD (Meena et al. 2020). With high SOM content, organic amendments are rich sources of different macro- and micronutrients (Stewart et al. 2009). Organic amendments also improve soil's physical, chemical, and biological properties (Meena et al. 2020). Thus, the integrated application of nutrients increases crop productivity and soil sustainability of the rain-fed rice production systems. Application of organic amendments, either sole or in combination with inorganic fertilizer, increases the addition of C inputs into the soil and leads to higher SOC content than inorganic fertilizers. The integrated application of fertilizer and manures increases C inputs and increases the availability of NPK, which promotes the growth of rain-fed rice and, subsequently, greater yield and biomass.

The greater yield under INM, rather than organic and inorganic, further added more C inputs by incorporating more crop residues, litter, and root biomass. Overall, the results of the analysis demonstrate that INM decreases BD, yields more SOC content and NPK stocks than those under organic and inorganic treatments. Many other researchers have also reported improvements in soil BD and an increase in SOC content and available NPK stocks under rice and other agricultural

FIGURE 8.1 Upland (A) and lowland (B) rain-fed rice farms in India.

production systems. The study results show that C content and NPK stocks were higher in soil under INM treatments than those under applied fertilizer treatment. Such a trend may be due to the probability that use of INM amendments reduces mineralization of biomass. Thus, slow release of NPK improves long-term effects on soil fertility by increasing the status of soil fulvic/humic acid ratio, which has positive impacts on sustainability (Weber et al. 2007). While applied fertilizers rapidly release NPK, mineralized uptake emitted to the environment and leached down below the root zone aggravate water and soil contamination. In contrast, INM decreases losses of nutrients and increases their retention in soil over a longer period.

The results also indicate significant changes in soil BD, SOC stocks, and plant available NPK stocks with nutrient management practices compared with those under "no-input control." These changes are attributed to a long-term balance between C inputs and decomposition (which is influenced by fertilization and other management practices), mineralization, and release of nutrients from fertilizers and organic manures (Waqas et al. 2020). The SOC storage and NPK stocks differed significantly probably due to differences in C input into the soil from organic and inorganic inputs and through INM. Soil under organic and INM treatments had 20 and 17% higher SOC stocks over those under control, respectively, than inorganic treatment over control (3.6%). A direct source of C through organic fertilizers increases SOC stocks (Li et al. 2018) and provides additional nutrients (N, P, S, etc.) to the soil. These nutrients are essential to convert biomass-C into SOM, leading to higher SOC stocks under organic treatment. However, a higher increase in SOC stocks might also be due to the addition of a higher input of biomass-C under INM through the combined application of fertilizers and organic manure; production and the attendant inputs into soil; and more crop residues, litter, and root-associated C. Combined long-term use of organic and inorganic amendments improves SOC stocks, increases total soil porosity, enhances SOC concentration by acting as a bacterial glue, cementing soil particles into stable and larger aggregates (Carter et al. 2003), and decreases soil BD. The results presented herein are in agreement with those from other studies involving long-term use of compost leading to a relatively lower BD of the soil (Jagadamma et al. 2009; Srinivasan et al. 2012).

The percentage change over control under different nutrient management practices is summarized in Table 8.3. Compared to inorganic and INM, a reduction in BD over its control was observed in soil under organic treatment. In the case of soil under inorganic treatments, BD tends to increase, leading to densification of soil over time. In contrast, use of INM and organic inputs increased SOC content by 17 and 20% over the respective no-input control. The highest increase in N content (relative to the control) was recorded with inorganic inputs, while the lowest was with that in soil under organic inputs. The increase in soil P content was higher with inorganic inputs and INM compared to that under organic inputs. Unlike N and P, an increase in soil K content was achieved with inorganic inputs, while the reverse trend was observed in soil under INM and organic treatments.

8.3.2 Variation in Yield and Yield Components with Nutrient Management Practices

Results of rice grain and straw yields under different nutrient management practices are summarized in Table 8.4. The analysis revealed no significant variation in rice grain and straw yields with

TABLE 8.3
Percent Change (δ) in the Treatment over Control

Variable	Type	% Change over Control (SE)	*P* Value
BD_δ	INM	−1.9 (1.4)	0.0182
	Inorganic	1.1 (1.4)	
	Organic	−6.5 (1.7)	
SOC_δ	INM	17.0 (6.7)	0.0131
	Inorganic	3.6 (6.7)	
	Organic	20.0 (6.7)	
N_δ	INM	8.3 (14.6)	0.2052
	Inorganic	25.5 (10.0)	
	Organic	7.5 (11.7)	
P_δ	INM	64.5 (73.1)	0.7477
	Inorganic	67.8 (40.5)	
	Organic	29.7 (53.6)	
K_δ	INM	−2.1 (11.2)	0.0346
	Inorganic	19.9 (6.8)	
	Organic	−3.1 (8.6)	
Grain yield$_\delta$	INM	80.9 (23.6)	0.0004
	Inorganic	59.9 (23.6)	
	Organic	33.4 (25.4)	

Note: Values within parenthesis are the standard errors of the estimates.

TABLE 8.4
Influence of INM, Inorganic, and Organic Nutriment Management Systems on Yield Attributes Under Rain-fed Rice Cultivation in India

Variable	Type	Estimate (SE)	*P* Value
Grain yield (Mg ha^{-1})	INM	3.3 (0.4)	0.4281
	Inorganic	3.8 (0.4)	
	Organic	3.6 (0.5)	
Straw yield (Mg ha^{-1})	INM	5.0 (0.6)	0.9938
	Inorganic	5.0 (0.6)	
	Organic	4.9 (0.6)	
RR (grain yield)	INM	1.8 (0.2)	0.0004
	Inorganic	1.3 (0.2)	
	Organic	1.6 (0.3)	

Note: Values within parenthesis are the standard errors of the estimates.

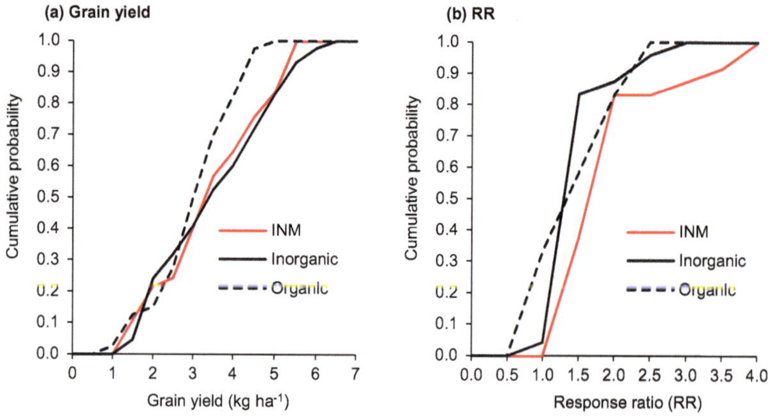

FIGURE 8.2 The cumulative probability distribution of grain yields and the response ratios.

nutrient management practices. The highest rice grain yield (3.8 Mg ha^{-1}) in rain-fed conditions was observed with inorganic inputs, but this was not significantly different from those achieved with organic (3.6 Mg ha^{-1}) or INM (3.3 Mg ha^{-1}) inputs. Similarly, there were no significant differences in straw yield among INM inorganic and organic treatments (Table 8.4). However, the RR significantly differed among nutriment management treatments. Significantly, higher RR was achieved with INM than that in soil under inorganic inputs (Figure 8.2). The increase in rice grain yield (over the control) was higher with INM and inorganic inputs than that with organic inputs. The cumulative probability distributions of yields, and RRs did not reveal significant differences among various nutrient management practices.

8.3.3 Implications for Management

The results demonstrate that soils under rain-fed rice can sustain crop productivity and enhance soil C and N concentration. This process is set in motion by (1) storage of nutrients in SOM, (2) mineralization of the nutrients from organic to available forms for uptake by plants, and (3) physical and chemical processes that control nutrient availability, displacement, sorption, and ultimate losses to the environment or water resources. Managed soils tend to transform toward an active system, and it is this very dynamic nature that enhances soils' functionality and increases provision of ecosystem services (Lal 2015). Overall, soil C and N stocks and functioning of soils depend on the interaction between soil properties, mineral matrixes, agricultural inputs, and management practices (Montgomery 2007). It is these interactions that lead to decomposition of biomass and buildup of SOM, availability and preservation of the available nutrients, and plant growth and development. The greater SOC content and NPK stocks under INM explain the higher yield increase under INM than those under other treatments. Unbalanced application of plant nutrients is also causing yield reductions in other parts of the world (Waqas et al. 2020). However, balanced fertilization, or INM, leads to more yield benefits (Yadav et al. 2019) than indiscriminate and unbalanced use of fertilization. An increase in agronomic yield under INM is primarily attributed to the improved availability of plant nutrients with the application of manure (Dai et al. 2019). Use of INM produces higher grain yields and leads to buildup of SOC stock. On the other hand, using organic manures showed lesser impact on rice yield but a larger impact on SOC content and stocks. However, inorganic fertilization has a high impact on rice yield but little effect on SOC stock. Many global analyses have also demonstrated that organic agriculture could reduce crop yield by 5–34%, compared with that of the mineral fertilizers (Knapp and Heijden 2018).

8.4 CONCLUSIONS

The results presented herein indicate that nutrient management practices significantly influenced soil BD, with INM maintaining an optimum BD level, whereas inorganic inputs increased soil compaction. Organic inputs resulted in higher SOC content (1.54%) than that in soil under inorganic (1.38%) or INM (1.46%) treatment. Available N stocks were significantly higher with INM (244 kg ha^{-1}) compared with organic (165 kg ha^{-1}) and inorganic (133 kg ha^{-1}) nutrient management. K stocks were significantly highest under INM (565 kgha^{-1}) than organic (304 kg ha^{-1}) and inorganic (131 kg ha^{-1}) nutrient management. Available P stocks did not vary significantly with nutrient management practices. Rice grain yield increased over the control at a higher rate with INM and inorganic inputs than that with organic inputs. The data support the conclusion that:

1. INM maintains optimum BD levels, SOC contents, N, and K stocks in rain-fed rice cultivation.
2. INM achieved lower rice grain yields although the change over the control was much higher with INM than with organic inputs alone.
3. Higher rice yields, and SOC buildup under rain-fed conditions over the years can be achieved with the balanced application of plant nutrients through an INM management system.

REFERENCES

Antil, R. S. 2012. Integrated plant nutrient supply for sustainable soil health and crop productivity. In: A Kumar (ed.), *Focus Global Reporter*, India Water Foundation, New Delhi, 3(2).

Carter, M. R., D. A. Angers, E. G. Gregorich, and M. A. Bolinder. 2003. Characterizing organic matter retention for surface soils in eastern Canada using density and particle size fractions. *Can. J. Soil Sci.* 83(1):11–23.

Chandrasekaran, B. 2008. Rice research programme in Tamil Nadu. In: Souvenir of 3rd National Symposium on System of Rice Intensification in India: Policies, Institutions and Strategies for Scaling up. Coimbatore, India.

Dai, H., Y. Chen, K. Liu, et al. 2019. Water-stable aggregates and carbon accumulation in barren sandy soil depend on organic amendment method: A three-year field study. *J. Clean. Prod.* 212:393–400.

Devi, T. P., A. A. Durai,T. A. Singh, S. Gupta, J. Mitra, A. Pattanayak, B. K. Sarma, and A. Das. 2008. Preliminary studies on physical and nutritional qualities of some indigenous and important rice cultivars of northeastern hill region of India. *J. Food Qual.* 31(6):686–700. https://doi.org/10.1111/J.1745-4557.2008.00228.X.

FAO. 2003. *FAOSTAT.* Food and Agriculture Organization of the United Nations, Rome.

Fuhrmann, I., S. Maarastawi, J. Neumann, et al. 2019. Preferential flow pathways in paddy rice soils as hot spots for nutrient cycling. *Geoderma* 337:594–606.

Hedges, L. V., J. Gurevitch, and P. S. Curtis. 1999. The meta-analysis of response ratios in experimental ecology. *Ecology* 80(4):1150–1156.

Jagadamma, S., R. Lal, and B. K. Rimal. 2009. Effects of topsoil depth and soil amendments on corn yield and properties of two alfisols in central Ohio. *J. Soil Water Conserv.* 64(1):70–80.

Knapp, S., and M. G. A. van der Heijden. 2018. A global meta-analysis of yield stability in organic and conservation agriculture. *Nat. Commun.* 9(1):3632. https://doi.org/10.1038/s41467-018-05956-1.

Ladha, J. K., V. Kumar, M. M. Alam, et al. 2009. Integrating crop and resource management technologies for enhanced productivity, profitability, and sustainability of the rice-wheat system in South Asia. In: *Integrated Crop and Resource Management in the Rice: Wheat System of South Asia*, International Rice Research Institute (IRRI), Philippines, 69–108.

Lal, R. 2015. Soil carbon sequestration in agro ecosystems of India. *J. Indian Soil Sci. Soc.* 63(2):125–143.

Li, Y. E., S. Shi, M. A. Waqas, et al. 2018. Long-term (20 years) application of fertilizers and straw return enhances soil carbon storage: A meta-analysis. *Mitigation Adapt. Strategies Global Change* 23(4):603–619.

Mahajan, G., V. Kumar, and S. Chauhan. 2017. Rice production in India. In: B. S. Chauhan, et al. (eds.), *Rice Production Worldwide.* Springer, New York. https://doi.org/10.1007/978-3-319-47516-5_3.

Meena, R. S., R. Lal, and G. S. Yadav. 2020. Long-term impacts of topsoil depth and amendments on soil physical and hydrological properties of an alfisol in central Ohio, USA. *Geoderma* 363: 114164.

Meetei, T. T., M. C. Kundu, and Y. B. Dev. 2020. Long-term effect of rice-based cropping systems on pools of soil organic carbon in farmer's field in hilly agroecosystem of Manipur, India. *Environ. Monit. Assess.* 192(4):209. https://doi.org/10.1007/s10661-020-8165-x.

Montgomery, D. R. 2007. Soil erosion and agricultural sustainability. *Proc. Natl Acad. Sci. U. S. A.* 104(33):13268–13272.

Nath, A. J., T. Bhattacharyya, S. K. Ray, et al. 2016. Assessment of rice farming management practices based on soil organic carbon pool analysis. *Trop. Ecol.* 57:607–611.

National Rice Research Institute (NRRI). 2020. *NRRI Research Bulletin No. 22 on Eco-regional Rice Farming for Enhancing Productivity, Profitability and Sustainability* (edited by H Pathak, R Tripathi, N N Jambhulkar, J P Bisen and B B Panda). ICAR-National Rice Research Institute, Cuttack, Odisha, India. p. 28.

Pang, X. P., and J. Letey. 2000. Organic farming: Challenge of timing nitrogen availability to crop nitrogen requirements. *Soil Sci. Soc. Am. J.* 65:247–253.

Rajkishore, S. K., S. K. Natarajan, A. Manikandan, N. S. Vignesh, and A. Balusamy. 2015. Carbon sequestration in rice soils: A review. *Ecoscane* 9(1):427–433.

Rao, K. S., S. K. Sharma, S. P. Singh, N. K. Sharma, G. N. Mishra, and S. Saha, 2008. Sustainable cropping systems for different rice ecologies. In: D. P. Singh, R. C. Dani, K. S. Rao, et al. (eds.), Proceedings of the National Symposium on 'Research Priorities and Strategies in Rice Production System for Second Green Revolution. Central Rice Research Institute, Cuttack. 165–188.

Srinivasan, V., H. P. Maheswarappa, and R. Lal. 2012. Long term effects of topsoil depth and amendments on particulate and non-particulate carbon fractions in a Miamian soil of Central Ohio. *Soil Till. Res.* 121:10–17.

Srinivasarao, Ch., R. Lal, J. V. N. S. Prasad, et al. 2015. Potential and challenges of rain-fed farming in India. In: Sparks, D. L. (ed.), *Advances in Agronomy*, Elsevier, vol. 133, 113–181.

Stewart, C. E., K. Paustian, R. T. Conant, A. F. Plante, and J. Six. 2009. Soil carbon saturation: Implications for measurable carbon pool dynamics in long-term incubations. *Soil Biol. Biochem.* 41(2):357–366.

United Nations. 2020. *World Population Prospects: The 2015 Revision, Key Findings and Advance Tables.* Department of economic and social affairs, population division, New York.

Waqas, M. A., P. Smith, X. Wang, et al. 2020. The influence of nutrient management on soil organic carbon storage, crop production, and yield stability varies under different climates. *J. Clean. Prod.* 268: 121922.

Weber, J., A. Karczewska, J. Drozd, M. Licznar, S. Licznar, E. Jamroz, and A. Kocowicz. 2007. Agricultural and ecological aspects of a sandy soil as affected by the application of municipal solid waste composts. *Soil Biol. Biochem.* 39(6):1294–1302.

Yadav, G. S., R. Lal, R. S. Meena et al. 2019. Conservation tillage and nutrient management effects on productivity and soil carbon sequestration under double cropping of rice in north eastern region of India. *Ecol. Indic.* 105:303–315.

Yimer, F., S. Ledin, A. Abdelkadir. 2006. Soil property variations in relation to topographic aspect and vegetation community in the south-eastern highlands of Ethiopia. *For. Ecol. Manag.* 232(1–3):90–99.

9 Soil Organic Matter
Bioavailability and Biofortification of Essential Micronutrients

*Sunil Mandi, Somanath Nayak, Yashbir Singh Shivay, and Bal Ram Singh**

CONTENTS

DOI: 10.1201/9781003102762-9

LIST OF ABBREVIATIONS

AR: average requirements
CA: conservation agriculture
CEC: cation exchange capacity
DOM: dissolved organic matter
DTPA: diethylenetriaminepentaacetic acid
FA: fulvic acid
FYM: farmyard manure
GM: green manuring
OM: organic matter
RDA: recommended dietary allowance
SOC: soil organic carbon
SOM: soil organic matter
UL: upper limit
ZnS: sphalerite
PA : Zn: phytic-acid-to-Zn
Zn-AML: Zn-amino lignosulfonate

9.1 INTRODUCTION

Soil organic matter (SOM) is an essential source of plant nutrients, both macro and micronutrients, and a major food source for microbes living in the soil. In a nutshell, it is the only material that plays a triple role, improving physical, chemical, and biological soil properties. The application of organic materials to the soil, such as manures, green manure, crop residues, etc., is an effective agronomic strategy for enhancing nutrient use efficiency, improving soil physicochemical and biological properties, and adding some micronutrients (Rengel et al. 1999) which sustain high crop productivity and maintain soil health. The beneficial effect of SOM on soil micronutrients and their uptake are well documented (Kirkby et al. 2011). Besides, organic material application may mitigate soil problems, such as salinity and alkalinity, acidity, and physical degradation (Frutos et al. 2010; Aydin et al. 2012). The application of these organic materials to soil may build up SOM and sequester organic carbon and thus mitigate climate change by reducing CO_2 emissions (Han et al. 2012). Worldwide, especially in Europe, policies are proposed to maintain or increase SOM content in agricultural fields (EC 2013; UNFCCC 2015). The main focus areas are those at a higher risk of SOM decline, adverse climatic conditions, and specialized cropping systems. To achieve this, COP21 proposed 4 per 1000 or 0.4% per year to increase SOM to a 40-cm depth (UNFCCC 2015).

The essential micronutrients for field crops are B, Cu, Fe, Mn, Mo, Zn, Cl, and Ni (Teklic et al. 2013). Based on physicochemical properties, the essential plant micronutrients are metals, except for B and Cl.

The accumulation of micronutrients by plants generally follows the order of Mn >Fe >Zn >B >Cu >Mo, but these orders may change in flooded rice (Fageria et al. 2002). Iron, Mn, Cu, and Mo are prosthetic groups that catalyze redox processes by electron transfer (Fageria et al. 2002). Moreover, Fe and Zn form enzyme-substrate complexes, enhancing enzyme reactions by influencing molecular configurations between enzymes and substrates (Zn) (Romheld and Marschner 1991). Co is not considered an essential micronutrient for plants, as it does not fulfill the definition of essentiality it does stimulate some plants' growth. It is well documented that Co is essential for the fixation of N_2 by bacteria or microorganisms, whether growing inside or outside root nodules (Marschner 1995). Even so, Co is essential for animal nutrition as a component of vitamin B12 (Needham 1983). Similarly, Se is not considered essential, though it has been found to promote growth and provide other beneficial effects (Marschner 2012).

Micronutrients present in soil constituents are not immediately available to plants. It has been shown that approximately 10% of micronutrients in the soil are bioavailable (Lake et al. 1984). Micronutrient content and soil mobility are governed by several factors, such as external application, SOM, pH, soil–plant or soil–microbe interactions, and plant genotype (Rengel 2015; Agrawal et al. 2016). Most metallic micronutrients in the soil form complexes with organic ligands. Organic compounds like carboxyl and phenolic groups form complexes with Cu, Zn, and Mn, making them less available to plants. Moreover, SOM contains many functional groups i.e., acidic group (carboxyl, enol, phenolic OH, quinone), neutral groups (alcoholic OH, ether, ketone, aldehyde, ester), and basic groups (amine, amide) which act to form complexes with micronutrients (Sparks 2003). Further, root exudation of organic acids (citric, malic, tartaric, oxalic, phenolic), sugars, and nonproteinogenic amino acids (phytosiderophores) and secretion of enzymes (phosphatases) have profound influences on plants' ability to absorb and utilize micronutrients from the soil (Baligar and Fageria 1997; Marschner 1995). The relationship of cations and anions with SOM have two directions i.e., mobilization or immobilization, depending primarily on soil pH. Typically, adsorption of cations is negligible when the soil solution is acidic, and basic soil solution anions adsorption is negligible when pH is in the alkaline range. Bleam (2012) demonstrated that the maximum adsorption edge reached for Co^{2+} and Cu^{2+} by imogolite is pH 6.0 and pH 7.7, respectively; whereas, for selenate (SeO_4^{2-}) and selenite (SeO_3^{2-}) by goethite are pH 4.4 and pH 9.8, respectively. The application of fresh organic manure can increase micronutrient bioavailability by forming chelation (De la Fuente et al. 2011), and mineralization may release free metals in soil solution (Bernal et al. 2002). Therefore, the availability and uptake of micronutrients are affected by SOM content. Besides, SOM plays a decisive role in toxic heavy-metal availability by changing soil properties and forming metal chelation and by forming a stable complex (Paré et al 1999). Barker and Bryson (2002) suggested that applied compost enhances organic pollutant degradation and a stable complex with toxic metals. Li et al. (2000) suggested that compost applications are an excellent organic amendment for bioremediation of heavy metal-polluted soils.

Micronutrient deficiency is a global problem as more than two billion people are affected (FAO 2015). Children and pregnant women are risk groups. Approximately 340 million children suffer from micronutrient deficiencies. For example, in Lagos (Nigeria), 20, 21, and 32% of school children are suffering from Fe, Zn, and Cu deficiency, respectively (Akeredolu et al. 2011). Indeed, in Sub-Saharan Africa, 50% of people are affected by Zn deficiency, and it is projected that its deficiency may increase by 450,000 persons globally per annum (McClafferty and Zuckermann 2015). In the human system, Fe, Cu, Mn, Co, Mo, Zn, I, and Se are considered essential micronutrients, but Cl as macronutrient is required for good human health (Garg et al. 2018). These micronutrients play crucial roles in human physical and mental development (Table 9.1). Indeed, without micronutrients, metabolic activities cannot be accomplished in the human body system. Lack of sufficient or optimum intake causes many physiological disorders. Moreover, immune system development in the human body can be achieved only by micronutrient involvement. For example, a worldwide ongoing pandemic (COVID-19) can be mitigated by developing the immune system in humans. For these cases, essential nutrients like vitamin D, vitamin A, B vitamins (folate, vitamin B6, and vitamin B12), vitamin C, and minerals (iron, copper, selenium, and zinc) are all necessary for a well-functioning immune system against COVID-19 (Alexander et al. 2020; Gasmi et al. 2020; McAuliffe et al. 2020).

Micronutrient deficiency is closely linked with their deficiency in soils and crops having low use efficiency of micronutrients in the staple food crops (Oliver and Gregory 2015). People relying primarily on micronutrient-deficient food can suffer from malnutrition or micronutrient deficiency. In southeast Asia, Fe, Zn, and I deficiency is a severe problem in all groups of people (Akhtar et al. 2013) because their dietary intake of micronutrients relies primarily on staple foods, i.e., cereal-based foods, especially in rural areas. Crops cultivated under these micronutrient-deficient soils contain less than the minimum standard requirement. For example, wheat, grown in Zn-deficient soil, contains <10 mg Zn kg^{-1}, whereas in Zn-sufficient soil, the grain Zn concentration was >20 mg kg^{-1} (Cakmak et al. 2010).

TABLE 9.1
Micronutrient Requirements for Humans

Micronutrients	ARs	RDAs	UL	Functions*
Fe (mg day^{-1})	22.4	22.4	45	It is a component of many proteins, including enzymes and hemoglobin. The hemoglobin is essential for transporting oxygen to tissues throughout the body for metabolism.
Zn (mg day^{-1})	8.9	10.2	25	It is a constituent of various enzymes, maintains the structural integrity of proteins, and regulates gene expression.
Cu (mg day^{-1})	0.7	0.9	5	It is a component of metalloenzymes that act as oxidase for the reduction of molecular oxygen.
Se (mg day^{-1})	45	55	300	It is a constituent of many selenoproteins that play vital roles in reproduction, thyroid hormone metabolism, DNA synthesis, and protection from oxidative infection and damage.
Mn (mg day^{-1})	2.4	2.4	11	It aids in the formation of bone and metabolism of amino acid, lipids, and carbohydrates.
Vitamin 12 (mg day^{-1})	2	2.4	-	Vitamin B12 is required for proper red blood cell formation, neurological function, and DNA synthesis. It functions as a cofactor for methionine synthase and L-methylmalonyl-CoA mutase.
I (μg day^{-1})	-	150	1100	Iodine functions as a component of various enzymes in the maintenance of the structural integrity of proteins and in the regulation of gene expression.
Mo (μg day^{-1})	-	45	2000	Cofactor of enzymes in humans: sulfite oxidase, xanthine oxidase, and aldehyde oxidase.
Cr (μg day^{-1})	-	25–35	-	Chromium potentiates the action of insulin in vivo and in vitro.

Source: FAO, IFAD, UNICEF, WFP and WHO 2020; US National Institute of Health 2020.
Abbreviations: AR, average requirements; RDA, recommended dietary allowance; UL, upper limit.
*Institute of Medicine 2001; US National Institute of Health 2020.

The build-up of SOM in soil converts fixed forms of micronutrients into available forms. The addition of SOM can increase the soluble and exchangeable form of micronutrients in the soil. Moreover, high amounts of SOM addition may result in more stable complexes of micronutrient formation. SOM can bind more with Zn, Cu, B, and Mo than with Fe and Mn due to its resistance to redox changes (Dhaliwal et al. 2019). Chelation of Zn and Fe with organic matter can increase phyto-availability and prevent insoluble forms with carbonates and oxides in the soil (Schulin et al. 2009). However, maintained SOM and organic amendment or fertilizer application increases micronutrient uptake and could decrease their deficiency in humans (Cakmak 2008). For example, in Finland, the application of Se-fortified NPK increased crop yields and increased daily Se intake and serum Se concentration in humans (Ekholm et al. 2005). This book chapter reviews the role of SOM in micronutrient availability, plant and human nutrition, detoxification of toxic metals, and, further, its critical roles for biofortification of food crops for human and animal consumption.

9.2 MICRONUTRIENT FORMS, STATUS, AND AVAILABILITY IN THE SOIL

9.2.1 Iron (Fe)

Iron is considered the fourth principal constituent in the lithosphere and comprising about 5%, with an average concentration in the soil of 40 g kg^{-1} (Masunaga and Fong 2018). Iron ions (Fe^{3+}) oxides mainly govern the concentration of Fe in soil solution. The solubility of the Fe oxides, such as hematite (Fe_2O_3), goethite (FeO[OH]), and magnetite (Fe_3O_4) are much lower than other micronutrient oxides. The formation of minerals mainly depends on soil aeration. Minerals such as goethite (α-FeOOH) and hematite (α-Fe_2O_3 are dominant in the aerobic soil. However, in submerged soil, Fe is present as Fe oxides in crystalline minerals, such as magnetite, lepidocrocite, maghemite, ferrocite, ferrihydrite, or non-crystalline precipitates (Cornell and Schwertmann 2003). Fe behavior in soil is mainly governed by the redox potential (i.e., oxidizing or reducing conditions) and pH. In aerated soil, Fe^{3+} is predominant in the soil solution with a very low concentration (10^{-9} to 10^{-20} M Fe^{3+}) at pH from 4–8. Iron solubility decreases as pH increases and minimum solubility is observed at pH ranges of 7.4 to 8.4 (Masunaga and Fong 2018), and under this soil pH range, Fe deficiency is common. Moreover, redox reaction can alter the Fe^{3+} to Fe^{2+} concentration in the soil solution. Lower redox potential favors the increase in bioavailability of Fe. Generally, redox potential below −200 mV increases the Fe^{2+} by reducing Fe^{3+}. Also, the application of organic matter decreases soil pH and reduces redox potential. Organic matter (OM) during decomposition releases organic acids and increases the population of Fe-reducing bacteria.

In submerged soil, Fe^{2+} is dominant. Neutral soil pH enhances the precipitation of Fe as ferrihydrite, whereas submerged and acid conditions increase the solubility of Fe minerals (Colombo et al. 2014). Decomposition of organic matter in submerged soil lowers the redox potential of soil, resulting in increased solubility of Fe (Rengel et al. 1999).

9.2.2 Zinc (Zn)

In soil, Zn content ranges from 10 to 100 mg kg^{-1}. Average Zn concentration in a solution is $10^{-2.21}$ M at 10% soil moisture content (Masunaga and Fong 2018). The most abundant Zn mineral is sphalerite (ZnS); other Zn minerals, such as zinc carbonates (smithsonite) and Zn silicates (hemimorphite) are present in lesser quantities (Mertens and Smolders 2013).

The global average Zn concentration in soil is 55 mg Zn kg^{-1}, but it may vary greatly. In general, soils with high clay content are associated with higher Zn concentrations than sandy soils, due to their higher native Zn concentration and their higher capacity to absorb and retain Zn in the soil matrix. Zn concentration ranges from 17 to 45 mg kg^{-1} in sandy soil, 55–140 mg kg^{-1} in peat soil, and 70–150 mg kg^{-1} in clay soil (Cleven et al. 1993). A low concentration of Zn in the soil causes deficiency to the plant and reduced uptake by plant products, leading to a deficiency for humans. About one-third of agricultural land and human population globally are Zn-deficient. The availability of Zn in soils is governed by the solubility of $ZnFe_2O_4$ (franklinite) depending on the activity of Fe^{3+}. Presence of amorphous iron $Fe(OH)_3$ decreases the franklinite solubility by higher concentration Fe^{3+}, whereas crystalline Fe(III) oxides (i.e., magnetite or goethite) permit higher equilibrium levels of Zn^{2+} in soils by lowering Fe^{3+}. Zn^{2+} is dominant in soils below pH 7.7, and $ZnOH^+$ is prevalent in soil above pH 7.7. Zinc can form complexes with sulfate, nitrate, phosphate, and chloride. $ZnSO_4$ complex is significant for the presence of Zn^{2+} in soil solution. The activity of $ZnSO_4$ equals that of Zn^{2+} when $SO4^{2-}$ is $10^{-2.33}$ M (Masunaga and Fong 2018).

Phosphate activity governs Zn^{2+} in soil solution in neutral-to-alkaline soils. For example, when $HPO_2^{4-} = 10^{-4}$ M, the activity of $ZnHPO_4$ is approximately equal to that of $ZnOH^+$. Therefore, in neutral and alkaline soils, Zn bioavailability decreases by precipitation with H PO_2^{4-} as $ZnHPO_4$. In these conditions, applying organic matter lowers the soil pH and increases Zn availability in soil solution (Mertens and Smolders 2013; Gramlich et al. 2013)

9.2.3 Copper (Cu)

Natural copper content in soils ranges from 2 to 250 mg kg^{-1} with an average of 30 mg kg^{-1} (Adriano 2013). The maximum concentration of Cu in a soil solution would be $10^{-2.33}$ M at 10% moisture content (Masunaga and Fong 2018). The most abundant copper mineral is chalcopyrite, $CuFeS_2$, and copper may substitute for Mg^{2+} or Fe^{2+} in ferromagnesian silicates. The lowering of the pH and the slightly reducing conditions that develop upon the oxidation of Fe^{2+} and Mn^{2+} and the precipitation of the (hydrous) oxides increase the mobility of Cu and Zn elements. Some Fe and Mn may remain in soil solution, either as Fe^{2+} and Mn^{2+}, as hydrolysis products, or as complexes (e.g., $FeSO_4$), and can be transported over considerable distances with subsurface waters. Upon exposure to the atmosphere, most of the Fe^{2+} and Mn^{2+} would precipitate as hydrous oxides and thereby lower the solubility of Cu^{2+} and Zn^{2+} as well, through precipitation, occlusion, or adsorption (Selbig et al. 2013) and in SOM and living organisms (Adriano 2013). The concentration of copper ions in the soil solution is relatively low; the central part is bound on soil colloids: clay minerals (Al-Qunaibit et al. 2004), SOM, aluminum and/or iron oxides (Boujelben et al. 2009), natural zeolites (Erdem et al. 2004), etc. The sorption/desorption of heavy metals in soil depends on SOM content and those of other Fe and Mn oxides. Copper has a very high affinity to organic material, and the bond between the organic material and copper is much stronger than that with other heavy metals (Adriano 2013). According to Adriano (2013), the affinity of copper additionally decreases in the chain Mn oxides >organic matter >Fe oxides >clay minerals. Copper deficiency can be expected in arid and semi-arid regions of the Indo-Gangetic plains of India, particularly for crops grown on soils with a coarse texture, high pH, high calcium carbonate, and low retention of water and nutrients (Sharma et al. 2015).

9.2.4 Manganese (Mn)

Mn is a dominant element in the earth's crust. It is approximately 1000 mg kg^{-1} in the soil (Masunaga and Fong 2018). In soil, Mn is present in the form of rhodonite, $MnSiO_3$, or as a trace constituent of ferromagnesian silicates. In soil solution, Mn concentration ranges from 0.01 to 1.0 mg L^{-1}. Mn behaves like Fe in soil, and the Mn^{2+} concentration increases under low pH and redox potential below –200 mV under poorly drained conditions. Therefore, Fe and Mn toxicity is prevalent in acid soils. However, as pH increases, Mn precipitates as MnO_2 and decreases solution and exchangeable Mn^{2+}.

A lesser supply of Mn may cause deficiency mainly in high pH and aerobic soils. On the contrary, its toxicity may occur in soils with over-supplied Mn or with Mn contained in irrigation water. Mn toxicity can be mitigated by soil liming and drainage and the use of Mn-tolerant plant species and varieties (Hayes et al. 2012; Hernandez-Soriano et al. 2012), but Mn concentrations can fluctuate widely and rapidly with changes in soil temperature, water regime and SOM content (Uren 2013). High SOM in peat and muck soil lowers Mn^{2+} availability by the formation of a stable complex.

9.2.5 Cobalt (Co)

Cobalt (Co) in the earth's crust is present with an average of 25 mg kg^{-1} and in soil from 1–40 mg kg^{-1} (Uren 2013). During the weathering of rocks, Co^{2+} is released from the ferromagnesian minerals and is retained in soil in association with an organic ligand, silicate layers. The behavior of Co in soil is the same as that of Mn. $Co(H_2O)_6^{2+}$ is the primary concern for controlling the solubility of Co in soil. As pH increases, the dominant form of Co is $CoOH^+$. Co present in organic matter is released during decomposition by microbial activities. Plants take up cobalt either as Co^{2+} or as an organic complex. Co concentration may be limited by the Fe, Ca, Mg, and Mn uptake (Lopez-Moreno et al. 2016).

9.2.6 Molybdenum (Mo)

Molybdenum (Mo), a transition element, ranges from 0.2 to 6.0 mg kg^{-1} in soils (He et al. 2006), and it is involved in cycling of N and C in the ecosystem. Mo, being a constituent of the nitrogenase enzyme (Wurzburger et al. 2012), affects N_2 fixation by legumes (Jean et al. 2012).

Mo has oxidation states varying from 2^- to 6^+ in its compounds (Xue et al. 2013). In soil, the metal tetrahedral layer and hydroxyl functional groups play a significant role in its adsorption (Donald 1998). The solubility and absorption of Mo in soil is governed by the presence of clay type, Fe, Al, and Mn oxides, Fe-sulfides, and SOM content (Xue et al. 2013). Mo can form a strong complex with HA and FA (Wichard et al. 2009). The soluble form of Mo in soils is anion molybdate (MoO_4^{2-}), and it is this form that is absorbed by plants. The highest Mo concentration in soil solution is $10^{-3.68}$ M at 10% moisture content (Masunaga and Fong 2018).

9.2.7 Selenium (Se)

Selenium (Se) is a semi-metallic trace element that exhibits both non-metals and metal's physical and chemical properties (Boyd 2011). In soils, Se ranges from 0.01 to 2 mg kg^{-1} with a world average mean value of 0.4 mg kg^{-1} (Fordyce 2013). In soils of the tropics, Se concentration can be as high as 2–4.5 mg Se kg^{-1} (Mehdi et al. 2013). Among factors that govern Se content in soil are soil type, texture, rainfall, parent rocks, weathering, soil pH, redox potential, and SOM content (Mehdi et al. 2013).

Se is present in soil in both inorganic and organic forms with oxidation states ranging from −2 to +6. (Jones et al. 2017). Selenate (SeO_4^{2-}) is a water-soluble, mobile, and bioavailable inorganic species in aerobic soils, with low adsorption affinity to oxide surfaces (Hartikainen 2005). The solubility and mobility of Se increase with high redox potential conditions. In anaerobic soils with low redox potential conditions, selenate may be reduced to selenite (SeO_3^{2-}) (Shahid et al. 2018). Selenite is less mobile and bioavailable than selenate (Hartikainen 2005). Se solubility and mobility are also determined by soil pH and SOM content. As soil pH increases, Se availability and solubility also increase. SOM is another crucial factor that determines the availability and solubility of Se in soil. SOM-rich soils contain high Se, but SOM may reduce Se availability by forming SOM-Se complexes (Winkel et al. 2015).

9.2.8 Iodine (I)

Iodine (I) is an element with multiple oxidation states −1, 0, +1, +3, +5, and +7, and forming different chemical compounds depending on Eh-pH conditions (Hou et al. 2009). Under natural conditions, I is found in both organic and inorganic compounds like iodide, iodate, elemental iodine, and organic iodine (Gilfedder et al. 2007; Yoshida et al. 2007). Iodine may be retained in an acid soil by sorption on positively charged hydrous iron and aluminum oxides and up to pH 8 by specific adsorption of iodate (Yoshida et al. 1992).

Organic substances limit iodate oxidation as iodine reacts with organic iodine at lower pH (Gallard et al. 2009). Moreover, iodine may form a strong complex with SOM (Steinberg et al. 2008).

9.3 MICRONUTRIENT COMPLEXES WITH ORGANIC MATTER: NATURE AND PROPERTIES

SOM is "a mixture of plant and animal residues in different stages of decomposition, substances synthesized microbiologically and/or chemically from the breakdown products, and the bodies of living and dead microorganisms and their decomposing remains" (Schnitzer 1978) and it can range from 0.5–5% in mineral soils but can be as high as 100% in organic soils (e.g., Histosols). Due to its high specific surface area (800–900 m^2/g) and high cation exchange capacity (150–300 cmol kg^{-1}) (Sparks 2003), SOM plays a crucial role in macro- and micronutrient retention in the soil. In alkaline soils, addition of manure may increase micronutrient availability.

SOM retains metal cations by complex formation (a metal–ligand interaction) and chelation. Metal complex formation can be divided into two groups: metal–ligand coordination complexes and supramolecular complexes (Tan 2010). In the metal–ligand complex, metal ions act as electron-pair acceptors and ligand as an electron-pair donor. However, in chelation, the organic ligand can bind the metal ion in more than one donor functional groups. In contrast, complexations are formed by only one donor atom of the ligand. Moreover, Tan (2010) believed that formation of the metal complex with ligand is either hard or soft base. The interaction of hard acid-hard base is mostly due to electrostatic forces resulting in strong complex formation. Whereas, soft acid-soft base interaction by covalent forces form a weak complex. These ligands can be negatively or positively or neutrally charged.

The micronutrient cations are present in the soil in many forms, such as water-soluble (free and complexed), exchangeable, specifically adsorbed, and organically complexed. The water-soluble and exchangeable are readily available to plants and, to some extent, specifically, the adsorbed pool may also be taken up by plants.

9.4 INTERACTION EFFECT OF SOIL ORGANIC MATTER ON THE BIOAVAILABILITY OF MICRONUTRIENTS

SOM can control the distribution of metal ions between soil matrix and solution (Shi et al. 2012) by changing binding energies and mean Gibbs-free energies for metals adsorption (Fan et al. 2015). However, OM sorptions of metals are stronger for Cu and Zn (He et al. 2017). In contrast, dissolved organic matter (DOM) enhances metal's solubility (Weng et al. 2002) because DOM could reduce the Cu and Zn adsorption to soil particles (Mesquita and Carranca 2005). Shi et al. (2017) reported that redistribution and reduced availability of Pb, Cd, and Cu in soils is improved by SOM. Furthermore, decomposition of organic matter releases micronutrients and the decomposed products form complex with both cations and anions (Figure 9.1). Thus, SOM has multiple roles for plant nutrition.

9.4.1 IRON

The SOM solubilizes Fe in four ways (Lindsay 1991): (1) complex formation of iron with organic acid ligand and functional groups, (2) siderophore production through microbial decomposition, (3)

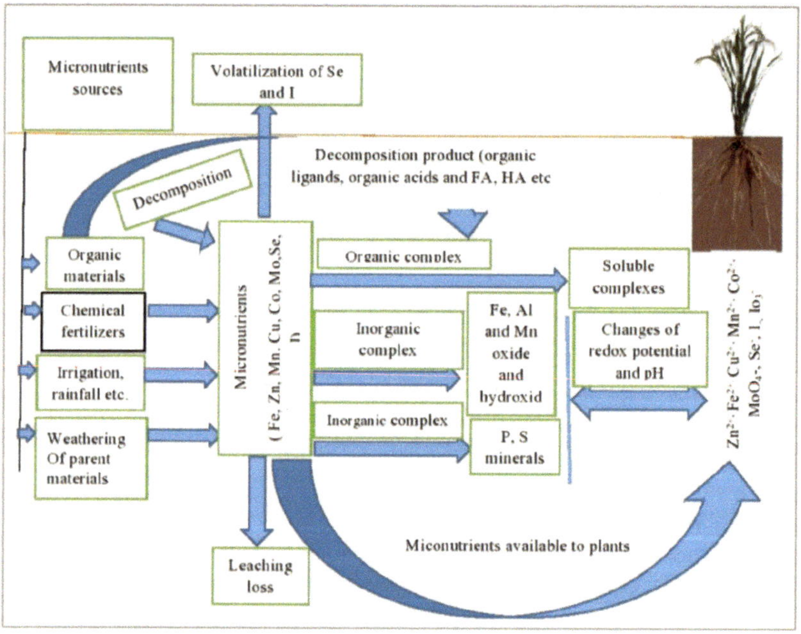

FIGURE 9.1 Micronutrients released from different sources and interactions with organic substances (The figure is drawn by the authors).

release of an electron and other reducing agents during decomposition, and (4) creation of reduction microsites.

Organic acids like citrate, oxalate, and malate can form a complex with Fe and solubilize Fe in acid soil but not in alkaline soils. Moreover, organic substances, such as cellulose, lignocellulose, sugars, organic acids, amino acids, pectin, lignin, tannin, etc., are used to nourish microorganisms and produce organic substances like fulvic and humic acids. These organic substances can solubilize Fe in soils (Colombo et al. 2012). Siderophores are low molecular weight Fe chelating agents produced by various organisms during organic matter decomposition and have a high affinity towards Fe (III) (Garg 2018). A siderophore is an organic compound that increases solubility and enhances the availability of Fe under Fe deficient conditions (Hayat et al. 2012).

Moreover, microbial decomposition of SOM releases electrons, and other reducing agents, that can reduce Fe in soil and increase its solubility. Another role of SOM for increasing Fe availability to plants is by reduction of microsites. The respiration process adjacent to the rhizosphere creates reduced conditions in that particular microsite. In this site, the oxygen entry in a water saturation condition is restricted and the sites become partially anaerobic, resulting in both dissolution and precipitation of Fe. The solubility of Fe under reduced microsites increases more than that of the Fe hydrolysis species.

9.4.2 Zinc

In developing countries, Zn deficiency is a severe problem as people consume low Zn content in staple foods, such as rice, wheat, maize, etc. (Cakmak 2008). Therefore, strategies for biofortification of Zn in grains may combat Zn deficiency in human nutrition (Cakmak 2008; White and Broadley 2011; Cakmak and Kutman 2018).

The availability of Zn in soil is governed by pH, cation exchange capacity (CEC) and the solubility of parent materials by which Zn adsorbed on clay surfaces, the formation of organo-Zn complex and formation of precipitates (Smolders and Mertens 2013; Fan et al. 2016). SOM plays a crucial role in Zn partitioning in soils. Moreover, SOM turnover is an important process that can affect Zn solubility. As Zn is released during litter decomposition and formation of the complex with dissolved organic compounds, its solubility and mobility increases (Scheid et al. 2009; Houben and Sonnet 2012). Also, SOM decreases the solubility of Zn by sorption onto the surface functional groups (Boguta and Sokolowska 2016), but further decomposition by microbial activities, the release of chelating agents, and functional groups improve Zn availability by forming a soluble complex with inorganic Zn (Baruah 2018). Both positive (Catlett et al. 2002) and negative (Eyupoglu et al. 1994) relations between SOM and Zn availability were found.

The availability of Zn in soil is directly affected whether the applied organic matter is matured, decomposable, or fresh because mature organic matter forms stable complexes with humic substances that restrict Zn availability (Smith 2009). The addition of decomposable organic matter results in the formation of strong chelation of Zn with labile organic compounds and functional groups, such as phenolic, carboxyl, amino, etc. (De la Fuente et al. 2011) and, thus, make it bioavailable. Moreover, roots excrete labile organic carbon compounds, enhancing Zn diffusion and its uptake by crops (Degryse et al. 2008).

Furthermore, SOM is an important reserve source for Zn. The addition of SOM undergoes decomposition and liberates Zn, which is sorbed on the surface of functional groups and decreases its solubility (Boguta and Sokolowska 2016), while the dissolved organic compounds forming complexes with Zn increases its solubility (Houben and Sonnet 2012).

Fan et al. (2016) showed a significant SOM effect on reducing non-specific adsorption of Zn in black soil. A 2–4-fold increase in Zn retention was found after the removal of OM (Hinz and Selim 1999). The complex of Zn-HA is better for plant growth, as it releases Zn slowly and gradually due to its partial solubility (Bednarek et al. 2011). This suggests that applying well-decomposed FYM or compost is good for reclamation of Zn-polluted soils, while the application of decomposable or fresh organic matter is good for biofortification of crops.

9.4.3 Copper

Cu is an essential component of numerous enzymes, proteins, and plastocyanin. It is also involved in the electron transport system. Cu is absorbed by plants in the form of Cu^{2+} ion and it enters from the soil into the plant system by binding itself with transporter present on the plasmalemma of root cells (Jiang et al. 2001). The concentration of Cu^{2+} ions in the soil system is strongly pH-dependent (Wu 2000) but also on total soluble Cu present in soil. SOM content plays an important role in its availability to plants and Cu concentration in soil. The Cu-organic complex makes Cu available by forming an inner-sphere complex with SOM (Boudesocque et al. 2007). Many researchers have found an increase of Cu fractions in soil with the application of SOM (Matijevic et al. 2014; Roussos et al. 2017). This trend might be attributed to release of micronutrients by SOM added to the soil, which lowers the redox potential in soil and increases its available form in soil. Increments of Diethylenetriaminepentaacetic acid (DTPA)-extractable Cu might be associated with its chelation with organic compounds that are liberated during the decomposition of applied SOM. Organic acids with low molecular weight compounds, such as oxalic, succinic, malic, citric, and acetic acids have been found to be effective for Cu adsorption on hydroxyapatite of clay-sized fractions (Wang et al. 2009). Nevertheless, SOM may also reduce Cu availability by a strong complex formation with it (Rengel et al. 1999). This may happen when high amounts of humified organic matter are applied (Clemente and Bernal 2006). Therefore, the less humified SOM with optimum quantity should be applied for increasing bioavailability and crop uptake.

9.4.4 Manganese

The literature on the interaction of SOM and Mn is limited. Mn availability depends on various factors like aeration, pH, redox conditions, $CaCO_3$, and SOM content. In anaerobic soil, Mn^{4+} is reduced to the Mn^{2+} form (Zhu et al. 2002). In submerged conditions, Mn^{4+} is reduced to Mn^{2+} and thus increases its solubility (Walia et al. 2010). The addition of organic matter releases organic compounds, reduces soil pH, and increases the intensity of reduction in soils by decomposition (Ojha et al. 2018). Therefore, SOM may increase the solubility of Mn by reducing higher valent Mn. A green manure application may have a substantial effect on the availability of Mn in soil. Dhaliwal and Manchanda (2008) found that green manuring in rice soil increased Mn availability and decreased its fixation in organic sites and oxide surfaces.

9.4.5 Cobalt

Plant roots absorbed Co as Co^{2+}. Co is a constituent of vitamin B12 and essential for nodule bacteria for free atmospheric N_2 fixation (Alloway 1990). Moreover, it has a beneficial effect on shoot and root growth (Aery and Jagetiya 2000). Bahnasawy et al. (2019) observed that significant fractions of Co are not bioavailable and recorded that Co is usually distributed in a sequence of residual-Co >exchangeable-Co >organic bound-Co >Fe-Mn-bound-Co >carbonate-bound-Co >soluble-Co in some soils of Brazil. Co bound or sorbed with humic acid is more easily exchangeable and available than that sorbed on metals oxide (McLaren et al. 1986). Besides, Li et al. (2001) reported that in some pumic soils of New Zealand, organically bound Co fractions range from 17 to 27% of total Co, and they also concluded that exchangeable and organic fractions of Co are bioavailable. Therefore, organic matter plays a crucial role in Co adsorption and release to soil solution (Anisimov et al. 2011). However, Aery and Jagetiya (2000) found that more cobalt was irreversibly bound in sandy loam soil with pH 7.7 than those with pH 6.4. The fulvic acid is responsible for binding Co both at low and high soil pH, whereas humic acid binds Co at low pH due to its low solubility in low pH. However, SOM forms complex with Co, but their low mobility in plants restricts its mobility from roots to shoots (Barysas et al. 2002; Bakkaus et al. 2005).

9.4.6 Molybdenum

Mo, an essential element, is required for both N_2-fixing legumes and non-fixing crops and plant growth development (Kaiser et al. 2005). It is a component of nitrogen-fixing enzyme nitrogenase (Stiefel 2002) and is required in very low concentrations (<5 mg kg^{-1}) for plants and animals (Xu et al. 2013). Reddy et al. (1997) reported that Mo in soil occurs in four major fractions, e.g., water-soluble, occluded with Al, Fe, and Mn oxides; solid-phase (e.g., molybdenite (MoS_2); powellite ($CaMoO_4$); ferrimolybdite ($Fe_2(MoO_4)_3$); wulfenite ($PbMoO_4$); and organically bound Mo. In soils, Mo is present in anionic forms of molybdate (MoO_4^{2-}) and/or tetrathiomolybdate (MoS_4^{2-}), which are highly soluble and susceptible to leaching (Wedepohl 1995). The main factors that govern Mo solubility in the soil are soil pH and oxides of Fe and Al; low pH favors fixation with Fe and Al oxides. (Xu et al. 2006). It is reported that organic matter ligands could form complexes with Mo in the soil (Wurzburger et al. 2012) by humic acid, and adsorbed Mo is the maximum at soil pH of 3.5 and 4, and adsorption decreases exponentially as soil pH increases (Bibak and Borggard 1994). Also at low pH, exchangeable Mo is higher and positively correlated with SOM content (Marks et al. 2015). X-ray spectroscopy studies by Wichard et al. (2009) revealed that organic substances like tannins and tannin-like compounds could form a strong complex with Mo in a wide range of soil pH and thus addition of natural organic matter could prevent leaching loss of Mo from the soil.

9.4.7 Selenium

Selenium plays multiple roles for normal physiological functions in humans and animals (Lenz and Lens 2009; Mao et al. 2016). Selenium acts as a cofactor of the enzyme iodothyronine deiodinase, which is required to convert hormone T4 to T3 (Schomburg and Köhrle 2008). Se deficiency in soil is a global problem (Williams et al. 2009). Consumption of food stuffs grown in Se-deficient soils results in low intake of Se, causing Se deficiency (Ngigi et al. 2019). About 15% of the global population suffers from Se deficiency (Dinh et al. 2018). Selenium content in soils range from 0.1 to 5 mg Se kg^{-1} (Bitterli et al. 2010). In many European countries, Se content in the arable lands is generally low from 0.11 to 0.64 mg Se kg^{-1} (Shand et al. 2012). Se availability in soils is governed by SOM and Fe and Al minerals (Ming-Ming et al. 2015). The role of SOM in controlling the Se availability in soils is established because increase in SOM causes a decrease in Se availability (Eich-Greatorex et al. 2007). Selenium oxyanions may be retained with SOM matter in several ways such as (1) the formation of ternary complexes of Se, OM, and Fe and Al oxy-hydroxides (Jordan et al. 2009; Tolu et al. 2014); (2) creation of anoxic zone that facilitate the reduction of Se (Di Tullo et al. 2016; Floor et al. 2011; Kausch et al. 2012); and (3) complex formation between Se and OM in SOM sorption sites (Floor et al. 2011; Winkel et al. 2015). Moreover, OM functional groups and Se anion are negatively charged and hence positively charged bridge-like metal forms a ternary complex (Se-metal-OM) (Dinh et al. 2017; Martin et al. 2017). Several researchers have reported that SOM retains Se by ternary complexes of Se, OM, and Fe oxy-hydroxides complex (Catrouillet et al. 2016; Fakour et al. 2016).

Tao et al. (2000) observed that the presence of fulvic acid (FA) could increase the sorption of Se on amorphous Fe oxides. However, Supriatin et al. (2016) recorded that OM fractions, such as hydrophobic neutral fractions, can retain higher Se than the hydrophilic and HA fractions. Qin et al. (2012) showed that the FA-Se complex is dominant and increases Se availability to plants. Many researchers also believe that positively charged functional groups of SOM can retain Se (Winkel et al. 2015; Dinh et al. 2017), and, thus, the Se-SOM complex may slowly release Se to soil solution (Šípková et al. 2016) and can act as a latent source of bioavailable Se in soil. (Li et al. 2017).

9.4.8 Iodine

Iodine is a trace element, a key constituent of thyroid hormones, and essential for human physical and mental development. Iodine deficiency disorders include such as "goitre, cretinism, hypothyroidism, abortion, stillbirth, brain damage, learning disabilities, mental retardation, psychomotor defects, and hearing and speech impairment" (Hetzel 2004). Iodine deficiency hits both developed and developing countries. Globally, 241 million people are estimated to have low iodine intake (Andersson et al. 2012). Iodine in soils is present mainly in the form of iodide (I^-), and iodate ($IO3^-$), and organic iodine species and added through atmospheric deposition and rock weathering (Johnson 2003). In inorganic species, iodine is covalently bound to aromatic moieties (Duborska et al. 2020). In soil, iodine concentration ranges from <0.1–150 mg kg^{-1} (Fuge 2013). The high content of iodine in coastal areas (Muramatsu and Wedepohl 1998) is caused by seawater, being rich in this element. Furthermore, high SOM content leads to high iodine as humus is the primary reservoir of soil iodine (Smyth and Johnson 2011), but high SOM content can reduce the bioavailability of iodine and plant uptake (Xu et al. 2012). The humus substances (i.e., humic acid) determines the iodine solubility in soil (Hansen et al. 2011). Humic acid contains both aliphatic and aromatic moieties depending upon the degree of humification and types of vegetation. Christiansen and Carlsen (1991) proposed HA presents three reaction sites: (1) weakly bonded by primary sites, (2) secondary sites susceptible to nucleophilic iodide-iodide substitution, and (3) tertiary electrophilic sites.

Distribution of iodine content in soil also depends also on metal oxides and hydroxides (Englund et al. 2010). Duborska et al. (2020) reported that SOM, amorphous and weakly crystalline forms of aluminum, clay fraction content, and aluminum and iron minerals have important roles in soil iodine retention.

9.5 EFFECT OF SOM ON CROP GROWTH AND YIELD

SOM content in soil can improve soil quality (Reeves 1997), soil health (Lal 2016), crop yield (Menšík et al. 2019), and, therefore, productivity. Yet, it is not easy to establish the direct effect of SOM on crop growth and yield, as numbers of other factors govern crop growth and yield. Furthermore, SOM dynamics and nutrients supply are also controlled by many factors. However, many studies have established that SOM can enhance crop yield (Lal 2020; Oldfield 2019; Ghaley et al. 2018). Ghaley et al. (2018) showed a strong positive relation between organic carbon in soil and wheat yield (Figure 9.2). In contrast, other studies reported that SOM content has no effects on crop yield (Oelofse et al. 2015; Hijbeek et al. 2017) because it is difficult to separate the effect of SOM, nutrients, and other parameters of crop yield (Murphy 2015). Ghaley et al. (2018) documented that increasing the SOC in soil increases the winter wheat yield (Figure 9.2). It is clear that application of manure or compost increased crop yields (Figure 9.3). Wood et al. (2016) found that particulate organic matter fraction was positively related to maize yield. Lal et al. (2020) showed a strong positive relationship between soil organic carbon (SOC) and cotton yield. Many studies have shown that

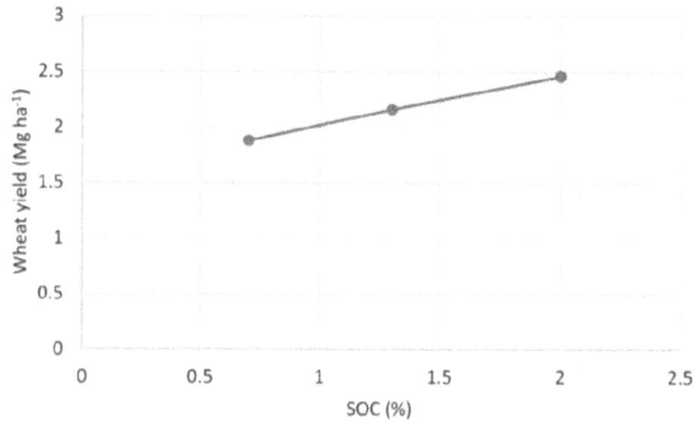

FIGURE 9.2 Relation between soil organic carbon (SOC) content and wheat yield (Data taken from Ghaley et al. (2018).

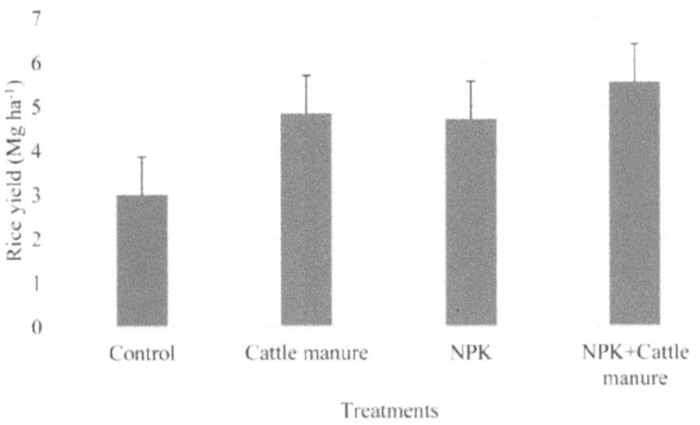

FIGURE 9.3 Impact of manure on rice yield (Data taken from Bi et al. (2009) and Shahid et al. (2016); bar indicates LSD ($P<0.05$)).

organic matter application significantly increased cereal crop growth and yield by the rational supply of both macro and micronutrients (Agbede et al. 2008; Wang et al. 2012). Hijbeek et al. (2017) analyzed many experiments across the world and showed that organic matter positively influences crop yields (Jeffery et al. 2017; Luo et al. 2018). Moreover, they found significant yield increases in crops grown in sandy soil or wet climate especially, root and tuber crops, and spring-sown cereals. In a meta-analysis, Oldfield et al. (2019) developed a quantitative model to explore the relationship between SOM and maize and wheat yield. They showed greater crop yield with increased SOC concentration, but yield increase leveled off at ~2%. They further estimated that reaching the target SOC concentration of 2% reduced the global N fertilizers inputs by 7% in maize and by 5% in wheat, whereas the potential yield increases were 10±11% (Mean±SD) for maize and 23±37% (Mean±SD) for wheat.

Ding et al. (2012) concluded that mixed use of inorganic and organic fertilizers sequestered higher SOC. Application of manure at 22.5 Mg ha^{-1} $year^{-1}$ improved soil productivity and C storage and crop yield, but the application of 15 Mg ha^{-1} $year^{-1}$ may be more efficient considering environmental factors. Hijbeek et al. (2017) suggested that the beneficial effect of SOM on crop growth by improving soil properties, increasing nutrients supply, and improving soil health.

9.6 EFFECT OF SOM ON THE UPTAKE AND NUTRITIONAL QUALITY OF MICRONUTRIENTS IN CROPS

SOM enhances the nutritional quality of crops by increasing the uptake of nutrients. Saha et al. (2019) showed that SOC is strongly related to B, Fe, Zn, Cu, and Mn uptake in rice and wheat. Shahid et al. (2016) reported that manure application increased micronutrient availability in soil, leading to their higher plant uptake. The external addition of organic matter results in SOM build-up upon decomposition and mineralization (Khaliq et al. 2017) leading to release of both macro and micronutrients in soil solution and increase in SOC stock. The release of micronutrients to soil follows different mechanisms (Wang et al. 2012). The stored SOC act as a storehouse of nutrients as it holds nutrients by its negative and positive charges. The uptake of micronutrients depends on crop type, crop ecosystem, and amount of organic matter added (Li et al. 2007; Mishra et al. 2006). Mishra et al. (2006) showed that the same amount of farm yard manure (FYM) application to rice and wheat resulted in different Zn and Cu uptake, being 33.6 mg kg^{-1} for rice and 47 mg kg^{-1} for wheat. They also showed that the application of green manuring, straw, and FYM to rice did not differ much for Zn and Cu uptake by the crop. This trend indicates that soil moisture, i.e., aeration and flooding, may control the uptake of micronutrients. Aghili et al. (2014) reported that green manure application of sunflower achieved 31 mg kg^{-1} of Zn in wheat grain but the uptake was higher when

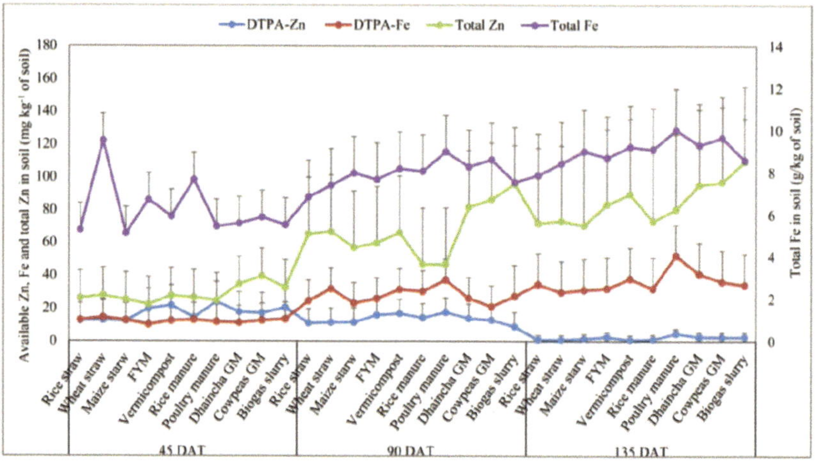

FIGURE 9.4 Effect of organic matter application on available and total status of Fe and Zn at different growth stages (Data taken from Dhaliwal et al. 2013); bar indicates LSD ($P<0.05$). *Abbreviations:* Cowpeas GM, Cowpeas green manure; Dhaincha GM, dhaincha green manure; DTPA-Fe, diethylenetriaminepentaacetic acid iron; DTPA-Zn, diethylenetriaminepentaacetic acid zinc; FYM, farmyard manure.

combined with chemical fertilizers (Aghili et al. 2014; Patil et al. 2017). SOM build-up in soil by applied organic materials resulted in significant decrease in the Cu concentration in the fababean pod (Matijevic et al. 2014). The higher uptake of micronutrients seems to be caused by higher bioavailability of micronutrients in organic manure applied fields (Figure 9.4).

9.7 EFFECT OF SOM ON TOXIC METAL UPTAKE BY CROPS

Cadmium (Cd), lead (Pb), arsenic (As), mercury (Hg), and chromium (Cr) are the toxic elements that are related to agricultural production system. Whereas, Cd, Pb, As, and Hg are cation, Cr is an anion. However, essential micronutrients such as Cu, Zn, Mn, Se, Ni, etc., are considered toxic when their high content in food crops may cause a toxicological effect. Nevertheless, in the present scenario, these essential micronutrient problems are not reported in the agriculture production system except for crops cultivated in heavily polluted soils. Therefore, careful management practices are required to avoid the toxic effect of essential micronutrients. The primary concern in this section is uptake of toxic metals such as Cd, Pb, As, and Hg related to SOM content. Uptake of these toxic metals by plants depends on the bioavailability or mobility of ions in the soil. The bioavailability or mobility in soil depends on metal's concentration, SOM status, soil reaction, and management practices.

These toxic metals have higher bioavailability and mobility under acidic soil reaction (Park et al. 2011; Branzini and Zubillaga 2012). Therefore, the alteration of soil reaction can control its availability and mobility, which can be achieved by adding organic and inorganic amendments. Generally, additions of organic matter lower soil pH, resulting in increased solubility and mobility. Therefore, organic matter addition, followed by leaching, can reduce toxic metal's problems in soil. Furthermore, added organic matter upon decomposition produces organic substances, which may make complexes with toxic metals. For example, organic substances, such as glycine, citric acid, tartaric acid, and gluconic acid, can chelate heavy metals, and humic acids can form a stable complex which is not available to plants. In contrast, toxic metals complex with acetic acid is more mobile and available to crop plants (Tate 1987).

Toxic metal ions have a strong affinity towards ligands and functional groups of organic substances and can form organo-metals complex (Bolan et al. 2003; Bradl 2004). The order of affinity of metal ions is given as follows: Hg^{2+}>Cd^{2+} >Pb^{2+}>As (V)>As (Sharma and Nagpal 2018). Thus, toxic metal cations bind with functional groups and ligands and are removed from the soil solution, leading to reduced bioavailability and plant uptake of toxic metals.

The behavior of anions is different from cations. Generally, soil particles and organic substances possess negative charges. Hence, anions present in soil are more prone to leaching, as it presents a free ion soil solution. Cr in soil is present in two forms: trivalent, Cr (III), and hexavalent, Cr (VI). The latter is a highly toxic, human mutagen and shows high solubility in the aqueous system. In contrast, Cr (III) is less toxic and relatively insoluble in the aqueous system. Therefore, strategies that can reduce hexavalent to low trivalent are desirable. A number of studies showed that organic matter constituents humic and non-humic substances (carbohydrate and protein) can reduce Cr (VI) to Cr (III) (Bolan et al. 2003; Bolan and Thiagarajan 2001) and a higher organic matter addition results in a greater reduction of Cr (VI) to Cr (III) (Enya et al. 2020) caused by retention of total Cr by SOM and reduction in the ratios of Cr (VI): total Cr. This trend indicates that the higher the organic matter application the greater is the reduction of Cr (VI) to Cr (III). Banks et al. (2006) recorded that high organic matter addition reduces the Cr uptake in the shoot biomass of tall fescue.

Organic matter application not only enhances the vegetative and crop productivity (Beesley et al. 2010; Park et al. 2011), but it also acts by altering soil chemistry such as pH, redox potential, and formation of organo-metal complexes (Bernal et al. 2007). The organic matter ligands form a stable complex with toxic metals and reduce their mobility. However, the retention of toxic metals depends on the type of organic ligands and surface area. The organic matter with high humic acid and humic substances can reduce bioavailability by tight adsorption on the surface area or stable complexation with humic acids (Varrault and Bermond 2011; Clemente et al. 2015) and, therefore, prevent leaching. Humic acids have a large molecular pore size than other organic substances and soil particles.

Application of biosolids mixed with iron oxides and phosphoric acid significantly reduces uptake of Pb (Brown et al. 2004). Nevertheless, the use of biosolids originated from municipal or industrial waste may be restricted due to high toxic metal contents (Clemente et al. 2015), other pollutants, and pathogens. Therefore, proper treatment and purification are needed before application to crop fields. Hence, alternative organic matter sources originated from the agriculture production systems must be utilized. Many studies have shown that the application of organic matter originated from agriculture systems, such as green manuring, compost, crop residues, and animal manures, can reduce toxic metals' uptake by various crops in diverse climatic regions (Table 9.2). Biochar is the other effective material for remediation of polluted soils in addition to supply of some essential nutrients. Biochar has a larger pore size and can hold toxic metals tightly, thus making them non-available to plants. Kalsi et al. (2016) showed that the addition of organic amendments significantly increased plant dry matter yield and micronutrient concentration in Indian mustard and significantly decreased Pb uptake when compared with no applied organic plots. Thus, increasing SOM content in soil can control the toxic metals solubility and crop uptakes.

9.8 SOIL ORGANIC AMENDMENTS FOR BIOFORTIFICATION OF CROPS WITH MICRONUTRIENTS

Biofortification of micronutrients, either agronomic fertilizer application (Dimkpa and Bindraban 2016; Cakmak and Kutman 2018) or breeding of crop species and varieties (Cakmak 2008; Zou et al. 2019; Kumar et al. 2019), are considered practical, efficient, and cost-effective approaches. However, SOM through its supply of nutrients also plays a vital role. The organic amendments (e.g., FYM, compost, crop residues, etc.) are a source of micronutrients such as Fe, Zn, Cu, Mn, Mo, etc. (Rutkowska et al. 2014; Masunaga and Fong 2018). Furthermore, the growing of efficient genotypes, intercropping, and micronutrient enriched materials increase the density of micronutrients in edible parts. Izydorczyk et al. (2021) suggested that foliar spray of Se and I with increased doses is the most effective way for Se and I biofortification. Manojlović et al. (2019) reported that agronomic

TABLE 9.2
Effect of Organic Amendments on Toxic Metals Uptake by Different Crops

Organic Amendments	Toxic Metals	Crops	Uptake of Toxic Metals	References
Sheep manure, sewage and sludge, municipal waste compost	Cd, Cr, Pb	*Citrus Sinensis* var *Valencia*	There was no significant increase in uptake of metals content in fruits and leaves except Pb in leaves	Canet et al. (1997)
OM added through sphagnum peat, farmyard manure, compost mixture	Cd	Ryegrass and wheat	Concentration of Cd decreased with increasing OM addition in ryegrass. In wheat, heavy metal content in shoots was reduced	Abd Elrahman et al. (2012)
Compost	Cd and Pb	Sunflower	Uptake of Cd and Pb accumulated more in leaves	Liphadzi and Kirkham (2006)
Rapeseed residue as a green manure, conocarpusbiochar, peat, compost, vermicompost	Cd and Pb	Rice, Maize, and Potato	Reduced uptake	Ok et al. (2011); Al-Wabel et al. (2014)
Municipal solid waste compost, cow dung manure	Cr	Rice	Enhanced uptake	Bhattacharyya et al. (2005)
Wheat straw biochar	Cd and Pb	Rice	Reduced uptake	Bian et al. (2014)
Biochar	Cd, Cr	Rice, wheat, and Tomato	Reduced uptake of metals. Rice had reduced Cd uptake (>50%) and tomato had reduced Cr uptake (>97%)	Cui et al. (2021); Herath et al. (2015)
Fresh ramial chipped wood, sewage sludge compost	Cr and As	*Phaseolus vulgaris*	Reduced uptake	Hattab et al. (2015)
Chicken manure biochar, Green waste biochars	Cd and Pb	*Brassica juncea*	Reduced uptake	Park et al. (2011)
Poultry manure, Cow dung, sludge	Cr and As	Maize	With increased OM, toxic metals uptake decreased	Okieimen and Ikhuoria (2011)
Rice straw derived-biochar	Cd		The addition reduced the uptake of biochar in Cd-polluted soil	Zhang et al. (2016)

biofortification of winter wheat can be effective in increasing Se grain concentration, where the efficiency depends on the rate of Se applied, application method, and local environmental conditions rather than on cultivar differences.

However, in Se- and I-deficient soils, growing efficient genotypes (White 2017) or application of Se and I-enriched OM is another way for biofortification (Li et al. 2017). In this section, the contribution of organic amendments, like green manuring, manure and compost, and micronutrient enriched organic materials are discussed. Moreover, efficient genotypes and intercropping are also considered due to their eco-friendly nature for biofortification of crops.

9.8.1 Green Manuring

Green manuring (GM) plays a positive role in increasing the density of micronutrients in edible portions of food crops. It has multiple benefits for improving soil properties and maintaining soil fertility (Singh and Shivay 2017). Grüter et al. (2017) and Kutman et al. (2011) suggested that leguminous green manuring may be used as a substitute for mineral fertilization in wheat for Zn biofortification. Green manuring with dhaincha (*Sesbania aculeata)* recorded higher grain yield and accumulated Fe concentration by 83.7 mg kg^{-1}, Zn by 21.7 mg kg^{-1}, Mn by 22.5 mg kg^{-1}, and Cu by 13.0 mg kg^{-1} in basmati rice grain, which was equally effective as other green manure crops like cowpea and mungbean (Pooniya and Shivay 2013). Summer green manuring with *sesbania* or *cowpea* produced higher rice and wheat yield than that under other treatments (Figure 9.5). This trend may be due to higher availability of micronutrients in green manure applied plots (Figure 9.4). Singh and Shivay (2013) recorded the residual effect of summer green manuring on increased grain yield of durum wheat and on improving quality like sedimentation value, hardiness, and beta carotene content. They also found that the residual effect of summer green manuring with dhaincha (*Sesbania aculeate)* accumulated significantly higher Zn concentration (43.2 mg kg^{-1}) in durum wheat grain and similar effect of summer green manuring with sunhemp (*Crotalaria juncea*) and cowpea was observed. Moreover, Dhaliwal and Manchanda (2008) showed that the residual effect of green manure applied to rice was equal to the two foliar sprays of manganese for increasing grain yield. Also, Baghaie and Aghilizefreei (2020) recorded that the incorporation of Fe-enriched green manuring increased the Fe concentration in durum wheat ranging from 40 to 119 mg kg^{-1}. The increasing grain yield and micronutrient density in grain may have been caused by the addition of green manuring crop containing substantial amounts of micronutrients (Singh and Shivay 2017). Indeed, applied green manuring produced organic acids resulting in lower soil pH and increased solubility of micronutrients. Leguminous green manuring with a lower C/N ratio may quickly release nutrients by decomposition to enhance the uptake of nutrients. Furthermore, green manuring enhances microbial activities (Aghili et al. 2014; Ok et al. 2011; Dhaliwal et al. 2019), and decomposition products like organic acids (amino acids), dissolved organic carbon contributing to chelation with micronutrients (Franchini et al. 2001; Soltani et al. 2014), which increase the mobility and uptake of micronutrients.

9.8.2 Manures and Compost

Application of FYM or compost can increase the bioavailability of metal micronutrients such as Zn, Fe, Mn, and Cu (Mann et al. 1978) and could supply the same amount of micronutrients as inorganic fertilizers (Yadav and Kumar 2000). However, Ojha et al. (2018) showed that after application of organic amendments, the availability of Fe and Mn was higher initially but decreased later after

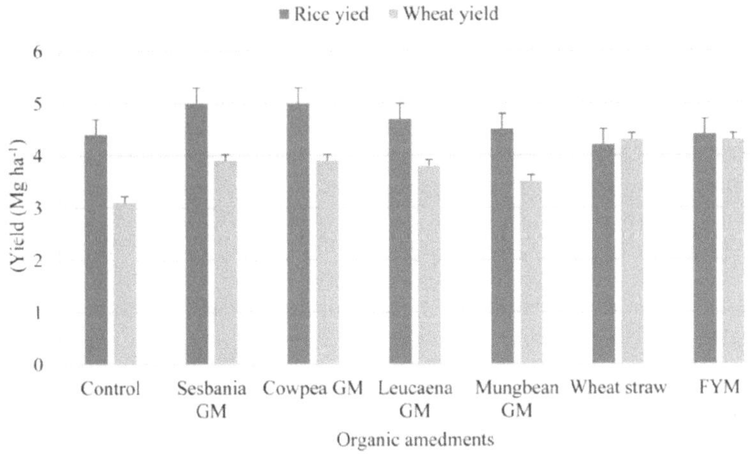

FIGURE 9.5 Increased rice and wheat yield by green manuring. (Data taken from Mishra et al. (2006); bar indicates LSD ($P<0.05$) *Abbreviations:* Cowpea GM, cowpea green manuring; FYM, farmyard manure; *Leucaena* GM, *Leucaena leucochephala* green manuring; Mungbean GM, mungbean green manuring; *Sesbania* GM, *Sesbania aculeata* green manuring.

reaching a peak, while the bioavailability of Cu decreased with time. They also found that Zn bioavailability increased throughout the incubation period. These differences in availability of different micronutrients may be due to their affinity of binding with organic substances and the stability of organo-metals complexes.

Furthermore, input of SOM improves soil physical, chemical, and biological properties. Therefore, applying FYM and compost may increase crop growth, yield, and density of metal micronutrients. On the other hand, application of well-decomposed or highly humified materials may reduce the availability of micronutrients by forming stable complexes.

Many researchers are interested in the biofortification of Fe and Zn in food crops due to its global concern. Shivay et al. (2010) showed that organically grown wheat contained a significant Fe and Zn amount in grain. Grüter et al. (2017) reported that wheat grain Zn concentration ranged from 45 to 50 mg kg^{-1} in FYM applied against that of 35–39 mg kg^{-1} in NPK-applied plots. However, many authors have found that application of FYM with inorganic fertilizers can increase micronutrient density in grains instead of FYM alone (Ram and Singh 2005; Ayeni et al. 2012).

For biofortification of Mo and I through application of compost and FYM is rather scarce. Li et al. (2017) showed that the application of manures such as compost, FYM, cattle manure, poultry manure, press mud increased the Se concentration in the soil but its uptake by different crops in seleniferous soil was decreased.

9.8.3 Crop Residue Retention or Incorporation

Crop residues are rich sources of both macro and micronutrients. In conservation agriculture (CA), maintaining the crop residues in the field is one of the principles. Retention of crop residues has many-fold benefits, such as mulching, nutrient addition, and conservation of soil against water and wind erosion. Crop residue retention can significantly increase rice grain yield and can act as a substitute for inorganic fertilizers (Huang et al. 2013). However, adding crop residues having a wide C/N ratio causes immobilization of nutrients. To avoid this, an initial additional chemical fertilizer application is required. Another option for supplying nutrients for crop requirements is the incorporation of legume residues with a narrow C/N ratio. Growing of wheat after leguminous crops enhances soil fertility, increases crop yield (Galantini et al. 2000), improves protein concentration and density of Zn in grain (Zentner et al. 1990; Lopez-Bellido et al. 1998), and enhances quality of wheat grain (Khoshgoftarmanesh et al. (2017). Aghili et al. (2014) reported that either clover or sunflower residues incorporation significantly increased protein content and grain yield of wheat and significantly increased grain Zn concentration but decreased the phytic-acid-to-Zn (PA:Zn) molar ratio.

In calcareous soil with high soil pH, soil application of straw may enhance the availability of micronutrients (Chen et al. 2017). Singh and Ram (2005) found that incorporation of rice residues increased Zn and Fe bioavailability in sodic calcareous soil, by lowering soil pH and, consequently, reducing their sorption to minerals by forming soluble complexes with micronutrients and increasing solubility of micronutrients. In China, Han et al. (2018) showed, based on 68 experimental studies, that incorporation of straw sequestered C at the rate of 0.35 Mg ha^{-1} $year^{-1}$ increased the grain yield by 9.3–18.4% However, they suggested that the combined application of straw and chemical fertilizers is better for both yield and grain quality.

When residues and other organic amendments (humified organic materials) are applied together, it may outperform chemical fertilizers. Walia and Kler (2010) showed that with combined application of FYM and residues incorporation to maize and wheat, total uptake of Fe, Zn, Cu, and Mn was significantly higher than that with the recommended dose of chemical fertilizers. Thomas et al. (2019) studied four types of organic amendments like vegetable waste compost, FYM, compost, and straw incorporation for five years. They reported that immediately after incorporating of wheat and barley straw, there was no significant uptake of Fe and Zn in grain, but after consecutive incorporation for 5 years, Zn and Fe uptake was significantly increased compared to control treatment.

9.8.4 Biochar

Biochar is the carbon-rich product produced by pyrolysis of crop residues or other waste products from agriculture and forestry (Xu et al. 2013; He et al. 2017). For long-term C sequestration, biochar may have a high potential due to its aromatic structure and recalcitrance in the environment (Lehmann 2007). Generally, biochar has a high pH value and cation exchange capacity, which can enhance soil productivity and bioavailability of micronutrients (Jeffery et al. 2011; Kookana 2010) due to its porous nature. Biochar application reduces the bioavailability of toxic metals (Zhang et al. 2013; Yuan et al. 2019).

9.8.5 Micronutrient Enriched Organic Matter

The addition of organic matter enriched with micronutrients is another effective strategy to increase micronutrient density in edible parts because organic substances act as chelators, forming a soluble complex with micronutrients and thereby increasing the bioavailability of micronutrients. Micronutrient-organic complexes such as polyflavonoids and lignosulfonates can increase crop yield and micronutrient uptake. Fe-organo complexes showed positive response on crop yield and Fe uptake (Salardini and Murphy 1978). Almendros et al. (2015) recommended the application of Zn-amino lignosulfonate (Zn-AML) at 10 mg Zn/kg in acid soils and Zn-DTPA-HEDTA-EDTA at the rate of 10 mg Zn kg^{-1} in calcareous soil for improving Zn uptake by plants.

In Switzerland, Gogo et al. (2013) recorded that hydrolyzed wool application increased grain yield 2-fold and protein content 1.5-fold, compared to the control with grain Zn concentration of 54.1 mg kg^{-1} and Fe concentration of 36.6 mg kg^{-1}, and compared well with mineral fertilizers. Veeranagappa et al. (2010) found that Zn enriched with compost enhances Zn uptake.

Se contamination in water is a major concern (Edelstein et al. 2016; Eich-Greatorex et al. 2007). The sources of Se may be the application of excessive amounts of inorganic fertilizers to the field or from parent materials enriched with Se. Therefore, Se enriched with organic materials may slowly release Se to soil solution by decomposition of organic matters. Wang and Gao (2001) suggested application of Se fertilizers with organic materials to reduce Se leaching because organic substances form complexes with Se.

Besides, Se-enriched crop residues can also be applied for Se biofortification (Ajwa et al. 1998; Stavridou et al. 2011). Bañuelos et al. (2016) showed that incorporation of desert prince's plume (*Stanleya pinnata*) in the soil enriched organic Se, increased SeCys (selenocysteine), MeSeCys (methlyselenocysteine), and SeMet (selenomethionine) in carrot roots. The desert prince's plume *(Stanleya pinnata)* is a rich source of Se, as it can accumulate more Se from soil containing sufficient amounts of Se. Furthermore, organo-selenium compounds showed higher biological functions in humans than inorganic Se (Rayman et al. 2008; Irons et al. 2006; Khanam and Platel 2016). However, in the short term, organic Se use efficiency is lower than inorganic Se fertilizers (Ebrahimi et al. 2015). Little research information is available on I biofortification with I-enriched organic materials. However, Weng et al. (2013) reported biofortification of vegetable crops through I-enriched organic fertilizers.

9.8.6 Efficient Genotypes and Intercropping

Growing efficient crops for micronutrient acquisition is the most crucial technique for biofortification. Bilski et al. (2012) showed that the growing of crops such as barley, oats, rye, wheat, and ryegrass on coal-based growth media resulted in significant uptake of Fe, Zn, and Se. Nair et al. (2013) recorded that the mungbean has a wide genetic variability for micronutrient content with Fe concentration ranging from 30 to 60 mg kg^{-1} and Zn concentration from 20 to 40 mg kg^{-1}, which indicates that it is an efficient crop among pulses for micronutrient biofortification. For Se, none of the cereals or pulses can accumulate Se in their edible parts (White 2017). However,

Izydorczyk et al. (2021) noted that crops such as rice, wheat, pea, and lentil are efficient for Se uptake. Biofortification of micronutrients, mainly Fe, Zn, Cu, and Mn, can be achieved by intercropping of graminaceous species and leguminous species (Zuo et al. 2004; Dragicevic et al. 2015; Shivay and Mandi 2020). Leguminous species (dicot) followed Strategy I for micronutrient acquisition, where roots release H^+ through H^+-ATPase 2, which reduces higher valent cations to a bioavailable form. However, gramineous species, such as wheat, barley, oat, and maize, mobilize micronutrients by releasing phytosiderophore (acting as chelators); the mechanism is known as Strategy II. Details of the mechanism of the intercropping system for Fe and Zn biofortification have been explained by Xue et al. (2016). Unfortunately, most of the works carried out on Fe and a few works on Zn (Xue et al. 2016) are mainly on peanut intercropped with maize, wheat, barley, oat, etc. Further, in the intercropping system, both graminaceous and leguminous species benefited by releasing root exudates such as protons, carboxylates, phosphatases, and phytosiderophores, which enhanced soil microbial activity and increased the solubility of Fe, Zn, Cu, and Mn (Hansen et al. 2006). For example, Zuo and Zhang (2009) reported Fe deficiency in peanut by intercropping with gramineous species. The results showed that peanut accumulated Fe by 1.4 times higher in grain when intercropped with maize than when grown as a monocrop. Similarly, Zuo and Zhang (2009) showed that seed Fe concentration increased in wheat and chickpea in the intercropping system. Also, Karlen et al. (1994) observed that crop rotation and cover crops may increase Fe, Cu, and Zn availability.

The literature review shows that a little research has been done on Se for increasing bioavailability by intercropping and crop rotation among anionic micronutrients. Chilimba et al. (2014) observed that Se applied in an intercropping system resulted in more uptake of Se than that in the monoculture system. The reason behind this trend is that, under the intercropping system, more foliage covers the soil surface than monoculture which slows down the Se deposition on the soil surface, resulting in more absorption before leaching, immobilization, and soil adsorption (Chilimba et al. 2014).

9.9 CONCLUSION

Applying organic matter and maintaining SOM has multiple roles, such as improving soil physicochemical and biological properties, both soluble and stable complexes formation with cations and anions, reducing leaching of anions, and regulating micronutrients solubility by changing soil pH and redox potential. Therefore, maintaining soil health and sustaining crop productivity can be achieved by applying organic manures. Maintaining SOM in soil enhances crop productivity and improves the nutritional quality of crops. In heavily polluted soils, applied organic matter reduces the availability and solubility of heavy metals. Thus, for the biofortification of essential micronutrients, SOM plays a vital role. Further, our conclusion is summarized:

- In neutral and alkaline soil in reaction, Fe remained as precipitation, whereas submerged and acid conditions increase the solubility of Fe minerals. Applying organic matter reduces soil pH and lowers the redox potential of soil, resulting in the increased solubility of Fe.
- For biofortification of metals cations, like Fe, Zn, Cu, and Mn, applied decomposable organic matter and intercropping of cereal and legumes can increase bioavailability and uptake.
- In soils with low micronutrients, the application of micronutrient enriched organic materials can lead to higher micronutrient density in edible parts.
- However, in polluted soils, the application of mature organic matter or biochar may reduce availability, thus avoiding toxic effects to both plants and humans.
- Furthermore, applying well-decomposed FYM or compost is good for reclamation of Zn-polluted soils, while the application of decomposable or fresh organic matter is good for biofortification of crops.

- Organic substances, like tannins and tannin-like compounds, could form a strong complex with Mo in a wide range of soil pH, and thus the addition of natural organic matter could prevent leaching loss of Mo from the soil.
- In I-deficient soils, growing efficient genotypes or applying I-enriched OM is another way for biofortification apart from foliar spraying.
- For Se biofortification, growing efficient genotypes with applied Se-enriched OM in seleniferous—but Se-deficient—soils, the application of Se fertilizers with OM effectively avoids leaching losses, thereby enhancing Se biofortification.

REFERENCES

Abd Elrahman, S.H., M.AM. Mostafa, T.A. Taha, M.A.O. Elsharawy and M.A. Eid. 2012. Effect of different amendments on soil chemical characteristics, grain yield and elemental content of wheat plants grown on salt-affected soil irrigated with low quality water. *Ann Agric Sci* 57(2):175–182.

Adriano, D.C. 2013. *Trace Elements in Terrestrial Environments: Biogeochemistry, Bioavailability, and Risks of Metals*. SpringerVerlag, New York.

Aery, N.C. and B.L. Jagetiya. 2000. Effect of cobalt treatments on dry matter production of wheat and DTPA extractable cobalt content in soils. *Commun Soil Sci Plant Anal* 31(9–10):1275–1286. https://doi.org/10.1080/00103620009370512.

Agbede, T., S. Ojeniyi and A. Adeyemo. 2008. Effect of poultry manure on soil physical and chemical properties, growth and grain yield of sorghum in southwest, Nigeria. *Am J Sustain Agric* 2:72–77. https://doi.org/10.1016/j.still.2008.12.014

Aghili, F., H.A. Gamper, J. Eikenberg, A.H. Khoshgoftarmanesh, M. Afyuni, R. Schulin, J. Jansa and E. Frossard. 2014. Green manure addition to soil increases grain zinc concentration in bread wheat. *PLOS ONE* 9(7):e101487. http://doi.org/10.1371/journal.pone.0101487.

Agrawal, R., B. Kumar, K. Priyanka, C. Narayan, K. Shukla, J. Sarkar and J. Anshumali. 2016. Micronutrient fractionation in coal mine-affected agricultural soils, India. *Bull Environ Contam Toxicol* 96(4):449–457.

Ajwa, H.A., G.S. Bañuelos and H.F. Mayland. 1998. Selenium uptake by plants from soils amended with inorganic and organic materials. *J Environ Qual* 27(5):1218–1227.

Akeredolu, I., B.E. Oguntona, C. Okafor and O.J. Osisanya. 2011. Iron, zinc, and copper malnutrition among primary school children in Lagos, Nigeria. *Food Nutri Sci* 2:1063–1070. https://doi.org/10.4236/fns.2011. 210142.

Akhtar, S., T. Ismail, S. Atukorala and N. Arlappa. 2013. Micronutrient deficiencies in South Asia–Current status and strategies. *Trends Food Sci Technol* 31(1):55–62. http://doi.org/10.1016/j.tifs.2013.02.005.

Alexander, J., A. Tinkov, T.A. Strand, U. Alehagen, A. Skalny and J. Aaseth. 2020. Early nutritional interventions with zinc, selenium and vitamin D for raising anti-viral resistance against progressive COVID-19. *Nutrients* 12(8):2358. https://doi.org/10.3390/nu12082358.

Alloway, B.J. 1990. *Heavy Metals in Soils*. Blackie, Glasgow.

Almendros, P., A. Obrador, D. Gonzalez and J.M. Alvarez. 2015. Biofortification of zinc in onions (*Allium cepa* L.) and soil Zn status by the application of different organic Zn complexes. *Sci Hortic* 186:254–265. https://doi.org/10.1016/j.scienta.2015.02.023.

Al-Qunaibit, M.H., W.K. Mekhemer and A.A. Zaghloul. 2004. The adsorption of Cu(II) ions on bentonite: A kinetic study. *J Colloid Interface Sci* 283(2):316–321.

Al-Wabel, M.I., A.R.A. Usman, A.H. El-Naggar, A.A. Aly, H.M. Ibrahim, S. Elmaghraby and A. Al-Omran. 2014. Conocarpus biochar as a soil amendment for reducing heavy metal availability and uptake by maize plants. *Saudi J Biol Sci* 22(4):503–511.

Andersson, M., V. Karumbunathan and M.B. Zimmermann. 2012. Global iodine status in 2011 and trends over the past decade. *J Nutr* 142(4):744–750.https://doi.org/10.3945/ jn.111.149393.

Anisimov, V.S., I.V. Kochetkov, S.V. Kruglov and R.M. Aleksakhin. 2011. Effect of organic matter on the parameters of the selective sorption of cobalt and zinc by soils and their clay fractions. *Eurasian Soil Sci* 44(6):618–627. https://doi.org/10.1134/S1064229311060020.

Aydin, A., C. Kant and M. Turan. 2012. Humic acid application alleviate salinity stress of bean (*Phaseolus vulgaris* L.) plants decreasing membrane leakage. *Afric J Agric Res* 7(7):1073–1086.

Ayeni, L.S., E.O. Adeleye and J.O. Adejumo. 2012. Comparative effect of organic, organomineral and mineral fertilizers on soil properties, nutrient uptake, growth and yield of maize (*Zea mays*). *Int Res J Agric Sci Soil Sci* 2:493–497.

Baghaie, A.H. and A. Aghilizefreei. 2020. Iron enriched green manure can increase wheat Fe concentration in Pb-polluted soil in the presence of *Piriformospora indica (P. indica)*. *Soil Sediment Contam* 29(7):721–743. https://doi.org/10.1080/15320383.2020.1771274.

Bahnasawy, N., A. Elwa, L. Sedky and T. El-Essawi. 2019. Distribution and forms of cobalt and its relationship to mineralogical composition in soils of the 10th of Ramadan City, Egypt. *Egypt J Soil Sci* 59(3):223–239. https://doi.org/10.21608/ejss.2019.12895.1273.

Bakkaus, E., B. Gouget, J.P. Gallien, H. Khodja, H. Carrot, J.L. Morel and R. Collins. 2005. Concentration and distribution of cobalt in higher plants: The use of micro-PIXE spectroscopy. *Nucl Instrum Methods B* 231:350–356.

Baligar, V.C. and N.K. Fageria. 1997. Nutrient use efficiency in acid soils: Nutrient management and plant use efficiency. In: Moniz A.C., A.M.C. Furlani, R.E. Schaffert, N.K. Fageria, C.A. Rosolem, H. Cantarella (eds.), *Plant-Soil Interactions at Low pH: Sustainable Agriculture and Forestry Production*. Soil Sci Soc Campinas/Vi¸cosa, Brazil, pp. 75–95.

Banks, M.K., A.P. Schwab and C. Henderson. 2006. Leaching and reduction of chromium in soil as affected by soil organic content and plants. *Chemosphere* 62(2):255–264. https://doi.org/10.1016/j.chemosphere.2005.05.020.

Bañuelos, G.S., I.S. Arroyo, S.R. Dangi and M.C. Zambrano. 2016. Continued selenium biofortification of carrots and broccoli grown in soils once amended with Se-enriched S. pinnata. *Front Plant Sci* 7:1251. https://doi.org/10.3389/fpls.2016.01251.

Barker, A.V. and G.M. Bryson. 2002. Bioremediation of heavy metals and organic toxicants by composting. *Sci World J* 2:407–420.

Baruah, R. 2018. Towards the bioavailability of zinc in agricultural soils. In: Meena V. (ed.), *Role of Rhizospheric Microbes in Soil*. Springer, Singapore. https://doi.org/10.1007/978-981-13-0044-8_4.

Barysas, D., T. Cesniene, L. Balciuniene, V. Vaitkuniene and V. Rancelis. 2002. Genotoxicity of Co^{2+} in plants and other organisms. *Biologija* 2:58–63.

Bednarek, R., H. Dziadowiec, U. Pokojska and Z. Prusinkiewicz. 2011. *Ecological and Soil Science Studies*. WNPWN, Warsaw.

Beesley, L., E. Moreno-Jime´nez and J.L. Gomez-Eyles. 2010. Effects of biochar and greenwaste compost amendments on mobility, bioavailability and toxicity of inorganic and organic contaminants in a multi-element polluted soil. *Environ Pollut* 158(6):2282–2287.

Bernal, M.P., D.J. Walker, R. Clemente and A. Roig. 2002. *The Effect of Organic Amendments and Lime on the Bioavailability of Heavy Metals in Contaminated Soils*. In: Proceeding of the 17th World Congress of Soil Science, Bangkok, Thailand.

Bernal, M.P., R. Clemente and D.J. Walker. 2007. The role of organic amendments in the bioremediation of heavy metal-polluted soils. In: Gore R.W. (ed.), *Environmental Research at the Leading Edge*. Nova Science Publishers Inc, New York, pp. 1–57.

Bhattacharyya, P., A. Chakraborty, K. Chakrabarti, S. Tripathy and M. Powell. 2005. Chromium uptake by rice and accumulation in soil amended with municipal solid waste compost. *Chemosphere* 60(10):1481–1486.

Bi, L., B. Zhang, G. Liu, Z. Li, Y. Liu, C. Ye, X. Yu, T. Lai, J. Zhang, J. Yin and Y. Liang. 2009. Long-term effects of organic amendments on the rice yields for double rice cropping systems in subtropical China. *Agric Ecosyst Environ* 129(4):534–541.

Bian, R., S. Joseph, L. Cui, G. Pan, L. Li, X. Liu, A. Zhang, H. Rutlidge, S. Wong, C. Chia, C. Marjo, B. Gong, P. Munroe and S. Donne. 2014. A three-year experiment confirms continuous immobilization of cadmium and lead in contaminated paddy field with biochar amendment. *J Hazard Mater* 272:121–128.

Bibak, A. and O.K. Borggard. 1994. Molybdenum adsorption by aluminum and iron oxides and humic-acid. *Soil Sci* 158(5):323–328.

Bilski, J., D. Jacob, F. Soumaila, C. Kraft and A. Farnsworth. 2012. Agronomic biofortification of cereal crop plants with Fe, Zn, and Se, by the utilization of coal fly ash as plant growth media. *Adv Biores* 3(4):130–136.

Bitterli, C., G.S. Ba~nuelos and R. Schulin. 2010. Use of transfer factors to characterize uptake of selenium by plants. *J Geochem Explor* 107(2):206–216. https://doi.org/10.1016/j.gexplo.2010.09.009.

Bleam, W.F. 2012. Adsorption and surface chemistry. In: Bleam W.F. (ed.), *Soil and Environmental Chemistry*. Academic Press, Cambridge, MA, pp. 371–407. https://doi.org/10.1016/B978-0-12-415797-2.00009-1.

Boguta, P. and Z. Sokolowska. 2016. Interactions of Zn (II) Ions with humic acids isolated from various type of soils. *PLOS ONE* 11(4):e0153626. https://doi.org/10.1371/journal.pone.0153626.

Bolan, N.S. and S. Thiagarajan. 2001. Retention and plant availability of chromium in soils as affected by lime and organic matter amendments. *Aust J Soil Res* 39(5):1091–1103.

Bolan, N.S., D.C. Adriano and R. Naidu. 2003. Role of phosphorus in (im) mobilization and bioavailability of heavy metals in the soil-plant system. *Rev Environ Contam Toxicol* 177:1–44.
Boudesocque, S., E. Guillon, M. Aplincourt, E. Marceau and L. Stievano. 2007. Sorption of Cu(II) onto vineyard soils: Macroscopic and spectroscopic investigations. *J Colloid Interface Sci* 307(1):40–49.
Boujelben, N., J. Bouzid and Z. Elouear. 2009. Adsorption of nickel and copper onto natural iron oxide-coated sand from aqueous solutions: Study in single and binary systems. *J Hazard Mater* 163(1):376–382.
Boyd, R. 2011. Selenium stories. *Nat Chem* 3(7):570. https://doi.org/10.1038/nchem.1076.
Bradl, H.B. 2004. Adsorption of heavy metal ions on soils and soils constituents. *J Colloid Interface Sci* 277(1):1–18.
Branzini, A. and M.S. Zubillaga. 2012. Comparative use of soil organic and inorganic amendments in heavy metals stabilization. *Appl Environ Soil Sci* 2012:1–7.
Brown, S., R. Chaney, J. Hallfrisch, J.A. Ryan and W.R. Berti. 2004. In situ soil treatments to reduce the phyto- and bioavailability of lead, zinc, and cadmium. *J Environ Qual* 33(2):522–531.
Cakmak, I. 2008. Enrichment of cereal grains with zinc: Agronomic or genetic biofortification? *Plant Soil* 302(1–2):1–17.
Cakmak, I. and U.B. Kutman. 2018. Agronomic biofortification of cereals with zinc: A review. *Eur J Soil Sci* 69(1):172–180. https://doi.org/10.1111/ejss.12437.
Cakmak, I., M. Kalayci, Y. Kaya, A.A. Torun, N. Aydin, Y. Wang Y., et al. 2010. Biofortification and localization of zinc in wheat grain. *J Agricul Food Chem* 58:9092–9102.
Canet, R., F. Pomares. and F. Tarazona. 1997. Chemical extractability and availability of heavy metals after seven years application of organic wastes to a citrus soil. *Soil Use Manag* 13(3):117–121. https://doi.org/10.1111/j.1475-2743.1997.tb00569.x.
Catlett, K.M., D.M. Heil, W.L. Lindsay and M.H. Ebinger. 2002. Soil chemical properties controlling zinc (2+) activity in 18 Colorado soils. *Soil Sci Soc Am J* 66(4):1182–1189.
Catrouillet, C., M. Davranche, A. Dia, B.L. Coz, E. Demangeat and G. Gruau. 2016. Does As(III) interact with Fe(II), Fe(III) and organic matter through ternary complexes? *J Colloid Interface Sci* 470:153–161.
Chen, Y., J. Cui, X. Tian, A. Zhao, M. Li, S. Wang, X. Li, Z. Jia and K. Liu. 2017. Effect of straw amendment on soil Zn availability and ageing of exogenous water-soluble Zn applied to calcareous soil. *PLOS ONE* 12(1):e0169776. https://doi.org/10.1371/journal.pone.0169776.
Chilimba, A.D., S.D. Young and E.J. Joy. 2014. Agronomic biofortification of maize, soybean and groundnut with selenium in intercropping and sole cropping systems. *Afric J Agricul Res* 9(50):3620–3626.
Christiansen, J.V. and L. Carlsen. 1991. Enzymatically controlled iodination reactions in the terrestrial environment. *Radiochim Acta* 52e53:327.e333.
Clemente, R. and M.P. Bernal. 2006. Fractionation of heavy metals and distribution of organic carbon in two contaminated soils amended with humic acids. *Chemosphere* 64(8):1264–1273.
Clemente, R., T. Pardo, P. Madejo´n, E. Madejo´n and M.P. Bernal. 2015. Food byproducts as amendments in trace elements contaminated soils. *Food Res Int* 73:176–189.
Cleven, R.F.M.J., J.A. Janus, J.A. Annema and W. Slooff. 1993. *Integrated Criteria Document Zinc* (RIVM Report 710401028). National Institute of Public Health and the Environment, Bilthoven, 278 p. Available at http://www.rivm.nl/bibliotheek/rapporten/710401028.html.
Colombo, C., G. Palumbo, V.M. Sellitto, C. Rizzardo, N. Tomasi, R. Pinton and S. Cesco. 2012. Characteristics of insoluble, high molecular weight iron-humic substances used as plant iron sources. *Soil Sci Soc Am J* 76(4):1246–1256. https://doi.org/10.2136/sssaj2011.0393.
Colombo, C., G. Palumbo,J. Z. He, R. Pinton and S. Cesco. 2014. Review on iron availability in soil: Interaction of Fe minerals, plants, and microbes. *J Soils Sediments* 14(3):538–548. https://doi.org/10.1007/s11368-013-0814-z.
Cornell, R.M. and U. Schwertmann. 2003. *The Iron Oxides*, 2nd ed. Wiley-VCH, Weinheim, New York.
Cui, L., G. Pan, L. Li, J. Yan, A. Zhang, R. Bian and A. Chang. 2012.The reduction of wheat cd uptake in contaminated soil via biochar amendment: A two-year field experiment. *Bioresour* 7(4):5666–5676.
De la Fuente, C., R. Clemente, T.G. Martínez-Alcalá, M.P. Bernal and M.P. Bernal. 2011. Impact of fresh and composted solid olive husk and their water-soluble fractions on soil heavy metal fractionation; microbial biomass and plant uptake. *J Hazard Mater* 186(2–3):1283–1289.
Degryse, F., V.K. Verma and E. Smolders. 2008. Mobilization of Cu and Zn by root exudates of dicotyledonous plants in resin-buffered solutions and in soil. *Plant Soil* 306(1–2):69–84.
Dhaliwal, S.S. and J.S. Manchanda. 2008. Effect of green manure, submergence and soil applied Mn on yield and uptake of Mn under rice-wheat system. *An Asian J Soil Sci* 3(1):166–172.

Dhaliwal, S.S., J.S. Manchanda, S.S. Walia and M.K. Dhaliwal. 2013. Differential response of manures in transformation of DTPA and total zinc and iron in rice transplanted on light textured soils of Punjab. *Int J Sci Environ Technol* 2(3):300–312.

Dhaliwal, S.S., R.K. Naresh, A. Mandal, R. Singh and M.K. Dhaliwal. 2019. Dynamics and transformations of micronutrients in agricultural soils as influenced by organic matter build-up: A review. *Environ Sustain Indicat* 1–2: 100007. https://doi.org/10.1016/j.indic.2019.100007.

Di Tullo, P., F. Pannier, Y. Thiry, I. Le Hécho and M. Bueno. 2016. Field study of time dependent selenium partitioning in soils using isotopically enriched stable selenite tracer. *Sci Total Environ* 562:280–288.

Dimkpa, C.O. and P.S. Bindraban. 2016. Fortification of micronutrients for efficient agronomic production: A review. *Agron Sustainable Dev* 36(1):7. https://doi.org/10.1007/s13593-015-0346-6.

Ding, X., X. Han, Y. Liang, Y. Qiao, L. Li and N. Li. 2012. Changes in soil organic carbon pools after 10 years of continuous manuring combined with chemical fertilizer in a mollisol in China. *Soil Till Res* 122:36–41. https://doi.org/10.1016/j.still.2012.02.002.

Dinh, Q.T., Z. Li, T.A.T. Tran, D. Wang and D. Liang. 2017. Role of organic acids on the bioavailability of selenium in soil: A review. *Chemosphere* 184:618–635.

Dinh, Q.T., Z. Cui, J. Huang, T. Tran, D. Wang, W. Yang, D. Liang, M. Wang, D. Yu and D. Liang. 2018. Selenium distribution in the Chinese environment and its relationship with human health: A review. *Environ Int* 112:294–309. https://doi.org/10.1016/j.envint.2017.12.035.

Donald, L.S. 1998. *Environmental Soil Chemistry.* Academic Press, San Diego, CA.

Dragicevic, V., S. Oljaca, M. Stojiljkovic, M. Simic, Z. Dolijanovic and N. Kravic. 2015. Effect of the maize-soybean intercropping system on the potential bioavailability of magnesium, iron and zinc. *Crop Pasture Sci* 66(11):1118–1127. Https://doi.org/10.1071/Cp14211.

Duborská, E., M. Bujdoš, M. Urík and P. Matúš. 2020. Iodine fractionation in agricultural and forest soils using extraction methods. *Catena* 195: 104749. https://doi.org/10.1016/j.catena.2020.104749.

Ebrahimi, N., H. Hartikainen, A. Simojoki, R. Hajiboland and M.M. Seppänen. 2015. Dynamics of dry matter and selenium accumulation in oilseed rape (*Brassica napus* L.) in response to organic and inorganic selenium treatments. *Agric Food Sci* 24(2):104–117.

EC. December 2013. *Regulation(EU) No. 1305/2013 of the European Parliament and of the Council of 17 on Support for Rural Development by the European Agricultural Fund for Rural Development (EAFRD) and Repealing Council Regulation.* No. 1698/2005. EC, Brussels.

Edelstein, M., D. Berstein, M. Shenker, H. Azaizeh and M. Ben-Hur 2016. Effects of selenium on growth parameters of tomato and basil under fertigation management. *Hortscience* 51(8):1050–1056.

Eich-Greatorex, S., T.A. Sogn, A.F. Øgaard and I. Aasen. 2007. Plant availability of inorganic and organic selenium fertiliser as influenced by soil organic matter content and pH. *Nutr Cycl Agroecosyst* 79(3):221–231. https://doi.org/10.1007/s10705-007-9109-3.

Ekholm, P., M. Eurola and E.-R.Venalainen. 2005. Selenium content of foods and diets in Finland. In: Eurola, M. (ed.), Proceedings twenty years of selenium fertilization. *Agrifood Res Rep* 69:39–45.

Englund, E., A. Aldahan, X.L. Hou, R. Petersen and G. Possnert. 2010. Speciation of iodine (^{127}I and ^{129}I) in lake sediments. *Nucl Instrum Methods Phys Res* 268(7–8):1102–1105. https://doi.org/10.1016/j.nimb.2009.10.109.

Enya, O., N. Heaney, G. Iniama and C. Lin. 2020. Effects of heavy metals on organic matter decomposition in inundated soils: Microcosm experiment and field examination. *Sci Total Environ* 724: 138223. https://doi.org/10.1016/j.scitotenv.2020.138223.

Erdem, E., N. Karapinar and R. Donat. 2004. The removal of heavy metal cations by natural zeolites. *J Colloid Interface Sci* 280(2):309–314.

Eyupoglu, F., N. Kurucu and U. Sanisa. 1994. *Status of Plant Available Micronutrients in Turkish Soils* (in Turkish). Annual Report, Report No: R-118. Soil and Fertilizer Research Institute, Ankara, pp. 25–32.

Fageria, N.K., V.C. Baligar and R.B. Clark. 2002. Micronutrients in crop production. *Adv Agron* 77:185–250. https://doi.org/10.1016/S0065-2113(02)77015-6.

Fakour, H., T.F. Lin and S.L. Lo. 2016. Equilibrium modeling of arsenic adsorption in a ternary arsenic-iron oxide-natural organic matter system. *CLEAN Soil Air Water* 44(10):1287–1295. https://doi.org/10.1002/clen.201500962.

Fan, T.T., Y.J. Wang, C.B. Li, D.M. Zhou and S.P. Friedman. 2015. Effects of soil organic matter on sorption of metal ions on soil clay particles. *Soil Sci Soc Am J* 79(3):794–802.

Fan, T.T., Y.J. Wang, C.B. Li, J.Z. He, J. Gao, D. Zhou, S.P. Friedman and D.L. Sparks. 2016. Effect of organic matter on sorption of Zn on soil: Elucidation by Wien effect measurements and EXAFS spectroscopy. *Environ Sci Technol* 50(6):2931–2937. https://doi.org/10.1021/acs.est.5b05281.

FAO, IFAD and WFP 2015. *The State of Food Insecurity in the world 2015. Meeting the 2015 International Hunger Targets: Taking Stock of Uneven Progress*. FAO, Rome.

FAO, IFAD, UNICEF, WFP and WHO. 2020. *The State of Food Security and Nutrition in the World 2020. Transforming Food Systems for Affordable Healthy Diets*. FAO, Rome. https://doi.org/10.4060/ca9692en.

Floor, G.H., S. Calabrese, G. Román-Ross, A. Aiuppa and A. Aiuppa. 2011. Selenium mobilization in soils due to volcanic derived acid rain: An example from Mt Etna volcano, Sicily. *Chem Geol* 289(3):235–244.

Fordyce, F.M. 2013. Selenium deficiency and toxicity in the environment. In: Selinus O. (ed.), *Essentials of Medical Geology*. Springer, Dordrecht, pp. 375–416.

Franchini, J.C., F.J. Gonzalez-Vila, F. Cabrera, M. Miyazawa and M.A. Pavan. 2001. Rapid transformations of plant water-soluble organic compounds in relation to cation mobilization in an acid oxisol. *Plant Soil* 231(1):55–63. http://doi.org/10.1023/A:1010338917775.

Frutos, I., A. G´arate and E. Eymar. 2010. Applicability of spent mushroom compost (SMC) as organic amendment for remediation of polluted soils. *ISHS Acta Hortic* 852:261–268.

Fuge, R. 2013. Soils and iodine deficiency. In: Selinus, O. (ed.), *Essentials of Medical Geology*. Springer, Dordrecht. https://doi.org/10.1007/978-94-007-4375-5_17.

Galantini, M.R., J.O. Landriscini, A.M. Iglesias, R.A. Miglierina and R. Rosell. 2000. The effects of crop rotation and fertilization on wheat productivity in the Pampean semiarid region of Argentina 2. Nutrient balance, yield and grain quality. *Soil Till Res* 53(2):137–144.

Gallard, H., S. Allard, R. Nicolau, U. von Gunten and J.P. Croue. 2009. Formation of iodinated organic compounds by oxidation of iodide-containing waters with manganese dioxide. *Environ Sci Technol* 43(18):7003–7009.

Garg, M., N. Sharma, S. Sharma, P. Kapoor, A. Kumar, V. Chunduri and P. Arora. 2018. Biofortified crops generated by breeding, agronomy, and transgenic approaches are improving lives of millions of people around the world. *Front Nutr* 5:12. https://doi.org/10.3389/fnut.2018.00012.

Gasmi, A., T. Tippairote, P.K. Mujawdiya, M. Peana, A. Menzel, M. Dadar, A.G. Benahmed and G. Bjørklund. 2020. Micronutrients as immunomodulatory tools for COVID-19 management. *Clin Immunol* 220:108545. https://doi.org/10.1016/j.clim.2020.108545.

Ghaley, B.B., H. Wösten, J.E. Olesen, K. Schelde, S. Baby, Y.K. Karki, C.D. Børgesen, P. Smith, J. Yeluripati, R. Ferrise, M. Bindi, P. Kuikman, J.P. Lesschen and J.R. Porter. 2018. Simulation of soil organic carbon effects on long-term winter wheat (*Triticum aestivum*) production under varying fertilizer inputs. *Front Plant Sci* 9:1158.

Gilfedder, B.S., M. Petri and H. Biester. 2007. Iodine and bromine speciation in snow and the effect of orographically induced precipitation. *Atmos Chem Phys* 7(10):2661–2669.

Gogos, A., M.W.H. Evangelou, A. Schäffer and R. Schulin. 2013. Hydrolysed wool: A novel soil amendment for zinc and iron biofortification of wheat. *New Zeal J Agr Res* 56(2):130–141. https://doi.org/10.1080/00288233.2013.775165.

Gramlich, A., S. Tandy, E. Frossard, J. Eikenberg and R. Schulin. 2013. Availability of zinc and the ligands citrate and histidine to wheat: Does uptake of entire complexes play a role? *J Agric Food Chem* 61(44):10409–10417.

Grüter, R., B. Costerousse, A. Bertoni, J. Mayer, C. Thonar, E. Frossard, R. Schulin and S. Tandy. 2017. Green manure and long-term fertilization effects on soil zinc and cadmium availability and uptake by wheat (*Triticum aestivum* L.) at different growth stages. *Sci Total Environ* 599–600:1330–1343.

Han, G.Z., G.L. Zhang, Z.T. Gong and G.F. Wang. 2012. Pedotransfer functions for estimating soil bulk density in China. *Soil Sci* 177(3):158–164. https://doi.org/10.1097/SS.0b013e31823fd493.

Han, X., C. Xu, J.A.J. Dungait, R. Bol, X. Wang, W. Wu and F. Meng. 2018. Straw incorporation increases crop yield and soil organic carbon sequestration but varies under different natural conditions and farming practices in China: A system analysis. *Biogeosciences* 15(7):1933–1946. https://doi.org/10.5194/bg-15-1933-2018.

Hansen, N.C., B.G. Hopkins, J.W. Ellsworth and V.D. Jolley. 2006. Iron nutrition in field crops'. In: Barton L.L., J. Abadía (eds.), *Iron Nutrition in Plants and Rhizospheric Microorganisms*. Springer, Netherlands, pp. 23–59.

Hansen, V., P. Roos, A. Aldahan, X. Hou and G. Possnert. 2011. Partition of iodine (^{129}I and ^{127}I) isotopes in soils and marine sediments. *J Environ Radioact* 102(12):1096–1104.

Hartikainen, H. 2005. Biogeochemistry of selenium and its impact on food chain quality and human health. *J Trace Elem Med Biol* 18(4):309–318.

Hattab, N., M. Motelica-Heino, O. Faure and J. Bouchardon. 2015. Effect of fresh and mature organic amendments on the phytoremediation of technosols contaminated with high concentrations of trace elements. *J Environ Manag* 159:37–47. https://doi.org/10.1016/j.jenvman.2015.05.0.

Hayat, R.I. Ahmed and R.A. Sheirdil. 2012. An overview of plant growth promoting rhizobacteria (PGPR) for sustainable agriculture. In: Ashraf M., M. Öztürk, M. Ahmad, A. Aksoy (eds.), *Crop Production for Agricultural Improvement*. Springer, Dordrecht, pp. 557–579. https://doi.org/10.1007/978-94-007-4116-4_22.

Hayes, R.C., M.K. Conyers, G.D. Li, G.J. Poile, A. Price, B.J. McVittie, M.J. Gardner, G.A. Sandral and J.I. McCormick. 2012. Spatial and temporal variation in soil Mn^{2+} concentrations and the impact of manganese toxicity on lucerne and subterranean clover seedlings. *Crop Pasture Sci* 63(9):875–885. HTTPS://doi.org/10.1071/CP12138.

He, P.J., Q.F. Yu, H. Zhang, L.M. Shao and F. Lu. 2017. Removal of copper (II) by biochar mediated by dissolved organic matter by biochar mediated by dissolved organic matter. *Sci Rep* 7(1).

He, Z.L., M. Zhang, X.E. Yang and P.J. Stofella. 2006. Release behavior of copper and zinc in sandy soils. *Soil Sci Soc Am J* 70(5):1699–1707.

Herath, I., P. Kumarathilaka, A. Navaratne, N., Rajakaruna and M. Vithanage. 2015. Immobilization and phytotoxicity reduction of heavy metals in serpentine soil using biochar. *J Soils Sediments* 15(1):126–138.

Hernandez-Soriano, M.C., F. Degryse, E. Lombi and E. Smolders. 2012. Manganese toxicity in barley is controlled by solution manganese and soil manganese speciation. *Soil Sci Soc Am J* 76(2):399–407. https://doi.org/10.2136/sssaj2011.0193.

Hetzel, B.S., F. Delange, J.T. Dunn, J. Ling, V. Mannar and C. Pandav. 2004.*Towards the Global Elimination of Brain Damage Due to Iodine Deficiency*. Oxford University Press, New Delhi, pp. 10–12.

Hijbeek, R., M.K. van Ittersum, H.F.M. ten Berge, G. Gort, H. Spiegel and A.P. Whitmore. 2017. Do organic inputs matter: A meta-analysis of additional yield effects for arable crops in Europe. *Plant Soil* 411(1):293–303.

Hinz, C. and H. Selim. 1999. Kinetics of Zn sorption-desorption using a thin disk flow method. *Soil Sci* 164(2):92–100.

Hou, X., V. Hansen, A. Aldahan, G. Possnert, O.C. Lind and G. Lujaniene. 2009. A review on speciation of iodine-129 in the environmental and biological samples. *Anal Chim Acta* 632:181–196. https://doi.org/10.1016/j.aca.2008.11.013.

Houben, D. and P. Sonnet. 2012. Zinc mineral weathering as affected by plant roots. *Appl Geochem* 27(8):1587–1592.

Huang, S., Y. Zeng, J. Wu, Q. Shi and X. Pan. 2013. Effect of crop residue retention on rice yield in China. A meta-analysis. *Field Crops Res* 154:188–194. https://doi.org/10.1016/j.fcr.2013.08.013.

Institute of Medicine. 2001. *Dietary Reference Intakes for Vitamin A, Vitamin K, Arsenic, Boron, Chromium, Copper, Iodine, Iron, Manganese, Molybdenum, Nickel, Silicon, Vanadium and Zinc*. The National Academics Press, Washington, DC. https://doi.org/10.17226/10026.

Irons, R., B.A. Carlson, D.L. Hatfield and C.D. Davis. 2006. Both selenoproteins and low molecular weight selenocompounds reduce colon cancer risk in mice with genetically impaired selenoprotein expression. *J Nutr* 136(5):1311–1317.

Izydorczyk, G., B. Ligas, K. Mikula, A. Witek-Krowiak, K. Moustakas and K. Chojnacka. 2021. Biofortification of edible plants with selenium and iodine: A systematic literature review. *Sci Total Environ* 754: 141983. https://doi.org/10.1016/j.scitotenv.2020.141983.

Jean, M.E., N. Cassar, C. Setzer and J.P. Bellenger. 2012. Short-term N_2 fixation kinetics in a moss-associated cyanobacteria. *Environ Sci Technol* 46(16):8667–8671.

Jeffery, S., F.G.A. Verheijen, M. van der Veldea and A.C. Bastos. 2011. A quantitative review of the effects of biochar application to soils on crop productivity using meta-analysis. *Agric Ecos Environ* 144(1):175–187.

Jeffery, S., D. Abalos, M. Prodana, A.C. Bastos, J.W. Van Groenigen, B.A. Hungate and F. Verheijen. 2017. Biochar boosts tropical but not temperate crop yields. *Environ Res Lett* 12(5): 053001.

Jiang, W., D. Liu and X. Liu. 2001. Effects of copper on root growth, cell division, and nucleolus of *Zea mays*. *Biol Plant* 44(1):105–109.

Johnson, C.C. 2003. The geochemistry of iodine and its application to environmental strategies for reducing the risks from iodine deficiency disorders (IDD). *Brit Geol Surv* DFID Kar Project R7411, Report CR/03/057N.

Jones, G.D., B. Droz, P. Greve, P. Gottschalk, D. Poffet, S.P. McGrath, L.H.E. Winkel, P. Smith and L.H. Winkel. 2017. Selenium deficiency risk predicted to increase under future climate change. *Proc Natl Acad Sci U S A* 114(11):2848–2853. https://doi.org/10.1073/pnas.1611576114.

Jordan, N., N. Marmier, C. Lomenech, E. Giffaut and J.J. Ehrhardt. 2009. Competition between selenium (IV) and silicic acid on the hematite surface. *Chemosphere* 75(1):129–134.

Kaiser, B.N., K.L. Gridley, J. Ngaire Brady, T. Phillips and S.D. Tyerman. 2005. The role of molybdenum in agricultural plant production. *Ann Bot* 96(5):745–754.

Kalsi, A., R. Sikka and D. Singh. 2016. Influence of organic and inorganic amendments on the bioavailability of lead and micronutrient composition of Indian mustard (*Brassica juncea* (L.) *Czern*) in a lead-contaminated soil. *Environ Earth Sci* 75(18):1254. https://doi.org/10.1007/s12665-016-6050-2.

Karlen, D.L., G.E. Varvel, D.G. Bullock and R.M. Cruse. 1994. Crop rotations for the 21st century. *Adv Agron* 53:1–45.

Kausch, M., P. Ng, J. Ha and C. Pallud. 2012. Soil-aggregate-scale heterogeneity in microbial selenium reduction. *Vadose Zone J* 11(2):1539–1663.

Khaliq, A., M. Zafar, M.K. Abbasi and I. Hussain. 2017. Soil-plant micronutrients dynamics in response to integrated fertilization under wheat–soybean cropping system at Rawalakot, Pakistan. *Arch Agron Soil Sci* 64(5):640–653. https://doi.org/10.1080/03650340.2017.1376245.

Khanam, A. and K. Platel. 2016. Bioaccessibility of selenium, selenomethionine and selenocysteine from foods and influence of heat processing on the same. *Food Chem* 194:1293–1299.

Khoshgoftarmanesh, A.M., M. Norouzi, M. Afyuni and R. Schulin. 2017. Zinc biofortification of wheat through preceding crop residue incorporation into the soil. *Eur J Agron* 89:131–139. https://doi.org/10.1016/j.eja.2017.05.006.

Kirkby, C.A., J.A. Kirkegaard, A.E. Richardson, L.J. Wade, C. Blanchard and G. Batten. 2011. Stable soil organic matter: A comparison of C:N:P:S ratios in Australian and other world soils. *Geoderma* 163(3–4):197–208.http://doi.org/10.1016/j.geoderma.2011.04.010.

Kookana, R.S. 2010. The role of biochar in modifying the environmental fate, bioavailability, and efficacy of pesticides in soils: A review. *J Aust Soil Res* 48(7):627–637.

Kumar, S., A. Palve, C. Joshi, R.K. Srivastava and Rukhsar. 2019. Crop biofortification for iron (Fe), zinc (Zn) and vitamin A with transgenic approaches. *Heliyon* 5(6):01914. https://doi.org/10.1016/j.heliyon.2019.e01914.

Kutman, U.B., B. Yildiz and I. Cakmak. 2011. Effect of nitrogen on uptake, remobilization and partitioning of zinc and iron throughout the development of durum wheat. *Plant Soil* 342(1–2):149–164. http://doi.org/10.1007/s11104-010-0679-5.

Lake, D.L., P.W. Kirk and J.N. Lester. 1984. Fractionation, characterization, and speciation of heavy metals in sewage sludge and sludge-amended soils: A review. *J Environ Qual* 13(2):175–183.

Lal, R. 2016. Soil health and carbon management. *Food Energy Secur* 5(4):212–222.

Lal, R. 2020. Soil organic matter content and crop yield. *J Soil Water Conserv* 75(2):27A–32A. https://doi.org/10.2489/jswc.75.2.27A.

Lehmann, J. 2007. A handful of carbon. *Nature* 447(7141):143–144.

Lenz, M. and P.N. Lens. 2009. The essential toxin: The changing perception of selenium in environmental sciences. *Sci Total Environ* 407(12):3620–3633.

Li, B., D. Zhou, L. Cang, H. Zhang, X. Fan and S. Qin. 2007. Soil micronutrient availability to crops as affected by long-term inorganic and organic fertilizer applications. *Soil Till Res* 96(1–2):166–173.

Li, Y.M., R.L. Chaney, G. Siebielec and B.A. Kerschner. 2000. Response of four turfgrass cultivars to limestone and biosolids-compost amendment of a zinc and cadmium contaminated soil at Palmerton, Pennsylvania. *J Environ Qual* 29(5):1440–1447.

Li, Z., R.G. McLaren and A.K. Metherell. 2001. Fractionation of cobalt and manganese in New Zealand soils. *Soil Res* 39(5):951–967.https://doi.org/10.1071/SR00081.

Li, Z., D. Liang, Q. Peng, Z. Cui, J. Huang and Z. Lin. 2017. Interaction between selenium and soil organic matter and its impact on soil selenium bioavailability: A review. *Geoderma* 295:69–79. https://doi.org/10.1016/j.geoderma.2017.02.019.

Lindsay, W.L. 1991. Iron oxide solubilization by organic matter and its effect on iron availability. *Plant Soil* 130(1–2):27–34.

Liphadzi, M.S. and M.B. Kirkham. 2006. Availability and plant uptake of heavy metals in EDTA-assisted phytoremediation of soil and composted biosolids. *S Afr J Bot* 72(3):391–397. https://doi.org/10.1016/j.sajb.2005.10.010.

Lopez-Bellido, L., M. Fuentes, J.E. Castillo and F.J. Lopez-Garrido. 1998. Effects of tillage: Crop rotation and nitrogen fertilization on wheat-grain quality grown under rainfed Mediterranean conditions. *Field Crops Res* 57(3):265–276.

Lopez-Moreno, M.L., L.L. Aviles, N.G. Perez, B.T. Irizarry, O. Perales, Y. Cedeno-Mattei and F. Roman. 2016. Effect of cobalt-ferrite nanoparticles on the growth and development of *Solanum lycopersicum* L. (tomato plants). *Sci Total Environ* 550:45–52.

Luo, G., L. Li, V.P. Friman, J. Guo, S. Guo, Q. Shen and N. Ling. 2018. Organic amendments increase crop yields by improving microbe-mediated soil functioning of agroecosystems: A meta-analysis. *Soil Biol Biochem* 124:105–115.

Mann, M.S., P.N. Takkar, R.L. Bansal and N.S. Randhawa. 1978. Micronutrient status of soil and yield of maize and wheat as influenced by micronutrient and FYM application. *J Indian Soc Soil Sci* 26:208–214.

Manojlović, M.S., Z. Lončarić, R.R. Cabilovski, B. Popović, K. Karalić, V. Ivezić, A. Ademi and B.R. Singh. 2019. Biofortification of wheat cultivars with selenium. *Acta Agr Scand B* 69(8):715–724.

Mao, J., V.J. Pop, S.C. Bath, H.L. Vader, C.W. Redman and M.P. Rayman. 2016. Effect of low dose selenium on thyroid autoimmunity and thyroid function in UK pregnant women with mild-to-moderate iodine deficiency. *Eur J Nutr* 55(1):55–61.

Marks, J.A., S.S. Perakis, E.K. King and J. Pett-Ridge. 2015. Soil organic matter regulates molybdenum storage and mobility in forests. *Biogeochemistry* 125(2):167–183. https://doi.org/10.1007/s10533-015-0121-4.

Marschner, H. 1995. *Mineral Nutrition of Higher Plants.* Academic Press, San Diego, CA.

Marschner, P. 2012. *Marschner's Mineral Nutrition of Higher Plants*, 3rd ed. Elsevier, Oxford.

Martin, D.P., J.M. Seiter, B.J. Lafferty and A.J. Bednar. 2017. Exploring the ability of cations to facilitate binding between inorganic oxyanions and humic acid. *Chemosphere* 166:192–196. https://doi.org/10.1016/j.chemosphere.2016.09.084.

Masunaga, T. and J.D.M. Fong. 2018. Strategies for increasing micronutrient availability in soil for plant uptake. In: Hossain M.A., T. Kamiya, D.J. Burritt, L.P. Tran, T. Fujiwara (eds.), *Plant Micronutrient Use Efficiency.* Academic Press, San Diego, CA, pp. 195–208. https://doi.org/10.1016/B978-0-12-812104-7.00013-7.

Matijevic, L., D. Romic and M. Romic. 2014. Soil organic matter and salinity affect copper bioavailability in root zone and uptake by *Vicia faba* L. plants. *Environ Geochem Health* 36(5):883–896. https://doi.org/10.1007/s10653-014-9606-7.

McAuliffe, S., S. Ray, E. Fallon, J. Bradfield, T. Eden and M. Kohlmeier. 2020. Dietary micronutrients in the wake of COVID-19: An appraisal of evidence with a focus on high-risk groups and preventative healthcare. *BMJ Nutr Prev Health 2020* bmjnph-2020-000100. https://doi.org/10.1136/ bmjnph-2020-000100.

McClafferty, B. and J.C. Zuckermann. 2015. *Cultivating Nutritious Food Systems: A Snapshot Report.* Global Alliance for Improved Nutrition (GAIN), Washington, p. 47.

McLaren, R.G., D.M. Lawson and R.S. Swift. 1986. Sorption and desorption of cobalt by soils and soil components. *J Soil Sci* 37(3):413–426.

Mehdi, Y., J.L. Hornick, L. Istasse and I. Dufrasne. 2013. Selenium in the environment, metabolism and involvement in body functions. *Molecules* 18(3):3292–3311. https://doi.org/10.3390/molecules18033292.

Menšík, L., L. Hlisnikovsky and E. Kunzová. 2019. The state of the soil organic matter and nutrients in the long-term field experiments with application of organic and mineral fertilizers in different soil-climate conditions in the view of expecting climate change. In: Larramendy M., S. Soloneski (eds.), *Organic Fertilizers: History, Production and Applications.* IntechOpen, London, pp. 1–29.

Mertens, J. and E. Smolders. 2013. "Zinc." In: Alloway B.J. (ed.), *Heavy Metals in Soils.* Springer, Dordrecht, pp. 465–493. https://doi.org/10.1007/978-94-007-4470-7_17.

Mesquita, M. and C. Carranca. 2005. Effect of dissolved organic matter on copper-zinc competitive adsorption by a sandy soil at different pH values. *Environ Sci Technol* 26(9):1065–1072.

Ming-Ming, F., B. Huang, J. Meng-Meng, H. Wen-You, S. Wei-Xia, D.C. Weindorf and Q. Chang. 2015. Effect of intensive greenhouse vegetable cultivation on selenium availability in soil. *Pedosphere* 25(3):343–350.

Mishra, B.N., R. Prasad, B. Gangaiah and B.G. Shivakumar. 2006. Organic manures for increased productivity and sustained supply of micronutrients Zn and Cu in a rice-wheat cropping system. *J Sustain Agric* 28(1):55–66.

Muramatsu, Y. and K.H. Wedepohl. 1998. The distribution of iodine in the Earth's crust. *Chem Geol* 147(3–4):201–216.

Murphy, B.W. 2015. Impact of soil organic matter on soil properties—A review with emphasis on Australian soils. *Soil Res* 53(6):605–635.

Nair, R.M., R.Y. Yang, W.J. Easdown, D. Thavarajah, P. Thavarajah, Jd. Hughes and J.D. Keatinge. 2013. Biofortification of mungbean (*Vigna radiata*) as a whole food to enhance human health. *J Sci Food Agric* 93(8):1805–1813. https://doi.org/10.1002/jsfa.6110.

Needham, P. 1983. The occurrence and treatment of mineral disorders in the field. In: C. Bould, E.J. Hewitt, P. Needham (eds.), *Diagnosis of Mineral Disorders in Plants*, vol 1: Principles. Her Majesty's Stationery Office, London, pp. 131–170.

Ngigi, P.B., C. Lachat, P.W.Masinde and G.D. Laing. 2019. Agronomic biofortification of maize and beans in Kenya through selenium fertilization. *Environ Geochem Health* 41(6):2577–2591. https://doi.org/10.1007/s10653-019-00309-3.

Oelofse, M., B. Markussen, L. Knudsen, K. Schelde, J. Olesen, L. Jensen and S. Bruun. 2015. Do soil organic carbon levels affect potential yields and nitrogen use efficiency? An analysis of winter wheat and spring barley field trials. *Eur J Agron* 66:62–73.

Ojha, S., S. Sourabh, S. Dasgupta, D.K. Das and A. Sarkar. 2018. Influence of different organic amendments on Fe, Mn, Cu and Zn availability in Indian soils. *Int J Curr Microbiol Appl Sci* 7(5):2435–2445. https://doi.org/10.20546/ijcmas.2018.705.280.

Ok, Y.S., A.R.A. Usman, S.L. Sang, A.M. Samy, A. El-Azeem, B. Choi, Y. Hashimoto and J.E. Yang. 2011. Effects of rapeseed residue on lead and cadmium availability and uptake by rice plants in heavy metal contaminated paddy soil. *Chemosphere* 85(4):677–682. https://doi.org/10.1016/j.chemosphere.2011.06.073.

Okieimen, F.E. and E.U. Ikhuoria. 2011. Effect of organic amendments on metal accumulation by maize (*Zea mays* L.) in contaminated soil. *Int J Agri Sci* 1:366–372.

Oldfield, E., M. Bradford and S. Wood. 2019. Global meta-analysis of the relationship between soil organic matter and crop yields. *Soil* 5(1):15–32.

Oliver, M.A. and P.J. Gregory. 2015. Soil, food security and human health: A review. *Eur J Soil Sci* 66(2):257–276. https://doi.org/10.1111/ejss.12216.

Paré, T., H. Dinel and M. Schnitzer. 1999. Extractability of trace metals during co-composting of biosolids and municipal solid wastes. *Biol Fertil Soils* 29(1):31–37.

Park, J.H., D. Lamb, P. Paneerselvam, G. Choppala, N. Bolan and J.W. Chung. 2011. Role of organic amendments on enhanced bioremediation of heavy metal(loid) contaminated soils. *J Hazard Mater* 185(2–3):549–574.

Patil, S., G. Girijesh, K. Nandini, K. Kumar, L. Pradeep and T.R. Kumar. 2017. Effect of zinc application through soil and foliar means on bio-fortification of zinc in rainfed maize (*Zea mays* L.). *Int J Pure Appl Biol Sci* 5:246–253. https://doi.org/10.18782/2320-051.2471.

Pooniya, V. and Y.S. Shivay. 2013. Enrichment of basmati rice grain and straw with zinc and nitrogen through ferti-fortification and summer green manuring under Indo-Gangetic plains of India. *J Plant Nutr* 36(1):91–117. https://doi.org/10.1080/01904167.2012.733052.

Qin, H., J. Zhu and H. Su. 2012. Selenium fraction in organic matter from Se- rich soils and weathered stone coal in selenosis areas of China. *Chemosphere* 86(6):626–633.

Ram, N. and V. Singh. 2005. Effect of 25 years of continuous fertilizer use on response to applied nutrients and uptake of micronutrients by rice-wheat-cowpea system. *Cereal Res Commun* 3:589–594.

Rayman, M.P., H.G. Infante and M. Sargent. 2008. Food-chain selenium and human health: Spotlight on speciation. *Br J Nutr* 100(2):238–253.

Reddy, K.J., L.C. Munn and L. Wang. 1997. Chemistry and mineralogy of molybdenum in soils. In: Gupta U.C. (ed.), *Molybdenum in Agriculture*. Cambridge University Press, Cambridge.

Reeves, D.W. 1997. The role of soil organic matter in maintaining soil quality in continuous cropping systems. *Soil Till Res* 43(1–2):131–167.

Rengel, Z. 2015. Availability of Mn, Zn and Fe in the rhizosphere. *J Soil Sci Plant Nutr* 15(2):397–409.

Rengel, Z., G.D. Batten and D.E. Crowley. 1999. Agronomic approaches for improving the micronutrient density in edible portions of field crops. *Field Crops Res* 60(1–2):27–40. https://doi.org/10.1016/S0378-4290(98)00131-2.

Römheld, V. and H. Marschner. 1991. Function of micronutrients in plants. In: Mortvedt J.J., F.R. Cox, L.M. Shuman, R.M. Welch (eds.), *Micronutrients in Agriculture*. Soil Science Society of America, Madison, WI, pp. 297–328.

Roussos, P.A., D. Gasparatos, K. Kechrologoua, P. Katsenos and P. Bouchagier. 2017. Impact of organic fertilization on soil properties, plant physiology and yield in two newly planted olive (*Olea europaea* L.) cultivars under Mediterranean conditions. *Sci Hortic* 220:11–19.

Rutkowska, B., W. Szulc, T. Sosulski and W. Stępień. 2014. Soil micronutrient availability to crops affected by long-term inorganic and organic fertilizer applications. *Plant Soil Environ* 60(5):198–203.

Saha, S., B. Saha, T. Seth, S. Dasgupta, M. Ray, B. Pal, S. Pati, S.K. Mukhopadhyay and G. Hazra. 2019. Micronutrients availability in soil–plant system in response to long-term integrated nutrient management under rice–wheat cropping system. *J Soil Sci Plant Nutr* 19(4):712–724. https://doi.org/10.1007/s42729-019-00071-6.

Salardini, A.A. and L.S. Murphy. 1978. Grain sorghum (*Sorghum bicolor* Pers.) responses to organic iron on calcareous soils. *Plant Soil* 49(1):57–70.

Scheid, S., M.S. Günthardt-Goerg, R. Schulin and B. Nowack. 2009. Accumulation and solubility of metals during leaf litter decomposition in non-polluted and polluted soil. *Eur J Soil Sci* 60(4):613–621.

Schnitzer, M. 1978. Humic substances: Chemistry and reactions. In: Schnitzer M., S.U. Khan (eds.), *Soil Organic Matter*. Elsevier, Amsterdam, pp. 1–58.

Schomburg, L. and J. Köhrle. 2008. On the importance of selenium and iodine metabolism for thyroid hormone biosynthesis and human health. *Mol Nutr Food Res* 52(11):1235–1246.

Schulin, R., A.H. Khoshgoftarmanesh, M. Afyuni, B. Nowack and E. Frossard. 2009. Effects of soil management on zinc uptake and its bioavailability in plants. In: Banuelos G.S., Z.Q. Lin (eds.), *Development and Uses of Biofortified Agricultural Products*. CRC Press, Boca Raton, FL, pp. 95–114.

Selbig, W.R., R. Bannerman and S.R. Corsi. 2013. From streets to streams: Assessing the toxicity potential of urban sediment by particle size. *Sci Tot Environ* 444:381–391.

Shahid, M., A.K. Shukla, P. Bhattacharyya, R. Tripathi, S. Mohanty, A. Kumar, B. Lal, P. Gautam, R. Raja, B.B. Panda, B. Das and A.K. Nayak. 2016. Micronutrients (Fe, Mn, Zn and Cu) balance under long-term application of fertilizer and manure in a tropical rice-rice system. *J Soils Sediments* 16(3):737–747. https://doi.org/10.1007/s11368-015-1272-6.

Shahid, M., N.K. Niazi, S. Khalid, B. Murtaza, I. Bibi and M.I. Rashid. 2018. A critical review of selenium biogeochemical behaviour in soil-plant system with an inference to human health. *Environ Poll* 234:915–934. https://doi.org/10.1016/j.envpol.2017.12.019.

Shand, C.A., J. Eriksson, A.S. Dahlin and D.G. Lumsdon. 2012. Selenium concentrations in national inventory soils from Scotland and Sweden and their relationship with geochemical factors. *J Geochem Explor* 121:4–14.

Sharma, A. and A.K. Nagpal. 2018. Soil amendments: A tool to reduce heavy metal uptake in crops for production of safe food. *Rev Environ Sci Biotechnol* 17(1):187–203. https://doi.org/10.1007/s11157-017-9451-0.

Sharma, B.D., J.S. Brar, J.K. Chanay, P. Sharma and P.K. Singh. 2015. Distribution of forms of copper and their association with soil properties and uptake in major soil orders in semi-arid soils of Punjab, India. *Commun Soil Sci Plant Anal* 46(4):511–527. https://doi.org/10.1080/00103624.2014.997388.

Shi, W., Z.F. Jin, S.Y. Hu, X.M. Fang and F.L. Li. 2017. Dissolved organic matter affects the bioaccumulation of copper and lead in *Chlorella pyrenoidosa*: A case of long-term exposure. *Chemosphere* 174:447–455.

Shi, Z.Q., E. Peltier and D.L. Sparks. 2012. Kinetics of Ni sorption in soils: Roles of soil organic matter and Ni precipitation. *Environ Sci Technol* 46(4):2212–2219.

Shivay, Y.S. and S. Mandi. 2020. Advances in understanding iron cycling in soils, uptake/use by plants and ways of optimising iron-use efficiency in crop production. In: Rengel Z. (eds.), *Achieving Sustainable Crop Nutrition*. Burleigh Dodds Science Publishing, Cambridge, UK, pp. 1–31. http://doi.org/10.19103/AS.2019.0062.15.

Shivay, Y.S., R. Prasad and A. Rahal. 2010. Studies on some nutritional quality parameters of organically or conventionally grown wheat. *Cereal Res Commun* 38(3):345–352. https://doi.org/10.1556/CRC.38.2010.3.5.

Singh, A. and Y.S. Shivay. 2013. Residual effect of summer green manure crops and Zn fertilization on quality and Zn concentration of durum wheat (*Triticum durum* Desf.) under a *Basmati* rice–durum wheat cropping system. *Biol Agric Hortic* 29(4):271–287. https://doi.org/10.1080/01448765.2013.832381.

Singh, A. and Y.S. Shivay. 2017. Effect of green manures and zinc fertilizer sources on DTPA-extractable zinc in soil and zinc concentration in *Basmati* rice plants at different growth stages. *Pedosphere*. https://doi.org/10.1016/S1002-0160(17)60442-9.

Singh, V. and N. Ram. 2005. Effect of 25 years of continuous fertilizer use on response to applied nutrients and uptake of micronutrients by Rice-Wheat-Cowpea system. *Cereal Res Commun* 33(2–3):589–594. https://doi.org/10.1556/CRC.33.2005.2-3.124.

Šípková, A., J. Száková, A. Hanč and P. Tlustos. 2016. Mobility of mercury in soil as affected by soil physicochemical properties. *J Soils Sediments* 16(9):2234–2241. https://doi.org/10.1007/s11368-016-1420-7.

Smith, S.R. 2009. A critical review of the bioavailability and impacts of heavy metals in municipal solid waste composts compared to sewage sludge. *Environ Int* 35(1):142–156.

Smolders, E. and J. Mertens. 2013. Cadmium. In: Alloway B.J. (eds.), *Heavy Metals in Soils: Trace Metals and Metalloids in Soils and Their Bioavailability*, 3rd ed. Springer, Netherlands.

Smyth, D. and C.C. Johnson. 2011. Distribution of iodine in soils of Northern Ireland. *Geochem Explor Environ Anal* 11(1):25–39.

Soltani, S., A.H. Khoshgoftarmanesh, M. Afyuni, M. Shrivani and R. Schulin. 2014. The effect of preceding crop on wheat grain zinc concentration and its relationship to total amino acids and dissolved organic carbon in rhizosphere soil solution. *Biol Fertil Soils* 50(2):239–247.

Sparks, D.L. 2003. Chemistry of soil organic matter. In: Sparks D.L. (eds.), *Environmental Soil Chemistry*. Academic Press, Burlington, pp. 75–113.

Stavridou, E., K. Thorup-Kristensen and S.D. Young. 2011. Assessment of selenium mineralization and availability from catch crops. *Soil Use Manag* 27(3):305–311.

Steinberg, S.M., G.T. Schmett, G. Kimble, D.W. Emerson, M.F. Turner and M. Rudin. 2008. Immobilization of fission iodine by reaction with insoluble natural organic matter. *J Radioanal Nucl Chem* 277(1):175–183.

Stiefel, E.I. 2002. The biogeochemistry of molybdenum and tungsten. *Met Ions Biol Syst* 39:1–29.

Supriatin, S., L.P. Weng and R.N.J. Comans. 2016. Selenium-rich dissolved organic matter determines selenium uptake in wheat grown on low-selenium arable land soils. *Plant Soil* 408(1–2):73–94. https://doi.org/10.1007/s11104-016-900-7.

Tan, K. 2010. *Principles of Soil Chemistry*. CRC Press, Boca Raton, FL. https://doi.org/10.1201/9781439894606.

Tao, Z., T. Chu, J. Du, X.X. Dai and Y. Gu. 2000. Effect of fulvic acids on sorption of U(VI), Zn, Yb, I and Se(IV) onto oxides of aluminum, iron and silicon. *Appl Geochem* 15(2):133–139.

Tate, R.L. III . 1987. *Soil Organic Matter: Biological and Ecological Effects*. Wiley, New York.

Teklic, T., Z. Lončarić, V. Kovačevićand B.R. Singh. 2013. Metallic trace elements in cereal grain–a review: How much metal do we eat? *Food Energy Secur* 2(2):81–95. https://doi.org/10.1002/fes3.24.

Thomas, C.L., G.E. Acquah, A.P. Whitmore, S.P. McGrath and S.M. Haefele. 2019. The effect of different organic fertilizers on yield and soil and crop nutrient concentrations. *Agron 9*:776. https://doi.org/10.3390/agronomy9120776.

Tolu, J., Y. Thiry, M. Bueno, C. Jolivet, M. Potin-Gautier and I. Le Hécho. 2014. Distribution and speciation of ambient selenium in contrasted soils, from mineral to organic rich. *Sci Total Environ* 479:93–101.

United Nations Convention on Climate Change (UNFCCC). 2015. *Join the 4/1000 Initiative Soils for Food Security and Climate*. Lima-Paris Action Agenda. Bonn, Germany.

Uren, N.C. 2013. Cobalt and manganese. In: Alloway B. (eds.), *Heavy Metals in Soils*, vol 22: Environmental Pollution. Springer, Dordrecht. https://doi.org/10.1007/978-94-007-4470-7_12.

US National Institute of Health. 2020. Dietary supplement fact sheets. https://www.nih.gov/ https://ods.od.nih.gov/factsheets/list-all/.

Varrault, G. and A. Bermond. 2011. Kinetics as a tool to assess the immobilization of soil trace metals by binding phase amendments for in situ remediation purposes. *J Hazard Mater* 192(2):808–812.

Veeranagappa, P., H.C. Prakasha, M.K. Basavaraja and M.H. Saqeebulla. 2010. Effect of zinc enriched compost on yield and nutrient uptake of rice (*Oryza sativa* L.). *Eur J Biol Sci* 3:23–29.

Walia, S.S. and D.S. Kler. 2010. Effect of organic and inorganic sources of nutrition on growth, macro and micro nutrient uptake in maize under maize-wheat sequence. *Indian J Ecol* 37:27–29.

Wang, J., H. Mao, H. Zhao, D. Huang and Z. Wang. 2012. Different increases in maize and wheat grain zinc concentrations caused by soil and foliar applications of zinc in Loess Plateau, China. *Field Crops Res* 135:89–96. https://doi.org/10.1016/j.fcr.2012.07.010.

Wang, S.S., D.L. Liang, D. Wang, W. Wei, D.D. Fu and Z.Q. Lin. 2012. Selenium fractionation and speciation in agriculture soils and accumulation in corn (*Zea mays* L.) under field conditions in Shaanxi Province, China. *Sci Tot Environ* 427:159–164. https://doi.org/10.1016/j.scitotenv.2012.03.091.

Wang, Y.J., C.B. Li, W. Wang, J. Jiang, D.M. Zhou, R.K. Xu and S.P. Friedman. 2009. Negative Wien effect measurements for exploring polarization processes of cations interacting with negatively charged soil particles. *Soil Sci Soc Am J* 73(2):569–578. https://doi.org/10.2136/sssaj2008.0018.

Wang, Z.J. and Y.X. Gao. 2001. Biogeochemical cycling of selenium in Chinese environments. *Appl Geochem* 16(11–12):1345–1351.

Wedepohl, K.H. 1995. The composition of the continental crust. *Geochim Cosmochim Acta* 59(7):1217–1232.

Weng, H.X., C.L. Hong, T.H. Xia, L. Bao, H. Liu and D. Li. 2013. Iodine biofortification of vegetable plants—An innovative method for iodine supplementation. *Chin Sci Bull* 58(17):2066–2072.

Weng, L., E.J.M. Temminghoff, S. Lofts, E. Tipping and W.H. Van Riemsdijk. 2002. Complexation with dissolved organic matter and solubility control of heavy metals in a sandy soil. *Environ Sci Technol* 36(22):4804–4810.

White, P.J. 2017. The genetics of selenium accumulation by plants. In: Pilon-Smits E., L. Winkel, Z.Q. Lin (eds.), *Selenium in Plants*, vol 11: Plant Ecophysiology. Springer, Cham. https://doi.org/10.1007/978-3-319-56249-0_9.

White, P.J. and M. Broadley. 2011. Physiological limits to zinc biofortification of edible crops. *Front Plant Sci* 2:80. https://doi.org/10.3389/fpls.2011.00080.

Wichard, T., B. Mishra, S.C.B. Myneni, J.P. Bellenger and A.M.L. Kraepiel. 2009. Storage and bioavailability of molybdenum in soils increased by organic matter complexation. *Nat Geosci* 2(9):625–629.

Williams, P.N., E. Lombi, G.X. Sun, K. Scheckel, Y.G. Zhu, X. Feng, J. Zhu, A.M. Carey, E. Adomako, Y. Lawgali, C. Deacon and A.A. Meharg. 2009. Selenium characterization in the global rice supply chain. *Environ Sci Technol* 43(15):6024–6030.

Winkel, L.H., B. Vriens, G.D. Jones, L.S. Schneider, E. Pilon-Smits and G.S. Bañuelos. 2015. Selenium cycling across soil-plant atmosphere interfaces: A critical review. *Nutrients* 7(6):4199–4239.

Wood, S.A., N. Sokol, C.W. Bell, M.A. Bradford, S. Naeem, M.D. Wallenstein and C.A. Palm. 2016. Opposing effects of different soil organic matter fractions on crop yields. *Ecol Appl* 26(7):2072–2085.

Wu, Q., W.H. Hendershot, W.D. Marshall and Y. Ge. 2000. Speciation of cadmium, copper, lead, and zinc in contaminated soils. *Commun Soil Sci Plant Anal* 31(9–10):1129–1144. https://doi.org/10.1080/00103620009370502.

Wurzburger, N., J.P. Bellenger, A.M. Kraepiel and L.O. Hedin. 2012. Molybdenum and phosphorus interact to constrain asymbiotic nitrogen fixation in tropical forests. *PLOS ONE* 7(3):e33710.

Xu, C., J. Zhong, P.G. Hatcher, S. Zhang, H.P. Li, Y.F. Ho, K.A. Schwehr, D.I. Kaplan, K.A. Roberts, R. Brinkmeyer, C.M. Yeager and P.H. Santschi. 2012. Molecular environment of stable iodine and radioiodine (I-129) in natural organic matter: Evidence inferred from NMR and binding experiments at environmentally relevant concentrations. *Geochim Cosmochim Acta* 97:166–182.

Xu, N., C. Christodoulatos and W. Braida. 2006. Adsorption of molybdate and tetrathiomolybdate onto pyrite and goethite: Effect of pH and competitive anions. *Chemosphere* 62(10):1726–1735.

Xu, N., W. Braida, C. Christodoulatos and J. Chen. 2013. A review of molybdenum adsorption in soils/bed Sediments: Speciation, mechanism, and model applications. *Soil Sediment Contam* 22(8):912–929. https://doi.org/10.1080/15320383.2013.770438.

Xue, Y., H. Duan, L. Liu, Z. Wang, J. Yang and J. Zhang. 2013. An improved crop management increases grain yield and nitrogen and water use efficiency in rice. *Crop Sci* 53(1):271–284.

Xue, Y., H. Xia, P. Christie, Z. Zhang, L. Li and C. Tang. 2016. Crop acquisition of phosphorus, iron and zinc from soil in cereal/legume intercropping systems: A critical review. *Ann Bot* 117(3):363–377. https://doi.org/10.1093/aob/mcv182.

Yadav, D.S. and A. Kumar. 2000. Integrated nutrient management in rice-wheat cropping system under Uttar Pradesh conditions. *Indian Farming* 65:28–35.

Yoshida, S., Y. Muramatsu and S. Uchida. 1992. *Water Air Soil Pollut* 63(3–4):321–329.

Yoshida, S., Y. Muramatsu, S. Katou and H. Sekimoto. 2007. Determination of the chemical forms of iodine with IC-ICP-MS and its application to environmental samples. *J Radioanal Nucl Chem* 273(1):211–214.

Yuan, P., J. Wang, Y. Pan, B. Shen and C. Wu. 2019. Review of biochar for the management of contaminated soil: Preparation, application and prospect. *Sci Tot Environ* 659.473–490. https://doi.org/10.1016/j.scitotenv.2018.12.400.

Zentner, R.P., K.E. Bowren, W. Edwards and C.A. Campbell. 1990. Effects of crop rotations and fertilization on yields and quality of spring wheat grown on a black chemozem in north central Saskatchewan. *Can J Plant Sci* 70(2):383–397.

Zhang, P., H. Sun, L. Yu and T. Sun. 2013. Adsorption and catalytic hydrolysis of carbaryl and atrazine on pig manure-derived biochars: Impact of structural properties of biochars. *J Hazard Mater* 244–245:217–224.

Zhang, R.H., Z. Li, X. Liu, B. Wang, G. Zhou, X. Huang, C. Lin, A. Wang and M. Brooks. 2016. Immobilization and bioavailability of heavy metals in greenhouse soils amended with rice straw-derived biochar. *Ecol Eng* 98:183–188. http://doi.org/10.1016/j.ecoleng.2016.10.057.

Zhu, Y.G., F.A. Smith and S.E. Smith. 2002. Phosphorus efficiencies and their effects on Zn and Mn nutrition of different barley cultivars grown in sand culture. *Aust J Agric Res* 53(2):211–216.

Zou, C., Y. Du, A. Rashid, H. Ram, E. Savasli, P.J. Pieterse, I. Ortiz-Monasterio, A. Yazici, C. Kaur, K. Mahmood, S. Singh, M.R. Le Roux, W. Kuang, O. Onder, M. Kalayci and I. Cakmak. 2019. Simultaneous biofortification of wheat with zinc, iodine, selenium, and iron through foliar treatment of a micronutrient cocktail in six countries. *J Agric Food Chem* 67(29):8096–8106. https://doi.org/10.1021/acs.jafc.9b01829.

Zuo, Y. and F. Zhang. 2009. Iron and zinc biofortification strategies in dicot plants by intercropping with gramineous species. A review. *Agron Sustainable Dev* 29(1):63–71.

Zuo, Y., Y. Liu, F. Zhang and P. Christie. 2004. A study on the improvement of iron nutrition of peanut intercropping with maize on nitrogen fixation at early stages of growth of peanut on a calcareous soil. *Soil Sci Plant Nutr* 50(7):1071–1078.

10 Raising Soil Organic Matter to Improve Productivity and Nutritional Quality of Food Crops in India

R. S. Antil, R. P. Narwal, and B. R. Singh

CONTENTS

ABBREVIATIONS

AICRP-IFS: All India Coordinated Research Project on Integrated Farming Systems
B: boron
BD: bulk density
CT: conventional tillage
DOC: dissolved organic carbon
FAI: Fertilizer Association of India
Fe: iron

DOI: 10.1201/9781003102762-10

FYM: farmyard manure
GM: green manure
HC: hydraulic conductivity
INM: integrated nutrient management
IR: infiltration rate
K: potash
LTEs: long-term experiments
MBC: microbial biomass carbon
Mg: megagram
Mn: manganese
Mo: molybdenum
MPa: megapascal
MT: minimum tillage
Mt: million tonnes
MWD: mean weight diameter
N: nitrogen
P: phosphorous
RDF: recommended dose of fertilizer
S: sulfur
SOC: soil organic C
SOM: soil organic matter
SSSA: Soil Science Society of America
WSA: water stable aggregates
Zn: zinc
ZT: zero tillage

10.1 INTRODUCTION

Indian agriculture has progressed rapidly in recent years and the country now ranks as the second-largest food producer in the world, reaching $415 billion in 2019. The country's agricultural production is more than the US, which once supplied food grains to India to meet our domestic food shortage. Unknown to many, India's international trade in agricultural products fetches higher earnings for the country than trade in services or manufacturing. Food-grain production of India has increased from 50.8 (1950–51) to 292 million tonnes (Mt) (2019–20). A nearly 6-fold increase in food-grain production during the last 69 years combined with inadequate and unbalanced use of nutrients has led to soil mining of all the essential nutrients. Extra mining of nutrients will have to be controlled and the lost nutrients replaced in order to sustain soil health. The maintenance of soil health is of great importance to ensure food and nutritional security for the country. For efficient use of fertilizers, all nutrients must be applied in balanced proportions and in accordance with crop requirements. The nitrogen (N): phosphorous (P): potash (K) consumption ratio (2018-19) of India was 7.2:2.7:1 compared to the ideal ratio of 4:2:1 (FAI 2019). The distortion in the NPK consumption ratio is more pronounced among different zones and states of India. The problem of unbalanced fertilizer use for secondary and micronutrients is greater, where applied fertilizers do not meet the requirement of the crops. Therefore, there is a need to narrow down the NPK consumption ratio to sustain crop productivity and restore soil health. Continuous use of unbalanced fertilizers has resulted in deterioration of soil health and stagnation of crop productivity (Das et al. 2015). The green revolution technologies viz. higher uses of chemical fertilizers and pesticides with the adoption of nutrient-responsive and high-yielding varieties of crops have increased the productivity of almost all crops. However, during the last decades, the compounded growth rates for production and productivity of major crops have generally declined or stagnated compared to the 1990s (Table 10.1).

TABLE 10.1
Compound Growth Rates in Production and Productivity of Major Food Grains in India (1995–96 to 2015–16, percent/annum)

Crops	Period I (1995–96 to 1999–2000)	Period II (2000–01 to 2004–05)	Period III (2005–06 to 2009-10	Period IV (2010-11 to 2015-16)	Overall Period (1995-96 to 2015-16)
Production (total)					
Rice (*Oryza sativa L.)*	3.64^{***}	-0.96^{NS}	0.01^{NS-}	1.21^{NS}	1.49^{***}
Wheat *(Triticum aestivum* L.)	4.51^{**}	-0.38^{NS}	3.75**	0.47^{NS}	1.94^{***}
Total cereals	3.35^{**}	-0.41^{NS}	1.63^{NS}	0.38^{NS}	1.71^{***}
Total pulses	2.28^{NS}	4.59^{NS}	2.11^{NS}	-1.36^{NS}	1.88^{***}
Total foodgrains	3.28^{**}	0.17^{NS}	4.45^{***}	0.27^{NS}	1.73^{***}
Productivity (total production (hectare))					
Rice (*Oryza sativa L.)*	2.23^{***}	0.86^{NS}	0.44^{NS}	1.41^{NS}	1.51^{***}
Wheat *(Triticum aestivum* L.)	1.93^{NS}	-0.97^{NS}	2.35**	-0.62^{NS} s	1.00^{***}
Total cereals	2.63^{**}	0.35^{NS}	1.85^{NS}	0.56^{NS}	1.77^{***}
Total pulses	4.06^{NS}	1.64^{NS}	1.80^{NS}	-0.48^{NS}	1.31^{***}
Total foodgrains	2.79**	0.28^{NS}	1.80^{NS}	0.59^{NS}	1.65^{***}

Source: Extracted and modified from Kumari et al. (2020).
***, **, and * are significant at 1, 5, and 10%, respectively. NS: non-significant.
Production growth rate changes in total crop production, and productivity growth rate changes in yield per hectare.

There are reports that farmers have to add higher quantities of fertilizers every year to obtain the same yield level as obtained in previous years (Antil and Raj 2020). The declining trend may be due to decline in soil organic matter (SOM) content. SOM is a heterogeneous and dynamic soil component that varies in decomposition rate, and turnover time and exerts a major influence on soil quality and the global carbon (C) cycle. It plays a crucial role in maintaining sustainability of cropping systems by improving soil physical, chemical, and biological properties. The preservation of SOM is crucial to ensuring long-term sustainability of agricultural ecosystems. Soils of India contain about 0.1 to 1% of soil organic carbon (SOC). However, even in small amounts, SOM is very important. The maintenance of SOM in agricultural soils, particularly of semiarid and subtropical regions of India, is generally governed by annual temperature, precipitation, and many interacting factors, such as soil type, tillage, application of fertilizers, quality and quantity of organics returned to soil, and the method of residue management. In India, most residues are used as feed and fuel and one-third of crop residues is used as a source of organic matter. Continuous use of the same crops in the cropping system, unbalanced and inappropriate use of chemical fertilizer, and minimum or no use of organic matter (OM) year after year are major constraints to SOC buildup in soils. Improvement of SOM can be achieved by adopting appropriate soil and crop management practices. The major management practices that improve or stabilize SOM include conservation tillage, crop rotation, use of organic manures, balanced fertilization, incorporation of crop residues, and including pasture in the rotation cycle. The use of organic manures along with chemical fertilizers may be an effective and alternative approach for further sustaining or improving SOM levels (Lal 2020, Antil and Raj 2020). In addition to providing nutrients and habitat to living organisms, SOM also binds soil particles into aggregates and improves water holding capacity of soil. SOM can also adsorb heavy metals in the soils, which reduces the toxicity of these metals to plants and their leaching into groundwater. SOM also functions as a sink for C and thus facilitates mitigation of carbon dioxide increase in the atmosphere. Hence, SOM plays a critical role in the global C budget that is a major factor affecting global warming. Overall, the adequate amount of SOM maintains soil quality, preserves sustainability of cropping systems, and reduces environmental pollution. The objective of this review is to provide updated information on the role of SOM in improving physical, chemical, and biological soil properties, and crop productivity and sustainability of cropping systems on the one hand and to suggest management practices to improve/maintain SOM content on the other. The knowledge of these practices is essential to maintaining SOM content at an appropriate level for maximizing and/or sustaining soil health and crop productivity.

10.2 SOIL ORGANIC MATTER (SOM)

The SOM is the organic fraction of the soil exclusive of undecayed plant and animal residues. SOM refers to the solid, non-mineral portions of the soil that originate from plant and animal residues (Aust and Lea 1991). Hayes and Swift (1983) used the term to refer more specifically to the nonliving components, which are a heterogeneous mixture composed largely of products resulting from microbial and chemical transformations of organic debris.

10.3 SOM AND THE C CYCLE

SOM is composed of soil microbes including bacteria and fungi, decaying material from once-living organisms, such as plant and animal tissues, fecal material, and products formed from their decomposition. The C is the most important single element in biological systems and substances that serve as the corner stone of cell structure. Plant tissues and microbial cells contain large quantities of C; approximately 40 to 50% on a dry weight basis, the ultimate source is CO_2. The C is constantly being fixed into organic form by photosynthetic organisms under the influence of light, and once bound, the C becomes unavailable for use in the generation of new plant life. It is thus essential for the carbonaceous materials to be decomposed and returned to the atmosphere for the survival of the higher organisms. The C cycle revolves around CO_2 and its fixation and regeneration. Chlorophyll-containing plants utilize the gas as their sole C source, and the carbonaceous matter thus synthesized serves to supply the animal world with preformed SOC. Upon the death of the plant or animal, microbial metabolism assumes the dominant role in the cyclic sequence. The dead tissues undergo decay and are transformed into microbial cells and a vast heterogeneous body of carbonaceous compounds known collectively as humus or as the soil organic fraction. The cycle is completed and C made available with the final decomposition and production of CO_2 from humus and the rotting tissues. The C cycle is presented in Figure 10.1. The C cycle in soil is a dynamic balance between photosynthesis, the respiration of decomposing organisms, and the stabilization of carbon.

10.4 FERTILITY STATUS AND NUTRIENT-BALANCE IN THE SOILS OF INDIA

Inadequate and unbalanced fertilizer use has caused widespread nutrient deficiencies and deterioration in soil health in many parts of India. It has been estimated that, in India, 63, 42, 13, and 40% of soils are deficient in N, P, K, and sulfur (S), respectively. On an average, 37% of soils are deficient in zinc (Zn), 13% in iron (Fe), 4% in copper (Cu), 7% in manganese (Mn), 23% in boron (B) and 13% in molybdenum (Mo) (Shukla et al. 2018).

Nutrient-balance sheets in most soils of India have been negative and continue to be so primarily because nutrient removal by crops far exceeds the nutrient additions through manures and fertilizers. The balance sheet of NPK in Indian agriculture in 2010 at 241 Mt of food-grain production showed a negative balance of 10 Mt between nutrient removal and addition (Figure 10.2).

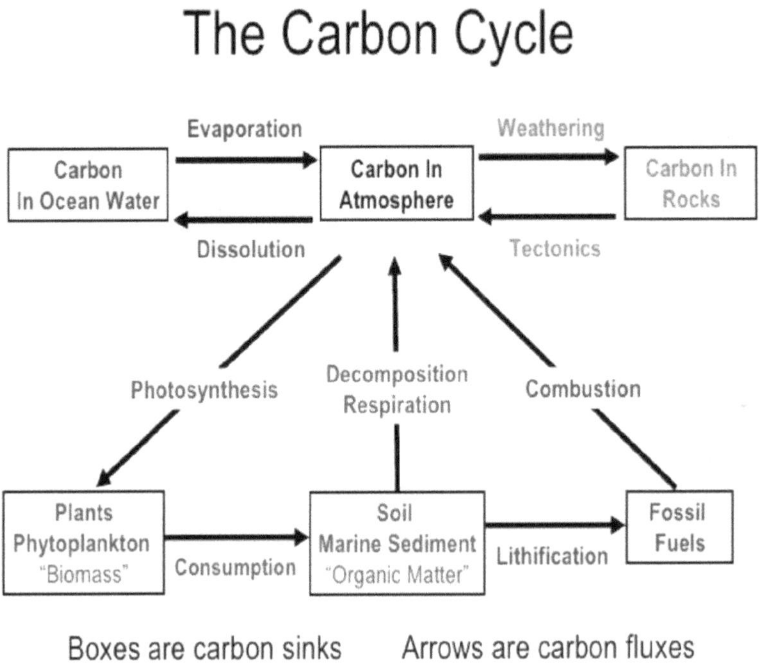

FIGURE 10.1 The C cycle (*Adapted from:* Berner 2004).

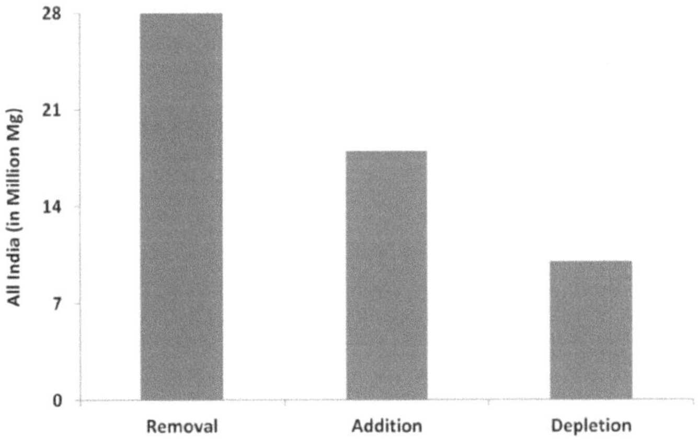

FIGURE 10.2 Balance sheet of NPK in Indian agriculture in 2010 (*Extracted and modified from:* Bhattacharjya et al. 2019).

Organic manures and bio-fertilizers contribute about 4 Mt, which means that about 6 Mt negative balance must be replenished by the soil. This negative balance is likely to increase to 15 Mt as the demand for food-grain production could be increased to 298 Mt in 2020–21. Recently, the government of India has announced the target of doubling food-grain production by 2050. It implies that for doubling the productivity, the nutrient removal would be more than double the present level. The demand for food grain is expected to be 405 Mt in 2050 to feed 1.7 billion people. Thus, the SOM/soil health would deteriorate further; this needs urgent attention.

10.5 NUTRIENT POTENTIALS OF ORGANIC RESOURCES

About 1688 Mt of total organic resources are available in India. From these organic resources, about 13 Mt of nutrients (N, P, and K) can be recycled in agriculture. But the utilization in 2007 was only 4 Mt for agricultural use (Bhattacharya 2007). It is possible that about 25% of the NPK requirement of Indian agriculture could be met by utilizing various organic resources (cattle dung, farmyard manure (FYM), crop residues, urban/rural wastes, and green manuring) which are readily available for agricultural use.

10.6 MANAGEMENT PRACTICES TO IMPROVE AND STABILIZE SOM

Adopting appropriate soil and crop management practices can improve or stabilize SOM content (Fageria 2012). The major management practices to improve or stabilize SOM are:

- Conservation tillage
- Balanced fertilization
- Use of organic manures
- Crop rotation
- Cover crops/green manuring
- Recycling crop residues
- Converting agriculturally marginal land to pasture

10.6.1 Conservation Tillage

Conservation tillage is defined as any tillage sequence with the objective of minimizing or reducing the loss of soil and water, and, operationally, a tillage and planting combination that leaves a 30% or greater cover of crop residues on the surface (SSSA 1997). Gosai et al. (2009) reported that concentration of SOM was higher under zero-tillage (ZT) and shallow-tilled plots than that under conventional tillage (CT) plots. Long-term studies show that conservation tillage minimizes soil disturbance, encourages buildup of organic material, preserves the soil structure, increases biodiversity, and conserves soil water (Sharma et al. 2016). Lower intensity of tillage leaves organic mulch at the soil surface, which reduces run-off, increases the SOM content in surface soil, and promotes greater aggregate stability, which reduces soil erosion risks. Sharma et al. (2016) reported that adoption of CT increased SOC content of soil by improving soil physical properties (Table 10.2).

The use of conservation tillage, including no tillage, is being considered as part of a strategy to reduce C loss from agricultural soils (Fageria and Moreira 2011). Conservation tillage plays an important role in the long-term C sequestration and buildup of SOM in intensive cropping systems. After four cropping cycles of wheat–mung bean (*Vigna radiate* L.)-rice, the highest SOM

TABLE 10.2
C Stock and C Sequestration Rate in Surface (0.4 m depth) Soils Under CT and ZT (Location: Dhiansar (J & K, India))

	C Stock (Mg ha^{-1})		C Sequestration Rate in ZT Over CT
Soil Type	ZT	CT	After 15 Years (Mg ha^{-1} year^{-1})
Sandy loam	18.8	22.3	0.24
Loam	19.8	26.7	0.46
Clay loam	23.8	33.1	0.62

Source: Extracted and modified from Sharma et al. (2016).

accumulation, the maximum root mass density (0–15 cm soil depth), and the improved physical and chemical properties were recorded in soil under conservation tillage practices (Alam et al. 2014).

10.6.2 Balanced Fertilization

Balanced fertilization means rational use of fertilizers and organic manures in such a manner that would increase crop yields, improve quality of crops and cost: benefit ratio, and have the least adverse environmental effects. The concept of balanced fertilization cannot be confined only to N, P, and K but must also be applied to all essential nutrients (which include primary, secondary, and micronutrients) in readily available form and in optimum quantities. Balanced fertilization must follow "4Rs" principles (Majumdar et al. 2013): using the right nutrient source, at the right rate, at the right time, and in the right place, so that fertilizer application can be managed to achieve economic, social, and environmental goals set by the stakeholders. Components of balanced fertilization include judicious use of chemical fertilizers based on deficient soil nutrients as established by soil testing in conjunction with other sources of plant nutrients such as organic manures and bio-fertilizers. The balanced fertilization could be achieved through application of a multi-nutrient in balanced proportion from fertilizer, organic, and biological sources. Fertilizer consumption is not only inadequate but also unbalanced. At present, the consumption ratio of the NPK (2018–19) is about 7.1:2.7:1 against the ideal ratio of 4:2:1 (FAI 2019). The NPK use ratio is quite wide whereas application of K, S, and micronutrients is often ignored. Results of long-term experiments (LTEs) have indicated that application of balanced fertilizer and integrated nutrient management (INM) were the best tools for obtaining the sustainability in SOM and crop yields of soybean and wheat (Tiwari 2008). Low or unbalanced fertilizer application is one of the important reasons for the decline in SOM (Singh et al. 2006, Antil and Raj 2020). Thirteen years of a rice-wheat cropping system without any fertilization decreased the SOC content of soil from its initial status of 4.2 g kg^{-1} to 3.0 g kg^{-1} soil. The SOM was either maintained or improved significantly with the balanced application of N, P, K and Zn fertilizers with and without organic amendments (Singh et al. 2012a). Main constraints in promoting balanced use of fertilizers include inadequate and ill-equipped soil testing facilities, lack of organic manures, inadequate extension services, wide gap in dissemination of knowledge between research institutions, soil testing laboratories and the extension machinery, and lack of awareness among farmers about benefits of balanced fertilization.

10.6.3 Use of Organic Manures

Uses of chemical fertilizers alone deteriorate soil fertility and create unfavorable soil physical, chemical and biological conditions in the intensive cropping system. Several studies have shown that the application of organic manure alone or in conjunction with fertilizers increased the SOC content and their fractions more effectively than the application of fertilizers alone (Singh et al. 2004, Bijay-Singh 2018, Antil and Raj 2020, Zhao et al. 2020). Continuous application of FYM for a period of 33 years in pearl millet (*Pennisetum glaucum* L.)—wheat cropping system increased SOC content (Antil et al. 2011). Long-term application of organic manures (15 Mg FYM or 7.5 Mg press mud or 5 Mg poultry manure ha^{-1} yr^{-1}) alone or in combination with NP fertilizers for 16 years in pearl millet–wheat cropping system increased the SOC significantly (Table 10.3).

Application of FYM as a component of INM increased SOC content of the soil and the increase ranged from 0.03 to 0.06% $year^{-1}$ under different cropping systems (mustard (*Brassica juncea* L.)-sorghum (*Sorghum bicolor* L.) (fodder), mustard–cotton (*Gossypium arboretum* L.), wheat–cotton, rice–wheat). The cumulative increase in SOC content under different cropping systems (Mustard–sorghum (fodder), mustard–cotton, wheat–cotton, rice–wheat) are presented in Figure 10.3. Thirty-one years of maize (*Zea mays* L.) wheat–cowpea (*Vigna unguiculata* L.) (fodder) cropping systems influenced the SOC significantly and the highest (6.5 g kg^{-1}) SOC (0–15 cm) was observed in the

TABLE 10.3
Effect of Long-Term Application of Organic Manures and Fertilizers on SOC Status After 16 Cycles of Pearl Millet–Wheat Cropping Sequence (Soil type: sandy loam; location: Hisar (Haryana, India))

Type of Manure	Dose (Mg ha^{-1})	Fertilizer (kg ha^{-1}) N	P_2O_5	SOC (%)
No manure	0	75	30	0.36
	0	150	60	0.43
FYM	15	0	0	1.01
	15	150	30	1.12
Poultry manure	5	0	0	0.59
	5	150	30	0.58
Pressmud	7.5	0	0	0.73
	7.5	150	30	0.75
Initial in 1995				0.39
LSD (P = 0.05)				0.05

Source: Extracted and modified from Kumara et al. (2013)

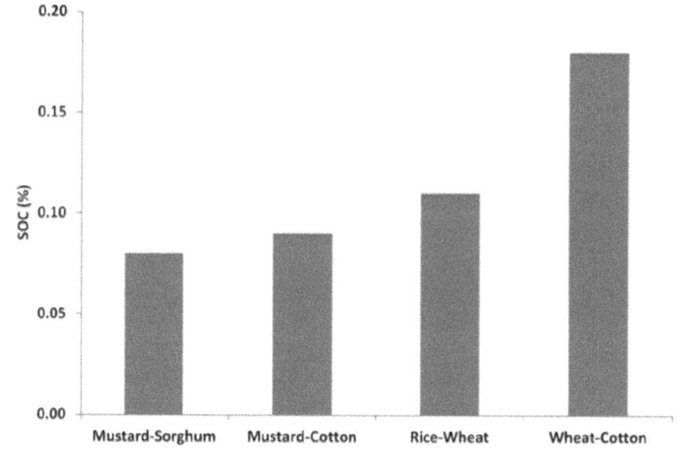

FIGURE 10.3 Influence of FYM on the build-up of SOC in soil as a component of INM under various cropping systems (*Extracted and modified from:* Antil and Raj 2020).

NPK + FYM treatment (Hati et al. 2006). Integrated use of NPK in conjunction with FYM and biofertilizers enhanced the SOM content in all types of soils in LTEs (Table 10.4).

The combined application of organic manures (FYM, vermicompost and rice straw compost) and recommended doses of fertilizers (RDF) and application of organic manures alone significantly increased the SOC content over control (Dhaliwal et al. 2020). The increase in SOC in INM treatments may be due to direct addition of C through manures and addition of more root biomass resulting from higher crop production under these treatments (Table 10.5).

TABLE 10.4
Effects of 45 Years of Soil Fertility Management on SOC Content of Some Soils in India*

Soil type	Location	SOC content (%)		
		Initial	NPK	NPK + FYM
Clay loam	Bhopal (MP) India	0.57	0.76	0.89
Sandy loam	Ludhiana (Punjab) India	0.44	0.46	0.52
Clay loam	Pantnagar (Uttrakhand) India	1.48	0.98	1.56
Silty clay loam	Palampur (HP) (India)	0.45	0.40	0.49

Source: Extracted and modified from Singh and Wanjari (2017).
*Long-term experiments were started in 1971–72.

TABLE 10.5
Effect of Different Organic Manures on Build-Up of SOC (g kg^{-1}) at Different Depths (Soil type: sandy loam; location: Ludhiana (Punjab, India))

Treatments	Depth (cm)	
	0–15	15–30
Control (no fertilizer and no manure)	0.48	0.41
Recommended fertilizer (120 kg N ha^{-1} + 60 kg P_2O_5 ha^{-1} + 30 kg K_2O ha^{-1})	1.55	1.47
100% recommended fertilizer +300 kg N ha^{-1} through FYM	6.37	2.81
100% recommended fertilizer + 400 kg N ha^{-1} through vermicompost	7.55	0.98
100% recommended fertilizer + 400 kg N ha^{-1} through rice straw compost	4.59	1.90
LSD (0.05)	0.61	0.80

Source: Extracted and modified from Dahaliwal et al. (2020)

Bijay-Singh (2018) conducted a meta-analysis of the data published on crop yields and soil parameters in maize–wheat, rice–rice, and rice–wheat cropping systems from LTEs. The buildup of SOC in upland system was less as compared to that under lowland systems. The application of RDF resulted in buildup of SOC over no-fertilizer application in all the three cropping systems (Table 10.6). Decrease in SOC content in the no-fertilizer application (control) from the initial values might be due to cultivation of the soil in maize–wheat cropping system.

10.6.4 Crop Rotation

The impacts of crop rotations are fundamental factors influencing changes in the SOC, and thus the sustainability of agricultural systems. Conventional monoculture agriculture systems can reduce the quality of soils by loss of SOM and structure because of low levels of organic inputs and frequent disturbances from tillage practices. Crop rotations have positive effects on soil properties related to the greater C inputs and diversity of plant residues to soils in comparison with continuous systems (Acosta-Martinez et al. 2004). Fundamental prerequisites of an appropriate crop rotation are listed in Table 10.7. The levels of SOC were greatest in the rotation maize–oats *(Avena sativa* L.)–clover (*Trifolium repens* L.) and least in the permanent maize rotation (Mengel et al. 2001). Crop rotations under conventional tillage that provide residues with low C/N ratios stimulate decomposition of native SOM to a greater extent than rotations providing residues with high C/N ratios (Wright and Hons 2004).

TABLE 10.6
Effect of Fertilizer Level Treatments on SOC Content of Long-Term Experiments on Maize–Wheat, Rice–Wheat, and Rice–Rice Cropping Systems in Different Locations in China (Soil type: clay loam to silt loam; location: different parts of China)

Cropping System	Number of Experiment	Cropping System Duration (years)	SOC ($g\ kg^{-1}$)		
			Initial	Control (No NPK fertilizer)	Optimum N, P, and K Fertilizer
Maize–wheat	12	6–25	6.4	5.8	6.8
Rice–wheat	10	9–27	14.3	14.9	16.3
Rice–rice	23	6–26	16.7	18.1	19.6

Source: Extracted and modified from Bijay-Singh (2018)

TABLE 10.7
Fundamental Prerequisite of an Appropriate Crop Rotation

Prerequisite	Advantage
Including legume crops in rotation	Supply N to succeeding crops.
Including crops of different root architectures	Improve soil physical and chemical properties, such as nutrient uptake from deeper soil layers, soil porosity, aeration, and drainage, and improve soil organic-matter content.
Including appropriate cover/green manure crops in the rotation	Protect soil from erosion, control weeds, supply SOM, conserve moisture, improve soil hydraulic conductivity, and control diseases and insects.
Including crops with different resistances to diseases and insects	Break cycle of diseases and insects, reduce host plant presence in rotation.
Including crops suited to a given agroecological region	Better economic return and ecological viability.

Source: Fageria (2012)

Shrestha et al. (2013) compared SOC amount and change (ΔSOC) for one continuous crop and four 3-yr fallow-containing crop rotations under zero tillage (ZT), and two fallow-containing crop rotations under minimum tillage (MT), from 1995 to 2005 in semiarid southwestern Saskatchewan, Canada. After 11 years, SOC (0–15 cm depth) was 0.2 Mg C ha^{-1} higher under continuous crop compared with fallow-containing systems. There were no significant differences in SOC and ΔSOC among fallow-containing rotations or between MT and ZT. Shrestha et al. (2015) observed that the SOC stock did not differ significantly because of rotation or fertilizer application types, even after 56 years. However, soil under the ZT system had a significantly higher SOC stock than that under the CT system at the 0–10 cm depth after 26 years of experimentation, but it was not significantly different at the 10–30 cm depth. Laamrani et al. (2020) observed that the combination of ZT and some specific cropping system practices increased total SOC concentrations in the topsoil after 36 years of long-term tillage and crop rotation. The highest total SOC concentrations were recorded in maize–maize–oats–barley (*Hordeum vulgare* L.) rotations with a red clover cover crop in both cereal phases combined with ZT, whereas the lowest total SOC concentrations were observed in the maize–maize–soybean (*Glycine max* L.)-soybean and maize–maize–oats–barley rotations without integrated cover crops, suggesting that the incorporation of cover crop is very important for total SOC storage in the topsoil.

10.6.5 Cover Crops/Green Manuring

Cover crops can increase SOC pools by adding both root and above ground biomass. Cover crops also reduce the risk of soil erosion and the resulting loss of C with soil particles. Cover crops also enhance nutrient cycling and increase soil health over time. The importance of green manuring in crop production has been known since ancient times (Fageria 2007). In recent years, cover crop importance has been increasing continuously because of the high cost of chemical fertilizers, increased risk of environmental pollution, and need for sustainable cropping systems. Cover crop/green manuring can improve the soil's physical, chemical, and biological properties and consequently improves sustainability of cropping systems. A production system that includes cover crops (legumes for N fixation, crop rotation, and ZT) can be adapted regionally and, therefore, contributes to the sustainability of soil management in the region. With improved land management, at least part of the OM lost can be restored. The increase in SOM in the absence of tillage can transform agricultural soils into C sinks. Cover crops or green manuring can increase cropping system sustainability by reducing soil erosion and increasing SOM content. Thirteen years of rice–wheat cropping without any fertilizers or green manuring significantly lowered the SOC content from its initial status of 4.2 g kg^{-1} soil to 3.0 g kg^{-1} soil (Table 10.8). There was significant buildup of SOC in green manured and RDF ($N_{150}P_{75}K_{75}Zn_{25}$) treatments. Application of 50% of RDF ($N_{75}P_{37.5}K_{37.5}Zn_{25}$) with *Sesbania* green manuring resulted in significant buildup of SOC. Continuous application of green manuring for 13 years increased SOC by 0.7 g kg^{-1} in surface soil over RDF ($N_{75}P_{37.5}K_{37.5}Zn_{25}$) only treatment (Singh et al. 2012a).

10.6.6 Recycling Crop Residues

Recycling or incorporating crop residues into the soil after harvest of grain is an important management practice to improve SOM. The incorporation of crop residue either partially or completely in the field depends upon the cultivation method (Kumari et al. 2019). The straw incorporation significantly increased the SOC concentration and storage levels compared to no incorporation of straw, where the increase was probably associated with the amount of crop residues incorporated into the soil (Malhi et al. 2011, Zhang et al. 2016). Crop residue management affects the biological and chemical processes that govern the conversion of C and N to SOM and the residual availability of N to succeeding crops (Bird et al. 2003). Bird et al. (2002) reported a sustained increase in SOM pools after four seasons of straw incorporation compared with straw burned. Crop residue is an important source of new C for building and maintaining SOM content. The crop residue management practices involving incorporation of crop residues alone or in combination with green manuring

TABLE 10.8
Effect of Different Treatments on SOC Content after 13 Cycles of Rice–Wheat Cropping System (Soil type: clay loam; location: Kaul (Haryana, India))

Treatment		
Rice	Wheat	SOC (g kg^{-1})
Control	Control	3.0
N_{150} $P_{75}K_{75}Zn_{25}$	N_{150} $P_{75}K_{75}$	5.0
N_{75} $P_{37.5}K_{37.5}Zn_{25}$ + 20 t GM ha^{-1}	N_{150} $P_{75}K_{75}$	5.7
Initial values		4.2
LSD (0.05)		0.9

Source: Extracted and modified from Singh et al. (2012a).

TABLE 10.9
Effect of Crop Residue Management in Rice–Wheat System on SOC (pooled data of 4 years) (Soil type: sandy clay; location: Katihar (Bihar), India)

Treatment	SOC (%)
Burned	0.38
Incorporated	0.58
Incorporated + *Sesbania* green manuring	0.62
LSD (0.05)	0.14

Source: Extracted and modified from Singh et al. (2019).

favorably and significantly improved the SOC in comparison to removal or burning practices of crop residues management (Singh et al. 2019). The SOC content increased by 63 and 53%, respectively, under incorporation of crop residues with green manuring and incorporation of crop residues alone in comparison to the practice of burning of crop residues (Table 10.9).

10.6.7 Converting Agriculturally Marginal Land to Pastures

Livestock grazing can have a major impact on SOC storage in grassland agroecosystems. Grasses in pasture generally maintain or increase SOM contents. The lower total soil C has been observed in cultivated cropland soils compared with that under pasture (Bronson et al. 2004). Moderate grazing of pastures may be the most effective strategy at storing SOC in pastures (Franzluebbers 2010). Hewins et al. (2018) suggested that current livestock grazing levels, which are typically moderate in intensity, on public land in these northern temperate grasslands maintain and even enhance SOC concentrations (0–30 cm) of mineral soil. Data from a meta-analysis indicated that improved grazing system practices (e.g., adjusted grazing intensity, fire management, legume or grass sowing, pasture fertilization) tend to lead to SOC sequestration by an average of 1.76 $MgCO_2$ ha $^{-1}$ yr $^{-1}$ (Godde et al. 2020).

10.7 SOM AND SOIL PROPERTIES

10.7.1 Soil Physical Properties

Physical properties of soils are those characteristics of a soil that are caused by physical forces and can be described by, or expressed in, physical terms or equations (SSSA 1997). Addition of SOM improves soil physical properties. Main soil physical properties that are influenced by SOM are discussed below.

Soil texture is the relative proportions of the various soil separates in a soil. The three soil separates that makes soil texture are sand (2 to 0.02 mm in diameter), silt (0.02 to 0.002 mm), and clay (0.002 mm or less in diameter). Soil texture is unchanged by cultural and management practices. SOM tends to increase as the clay content increases (Hartati and Sudarmadj 2016). Texture affects aggregation, and the extent of physical protection of SOM increases with increase in clay content (Vandenbygaart and Kay 2004). Soil compaction affects physical, chemical, and biological properties and is one of the main causes of agricultural soil degradation (Imhoff et al. 2004). No significant relationship was observed between soil texture and unprotected (coarse- and fine-particulate SOM) SOC (Plante et al. 2006).

Soil structure is an important property that mediates many physical and biological processes and controls SOM decomposition (Mikha and Rice 2004). From the agronomic standpoint, soil structure affects plant growth through its influence on water infiltration, percolation, and retention, soil aeration, and mechanical impedance to root growth. With better soil structure, water infiltration through the soil improves, which improves soil's ability to take up and hold water, infiltration, and stable soil aggregates. The major binding agents responsible for aggregate formation are the silicate clays, oxides of iron and aluminum, and OM and its biological decomposition products (Denef et al. 2004). The stability of aggregates is fundamental for the favorable physical conditions in the soil and sustaining crop productivity (Eynard et al. 2004). Iron and aluminum oxides and SOM increase the stability of soil aggregates. SOM compounds bind the primary particles into aggregate, physically and chemically, and this, in turn, increases the stability of the aggregates and limits their breakdown during the wetting process (Lado et al. 2004a, b). Six et al. (2000) reported that the quantity of clay- and silt-size particles is positively correlated with SOM in the same size fractions across a broad range of climates and soil types. Surface soils with high SOM contents have well granulated and stable structures. A marked reduction in the quantity and/or quality of SOM often leads to a deterioration of soil structure (Zaher et al. 2005). The number of correlations between SOM parameters and the soil structure decreased in the following order: a 94-year-old experiment with mineral fertilization > 41-year-old experiment with mineral fertilization > 25-year-old experiment with mineral fertilization + FYM (Simansky and Jonczak 2019).

Among different physical properties of soil, bulk density (BD) is an important parameter for assessment of soil health, mainly due to its relationships with other soil state (strength and porosity) and rate (moisture retention and flow characteristics) variables. Soil aggregation, a physical property related to soil structure, is greatly influenced with addition of organic resources. Addition of NPK along with FYM or green gram residue significantly improved soil aggregation, soil water retention, micro porosity, and available water capacity and reduced the BD of the soil at 0–30 cm depth (Hati et al. 2006, Das et al. 2014). The hydraulic conductivity (HC), water stable aggregates (WSA) >0.25 mm and mean weight diameter (MWD) increased significantly with the addition of FYM. However, the addition of FYM reduced the BD. Time of application of FYM can also influence soil physical properties. The increase in HC, WSA >0.25 mm, and MWD was higher when FYM was applied in the rabi (winter) season compared to the kharif (summer) season. This trend may be due to the fact that after application of FYM in kharif (summer), two crops have been taken i.e., pearl millet (summer) and wheat (winter). Whereas, when FYM is applied in the rabi (winter) season, only wheat crop has been taken. Soil sampling was carried out after harvesting of wheat. Application of FYM during both seasons gave the highest values of HC, WSA >0.25 mm and MWD as compared to the application of FYM only once in either of the seasons (Antil et al. 2011). The HC, moisture retention at 0, 0.01, 0.03, and 0.1 megapascal (MPa) and infiltration rate (IR) significantly increased with increasing levels of FYM after 23 cycles of pearl millet–wheat cropping system (Table 10.10). On the other hand, dispersion percentage and BD decreased significantly with the application of FYM. SOC was positively correlated with IR ($r = 0.97$), HC ($r = 0.89$), and moisture retention ($r = 0.94$) and negatively with BD ($r = -0.93$) and dispersion percentage ($r = -0.75$). Reductions in BD have been reported with the application of cattle manure (Nymangara et al. 2001), poultry manure (Tejada and Gonzales 2008), and FYM (Bandyopadhyay et al. 2011, Dubey et al. 2014) in LTEs.

The soil strength and IR increased significantly by incorporating green manure, wheat cut straw, and FYM in fertilization schedule (Table 10.11). However, reduction in BD was observed (Walia et al. 2010).

10.7.2 Soil Chemical Properties

Application of NP fertilizers for 16 years in the pearl millet–wheat cropping system decreased the SOC content from its initial level. However, the application of organic manures alone or in combination with NP fertilizers increased the SOC significantly (Table 10.12). Application of organic manures with or without NP fertilizers could not sustain the initial level of N. However, soil fertility

TABLE 10.10
Effect of FYM Levels on Some Physical Properties of Soil After 23 Cycles of Pearl Millet–Wheat Cropping System (Soil type: sandy loam; location: Hisar (Haryana, India))

					IR (cm h^{-1}) After		Moisture Retention (at Different MPa)%			
FYM Level (Mg ha^{-1})	SOC (%)	BD (Mg m^{-3})	Dispersion Percentage	HC (cm h^{-1})	30 minutes	105 minutes	0	0.01	0.03	0.1
0	0.94	1.48	56.94	0.28	0.6	0.3	42.7	24.8	16.9	10.6
15	1.22	1.44	53.24	0.41	0.9	0.6	46.5	27.4	18.0	12.5
30	1.30	1.32	50.16	0.41	1.8	1.2	48.0	28.4	18.3	13.4
45	1.36	1.31	48.47	0.45	2.1	1.4	49.6	28.8	19.4	13.8
LSD (0.05)	0.09	0.08	1.13	0.18	ND	ND	0.5	0.4	0.3	0.2

Source: Extracted and modified from Antil et al. (2011)

TABLE 10.11
Effect of Chemical Fertilizers and Organic Manures on Physical Properties of Soil Measured in 2007 After 23 Years of Cropping (Soil type: loamy sand; location Ludhiana (Punjab, India))

Fertilizer Rate (% of recommended NPK)					
Rice	Wheat	SOC (%)	BD (Mg m^{-3})	Soil Strength (MPa)	IR (cm h^{-1})
Control	Control	0.300	1.30	2.72	0.300
100	100	0.369	1.43	6.06	0.369
50+50% N (FYM)	100	0.510	0.97	4.48	0.510
50+50% N (wheat cut straw)	100	0.393	1.07	3.24	0.393
50+50% N (Sesbania green manuring)	100	0.396	1.03	4.24	0.396
LSD (0.05)			NS	0.17	0.114

Source: Extracted and modified from Walia et al. (2010)

with respect to P, K, Zn, Mn, Fe, and Cu can be maintained with the application of organic manures (15 Mg FYM or 7.5 Mg pressmud or 5 Mg poultry manure ha^{-1} yr^{-1}) in conjunction with RD of N under pearl millet–wheat cropping systems (Antil et al. 2011, Kumara et al. 2013).

The increase in SOM, available P, and K contents of soils ranged from 0.05 to 0.10% $year^{-1}$, 0.9 to 3.3 and 25 to 109 kg ha^{-1} $year^{-1}$, respectively, under different cropping systems (mustard–cotton, wheat–cotton, mustard–sorghum, and rice–wheat) with the application of 15 Mg FYM ha^{-1} as a component of INM (Antil and Narwal 2007). Thirteen years of a rice–wheat cropping system without any fertilizers or organic amendments had deleterious effects on soil health, resulting in significant decrease in SOC and available N, P, K, and DTPA-extractable micronutrients. However, soil health with respect to contents of SOC and available N, P, K, Zn, Mn, Fe, and Cu was either maintained or improved significantly with the balanced application of NPKZn alone or in combination with organic manures (Singh et al. 2012a).

10.7.3 Soil Biological Properties

Generally, enzyme activities in the soil are closely related to the SOM content. Application of balanced amounts of fertilizers and manures improves the SOM and microbial biomass C (MBC)

TABLE 10.12

Effect of Long-Term Application of Organic Manures and Fertilizers on SOC (%), Available N, P, K, Zn, Mn, Fe, and Cu (mg kg^{-1}) Status of Soil After 16 Cycles of Pearl Millet–Wheat Cropping Sequence (Soil type: sandy loam; location: Hisar (Haryana, India))

Type of Manure	Dose (Mg ha^{-1})	Fertilizer (kg ha^{-1})		SOC	N	P	K	Zn	Mn	Fe	Cu
		N	P_2O_5								
No manure		75	30	0.36	64.5	8.9	173.4	0.72	3.30	5.02	0.48
		150	60	0.43	67.5	9.6	178.4	0.76	3.55	5.03	0.50
FYM	15	0	0	1.01	82.6	26.0	294.4	1.84	4.29	7.29	0.78
	15	150	30	1.12	84.6	28.0	302.4	1.79	4.59	7.27	0.76
Poultry	5	0	0	0.59	78.4	28.6	247.4	1.65	3.81	6.82	0.72
manure	5	150	30	0.58	80.9	29.0	253.4	1.72	3.92	6.80	0.69
Pressmud	7.5	0	0	0.73	95.7	27.1	253.6	1.73	4.00	6.99	0.74
	7.5	150	30	0.75	98.3	27.3	267.4	1.81	4.14	8.09	0.77
Initial in 1995				0.39	98.0	12.6	217.0	-	-	-	-
LSD (0.05)				0.05	2.2	0.2	2.6	0.01	0.06	0.04	0.02

Source: Extracted and modified from Kumara et al. (2013)

content of soils, which corresponds with higher enzyme activity (Mandal et al. 2007). The dissolved OC (DOC), MBC, light C fraction and heavy C fraction content of soil increased significantly with increasing levels of FYM (Kumara 2013). The highest MBC (218.2 µg g^{-1}), urease activity (221.9 µg urea $g^{-1}h^{-1}$), alkaline phosphatase (121.5 µg PNP $g^{-1}h^{-1}$), and dehydrogenase activity (44.5 µg TPF $g^{-1}d^{-1}$), recorded with treatment 70% RDF with 30% N by FYM and pressmud equally. Significantly lowest value of urease (169.8 µg urea g^{-1} h^{-1}), alkaline phosphatase (81.83 µg PNP $g^{-1}h^{-1}$), and dehydrogenase activity (27.07 µg TPF $g^{-1}d^{-1}$) were observed with the unfertilized control (Jat and Singh 2017). Application of organic manures alone or in combination with N or NP fertilizers significantly increased dehydrogenase, alkaline phosphatase, and urease activity as compared to application of fertilizers alone (Table 10.13). However, combined application of organic manures and fertilizers significantly decreased dehydrogenase and urease activity compared with organic manures applied alone, while alkaline phosphatase activity showed the reverse trend (Sheoran et al. 2020).

10.7.4 Nutrient Use Efficiency

Long-term studies indicated that application of organic manures along with fertilizers enhanced the use efficiency of N at different locations (Dwivedi et al. 2016). In Inceptisols (Ludhiana), the N use efficiency in maize recorded under 100% N (alone) was 16.7% which increased to 23.5, 36.4, 40.2% on integration with P, PK, and FYM, respectively (Table 10.14). A similar trend was noted in Mollisols (Pantnagar) under a rice–wheat system. INM led to an increase in use efficiency of P and K at different locations (Ludhiana, Palampur, and Pantnagar) under AICRP-LTEs.

10.8 SOM AND CROP YIELD

Uses of chemical fertilizers alone deteriorate soil fertility and create unfavorable soil physical, chemical and biological conditions in the intensive cropping system. It can be overcome by use of

TABLE 10.13
Effect of Long–Term Application of Organic Manures and Fertilizers on Soil Enzyme Activity (Soil type: sandy loam; location: Hisar (Haryana, India))

Type of Manure	Dose (Mg ha^{-1})	Fertilizer (kg ha^{-1})		Dehydrogenase Activity (µg TPF g^{-1} $24h^{-1}$)	Alkaline Phosphatase Activity (µg PNP $g^{-1}h^{-1}$)	Urease Activity (µgNH_4^+–N $g^{-1}h^{-1}$)
		N	P_2O_5			
No manure	0	75	30	36.5	572	58.8
	0	150	60	32.3	581	65.0
FYM	15	0	0	63.7	684	87.7
	15	150	30	50.5	734	71.3
Poultry	5	0	0	48.6	705	85.2
manure	5	150	30	40.1	756	75.8
Pressmud	7.5	0	0	58.1	665	97.6
	7.5	150	30	39.4	673	71.3
LSD (0.05)				5.7	18	7.9

Source: Extracted and modified from Sheoran et al. (2020)

TABLE 10.14
N Use Efficiency in Different Crops as Affected by Long-Term Nutrient Supply with and Without FYM

Soil Type	Location	Crop	N Use Efficiency (%)			
			N	NP	NPK	NPK+FYM
Sandy loam	Ludhiana (Punjab, India)	Maize	16.7	23.5	36.4	40.2
		Wheat	32.0	50.6	63.1	67.8
Silty clay loam	Palampur (HP, India)	Maize	6.4	34.7	52.6	63.7
		Wheat	1.9	35.6	50.6	72.6
Clay loam	Pantnagar (Uttrakhand, India)	Rice	37.5	40.7	44.4	61.7
		Wheat	42.4	46.1	48.4	47.9

Source: Extracted and modified from Singh et al. (2012b)

organics along with fertilizers for sustaining crop production. The findings of LTEs carried out under AICRP-IFS (AICRP-CS Annual Reports 1992–93 to 2001–02) show that part of fertilizer N requirements of a monsoon crop can be met by adding FYM with annual production either at par with fertilizer application alone at recommended levels or slightly higher with an INM package (Table 10.15). It was further noticed that at a few locations, the fertilizer requirements of the winter wheat could be reduced to about 25% by substituting 25% N needs of preceding monsoon crop through FYM.

A similar advantageous effect of conjoint use of fertilizers and FYM was recorded in soybean/maize–wheat (Table 10.16) at different locations. At Ranchi, use of lime as an ingredient of INM also significantly improved the productivity of the system.

The LTEs at Pantnagar (Mollisols), Barrackpore (Inceptisols), and Raipur (Vertisols) revealed that incorporation of FYM along with NPK gave the highest production of a rice–wheat system (Singh et al. 2012b). Similar trends in the yield were also observed at other locations.

TABLE 10.15
Effect of Integrated Nutrient Supply Through Fertilizers and FYM on the Productivity of Crops Under AICRP-CS

Treatment		Grain Yield ($Mg\ ha^{-1}$)		
Monsoon	Winter	Monsoon	Winter	Total
Sorghum (*Sorghum bicolor* L.) -wheat av. of 7 yrs (Soil type: black soil; location: Parbhani (Gujarat) India)				
100% NPK	100% NPK	2.97	2.64	5.62
50% NPK+ 50% N (FYM)	100% NPK	2.85	2.78	6.53
Pearl millet (*Pennisetum glaucum* L.) -wheat) av. of 5 yrs (Soil type—loamy sand, location—Hanumangarh (Rajasthan) India)				
100% NPK	100% NPK	2.72	3.61	6.33
50% NPK+ 50% N (FYM)	100% NPK	2.56	2.40	6.38
Maize (*Zea mays* L.) -wheat av. of 8 yrs (Soil type—red soil, location—Ranchi (Bihar) India)				
100% NPK	100% NPK	2.92	2.71	5.63
75% NPK+ 25% N (FYM)	100% NPK	3.30	2.40	5.70
Rice—wheat av of 5 yrs (Soil type—clayey soil, location—Varansi (UP) India)				
100% NPK	100% NPK	4.33	3.67	8.00
50% NPK+ 50% N (FYM)	100% NPK	4.70	4.02	8.72

Source: Extracted and modidified from AICRP-IFS Reports (2005–2010 and Das et al. 2014)

TABLE 10.16
Effect of INM on Productivity ($Mg\ ha^{-1}$) of Soybean/Maize–Wheat System Under LTFE at Different Locations

Treatment	Ranchi (Bihar) India Soil Type—Red Soil 1972–2009[b]		Jabalpur (MP) India, Soil Type—Black Soil 1972–2009[a]		Palampur (MP) India, Soil Type—Red Soil 1999–2009[c]	
Unfertilized	0.61	0.69	0.81	1.24	0.29	0.38
100% NPK	1.50	2.80	1.82	4.42	3.24	2.29
NPK + FYM	1.87	3.33	2.00	4.85	4.66	3.10
NPK + Lime	1.80	3.17	-	-	4.11	2.85
LSD (0.05)	0.21	0.39	0.26	0.44	0.71	0.50

Source: Extracted and modified from [a](Dwivedi et al. 2007), [b] (Mahapatra et al.2007), and [c] (Sharma et al. 2005)

To study the contribution of FYM on the grain yield of pearl millet and wheat crops, linear regression was fitted between the intercept of linear model and the SOC content. It has been observed that each unit (0.1%) increase in the SOC content increased the productivity of pearl millet by 272.6 kg ha^{-1} and that of wheat by 1590.9 kg ha^{-1} (Figure 10.4).

Lowest grain yield of pearl millet and wheat crops was recorded when either 15 Mg FYM or 7.5 Mg pressmud or 5 Mg poultry manure ha^{-1} was applied alone. However, a significant increase in yield was obtained when organic manures were applied in combination with RD of N, and that was comparable with RD of applied NP alone or NP applied in combination with organic manures indicating that the amount of N released by organic manures was not good enough to meet the remaining N requirement of the crop (Kumara et al. 2013). Results of three years of field demonstrations at farmers' fields indicated that the application of 15 Mg FYM ha^{-1} $year^{-1}$ in conjunction with

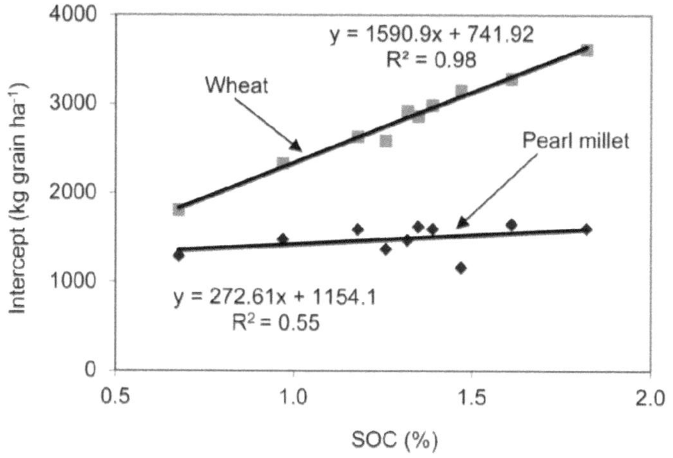

FIGURE 10.4 Influence of SOC on the intercept of linear model in pearl millet and wheat after 30 cycles of pearl millet–wheat cropping system (*Extracted and modified from:* Gupta et al. 2003).

RDF increased the productivity of different cropping systems (pearl millet–wheat, wheat–cotton, rice–wheat) compared to the application of RDF alone (Antil and Narwal 2007). The rice and wheat grain yield decreased significantly with the application of unbalanced fertilizers compared to balanced fertilization (Singh et al. 2012a).

10.9 SOM AND NUTRITIONAL QUALITY OF FOOD CROPS

Several factors can directly or indirectly affect the nutritional quality of crops. Different strategies may be applied for enhancing the nutritional quality of food crops. The building of SOM could be a tool in managing agriculture to produce more nutritious food. Because nutrients in crops come from soil, soil management such as building SOM content could be a tool in managing agriculture to produce more nutritious food. Wood, Baudron and Baudron (2018) found that protein content was related to SOM N, and Zn content to SOC. Increasing SOC by 1% was associated with an increase in Zn equivalent to the needs of 0.2 additional people per hectare; increasing SOM N by 1% was associated with an increase in protein equivalent to the daily needs of 0.1 additional people per hectare. The realistic gains in SOM could make human health-relevant increases in wheat nutrient content. SOM management can therefore be an additional tool for feeding the world population. Ali et al. (2020) reported that the application of organic manures (vermicompost, FYM) alone or in combination improved the quality of bread wheat, protein, starch, wet and dry gluten, and the gluten index of bread wheat as compared to the control.

Pooled data of protein and nutrient content revealed that the organic manures applied alone resulted in lower protein, P, and K contents in wheat grains. However, when organic manures were applied in conjunction with N or NP fertilizers, then only protein content improved significantly. The protein, P, and K contents in wheat grain ranged from 15.0 to 15.9, 0.44 to 0.53, and 0.25 to 0.30%, respectively, under different treatment combinations (Table 10.17). Among all treatments, the highest (15.9%) and the lowest (15.0%) protein content in wheat grain was observed with pressmud$_{7.5}$N$_{150}$P$_{30}$ and poultry manure$_5$, respectively. The P content in wheat grain was the highest (0.53%) and the lowest (0.44%) with application of FYM$_{15}$N$_{150}$P$_{30}$ and N$_{75}$P$_{30}$, respectively. The highest K content (0.30%) in wheat grain was observed in FYM$_{15}$N$_{150}$P$_{30}$ and poultry manure$_5$N$_{150}$P$_{30}$ treatments, while the lowest (0.25%) was observed in FYM$_{15}$N$_{150}$. The increasing levels of FYM from 0 to 15 Mg ha^{-1} increased the protein, P, and K content in wheat grain from 8.9 to 12.1%, 0.51 to 0.84%, and 0.92 to 1.44%, respectively. There was an increase in protein, P, and K contents in

TABLE 10.17
Effect of Long-Term Application of Organic Manures and Fertilizers on Protein, P, and K Content in Wheat Grain (pooled of years 2013 and 2014) (Soil type: sandy loam; location: Hisar (Haryana) India)

Type of Manure	Dose (Mg ha^{-1})	Fertilizer (kg ha^{-1})		Protein (%)	P (%)	K (%)
		N	P_2O_5			
No manure	0	75	30	15.3abc	0.44a	0.29bc
	0	150	60	15.6cd	0.50b	0.28abc
FYM	15	0	0	15.1a	0.51bc	0.27abc
	15	150	30	15.8d	0.53c	0.30c
Poultry manure	5	0	0	15.0a	0.51bc	0.27abc
	5	150	30	15.7d	0.52bc	0.30c
Pressmud	7.5	0	0	15.ab	0.50b	0.26ab
	7.5	150	30	15.9d	0.52bc	0.27abc
LSD (0.05)				0.38	0.02	0.03

Source: *Extracted and modified from* Sheoran et al. (2017).
Note: Different lower-case letters within columns indicate significance at $p < 0.05$.

wheat grain with the addition of 15 Mg FYM ha^{-1} followed by 10 and 5 Mg ha^{-1}, and values were 11.1, 11.6, and 12.1% protein; and 0.71, 0.76, and 0.84% P; and 1.28, 1.36, and 1.44% K, respectively. Higher content of protein, P, and K was observed when FYM was applied in both the seasons—kharif (summer) and rabi (winter)—followed by applications in rabi (winter) and kharif (summer) season (Kumara 2013).

10.10 CONSTRAINTS AND SUGGESTIONS FOR IMPROVING SOM

SOM contributes to a range of biological, chemical, and physical properties of soil and is essential for soil health. SOM plays a central role in maintaining tropical soil fertility. Its conservation and maintenance in tropical cultivation systems is imperative if soil degradation is to be halted and cropping made sustainable. The SOC content has decreased to 0.3–0.4% in India, but it should be between 1 to 1.5% (Krishna Kumar 2014). It is well below the acceptable limit and is a cause for concern. The possible reasons for declining SOC in tropical conditions are mainly increasing atmospheric temperature, over exploitation, extensive mining of soil fertility, soil degradation, inappropriate soil tillage, poor crop management, indiscriminate use of fertilizers, accelerated soil erosion, and enhanced decomposition and nutrient mineralization rates. The inherent factors affecting the rate of organic matter are soil texture, clay mineralogy, climatic conditions, such as rainfall and temperature; and soil moisture and aeration (oxygen levels). The effective ways to increase SOM are:

- Following conservation tillage practices
- Following ZT instead of conventional tillage
- Crop rotation (with fallow cropping)
- Growing perennial pasture
- Using cover crops
- Including legumes in the system
- Growing green manure crops

- Using organic amendments
- Incorporating plant residues regularly
- Adding SOM inputs regularly

10.11 CONCLUSIONS

SOM has long been recognized as an important indicator of soil productivity. Its role in improving crop productivity and sustainability of agricultural systems is enormous. SOM modifies soil physical, chemical, and biological properties in favor of better soil quality and, consequently, greater crop yields. A substantial amount of N requirement of plants is satisfied from the mineralization of SOM. Plant and animal residues are major sources of SOM formation. SOM formation and accumulation is, however, highly dependent on management practices and the amount and placement of organic materials. For example, improving SOM with conventional tillage in crop production is difficult due to rapid decomposition of organic materials by microbial action and loss of SOC, especially under tropical conditions. However, soil characteristics are improved with the addition of organic materials to the soil, which are in favor of greater crop yields. Adopting appropriate soil and crop management practices can be helpful in improving and/or stabilizing SOM of soils. These practices are appropriate crop rotation, adopting conservation or minimum tillage, applying farmyard and compost manures, and liming acidic soils. In addition, use of fertilizers at an adequate rate can also improve organic-matter content of soils by increasing shoot and root biomass. SOM slowly releases essential plant nutrients and also reduces their leaching to groundwater. SOM stores large amounts of SOC and avoids CO_2 escape to the atmosphere. Hence, SOM plays an important role in reducing global warming and environmental pollution.

REFERENCES

Acosta-Martinez, V., T.M. Zobeck and V. Allen. 2004. Soil microbial, chemical, and physical properties in continuous cotton and integrated crop–livestock systems. *Soil Science Society of America Journal* 68(6): 1875 1881.

AICRP-IFS Reports. 2005–2010. *AICRP on Integrated Farming Systems*. PDFSR, Modipuram, India.

Alam, M.K., M.M. Islam, N. Salahin and M. Hasanuzzaman. 2014. Effect of tillage practices on soil properties and crop productivity in wheat-mungbean-rice cropping system under subtropical climatic conditions. *The Scientific World Journal* 2014: 437283. https://doi.org/10.1155/2014/437283.

Ali, N., M.N. Khan, M.S. Ashraf et al. 2020. Influence of different organic manures and their combinations on productivity and quality of bread wheat. *Journal of Soil Science and Plant Nutrition* 20(4): 1949–1960.

Antil, R.S. and R.P. Narwal. 2007. Integrated nutrient management for sustainable soil health and crop productivity. *Indian Journal of Fertilisers* 3: 111–112.

Antil, R.S. and Dev Raj. 2020. Integrated nutrient management for sustainable crop production and improving soil health. In: *Nutrient Dynamics for Sustainable Crop Production*, ed. R.S. Meena et al. Springer Nature Singapore Pte Ltd, Singapore, 67–102.

Antil, R.S., R.P. Narwal, B. Singh and J.P. Singh. 2011. Long-term effects of FYM and N application on soil health and productivity under pearl millet – Wheat cropping system. *Indian Journal of Fertilisers* 7: 14–32.

Aust, W.M. and R. Lea. 1991. Soil temperature and organic matter in a disturbed forested wetland. *Soil Science Society of America Journal* 55(6): 1741–1746.

Bandyopadhyay, P.K., S. Saha and S. Mallick. 2011. Comparison of soil physical properties between a permanent fallow and a long-term rice–wheat cropping with inorganic and organic inputs in the humid subtropics of eastern India. *Communication in Soil Science and Plant Analyses* 42(4): 435–449.

Berner, R.A. 2004. *The Pharnerozoic Carbon Cycle*. Oxford University press, Oxford.

Bhattacharya, P. 2007. Prospect of organic nutrient resources utilization in India. *Indian Journal of Fertilisers* 3: 93–107.

Bhattacharjya, S., A. Sahu, M.C. Manna and A.K. Patra. 2019. Potential of surplus crop residues, horticultural waste and animal excreta as a nutrient source in the central and western regions of India. *Current Science* 116(8): 1314–1323.

Bijay-Singh. 2018. Optimizing fertilizer use to maintain soil Health. In: *Managing Soil Health for Sustainable Agriculture* Vol. 2: Monitoring and management, ed. D. Reicosky. Burleigh Dodds Science Publishing, Cambridge, UK (ISBN: 9781786761927, www.bdspublishing.com).

Bird, J.A., C.V. Kessel and W.R. Horwath. 2002. Nitrogen dynamics in humic fractions in temperate rice. *Soil Science Society of America Journal* 66: 478–488.

Bird, J.A., C.V. Kessel and W.R. Horwath. 2003. Stabilization of ^{13}C-carbon and immobilization of ^{15}N-nitrogen from rice straw in humic fractions. *Soil Science Society of America Journal* 67(3): 806–816.

Bronson, K.F., T.M. Zobeck, T.T. Chua, V. Acosta-Martinez, R.S.V. Pelt and J.D. Booker. 2004. Carbon and nitrogen pools of southern high plains cropland and grassland soils. *Soil Science Society of America Journal* 68: 695–1704.

Das, B., D. Chakraborty, V.K. Singh, P. Aggarwal, R. Singh, B.S. Dwivedi and R.P. Mishra. 2014. Effect of integrated nutrient management practice on soil aggregate properties, its stability and aggregate-associated carbon content in an intensive rice-wheat system. *Soil and Tillage Research* 136: 9–18.

Das, D., B.S. Dwivedi, M.C. Meena and V.K. Singh. 2015. Integrated nutrient management for improving soil health and crop productivity. *Indian Journal of Fertilisers* 11: 64–83.

Denef, K., J. Six, R. Merckx and K. Paustian. 2004. Carbon sequestration in microaggregates of no-tillage soils with different clay mineralogy. *Soil Science Society of America Journal* 68(6): 1935–1944.

Dhaliwal, S.S., J.K. Dhaliwal, A.K. Shukla, V. Sharma, M.K. Dhaliwal. 2020. Effect of different organic manures on build-up of soil organic carbon and DTPA-extractable micronutrients in soil profile under basmati rice-wheat system. *Journal of the Indian Society of Soil Science* 68(1): 91–99.

Dubey, R., R.S. Sharma and D.P. Dubey. 2014. Effect of organic, inorganic and integrated nutrient management on crop productivity, water productivity and soil properties under various rice-based cropping systems in Madhya Pradesh, India. *International Journal of Current Microbiology and Applied Sciences* 3: 381–389.

Dwivedi, A.K., M. Singh, D.L. Kauraw, R.H. Wanjari and S.S. Chauhan. 2007. *Impact of Fertilizer and Manure Use for Three Decades on Crop Productivity and Sustainability and Soil Quality under Soybean-Wheat System on Vertisols in Central India*. Indian Institute of Soil Science, Bhopal, India, pp. 51.

Dwivedi, B.S., V.K. Singh, M.C. Meena, A. Dey and S.P. Datta. 2016. Integrated nutrient management for enhancing nutrient use efficiency. *Indian Journal of Fertilisers* 12: 62–71.

Eynard, A., T.E. Schumacher, M.J. Lindstrom and D.D. Malo. 2004. Porosity and poresize distribution in cultivated Ustolls and Usterts. *Soil Science Society of America Journal* 68(6): 1927–1934.

Fageria, N.K. 2007. Green manuring in crop production. *Journal of Plant Nutrition* 30(5): 691–719.

Fageria, N.K. 2012. Role of soil organic matter in maintaining sustainability of cropping systems. *Communications in Soil Science and Plant Analysis* 43(16): 2063–2113.

Fageria, N.K. and A. Moreira. 2011. The role of mineral nutrition on root growth of crop plants. *Advances in Agronomy* 110: 251–331.

FAI. 2019. *Fertilizer and Agriculture Statistics (Northern Region). 2018-19.* The Fertilizer Association of India, New Delhi, 105–106.

Franzluebbers, A.L. 2010. Soil organic carbon in managed pastures of the southeastern United States of America. *Grassland Carbon Sequestration: Management, Policy and Economics* 11: 163–175.

Godde, C.M., I.J.M. de Boer, E.Z. Ermgassen et al. 2020. Soil carbon sequestration in grazing systems: Managing expectations. *Climatic Change* 161(3): 385–391.

Gosai, K., A. Arunachalam and B.K. Dutta. 2009. Influence of conservation tillage on soil physicochemical properties in a tropical rainfed agricultural system of northeast India. *Soil and Tillage Research* 105(1): 63–71.

Gupta, A.P., R.P. Narwal and R.S. Antil. 2003. Influence of soil organic matter on the productivity of pearl millet – Wheat cropping sequence. *Archives of Agronomy and Soil Science* 49(3): 325–332.

Hartati, W. and T. Sudarmadji. 2016. Relationship between soil texture and soil organic matter content on mined-out lands in Berau, East Kalimantan, Indonesia. *Nusantara Bioscience* 8: 83–88.

Hati, K.M., A. Swarup, D. Singh, A.K. Misra and P.K. Ghosh. 2006. Long-term continuous cropping, fertilization and manuring effects on physical properties and organic carbon content of a sandy loam soil. *Australian Journal of Soil Research* 44(5): 487–495.

Hayes, M.H.B. and R.S. Swift. 1983. The chemistry of soil organic colloids. In: *The Chemistry of Soil Constituents*, eds. D.J. Greenland and M.H.B. Hayes. Wiley, New York, 179–320.

Hewins, D.B., M.P. Lyseng, D.F. Schoderbek et al. 2018. Grazing and climate effects on soil organic carbon concentration and particle-size association in northern grasslands. *Scientific Reports* 8(1): 1336. 10.1038/s41598-018-19785-1 (www.nature.com/scientific reports/).

Imhoff, S., A.V. Da Silva and D. Fallow. 2004. Susceptibility to compaction, load support capacity, and soil compressibility of hapludox. *Soil Science Society of America Journal* 68(1): 17–24.

Jat, L.K. and Y.V. Singh. 2017. Short-term effects of organic and inorganic fertilizers on soil properties and enzyme activities in rice production. *International Journal of Current Microbiology and Applied Sciences* 6(2): 185–194.

Krishna Kumar, N.K. 2014. Decline in soil organic carbon content worries ICAR. *Hindu News Paper*, Dec. 10, 2014.

Kumara, B.H. 2013. *Long-Term Effects of Organic Manures and Fertilizers on Soil Organic Carbon Pools and Nitrogen Fractions*. Ph.D. thesis. Department of Soil Science, CCS Haryana Agricultural University, Hisar, India.

Kumara, B.H., R.S. Antil and Dev Raj. 2013. Long-term effects of nutrient management on soil health and crop productivity under pearl millet: Wheat cropping system. *Indian Journal of Fertilisers* 9(12): 86–97.

Kumari, R., R. Singh and N. Kumar. 2019. Effect of crop residue management on soil organic carbon, soil organic matter and crop yield: An over-view. *Journal of Applied and Natural Science* 11(3): 712–717.

Kumari, N., V.P. Mehta and J.K. Bhatia. 2020. Foodgrains production in India: Trend and decompositions analysis. *Economic Affairs* 65(3): 333–342.

Laamrani, A., P.R. Voroney, A.A. Berg, A.W. Gillespie, M. March, B. Deen and R.C. Martin. 2020. Temporal change of soil carbon on a long-term experimental site with variable crop rotations and tillage systems. *Agronomy* 10(6): 840. 10.3390/agronomy10060840.

Lado, M., A. Paz and M. Ben-Hur 2004a. Organic matter and aggregate size interactions in saturated hydraulic conductivity. *Soil Science Society of America Journal* 68(1): 234–242.

Lado, M., A. Paz and M. Ben-Hur 2004b. Organic matter and aggregate size interactions in infiltration, seal formation, and soil loss. *Soil Science Society of America Journal* 68(3): 935–942.

Lal, R. 2020. Managing organic matter content for restoring health and ecosystem services of soils of India. *Journal of the Indian Society of Soil Science* 68(1): 1–16.

Mahapatra, P., R.P. Singh, B.P. Singh, B. Mishra and A.K. Sarkar. 2007 *SSAC (BAU) Technical Bulletin No. 4*, BAU, 1–75.

Majumdar, K., A.M. Johnston, S. Dutt, T. Satyanarayana and T.L. Roberts. 2013. Fertiliser best management practices concept, global perspectives and application. *Indian Journal of Fertilisers* 9(4): 14–31.

Malhi, S.S., M. Nyborg, T. Goddard and D. Puurveen. 2011. Long-term tillage, straw management and N fertilization effects on quantity and quality of organic C and N in a Black chernozem soil. *Nutrient Cycling in Agroecosystems* 90(2): 227–241.

Mandal, A., A.K. Patra, D. Singh, A. Swarup and R.E. Masto. 2007. Effect of long-term application of manure and fertilizer on biological and biochemical activities in soil during crop development stages. *Bioresource Technology* 98(18): 3585–3592.

Mengel, K., E.A. Kirkby, H. Kosegarten and T. Appel. 2001. *Principles of Plant Nutrition*, 5th ed. Kluwer Academic, Dordrecht, the Netherlands.

Mikha, M.M. and C.W. Rice. 2004. Tillage and manure effects on soil and aggregate associated carbon and nitrogen. *Soil Science Society of America Journal* 68(3): 809–816.

Nymangara, J., J. Gotosa and S.E. Mpofu. 2001. Cattle manure effects on structural stability and water retention capacity of a granite soil in Zimbabwe. *Soil and Tillage Research* 62(3–4): 157–162.

Plante, A.F., R.T. Conant, C.E. Stewart, K. Paustian and J. Six. 2006. Impact of soil texture on the distribution of soil organic matter in physical and chemical fractions. *Soil Science Society of America Journal* 70(1): 287–296.

Sharma, P., V. Abrol, K.R. Sharma, N. Sharma, V.K. Phogat and V. Vikas. 2016. Impact of conservation tillage on soil organic carbon and physical properties: A review. *International Journal of Bio-Resource and Stress Management* 7(1): 151–161.

Sharma, S.P., M.V. Singh, S.K. Subehia, P.K. Jain, V. Kaushal and T.S. Verma. 2005. *Research Bulletin No. 2*. AICRP-LTFE, IISS, Bhopal, MP and Department of Soils, CSK HPKV, Palampur, HP, 1–88.

Sheoran, Sunita, Dev Raj, R.S. Antil, V.S. Mor and D.S. Dahiya. 2017. Productivity, seed quality and nutrient use efficiency of wheat (*Triticum aestivum*) under organic, inorganic and integrated nutrient management practices after twenty years of fertilization. *Cereal Research Communications* 45(2): 315–325.

Sheoran, Sunita, Dev Raj, R.S. Antil, V.S. Mor, K.S. Grewal. 2020. Soil microbial properties influenced with long term application of manures and fertilizers. *Indian Journal of Agricultural Sciences* 90: 178–182.

Shrestha, B.M., B.G. McConkey, W.N. Smith, R.L. Desjardins, C.A. Campbell, B.B. Grant and P.R. Miller. 2013. Effects of crop rotation, crop type and tillage on soil organic carbon in a semiarid climate. *Canadian Journal of Soil Science* 93(1): 137–146.

Shrestha, B.M., B.R. Singh, C. Forte and G. Certini. 2015. Long-term effects of tillage, nutrient application and crop rotation on soil organic matter quality assessed by NMR spectroscopy. *Soil Use and Management* 31(5): 358–366.

Shukla, A.K., S.K. Behera, A. Pakhre and S.K. Chaudhari. 2018. Micronutrients in soils, plants, animals and humans. *Indian Journal of Fertilisers* 14(4): 30–54.

Simansky, V. and J. Jonczak. 2019. Sorption capacity of sandy soil under long-term fertilisation. *Agriculture. (Poľnohospodárstvo)* 65: 164–171.

Singh, J.P., J. Kaur, D.S. Mehla and R.P. Narwal. 2012a. Long-term effects of nutrient management on soil health and crop productivity under rice-wheat cropping system. *Indian Journal of Fertilisers* 8(8): 28–48.

Singh, M. and R.H. Wanjari. 2017. Annual Report 2016-17. All India Coordinated Research Project on Long-Term Fertilizer Experiments to study changes. In: *Soil Quality, Crop Productivity and Sustainability.* Indian Institute of Soil Science, Nabibagh, Bhopal.

Singh, M., B.S. Dwivedi and S.P. Datta. 2012b. Integrated nutrient management for enhancing productivity, nutrient use efficiency and environmental quality. In: *Soil Science in the Service of Nation.* ISSS, New Delhi, 55–67.

Singh, R.K., P.K. Ghosh, K.K. Bandyopadhyay, A.K. Mishra, K.G. Mandal and K.M. Hati. 2006. Integrated plant nutrient supply for sustainable production in soybean-based cropping system. *Indian Journal of Fertilisers* 1: 25–32.

Singh, R.K., G.K. Sharma, P. Kumar, S.K. Singh and R. Singh. 2019. Effect of crop residues management on soil properties and crop productivity of rice-wheat system in inceptisols of Seemanchal *r*egion of Bihar. *Current Journal of Applied Science and Technology* 37(6): 1–6.

Singh, Y., B. Singh, J.K. Ladha, C.S. Khind, R.K. Gupta, O.P. Meelu and E. Pasuquin. 2004. Long-term effects of organic inputs on yield and soil fertility in the rice–wheat rotation. *Soil Science Society of America Journal* 68(3): 845–853.

Six, J., E.T. Elliott and K. Paustian. 2000. Soil microaggregate turnover and microaggregate formation: A mechanism for C sequestration under no-tillage agriculture. *Soil Biology and Biochemistry* 32(14): 2099–2013.

Tejada, M. and J.L. Gonzales. 2008. Influence of two organic amendments on the soil physical properties, soil losses, sediments and runoff water quality. *Geoderma* 145(3–4): 325–334.

Tiwari, K.N. 2008. Future plant nutrition research in India. *Journal of the Indian Society of Soil Science* 56: 327–336.

Vandenbygaart, A.J. and B.D. Kay. 2004. Persistence of soil organic carbon after plowing a long-term no-till field in southern Ontario, Canada. *Soil Science Society of America Journal* 68(4): 1394–1402.

Walia, M.K., S.S. Walia and S.S. Dhaliwal. 2010. Long-term effect of integrated nutrient management on properties of Typic Ustochrept after 23 cycles of an irrigated rice-wheat system. *Journal of Sustainable Agriculture* 34: 724–743.

Wood, S.A., F. Baudron and F. Baudron. 2018. Soil organic matter underlies crop nutritional quality and productivity in smallholder agriculture. *Agriculture, Ecosystems and Environment* 266: 100–108.

Wright, A.L. and F.M. Hons. 2004. Soil aggregation and carbon and nitrogen storage under soybean cropping sequences. *Soil Science Society of America Journal* 68(2): 507–513.

Zaher, H., J. Caron and B. Ouaki. 2005. Modeling aggregate internal pressure evolution following immersion to quantify mechanisms of structural stability. *Soil Science Society of America Journal* 69(1): 1–12.

Zhang, P., X. Chen, T. Wei et al. 2016. Effects of straw incorporation on the soil nutrient contents, enzyme activities, and crop yield in a semiarid region of China. *Soil and Tillage Research* 160: 65–72. 10.1016/j.still.2016.02.006.

Zhao, Z., C. Zhang, F. Li, S. Gao and J. Zhang. 2020. Effect of compost and inorganic fertilizer on organic carbon and activities of carbon cycle enzymes in aggregates of an intensively cultivated Vertisol. *PLOS ONE* 15(3): e0229644. 10.1371/journal.pone.0229644.

11 Role of Legumes in Managing Soil Organic Matter and Improving Crop Yield

Ahmad Latif Virk, Gulab Singh Yadav, Xin Zhao, Zheng-Rong Kan, Jian-Ying Qi, Naeem Ahmad, Rattan Lal, and Hai-Lin Zhang

CONTENTS

11.1 INTRODUCTION

Approximately 2344 Gt organic carbon (C) is present (~1 m) in the form of soil organic matter (SOM) (Stockmann et al. 2013, Lal 2018). Restoration and maintenance of soil organic matter (SOM) are difficult in agricultural soils. SOM is highly sensitive to multiple agronomic practices, i.e., tillage, fertilization, crop types, land use change, and pedo-climatic conditions. Agricultural production systems contribute almost 37% of global emissions (Tubiello et al. 2015). With better management and under site-specific conditions, soil is a sink of atmospheric CO_2 (Paustian et al. 2016). However, SOM pools are vulnerable to several factors, i.e., climate, crop type, tillage, and these changes could aggravate emission of CO_2 (Lal 2004). Crop growth and yield are adversely affected under global warming that could exacerbate global food insecurity (Meena and Lal 2018). Thus, management of SOM can reduce atmospheric emissions, increase crop yield, and alleviate global food insecurity (Meena et al. 2018). But maintaining SOM is not a one-step approach; it includes a combination of management practices in different agroecosystems (Barton et al. 2016). Whereas no-till (NT) is a promising option to maintain SOM, climate (especially temperature and precipitation) is another determinant that may cause slight to severe changes in SOM (Xue et al.

DOI: 10.1201/9781003102762-11

2018). The detrimental effects of climate on SOM loss cannot be eliminated, but these can be reduced (Zhao et al. 2020a). In this perspective, inclusion of legumes in agroecosystem may maintain SOM pool, and also improve soil fertility and agronomic yield of subsequent crops (Meena and Lal 2018, Meena et al. 2018).

An adequate concentration of SOM improves soil fertility level and enhances crop growth and yield (Lal 2010). SOM has multiple functions in soil: providing habitat and nutrients to microorganisms (Allison et al. 2010), binding soil particles into aggregates that improve water-holding capacity, and aeration (Cates et al. 2016). Thus, depletion of SOM can degrade soil structure stability due to clay dispersion (Schjønning et al. 2012). Therefore, a low level of SOM leads to soil fertility decline and decreased crop productivity. SOM comes from decomposition of litter and root biomass (Legay et al. 2014), and the decomposition process is mostly influenced by tillage type (Kubar et al. 2019), fertilization (Zhou et al. 2019), C/N ratio, and microbial community (Allison et al. 2010). These management practices can influence the decomposition process and affect stability of SOM (Wang et al. 2019a). Changes in the degree of decomposition of SOM can alter soil functions and atmospheric emissions (Lal 2004). Therefore, proper maintenance of SOM is necessary to managing soil fertility, crop productivity, and environmental sustainability (Kan et al. 2020a). In this perspective, legumes can play a vital role in maintaining SOM (Meena and Lal 2018) because nitrogen (N) addition through biological nitrogen fixation (BNF) can restore optimum C/N ratio and stabilize SOM (Veloso et al. 2019). In addition, microorganisms have a major role in SOM stabilization (Lehman et al. 2015, Kan et al. 2020b), and legumes are important for microbial community development (Stagnari et al. 2017). Stability of soil structure is a critical factor in SOM stabilization, and legumes play an important role in improving soil structure due to root exudation (Ayres et al. 2007). Mineralization of SOM is a natural process, and the amount of SOM is decreased over time (Kan et al. 2020c). Thus, consistent addition of organic matter is necessary to sustain the SOM level (Koenig and Cochran 1994). Whereas the decomposition of SOM depends on soil microbes, the process is mediated by C/N ratio. A higher C/N ratio delays the decomposition process and leads to less accrual of SOM (Sprunger et al. 2019). Contrarily, a lower C/N ratio may accentuate the decomposition process and decrease the rate of SOM accrual (loss $\leq$ addition) (Zhou et al. 2019). A lower C/N ratio of roots and residues of legumes than those of cereals can increase the rate of SOM accrual (Lupwayi et al. 2011). So, the positive effects of legumes for SOM maintenance cannot be neglected. Furthermore, combined effects of legume and SOM addition improve soil fertility (Sánchez-Navarro et al. 2019).

Soil fertility denotes the concentration of essential nutrients that are required for crop growth and development (Pu et al. 2019). Most soils are nutrient deficient due to the continuous cultivation of exhaustive crops (Chauhan et al. 2012), low level of fertilization (Kitonyo et al. 2018), and higher emissions (mainly C and N) (Badagliacca et al. 2018). Some soils have nutrients in a form that is non-available to plants, and the availability of these nutrients is essential for plant growth, e.g., phosphorus (Hazra et al. 2014). It is well documented that the addition of SOM significantly increases microbial activities (Yusuf et al. 2009), and some of the microbes (e.g., mycorrhizae) make available to plants those nutrients which are otherwise fixed (Johansson et al. 2004). Other growing conditions (e.g., water content and oxygen level in soil pores) determine crop growth and development (Kan et al. 2020a). Thus, soil physical and chemical properties are the key drivers for plant growth and development, and these properties are significantly improved by SOM (Qi et al. 2020), and the latter impacts crop yield indirectly through changes in soil functions, fertility, and eco-system functions that affect crop growth and yield.

Improved physico-chemical properties facilitate crop root development, nutrient uptake, plant physiology, and crop yield (Kan et al. 2020a). Hence, soil physico-chemical properties and crop growth is an integrated system, that could be sustained with a constant supply of SOM (Douglas et al. 1980). Additionally, a legume-based cropping system is one of the promising options to maintain and sequester SOM, which can contribute to increase yield of subsequent crops (Sainju et al. 2003). Therefore, this review is aimed at understanding the (1) underlying mechanisms of SOM-induced improvement of soil physical and chemical properties, (2) role of legumes in maintaining SOM in agroecosystems, and (3) interplay of SOM and soil properties for the improvement of crop yield under legume-based cropping systems.

11.2 ROLE OF LEGUMES IN SOM MANAGEMENT

SOM is a strong soil quality indicator under climate anomalies (Lal 2004), tillage practices (Xue et al. 2018), and crop type (Sánchez-Navarro et al. 2019). SOM content can be improved by soil conservation approaches (e.g., NT, residue retention) (Zhao et al. 2020a) and with the inclusion of legumes in the rotation cycle (Sánchez-Navarro et al. 2019). The addition of legumes affects SOM in multiple ways as discussed below.

11.2.1 Soil Structure

SOM improves soil structure and aggregation (Lal 1989), which protects SOM through encapsulation in aggregates (Sainju et al. 2003) and enhances soil fertility and quality (Lal 1989). Legumes have a distinct role in improving soil structure and fertility not only for medium fertile land but also for marginal lands (Yusuf et al. 2009). Legumes differ from other crops regarding their effects in improving soil structure and maintaining SOM. For example, legume cultivation improves the quantity of organic acids, amino acids, mucilage, and vitamins in the soil (Sainju et al. 2003). These organic substances increase binding of soil particles and stabilize aggregates (Grandy and Robertson 2007). Therefore, increased aggregation encapsulates a higher amount of SOM, protects it against microbial processes (Kan et al. 2020b), and also improves soil physical and biological functioning (Figure 11.1). Nyawade et al. (2019) reported that intercropping of potatoes (*Solanum tuberosum*) with lima bean (*Phaseolus lunatus*) and dolichos (*Lablab purpureus*) increased light organic matter by 12–28%, dissolved organic matter by 7–21%, and microbial biomass by 15–38 % in some soils of Kenya. However, the maintenance of SOM depends upon multiple factors, such as agronomic practices, climate, legume type, and water availability. But continuous cultivation of legumes can significantly enhance soil structure and porosity over time. Udom and Omovbude (2019) reported that growing of legumes and the addition of legume residues (*Calopogonium mucunoides* and *Centrosema pubescens)* increased total soil porosity and macro-aggregate formation and reduced soil bulk density in some soils of Nigeria. Thus, legume cultivation can stabilize SOM

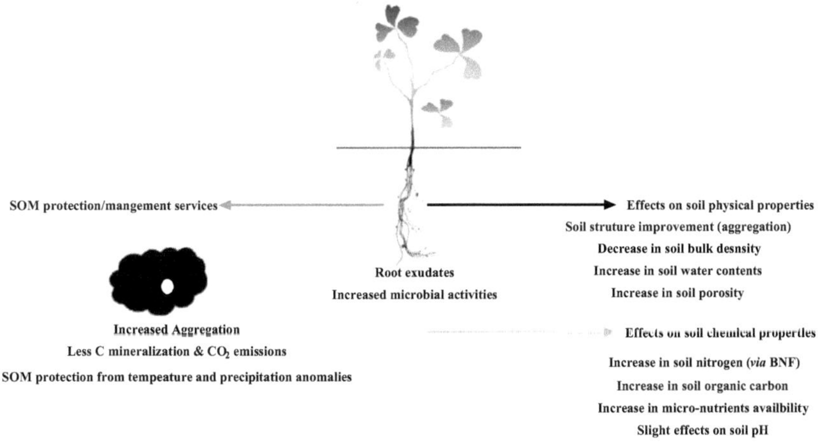

FIGURE 11.1 Role of legumes for improving soil physico-chemical properties and SOM protection.

by improving aggregation, soil physical and biochemical properties (Stagnari et al. 2017), and crop yields (Zhou et al. 2019).

11.2.2 Soil Microbial Functions

Soil microorganisms have a primary role in the utilization and transformation of SOM. More specifically, soil microbes are associated with specific plant types (due to root exudation) (Zak et al. 2003) and influence nutrient transformation in different soil layers (Tang et al. 2014). Legumes nurture a large group of microbes that, in turn, provide essential services, such as increased nutrient availability and protection from pathogens (Bais et al. 2006). The structure and community of microbes are changed by basic soil properties, management practices, and soil micro- and macro-environment (Legay et al. 2014). Continuous cultivation of legume accumulates a large quantity of root exudates, including some allelochemicals that can alter the composition of the microbial community in the rhizospheric soil (Duchene et al. 2017). Li et al. (2012) reported that the population of bacteria and actinomycetes declined, and fungal abundance increased under the monoculture of peanut (*Arachis hypogaea* L.). Badawi et al. (2011) also reported that co-inoculation with *Bradyrhizobium* increased the growth and yield of peanut. In addition, legume cultivation plays a vital role in increasing soil microbial biomass (SMB), which moderates nutrient cycling and SOM decomposition (Stockmann et al. 2013) that can improve crop productivity (Yadav et al. 2019b). Legume cultivation significantly improves SMB due to specific association with roots (Rejili et al. 2012), which further improves soil physical properties, nutrient availability, plant growth, and productivity (Mhango et al. 2013).

11.2.3 C Mineralization and CO_2 Emissions

SOM losses in the form of mineralization and CO_2 emissions are a natural process that is primarily driven by soil microbes (Kan et al. 2020b). These emissions are affected by multiple factors, such as nitrogen availability, soil aeration, water contents, and other soil biochemical attributes due to positive and negative impacts on microbial functions (Barton et al. 2016). Several crop management practices also affect SOM mineralization to a different extent, and a particular crop is also one of the most important factors (Barton et al. 2011). Crop types have a major role in microbial community development (Xiao et al. 2019), and most of (bacterial) microbes are responsible for SOM mineralization (Li et al. 2010). Higher mineralization leads to rapid loss of SOM and degradation of soil quality (Mehra et al. 2019). Specifically, legumes have a pertinent role in reducing SOM mineralization through increased C addition (via rapid residue decomposition), aggregation (C encapsulation and protection), and proper supply of economical N addition through BNF (Sainju et al. 2003). Inorganic N rapidly decreases over time, and microbes are in search of another N source to obtain energy (Xiao et al. 2019), which aggravates SOM loss (Lal 2018) and increases gaseous emissions. Wang et al. (2020) reported that the residue addition of maize and wheat under NT decreased C emissions by 14.5% compared with that under conventional tillage in some soils of northwest China. Barton et al. (2014) reported that the inclusion of lupin (*Lupinus angustifolius*) in double-cropping wheat systems reduced C emissions by 56% in some soils of southwest Australia.

11.2.4 SOM Protection from Temperature and Precipitation Driven Changes

SOM dynamic depends on abiotic factors (i.e., sunshine, temperature, precipitation). Thus, changes in abiotic factors directly affect microbial functions (Kan et al. 2020b) and the stability of SOM. Some legume cover crops significantly impact SOM protection through ground cover (Franke et al. 2008) that reduces the hazardous effects of temperature on soil and SOM (Curtin et al. 2000). Legume crops significantly impact SOM protection under climate anomalies. For example, root exudates and mycorrhizae association increase soil aggregation that encapsulates SOM and protects it from further exposure to oxidation (Lehman et al. 2015). Most legume crops are also grown as cover crops, which protect the soil surface from direct solar insolation (Duchene et al. 2017, Veloso et al. 2019) and decrease oxidation (Zhong et al. 2018). It is widely known that the rewetting of soil aggravates SOM loss due to the rapid increase in microbial activities (Bell et al. 2017). But continuous addition of residues (legume or non-legume) significantly improves soil water-holding capacity that may protect SOM through less fluctuations in soil water content (Sainju et al. 2007, Li et al. 2016). In addition, better aggregation significantly contributes to soil water retention and decreases water fluctuations and their effects on microbial activity (Muhammad et al. 2019).

11.2.5 SOM Stabilization and Temporal Variability

Long-term persistence and least variations over time in SOM is a priority in modern agriculture (Xue et al. 2018). The persistence of SOM improves soil biological activity that improves crop growth and development (Yadav et al. 2019b). Legume crops may increase C storage by enhancing the microbial efficiency and stabilizing SOM through plant residue decomposition that is enriched in N and labile C substances (Oelbermann et al. 2015). In addition, labile C sources in residues are responsible for enhancing microbial biomass, which facilitates accumulation of C through mineral-SOM associations (Veloso et al. 2019). The mineral-SOM association is established with the binding of SOM to soil clay and silt particles (Veloso et al. 2018), and the main binding agents come from the decomposed organic residues (Meena and Lal 2018). These associations come into action by microbes (Xing et al. 2019) because microbes are one of the main identified facilitators for the SOM–mineral association, especially under residue addition (Gao et al. 2017). Moreover, legume roots are also responsible for increasing microbial abundance and diversity (Duchene et al. 2017), which significantly increases the macro-aggregate formation, thus storing a higher amount of C (Veloso et al. 2019). The labile SOM fractions are susceptible to a range of factors, such as temperature, precipitation, microbes, and other soil chemical properties (Xue et al. 2018). The mineral-associated SOM is relatively resistant and less affected by climate anomalies and management practices (Koenig and Cochran 1994). Cultivation of legumes can increase mineral-SOM association due to increased microbial abundance and root exudation, can stabilize SOM, and can reduce CO_2 emissions (Rasmussen et al. 2005, Ågren and Wetterstedt 2007, Jones and Singh 2014).

11.3 EFFECTS OF SOM ON SOIL PHYSICO-CHEMICAL PROPERTIES UNDER LEGUMES

Legumes are the renowned facilitator of SOM increment. Previous studies have reported that SOM content declines under conventional cropping systems (Wang et al. 2019a, Yadav et al. 2019a). Legumes improve soil physical properties (Figure 11.1), including bulk density, aggregate stability, and porosity, by increasing SOM and enhancing microbial processes (Mhango et al. 2013, Lira Junior et al. 2020). Legumes can also improve chemical properties such as pH, nutrient cycling, SOC stock, and N-fixation (Zhong et al. 2018) and release some organic acids (Meena and Lal 2018). Pikul et al. (2008) reported that the inclusion of leguminous crops and their residues increases the nutrient availability, such as N, P, and K, and trace elements to the subsequent crop by moderating soil pH; and the decomposition of leguminous crops have positive effects on soil fertility, SOM, and nutrient status (Romaniuk et al. 2018), especially available N-stock for succeeding crops (Hauggaard-Nielsen and Jensen 2005).

11.3.1 Soil Bulk Density and Aeration

A high soil bulk density increases compaction, reduces water infiltration rate, decreases nutrient flow in soil layers, inhibits root system development, and adversely affects plant growth (Torabian et al. 2019). Some legume crops can reduce soil compaction through systematic root development, called "biodrilling." For example, alfalfa (*Medicago sativa* L.) roots are large in diameter, and are long and efficient in creating macrospores in compacted soils (Angers and Caron 1998). Specific root penetration of legumes in compacted soils decreases bulk density and increases macro-aggregate formation and porosity (Rücknagel et al. 2016). Sultani et al. (2007) reported that cultivation and incorporation of sesbania (*Sesbania sesban*) decreased soil bulk density by 7%, increased porosity by 11%, water availability by 17%, and meso/macro-pores by 41% as compared to that of fallow treatment. Moreover, the stability of macro-aggregates is also one of the major factors of reducing soil bulk density (Martins et al. 2012). Legume root exudates and associated microbes play a vital role in aggregate formation and increasing the stability that can significantly decrease soil bulk density (Sainju et al. 2003). McVay et al. (1989) reported that crimson clover (*Trifolium incarnatum*) decreased soil bulk density and increased water infiltration rate compared with soil under wheat and fallow treatments. Legume cultivation rapidly increased SOM (due to rapid decomposition of roots) and maintained the SOM level over time that significantly reduced soil bulk density (Oliveira et al. 2019). Other studies also reported that SOM contents, legume root numbers, and an abundance of earthworms and microbes are significantly correlated with soil bulk density (Wick et al. 2017).

Soil aeration is an important indicator that affects root growth, microbial activity, and biochemical reactions (Curtin et al. 2000). The aeration status can be determined by assessing composition of soil air, O_2 uptake and diffusion rate, and oxidation-reduction potential (Lehman et al. 2015). All these activities are significantly influenced by living roots and growing plants (Lal 2004). Particularly, legumes can improve soil aeration by increasing macropores and water-stable macro-aggregates (Sainju et al. 2003, Ashagrie et al. 2007, Oliveira et al. 2019).

11.3.2 Soil Moisture

Soil moisture is a crop growth-limiting factor (Ellert and Janzen 2008), and water storage capacity is affected by crop, management practice, and environmental conditions (Kan et al. 2020a). Some legumes can be used for soil rehabilitation to improve soil water content through SOM restoration (Franke et al. 2008). Legume roots decompose more quickly than those of cereals due to the low C/N ratio (Wang and Sainju 2014). Therefore, maintenance of SOM improves water-holding capacity (WHC) and water storage (Franzluebbers et al. 1994). Legume cultivation can improve aggregation even in sandy loam soils, by increased microbial abundance, sloughed roots, and other aromatic secretions which reduce water infiltration rate and retain water in the soil for a long period of time (Sultani et al. 2007). Furthermore, addition or incorporation of legume residue significantly improves soil water storage. Sajjad et al. (2019) reported that cowpea (*Vigna unguiculata* L.) residue incorporation increased moisture content by 3.6% and 11.8% at 0–15 cm and 15–30 cm soil layers, respectively. Some data on effects of legumes on soil physical characteristics are presented in Table 11.1.

11.3.3 Soil Nutrient Contents

Soil fertility affects crop growth and development (Xue et al. 2015). Most cereals are nutrient-exhaustive crops (Kumar et al. 2019), but legumes can substantially improve soil nutrients through BNF and other microbial functions (Oelbermann et al. 2015). Rochester et al. (2001) reported that fababean (*Vicia faba* L.) and soybean (*Glycine max* L.) could fix 244–288 kg N ha^{-1}. Root exudates of legumes can improve microbial activities and enhance phosphorus (P) availability (Tang et al. 2014). Legume roots can facilitate plant growth-promoting rhizobacteria (PGPR) and mycorrhizae linkages, which could enhance P availability (Hayman 1986). Legume residue addition significantly enhances the SOM, and the latter enhances soil nutrient concentrations through increased microbial bioturbation (Curtin et al. 2000). Sánchez-Navarro et al. (2019) reported that cowpea inclusion in a vegetable cropping system significantly enhances SOM, N, and P availability compared with mono-cropping in some soils of Brazil. Hossain et al. (2016) reported that the inclusion of legumes in the rice-wheat cropping system with poultry manure addition significantly enhanced soil N, P, K, and

TABLE 11.1
Effects of Legumes on Soil Physical Properties

Crop/Treatment	Soil Depth (cm)	Δ BD (%)[a]	Δ SP (%)[b]	Δ AWC (%)[c]	Reference
Cluster bean vs fallow	0–15	−7.1	~7	12.5	(Sultani et al. 2007)
	15–30	−3.9	10.81	1.2	
Rice bean vs fallow	0–15	−1.7	5.10	0.51	(Sultani et al. 2007)
	15–30	−1.1	4.98	−0.79	
Sesbania vs fallow	0–15	−8.38	13.9	36.36	(Sultani et al. 2007)
	15–30	−6.7	10.81	1.56	
Cowpea vs fallow	-	−1.62	-	11.42	(Rahim et al., 2018)
Mungbean vs fallow	-	−3.25	-	7.79	(Rahim et al., 2018)
Sesbania vs fallow	-	−8.13	-	18.5	(Rahim et al., 2018)

a: soil bulk density; b: soil porosity; c: available water contents.

zinc in some soils of Bangladesh. Similarly, N'Dayegamiye et al. (2015) reported that the inclusion of legumes, in combination with wheat and maize, could reduce the inorganic fertilizer application rate by about 37 kg N ha^{-1} and 51–77 kg N ha^{-1}, respectively, in the soils of eastern Canada through BNF.

Some other aspects of legumes for SOM improvement are also crucial (Meena and Lal 2018). SOM sequestration and improvement are driven by microbes from crop residues, plant roots, and other plant remains (Urquiaga et al. 1998). But the efficient utilization and transformation of crop residues into SOM is mainly done by microbes, and these microbes need N to maintain decomposition services (Zak et al. 2017). In this perspective, legume cultivation is the only economical option to add N into the soil to restore and maintain biological services and SOM (Meena and Lal 2018). As a result of increased SOM content, soil biological and physical properties are enhanced, which ultimately facilitates plant growth (Figure 11.1) (Lal 1989). Latati et al. (2017) reported that intercropping of common bean with maize significantly enhanced SOM stock in some Algerian soils, and increase in SOM stock was positively correlated with nodule carbon and microbial biomass carbon. Thus, a significant increase in soil quality improves nutrient reserves and boosts plant growth by providing sufficient water and nutrients.

11.4 ROLE OF LEGUMES FOR YIELD IMPROVEMENT THROUGH IMPROVED SOIL CONDITIONS

11.4.1 Crop Growth under Improved Soil Conditions

Soil conditions (physical or chemical) affect plant growth in many ways, such as through growth and expansion of roots (Passioura 1991). Some soils are compacted and suppress growth of roots and root hairs thereby limiting uptake of nutrients and water (Kan et al. 2020a). Improved soil structure and macropores enhance contact of roots with solids and liquids, and increase uptake of nutrients and water (Wang et al. 2019b). Growing crops also affect soil conditions either positively or negatively (Talgre et al. 2017). But legumes are known to improve soil structure and chemical properties (Oelbermann et al. 2015). Hauggaard-Nielsen et al. (2009) reported that legume intercropping significantly improves soil nutrient reserves and SOM that positively affect the growth of the subsequent crop. High soil compaction can lead to abnormal root growth even in fertile soil. Roots are unable to absorb nutrients and water efficiently and, thus, adversely affect plant growth (Oelbermann et al. 2015). Well aerated and fertile soils have vigorous root development

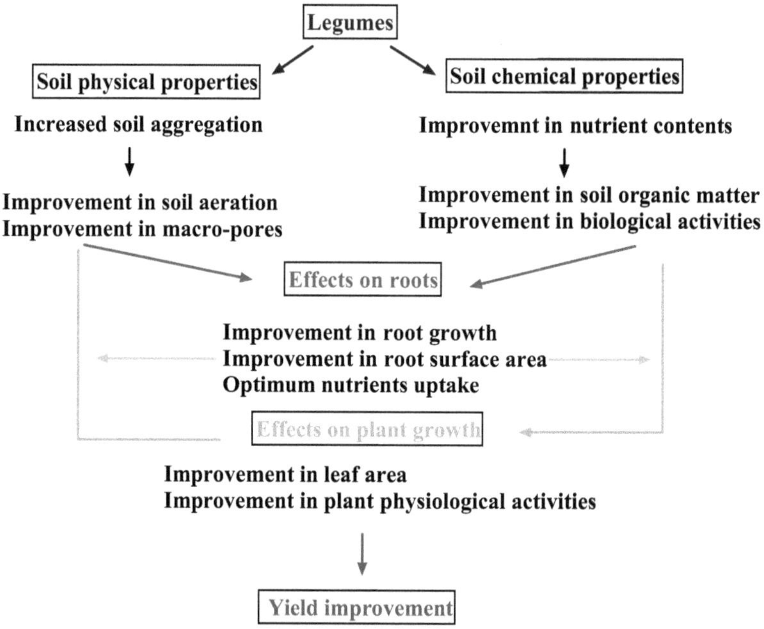

FIGURE 11.2 Role of legumes for improving crop yield through different supportive services.

with a higher root surface area that allows uptake and transport of nutrients to the upper parts of the plant (Kan et al. 2020a). The growing pattern of roots of legumes differs from that of cereals, and it improves soil structure through aggregation and development of macropores (Mhango et al. 2013) and chemical attributes through the increase in SOM, microbial activities, WHC, and nutrient concentrations (Hargrove 1986). Such improvements in soil attributes create a favorable environment for roots of the following crop to absorb nutrients and water and promote respiration that enhances plant growth and other functions (Figure 11.2) (Drinkwater et al. 1998, Alvey et al. 2001). Other crop management practices may also affect soil physical and chemical properties, but have different influences depending upon environmental conditions (Wang et al. 2016). In addition, negative effects of intensive agriculture practices can also be remediated through legume cultivation because legume roots and nodules have higher biomass production with a low C/N ratio that decomposes quickly and restores SOM content and biological activities (Sprunger et al. 2019). Hupe et al. (2018) reported that rhizodeposition from pea (*Pisum sativum* L.) cultivation significantly affects the development of mycorrhizae and other microbial biomasses that could improve soil biological functions and growth of the subsequent crop. Thus, maintenance of SOM for a long period of time improves soil physical and chemical properties that facilitate plant growth by providing a favorable underground environment. Improved soil properties help plant roots to grow faster, with a higher leaf surface area, and provide an optimum amount of nutrients that enhance plant growth and yield.

11.4.2 Relationship between SOM and Crop Yield under Legume Cultivation

Incorporation of legumes in the existing cropping system can increase SOM content, and the restored SOM has several advantages to improve soil fertility and productivity (Sainju et al. 2003). Whereas there are not many reports of direct positive effects of SOM content on crop yields (Zeng et al. 2016, Stagnari et al. 2017, Sánchez-Navarro et al. 2019), several studies have reported indirect influences of SOM to improvement in crop yield (Danga et al. 2009, Zeng et al. 2016). For example, restoration of SOM improves soil physical properties, water retention, and nutrient concentration, while minimizing the risks of topsoil loss via erosion (Jensen et al. 2012). Further, improvement in soil physical properties facilitates crop growth by providing favorable edaphic conditions (Figure 11.2) (Rochester et al. 2001). Maintenance of SOM for a long period of time decreases reliance on inorganic fertilizer and leads to sustainable crop production (Foley et al. 2011). Some studies have reported a positive correlation between SOM content and agronomic yield (Danga et al. 2009, Yusuf et al. 2009, Mucheru-Muna et al. 2010, N'Dayegamiye et al. 2015, Uzoh et al. 2019). Nonetheless, it is unclear how much yield could be improved with unit change in SOM content (Herrick 2000). Because the efficiency of SOM for soil fertility improvement is also affected by management practices, cropping patterns, regional climate, and soil types (Xue et al. 2018). Huang et al. (2013) reported that straw addition could increase SOM and rice yield by 5% in some regions of China.

Mucheru-Muna et al. (2010) reported that inclusion of common bean (*Phaseolus vulgaris* L.), cowpea (*Vigna unguiculata* L.), and groundnut (*Arachis hypogaea* L.) as an intercrop significantly enhanced maize yield (almost 14%) and net profits by 40% in some soils of Kenya. Sánchez-Navarro et al. (2019) hypothesized that increases in the yield of a crop after legume cultivation is mainly due to improved soil fertility, better root distribution, enhanced SOM content, and favorable soil water regime during the growing season. Johansen et al. (1991) reported that maize yield was increased by 17% in maize-soybean rotation compared to that for a continuous maize system, and improvement in maize yield was mainly due to increased soil N content as a result of BNF by soybean. Similarly, Danga et al. (2009) reported that wheat yield was increased about 37% when chickpea (*Cicer arietinum* L.) was incorporated prior to wheat sowing compared to the fallow-wheat rotation system. Njunie et al. (2004) observed that an ample amount of N is added into the soil by legume cultivation or legume residue addition, and an increase in soil N content favors rapid SOM sequestration that improves soil physical, chemical, and biological activities. Thus, the positive effects on subsequent crop yield after legume cultivation are the integration of the whole system rather than a single factor. Other results regarding the impact of legumes on subsequent crop yields are presented in Table 11.2.

More importantly, large variations in regional climatic conditions, topography, and soil types impact plant growth and yield (Zhao et al. 2015). Differences in yield variations might be linked with differences in crop management practices, experimental duration, and legume species (Zeng et al. 2016).

In general, significant impacts of legumes are observed on soil organic C and aggregation after 1–6 years experiment duration (Table 11.3). Data from prior studies indicate that legume inclusion in the existing cropping system significantly improves soil quality and nutrient concentration even over a short time (Uzoh et al. 2019), but their effects on plant growth and yield are observed over a long time period (Zeng et al. 2016).

11.5 LIMITATIONS AND FUTURE PROSPECTS

Maintaining the optimum amount of SOM is difficult in intensive farming practices (Zhao et al. 2020b) because higher inorganic nitrogen application and repeated tillage operations increase oxidation of SOM (Plaza-Bonilla et al. 2014, Li et al. 2018). Moreover, climate anomalies also increase SOM oxidation, especially under higher temperatures and rainfall (Zhao et al. 2020b). Therefore, a severe depletion of SOM adversely affects soil structure, chemical properties, and microbial activities (Chen et al. 2014) and exacerbates negative impacts on yield stability and sustainability (Zhao et al. 2015). In this context, incorporation of legume in the rotation cycle can protect SOM from oxidation and microbial processes through increased aggregation, SOM–mineral association, and rapid addition of SOM via decomposition of root/crop residues (Veloso et al. 2018, 2019). The intensity of SOM accrual under different legume species has not been precisely quantified (Stagnari et al. 2017). The root exudation capacity of legume could be changed with respect to the cultivated environment because of differences in root development and growth in different soils (Graham 1991) which affect aggregation and microbial community development and, thus, change plant growth and yield (Xiao et al. 2019). Whereas quantification of root exudates from legumes under field conditions is

TABLE 11.2
Effects of Legume Inclusion on Subsequent Crop Yields

Legume Species	Compared Treatment/ Cropping System	Agronomic Approach	Effects on Yield	Regions	Experiment Duration (Years)	Reference
Wheat-mung bean	Wheat-fallow	CR[a]	−27%	China Loess Plateau	1	(Li et al. 2011)
Wheat-soybean	Wheat-fallow	CR	−12%	China Loess Plateau	1	(Li et al. 2011)
Wheat-qiubean	Wheat-fallow	CR	−10%	China Loess Plateau	1	(Li et al. 2011)
Wheat-mung bean	Wheat-fallow	CR	−4%	China Loess Plateau	2	(Yang et al. 2012)
Wheat-soybean	Wheat-fallow	CR	23%	China Loess Plateau	2	(Yang et al. 2012)
Wheat-qiubean	Wheat-fallow	CR	-7%	China Loess Plateau	2	(Yang et al. 2012)
Wheat-changwuhuai bean	Wheat-fallow	CR	30%	China Loess Plateau	3	(Yao et al. 2014)
Wheat-soybean	Wheat-fallow	CR	23%	China Loess Plateau	3	Yao et al. 2014)
Wheat-mung bean	Wheat-fallow	CR	16%	China Loess Plateau	3	Yao et al. 2014)
Oat-soybean	Sole oat	IC[b]	−51%	Northeast China	1	(Feng et al. 2015)
Oat-peanut	Sole oat	IC	−41%	Northeast China	1	(Feng et al. 2015)
Oat-soybean	Sole oat	IC	−43%	Northeast China	2	(Feng et al. 2015)
Oat-peanut	Sole oat	IC	3%	Northeast China	2	(Feng et al. 2015)
Tomato-kidney bean	Sole tomato	IC	16%	East China	1	(Dai et al. 2015)
Tomato-common bean	Sole tomato	IC	4%	East China	1	(Dai et al. 2015)
Maize-soybean	Sole maize	IC	−30%	Southwest China	1	(Liu and Li 2017)
Millet-mung bean	Sloe millet	IC	−3%	Northeast China	1	(Xue et al. 2019)
Maize-soybean	Sole maize	IC	−11%	Southwest China	1	(Zhao et al. 2020a)
Maize-soybean	Sole maize	IC	3–5%			(Sekamatte et al. 2003)
Wheat-Mungbean	Wheat-sorghum	CR	8%	Northwest Frontier Province (NWFP), Pakistan	4	(Shah et al. 2003)
Wheat-mung bean/ mash bean	Wheat-sorghum	CR	20%	Ugnada (East Africa)	1	(Hayat et al. 2008)
Wheat-mung bean	Wheat-rice	RA[c]	21%	Punjab Pakistan	1	(Yaqub et al. 2010)
Maize-cowpea	Sole maize	CR	17%	Kenya	5	(Rao and Mathuva 2000)
Maize-pigeon pea	Sole maize	CR	24%	Kenya	5	(Rao and Mathuva 2000)
Wheat-Sunn-hemp	Wheat-Fallow	RA	21%	Central Alabama	2	(Balkcom and Reeves 2005)
Sugarcane-cowpea	sugarcane-Fallow	RA	26%	Lakeland USA	2	(White et al. 2020)
Maize-hairy vetch	no cover crop	CR	14%	Carbondale USA	4	(Singh et al. 2020)
Maize-Velvet bean	Maize-maize	CR	>100%	Oyo state Nigeria	2	(Uzoh et al. 2019)
Maize-cowpea	Maize-maize	CR	36%	Oyo state Nigeria	2	(Uzoh et al. 2019)
Maize-soybean	Maize-maize	CR	24%	Oyo state Nigeria	2	(Uzoh et al. 2019)

a: crop rotation; b: intercropping; c: residue application.

TABLE 11.3
Effects of Legume Inclusion on Soil Aggregation and Organic Carbon in Different Regions

Legume Species	Compared Treatment/ Cropping System	Agronomic Approach	Soil Depth	Effects on Aggregation and Soil Organic Carbon	Regions	Experiment Duration (Years)	Reference
Maize-soybean	Sole soybean	Intercropping	0–20 cm	16% $AS^{a}_{>2mm}$ 28% AS_{1-2mm} 15% $AS_{0.5-1mm}$ 12% $AS_{0.25-0.5mm}$ 5% $AS_{0.106-0.25}$	Southwest China	1	(Wang et al. 2020)
Maize-soybean	Sole soybean	Intercropping	20–40 cm	31% $AS_{>2mm}$ 2% AS_{1-2mm} 14% $AS_{0.5-1mm}$ 29% $AS_{0.25-0.5mm}$ 34% $AS_{0.106-0.26}$	Southwest China	1	(Wang et al. 2020)
Apple-*Trifolium repens*	Sole apple	Intercropping	0–5 cm	15% $AS_{>2mm}$ –2% AS_{1-2mm} 14% $AS_{0.5-1mm}$ 36% $AS_{0.25-0.5mm}$	China Loess Plateau	6	(Wang et al. 2013)
Apple-*Trifolium repens*	Sole apple	Intercropping	5–10 cm	88% $AS_{>2mm}$ 104% AS_{1-2mm} 69% $AS_{0.5-1mm}$ 25% $AS_{0.25-0.5mm}$	China Loess Plateau	6	(Wang et al. 2013)
Apple-*Trifolium repens*	Sole apple	Intercropping	10–20 cm	15% $AS_{>2mm}$ 22% AS_{1-2mm} –9% $AS_{0.5-1mm}$ 41% $AS_{0.25-0.5mm}$	China Loess Plateau	6	(Wang et al. 2013)
Apple-*Trifolium repens*	Sole apple	Intercropping	20–40 cm	15% $AS_{>2mm}$ 88% AS_{1-2mm} 15% $AS_{0.5-1mm}$ –4% $AS_{0.25-0.5mm}$	China Loess Plateau	6	(Wang et al. 2013)
Maize-pea	Maize-fallow	Crop rotation	0–20 cm	5%[b]	China Loess Plateau	1	(Wei et al. 2019)
Maize-pea	Maize-fallow	Crop rotation	20–40 cm	12%	China Loess Plateau	1	(Wei et al. 2019)
Maize-broad bean	Maize-fallow	Crop rotation	0–20 cm	2%	China Loess Plateau	1	(Wei et al. 2019)
Maize-broad bean	Maize-fallow	Crop rotation	20–40 cm	6%	China Loess Plateau	1	(Wei et al. 2019)
Maize-soybean	Maize-fallow	Crop rotation	0–20 cm	6%	China Loess Plateau	1	(Wei et al. 2019)
Maize-soybean	Maize-fallow	Crop rotation	20–40 cm	–4%	China Loess Plateau	1	(Wei et al. 2019)
Maize-alfalfa	Maize-fallow	Crop rotation	0–20 cm	5%	China Loess Plateau	1	(Wei et al. 2019)
Maize-alfalfa	Maize-fallow	Crop rotation	20–40 cm	12%	China Loess Plateau	1	(Wei et al. 2019)
Maize-soybean	Maize-maize	Crop rotation	0–20 cm	2%	Northeast China	2	(Yang et al. 2019)

a: aggregate size; b: soil organic carbon.

a daunting challenge, it can be determined in a pot experiment to understand the behavior of root exudates on soil structure and stability as well as on the microbial community. Climate change has severe effects on global agriculture, but the role of legumes on SOM protection under different climates (especially under temperature extremes) has not been widely assessed. These knowledge gaps should be addressed with regard to yield stability. Indeed, SOM stabilization leads to yield stabilization, which could play an important role in mitigating food insecurity (Lal 2010, 2018). Impacts of legume intercropping in different crop combinations with various soil characteristics are still scarce; this should be addressed in future research to know the specific role of legume species in combination with other crops on soil physico-chemical characteristics.

11.6 CONCLUSIONS

The available information in the literature supports the conclusion that legume cultivation stabilizes SOM content and has positive effects on subsequent crop yield.

- Legume cultivation improves soil aggregation through root exudates and leads to a rapid addition of SOM from root/shoot decomposition.
- Legume plants increase SOM–mineral association and, thus, protect SOM from oxidation.
- Increased aggregation and SOM–mineral association reduces C mineralization and CO_2 emissions.
- Some legume species, such as alfalfa, reduce soil compaction through increased root penetration.
- Increment in SOM improves soil physical (aggregation and stability, WHC, aeration), biological attributes (microbial abundance and diversity), and chemical properties (N addition via BNF) which enhance plant growth and yield.
- Further strategic inclusion of legumes in existing cropping systems can improve soil health, crop yield, and mitigate food insecurity.

REFERENCES

Ågren, G. I. and J. Å. M. Wetterstedt. 2007. What determines the temperature response of soil organic matter decomposition? *Soil Biology and Biochemistry* 39(7):1794–1798.

Allison, S. D., M. D. Wallenstein, and M. A. Bradford. 2010. Soil-carbon response to warming dependent on microbial physiology. *Nature Geoscience* 3(5):336–340.

Alvey, S., M. Bagayoko, G. Neumann, and A. Buerkert. 2001. Cereal/legume rotations affect chemical properties and biological activities in two West African soils. *Plant and Soil* 231(1):45–54.

Angers, D. A. and J. Caron. 1998. Plant-induced changes in soil structure: Processes and feedbacks. *Biogeochemistry* 42(1/2):55–72.

Ashagrie, Y., W. Zech, G. Guggenberger, and T. Mamo. 2007. Soil aggregation, and total and particulate organic matter following conversion of native forests to continuous cultivation in Ethiopia. *Soil and Tillage Research* 94(1):101–108.

Ayres, E., K. M. Dromph, R. Cook, N. Ostle, and R. D. Bardgett. 2007. The influence of below-bround herbivory and fefoliation of a legume on nitrogen transfer to neighbouring plants. *Functional Ecology* 21(2):256–263.

Badagliacca, G., E. Benítez, G. Amato, L. Badalucco, D. Giambalvo, V. A. Laudicina, and P. Ruisi. 2018. Long-term no-tillage application increases soil organic carbon, nitrous oxide emissions and faba bean (Vicia faba L.) yields under rain-fed Mediterranean conditions. *Science of the Total Environment* 639:350–359.

Badawi, F. S. F., A. M. M. Biomy, and A. H. Desoky. 2011. Peanut plant growth and yield as influenced by co-inoculation with Bradyrhizobium and some rhizo-microorganisms under sandy loam soil conditions. *Annals of Agricultural Sciences* 56(1):17–25.

Bais, H. P., T. L. Weir, L. G. Perry, S. Gilroy, and J. M. Vivanco. 2006. The role of root exudates in rhizosphere interactions with plants and other organisms. *Annual Review of Plant Biology* 57:233–266.

Balkcom, K. S. and D. W. Reeves. 2005. Sunn-hemp utilized as a legume cover crop for corn production. *Agronomy Journal* 97(1):26–31.

Barton, L., K. Butterbach-Bahl, R. Kiese, and D. V. Murphy. 2011. Nitrous oxide fluxes from a grain-legume crop (narrow-leafed lupin) grown in a semiarid climate. *Global Change Biology* 17(2):1153–1166.

Barton, L., T. Thamo, D. Engelbrecht, and W. K. Biswas. 2014. Does growing grain legumes or applying lime cost effectively lower greenhouse gas emissions from wheat production in a semi-arid climate? *Journal of Cleaner Production* 83:194–203.

Barton, L., F. C. Hoyle, K. T. Stefanova, and D. V. Murphy. 2016. Incorporating organic matter alters soil greenhouse gas emissions and increases grain yield in a semi-arid climate. *Agriculture, Ecosystems and Environment* 231:320–330.

Bell, L. W., J. Lawrence, B. Johnson, and M. B. Peoples. 2017. New ley legumes increase nitrogen fixation and availability and grain crop yields in subtropical cropping systems. *Crop and Pasture Science* 68(1):11–26.

Cates, A. M., M. D. Ruark, J. L. Hedtcke, and J. L. Posner. 2016. Long-term tillage, rotation and perennialization effects on particulate and aggregate soil organic matter. *Soil and Tillage Research* 155:371–380.

Chauhan, B. S., G. Mahajan, V. Sardana, J. Timsina, and M. L. Jat. 2012. Productivity and sustainability of the rice–wheat cropping system in the Indo-Gangetic Plains of the Indian subcontinent: Problems, opportunities, and strategies. *Advances in Agronomy* 117:315–369.

Chen, L., J. Zhang, B. Zhao, P. Yan, G. Zhou, and X. Xin. 2014. Effects of straw amendment and moisture on microbial communities in Chinese fluvo-aquic soil. *Journal of Soils and Sediments* 14(11):1829–1840.

Curtin, D., H. Wang, F. Selles, R. P. Zentner, V. O. Biederbeck, and C. A. Campbell. 2000. Legume green manure as partial fallow replacement in semiarid Saskatchewan: Effect on carbon fluxes. *Canadian Journal of Soil Science* 80(3):499–505.

Dai, H. H., X. F. Hu, M. Y. Cao, M. Y. Yang, and J. Wang. 2015. Effects of intercropping with leguminous crops on tomato yield, soil nutrient and enzyme activity. *Acta Pedologica Sinica* 52:911–918.

Danga, B. O., J. P. Ouma, I. I. C. Wakindiki, and A. Bar-Tal. 2009. Legume-wheat rotation effects on residual soil moisture, nitrogen and wheat yield in tropical regions. *Advances in Agronomy* 101:315–349.

Douglas, C. L., R. R. Allmaras, P. E. Rasmussen, R. E. Ramig, and N. C. Roager. 1980. Wheat straw composition and placement effects on decomposition in dryland agriculture of the Pacific Northwest. *Soil Science Society of America Journal* 44(4):833–837.

Drinkwater, L. E., P. Wagoner, and M. Sarrantonio. 1998. Legume-based cropping systems have reduced carbon and nitrogen losses. *Nature* 396(6708):262–265.

Duchene, O., J. F. Vian, and F. Celette. 2017. Intercropping with legume for agroecological cropping systems: Complementarity and facilitation processes and the importance of soil microorganisms. A review. *Agriculture, Ecosystems and Environment* 240:148–161.

Ellert, B. H. and H. H. Janzen. 2008. Nitrous oxide, carbon dioxide and methane emissions from irrigated cropping systems as influenced by legumes, manure and fertilizer. *Canadian Journal of Soil Science* 88(2):207–217.

Feng, X. M., Y. Yang, C. Z. Ren, Y. G. Hu, and Z. H. Zeng. 2015. Effects of legumes intercropping with oat on photosynthesis characteristics and Grain Yield. *Acta Agronomica Sinica* 41(9):1426–1434.

Foley, J. A., N. Ramankutty, K. A. Brauman, E. S. Cassidy, J. S. Gerber, M. Johnston, N. D. Mueller, C. O'Connell, D. K. Ray, P. C. West, C. Balzer, E. M. Bennett, S. R. Carpenter, J. Hill, C. Monfreda, S. Polasky, J. Rockström, J. Sheehan, S. Siebert, D. Tilman, and P. M. Zaks. 2011. Solutions for a cultivated planet. *Nature* 478(7369):337–342.

Franke, A. C., G. Laberge, B. D. Oyewole, and S. Schulz. 2008. A comparison between legume technologies and fallow, and their effects on maize and soil traits, in two distinct environments of the West African savannah. *Nutrient Cycling in Agroecosystems* 82(2):117–135.

Franzluebbers, K., R. W. Weaver, A. S. R. Juo, and A. J. Franzluebbers. 1994. Carbon and nitrogen mineralization from cowpea plants part decomposing in moist and in repeatedly dried and wetted soil. *Soil Biology and Biochemistry* 26(10):1379–1387.

Gao, D., X. Wang, S. Fu, and J. Zhao. 2017. Legume plants enhance the resistance of soil to ecosystem disturbance. *Frontiers in Plant Science* 8:1295.

Graham, T. L. 1991. Flavonoid and isoflavonoid distribution in developing soybean seedling tissues and in seed and root exudates. *Plant Physiology* 95(2):594–603.

Grandy, A. S. and G. P. Robertson. 2007. Land-use intensity effects on soil organic carbon accumulation rates and mechanisms. *Ecosystems* 10:58–73.

Hargrove, W. L. 1986. Winter Legumes as a nitrogen source for no-till grain sorghum 1. *Agronomy Journal* 78(1):70–74.

Hauggaard-Nielsen, H. and E. S. Jensen. 2005. Facilitative root interactions in intercrops. In *Root Physiology: From Gene to Function*, ed. H. Lambers and T. D. Colmers, 237–250. Springer: Dordrecht.

Hauggaard-Nielsen, H., M. Gooding, P. Ambus, G. Corre-Hellou, Y. Crozat, C. Dahlmann, A. Dibet, P. von Fragstein, A. Pristeri, M. Monti, and E. S. Jensen. 2009. Pea-barley intercropping for efficient symbiotic N2-fixation, soil N acquisition and use of other nutrients in European organic cropping systems. *Field Crops Research* 113(1):64–71.

Hayat, R., S. Ali, M. T. Siddique, and T. H. Chatha. 2008. Biological nitrogen fixation of summer legumes and their residual effects on subsequent rainfed wheat yield. *Pakistan Journal of Botany* 40:711–722.

Hayman, D. S. 1986. Mycorrhizae of nitrogen-fixing legumes. *Mircen: Journal of Applied Microbiology and Biotechnology* 2(1):121–145.

Hazra, K. K., M. S. Venkatesh, P. K. Ghosh, A. N. Ganeshamurthy, N. Kumar, N. Nadarajan, and A. B. Singh. 2014. Long-term effect of pulse crops inclusion on soil–plant nutrient dynamics in puddled rice (*Oryza sativa* L.)-wheat (*Triticum aestivum* L.) cropping system on an inceptisol of Indo-Gangetic plain zone of India. *Nutrient Cycling in Agroecosystems* 100(1):95–110.

Herrick, J. E. 2000. Soil quality: An indicator of sustainable land management? *Applied Soil Ecology* 15(1):75–83.

Hossain, M. S., A. Hossain, M. A. R. Sarkar, M. Jahiruddin, J. A. Teixeira da Silva, and M. I. Hossain. 2016. Productivity and soil fertility of the rice-wheat system in the high Ganges river floodplain of Bangladesh is influenced by the inclusion of legumes and manure. *Agriculture, Ecosystems and Environment* 218:40–52.

Huang, S., Y. Zeng, J. Wu, Q. Shi, and X. Pan. 2013. Effect of crop residue retention on rice yield in China: A meta-analysis. *Field Crops Research* 154:188–194.

Hupe, A., H. Schulz, C. Bruns, T. Haase, J. Heß, R. G. Joergensen, and F. Wichern. 2018. Even flow? Changes of carbon and nitrogen release from pea roots over time. *Plant and Soil* 431(1–2):143–157.

Jensen, E. S., M. B. Peoples, R. M. Boddey, P. M. Gresshoff, H. Hauggaard-Nielsen, B. J. R. Alves, and M. J. Morrison. 2012. Legumes for mitigation of climate change and the provision of feedstock for biofuels and biorefineries. A review. *Agronomy for Sustainable Development* 32(2):329–364.

Johansen, C., K. K. Lee, and K. L. Sahrawat. 1991. *Phosphorus Nutrition of Grain Legumes in the Semi-Arid Tropics*. International Crops Research Institute for the Semi-Arid Tropics: Patancheru, Andhra Pradesh, India.

Johansson, J. F., L. R. Paul, and R. D. Finlay. 2004. Microbial interactions in the mycorrhizosphere and their significance for sustainable agriculture. *FEMS Microbiology Ecology* 48(1):1–13.

Jones, E. and B. Singh. 2014. Organo-mineral interactions in contrasting soils under natural vegetation. *Frontiers in Environmental Science* 2:2.

Kan, Z. R., A. L. Virk, G. Wu, J. Y. Qi, S. T. Ma, X. Wang, X. Zhao, R. Lal, and H. L. Zhang. 2020a. Priming effect intensity of soil organic carbon mineralization under no-till and residue retention. *Applied Soil Ecology* 147:103–115.

Kan, Z. R., Q. Y. Liu, C. He, Z. H. Jing, A. L. Virk, J. Y. Qi, X. Zhao, and H. L. Zhang. 2020b. Responses of grain yield and water use efficiency of winter wheat to tillage in the North China Plain. *Field Crops Research* 249:107–116.

Kan, Z. R., Q. Y. Liu, G. Wu, S. T. Ma, A. L. Virk, J. Y. Qi, X. Zhao, and H. L. Zhang. 2020c. Temperature and moisture driven changes in soil carbon sequestration and mineralization under biochar addition. *Journal of Cleaner Production* 265:1–11.

Kitonyo, O. M., V. O. Sadras, Y. Zhou, and M. D. Denton. 2018. Nitrogen fertilization modifies maize yield response to tillage and stubble in a sub-humid tropical environment. *Field Crops Research* 223:113–124.

Koenig, R. T. and V. L. Cochran. 1994. Decomposition and nitrogen mineralization from legume and non-legume crop residues in a subarctic agricultural soil. *Biology and Fertility of Soils* 17(4):269–275.

Kubar, K. A., L. Huang, J. Lu, X. Li, B. Xue, and Z. Yin. 2019. Long-term tillage and straw returning effects on organic C fractions and chemical composition of SOC in rice-rape cropping system. *Archives of Agronomy and Soil Science* 65(1):125–137.

Kumar, N., C. P. Nath, K. K. Hazra, K. Das, M. S. Venkatesh, M. K. Singh, S. S. Singh, C. S. Praharaj, and N. P. Singh. 2019. Impact of zero-till residue management and crop diversification with legumes on soil aggregation and carbon sequestration. *Soil and Tillage Research* 189:158–167.

Lal, R. 1989. Conservation tillage for sustainable agriculture: Tropics Versus temperate environments. *Advances in Agronomy* 42:85–197.

Lal, R. 2004. Soil carbon sequestration to mitigate climate change. *Geoderma* 123(1–2):1–22.

Lal, R. 2010. Beyond Copenhagen: Mitigating climate change and achieving food security through soil carbon sequestration. *Food Security* 2(2):169–177.

Lal, R. 2018. Digging deeper: A holistic perspective of factors affecting soil organic carbon sequestration in agroecosystems. *Global Change Biology* 24(8):3285–3301.

Latati, M., A. Aouiche, S. Tellah, A. Laribi, S. Benlahrech, G. Kaci, F. Ouarem, and S. M. Ounane. 2017. Intercropping maize and common bean enhances microbial carbon and nitrogen availability in low phosphorus soil under Mediterranean conditions. *European Journal of Soil Biology* 80:9–18.

Legay, N., C. Baxendale, K. Grigulis, U. Krainer, E. Kastl, M. Schloter, R. D. Bardgett, C. Arnoldi, M. Bahn, M. Dumont, F. Poly, T. Pommier, J. C. Clément, and S. Lavorel. 2014. Contribution of above- and below-ground plant traits to the structure and function of grassland soil microbial communities. *Annals of Botany* 114(5):1011–1021.

Lehman, R., C. Cambardella, D. Stott, V. Acosta-Martinez, D. Manter, J. Buyer, J. Maul, J. Smith, H. Collins, J. Halvorson, R. Kremer, J. Lundgren, T. Ducey, V. Jin, and D. Karlen. 2015. Understanding and enhancing soil biological health: The solution for reversing soil degradation. *Sustainability* 7(1):988–1027.

Li, H., M. Dai, S. Dai, and X. Dong. 2018. Current status and environment impact of direct straw return in China's cropland: A review. *Ecotoxicology and Environmental Safety* 159:293–300.

Li, K. Y., Z. H. Wang, H. B. Zhao, N. Zhao, Y. J. Gao, and L. Graham. 2011. Effect of rotation with legumes and N fertilization on yield and grain nutrient contents of wheat in dryland of the Loess Plateau. *Agricultural Research in the Arid Area* 29:110–123.

Li, P., C. Dai, X. Wang, T. Zhang, and Y. Chen. 2012. Variation of soil enzyme activities and microbial community structure in peanut monocropping system in subtropical China. *African Journal of Agricultural Research* 7:1870–1879.

Li, Q., P. Yu, G. Li, and D. Zhou. 2016. Grass–legume ratio can change soil carbon and nitrogen storage in a temperate steppe grassland. *Soil and Tillage Research* 157:23–31.

Li, Z. P., C. W. Han, and F. X. Han. 2010. Organic C and N mineralization as affected by dissolved organic matter in paddy soils of subtropical China. *Geoderma* 157(3–4):206–213.

Lira, M. A., Junior, F. J. C. Fracetto, J. da S. Ferreira, M. B. Silva, and G. G. M. Fracetto. 2020. Legume-based silvopastoral systems drive C and N soil stocks in a subhumid tropical environment. *Catena* 189 104508.

Liu, C. M. and C. Y. Li. 2017. Effects of maize intercropping with soybean and potato on maize leaf senescence, yield, and disease control. *Journal of Jiangsu Agricultural Science* 45:75–78.

Lupwayi, N. Z., A. C. Kennedy, and R. M. Chirwa. 2011. Grain legume impacts on soil biological processes in sub-Saharan Africa. *African Journal of Plant Science* 5:1–7.

Martins, M. dos R., D. A. Angers, J. E. Corá. 2012. Carbohydrate composition and water-stable aggregation of an oxisol as affected by crop sequence under no-till. *Soil Science Society of America Journal* 76(2):475–484.

McVay, K. A., D. E. Radcliffe, and W. L. Hargrove. 1989. Winter legume effects on soil properties and nitrogen fertilizer requirements. *Soil Science Society of America Journal* 53(6):1856–1862.

Meena, B. L., R. K. Fagodiya, K. Prajapat, M. L. Dotaniya, M. J. Kaledhonkar, P. C. Sharma, R. S. Meena, T. Mitran, and S. Kumar. 2018. Legume green manuring: An option for soil sustainability. In *Legumes for Soil Health and Sustainable Management*, ed. R. Meena, A. Das, G. Yadav, and R. Lal, 387–408. Springer: Singapore.

Meena, R. S. and R. Lal. 2018. Legumes for soil health and sustainable management. In *Legumes for Soil Health and Sustainable Management*, ed. R. Meena, A. Das, G. Yadav, and R. Lal, 3–31. Springer: Singapore.

Mehra, P., B. Sarkar, N. Bolan, S. Chowdhury, and J. Desbiolles. 2019. Impact of carbonates on the mineralisation of surface soil organic carbon in response to shift in tillage practice. *Geoderma* 339:94–105.

Mhango, W. G., S. S. Snapp, and G. Y. K. Phiri. 2013. Opportunities and constraints to legume diversification for sustainable maize production on smallholder farms in Malawi. *Renewable Agriculture and Food Systems* 28(3):234–244.

Mucheru-Muna, M., P. Pypers, D. Mugendi, J. Kung'u, J. Mugwe, R. Merckx, B. Vanlauwe. 2010. A staggered maize-legume intercrop arrangement robustly increases crop yields and economic returns in the highlands of Central Kenya. *Field Crops Research* 115(2):132–139.

Muhammad, I., U. M. Sainju, F. Zhao, A. Khan, R. Ghimire, X. Fu, and J. Wang. 2019. Regulation of soil CO_2 and N_2O emissions by cover crops: A meta-analysis. *Soil and Tillage Research* 192:103–112.

N'Dayegamiye, A., J. K. Whalen, G. Tremblay, J. Nyiraneza, M. Grenier, A. Drapeau, and M. Bipfubusa. 2015. The benefits of legume crops on corn and wheat yield, nitrogen nutrition, and soil properties improvement. *Agronomy Journal* 107(5):1653–1665.

Njunie, M. N., M. G. Wagger, and P. Luna-Orea. 2004. Residue decomposition and nutrient release dynamics from two tropical forage legumes in a Kenyan environment. *Agronomy Journal* 96(4):1073–1081.

Nyawade, S. O., N. N. Karanja, C. K. K. Gachene, H. I. Gitari, E. Schulte-Geldermann, and M. L. Parker. 2019. Short-term dynamics of soil organic matter fractions and microbial activity in smallholder potato-legume intercropping systems. *Applied Soil Ecology* 142:123–135.

Oelbermann, M., A. Regehr, and L. Echarte. 2015. Changes in soil characteristics after six seasons of cereal-legume intercropping in the Southern Pampa. *Geoderma Regional* 4:100–107.

Oliveira, M., P. Barré, H. Trindade, and I. Virto. 2019. Different efficiencies of grain legumes in crop rotations to improve soil aggregation and organic carbon in the short-term in a sandy cambisol. *Soil and Tillage Research* 186:23–35.

Passioura, J. B. 1991. Soil structure and plant growth. *Australian Journal of Soil Research* 29(6):717–728.

Paustian, K., J. Lehmann, S. Ogle, D. Reay, G. P. Robertson, and P. Smith. 2016. Climate-smart soils. *Nature* 532(7597) :49–57.

Pikul, J. L., J. M. F. Johnson, T. E. Schumacher, M. Vigil, and W. E. Riedell. 2008. Change in surface soil carbon under rotated corn in Eastern South Dakota. *Soil Science Society of America Journal* 72(6):1738–1744.

Plaza-Bonilla, D., J. Álvaro-Fuentes, and C. Cantero-Martínez. 2014. Identifying soil organic carbon fractions sensitive to agricultural management practices. *Soil and Tillage Research* 139:19–22.

Pu, C., Z. R. Kan, P. Liu, S. T. Ma, J. Y. Qi, X. Zhao, and H. L. Zhang. 2019. Residue management induced changes in soil organic carbon and total nitrogen under different tillage practices in the North China Plain. *Journal of Integrative Agriculture* 18(6):1337–1347.

Qi, J. Y., X. Z. Zhang, S. S. Li, A. L. Virk, X. Wang, Z. R. Kan, X. Zhao, X. P. Xiao, and H. L. Zhang. 2020. Effects of different tillage practices on the distribution of soil total nitrogen and carbon/nitrogen ratio at different soil depths in a double rice cropping system. *Archives of Agronomy and Soil Science* 67(5):714–725. DOI: 10.1080/03650340.2020.1752369.

Rahim, H., M. I. Ahmad, M. Arif, Z. Rahim, S. Ahmad, Z. Khan, K. Z. Laiq, M. Ayoub, and M. Haris. 2018. Residual effect of biochar and summer legumes on soil physical properties and wheat growth. *Pure and Applied Biology* 7(4):16–26.

Rao, M. R. and M. N. Mathuva. 2000. Legumes for improving maize yields and income in semi-arid Kenya. *Agriculture, Ecosystems and Environment* 78(2):123–137.

Rasmussen, C., M. S. Torn, and R. J. Southard. 2005. Mineral assemblage and aggregates control carbon dynamics in a California conifer forest. *Soil Science Society of America Journal* 69(6):1711–1721.

Rejili, M., M. Mahdhi, A. Fterich, S. Dhaoui, I. Guefrachi, R. Abdeddayem, and M. Mars. 2012. Symbiotic nitrogen fixation of wild legumes in Tunisia: Soil fertility dynamics, field nodulation and nodules effectiveness. *Agriculture, Ecosystems and Environment* 157:60–69.

Rochester, I. J., M. B. Peoples, N. R. Hulugalle, R. R. Gault, and G. A. Constable. 2001. Using legumes to enhance nitrogen fertility and improve soil condition in cotton cropping systems. *Field Crops Research* 70(1):27–41.

Romaniuk, R., M. Beltrán, L. Brutti, A. Costantini, S. Bacigaluppo, H. Sainz-Rozas, and F. Salvagiotti. 2018. Soil organic carbon, macro- and micronutrient changes in soil fractions with different lability in response to crop intensification. *Soil and Tillage Research* 181:136–143.

Rücknagel, J., P. Götze, B. Koblenz, N. Bachmann, S. Löbner, S. Lindner, J. Bischoff, and O. Christen. 2016. Impact on soil physical properties of using large-grain legumes for catch crop cultivation under different tillage conditions. *European Journal of Agronomy* 77:28–37.

Sainju, U. M., W. F. Whitehead, and B. P. Singh. 2003. Cover crops and nitrogen fertilization effects on soil aggregation and carbon and nitrogen pools. *Canadian Journal of Soil Science* 83(2):155–165.

Sainju, U. M., H. H. Schomberg, B. P. Singh, W. F. Whitehead, P. G. Tillman, and S. L. Lachnicht-weyers. 2007. Cover crop effect on soil carbon fractions under conservation tillage cotton. *Soil and Tillage Research* 96(1–2):205–218.

Sajjad, M. R., R. Bibi, U. Adnan, A. Afzal, A. Ali Arshad, and T. Rafique. 2019. Performance of green manuring for soil health and crop yield improvement. *Pure and Applied Biology* 8(2):1543–1553.

Sánchez-Navarro, V., R. Zornoza, Á. Faz, and J. A. Fernández. 2019. Comparing legumes for use in multiple cropping to enhance soil organic carbon, soil fertility, aggregates stability and vegetables yields under semi-arid conditions. *Scientia Horticulturae* 246:835–841.

Schjønning, P., L. W. de Jonge, L. J. Munkholm, P. Moldrup, B. T. Christensen, and J. E. Olesen. 2012. Clay dispersibility and soil friability-testing the soil clay-to-carbon saturation concept. *Vadose Zone Journal* 11(1):67.

Sekamatte, B. M., M. Ogenga-Latigo, and A. Russell-Smith. 2003. Effects of maize-legume intercrops on termite damage to maize, activity of predatory ants and maize yields in Uganda. *Crop Protection* 22(1):87–93.

Shah, Z., S. H. Shah, M. B. Peoples, G. D. Schwenke, and D. F. Herridge. 2003. Crop residue and fertiliser N effects on nitrogen fixation and yields of legume-cereal rotations and soil organic fertility. *Field Crops Research* 83(1):1–11.

Singh, G., A. D. G. M. Thilakarathne, K. W. J. Williard, J. E. Schoonover, R. L. Cook, K. L. Gage, and R. McElroy. 2020. Tillage and legume non-legume cover cropping effects on corn–soybean production. *Agronomy Journal* 112(4):2636–2648.

Sprunger, C. D., S. W. Culman, C. A. Palm, M. Thuita, and B. Vanlauwe. 2019. Long-term application of low C:N residues enhances maize yield and soil nutrient pools across Kenya. *Nutrient Cycling in Agroecosystems* 114(3):261–276.

Stagnari, F., A. Maggio, A. Galieni, and M. Pisante. 2017. Multiple benefits of legumes for agriculture sustainability: An overview. *Chemical and Biological Technologies in Agriculture* 4(1):2.

Stockmann, U., M. A. Adams, J. W. Crawford, D. J. Field, N. Henakaarchchi, M. Jenkins, B. Minasny, A. B. McBratney, V. de R. de Courcelles, K. Singh, I. Wheeler, L. Abbott, D. A. Angers, J. Baldock, M. Bird, P. C. Brookes, C. Chenu, J. D. Jastrow, R. Lal, J. Lehmann, A. G. O'Donnell, W. J. Parton, D. Whitehead, and M. Zimmermann. 2013. The knowns, known unknowns and unknowns of sequestration of soil organic carbon. *Agriculture, Ecosystems and Environment* 164 :80–99.

Sultani, M. I., M. A. Gill, M. M. Anwar, and M. Athar. 2007. Evaluation of soil physical properties as influenced by various green manuring legumes and phosphorus fertilization under rain fed conditions. *International Journal of Environmental Science and Technology* 4(1):109–118.

Talgre, L., H. Roostalu, E. Mäeorg, and E. Lauringson. 2017. Nitrogen and carbon release during decomposition of roots and shoots of leguminous green manure crops. *Agronomy Research* 15:594–601.

Tang, X., L. Bernard, A. Brauman, T. Daufresne, P. Deleporte, D. Desclaux, G. Souche, S. A. Placella, and P. Hinsinger. 2014. Increase in microbial biomass and phosphorus availability in the rhizosphere of intercropped cereal and legumes under field conditions. *Soil Biology and Biochemistry* 75:86–93.

Torabian, S., S. Farhangi-Abriz, and M. D. Denton. 2019. Do tillage systems influence nitrogen fixation in legumes? A review. *Soil and Tillage Research* 185:113–121.

Tubiello, F. N., M. Salvatore, A. F. Ferrara, J. House, S. Federici, S. Rossi, R. Biancalani, R. D. Condor Golec, H. Jacobs, A. Flammini, P. Prosperi, P. Cardenas-Galindo, J. Schmidhuber, M. J. Sanz Sanchez, N. Srivastava, and P. Smith. 2015. The contribution of agriculture, forestry and other land use activities to global warming, 1990-2012. *Global Change Biology* 21:2655–2660.

Udom, B. E. and S. Omovbude. 2019. Soil physical properties and carbon/nitrogen relationships in stable aggregates under legume and grass fallow. *Acta Ecologica Sinica* 39(1):56–62.

Urquiaga, S., G. Cadisch, B. J. R. Alves, R. M. Boddey, and K. E. Giller. 1998. Influence of decomposition of roots of tropical forage species on the availability of soil nitrogen. *Soil Biology and Biochemistry* 30(14):2099–2106.

Uzoh, I. M., C. A. Igwe, C. B. Okebalama, and O. O. Babalola. 2019. Legume-maize rotation effect on maize productivity and soil fertility parameters under selected agronomic practices in a sandy loam soil. *Scientific Reports* 9(1):8539.

Veloso, M. G., D. A. Angers, T. Tiecher, S. Giacomini, J. Dieckow, and C. Bayer. 2018. High carbon storage in a previously degraded subtropical soil under no-tillage with legume cover crops. *Agriculture, Ecosystems and Environment* 268:15–23.

Veloso, M. G., D. Cecagno, and C. Bayer. 2019. Legume cover crops under no-tillage favor organomineral association in microaggregates and soil C accumulation. *Soil and Tillage Research* 190:139–146.

Wang, D., M. R. C. Laza, K. G. Cassman, J. Huang, L. Nie, X. Ling, G. S. Centeno, K. Cui, F. Wang, Y. Li, and S. Peng. 2016. Temperature explains the yield difference of double-season rice between tropical and subtropical environments. *Field Crops Research* 198:303–311.

Wang, H., S. Wang, Q. Yu, Y. Zhang, R. Wang, J. Li, and X. Wang. 2020. No tillage increases soil organic carbon storage and decreases carbon dioxide emission in the crop residue-returned farming system. *Journal of Environmental Management* 261:2–8.

Wang, J. and U. M. Sainju. 2014. Soil carbon and nitrogen fractions and crop yields affected by residue placement and crop types. *PLOS ONE* 9(8):e105039.

Wang, M. X., Y. M. Li, J. J. Jiu, J. X. Zhao, Z. L.Wang, and M. P. Fan. 2020. Effects of planting patterns on stability and organic carbon content of soil aggregate. *Guangdong Agricultural Sciences* 47:52–59.

Wang, X., J. Y. Qi, X. Z. Zhang, S. S. Li, A. L. Virk, X. Zhao, X. P. Xiao, and H. L. Zhang. 2019a. Effects of tillage and residue management on soil aggregates and associated carbon storage in a double paddy cropping system. *Soil and Tillage Research* 104:39–43.

Wang, X., N. Samo, C. Zhao, H. Wang, G. Yang, Y. Hu, Y. Peng, and F. Rasul. 2019b. Negative and positive impacts of rape straw returning on the roots growth of hybrid rice in the Sichuan Basin area. *Agronomy* 9(11):690.

Wang, Y., L. I. Tong-Chuan, D. Y. Zhang, M. L. Jia, and H. K. Li, and W. D. Cao. 2013. Effects of intercropping white clover on soil aggregates and soil organic carbon of aggregates in apple-white clover intercropping system. *Acta Agrestia Sinca* 21:485–493.

Wei, F., M. M. Hao, and L. L. Nan. 2019. Effects of planting different leguminous green manure on soil chemical properties in the arid region of central Gansu. *Grassland and Turf* 39:34–37.

White, P. M., G. Williams, H. P. Viator, R. P. Viator, and C. L. Webber. 2020. Legume cover crop effects on temperate sugarcane yields and their decomposition in soil. *Agronomy* 10(5):703.

Wick, A., M. Berti, Y. Lawley, and M. A. Liebig. 2017. Integration of annual and perennial cover crops for improving soil health. In *Soil Health and Intensification of Agroecosystems*, ed. M. Al-Kaisi and B. Lowery, 127–150. Academic Press: San Diego, CA.

Xiao, D., Y. Tan, X. Liu, R. Yang, W. Zhang, X. He, and K. Wang. 2019. Effects of different legume species and densities on arbuscular mycorrhizal fungal communities in a karst grassland ecosystem. *Science of the Total Environment* 678:551–558.

Xing, W., X. Lu, F. Xu, J. Ying, D. Chen, and Y. Bai. 2019. Linking microbial community structure to carbon substrate chemistry in soils following aboveground and belowground litter additions. *Applied Soil Ecology* 141:18–25.

Xue, J. F., C. Pu, S. L. Liu, Z. Du Chen, F. Chen, X. P. Xiao, R. Lal, and H. L. Zhang. 2015. Effects of tillage systems on soil organic carbon and total nitrogen in a double paddy cropping system in Southern China. *Soil and Tillage Research* 153:161–168.

Xue, J. F., C. Pu, X. Zhao, Y. H. Wei, Y. L. Zhai, X. Q. Zhang, R. Lal, and H. L. Zhang. 2018. Changes in soil organic carbon fractions in response to different tillage practices under a wheat-maize double cropping system. *Land Degradation and Development* 29(6):1555–1564.

Xue, R. F., Y.Zhao, Y. J.Wang, J.Chen, and W. Ge. 2019. Effect of intercropping modes between millet and mung bean on their growth and yields. *Journal of Henan Agricultural Sciences* 48:37–40.

Yadav, G. S., A. Das, R. Lal, S. Babu, M. Datta, R. S. Meena, S. B. Patil, and R. Singh. 2019a. Impact of no-till and mulching on soil carbon sequestration under rice (Oryza sativa L.)-rapeseed (Brassica campestris L. var. rapeseed) cropping system in hilly agro-ecosystem of the Eastern Himalayas, India. *Agriculture, Ecosystems and Environment* 275:81–92.

Yadav, G. S., R. Lal, R. S. Meena, S. Babu, A. Das, S. N. Bhowmik, M. Datta, J. Layak, and P. Saha. 2019b. Conservation tillage and nutrient management effects on productivity and soil carbon sequestration under double cropping of rice in north eastern region of India. *Ecological Indicators* 105:303–315.

Yang, D. G., Y. Wu, X. L. Song, B. Tao, J. L. Gu, L. M. Dong, S. D. Ji, and Y. H. Han. 2019. Effects of crop rotation on soil fertility and growth and development of maize. *Journal of Maize Sciences* 27:127–133.

Yang, N., H. Zhao, Z. H.Wang, D. B. Zhang, and Y. J. Gao. 2012. Accumulation and translocation of dry matter and nutrients of wheat rotated with legumes and its relation to grain yield in a dryland area. *Acta Ecologica Sinca* 32(15):4827–4835.

Yao, P. W., D. B. Zhang, Z. Wang, Q. H. Cao, C. W. Yu. , J. Li, W. D. Cao, and Y. J. Gao. 2014. Nutrients accumulation of leguminous green manures and the effects on growth and yield of winter wheat. *Journal of Northwest A&F University (Natural Science)* 42:111–117.

Yaqub, M., T. Mahmood, M. Akhtar, M. M. Iqbal, and S. Ali. 2010. Induction of mungbean [Vigna radiata (L.) wilczek] as a grain legume in the annual rice-wheat double cropping system. *Pakistan Journal of Botany* 42:3125–3135.

Yusuf, A. A., R. C. Abaidoo, E. N. O. Iwuafor, O. O. Olufajo, and N. Sanginga. 2009. Rotation effects of grain legumes and fallow on maize yield, microbial biomass and chemical properties of an alfisol in the Nigerian savanna. *Agriculture, Ecosystems and Environment* 129(1–3):325–331.

Zak, D. R., W. E. Holmes, D. C. White, A. D. Peacock, and D. Tilman. 2003. Plant diversity, soil microbial communities, and ecosystem function: Are there any links? *Ecology* 84(8):2042–2050.

Zak, D. R., Z. B. Freedman, R. A. Upchurch, M. Steffens, and I. Kögel-Knabner. 2017. Anthropogenic N deposition increases soil organic matter accumulation without altering its biochemical composition. *Global Change Biology* 23(2):933–944.

Zeng, Z. H., Z. Y. Lu, Y. Jiang, K. Zhang, Y. D. Yang, and P. Y. Zhao. 2016. Legume-cereal crop rotation systems in China. In *Crop Rotations: Farming Practices, Monitoring and Environmental Benefits*, ed. B. Ma, 52–69. Nova Science Publishers, Inc.: New York.

Zhao, D. Q., T. Li, Y. T. Hou, J. C.Yuan, and Y. C. Liao. 2020a. Benefits and marginal effect of dry matter accumulation and yield in maize and soybean intercropping patterns. *Scientia Agricultura Sinica* 53:1971–1985.

Zhao, X., S. Wang, and G. Xing. 2015. Maintaining rice yield and reducing N pollution by substituting winter legume for wheat in a heavily-fertilized rice-based cropping system of southeast China. *Agriculture, Ecosystems and Environment* 202:79–89.

Zhao, X., A. L. Virk, S. T. Ma, Z. R. Kan, J. Y. Qi, C. Pu, X. G. Yang, and H. L. Zhang. 2020b. Dynamics in soil organic carbon of wheat-maize dominant cropping system in the North China Plain under tillage and residue management. *Journal of Environmental Management* 265:110549.

Zhong, Z., X. D. Huang, F. S. Xing, and B. Weng. 2018. Long-term effects of legume mulching on soil chemical properties and bacterial community composition and structure. *Agriculture, Ecosystems and Environment* 268:24–33.

Zhou, X., Y. H. Lu, Y. L. Liao, Q. D. Zhu, H. D. Cheng, X. Nie, W. Cao, and J. Nie. 2019. Substitution of chemical fertilizer by Chinese milk vetch improves the sustainability of yield and accumulation of soil organic carbon in a double-rice cropping system. *Journal of Integrative Agriculture* 18:381–392.

12 Managing Soil Organic Carbon in Croplands of the Eastern Himalayas, India

Anup Das, J. Layek, Gulab Singh Yadav, R. Lal, S. Saha, R. Singh, M. A. Ansari, S. Babu, K. P. Mohapatra, S. Devi, B. K. Kandpal, and P. K. Ghosh*

CONTENTS

* Corresponding author: jayanta.icar@gmail.com

DOI: 10.1201/9781003102762-12

ABBREVIATIONS

AFSs: agroforestry systems
BMPs: best management practices
C: carbon
CA: conservation agriculture
CO: carbon monoxide
CT: conventional tillage
EHR: Eastern Himalayan region
FYM: farmyard manure
GHGs: greenhouse gases
GLM: green leaf manure
GM: green manure
ICAR: Indian Council of Agricultural Research
IFS: integrated farming system
IHR: Indian Himalayan region
INM: integrated nutrient management
LC: leaf compost
MPTs: multipurpose trees
MT: minimum tillage
NEH: North Eastern Hill
NER: North Eastern Region
NT: no-till
RI: residue incorporation
RR: residue retention
SOC: soil organic carbon
TOC: total organic carbon
VC: vermicompost
WB: weed biomass

12.1 INTRODUCTION

The Indian Himalayan region (IHR) comprises 11 states (Uttarakhand, Himachal Pradesh, Sikkim, Meghalaya, Manipur, Mizoram, Tripura, Arunachal Pradesh, Nagaland, and the hill regions of Assam & West Bengal) and two union territories (Ladakh and Jammu & Kashmir). The IHR covers a geographical area of about 53.7 million ha (M ha) with a population of about 52.7 million (Apollo, 2017). The Eastern Himalayan region (EHR) of India lies between the latitudes 26°40'–29° 30' N and longitudes 88° 5'–97° 5' E and comprises the state of Sikkim, the northern extent of West Bengal including Darjeeling, and the northeastern Indian states of Assam, Arunachal Pradesh, Manipur, Mizoram, Tripura, Meghalaya and Nagaland (CEPF, 2005). The EHR has a geographical area of 20.2 M ha or about 37.9% of the total IHR (53.4 M ha) and 6.0% of India's total geographical area (328.7 M ha) (Nandy et al. 2006). Arunachal Pradesh (15.7%) encompasses the major share of the area in IHR. The average agricultural land to the total geographical area of the IHR is about 11.2% as against about 43% of the national average. The landholding in the hill area is smaller (700–900 m^2) than the national average (1370 m^2). The EHR has varied agro-climatic conditions ranging from sub-tropical to mild tropical, temperate, and alpine zones. The IHR has the world's highest rainfall area in Mawsynram-Cherrapunjee of Meghalaya. The NEH of India, along with the Assam Valley, is known as North Eastern Region (26.3 M ha). In many cases it's difficult to segregate NEH from NER, as both are similar in socio-cultural life and agricultural practices. The mean soil organic carbon or SOC (%) in Eastern Himalaya Zone in northeast India is 1.88% (0.85–3.56% at 0–30 cm depth), one of the highest recorded SOC values across all the agro-climatic regions of India (Bhattacharyya et al. 2008). However, the EHR of India is also highly variable in climatic conditions (tropical to alpine climate), topography, rainfall pattern, vegetation, land use, and cultural diversity and ethnicity. Thus, the soil organic carbon (SOC) concentration is also highly variable across the entire region (Das et al. 2017). Extreme rainfall erosivity (1,894.6 MJ mm ha^{-1} h^{-1} $year^{-1}$) and high soil erodibility (up to 0.055 Mg h MJ^{-1} mm^{-1}) increases the risk of top soil erosion (29–79 Mg $ha^{-1}yr^{-1}$; Poręba and Prokop, 2011), and the SOC loss particularly from the top soil (0–15 cm) in the EHR (Dabral et al. 2008). Practicing suitable soil conservation measures are essential for gradual soil C buildup and its long-term maintenance in the fragile ecologically sensitive agroecosystems of the EHR (Saha et al. 2015). A soil rich in SOC holds moisture, nutrients and promotes microbial activities resulting in better nutrient cycling and crop growth. The focus of this chapter is on the status of SOC variability and potential C management options across the EHR, maintaining long-term sustainability in agroecosystem productivity.

12.2 STATUS OF FARMING IN THE EASTERN HIMALAYAN REGION (EHR)

Agriculture in EHR is characterized by fragility, marginality and inaccessibility. About 70% of the geographical area of the region is under hills/mountains and rolling topography. Traditionally, farmers both at upland terrace and valley land follow mono-cropping under rain-fed conditions. Rice (*Oryza sativa* L) is the major crop occupying more than 80% of the cultivated area followed by maize (*Zea mays*). Despite the fact that the region receives high rainfall (>2000 mm), there is severe water scarcity during the post-rainy season (November to April). Indeed, cultivation of winter and summer crops is difficult in the absence of soil moisture conservation measures and irrigation facilities. Thus, cropping intensity of the region is only 134%. Use of fertilizer, manure, and pesticides in agriculture in the region is the lowest in the country. The use of N, P, and K fertilizer in the NER is only 13.4, 11.1, and 11.1%, respectively, of the crop removal, leading to the perpetual depletion of soil fertility and low productivity (Yadav et al. 2019a). Agronomic yield of most crops is much less than India's national average. For example, yield of rice and maize in the NER is 2.3 and 2.0 Mg ha^{-1}, respectively, as against India's national average of 2.7 and 3.0 Mg ha^{-1} (Agricultural Statistics at a Glance, 2019). About 85% of the farmers of this region are small and marginal in nature. The average income of farming family ranges from US$ 1400–2000 $annum^{-1}$. Though the region is comfortable in food grain production at present with marginal deficiency, it is expected to be 32% deficient in food grains by 2050 (Vision 2050, 2013). Thus, the immediate focus in the region is on crop yield enhancement, farm diversification, and increasing farmers' income while conserving natural resources for posterity.

12.3 STATUS OF NATURAL RESOURCES IN THE REGION

Intensive agriculture and excessive use of external inputs lead to degradation of soil, water, and genetic resources. Widespread soil erosion, nutrient mining, depleting ground water table and dwindling biodiversity are major concerns which are threatening the food security and livelihood opportunities of farmers, especially of the poor and underprivileged. Ecological restoration of degraded hill lands would trigger the process of soil/terrestrial carbon (C) sequestration, enhancements in productivity and use efficiency of inputs, mitigation and adaptation of climate change, provisioning of essential ecosystem services (ESs), increase in biodiversity by restoration of wildlife habitat, and increase in human wellbeing (Lal, 2015b). Historic land misuse and soil mismanagement in hill agriculture in India have strongly depleted the soil/terrestrial C reserves. *Jhum* /shifting cultivation (practice of rotational farming) in which land is cleared for cultivation, normally by fire, and then left to regenerate after a few years and other extractive practices of hill farming in NER have adversely impacted biodiversity and jeopardized the stability of ecosystems in these fragile environments. A major land mass in the NER (~15%) falls under sloping, undulating, eroded, and unproductive land (Saha et al. 2012). Loss by soil erosion of 46 Mg $ha^{-1}yr^{-1}$have been reported in area under shifting cultivation as against the national average loss of 16.4 Mg $ha^{-1}yr^{-1}$ (Bhattacharyya et al. 2015). Intensive tillage, residue burning and along the slope cultivation are the major threats to environment and food security in the hills. Excessive tillage disrupts soil structure and breaks pore continuity which exacerbate displacement and transport of soil particles and lead to soil and nutrient erosion and loss of SOC during heavy rains (Saha et al. 2007; Choudhury et al. 2016). Burning of biomass releases carbon monoxide (CO), carbon dioxide (CO_2), nitrous oxides (NO_x) and other greenhouse gases (GHGs) into the atmosphere. It is estimated that about 10 Mg ha^{-1} dry biomass is burned in the shifting cultivation amounting to burning of about 9 Tg (10^{12}g) biomass annually in NER (Das et al. 2011). Thus, it is important to identify and employ efficient technologies to use scarce natural resources to attain food security, sustain soil health, and mitigate the impact of climate change. Lal (2015b) proposed a model for restoring hill ecosystems and advancing food security. He emphasized the need for erosion control, nutrient recycling, restoring SOC stock, biomass production and ecosystem stability for restoring degraded hill regions.

Soil acidity is another important aspect of natural resource management in the EHR. About 85% soils of the NEH region are acidic in nature with majority of areas having pH below 5.5. About 47.5% of soils of the region are subject to degradation of various kinds, among which, soil acidity and water erosion are the most dominant processes. The average productivity of the red lateritic sandy to clay loam-textured soils of the EHR is inherently low due to deficiency of some essential nutrients like P, Zn, B and toxicity of Al, Fe and Mn with high to moderate acidity (Hmar and Ramanujam, 2014). Inadequate and unbalanced use of fertilizers to the crops is also responsible for low system productivity and decline in SOC and fertility (Das et al. 2020).

12.4 CROP PRODUCTION, SOIL FERTILITY AND SOC INTER-LINKAGE

Stagnating and declining crop yields, especially in South Asia's cereal-based cropping system (Bhowmick, 2005; Das et al. 2014) are attributed to decline in soil quality caused by removal or burning of crop residues, low rate or unbalanced application of fertilizers, indiscriminate/excessive grazing of stubbles, and inappropriate use of irrigation water. Most crop land soils of South Asia, similar to those of sub-Saharan Africa (IFDC, 2006), have a negative nutrient budget (Sahu et al. 2020). The problems of soil degradation, water pollution and contamination by fertilizers and pesticides, low and declining crop yields, and poor environment quality can be partly addressed by restoring degraded ecosystems by restoring SOC stock (Lal, 2008d). However, land managers must be compensated for the ecosystem services (ESs) of value to humans and nature (Lal, 2008e). Sequestration of C in soils and ecosystems depends on the use of precious inputs (e.g., crop residues, fertilizers, animal manure and compost, irrigation) for which farmers must be compensated. Continuous use of unbalanced fertilizers under intensive cultivation has adverse impacts on soil (Das et al. 2014). Adoption of those agricultural practices which involve application of off-farm or external chemical inputs release CO and other GHGs into the atmosphere during production, transportation, and storage (Saha et al. 2018). Industrial agriculture inputs contribute to numerous forms of environmental degradation, including air and water pollution, soil depletion, and diminishing of biodiversity and soil quality (Verma et al. 2015). Hence, sustainability of agricultural systems with low or no emission of GHGs has become an important issue all over the world. Intensive agronomic practices have a great impact on SOC characteristics that often lead to decreased levels of SOC and, as a consequence, to deterioration of soil quality (Haynes, 1999). Nutritional management usually affects the input of biomass C to the soil and, therefore, can affect the quantity and quality of SOC stock. Again, the increase in extreme climatic variability in the IHR (Pandey et al. 2009) is aggravating the uncertainty of agricultural productivity. Hence, sustainability of agricultural systems with low or without emission of GHGs is a prime consideration. Long-term application of mineral fertilizers alone is insufficient to maintain SOC levels unless it is combined with a high return of organic substances (Layese et al. 2002; Babu et al. 2015). Crop residues, green manures, and composts are considered to be important organic inputs in maintaining SOC concentration. Continuous input of these organic resources modifies the activity of microbial communities and consequently the SOC turnover (Kautz et al. 2004; Das et al. 2014). However, the magnitude of response may depend upon climatic conditions, soil indigenous attributes, amount and composition of residues returned to the soil, and intensity of tillage disturbances. Hence, monitoring qualitative and quantitative changes in the SOC is essential to keeping the soil viable and productive for long-term sustainability and enhancing environmental quality (Lal, 2004a). Soil C sequestration (SCS) is based on the natural process of photosynthesis and conversion of biomass C into soil C stock. Therefore, the rate of SCS depends on the rate of soil application of biomass C along with those of N, P, K, S, and other components essential to conversion of biomass into stable SOC with a long mean residence time (Lal, 2008b).

12.5 LAND USE AND SOC

Land-use change can affect soil quality and productivity by altering structure and functions of ecosystems and the biogeochemical cycle. Conversion of natural forest or grasslands to crop production can decrease SOC levels (Grogan et al. 2012; Wang et al. 2017). Conversely, conversion of fallow and degraded lands to permanent vegetation cover (by afforestation) could buildup long-term fertility through the increase in input of organic matter. Continuous and intensive tillage practices, removal or burning of residues, inadequate nutrient management, and improper crop rotations are some of the reasons for decreasing SOC levels in cultivated soils. Land-use change and management dictate whether the soil is a source or sink of atmospheric C (Lal, 2004b). As a rule of thumb, land management practices which cause a minimal soil disturbance, increase SOC stock. The SOC loss occurs when natural forest ecosystems are converted to cultivated system (Boerner et al. 2005). Crop rotation, minimum tillage, no-till (NT), intercropping, mulching, organic farming, cover cropping and management of crop residues contribute to SOC buildup. Lal, 2003 reported that approximately 1.14 Pg (10^{15} g) of C is released into atmosphere through global water erosion processes.

12.6 SOC AND POTENTIAL OF SOIL AS C-RESERVE

The organic residue input is primarily sourced from different plant/animal byproducts that facilitates SOC buildup, improves soil hydro-physical and chemical properties, conserves soil moisture, and enhances system productivity after 3–5 years adaptation of conservation tillage against conventional tillage practiced in the hill ecosystems of EHR (Saha and Ghosh, 2013). Land management practices followed in a particular system can influence the amount of SOC concentration in different parts of EHR (Arunachalam, 2002; Pathak et al. 2018; Nath et al. 2018). Extensive soil tillage increases aeration, increases the rate of decomposition of organic matter and decreases the SOC concentration (Chakraborty et al. 2017). Increased fallowing in crop production also decreases SOC, because of reduction in amount of non-harvested plant residues addition to the soils (Sainju et al. 2009). Application of fertilizer may increase SOC concentration because it increases biomass (above and below grounds production) and therefore, increases productivity (Zang et al. 2016). Increased use of manure and other soil amendments has similar effects (Saha et al. 2012). Increased use of legumes and forages in crop rotation also increases SOC as forage crops develop extensive root systems and deposit more residues than other crops (Gentile et al. 2005). Irrigation has similar effects of enhancing the SOC by improving productivity and increasing soil moisture content (Condron et al. 2014).

About 20% of global CO_2 emissions are of soil origin (Rastogi et al. 2002; Barnard et al. 2015). Further, about one-third of global CH_4 emissions and two-thirds of N_2O emissions are also contributed by soil (Smith et al. 2003). Therefore, carbon sequestration in soils for longer period and deeper layers is considered is an important strategy to mitigate the increasing level of CO_2 concentration in the atmosphere and to improve soil resilience (Lal, 2018). Globally, soils store 1,500 and 2,400 Pg of organic C up to 1 m depth of mineral soil layer that is 3–4 times more than the C content in vegetation (450–650 Pg C) and 2–3 times more as in the atmosphere (~829 Gt C) (Stockmann et al. 2013; Ciais et al. 2014). Even a minor change in terrestrial C stock can impact global C cycle significantly with profound effect on climate change and global warming. The SOC is the key indicator of soil health and quality (Sharma et al. 2005). Thus sustaining and improving its quality and quantity is of paramount importance for long-term soil fertility and crop production. The SOC concentration directly influences physical, chemical and biological properties of soil (Esmaeilzadeh and Ahangar, 2014). Hence, sustenance and maintenance of SOC is an important aspect of soil health management toward sustainable crop production (Saha et al. 2012). The SOC comprises of various pools and fractions of SOC with varying degree of decomposition and stability. These fractions and pools are useful to assessing the impact of management practices on SOC dynamics. The amount of C stored in a given soil is influenced by the input and output of C substances and environmental conditions. The C inputs may be from any type of organic biomass and the output may be CO_2 emission and crop harvest. Temperature also influences the SOC dynamics by regulating the degree of C buildup and SOC disintegration (Chatterjee et al. 2019).

12.7 WHY ORGANIC MATTER MANAGEMENT IS IMPORTANT IN CROPLANDS OF THE EHR

The organic matter in soil exists as 4 distinct fractions viz., dissolved organic matter, particulate organic matter, humus, and resistant organic matter which vary widely in size, turnover time, and composition in the soil. Maintaining and improving SOC levels are preconditions to ensuring soil fertility, productivity, and sustainability of agricultural ecosystems (Ghosh et al. 2009). Increase in SOC concentration by 0.01% can substantially reduce the adverse consequences of annual increase in atmospheric CO_2 concentration (Lal et al. 1998). The amount of SOC is a net balance between C input (quantity & quality) and output from soil (Regnier et al. 2013). The SOC stocks depend on diverse factors which influence C input and decompositions such as vegetation type and its net primary productivity (Ren et al. 2012), soil type, moisture content and land use practices. Climatic factors especially temperature and moisture regimes have been regarded as the primary control of SOC decomposition. The leaf and root biomass C are recycled within ecosystem through decomposition, erosion and leaching. Since soil carbon is considered one of the most important keys to soil fertility, productivity and quality, decline in SOC concentration not only affects sustainability of agricultural ecosystems, but also overall quality of environment, as soil contains a significant part of global carbon. The extent of variation in SOC concentration and stock relies on the land-use type and land-use change and post-conversion land management (West et al. 2010). Besides enhancing food security, carbon sequestration has the potential to offset fossil-fuel emissions by 0.4–1.2 Pg of carbon (10^9 metric ton) per year, or 5–15% of the global fossil-fuel emissions (Lal, 2004a). Thus, carbon management is the key to environment management and sustainability of soil health vis-a-vis agricultural productivity.

The SOC in the humid sub-tropical climate of the EHR becomes more important in view of undulating hilly terrains (Yadav et al. 2017). Halting land-use conversion would be an effective mechanism to reduce soil C losses, but with a growing population and changing dietary preferences in the developing world, more land is likely to be required for agriculture. However, with judicious management of existing lands and narrowing the yield gap, some land can and must be returned back to nature (Lal, 2016, 2018). Thus, understanding the SOC dynamics under diverse land use system for climate resilient agriculture and for maximizing the productivity of existing agricultural land and applicability of best management practices are among the highest priorities to set aside agriculturally marginal lands in ecologically sensitive eco-regions of the EHR for nature.

12.8 STATUS OF SOC IN THE EHR

The SOC in soils of the EHR are in general much higher than those observed in other parts of India like Indo-genetic plains where SOC ranges from 5–6 g kg^{-1}. Soils in the valleys of the EHR are rich in SOC but are mostly degraded. The SOC concentration is low to medium in arable lands on steep slopes because of accelerated soil erosion (Das et al. 2017).The controlling factor for SOC in different agroecosystems of EHR may be climate, topography, soil type, management practices and interaction among different factors (Singh et al. 2014; Choudhury et al. 2016). Among the NER states, Tripura has the highest land area (>89%)under relatively low SOC concentration (10–15 g kg^{-1}) followed by Assam (62.83%) and Meghalaya (19.9%). The SOC concentrations in mid and high attitude areas of the NER ranges from 10–35 g kg^{-1}. The general trend is that SOC concentrations increase with increase in altitude. In Arunachal Pradesh, SOC concentration ranges from 14 g kg^{-1} (mid altitude) to 59 g kg^{-1} (high altitudes). In Manipur, SOC concentration ranges from 5.1–36.6 g kg^{-1} in valley districts (Bishnupur and Imphal) and 14.4–51 g kg^{-1} in hill districts (Ukhrul and Tamenglong). About 13% soils in the valleys and 5.5% in the high altitude (>1700 m MSL) are low in SOC in Meghalaya. The soils of Mizoram are low to medium in SOC (4.2–13.4 g kg^{-1}) concentration. In Sikkim, the SOC concentration is high to very high in most of the soils ranging from 23.5 g kg^{-1} to 44.5 g kg^{-1} (Sharma et al. 2005; Das et al. 2011). Sikkim has the maximum area (82%) under high SOC density (40–50 Mg/ha) among all the NEH states. Similarly, Sikkim is the only state having considerable area (>10%) under very high SOC density (50–60 Mg ha^{-1}) (Choudhury et al. 2013). Under minimum disturbance, the soils of primary and secondary forests (regenerated by *jhum* fallows) in Eastern Himalaya may be a chief source or sink of carbon to play an important role in both climate change adaptation and mitigation (Vashum et al. 2016). However, over the years due to non-adoption of scientific conservation practices and soil fertility–building measures, soil erosion, continuous cropping without addition of sufficient organic manures etc., the SOC concentration in most part of the EHR is showing signs of decline with some areas like the Shillong plateau becoming unfit for cultivation (Behera et al. 2016). Agriculture, horticulture, and shifting cultivation are the three prominent land uses practiced by farmers of the EHR. Soil properties and SOC status of land under these systems are varied. For example, with respect to Tripura, the SOC is the highest under agriculture land use (0.82%) than those under shifting cultivation (0.80%) and horticulture (0.67%) land use systems. Further, SOC concentration may vary according to land topography among land use systems (Bhattacharyya et al. 1998).

12.9 SOC MANAGEMENT PRACTICES IN THE EHR

12.9.1 Indigenous SOC Management Practices

A large number of indigenous farming systems are being practiced in NER and production is maintained only through organic nutrition. There are "*Zabo*" systems practiced in Phek district of Nagaland which have a combination of forest, agriculture, livestock, and fisheries. Rice-based farming systems of Apatani plateau occupying a stretch of 26 km^2 area in Subansiri district of Arunachal Pradesh is inhabited by Apatani Tribe. Bamboo drip irrigation system of Jaintia and Khasi Hills of Meghalaya, agriculture with alder in Nagaland, aerobic rice with dragon fruit in the steep slopes of *Lusai* hills, rice cultivation on terraces in Nagaland, Manipur and Sikkim, "*Taungya System*"(method of establishing forest species in temporary combination with field crops) are some of the systems which have inbuilt mechanisms of resource conservation. Homestead agroforestry where farmers of Assam and Tripura grow number of tree species along with livestock, poultry and fish mainly for the purpose of meeting their own needs is also a low input popular system (Datta et al. 2012). However, these systems have changed over the years due to population pressure necessitating need for adoption of modern management practices to increase system productivity (Das et al. 2012). The seasonal variation in SOC stock (0–45 cm) ranges from 89.9 ± 2.16 (January) to 117.9 ± 3.12 Mg C ha^{-1} (August) in traditional cardamom-based agroforestry system of Sikkim Himalayas, and it exhibits considerable potential for large scale adaption toward the climate change mitigation (Lepcha and Devi, 2020). The rate of aboveground C sequestration for traditional village level bamboo cultivation and bamboo agroforestry systems ranges between 18.93 and 23.55 Mg ha^{-1} yr^{-1} that facilitates soil C sequestration @ 0.28–0.59 Mg C ha^{-1} $year^{-1}$ with the estimated soil carbon stock of 26–35 Mg C ha^{-1} (30 cm depth) in the northeast Indian state of Assam (Nath and Das, 2012; Nath et al. 2015).

12.9.2 Improved SOC Management Practices

Agriculture in the EHR is practiced in valley lowlands, well drained uplands, terraced lands, and sloping lands. Shifting cultivation is practiced in hill slopes across the EHR excluding Sikkim and Darjeeling Hills. The strategy of agriculture and SOC management varies from location to location as per soil conditions, climate, land-use change, fertility status, existing land management practices, crops and cropping systems, and resources available with the farmers. Based on these factors, different conservation-effective best management practices (BMPs) have been developed/standardized and recommended to the stakeholders (Table 12.1).

12.9.2.1 Impact of Organic Manure on Soil Carbon

Organic amendments cover a wide range of inputs, from animal manure to solid wastes, oilcakes, biomass from shrubs and crop residues, various composts and especially grown cover crops, often legumes, which are plowed under as green manure (Callow, 1997). The main advantage of organic amendments is that they are rich in labile carbon fractions, which are a source of energy for microorganisms. Organic amendments improve soil quality by reducing compaction, decreasing soil bulk density, improving soil aggregation status, thus promoting drainage and soil water-holding capacity (Singh et al. 2019). Application of manures and other organic amendments is another important strategy of SOC sequestration (Indoria et al. 2018).

12.9.2.1.1 Impact of Organic Manure in Vegetable-Based Systems

Long-term applications of organic amendments can improve SOC concentration and available nutrient status in soil, thereby sustaining soil health (Tiwari et al. 2002; Panwar et al. 2010). The differences in SOC with the application of different sources of nutrients might be due to the results of the differential rate of organic matter oxidation by microbes (Trehan, 1997). In a five-year study

TABLE 12.1
Ecosystem Specific Soil Health Management Practices for Improving SOC

Valley Lands/Lowlands	Uplands/Terraced Lands	Sloping Lands	Shifting Cultivation
Conservation tillage (MT/NT)	Conservation tillage (MT/NT)	Agroforestry systems	Strip/intercropping
Crop residue recycling	Mulching/residue retention	Hedgerows on contours	Fast-growing Multipurpose tree species (MPTs) like *Alnus nepalensis*
Organic manure	Cover crops	Cover crops	Improved varieties
Weed biomass	Green manuring	Perennial fodder crops	Foliar nutrition
Green leaf manure (GLM)/Green manure	Integrated nutrient management (INM)	Medicinal and aromatic grasses	Biofertilizer
INM	Perennial fodder crops	Vegetative barrier through aromatic and medicinal grasses	Seed treatments
Crop diversification with pulses/oilseeds	Crop diversification	Permanent soil cover	Hedgerow barriers on contours
Integrated Farming system (IFS)	Rotation with legumes	Protected cultivation of high value organic crops	Residue retention/Mulching
Intensification of rice–fallow areas	Biochar application	Across the slope cultivation	MT/NT
	Hedgerows on terrace risers/boundaries		
	MPTs on fences		
	Watershed based IFS		

at mid-altitude of Meghalaya, a maize + soybean–French bean cropping system resulted in higher SOC (22. 0 g kg^{-1}) than that under other maize-based cropping systems. The increase in SOC was 22.2% compared to that under control after five cropping cycles. Integrated application of 50% N through FYM and remaining 50% through vermicompost (VC) resulted in the maximum SOC concentration (23.6 g kg^{-1}) compared to other amendments. The maximum soil microbial biomass carbon (SMBC) and water-holding capacity were recorded with integrated application of FYM + VC (50% N from each source) as sources of organic amendments followed by FYM and VC, which were on par with each other (Patel et al. 2015).

12.9.2.1.2 Impact of Organic Manure in Field Crop-Based Systems

Increase in the SOC due to organic farming has been reported by several researchers (Babhulkar et al. 2000; Stolze et al. 2000). In a study on rice and maize-based cropping systems at mid-altitude of Meghalaya, integrated application of recommended dose of N through 1/3rd each of FYM, VC, and leaf compost (LC) resulted in the maximum SOC (25 g kg^{-1}), available N (247.2 kg ha^{-1}), P (22.7 kg ha^{-1}), and K (247.6 kg ha^{-1}) after three cropping cycles compared to other sources, which were 47.2, 24.9, 54.4, and 33.3%, respectively, higher compared to the control. The minimum bulk density (BD) was recorded with FYM (1.11 Mg m^{-3}) followed by that through 1/3rd each of FYM, VC, and LC (Das et al. 2014). Mixing manures of different origin has the potential to synchronize mineralization with crop demands and enhance productivity (Handayanto et al. 1997), although the practicalities of this on a farm scale are questionable (Efthimiadou et al. 2010). Lowland rice productivity and soil fertility with incorporation (10 Mg of fresh biomass ha^{-1}) of Indian coral tree (*Erythrina indica*) leaves have been reported significantly higher than the control but remained at par with northern black wattle (*Acacia auriculiformis*), tree bean, and yellow casia (*Casia siamea*) leaves (Das et al. 2010). The SOC concentration under northern black wattle and Indian coral tree biomass have been reported 10.3 and 9.2% higher than that under recommended NPK and 15.2 and 14% higher than the control, respectively. Continuous applications of organic amendments were reported to improve the SOC, available P, and K in soil, thereby sustaining the soil health (Panwar et al. 2010). Ding et al. (2012) showed that organic fertilization has a great positive impact on soil C storage and sequestration, and has been attributed to greater C inputs through organic manure and root biomass due to better crop growth.

12.9.2.2 Integrated Nutrient Management

Integrated nutrient management (INM) practice is a viable option for restoring the soil physical structure and chemical fertility, improving SOC concentration and therefore, sustaining the system productivity (Aulakh, 2010). An integrated approach recognizes that soils are the storehouse for most of the essential plant nutrients. The way in which nutrients are managed has a major impact on plant growth, soil fertility, C-sequestration and agricultural sustainability (Peter et al. 2020). Application of the recommended dose of nutrients through the sole source of synthetic fertilizer is neither cost-effective nor economically and environmentally viable. The hill ecosystems are more prone to erosion, and thus, soil, nutrient, and carbon are lost by runoff and erosion. Also, the farmers in the region are resource-poor and can't afford to apply the full amounts of nutrients through chemical fertilizers. Considering the huge potential of organic manure in the region (i.e., 47 Tg reported by Bujarbaruah, 2004), a large portion of the nutrient requirement in agriculture can be met through organic sources. Crop residues, animal excreta-based manure, forest litter, weed biomass, green manuring, cover crops, etc. are important sources of organic manure in the EHR (Das et al. 2017). Biofertilizer, lime, rock phosphate, etc. also forms an integral component of INM in the EHR, which is essential to SOC sequestration. The humification process can be severely constrained by the lack of N, P, S, and other building blocks of soil humus (Himes, 1998). The furrow application of lime 250–500 kg ha^{-1}every year is reported to substantially enhance crop yield in the EHR (Das et al. 2017). The use of acid-tolerant varieties and the application of organic manure also improve the productivity of such soils. Significantly higher SOC and SMBC of soils under 50% NPK + GLM as compared to 50% NPK alone at 0–15 cm soil depth have been reported from a 6-year study at mid-altitude of the EHR (Das et al. 2020). The C inputs under INM are generally higher due to high biomass yields and the external organic C inputs through various organic manures such as FYM, GM, CR, etc. The strong influence from increasing C stock through long-term INM under a rice–wheat system has been widely reported (Majumder et al. 2008; Sodhi et al. 2009; Nayek et al. 2012). SOC concentrations under rice–rice system in paddies of Tripura have been reported to increase at a rate of 428 kg $ha^{-1}year^{-1}$ with the adoption of MT along with improved INM. The improved INM comprised 25% N (20 kg N) through green leaf manuring (GLM) + 60 kg N, 9 kg phosphorus (P), 17 kg potassium (K), 2 kg Boron (B), and 5 kg zinc (Zn) ha^{-1} through fertilizer + cellulose decomposition microorganism along with 30% residue retention in rainy season rice. The tillage and INM for winter rice were CT along with 25% N through FYM and 75% N and remaining P and K (after deducting quantity supplied by FYM) through inorganic fertilizer + 30% residue incorporation (RI). This combination of tillage and nutrient management also improves SOC concentration, stock, accumulation, sequestration, C-retention efficiency, soil microbial biomass C and dehydrogenase activities in paddy soils of Tripura (Yadav et al. 2017c).

Organic substitution of N significantly contributes to building SOC stock vis-a-vis improving the soil structure. However, improvement in soil structure or soil C may not proportionately improve crop yields. Nevertheless, organic–inorganic combinations provide greater scope in improving soil quality and long-term C accumulation, while sustaining the net system productivity (Datta et al. 2012). Ladha et al. (2011) showed that long-term use of synthetic fertilizer N leads to a smaller decrease of SOC as compared with no use of synthetic N. In inorganic fertilized and unfertilized control treatments, most of the SOC is supplied through rice root biomass, since the aboveground part is removed at the harvesting stage. Through the years, rice grain yield increased in the control and inorganic treatments, but total SOC concentrations significantly decreased over time, irrespective of the increase in root biomass. This trend shows that continuous rice cropping without organic matter addition might significantly decrease SOC concentration, which may then result in soil degradation. There have been many reports about the degradation of soil properties due to continuous rice cropping with conventional management practices (Wright et al. 2007; Lee et al. 2009).

12.9.2.3 Conservation Agriculture as a Means of Managing SOC in the EHR

The resource conservation technology measures (e.g., laser land leveling, crop residue retention, and NT farming) are popular in the Indo Gangetic Plains and reported to cover an area of about 3 M ha in India. However, meager attempts have been made to promote CA technologies suitable for the low input hill agriculture, where such practices are in high need due to the fragility of the eco-regions. In the rain-fed hill zones, mechanization is meager mainly due to difficult terrains, small holdings, and the poor economic condition of the farmers. High rainfall and excessive disturbance to the soil along with faulty agricultural practices are causing serious soil degradation in terms of erosion and nutrient and SOC loss in the Eastern Himalayas (Mandal and Sharda, 2013). With the use of an energy-intensive conventional tillage system, there is a gradual decline in SOC through accelerated oxidation and burning of crop residues causing pollution, GHG emissions, and loss of valuable plant nutrients. Retention of crop residues on the soil surface in combination with NT/MT led to soil quality improvement and an overall increase in resource use efficiency. Retention of plant biomass enhances the microbial population that increases the decomposition rate of crop residues/other biomass and accretion of SOC buildup in surface soils (Lal et al. 2000). Thus, conservation agriculture (CA) may lead to sustainable agriculture through the efficient use of water and nutrients by improving nutrient balances and availability, water infiltration and retention by soils, and minimizing evaporation loss (Das et al. 2013). Reduced tillage coupled with residue management would reverse the trend of soil degradation. Crop rotations increase SOC concentration, more so with the introduction of leguminous crops. The SOC status is poor to medium on steep slopes because of continuous loss of topsoil by erosional processes.

About 60% of the cultivated area in the EHR is under fallow after the rainy season of rice. Retention or external application of crop residues during the rice fallows increases the soil's hydraulic conductivity and water infiltration rate by improving soil structure, proportion of macro pores, and aggregate stability. Thus, if crop residues are retained on the soil surface in combination with suitable planting techniques, it can conserve soil moisture and alleviate drought conditions in pulses. Further, plowing and pulverizing soil on steep slopes aggravates erosion hazards during high intensity rains and lead to loss of plants nutrient and SOC. Ghosh et al. (2010) reported that double NT practice in rice-based systems was cost-effective, restored SOC stock, favored biological activity, conserved water, and produced better yield, which were 70.8, 46.7, and 49.0% higher compared to CT. Adoption of NT, along with retention of maize stalks and application of *Ambrosia* spp. mulch at 5 Mg ha^{-1}, resulted in the maximum improvement in soil quality parameters and enhanced rapeseed (*Brassica napus* subsp. napus) yield in maize–rapeseed cropping systems in the Eastern Himalayas (Das et al. 2014a). Mulching treatment, comprised of the retention of maize stock cover along with application of *Ambrosia* spp. mulch at 5 Mg $ha^{-1}$and poultry manure at 5 Mg ha^{-1}, enhanced the mean SOC concentration by 30% and mean weight diameter of aggregates by 100% as compared with those under control. Thus, NT and mulching are recommended measures for protecting surface soil and improving soil quality in the EHR.

The average grain yield of rice under 50% NPK + weed biomass (WB) was 16.7% and 9.10% higher than that of under 50% NPK and 100% NPK, respectively (Das et al. 2020). The residual effect of tillage and nutrient management practices applied to rice can have a significant effect on green pod yield of pea in rice fallows. The pooled green pod yield of pea was the highest under MT followed by CT and the lowest was under NT (Das et al. 2020). Among different NM practices followed in rice, productivity of pea was recorded significantly higher under, 50% NPK+WB than that of 50% NPK alone. The increase in SOC and soil nutrient status due to residue management under the groundnut (*Arachis hypogaea*)–rapeseed system is also reported at Umaim, Meghalaya (Kuotsu et al. 2014). The NT with residue retention may increase carbon efficiency. Further, the C sustainability index [i.e., (C output-C input)/C input; 9.36] were higher under NT than those under conventional tillage (CT) with residue incorporation (Yadav et al. 2018a). Retention of residues under NT can store more moisture in soils and promote better crop growth and development, leading to higher SOC sequestration than those under CT soils (Yadav et al. 2018b). The SOC concentration in the EHR is different from location to location based on altitude, climate, land use pattern, crop management practices, cropping systems and other factors (West et al. 2010). The SOC concentration and soil biological activities were significantly higher in 0–15 cm soil depth under NT–NT and MT–NT as compared to CT–NT systems (Table 12.1). On average, significantly higher soil biological activities at harvest of lentil were observed under mulch followed by 40 cm standing stubble as compared to residue management (Das et al. 2019). In another study, adoption of conservation tillage (minimum tillage with 30% residue on surface) in a rice–rapeseed system improved SOC concentration significantly over CT (Das et al. 2020). Such a trend can be attributed to the addition of biomass-C input in the soil surface and reduced mineralization due to less soil disturbance and reduced oxidation (Ghosh et al. 2010). In a maize–rapeseed system, retention of maize stover in-situ and application of *Ambrosia* spp. biomass 10 Mg ha^{-1} as mulch enhanced productivity and SOC concentration significantly over residue removal (Das et al. 2017). A rice–rice system with reduced tillage (RT) + INM comprising 25% N through FYM and 75% N with remaining P and K through inorganic fertilizer + 30% residue retention (RR) in wet season rice and CT + INM + 30% RI in dry season rice increased SOC stocks by 4.14% (Table 12.2) over farmers practices in the paddy soils of Tripura (Yadav et al. 2019a). However, NT with residue retention was found to be better for increasing SOC concentration and stocks in the upland of Tripura (Yadav et al. 2019b). Adaption of NT with 100% residue retention practices in a rice–mustard system in Tripura added more C to the soil and increased soil C stocks than that of CT with 100% residue incorporation (Yadav et al. 2019b). Under the high-altitude conditions of Sikkim, NT in a maize–pea system under recommended organic nutrition had a positive impact on SOC after 5 consecutive years. Use of NT

TABLE 12.2
Location-Specific Management Options to Increase SOC Concentration in Crop Production

Crop/Cropping System	Improved Management Options	SOC Concentration/ SOC Stock	References
Location: Umiam, Meghalaya, 950–1000 m ASL			
Rice (upland) - lentil	Mulching 5 Mg ha^{-1}/40 cm standing stubble in lentil, 0–15 cm (soil depth)	17.5 g kg^{-1} compared to 16 g kg^{-1} under control	Das et al. (2019)
Rice (upland) - mustard	Conservation tillage, 0–15 cm	17.5 g kg^{-1} compared to 16.4 g kg^{-1} under CT	Das et al. (2020)
	50% NPK + WB/in-situ rice residue retention (ISRR), 0–15 cm	32 Mg ha^{-1} relative to 29.1 Mg ha^{-1}under 50% NPK only	
Maize–rapeseed	Retention of maize stover in-situ + Ambrosia biomass 10 Mg ha^{-1} in rapeseed, 0–15 cm	26.6 g kg^{-1} compared to 20.8 g kg^{-1} under residue removal	Das et al. (2017)
Location: Tripura, Altitude: 51 m ASL			
Rice–rice system	Reduce tillage (RT) + integrated nutrient management (INM) comprising 25% N through FYM and 75% N & rest of P & K (after deducting quantity supplied through FYM) through inorganic fertilizer + 30% residue retention (RR) in WR and CT + INM + 30% RI in DR, 0–20 cm	21.63 Mg ha^{-1} relative to 20.77 Mg ha^{-1}under farmers practice (FP)	Yadav et al. (2019a)
Rice (upland) - mustard	No-till with 100% residue retention, 0–30 cm	29.92 Mg ha^{-1} relative to 29.18 Mg ha^{-1}under CT	Yadav et al. (2019b)
Location: Sikkim, Altitude 1300 m			
Rice–pea	Double NT + 50% FYM + 50% VC+ biofertilizers, 0–20 cm	53.46 Mg ha^{-1} relative to 50.76 Mg ha^{-1}under rice–fallow	Singh et al. (2018)
Rice–maize	Rice–broccoli–*Sesbania* + recommended organic nutrition, 0–40 cm	69.2 Mg ha^{-1} relative to 66.7 Mg ha^{-1}under FP	Babu et al. (2020)

recorded 8.1% higher SOC concentration at 0–10 cm soil depth compared to CT after 5 years of experimentation (unpublished data). The enhancement in SOC was mainly due to retention of crop residues and less soil disturbance under NT compared to CT. Double NT along with application of recommended dose of N through 50% each of FYM and VC + biofertilizers resulted in the highest SOC stock (Babu et al. 2020).

12.9.2.4 SOC under Intensified Maize and Rice-Based Cropping Sequences in High Altitudes

Maize–fallow, rice–fallow, and maize–rice–fallow are the primary cropping systems in the Sikkim Himalayas (Singh et al. 2017). Maize is a nutrient-exhaustive crop and, hence, its continuous cultivation has depleted inherent soil fertility, causing deficiency of several nutrients. Thus, soil quality improvement through cropping sequences is important for sustaining crop productivity and soil health. Inclusion of leguminous crops, like soybean, black gram, kidney beans, and vegetable pea can restore the soil fertility of maize and rice-based cropping systems in Sikkim (Singh et al. 2017; Babu et al. 2020). The rice–vegetable pea system had the maximum SOC (1.25%) content in 0–15 cm depth compared to that under rice–potato (1.18%). The enhancement in SOC concentration in rice–vegetable pea was 6.83% higher compared to initial SOC concentration (1.17%) after three years of experimentation (Avasthe et al. 2020). In a long-term study, Babu et al. (2020) reported that SOC stock varied from 18.4 Mg ha^{-1} to 69.2 Mg ha^{-1} in 0–40 cm depth, across six cropping sequences. Rice–buckwheat–sesbania (green manure) and the rice–coriander (*Coriandrum sativum*)-cowpea (*Vigna unguiculata* L.) system had the highest SOC stock (68.9–69.2 Mg ha^{-1}) at 0–40 cm soil depth. The total SOC concentration was 4.2–5.0% higher under the rice–buckwheat (*Fagopyrum* spp.)–sesbania (green manure) system over rice (*Trigonella foenum-graecum* L.)–fenugreek–baby corn cropping sequences. Hence, the addition of GM crops to the cropping system can accumulate higher SOC concentration relative to the conventional cropping sequences (rice–maize) without GM crops in an organic production system (Babu et al. 2020). Similarly, carbon retention efficiency (CRE) is higher under rice–coriander–cowpea (15.1%) and rice–buckwheat-*sesbania* (14.9%) cropping sequences than under others. The lowest CRE is observed under the rice–fenugreek–baby corn cropping sequence (Babu et al. 2020).

12.9.2.5 Integrated Farming Systems (IFS) for Managing SOC

Considering the small and marginal nature of more than 85% of farmers of the EHR, the recommended systems for the regions are IFS with a compatible combination of crop, livestock, fishery, MPTs, and other components. Under IFS, there is minimal dependence on external resources, and about 70% of the input requirements are met within the model thus reducing dependence on external resources. Eight farming system models have been established at ICAR Research Complex for NEH Region, Umiam, Meghalaya during 1983 with various farming enterprises and conservation measures for ensuring food and nutritional security of hill farmers and conservation of natural resources. Assessment of soil fertility after 35 years of establishment revealed that the SOC buildup in IFS followed the trend of natural forest block> timber-based system> agri-horti-silvi-pastoral> agro-pastoral system (Figure 12.1) >livestock-based farming system (Das et al. 2019). There are numerous reports of improvement in overall soil fertility due to various farming systems over shifting cultivation-based land use (Das et al. 2014; Das et al. 2020). Alder tree (*A. nepalensis*) and *Michelia oblonga* improve the accumulation of SOC and available nutrients compared to other tree species under the mid-hill conditions of Meghalaya (Ramesh et al. 2013). The agri-horti-silvi-pastoral land-use models increase SOC by ~45%, mean weight diameter of soil aggregates by 29%, and in-situ soil moisture content by 21%. Further, these systems decrease clay dispersion by 53%,

FIGURE 12.1 Agro-pastoral based farming system model for sloping lands of the EHR (ICAR Research Complex for NEH Region, Umiam, Meghalaya).

soil loss by ~99%, and soil erosion ratio by 45.9% under moderate to steep slopes. Multipurpose trees like *Michelia oblonga* are identified as a better bio-ameliorant for these soils because continuous leaf litter and root exudates improve soil physical behavior and SOC concentration (Ghosh et al. 2009). Das et al. (2010) reported that continuous application of leguminous tree leaves (at 10 Mg ha^{-1} on fresh weight basis) in lowland rice fields sustain soil and crop productivity. Under the mild tropical conditions of Tripura, IFS comprising agriculture and horticultural crops, duckery, fishery, and vermicomposting registered maximum SOC concentration and available nutrient status followed by those of IFS comprising agri-horti crops-fishery and duckery (Debnath et al. 2019). The IFS models improve soil fertility and SOC status through efficient recycling of on-farm biomass and improvement in soil physicochemical and biological properties.

12.9.2.6 Cropping Systems for the Enhancement of SOC

Mono-cropping, cultivation of exhaustive crops, and non-inclusion of leguminous and soil fertility building crops are major contributors of SOC decline in the EHR. Adoption of location-specific suitable cropping sequences and conservation-effective management practices may positively affect SOC concentration and improve quality of degraded hilly soils (Das et al. 2014; Lal, 2015a). In lowlands and valleys, rice–pea, rice–lentil, rice–mustard/rapeseed, and, in terraces and uplands rice–mustard, maize–kidney bean, maize–black gram/green gram, maize–groundnut, vegetable–maize, etc., are identified as some of the cropping systems with the most potential for enhancing productivity and income of the farmers. Most of these cropping systems also enhance SOC concentration due to high biomass turnover per unit area than mono-cropping. Higher SOC and total N contents under NT and residue retention in rice–pea (*Pisum sativum*) (Ghosh et al. 2010), rice–lentil (*Lens culinaris*), (Das et al. 2014b), maize–rapeseed (*Brassica campestris* var. toria), (Das et al. 2011), and maize–kidney bean (*Phaseolus vulgaris*) (Ngangom et al. 2020) have been reported from the diverse ecosystems of the NER. Intercropping of kidney bean, groundnut, and cowpea with maize and soybean and groundnut with rice also improve soil fertility and enhance SOC in the EHR (Layek et al. 2018).

12.9.2.7 Perennial Fodder Grass for Restoring SOC Concentration

Cultivation of perennial fodder crops, with a strong fibrous root system even in degraded hilly soils, can improve soil quality. Perennial grasses provide greater amounts of leaf and root biomass litter for C storage than annual cereal crops, and, hence, grasslands generally have higher SOC levels than croplands (Sarkar et al. 2018). In a study conducted in a degraded land at mid-altitude of Meghalaya, the SOC stock (0–15 cm) after the third year was 5.4–7.5% higher under forages and 2.3–10.4% higher under different fertilizer types compared to the antecedent stock. Among forages, the highest SOC stock was observed under napier (*Pennisetum purpureum*) followed by that under congo signal (*Brachiaria rosenesis*) grass. Among the fertilizers, the maximum SOC stock in 0–30 cm soil depth was observed under organic followed by that under inorganic fertilizers. The SOC stock (0–30 cm soil depth) in the third year under organic fertilizer was 8.1 and 2.1 Mg ha^{-1} higher than that under the control and inorganic fertilizers, respectively (Das et al. 2016). On sloping lands (35%) in mid-altitude of Meghalaya, a fodder crop-based system recorded the maximum SOC concentration (1.80%) and SOC stock (29.7 Mg ha^{-1}) followed by that under a cover crop-based system (1.61%, 26.8 Mg ha^{-1}) at the end of three cropping cycles in 0–15 cm soil depth (Das et al. 2017). Perennial fodder crops increased SOC concentration by 19.9% to 39.6% compared to the soil under conventional maize cultivation. The relative efficacy of the fodder species to SOC accumulation was found most effective for African bristle grass (*Setaria sphacelat*), congo signal (*Brachieria rosenesis*), and Guinea grass (*Panicum maximum*) vs. Makunia in increasing SOC (Sarkar et al. 2018). A geospatial study in the NEH region of India, covering about 10.1 M ha, reported high SOC concentration from grassland that is second only to that under forest land use (Choudhury et al. 2013). The SOC stock (0–15 cm depth) under perennial grassland was 47 Mg ha^{-1} and 23 Mg ha^{-1} under degraded and abandoned *jhum* land, respectively.

The information on SOC stocks and its fractionation into different pools under perennial forage grasses and arable lands in the low hills of northeast India are lacking. Plantation of fodder grasses, like hybrid napier (*Pennisetum glaucum* × *P. purpureum*), congo signal, and combo napier (*Pennisetum purpureum*) and natural grasses, increased the total organic carbon (TOC) stock by 41.5–64.4% at 0–30 cm soil depth compared to cultivated field (59.5 Mg ha^{-1}) over 15 years (Das et al. 2017). Soils under natural grasses had higher TOC stock (97.8 Mg ha^{-1}) and Walkey and Black carbon stock (69.6 Mg ha^{-1}) than those in soils under hybrid napier (84.6 and 61.5 Mg ha^{-1}), congo signal (84.8 and 66.3 Mg ha^{-1}), and combo napier (84.2 and 64.1 Mg ha^{-1}), respectively. The passive carbon pools in soils under natural grass (46.1 Mg ha^{-1}) was significantly higher than those in soil under hybrid napier (39.4 Mg ha^{-1}), combo napier (38.2 Mg ha^{-1}), congo signal (34.9 Mg ha^{-1}), and also in arable land (26.8 Mg ha^{-1}). In general, all the C fractions were higher in soils (0–30 cm depth) under grasses than those in soils under arable crops. The accumulation of greater proportion of TOC pools in soils under planted fodder grasses and natural grass over longer period of time and year-round addition of C through root decomposition of grasses helps in the rehabilitation of degraded croplands and could effectively enhance the soil C storage and stability and reduce the decomposability of C pools present in soils.

12.9.2.8 Cover Crops to Sustain SOC

Heavy rainfall (>2000 mm), cultivation in gentle to steep slopes, along the slope cultivation, loose soil, barren surface, and removal of crop residues are some of the factors which contribute to land degradation, decline in SOC, and poor productivity. A permanent soil cover is therefore recommended by the researchers globally (Lal, 2015) including the EHR for conserving soil resources, building fertility, and SOC sequestration (Das et al. 2017). Ricebean (*Vigna umbellata*) (Figure 12.2), cowpea (*Vigna unguiculata*), groundnut (*Arachis hypogea*), soybean (*Glycine max*), and velvet bean are some of the prominent cover crops recommended for the region (Das et al. 2017). Since most of the farmers in the EHR are resource-poor, a dual-purpose cover crop i.e., which serves the purpose of soil conservation and provides some grains or pods to farmers, are found most feasible for the EHR.

12.9.2.9 Agroforestry Systems (AFS) as an Option for Restoring SOC

In the hills, AFS play an important role in sustainability, resource conservation, and food security (Bhatt et al. 2006). Numerous studies have described the beneficial effects of AFS in long-term soil

FIGURE 12.2 Ricebean as a cover crop in orchards (ICAR Research Complex for NEH Region, Umiam, Meghalaya).

productivity and sustainability, but the magnitude of the beneficial effects may vary with a number of site-specific factors and attributes of associated tree species (Das et al. 2019). Agri-horticulture, agri-silviculture, agri-horti-silvi-pasture, agri-pisciculture, and multi-tier systems are some of the important AFS in the NER (Das et al. 2017). Multipurpose trees (MPTs), like *M. oblonga* were identified as a better bio-ameliorant for these soils because continuous leaf litter and root exudates improved soil physical behavior and SOC (Ghosh et al. 2009). Growing hedgerow (Figure 12.3) species on terrace risers or on contours in slopes conserves soil and water and enriches soil fertility and SOC through leaf biomass addition. Leguminous hedgerow species, like pigeon pea (*Cajanus Cajan*), rattlepods (*Crotolaria tertragona*), tickclover (*Desmodium rensonii*), true indigo (*Indigofera tinctoria*), Mexican lilac (*Glyricidia* spp.), and white hoary pea (*Tephrosia candiada*), produce 9–19.5 Mg pruned biomass annually and contain 3–3.5% N in leaf biomass; they have also been found suitable for the EHR (Tomar et al. 2007). Laxminarayan et al. (2006) reported significant improvement in soil fertility and SOC concentration through the addition of prunings of rattlepods in the mid-altitude of Meghalaya. Agroforestry tree species, like *Alder*, has an annual potential litter production of 473.7 g m^{-2} and fine root biomass of 496 g m^{-2}. The tree bean (*Parkia roxburghii*) and white teak (*Gmelina arborea)* also have very good litter and fine root production potential which enriches SOC and productivity of associated crops (Tomar et al. 2007). The SOC under different agroforestry systems e.g., neem (*Azadirachta indica* A. Juss) + pineapple (*Ananas comosus* L. merr.), teak (*Tectona grandis* Linn) + pineapple, eucalyptus (*Eucalyptus globulus* L.) + pineapple, sisso (*Dalbergia sissoo* Roxb. ex DC.) + pineapple, established in 1989 were monitored in the low altitude of Tripura at *Cocotilla* farm, ICAR Research Complex for the NEH region, Tripura Center and compared to adjacent cultivated land. The SOC concentration was higher under a sisso–pineapple-based system followed by the teak–pineapple system; whereas cultivated land had the lowest concentration of SOC. The SOC concentration in soils in all the agroforestry systems was significantly higher over adjacent cultivated fields.

12.9.2.10 Prospects of Biochar in Soil Health Management in the Hills

Biochar is a carbon-rich, fined-grained, porous by-product of plant biomass that has been subjected to a thermochemical conversion process (pyrolysis) at low temperatures (~350–600°C) in an environment with little or no oxygen (Amonette and Joseph, 2009). It is an important and popular carbon sequestration method to mitigate climate change during modern times since it plays a significant role in sequestering carbon (Lal, 2008a). Biochar can sequester more than 50% of the C in

FIGURE 12.3 Leguminous hedgerow for soil and water conservation.

a highly stable form in soil (Lehmann et al. 2002) and can recover the atmospheric C of 190–213 Tg (Srinivasarao et al. 2013). Moreover, its application can improve soil's physical, chemical, and biological properties (Downie and Van Zwieten, 2013) and serve as a low-cost and effective tool for reducing soil and nutrient loss on sloping lands. Biochar can also sequester carbon from the atmosphere in the hill region, where farmers can adapt the "slash and char" instead of the "slash and burn" practice (Lehmann et al. 2002). There is an ample scope of production and utilization of biochar in the hill region because of abundant availability of crop residues, animal dung, and weed biomass in these areas that can be converted to biochar for soil application (Das et al. 2018d). Biochar productivity from different weed biomasses from the hills of North Eastern India ranged between 35.9% and 48.3% on a dry mass basis (Mandal et al. 2015). Irrespective of sources of biochar (weeds maize stalk and pine needles), its application improves soil pH by 0.26–0.30 units and SOC by 1.5–1.8.18 g kg^{-1} (Mandal et al. 2015). Biochar can significantly increase the soil's water retention and water-holding capacity at 0–15 cm soil depth which helps in growing winter season crops like kidney bean, mustard, etc. in the upland hill ecosystem of the EHR under residual soil moisture (Layek et al. 2020). While the application of weed and maize biochar improves soil pH significantly, irrespective of the source of biochar, its application also enhances SOC concentration, SMBC, and the dehydrogenase enzyme of surface soil (Layek et al. 2020).

12.9.2.11 Improved Shifting Cultivation Practices

Layek et al. (2018) stated that shifting cultivation (locally called *jhum*) causes mass scale destruction of forest resources, degradation of land, and losses of valuable flora, fauna, and animal resources besides the creation of an ecological environment that is non-conducive for crop production. The burning of biomass on *jhum* land releases a large amount of CO_2 into the atmosphere which is the major component in GHGs and cause of climate change (Rastogi et al. 2002). Burning of aboveground vegetation can increase soil pH and cations but decrease carbon and nitrogen contents in the surface soil (Ram and Ramakrishnan, 1988). The maximum soil loss of 30.2–70.2 Mg ha^{-1} y^{-1} is reported under shifting cultivation and the minimum of 0.04–0.52 Mg ha^{-1} y^{-1}under the natural forest system (Saha et al. 2012). The estimated loss of 702.9 kg of SOC is associated with the supportive loss of 63.5 kg of phosphorus (P) and 5.9 kg of potassium (K) ha^{-1} from the 44–53% slopes under shifting cultivation of the NEH region (Ram and Singh, 1993). The length of secondary fallow period has a marked influence on conserving SOC and soil available nutrients (Sarkar et al. 2015). The same study also reported that the oxidizable SOC, which constitutes 71.9% of the total SOC, increases with fallowing duration of *jhum* after forest clearing and successive cropping for 2–3 years. In the NER, increasing human population density has resulted in the practice of slash-and-burn, which includes shortening of the fallow period (Yadav, 2013) as well as permanent conversion of forest to agriculture (Yadav et al. 2012). Perhaps the most important and universal impact of slashing and felling in regard to sustainability is the disruption of natural nutrient cycling and the acceleration of nutrient flow out of the agroecosystem. Pressure due to the increase in population has resulted in more extensive *jhumming* and a subsequent significant reduction in secondary *jhum* fallow period from 25–30 years to 5–6 years (Saha et al. 2015; Punitha et al. 2018). Several experimental evidences clearly showed that the increase in the length of secondary fallow periods from 3 years to 23 years decreases availability of soil nutrients, SOC stock, SMBC, and associated enzyme activities with a concomitant increase in SOC stocks in *jhum* agroecosystems of Mizoram (Sarkar et al. 2015; Lungmuana et al. 2017). Any shortening of fallow periods increases soil erosion rates and SOC decomposition that depletes C storage potential. Under the sub-tropical humid environment of Mizoram, the reduction of SOC stock by 18–38% in 0–10 cm layer and 7–15% in subsurface soil has exhausted forest litter (labile soil C fractions under extensive burning) followed by post-burn crop cultivation. Maintenance of secondary forest fallow period >10 years is congenial for conserving soil health for the first and second year under shifting cultivation in Mizoram (Lungmuana et al. 2017). The inclusion of leguminous crops in *jhum* (Figure 12.4) enhanced the system productivity and profitability under *jhum* cultivation (Ansari et al. 2017). The maize + groundnut intercropping

FIGURE 12.4 Strip cropping of rice and soybean in *jhum* fields for soil and water conservation (Ri-bhoi District, Meghalaya).

under 50% RDF + FYM 2 Mg ha^{-1} + 500 kg lime ha^{-1} resulted in significantly higher available P, soil pH, and SOC concentration of soil as compared to the conventional practice of mono-cropping and no fertilization. Since the N fertilizer speeds up the mineralization of organic matter and facilitates slow but steady release of nutrients to crops, integrated application of chemical fertilizers and organic manure could enhance nutrient availability and increase the crop yield (Das et al. 2020). Similarly, in respect of nutrient management in rice, application of the minimum amount (50% RDF) of fertilizer under slash-and-burn agriculture could supplement the nutrient shortfall when lands are used more than once at the same site. Improved rice varieties like Bhalum-2, Bhalum-3, and IURON-514 can produce significantly higher yields than local varieties. Soybean and groundnut as intercrops with rice (Figure 12.4) enhance productivity and soil health. *Tephrosia* spp. along the contour at suitable intervals checks soil loss and improves soil fertility (Layek et al. 2018).

Developing improved methods of *jhum* cultivation, like proper bunding, residue recycling, use of high yielding varieties, etc. are a few short-term options. However, long-term viable options would be proper land configuration like terraces, water harvesting, agroforestry, adoption of integrated farming system approaches, etc. Proper fallow management with seeding of leguminous crops like rice bean, velvet bean (*Mucuna spp*), wild sunflower, cowpea (*Vigna sinensis*), etc. helps in early fertility-building of *jhum* fallow areas. The rate of rehabilitation of degraded *jhum* land can be further enhanced by maintaining alder tree (*Alnus nepalensis*). Besides fixing atmospheric N, the litters from lopping of alder trees enriches soil with P, K, and C (Das et al. 2018b).

Permanent changes in existing fallow period based *jhum* land use and cover also has the potential for the restoration of SOC dynamics, SOC, and microbial activities (Gispert et al. 2013). Technological alternatives through the replenishment of secondary *jhum* fallows with the perennial plantation cropping options viz. oil palm, rubber, teak, coffee, banana, mandarin, etc. have the potential for sustainable management of *jhum* fallow toward SOC conservation on the steep slopes of the NEH region (Grogan et al. 2012). The identification of promising and sustainable land use in the fragile degraded ecosystems of the north is of utmost importance for the conversion of degraded *jhum* fallows in the sub-tropical EHR. Conversion of secondary *jhum* fallows into plantations crop [arecanut (Areca catechu) and rubber (*Hevea brasiliensis*)] conserves SOC, increases the soil's water-holding capacity (WHC), and improves soil health under the sub-tropical warm humid environment of Mizoram (Lungmuana et al. 2019). Further increase in temperatures from global climate change has the clear potential for increasing available nitrogen content (by 1.85–49.32%) along with the soil C mobilizing enzyme activities viz. dehydrogenase, fluorescein diacetate hydrolase, β–glucosidase etc.; thus, reducing soil health under the established orchards and agricultural land-use systems in EHR (Saha et al. 2015). With minimum disturbance, soils of primary and secondary forests (regenerated in *jhum* fallows) in EHR are sinks of carbon that play an important role in both climate change adaptation and mitigation (Vashum et al. 2016).

12.10 CONCLUSIONS

Improving SOC through the adoption of location-specific conservation-effective management practices in croplands of the EHR offers an opportunity for soil quality improvement, SOC buildup for sustainable agriculture, and environmental security. For valleys, conservation tillage, residue management, INM, crop diversification, and IFS are recommended to improve SOC and soil quality. Whereas, in uplands and terrace lands, in-situ soil moisture conservation practices comprising mulching, no-till, efficient cropping system, cover crops, intercropping, and IFS (including multipurpose trees) are found suitable for enhancing SOC. Hedgerow on terrace risers and contours on sloping lands, agroforestry in sloping and degraded lands, cover crops, strip cropping, and perennial forage crops are recommended strategies for sloping and undulating lands in the EHR of India. *Jhum* (Shifting cultivation) rehabilitation through conservation measures, like residue retention, cover cropping, biofertilizer, fast-growing trees, and effective fallow management with leguminous crops are found effective for reducing the *jhum* cycle and improving the SOC status. Improved shifting cultivation practices and the application of biochar to enhance SOC and productivity should also be included in the strategy for SOC sustenance and building in the EHR of India.

REFERENCES

Agricultural Statistics at a Glance. 2019. *Directorate of Economics and Statistics*. GOI.

Amonette, J. E., and S. Joseph. 2009. Physical properties of biochar. In: *Biochar for Environmental Management. Science and Technology*, eds. J. Lehmann, and S. Joseph, 33–53. London: Earthscan.

Ansari, M. A., P. K. Saraswat, S. S. Sharma, P. Punitha, M. H. Ansari, N. Prakash, R. K. Mishra, N. Lal, and Y. Ramakrishna. 2017. Significance and strategies of legume production for achieving nutritional security in North East Indian Himalayan Region. *Journal of Food Legumes* 30(3):56–63.

Apollo, M. 2017. The population of Himalayan regions – By the numbers: Past, present and future. In: *Contemporary Studies in Environment and Tourism*, eds. R. Efe, and M. Ozturk, 143–159. Cambridge Scholars Publishing, Newcastle upon Tyne, UK .

Arunachalam, A. 2002. Dynamics of soil nutrients and microbial biomass during first year cropping in an 8-year *jhum* cycle. *Nutrient Cycling in Agroecosystems* 64(3):283–291.

Aulakh, M. S. 2010. *World Congress of Soil Science, Soil Solutions for a Changing World* (August):1–6. Brisbane, Australia.

Avasthe, R. K., S. Babu, R. Singh, G. S. Yadav, and A. Kumar. 2020. Productivity and profitability assessment of organically grown vegetables embedded in rice based cropping sequences in Sikkim Himalayas, North East India. *Journal of Environmental Biology* 41(1):111–117.

Babhulkar, P. S., R. M. Wandile, W. P. Badole, and S. S. Balpande. 2000. Residual effect of long-term application of FYM and fertilizers on soil properties (vertisols) and yield of soybean. *Journal of the Indian Society of Soil Science* 48(1):89–92.

Babu, S., R. Singh, R. K. Avasthe, and G. S. Yadav. 2015. Strategies for enhancing productivity of organic agriculture in North Eastern Region of India. In: Souvenir and Conference Book of National Conference on Global Research Initiatives for Sustainable Agriculture and Allied Sciences, 55–58. Gwalior, Madhya Pradesh, India.

Babu, S., R. Singh, R. K. Avasthe, G. S. Yadav, K. P. Mohapatra, T. Selvan, A. Das, V. K. Singh, D. Valente, and I. Petrosillo. 2020. Soil carbon dynamics in Indian Himalayan intensified organic rice-based cropping sequences. *Ecological Indicators* 114:106292.

Barnard, R. L., C. A. Osborne, and M. K. Firestone. 2015. Changing precipitation pattern alters soil microbial community response to wet-up under a Mediterranean-type climate. *The ISME Journal* 9(4):946–957.

Behera, R. N., D. K. Nayak, P. Andersen, and I. E. Maren. 2016. From jhum to broom: Agricultural land-use change and food security implications on the Meghalaya Plateau, India. *Ambio – A Journal of the Human Environment* 45(1):63–77.

Bhatt, B. P., K. M. Bujarbaruah, and Y. P. Sharma. 2006. Integrated farming system: A sustainable alternative for the benefit of small scale farmers and the environment. In: *Agroforestry in North East India: Opportunities and Challenges*, eds. B. P. Bhatt, and K. M. Bujarbaruah, 537–555. Umiam: ICAR Research Complex for NEH Region.

Bhattacharya, T., D. K. Pal, P. Chandran, S. K. Ray, C. Mandal, and B. Telpande. 2008. Soil carbon storage capacity as a tool to prioritize areas for carbon sequestration. *Current Science* 95:482–494.

Bhattacharyya, R., B. N. Ghosh, P. K. Mishra, B. Mandal, S. Ch, D. Sarkar, K. Das, K. K. Anil, M. Lalitha, K. M. Hati, and A. J. Franzluebbers. 2015. Soil Degradation in India: Challenges and Potential Solutions. *Sustainability* 7:3528–3570.

Bhattacharyya, T., D. Sarkar, S. K. Gangopandhyay, P. N. Dubey, U. Baruah, G. S. Chamuah, S. Mukhopadhyay, D. C. Nayak, A. K. Maji, R. K. Saxena, A. K. Barthwar, N. D. R. Krishna, C. Mandar, and J. Sehgal. 1998. Soils of Tripura. 1-Characterisation and classification. *Agropedology* 8:47–54.

Bhowmick, B. C. 2005. *Changing Pattern of Rice Production Systems and Technology in Assam: A Spatio-temporal Analysis of Performance and Prospects.* NCAP (ICAR), New Delhi.

Boerner, R. E. J., J. A. Brinkman, and A. Smith. 2005. Seasonal variations in enzyme activity and organic carbon in soil of a burned and unburned hardwood forest. *Soil Biology and Biochemistry* 37(8):1419–1426.

Bujarbaruah, K. M. 2004. Organic farming: Opportunities and challenges in north eastern region of India. In: International Conference on Organic Food, 13–23. Barapani: ICAR Research Complex for NEH Region.

Callow, J. A. 1997. *Advances in Botanical Research*, 26. Academic Press, Harcourt Brace & Company, Publishers, San Diego, London, Boston, New York, Sydney, Tokyoand Toronto.

CEPF (Profile, C.E.). 2005. *Ecosystem Profile: Eastern Himalayas Region.* http://www.cepf.net/Documents/final.ehimalayas.ep.pdf.

Chakraborty, D., J. K. Ladha, D. S. Rana, M. L. Jat, M. K. Gathala, S. Yadav, A. N. Rao, M. S. Ramesha, and A. Raman. 2017. A global analysis of alternative tillage and crop establishment practices for economically and environmentally efficient rice production. *Scientific Reports* 7(1):9342.

Chatterjee, D., R. Kuotsu, M. Ao, S. Saha, S. K. Ray, and S. V. Ngachan. 2019. Does rise in temperature adversely affect soil fertility, carbon fractions, and microbial biomass and enzyme activities under different land uses? *Current Science* 116(12):2044.

Choudhury, B. U., K. P. Mohapatra, A. Das, P. T. Das, L. Nongkhlaw, R. A. Fiyaz, S. V. Ngachan, S. Hazarika, D. J. Rajkhowa, and G. C. Munda. 2013. Spatial variability in distribution of organic carbon stocks in the soils of North East India. *Current Science* 104:5.

Choudhury, B. U., A. R. Fiyaz, K. P. Mohapatra, and S. Ngachan. 2016. Impact of land uses, agrophysical variables and altitudinal gradients on soil organic carbon concentration of North-Eastern Himalayan Region of India. *Land Degradation and Development* 27(4):1163–1174.

Ciais, P., C. Sabine, G. Bala, L. Bopp, V. Brovkin, J. Canadell, and C. Jones. 2014. Carbon and other biogeochemical cycles. In: *Climate Change 2013: The Physical Science Basis. Contribution of Working Group I to the Fifth Assessment Report of the Intergovernmental Panel on Climate Change.* Cambridge University Press, 465–570.

Condron, L. M., D. W. Hopkins, E. G. Gregorich, A. Black, and S. A. Wakelin. 2014. Long-term irrigation effects on soil organic matter under temperate grazed pasture. *European Journal of Soil Science* 65(5):741–750.

Dabral, P. P., N. Baithuri, and A. Pandey. 2008. Soil erosion assessment in a hilly catchment of North Eastern India using USLE, GIS and remote sensing. *Water Resources Management* 22(12):1783–1798.

Das, A., S. Basavaraj, J. Layek, G. I. Ramkrushna, R. Lal, K. Rangappa, G. S. Yadav, S. Babu, P. K. Ghosh, and S. Ngachan. 2020. Can conservation tillage and residue management enhance energy use efficiency and sustainability of rice-pea system in the Eastern Himalayas? *Archives of Agronomy and Soil Science* 66(6):830–846.

Das, A., D. P. Patel, M. Kumar, G. I. Ramkrushna, S. V. Ngachan, and M. Lyngdoh. 2014b. Influence of cropping systems and organic amendments on productivity and soil health at mid altitude of North East India. *Indian Journal of Agricultural Sciences* 84(12):1525–1530.

Das, A., D. P. Patel, R. Lal, M. Kumar, G. I. Ramkrushna, J. Layek, J. Buragohain, S. V. Ngachan, P. K. Ghosh, B. U. Choudhury, K. P. Mohapatra, and B. G. Shivakumar. 2016. Impact of fodder grasses and organic amendments on productivity and soil and crop quality in a subtropical region of eastern Himalayas, India. *Agriculture, Ecosystems and Environment* 216:274–282.

Das, A., G. C. Munda, P. K. Ghosh, G. I. Ramkrushna, D. P. Patel, B. U. Choudhury, S. Ngachan, and J. Layek. 2011. Managing degraded land and soil organic carbon for climate resilient agriculture in North East India. In: *Soil Carbon Sequestration for Climate Change Mitigation and Food Security*, ed. C. H. Srinivasa Rao, B. Venkateswarlu, K. Srinivas, S. Kundu, and A. K. Singh, 241–265. Hyderabad: Central Research Institute for Dry land Agriculture.

Das, A., G. I. Ramkrushna, B. Makdoh, J. Layek, S. Mandal, and R. Lal. 2017. *Lower Himalayas: Soil Management. Encyclopedia of Soil Science*, 3rd Edition. Taylor & Francis, Boca Raton, 1382–1387.

Das, A., G. I. Ramkrushna, B. U. Choudhury, G. C. Munda, D. P. Patel, S. V. Ngachan, P. K. Ghosh, A. K. Tripathi, S. Das, and M. Kumar. 2012. Natural resource conservation through indigenous farming systems: Wisdom alive in North East India. *Indian Journal of Traditional Knowledge* 11(3):505–513.

Das, A., G. S. Yadav, S. Babu, D. Sarkar, R. S. Meena, and R. Lal. 2018b. Managing nitrogen in small land holder hill farms of north eastern Indian Himalayas. In: *Soil Nitrogen Uses and Environmental Impacts Series: Advances in Soil Science*, eds. R. Lal and B. A. Stewart. CRC Press, Taylor & Francis, 257–282.

Das, A., J. Layek, G. I. Ramkrushna, S. Basavaraj, R. Lal, R. Krishnappa, G. S. Yadav, and S. Babu. 2020. Conservation tillage and residue management improves soil properties under upland rice–rapeseed system in subtropical eastern Himalayas. *Land Degradation and Development*. DOI: 10.1002/ldr.3568.

Das, A., J. M. S. Tomar, T. Ramesh, G. C. Munda, P. K. Ghosh, and D. P. Patel. 2010. Productivity and economics of lowland rice as influenced by incorporation of N-fixing tree biomass in mid-altitude subtropical Meghalaya, North East India. *Nutrient Cycling in Agroecosystems* 87(1):9–19.

Das, A., R. Lal, D. P. Patel, G. I. Ramkrushna, J. Layek, S. V. Ngachan, P. K. Ghosh, J. Bordoloi, M. Kumar, J. Bordoloi, and M. Kumar. 2014a. Effects of tillage and biomass on soil quality and productivity of lowland rice cultivation by small scale farmers in North Eastern India. *Soil and Tillage Research* 143:50–58.

Das, A., S. Babu, A. S. Panwar, S. V. Ngachan, D. J. Rajkhowa, N. Prakash, and M. Tahashildar. 2019. Micro-watershed based farming system models for natural resource conservation and livelihood of hill farmers, 117. ICAR Research Complex for NEH Region, Umiam-793 103. Meghalaya.

Das, S. K., R. K. Avasthe, M. Singh, and A. Yadav. 2018d. Soil health improvement using biochar application in Sikkim: A success story. *Innovative Farming* 3(1):48–50.

Das, S., D. Chakraborty, V. K. Singh, P. Aggarwal, R. Singh, R. P. Mishra, and B. S. Dwivedi. 2014. Effect of integrated nutrient management practice on soil aggregate properties, its stability and aggregate-associated carbon content in an intensive rice–wheat system. *Soil and Tillage Research* 136:9–18.

Das, T. K., R. Bhattacharya, A. R. Sharma, S. Das, A. A. Saad, and H. Pathak. 2013. Impacts of conservation agriculture on total soil organic carbon retention potential under an irrigated agro-ecosystem of the western Indo-Gangetic Plains. *European Journal of Agronomy* 51:34–42.

Datta, M., A. Das, and S. V. Ngachan. 2012. Soil carbon management in hill agriculture options and opportunities in Northeast India. In: *Carbon Management in Agriculture for Mitigating Greenhouse Effect*, eds. A. K. Singh, S. V. Ngachan, G. C. Munda, K. P. Mohapatra, B. U. Choudhury, A. Das, Ch. SrinivasaRao, D. P. Patel, D. J. Rajkhowa, G. I. Ramkrushna,and A. S. Panwar, 102–114. Umiam-793 103. Meghalaya, India: ICAR Research Complex for NEH Region.

Debnath, C., G. S. Yadav, L. Sahoo, H. L. Devi, and V. Singh. 2019. Production and profit from different farming system approaches in Tripura. *Indian Journal of Hill Farming* 32(1):10–15.

Ding, X., X. Han, Y. Liang, Y. Qiao, L. Li, and N. Li. 2012. Changes in soil organic carbon pools after 10 years of continuous manuring combined with chemical fertilizer in a mollisol in China. *Soil and Tillage Research* 122:36–41.

Downie, A., and L. Van Zwieten. 2013. Biochar: A co-product to bioenergy from slow-pyrolysis technology. In: *Advanced Biofuels and Bioproducts*, ed. J. W. Lee, 97–117. New York: Springer.

Efthimiadou, A., D. Bilalis, A. Karkanis, and B. Fround-Williams. 2010. Combined organic/inorganic fertilization enhances soil quality and increased yield, photosynthesis and sustainability of sweet maize crop. *Australian Journal of Crop Science* 4(9):722–729.

Esmaeilzadeh, J., and A. G. Ahangar. 2014. Influence of soil organic matter content on soil physical, chemical and biological properties. *International Journal of Plant, Animal and Environmental Sciences* 4(4):244–252.

Gentile, R. M., D. L. Martino, and M. H. Entz. 2005. Influence of perennial forages on subsoil organic carbon in a long-term rotation study in Uruguay. *Agriculture, Ecosystems and Environment* 105(1–2):419–423.

Ghosh, P. K., A. Das, R. Saha, K. Enboklang, A. K. Tripathi, G. C. munda, and S. V. Ngachan. 2010. Conservation agriculture toward achieving food security in North East India. *Current Science* 99:915–921.

Ghosh, P. K., R. Saha, J. J. Gupta, T. Ramesh, A. Das, T. D. Lama, G. C. Munda, J. S. Bordoloi, M. R. Verma, and S. V. Ngachan. 2009. Long-term effect of pastures on soil quality in acid soil of North – East India. *Australian Journal of Soil Research* 47(4):372–379.

Gispert, M., M. Emran, G. Pardini, S. Doni, and B. Ceccanti. 2013. The impact of land management and abandonment on soil enzymatic activity, glomalin content and aggregate stability. *Geoderma* 202–203:51–61.

Grogan, P., F. Lalnunmawai, and S. K. Tripathi. 2012. Shifting cultivation in steeply sloped regions: A review of management options and research priorities for Mizoram state, Northeast India. *Agroforestry Systems* 84(2):163–177.

Handayanto, E., K. E. Giller, and G. Cadisch. 1997. Regulating N release from legume tree prunings by mixing residues of different quality. *Soil Biology and Biochemistry* 29(9–10):1417–1426.

Haynes, R. J. 1999. Size and activity of the soil microbial biomass under grass and arable management. *Biology and Fertility of Soils* 30(3):210–216.

Himes, F. L. 1998. Nitrogen, sulfur and phosphorus and the sequestering of carbon. In: *Soil Processes and the Carbon Cycle*, 315–319, eds. R. Lal, J. M. Kimble, R. F. Follett, and B. A. Stewar. Boca Raton, FL: CRC Press.

Hmar, L., and S. N. Ramanujam. 2014. Earthworm cast production and physico-chemical properties in two agroforestry systems of Mizoram (India). *Tropical Ecology* 55(1):75–84.

Indoria, A. K., K. L. Sharma, K. S. Reddy, C. Srinivasarao, K. Srinivas, S. S. Balloli, M. Osman, G. Pratibha, and N. S. Raju. 2018. Alternative sources of soil organic amendments for sustaining soil health and crop productivity in India–impacts, potential availability, constraints and future strategies. *Current Science* 115(11):2052.

IFDC (International Fertilizer Development Center). 2006. Africa Fertilizer Summit Proceedings, Abuja, Nigeria, June 9–13.

Kautz, T., S. Wirth, and F. Ellmer. 2004. Microbial activity in a sandy arable soil is governed by the fertilization regime. *European Journal of Soil Biology* 40(2):87–94.

Kuotsu, K., A. Das, R. Lal, G. C. Munda, P. Ghosh, and S. V. Ngachan. 2014. Land forming and tillage effects on soil properties and productivity of rain-fed groundnut (*Arachishypogaea* L.) – Rapeseed (*Brassica campestris* L.) cropping system in northeastern India. *Soil and Tillage Research* 142:15–24.

Ladha, J., C. K. Reddy, A. T. Padre, and C. Kessel. 2011. Role of nitrogen fertilization in sustaining organic matter in cultivated soils. *Journal of Environmental Quality* 40(6):1756–1766.

Lal, J. K., B. Mishra, and A. K. Sarkar. 2000. Effect of plant residues incorporation on specific microbial groups and availability of some plant nutrients in soil. *Journal of Indian Society of Soil Science* 48:67–77.

Lal, R. 2003. Soil erosion and the global carbon budget. *Environment International* 29(4):437–450.

Lal, R. 2004a. Soil carbon sequestration impacts on global climate change and food security. *Science* 304(5677):1623–1627.

Lal, R. 2004b. Carbon emission from farm operations. *Environment International* 30(7):981– 990.

Lal, R. 2008a. Carbon sequestration. *Philosophical Transactions of the Royal Society of London Series B* 363(1492):815–830.

Lal, R. 2008b. Carbon sequestration in soil. *CAB Reviews: Perspectives in Agriculture, Veterinary Science, Nutrition and Natural Resources*. http://www.cababstracts.org/cabreviews

Lal, R. 2008e. Soil carbon stocks under present and future climate with specific reference to European ecoregions. *Nutrient Cycling in Agroecosystems* 81:113–127. DOI: 10.1007/ s10705-007-9147-x.

Lal, R. 2008d. Sequestration of atmospheric CO2 in global carbon pools. *Energy & Environmental Science* 1:86–100. DOI: 10. 1039/b809492f.

Lal, R. 2015a. Soil carbon sequestration and aggregation by cover cropping. *Journal of Soil and Water Conservation* 70(6):329–339.

Lal, R. 2015b. International year of soil and the sustainable development goals of the united Nation. In: *Souvenir of Sustaining Hill Agriculture in Changing Climate*. Agartala, Tripura, 5–7 Dec.

Lal,R. 2016. Feeding 11 billion on 0.5 billion hectare of cropland. *Food and Energy Security Journal* 5(4):239–251.

Lal, R. 2018b. Digging deeper: A holistic perspective of factors affecting soil organic carbon sequestration in agroecosystems. *Global Change Biology*. DOI: 10.1111/gcb.14054.

Lal, R., J. M. Kimble, R. F. Follett, and C. V. Cole. 1998. *The Potential of U.S. Cropland to Sequester Carbon and Mitigate the Greenhouse Effect*, 128. Boca Raton, FL: Lewis Publishers.

Laxminarayana, K., B. P. Bhatt, and T. Rai. 2006. Soil fertility buildup through Hedgerow intercropping in integrated Farming system; a case study. In: *Agroforestry in North East India, Opportunities and Challenges*, eds. B. P. Bhatt, and K. M. Bujarbaruah, 479–490. Umiam, Meghalaya, India: ICAR Research Complex for NEH Region.

Layek, J., A. Balusamy, S. Hazarika, S. Devi, R. Krishnappa, K. P. Mohapatra, S. Mandal, and B. U. Choudhary. 2020. *Prospects of Bio-Char for Sustainable Agriculture and Carbon Sequestration in*

North Eastern Hill Region of India, 46. Umiam – 793 103, Meghalaya, India: ICAR Research Complex for North Eastern Hill (NEH) Region.

Layek, J., A. Das, G. I. Ramkrushna, A. S. Panwar, B. C. Verma, and A. Roy. 2018. Improving rice production under shifting cultivation: A case study. In: *Conservation Agriculture for Advancing Food Security in Changing Climate*, eds. A. Das, K. P. Mohapatra, S. V. Ngachan, A. S. Panwar, D. J. Rajkhowa, G. I. Ramkrushna, and J. Layek, 143–153. New Delhi - 110 002. India: Today & Tomorrow's Printers and Publishers.

Layek, J., A. Das, T. Mitran, C. Nath, R. S. Meena, G. S. Yadav, B. G. Shivakumar, S. Kumar, and R. Lal. 2018. Cereal + legume intercropping: An option for improving productivity and sustaining soil health. In: *Legumes for Soil Health and Sustainable Management,* eds. Meena, R. S., A. Das, R. Lal, and G. S. Yadav, 347–386. Singapore: Springer.

Layese, M. F., C. E. Clapp, R. R. Allmaras, D. R. Linden, S. M. Copeland, J. A. E. Molina, and R. H. Dowdy. 2002. Current and relic carbon using natural abundance carbon-13. *Soil Science* 167(5):315–326.

Lee, S. B., C. H. Lee, K. Y. Jung, K. D. Park, D. P. Lee, and J. Kim. 2009. Changes of soil organic carbon and its fractions in relation to soil physical properties in a long-term fertilized paddy. *Soil and Tillage Research* 104(2):227–232.

Lehmann, J., J. P. da Silva Jr. , M. Cravo, J. Greenwood, T. Nehls, C. Steiner, and B. Glaser. 2002. Slash-and-char – A feasible alternative for soil fertility management in the central Amazon? In: Proceedings of the 17th World Congress of Soil Science, 1–12. Bangkok: Soil and Fertility Society of Thailand.

Lepcha, N. T., and B. N. Devi. 2020. Carbon cycling and balance in a traditional cardamom based agroforestry system of Sikkim Himalayas. *Tropical Ecology*:61; 527-537–.

Lungmuana, S. B., S. S. Vanthawmliana, A. R. Rambuatsaiha, S. Saha, S. K. Dutta Rambuatsaiha, A. R. Singh, and T. Boopathi. 2017. Impact of secondary forest fallow period on soil microbial biomass carbon and enzyme activity dynamics under shifting cultivation in North Eastern Hill region, India.

Lungmuana, S. B. Singh, B. U. Choudhury, Vanthawmliana, S. Saha, and V. Hnamte. 2019. Transforming jhum to plantations: Effect on soil microbiological and biochemical properties in the foot hills of North Eastern Himalayas, India. *CATENA* 177:84–91.

Majumder, B., B. Mandal, P. K. Bandyopadhyay, A. Gangopadhyay, P. K. Mani, A. L. Kundu, and D. Mazumdar. 2008. Organic amendments influence soil organic carbon pools and rice–wheat productivity. *Soil Science Society of America Journal* 72(3):775–785.

Mandal, D., and V. N. Sharda. 2013. Appraisal of soil erosion risk in the Eastern Himalayan region of India for soil conservation planning. *Land Degradation and Development* 24:430–437.

Mandal, S., B. C. Verma, G. I. Ramkrushna, R. K. Singh, and D. J. Rajkhowa. 2015. Characterization of biochar obtained from weeds and its effect on soil properties of North Eastern Region of India. *Journal of Environmental Biology* 36(2):499.

Nandy, S. N., P. P. Dhyani, and P. K. Samal. 2006. Resource information database of the Indian Himalaya. In: *Gb Pant Institute of Himalayan Environment & Development.* Almora, 96.

Nath, A. J., and A. K. Das. 2012. Carbon pool and sequestration potential of village bamboos in the agroforestry system of northeast India. *Tropical Ecology* 53(3):287–293.

Nath, A. J., B. Brahma, G. W. Sileshi, and A. K. Das. 2018. Impact of land use changes on the storage of soil organic carbon in active and recalcitrant pools in a humid tropical region of India. *Science of the Total Environment* 624:908–917.

Nath, A. J., R. Lal, and A. K. Das. 2015. Ethnopedology and soil properties in bamboo (*Bambusa*sp.) based agroforestry system in North East India. *CATENA* 135:92–99.

Nayek, A. K., B. Gangwara, K. Arvind, S. Shukla, P. Mazumdar, A. Kumar, R. Rajab, A. Kumar, V. Kumar, P. K. Rai, and U. Mohan. 2012. Long-term effect of different integrated nutrient management on soil organic carbon and its fractions and sustainability of rice–wheat system in Indo Gangetic Plains of India. *Field Crops Research* 127:129–139.

Ngangom, B., A. Das, R. Lal, R. G. Idapuganti, J. Layek, S. Basavaraj, S. Babu, G. S. Yadav, and P. K. Ghosh. 2020. Double mulching improves soil properties and productivity of maize-based cropping system in eastern Indian Himalayas. *International Soil and Water Conservation Research* 8(3):308–320.

Pandey, A., A. Mathur, S. K. Mishra, and B. C. Mal. 2009. Soil erosion modeling of a Himalayan watershed using RS and GIS. *Environmental Earth Sciences* 59(2):399–410.

Panwar, N. R., P. Ramesh, A. B. Singh, and S. Ramana. 2010. Influences of organic, chemical and integrated management practices on soil organic carbon and soil nutrient status under semi-arid tropical conditions in central India. *Communications in Soil Science and Plant Analysis* 41(9):1073–1083.

Patel, D. P., A. Das, M. Kumar, G. C. Munda, S. V. Ngachan, G. I. Ramkrushna, J. Layek, J. Naropongla, J. Buragohain, and U. Somireddy. 2015. Continuous application of organic amendments enhances soil health, produce quality and system productivity of vegetable based cropping systems at subtropical eastern Himalayas. *Experimental Agriculture* 51(1):85–106.

Pathak, K., A. J. Nath, A. K. Das, and M. Bhar. 2018. Litter dynamics in *Imperata cylindrica* grassland under culturally managed system in North East India. *Journal of Tropical Agriculture* 56(2):99–106.

Peter, G., G. Francesco, and Y. Montague. 2020. Integrated nutrient management, soil fertility, and sustainable agriculture: Current issues and future challenges. A 2020 Vision for food, agriculture, and the environment. Brief 67. International Food Policy Research Institute.

Poręba, G., and P. Prokop. 2011. Estimation of soil erosion on cultivated fields on the hilly Meghalaya Plateau, North-East India. *Geochronometria* 38(1):77–84.

Punitha, P., M. A. Ansari, D. K. Pandey, D. Ram, S. Datt, P. K. Sharma, M. Aheibam, S. S. P. Jyothi, and N. Prakash. 2018. Shifting cultivation in North East India: Social dimension, cross cultural reflection and strategies for improvement. *Indian Journal of Agricultural Sciences* 88(6):811–819.

Ram, M., and B. P. Singh. 1993. Soil fertility management in farming systems. In: *Lectures Notes, off Campus Training on Farming System*. Aizawl, India.

Ram, S. C., and P. S. Ramakrishnan. 1988. Hydrology and soil fertility of degraded grasslands at Cherrapunji in northeastern India. *Environmental Conservation* 15(1):29–35.

Ramesh, T., K. M. Manjaiah, J. M. S. Tomar, and S. V. Ngachan. 2013. Effect of multipurpose tree species on soil fertility and CO_2 efflux under hilly ecosystems of Northeast India. *Agroforestry Systems* 87(6):1377–1388.

Rastogi, M., S. Singh, and H. Pathak. 2002. Emission of carbon dioxide from soil. *Current Science* 82(5):510–518.

Regnier, P., P. Friedlingstein, P. Ciais, F. T. Mackenzie, N. Gruber, I. A. Janssens, et al. 2013. Anthropogenic perturbation of the carbon fluxes from land to ocean. *Nature Geoscience* 6(8):597–607.

Ren, W., H. Q. Tian, B. Tao, Y. Huang, and S. F. Pan. 2012. China's crop productivity and soil carbon storage as influenced by multifactor global change. *Global Change Biology* 18(9):2945–2957.

Saha, R., and P. K. Ghosh. 2013. Soil organic carbon stock, moisture availability and crop yield as influenced by residue management and tillage practices in maize–mustard cropping system under hill agro-ecosystem. *National Academy Science Letters* 36(5):461–468.

Saha, R., J. M. S. Tomar, and P. K. Ghosh. 2007. Evaluation and selection of multipurpose tree for improving soil hydro-physical behaviour under hilly ecosystem of North East India. *Agroforestry Systems* 69(3):239–247.

Saha, R., R. S. Chaudhary, and J. Somasundaram. 2012. Soil health management under hill agroecosystem of North East India. *Applied and Environmental Soil Science*. 9 pages: Article ID 696174.DOI: 10.1155/2012/696174.

Saha, S., D. Chakraborthy, B. U. Choudhury, S. B. Singh, N. Chinza, C. Lalzarliana, A. R. Singh, S. Chowdhury, T. Boopathi, Lungmuana A. R. Singh, and S. V. Ngachan. 2015. Spatial variability in temporal trends of precipitation and its impact on the agricultural scenario of Mizoram. *Current Science* 109 (12):2278–2282.

Saha, S., D. Chakraborty, R. K. Paul, S. Samanta, and S. B. Singh. 2018. Disparity in rainfall trend and patterns among different regions: Analysis of 158 years' time series of rainfall dataset across India. *Theoretical and Applied Climatology* 134(1–2):381–395.

Sahu, G., S. Das, and S. Mohanty. 2020. Nutrient budgeting of primary nutrients and their use efficiency in India. *International Research Journal of Pure and Applied Chemistry*:21: 92–114.

Sainju, U. M., A. W. Lenssen, T. Caesar-Tonthat, and R. G. Evans. 2009. Dryland crop yields and soil organic matter as influenced by long-term tillage and cropping sequence. *Agronomy Journal* 101(2):243–251.

Sarkar, D., B. Meitei, Ch. , A. Das, P. K. Ghosh, and B. Mandal. 2018. Changes in soil organic carbon pools in a long-term trial with perennial fodder crops in acid soils of North-East India. *Grass and Forage Science* 73(2):473–481.

Sarkar, D., Ch. Meitei, L. K. Baishya, S. Ghosh, K. L. Chongloi, D. J. Rajkhowa, and D. Rajkhowa. 2015. Potential of fallow chrono sequence in shifting cultivation to conserve soil organic carbon in northeast India. *CATENA* 135:321–327.

Sharma, K. L., U. K. Mandal, K. Srinivas, K. P. R. Vittal, B. Mandal, J. K. Grace, and V. Ramesh. 2005. Long-term soil management effects on crop yields and soil quality in a dryland alfisol. *Soil and Tillage Research* 83(2):246–259.

Singh, A. K., L. J. Bordoloi, M. Kumar, S. Hazarika, and B. Panwar. 2014. Land use impact on soil quality in eastern Himalayan region of India. *Environmental 2014 Monitoring and Assessment* 186(4):2013–2024.

Singh, R., S. Babu, R. K. Avasthe, G. S. Yadav, and S. V. Ngachan. 2017. Organic production technology for alternative cropping systems in Sikkim. In: *ICAR – National Organic Farming Research Institute*, 34. Tadong, Gangtok.

Singh, S. B., B. U. Choudhury, S. Saha, V. Hnamte, S. Saha, and V. Hnamte. 2019. Transforming *jhum* to plantations: Effect on soil microbiological and biochemical properties in the foot hills of North Eastern Himalayas, India. *CATENA* 177:84–91.

Singh, R., B. Subhash, R. K. Avasthe, and G. S. Yadav. 2018. *Crop Diversification and Intensification for Enhancing Livelihood Security in Sikkim*. Tadong, Gangtok, Sikkim, India: ICAR-National Organic Farming Research Institute. 36 p.

Smith, K. A., T. Ball, F. Conen, K. E. Dobbie, J. Massheder, and A. Rey. 2003. Exchange of greenhouse gases between soil and atmosphere: Interactions of soil physical factors and biological processes. *European Journal of Soil Science* 54(4):779–791.

Sodhi, G. P. S., V. Beri, and D. K. Bendi. 2009. Soil aggregation and distribution of carbon and nitrogen in different fractions under long-term application of compost in rice–wheat system. *Soil and Tillage Research* 103(2):412–418.

Srinivasarao, Ch., B. Venkateswarlu, R. Lal, A. K. Singh, and S. Kundu. 2013. Sustainable management of soils of dryland ecosystems of India for enhancing agronomic productivity and sequestering carbon. In: *Advances in Agronomy*, 121, ed. D. L. Sparks, 253–329. Burlington: Academic Press.

Stockmann, U., M. A. Adams, J. W. Crawford, D. J. Field, N. Henakaarchchi, M. Jenkins, et al. 2013. The knowns, known unknowns and unknowns of sequestration of soil organic carbon. *Agriculture, Ecosystems and Environment* 164:80–99.

Stolze, M., A. Piorr, A. M. Haring, and S. Dabbert. 2000. *Environmental Impacts of Organic Farming in Europe*. Stuttgart-Hohenheim: UniversitätHohenheim.

Tiwari, A., A. K. Dwivedi, and P. R. Dikshit. 2002. Long-term influence of organic and inorganic fertilization on soil fertility and productivity of soybean-wheat system in a vertisol. *Journal of Indian Society of Soil Science* 50:472–475.

Tomar, J. M. S., A. Anna Durai, A. Das, and R. Singh. 2007. Agroforestry – A viable component of organic farming. In: *Advances in Organic Farming Technology in India*, eds. G. C. Munda, P. K. Ghosh, A. Das, S. V. Ngachan, and K. M. Bujarbaruah, 377–386. Umiam, Meghalaya, India: ICAR Research Complex for NEH Region.

Trehan, S. P. 1997. A rapid method for the estimation of potential immobilization of N after the addition of cattle slurry to soil. *Journal of the Indian Society of Soil Science* 45(1):14–19.

Vashum, K. T., and S. Jayakumar. 2016. Soil organic carbon status and sustainability of slash-and-burn cultivation in India. *Journal of Sustainable Forestry* 35(8):591–603.

Verma, J. P., V. S. Meena, A. Kumar., R. S. Meena, and R. S. Meena. 2015. Issues and challenges about sustainable agriculture production for management of natural resources to sustain soil fertility and health. *Journal of Cleaner Production* 107:793–794.

Vision 2050. 2013. ICAR Research Complex for NEH Region, Umiam, Meghalaya, India, 1–22.

Wang, Z., S. Liu, C. Huang, Y. Liu, and Z. Bu. 2017. Impact of land use change on profile distributions of organic carbon fractions in peat and mineral soils in Northeast China. *CATENA* 152:1–8.

West, P. C., H. K. Gibbs, M. Chad, J. Wagner, C. C. Barford, S. R. Carpenter, and J. A. Foley. 2010. Trading carbon for food: Global comparison of carbon stocks vs crop yields on agricultural land. *Proceedings of the National Academy of Sciences of the United States of America* 107(46):19645–19648.

Wright, A. L., F. Dou, and F. M. Hons. 2007. Crop species and tillage effects on carbon sequestration in subsurface soil. *Soil Science* 172(2):124–131.

Yadav, G. S., A. Das, R. Lal, S. Babu, R. S. Meena, P. Saha, R. Singh, and M. Datta. 2018a. Energy budget and carbon footprint in a no-till and mulch based rice–mustard cropping system. *Journal of Clearner Production* 191:144–157.

Yadav, G. S., A. Das, R. Lal, S. Babu, R. S. Meena, S. B. Patil, P. Saha, and M. Datta. 2018b. Conservation tillage and mulching effects on the adaptive capacity of direct-seeded upland rice (Oryza sativa L.) to alleviate weed and moisture stresses in the North Eastern Himalayan Region of India. *Archives of Agronomy and Soil Science* 64(9):1254 1267.

Yadav, G. S., R. Datta, S. I. Pathan, R. Lal, R. S. Babu, S. Babu, A. Das, S. N. Bhowmik, M. Datta, P. Saha, and P. K. Mishra. 2017. Effects of conservation tillage and nutrient management practices on soil fertility and productivity of rice (Oryza sativa L.) – Rice system in north eastern region of India. *Sustainability* 9(10):1816.

Yadav, G. S., R. Lal, S. Babu, M. Datta, R. S. Meena, S. B. Patil, R. Singh, and R. Singh. 2019b. Impact of no-till and mulching on soil carbon sequestration under rice (*Oryza sativa* L.)-rapeseed (*Brassica campestris L. var. rapeseed*) cropping system in hilly agro-ecosystem of the Eastern Himalayas, India. *Agriculture, Ecosystems and Environment* 275:81–92.

Yadav, G. S., R. S. Meena, S. Babu, A. Das, S. N. Bhowmik, M. Datta, J. Layek, P. Saha, and P. Saha. 2019a. Conservation tillage and nutrient management effects on productivity and soil carbon sequestration under double cropping of rice in north eastern region of India. *Ecological Indicators* 105:303–315.

Yadav, P. K. 2013. Slash-and-burn agriculture in North-East India. *Expert Opinion on Environmental Biology* 2(1):1–4.

Yadav, P. K., M. Kapoor, and K. Sarma. 2012. Impact of slash-and- burn agriculture on forest ecosystem in Garo Hills landscape of Meghalaya, North-East India. *Journal of Biodiversity Management and Forestry* 1:1. of 2, 2.

Zang, H., J. Wang, and Y. Kuzyakov. 2016. N fertilization decreases soil organic matter decomposition in the rhizosphere. *Applied Soil Ecology* 108:47–53.

13 Soil Organic Carbon Restoration in India

Programs, Policies, and Thrust Areas

Ram Swaroop Meena, Sandeep Kumar, Seema Sheoran, Manoj Kumar Jhariya, Rajan Bhatt, Gulab Singh Yadav, Kodigal A. Gopinath, Cherukumalli Srinivasa Rao, and Rattan Lal

CONTENTS

DOI: 10.1201/9781003102762-13

ABBREVIATIONS

AMF:	Arbuscular Mycorrhizal Fungi
BMPs:	Best Management Practices
BNF:	Biological Nitrogen Fixation
C:	Carbon
CA:	Conservation Agriculture
CO_2:	Carbon-dioxide
DAC&FW:	Department of Agriculture Cooperation & Farmer's Welfare
GHGs:	Greenhouse Gases
GOI:	Government of India
ICAR:	Indian Council of Agricultural Research
IIPR:	Indian Institute of Pulses Research
Mg:	Mega Grams
MGNREGA:	Mahatma Gandhi National Rural Employment Guarantee Act
Mt:	Metric Tons
NAPCC:	National Action Plan on Climate Change
NARS:	National Agricultural Research System
NFP:	National Forest Policy
NFSM:	National Food Security Mission
NMEEE:	National Mission for Enhanced Energy Efficiency
NMSA:	National Mission for Sustainable Agriculture
NPMCR:	National Policy for Management of Crop Residue
NPOF:	National Project on Organic Farming
NPOP:	National Programme for Organic Production
OVCDNERS:	Organic Value Chain Development in North Eastern Region Scheme
PAU:	Punjab Agricultural University
Pg:	Peta Grams
PKVY:	Paramparagat Krishi Vikas Yojana
PMKSY:	Pradhan Mantri Krishi Sinchai Yojana
REDD:	Reducing Emissions from Deforestation and Forest Degradation
RKVY:	Rashtriya Krishi Vikas Yojana
SDM:	Sustainable Development Mechanism
SHMS:	Soil Health Management Scheme
SOC:	Soil Organic Carbon
SOM:	Soil Organic Matter
TRFA:	Targeting Rice–Fallow Areas
UNFCCC:	United Nations Framework Convention on Climate Change

13.1 INTRODUCTION

India, with 328.7 million hectares (M ha) of the geographical area, is inhabited by 17.6% of the global population (1.37 billion out of 7.77 billion in 2020). Further, as far as land uses are concerned, it is estimated to have 162 M ha of arable land (~12% of the world) of which 57.0 M ha (~21% of the world) is irrigated, 68.5 M ha is forest and woodland (1.6% of the world), 11.05 M ha is permanent pasture (0.3% of the world), and 8 M ha is permanent cropland (6.0% of the world) (Lal 2004; Ramachandra and Shwetmala 2012). Generally, faulty management practices accelerate the production and emission of greenhouse gases (GHGs) in the atmosphere, which are further responsible for abrupt changes in our climate. The GHG emissions in India originate primarily from the combustion of fossil fuel and manufacturing of industrial products, agricultural practices, and associated energy use from them. Intensive cultivation of ~12 M ha under the rice–wheat (*Oryza sativa–Triticum aestivum*) system and mismanagement of finite natural resources are adversely affecting soil health and food quality and exacerbating the environmental problems. In general, inappropriate management practices seem to accelerate the production and thus the emission of GHGs in the atmosphere which are further responsible for the abrupt changes in the climate (Lal 2004a, 2004b). In several states (e.g., Punjab, Haryana, Uttar Pradesh, Madhya Pradesh, West Bengal, Tamil Nadu, etc.), soil organic C (SOC) concentration has declined to ~0.2–0.4% (MoA&FW 2019). So, now is the time to focus on the restoration of the SOC pool for advancing sustainability of food, nutritional, economic, and environmental security in India.

SOC is a key determinant in soil health and a novel weapon to fight against climate change, food and nutritional insecurities, and other environmental concerns. Through carbon (C) sequestration, soil can capture the atmospheric carbon-dioxide (CO_2). There are two types of soil C sequestration, 1) biomass-based organic C, which is a key part of the food chain in the ecosystem and 2) paedogenic calcium carbonate ($CaCO_3$), called an inorganic C sequestration which is often bane for farmers. It is in this context that the French government at the conference of parties (COP)-21, Paris Climate Summit in 2015, initiated a program called "4 per 1000." Lal (2004a) described five principal global C pools, out of which the oceanic pool is the largest; other pools include geologic, pedologic (soil), biotic, and atmospheric pools. The aim of "4 per 1000" initiative is to capture the atmospheric CO_2 and restore SOC in the agricultural soils at the rate of 0.4% yr^{-1} to promote food, nutritional, environmental, and economic security. Thus, best management practices (BMPs) should be adopted to restore the atmospheric C in terrestrial and agriculture ecosystems. Agricultural sustainability largely depends on soil C stock and its turnover. Thus, soil C sequestration is a viable option to restore the SOC pools in agroecosystems by reducing net greenhouse gas (GHG) emissions and capturing the atmospheric CO_2. Soil C sequestration also helps to improve soil health and mitigate climate change (Lal 2004b, 2011; Bronick and Lal 2005). India has a large potential to sequester C in the forest and agricultural soils (Figure 13.1).

There is a rapid increase in India's CO_2 emissions since 1970, and, particularly in recent decades, emissions of CO_2 flux into the atmosphere from different sectors have increased because of anthropogenic activities (Figure 13.2).

Furthermore, to meet the ever-increasing fertilizer demand, production of the nitrogenous fertilizers through the Haber-Bosch process results in the release of 0.52 Mg (Megagrams) of CO_2-C $Mg^{-1}$of ammonia (NH_3)-N (nitrogen) (Bakker et al. 2012). Thus, the emission of CO_2 into the atmosphere as a result of anthropogenic activities is a major concern. In most agroecosystems of India, the SOC ranges from ~ 0.25% to 1.15% in topsoil (30 cm), which is more than that in horizons beneath (Bhattacharyya et al. 2000). Small changes in SOC may cause significant alterations in atmospheric CO_2. Cropping systems can be a major source of CO_2 release to the atmosphere through unsustainable crop management practices in different cropping systems Figure 13.3.

The SOC content and stock depend on soil, cultivation practices, various inputs for crop production, and residue management in the cropping systems (Loveland and Webb 2003; Giller et al. 2011). In India, the loss of the different types of SOC pools is a big challenge to stopping the C fluxes for

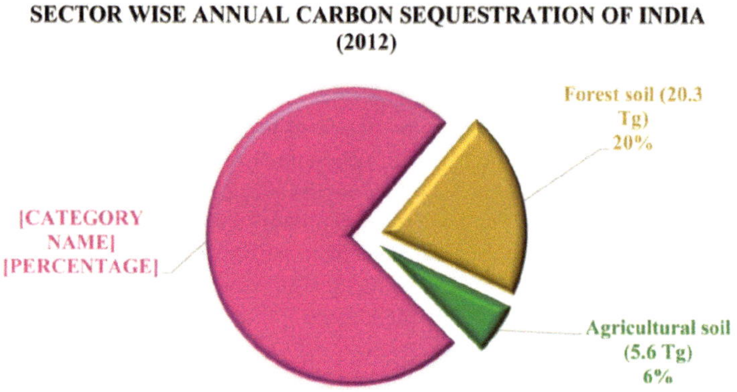

FIGURE 13.1 Annual carbon sequestration in forest and agricultural soils of India (Data source: Ramachandra and Shwetmala 2012).

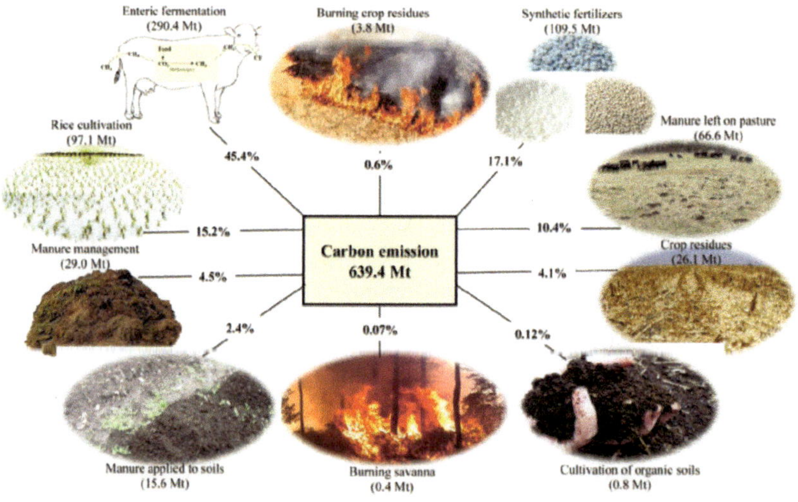

FIGURE 13.2 Carbon emission from the different agriculture sectors of Indian (Data source: FAOSTAT 2020).

anthropogenic activities. On the other hand, judicious management can be useful in managing the direct and indirect factors related to agricultural systems and climate change (Olivier et al. 2017; Meena et al. 2018).

Out of the total land area of 297.3 Mha in India, around 27.3%, 20.3%, 17.4%, 12.3%, 8.3%, 6.2%, 0.6%, and 0.27% area are covered by Alfisols, Vertisols, Inceptisols, Ultisols, Entisols, Aridisols, Mollisols, and Gelisols, respectively (Vermeulen et al. 2019). The SOC stocks in the soils of India, varying among different land uses and cropping systems, are shown in Tables 13.1 and 13.2.

Indo-Gangetic Plains (IGPs) (constitute about 60% of the total area of the Great Plains) have many sustainability issues *viz.*, the decline in underground water and SOC levels and, thus, poor soil health and low yields of reduced nutritional quality (Bhatt et al. 2016). Intensive tillage and open burning of crop residues are also responsible for the increased loss of the C stocks into the atmosphere. This is the reason nowadays that conservation agriculture (CA) practices (*viz.*, minimum tillage, crop rotation, and crop residues placement as mulch on to the bare soil) are recommended for regions of low SOC concentration (Meena et al. 2020a). Further, the practices, such as green

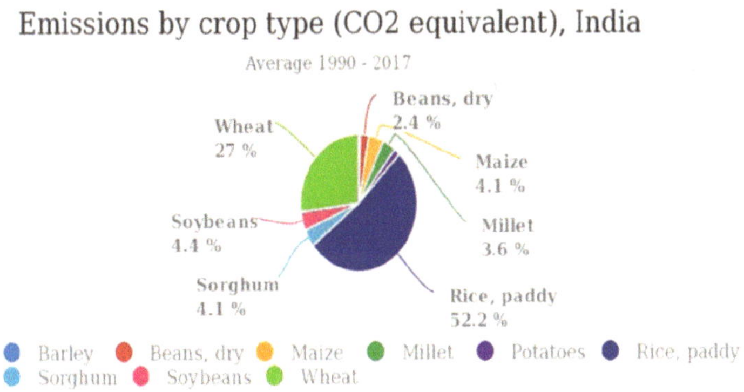

FIGURE 13.3 Emission of CO_2 equivalent in different cropping systems in India (Data source: FAOSTAT 2020).

TABLE 13.1
Mean Annual Carbon Inputs, Depletion Rate, and Input Requirement in Land Use and Cropping Systems (Data sources: Srinivasarao et al. 2011, 2012)

Location	Production System	Mean Annual C Inputs (Mg C ha^{-1} yr^{-1})	Mean C Depletion Rate (Mg C ha^{-1} yr^{-1})	Critical C Input Requirement (Mg ha^{-1} yr^{-1})
Anantapur	Groundnut (*Arachis hypogaea*)	0.5–3.5	0.18	1.12
Bangalore	Groundnut-finger millet (*Eleusine coracana*)	0.3–3.0	0.92	1.62
Bangalore	Finger millet	0.3–3.1	0.25	1.13
Solapur	Winter sorghum (*Sorghum bicolor*)	0.6–3.4	0.23	1.10
SK Nagar	Pearlmillet (*Pennisetum glaucum*)-cluster bean (*Cyamopsistetragonoloba*)-castor (*Ricinus communis*)	0.2–1.9	0.67	3.30
Indore	Soybean (*Glycine max*)-safflower (*Carthamus tinctorius*)	1.9–7.0	0.47	3.47
Varanasi	Upland rice -lentil (*Lens culinaris*)	1.1–5.6	0.15	2.47

manuring, application of farmyard manure, compost, vermicompost, biochar, etc., can enhance the SOC stock and improve soil properties and agronomic yields (Bhattacharrya et al. 2000). In general, SOC concentration increases with an increase of rainfall and decrease of the mean annual temperature (Table 13.3).

Soils of arid and semi-arid climatic regions, though covering around 33% of the country, have lower SOC contents due to prevailing light-textured soils and higher temperatures which further accelerate the decomposition rates of the organic matter. Out of the total area (160 M ha) in the Indian Peninsula, around 30 M ha area is occupied by the black cotton soils (Vertisols) which have inherently lower SOC content due to higher temperature and associated soil carbon oxidation.

From a policy point of view, the Indian government has taken a few initiatives to enhance C sequestration by launching different schemes. But on the practical aspect, still much more attention is required to successfully implement these plans and policies to achieve the required target of soil C restoration. Therefore, considering the importance of soil C, this chapter is mainly focused on

TABLE 13.2
Distribution of Labile Carbon (mg g^{-1}) Under Different Land-Use Systems (Data Source: Pandher et al. 2020)

Depths of Soil	Cropping Systems				
			Agroforestry		
	Maize (*Zea mays*)-wheat	Agro-horticulture	3 years	6 years	Mean
0–15	1.13	1.70	1.31	1.72	1.47
15–30	1.47	0.97	1.00	1.29	1.18
30–60	0.83	1.19	1.31	1.35	1.17
60–90	0.92	0.74	0.57	1.57	0.95
90–120	0.78	1.01	0.44	0.78	0.75
Mean	1.03	1.12	0.93	1.34	

TABLE 13.3
Soil Organic Carbon Concentrations of Soils of India Concerning the Rainfall Regime and Temperature (Data Source: Srinivasararao et al. 2009)

Rainfall (mm yr^{-1})	Mean Annual Temperature (°C)	Soil Organic C Content (g kg^{-1})	
		Surface	Sub-surface
Less than 500	25.9–26.7	1.20–8.0	1.2–4.0
500-1000	23.9–27.9	1.80–12.5	0.7–11.7
More than 1000	24.4–27.2	2.6–9.0	2.3–8.4

the effective plans and policies for soil management and C restoration in agricultural ecosystems that will eventually make the country more secure in the aspects of food, nutrition, environment, and economy. A roadmap on soil C policies is also discussed to encourage the implementation of BMPs for soil health and C stock restoration, and realize the "*Sustainable Development Goals*" of the United Nations for a better country and the planet in general.

13.2 SOIL ORGANIC CARBON IN AGRICULTURE

Being a large country, the agroecosystems of India are a key component of the global C cycle. It has diverse agricultural practices and cropping systems. Ten approaches which affect soil properties are (Lal 1998; Meena et al. 2020a): (1) Improving the nutrient use efficiency by using the slow-release fertilizers *viz.*, poly-coated or neem coated urea, (2) Avoiding in-field burning of crop residues, (3) Enhancing water use efficiency, (4) Minimizing soil tillage, (5) Adopting erosion control measures in hilly regions, (6) Strengthening soil biodiversity, (7) Increasing soil aggregation, (8) Reducing losses of nitrogen by drainage, erosion, and volatilization, (9) Increasing buffering capacity against sudden fluctuations in soil reactions, and (10) Moderating soil temperatures. The judicious management of SOC stock is a common strategy in all these ten factors.

Climate change and soil health are two interrelated factors which affect the food, nutritional, environmental, and economic status of a country. However, most soils of India are less fertile and have low SOC stock. So, it is the need of the time to focus on restoration of SOC and nutritional security in India. Now there is an urgency to take needful actions for the implementation of BMPs and different policies and planning for sustainable and eco-friendly agricultural systems in India.

The Indian Council of Agricultural Research (ICAR) has developed numerous strategies to enhance soil C restoration for different agriculture systems to ensure food, environmental, and economic security. Choice of BMPs (Figure 13.4) to be followed for sustainable soil management involve: (1) improvement of soil fertility maintenance for SOC sequestration e.g., judicious use of organic and inorganic source of nutrients for enhancing nutrient use efficiency, increasing biological nitrogen fixation (BNF), nutrient cycling through cover crops and fallow plantation, (2) use of organic inputs (animal manure, green manuring, recycling of organics), (3) recycling of plant litter and agricultural by-products, (4) recycling urban waste, (5) changing land use patterns, (6) adopting soil biological management, (7) using modern biofuel and other renewable sources (e.g., solar, wind, hydro, geo, tidal) for energy rather than crop residues and animal waste, (8) adopting ecological weed management (e.g., integrated weed management, minimal use of herbicides), (9) establishing land restoration measures (e.g., erosion control through runoff management with terraces, vegetative barriers, soil surface amendment, mulch farming, and fallow plantation), (10) promoting conservation agriculture (zero or reduced tillage, surface residue retention, mulch farming, legume-based crop rotation), (11) growing legumes, grain crops, grasses, green manure crops, shrubs, and trees, etc., (12) promoting eco-friendly farming systems with high diversity (e.g., mixed farming, agroforestry, silvo-pastoral systems, etc.), (13) encouraging crop breeding programs for deeper root systems, and (14) creating environmental awareness at the farmer levels.

Concerning C sequestration in the different land-use systems, Lal et al. (1998) addressed two focal points i.e., soil C management aspects as well as the policy instruments. From a management point of view, Lal (2002) suggested that the focus should be to promote suitable land-use and agricultural practices, soil conservation and BMPs, diversified crops and tree plantation, and proper management of agricultural waste for better soil, health, and eco-friendly agriculture systems.

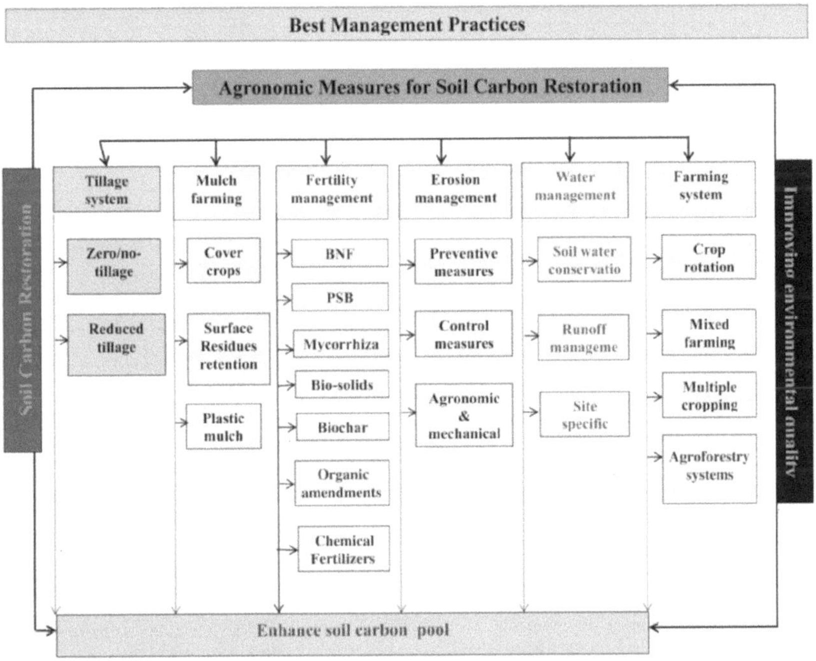

FIGURE 13.4 Best management practices (BMPs) for improving soil carbon pool.

13.3 NEED FOR SOIL ORGANIC CARBON RESTORATION IN THE SOILS OF INDIA

India has diverse ecological regions. Therefore, there is huge potential to improve the SOC pool in cultivated soils. The SOC concentration in cultivated soils is <0.05% compared to 0.15–0.20% in uncultivated soils (Sahoo et al. 2019). Soil C stock distribution by order in Indian soils is presented in Table 13.4.

It is important to formulate and implement policies for SOC restoration in order to strengthen the agricultural soil systems. Thus, the government has to work in collaboration with farmers to re-carbonize soil by promoting recycling of crops and animal wastes back into the soil profile. The low SOC content and stock may be due to intensive tillage practices, crop residue burning, and nutrient mining because of extractive practices. Accelerated soil erosion by water is also a big problem in the country, which can lead to an annual loss of ~00.7 Pg (Petagrams) of SOC (Lal 2004a). The SOC stock, has a positive relationship with soil fertility and productivity and with that of the country's agricultural production trend. Redesigned policies should be focused on soil, environmental, and human health through soil C restoration. Through planning to manage the agroecosystem, it is possible to establish a link between soil and human wellbeing toward achieving sustainability.

For that, planning on soil C must be prioritized at all levels by the governing bodies through providing adequate funding for soil C restoration and its research, better knowledge transfer along with extensive dialogue among farmers through demonstration techniques. In this context, chemical fertilizers have to be gradually replaced by biofertilizers including animal manures and composts wherever possible without yield reduction (Meena et al. 2020b). This can be accomplished by the inclusion of livestock in the existing farming systems to ensure adequate availability of manure and compost for application to the soil, and crop residues must also be recycled as a nutrient source. In this regard, the government has to support landowners and tenants for building up C stock in soil strata. Mixed farming can also be supported by giving a special scoring uplift to farmers for

TABLE 13.4
Carbon Stock (in Pg = 10^{15}g) Distribution by Order in Indian Soils (Modified from: Bhattacharyya et al. 2000; Pal et al. 2015)

	Soil Depth	Carbon Stock (Pg)		
Soil Order	Range in cm	SIC	SOC	TC
Vertisols	0–30	1.07(26)	2.59(27)	3.66(27)
	0–150	6.14(18)	8.77(29)	14.90(23)
Entisols	0–30	0.89(21)	0.62(6)	1.51(11)
	0–150	2.86(8)	2.56(8)	5.42(8)
Inceptisols	0–30	0.62(15)	2.17(23)	2.79(20)
	0–150	7.04(21)	5.81(19)	12.85(20)
Aridisols	0–30	1.40(34)	0.74(8)	2.14(16)
	0–150	13.40(39)	2.02(7)	14.42(24)
Alfisols	0–30	0.16(4)	3.14(33)	3.30(24)
	0–150	4.48(13)	9.72(32)	14.20(22)
Ultisols	0–30	0.00	0.20(2)	0.20(1)
	0–150	0.00	0.55(2)	0.55(1)
Mollisols	0–30	0.00	0.09(1)	0.09(1)
	0–150	0.07(0.2)	0.49(2)	0.56(1)
Total	0–30	4.14	9.55	13.69
	0–150	33.98	29.92	63.90

middle-tier agri-environmental agreements. Besides, the impact of contract farming and land rental agreements on soil C trends is still not explored.

13.4 INDIA'S PROGRESS AT THE INTERNATIONAL LEVEL

The Government of India (GOI), after considering the consequences of climate change, initiated a program in 2008 named "National Action Plan on Climate Change (NAPCC)" with eight missions. Under NAPCC, NMEEE (National Mission for Enhanced Energy Efficiency) was launched to identify mitigation strategies against climate change. Further, PAT (Perform, Achieve, and Trade) under the NMEEE is designed to reduce CO_2 emissions to about 30 million tonnes, which reduces CO_2. Article 6.4–7.7 of the Paris Agreement establishes a mechanism to contribute to the mitigation of GHGs and support the Sustainable Development Mechanism (SDM). The SDM is considered to be a successor to the Clean Development Mechanism (CDM), a flexible mechanism under the Kyoto Protocol by which parties could collaboratively pursue emission reductions for their intended nationally determined contributions.

The highlights of Indian contributions toward resolving the issue submitted to UNFCCC (United Nations Framework Convention on Climate Change) is as follows (Ministry of Power 2020):

- To promote a healthy and sustainable way of living based on traditions and values.
- To adopt a climate-friendly and a cleaner path for economic development.
- To reduce the emissions by 33–35% by 2030 compared with 2020.
- To transfer technology involving low-cost international finance, around 40% electric power to be generated from non-fossil fuel by 2030.
- To identify and realize an additional C sink of 2.5–3 Pg of CO_2 equivalent through afforestation.
- To propagate advanced climate-smart technologies for mitigating the adverse effects of climate change—particularly in agriculture, water resources, etc.
- To assemble funds from developed countries for developing a mechanism to cut down the emissions into the atmosphere in developing countries.
- To quickly transfer climate-smart technology in India's villages and remote places; and for joint international activities, collaborative research projects must be established to find out some more advanced ways to enhance the SOC and mitigate the consequences of climate change.

13.5 THE NEED TO REFORM THE GOVERNMENT POLICIES IN INDIAN AGRICULTURE FOR SOIL ORGANIC CARBON RESTORATION

In the present scenario of South Asia, India is a leading country in the agricultural sector of the developing world. Growing concern about SOC at the international platform does not yet match with the action and investment on soil C sequestration strategies at ground level in India. Many surveys and reports depict that most of the farmers are not bothered about soil health and its relation to food, nutrition, environmental, and economic conditions of the country. Likewise, farmers are not fully aware of the importance of the soil C sequestration (Conference Report 2017). India does not have any separate organization to work and research on the soil C sequestration. It is a big challenge to politicians, researchers, policymakers, and extension workers. Now there is an urgent need to address these challenges to solve the issue at the ground level for future sustainability because soil C is a key element on the earth. The important challenges and barriers in the way of soil C sequestration arc limited finance, lack of adequate knowledge, the governmental interest in a large business, existing subsidies on depleting or C-exhaustive farm practices, lack of policy agenda to safeguard rural communities and small farmers, and limited/no demonstration/policies to advertise stories of farmers who have successfully adopted soil health/C sequestration practices.

To bring comprehensive changes in the agricultural policies of the country, it is important to establish plans for soil C restoration. Such plans are essential to the transformation and establishment of the protocol to offset markets and incentive payments. Along with designing financial policies, there is a need for setting up guidelines to promote C management practices across different agroecoregions, landscapes, and cultural contexts to the farmers (Conference Report 2017).

The GOI has recently emphasized balanced plant nutrition including organic and inorganic fertilizers in the Union Budget 2019–20. Although, its implementation is still challenging due to the lack of seriousness by the government to promote the use of organic inputs in farming, and policies favoring unbalanced and excessive use of chemical fertilizers. The government should reduce subsidies on the most harmful agricultural inputs (e.g., chemical fertilizers, pesticides), machines (e.g., extensive tillage equipment), and promote eco-friendly inputs and reduce the energy consumption in agriculture. The incentive payments and subsidies should be reduced for C-exhaustive inputs, which will help to promote C-rich strategies. The government efforts must be on reforming the policies that sub-optimize the subsidies on fertilizers and extensive tillage implements so that their use can be minimized. Removing such subsidies would promote more efficient use of agricultural crop and industrial waste after conversion to safe nutrient sources. Payments should be given to farmers as an incentive for planting trees/shrubs in the agroforestry system or transforming tillage practices from conventional to no-till that ensure a positive impact on soil C build-up (OECD/IEA/NEA/ITF 2015). This will enhance the SOC stock and restore soil health in India. At the same time, the government has to provide financial assistance to promote the machinery, implements, practices, infrastructure, and skills that reduce C loss and enhance C sequestration in the soil in a way that is compatible with continuous productivity improvement. A successful policy puts efforts to measure results, think for farmers' direct benefits beyond soil C sequestration, reinstate C exhausted lands, ensures active participation of farmers, and availability of tools for education and C sequestration. The Healthy Soil Initiative scheme in California is a good example of a farmer's involvement in the C management system (Conference Report 2017). The government should also emphasize on knowledge sharing among farmers, motivation, education, rewards, and monetary support to the farmers and land managers to promote such activities for efficient soil C sequestration.

The protagonists across civil society and business governments must work together to restore SOC stock by identifying and resolving barriers beyond their benefits that urgently calls for a cross-sectoral agenda (Figure 13.5) (Vermeulen et al. 2019). All the necessary need-based information related to soil C importance and practices should be transmitted at the farmer's level in local values, languages, relationships, and the possible diversification in the traditional approaches.

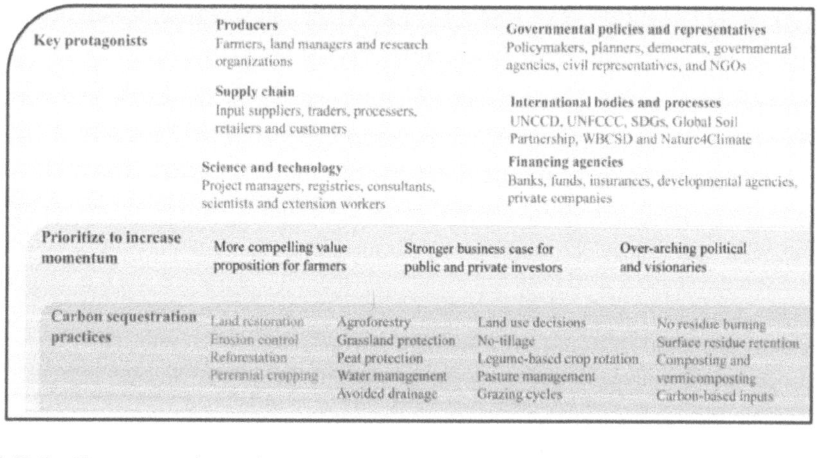

FIGURE 13.5 Key protagonists and approaches for soil carbon restoration (Modified from: Vermeulen et al. 2019).

This calls for better coordination among scientists, farmers, and land managers for designing of efficient policy. It will help to generate a holistic approach for developing an action plan by taking advantage of available technical and scientific resources to restore the C pool in national soil and terrestrial ecosystems. The science and policy need to be brought together at a common platform that is useful to all the living entity including soil. The role of land managers and farmers is to identify such management practices with the help of scientists that are productive, C-rich, eco-friendly and have a wider-scale adaptability. In this series, six important steps have to be followed (Kimble et al. 2002): 1) Assessing the trends of C loss at regional and national scale, 2) Measuring the C sequestration potential of management practices in major soils of key ecological regions, 3) Analyzing the economics of soil organic C from a productivity and environmental point of view, 4) Developing policy plans that help to build soil C stock, 5) Classifying standards and index for the espousal of C sequestration practices, and 6) Forming policies for practical executions of identified strategies.

To restore the SOC in the agroecosystems of India, three important priorities are: 1) an overarching case and sight for action (e.g., it can involve a few countries that have already included soil in their national agenda and are considered future leaders, such as Canada, Australia, Bhutan, Uruguay, Ethiopia, etc.; 2) a robust business plan in the amalgamation of private and public investors that will help to produce finance level and improved soil C management; and 3) a more attractive value proportion for land managers and producers that will provide the financial support to restore soil C through daily management activities. The processes for soil C restoration awareness, policies, knowledge, finance, and on-farm implementation are of a critical need in India (Figure 13.6).

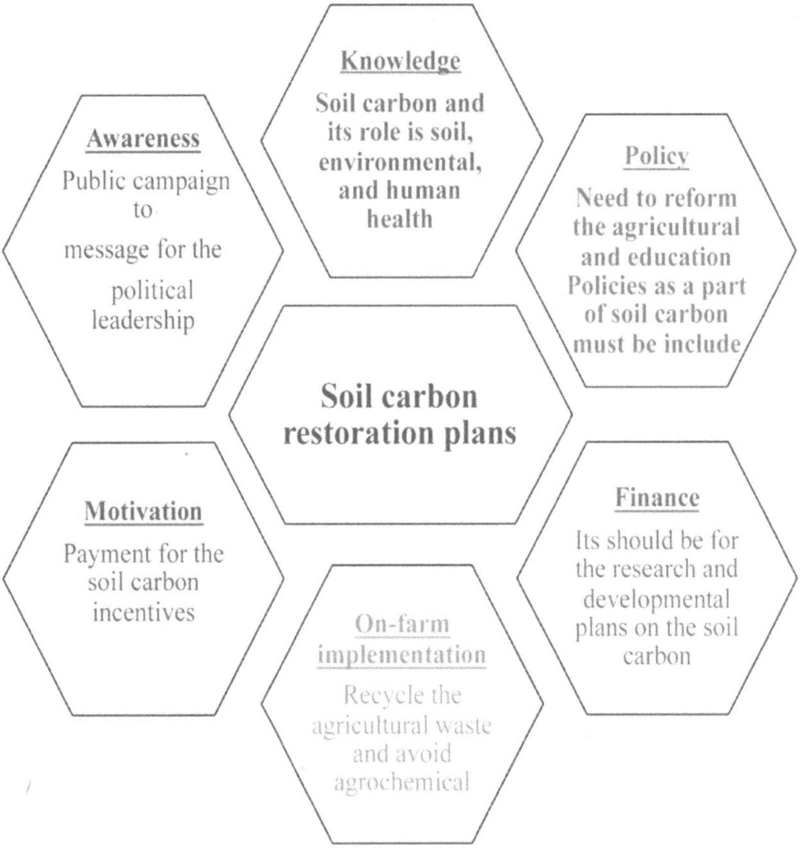

FIGURE 13.6 Steps and action plans for soil carbon restoration in soils of India.

13.6 INTERVENTIONS AS A CARBON OFFSETTING OPTION/STRATEGIES FOR SOIL CARBON SEQUESTRATION AND REDUCING THE FOOTPRINT

13.6.1 Reduce Food Wastage

As per the estimates of FAO (2019), one-third of all food produced in India is lost or wasted on the journey from farm to mouth, and hence, water, energy, nitrogen, C, and other factors are direct/indirectly adversely affected through the food wastage (Figure 13.7). Therefore, for reducing the footprints, food wastage must be reduced to sustainably reduce the C footprint. Most of the agricultural produce is lost during its production, post-harvest handling, storage, and consumption phases. This fact is particularly true in India for vegetables as the country leads globally in vegetable production and consumption of more than 50% (Scialabba and Lindenlauf 2010). Therefore, first food loss must be controlled in the field, then during packing and storing stages, and, lastly, at home. Further, proper distribution is also important as, at certain places, people are prone to hunger while on the other side, food is wasted because of poor storage. There is a need to focus on the new cold storage and post-harvest technologies to reduce spoilage of the agricultural products, particularly more perishable ones, like vegetables, fruits, raw materials, dairy products, etc. It is important for their wide-scale and demand-based utilization as it also reduces the pressure for more production and ultimately will help to reduce gaseous emissions that cooccur during production.

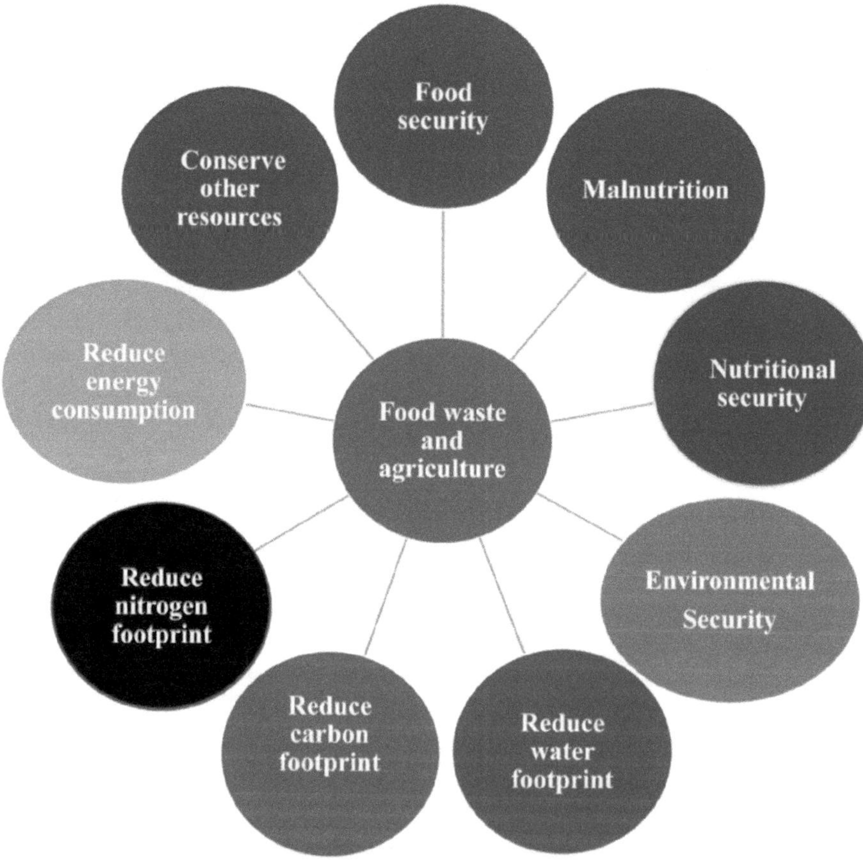

FIGURE 13.7 Food waste in relation to agriculture.

13.6.2 Industrial Waste as a Carbon Input Source for Cultivated Soils

A rapid industrialization in India has generated large amounts of industrial waste—both in terms of fuel gas and by-products and other wastes which have negative effects on the ecosystems (Yadav et al. 2019). However, these waste products can be judiciously utilized as a good C source to restore SOC content. Also, continuous increase of CO_2 concentration results in global warming which has its own adverse consequences on the agricultural sector, opening new land, which further leads to many ecological and economic disruptions (Saravanan et al. 2018). To reduce the CO_2 load in the atmosphere, we can't directly opt for the industry to shut down, as this will affect the country's economy. Hence, industrial waste can be converted to a C-based soil amendment (Yadav et al. 2019). Further, microalgae have a more rapid growth rate and CO_2 fixation efficiency (10–50 times more) than terrestrial plants and also can utilize concentrated amounts of CO_2 present in industrial wastes (Chiu et al. 2008; Tang et al. 2011). Microalgae can be a viable option for reducing C levels in the atmosphere with a long-term and sustainable solution to global warming. Some methods (*viz*., physical separation, chemical, and biological methods) can be applied for sequestering C from various industrial wastes as power plants, steel and cement manufacturing units, and refineries. In general, renewable and green sources of energy result in the least CO_2 emissions (Yadav and Sen 2017). Presently, biological methods are considered as the most effective techniques to capture and store CO_2 from flue gas and wastewater treated with microalgae (Yadav et al. 2007, Yadav et al. 2015). Hence, it will be a preferred choice for promoting CO_2 sequestration from industrial wastes throughout the country (Ugwu et al. 2008). Further, microalgal CO_2 fixation can be improved by reactor operation modes, CO_2 delivery systems, and genetic engineering applications. Fixation of atmospheric CO_2 in the biomass can be used sustainably for biodiesel, bioethanol, biohydrogen, or methane production (Tang et al. 2011; Kumar et al. 2011). India has still not implemented any policies to use industrial waste as a C source for soil. Therefore, there is a need to do brainstorming and come up with ideas/plans on the ground level to practically utilize and convert the industrial waste as a good and safe C source for the soil.

13.6.3 Soil Microbial-Based Carbon Sequestration

Being the hub of living and non-living matter, the soil ecosystem is highly complex with the multitude of interactions among soil microbes (Nielsen et al. 2011; Meena et al. 2020). The soil microbes play an important role in sequestered global C (Jansson 2011). Although higher plants contribute dominantly to the uptake of atmospheric CO_2, soil microbes also significantly impact the C budget being as plant symbionts, pathogens, or detritivores (Singh et al. 2010). There are several microbes in the soil subsurface (Blume et al. 2002; Fierer et al. 2003) that contribute to the CO_2 fixation (Table 13.5).

Thus, the microbial community plays a critical role in the mineralization of organic matter. A wide range of factors (e.g., soil profile, temperature, moisture, roots impacts, physical and chemical properties of soil) regulate the metabolism of soil's biotic components (bacteria, fungi, viruses,

TABLE 13.5
CO_2 Sequestration Capabilities of Microalgae Strain Under Different CO_2 Concentrations

Microalgal Species	CO_2 (%, v/ v)	Biomass Productivity (g L^{-1} d^{-1})	References
Dunaliella sp.	12	0.71	Hulatt and Thomas (2011)
Scenedesmus obliquus	20	0.38	Ho et al. (2010)
Chlorella sp.	70	0.118	Sung et al. (1999)
Nannochloris sp.	15	0.35	Negoro et al. (1991)
Chlorella sp.	50	0.95	Maedal et al. (1995)
Chlorella sp.	20	0.70	Sakai et al. (1995)

archaea, and fauna), and subsequently affect the mineralization or stabilization of soil C. As roots are in close association with soil microbial community and their biomass is the main harbinger of stabilized soil organic matter (SOM) (Kallenbach et al. 2016), it is a viable option to identify or develop plant cultivars that modulate the soil–plant microbiome to increase SOC sequestration in the rhizosphere. Soil microbes possess many traits like a biochemical constituent of cell walls and envelopes that directly affect the persistence of soil C (Kallenbach et al. 2016). Also, microbial population dynamics, growth rates, and mortality are likely to be linked with C turnover (Roller et al. 2016). There are several challenges to manipulate microbial community for improved C sequestration due to huge diversity, lack of cultivability, characterization of soil microbes, and limited understanding about their ecological roles (Singh et al. 2010; Uroz et al. 2013).

By utilizing the quantitative stable isotope probing (qSIP) (Hungate et al. 2015; Koch et al. 2018), C persistence traits of microbial strain genomes can be analyzed to establish robust mechanistic linkages between C cycling and soil microbial community (Blazewicz et al. 2014; Meena and Lal 2018). The application of "omics" techniques can also help understand the locking of soil C propelled by biological activities undertaken into the soil micro-niche (Trivedi et al. 2013). Among the fungi, arbuscular mycorrhizal fungi (AMF) in symbiotic association with crop roots play a significant role in C stabilization (Wilson et al. 2009; Lambers et al. 2009). Thus, plant roots, AMF, and SOC have a complex symbiotic interaction, beneficially affecting each other (Feeney et al. 2006; Hinsinger et al. 2009). The AMF especially of the genus Glomeromycota supply nutrients to plants, which in turn dispense about 20% of the fixed C in soil (Kell 2011). Microalgae are photosynthetic organisms using CO_2, water, sunlight, and inorganic salts for their growth and development (Razzak et al. 2013) and are superior to plants in sequestering C (Rahaman et al. 2011; Van et al. 2012). Further, around 50% of C in cell dry mass of microalgae can be sequestered (Sánchez Mirónet al. 2003) as up to 183 Mg of CO_2 could be sequestered by the production of 100 Mg of biomass (Chisti 2008) which can be converted into several products *viz*., pigments, cosmetics, food, and animal feed (Kumar et al. 2011; Yadav et al. 2015).

Thus, there is a need to focus on designing research experiments, plans, and policies for studying the role of *in-situ* microbes in C stabilization and capture in the soil. It is also important to know the role of microbiota in CO_2 capture and C locking up within soil aggregates and stabilize it for a long time.

13.7 IN-FIELD BURNING OF CROP RESIDUES IN INDIA

Agricultural cropping systems produce a lot of residues in India. Most of the crop residues are used for fodder, fuel, and other domestic purposes, but still, crop residues are burned annually without any adequate management that causes emissions of particulate matter into the atmosphere (Bhuvaneshwari et al. 2019). Regrettably, the quantity of crop residues being burned in India is much more than the agricultural wastes generated in Bangladesh (72 Mt—Metric tons), Indonesia (55 Mg), Myanmar (19 Mg), combined. It resulted in a loss of ~00.4 Pg of organic C (Kumar et al. 2018). One Mg of rice residue contains about 400 kg C which gets lost during the burning of crop residues. Likewise, burning of one-ton rice residue releases 1460 kg CO_2 into the atmosphere (Singh et al. 2018). This is the amount that can enrich the soil with C if instead of burning residues, any viable and sustainable scopes are generated to gain an advantage (The Economics Time 2019). Also, the crop residue burning adversely impacts soil properties, minerals, fertility, and microbial activities that indirectly reduce the C capturing, and retention capacity of the soil, and have an adverse effect on soil health and quality (Meena et al. 2019).

13.7.1 Biochar as an Option for Crop Residue Management

Biochar being prepared through pyrolysis, gasification, and hydrothermal carbonization, comprised of~35–60% C, which might be otherwise emitted into the atmosphere. The end products from biochar are bio-oil and syngas (Chowdhury et al. 2017). Application of biochar to soil enhances SOC pool, and therefore, improves soil physico-chemical and biological properties, and restore soil health (Day et al. 2005; Sohi et al. 2010; Srinivasarao et al. 2012, 2013). Biochar is a good option for long-term C storage and ranks high among options of soil C sequestration. Even small-scale production can be economical and sustainable (Pratt and Moran 2010). To avoid in-field burning of residues and to promote the SOC, Punjab Agricultural University (PAU, India) recommended biochar as an option for the better management of paddy straw. Biochar is also used in croplands for better use of fertilizers (PAU 2020). The concept of biochar is popularized among paddy farmers of Punjab state in India through different extension centers *viz.*, Krishi Vigyan Kendras (Farm Advisory Service Scheme). Being fine-grained, soft, C-rich, porous structure, and high surface area, biochar is ideal for reducing the C-footprints and for increasing SOC sequestration (Srinivasarao et al. 2013; PAU 2020). It can also help mitigate global warming consequences because of its effects on the recalcitrant C storage in soil. In addition to saving of nitrogen, application of biochar @ 5 Mg ha^{-1} increases the crop yield by 10% (PAU 2020)

Application of biochar can mitigate 12–50% of anthropogenic C emissions depending on factors like kind of material used, pyrolysis conditions and energetic performance of the biochar production system, etc. (Cayuela et al. 2014; Chowdhury et al. 2017).

13.7.2 Government Efforts to Promote *In-Situ* Management of Crop Residue

The GOI initiated a scheme to promote the mechanization in agriculture for *in-situ* crop residue management in Punjab, Haryana, Uttar Pradesh, and Delhi in 2018–19 and 2019–20 by releasing a fund of ₹ 1151.80 crores (~US$ 156.0 million) (MoA & FW 2019). Under this scheme, the government is providing up to 80% subsidy on the total cost of custom hiring of farm machinery for *in-situ* crop residue management through cooperative society, private entrepreneurs, or self-help groups. Besides, there is a provision of monetary support of up to 50% to the total outlay for the procurement of agricultural machinery and equipment to the individual farmer for managing the crop residues. The important machinery and equipment for *in-situ* residue management involved under this scheme are Super Straw Management System, Happy Seeder, Rotary Mulcher, Hydraulic Reversible Mould Board Plough, Paddy Straw Chopper/Shredder, Shrub Master, Rotavator, and Zero-Till Seed-Cum-Fertilizer Drill. As per the government report (GOI 2019), a total of 32,570 different types of machinery have been distributed to the farmers besides setting up of 7960 custom hiring centers in Punjab, Haryana, Uttar Pradesh, etc. The Department of Agriculture in the Punjab government is organizing training programs at village, block, and district level to create awareness among the farmers about the importance of crop residue incorporation in soil (Kumar et al. 2015a). As a result, during 2018, more than 4500 villages in Punjab and Haryana were declared as zero stubble burning villages as there was not a single crop burning incident from these villages. Besides this, the Department of Agriculture Cooperation & Farmer's Welfare (DAC&FW) is providing separate funds to the States under the scheme of Sub-Mission on Agriculture Mechanization for crop residue management. Under this mission, States have also been directed to conduct a demo of types of machinery for residue management at farmer's fields at the rate of ₹ 4000 (~ US$55) ha^{-1}. Similarly, in 2014, a National Policy for Management of Crop Residue (NPMCR) was developed by the Ministry of Agriculture & Farmer's Welfare to promote *in-situ* crop residue management, promote farm types of machinery to incorporate crop residues within the field on subsidy, and to provide financial aid to the researchers for innovative ideas and project proposal on the same issues (Datta et al. 2020).

The economic conditions of farmers do not allow them to adopt a long-term solution for residue management. Therefore, the local government, municipalities, or farmers' association should assist the farmers in education, awareness creation, and capacity building for effective incorporation of C-rich agricultural wastes into the soil body (Bhuvaneshwari et al. 2019). They should be made aware of the advantages of crop residues on soil health, and the adverse effects of chemical fertilization and residue burning.

There is also a need to give attention to three important policy-related issues to crop residues management: 1) to think of a self-running mechanism in a holistic way rather than in isolation, 2) to identify strategies to empower stakeholders, and 3) to lean in the direction of nexus thinking rather than sectorial thinking.

13.8 POLICIES FOR PROMOTION OF CONSERVATION AGRICULTURE

Conservation agriculture (CA) is not a new approach, but still, it has a need for wider adoption as a challenging task, especially in the rice–wheat cropping system in India. It is a matter of grave concern for policymakers and institutional leaders to transform the existing tillage practices into a system-based CA. To make this transformation practically successful, there is a need to support farmers by providing incentives and motivation for faster adoption of a system that will help to enhance the soil C pool. It can be achieved by a regulatory standard framework, research and development, wider training programs, legislations, incentive, and payback programs. Enhancement of SOC concentration with the adoption of CA is a direct investment into soil fertility and C stock, which is still not standardized by the government. Some cases have been observed in Africa and Europe where farmers plow the soil as they know that the decomposition of organic matter upon plowing will release the nutrients and the cost of fertilizer input will be reduced (Haugen-Kozyra and Goddard 2009; Kassam et al. 2014).

Therefore, a policy framework is necessary for India to hold the landowner responsible for ensuring the continuous addition of biomass-C into the soil. Furthermore, rules and regulations should be the same for farmers who have or do not have their land to adopt CA-based practices to improve the SOC pool. Policies should encourage and enable the unification and validation of CA protocol in a practical form so that farmers can get pecuniary assistance for providing certain ecosystem services. Likewise, the farmers get financial and technical help from the government for adopting the CA system to reduce soil erosion and subsequent C loss (Jangir et al. 2017). Such plans encourage producers to transform their existing system of crop management into CA-based systems.

13.8.1 Government Interventions for Conservation Agriculture and Needful Action Plans

The Indian government has to demonstrate and create awareness by conveying various CA promoting policies through private and public extension services. Farmers suffer from yield losses caused by the weed proliferation in CA systems initially. Therefore, governmental agencies have to ensure the adequate and timely availability of subsidized and safe herbicides to lessen the initial costs and credit lines for the procurement of CA tools and implements. Knowledge sharing must be promoted among farmers at each level for the implementation of CA tactics. This can be achieved by creating a strong network and task force at the local, regional, and national levels to strengthen mutual learning and capacity building. It will subsequently give a whole view of its benefits, constraints, awareness, and opportunities within the policy framework. Government has taken appreciable steps in the direction of providing up to 50% subsidy on the total cost of procurement of CA types of machinery and implements such as Super Straw Management System, Happy Seeder, Rotary Mulcher, Hydraulic Reversible Mould Board Plough, Paddy Straw Chopper/Shredder, Shrub Master, Rotavator, and Zero-Till Seed-Cum-Fertilizer Drill to the stakeholders. This provides opportunities and easiness for farmers to opt for the CA adoption by incorporating crop residues in soil or retained on the surface. At the farm level, the changes in concentration of SOC will determine the macro-level effects on soil health and crop productivity (Lal 2010b). Incentives should also be given to producers for the adoption of novel practices and technologies for risk management. For endorsing the CA system, the policies should be designed in a way that gives a complete understanding of micro-level problems. As we know, the production economics in CA improve overtime, and that larger farmers do not return back to conventional farming after conversion to CA practices.

13.8.2 Technological Interventions in the Cropping System

There is also a need to diversify the existing cropping systems before recommending any modern crop establishment techniques to enhance SOC stock in soils of India. Thus, outflows of C from the agricultural soils of India should be reduced by adopting climate-smart agricultural techniques which help to restore the SOC pool for overall improvement in agricultural sustainability in India (Lal 2011; Doddabasawa et al. 2019). At present, rice–wheat cropping system covering ~12 M ha is extensively practiced in India using conventional practices of both crop establishment and irrigation which have generated many sustainability issues (Meena et al. 2020; Kumar et al. 2021). Moreover, food preferences toward animal-based produces and food wastage are constantly developing pressure for producing more and more agricultural commodities, and that compels farmers to over-fertilization with large C-footprints. Under field conditions, excessive tillage is performed under standing water (puddling for rice) or without standing water (wheat), which results in the exposure of the hitherto protected SOM within aggregates to microbial processes and exacerbating release of GHGs into the atmosphere. Further, soil anaerobiosis in rice fields aggravate production of many GHGs (methane, nitrous oxide, etc.), which further negate sustainability. Therefore, there is an urgent need for intervention of diversified technologies in regions of rice-based farming systems and conversion of transplanted/flooded to direct-seeded/aerobic rice or other upland crops (e.g., cotton, vegetables). Conversion to aerobic rice or other upland crops may save inputs (e.g., water, energy for puddling) and other finite but precious resources.

Scientists from across the country suggested many technological interventions for enhancing C sequestration, and further reduce the C-footprints (Gan et al. 2014). Reduction in C-footprints of production system depends upon several factors *viz*., soil type, crop diversification, land cover, air temperature, and land management practices (Sainju et al. 2010; Pinheiro et al. 2015). Further, temperate regions have a better potential for C sequestration as compared to tropical regions (Lal 2004a; 2010a), and the former ranges from 50 kg C $ha^{-1}yr^{-1}$ to 1500 kg C $ha^{-1}yr^{-1}$ with the adoption of different climate-smart technologies. A linear relationship is observed for soils of India between C input and CO_2 output. An increase of 00.4 Pg CO_2 eq yr^{-1} of C input resulted in a corresponding increase in C output by ~0.2 Pg CO_2 eq yr^{-1} through an increase in system productivity (Maheswarappa et al. 2011). Usually, the landowners take care of maintenance and improvement of soil fertility. But it is likely that many tenant farmers take the land on lease and take crop without any C addition into the soil. So, there is a need to encourage farmers and landowners to diversify some high remunerative and biomass production crops with resource conservation technologies as per regional climate for betterment of a sustainable future.

13.8.3 Policies for Promotion of Crop Diversification

Adequate and proper policy planning is important to promoting crop diversification. The policy should cover the infrastructure availability, supporting price, technological interventions, and farmer's training to select alternative/ remunerative crops suitable to soil conditions. After analyzing the negative consequences of rice–wheat system on ecological conditions and soil health, the government is now looking for new scope through crop diversification that is remunerative and promotes C accretion. For instance, in Punjab, basmati rice, oilseeds, pulses, fruits, and vegetables are being encouraged as a substitute for rice and wheat monoculture. In 2002, a "multi-year contract farming" scheme was launched by the Punjab government to boost crop diversification to protect natural resources including soil health. The government of Punjab constituted Agricultural Diversification Fund (ADF) in 2005–06 to improved soil health under rice–wheat cropping system (Kumar et al. 2015a). Under ADF, the state government released a fund of ₹ 50.56 crores (~US$6.85 million) in 2006–07 to empower the agricultural infrastructure and promote crop diversification in the rice–wheat cropping system. Besides this, the Punjab government also released an additional fund of ₹ 10 crores (~US$ 1.35 million) in 2005–06 to establish Agricultural Research and Development

Fund (ARDF) for the screening of alternative crops and improved crop management strategies. Cultivation of other low water requiring traditional crops (i.e., pearl millet, maize, sorghum, millet, barley (*Hordeum vulgare*), pulses, etc.) should be reinforced by offering better support prices. The Crop Diversification Program of the Punjab Agriculture Department has been successful in achieving 0.5 M ha area under the "Crop Diversification Pilot Scheme" in 2020. There is a need to extend a similar in other states to enhance crop and soil productivity.

13.8.4 Fallow Periods and Their Management Plans

Sustainable management of soil C sequestration becomes more crucial when land is set aside and returned to nature. In India, there are three types of fallow periods *viz.*, summer fallow (April–June), rice fallow (November–February), and fallow period in drylands (*Kharif* fallow in absence of adequate rainfall or *rabi* fallow due to lack of residual moisture). In India, ~11M ha area remains fallow after rice, of which about 82% is in the Eastern States like West Bengal, Odisha, Jharkhand, Assam, Bihar, Chhattisgarh, Eastern Uttar Pradesh, etc. The remainder of fallow land is in peninsular India *viz.*, Karnataka, Tamil Nadu, and Andhra Pradesh (IIPR 2016). These are the areas where rice is taken during the *Kharif* season (June to October) while the land is fallowed between November and February. However, plowed bare fallowing has negative effects on the SOC storage capacity of agricultural soils (Rasmussen et al. 1998). During this period, SOC is lost due to its increased decomposition in absence of any input of biomass-C, and the increased temperature and mineralization with soil tillage (Eagle and Olander 2012). The C loss in summer fallowing is also accelerated through soil erosion. Thus, C capture and retention capacity of these soils can be greatly enhanced by elimination of summer fallow by cultivating a short-duration crop during the rice fallow period (Miglierina et al. 2000). In this context, there is a pertinent example from western Canada which has successfully transformed the wheat-fallow period to wheat–wheat in semi-arid regions, and this has improved the SOC content by 0.03 Mg ha^{-1} (Smith et al. 2001). Formulation and implementation of appropriate policies had great impacts on utilizing these fallow lands for the cultivation of pulses and oilseeds using proper management technologies. Recently, the Indian government has undertaken a good initiative for mapping of potential districts for pulses and oilseeds cultivation. The government, in coordination and collaboration with international institutions and organizations, has initiated the screening, identification of suitable crops and cultivars, agronomic management and other promotional technologies. In this direction, a scheme "Targeting Rice Fallow Areas" (TRFA) was implemented under Rashtriya Krishi Vikas Yojana (RKVY) in eastern India by DAC&FW during 2016–17 where 4.5 M ha under rice–fallow area was targeted to bring under *rabi* pulses and oilseed within the next three years by releasing a fund of ₹ 75 crores (~US$ 10 millions) (NAAS 2016). Initially, in 2016–17, the TRFA was initiated in 15 districts of 6 states (West Bengal, Odisha, Jharkhand, Chhattisgarh, Bihar, and Assam) to cover 0.78 M ha. In 2017–18, the area under the scheme was extended to 40 districts and 4000 villages to cover 0.6 M ha with support of Minikit distribution, cluster demonstration, and farmers' training (MoAFW 2018).

13.8.5 Potential to Promote Pulses in the Rice-Based System

Most of the rice fallows are cultivated to diverse pulses and oilseeds such as lentil, pea (*Pisum sativum*), green gram (*Vigna radiata*), chickpea (*Cicer arietinum*), black gram (*Vigna mungo*), pigeon pea (*Cajanus cajan*), lathyrus (*Lathyrus sativus*), mustard (*Brassica* spp.), sesame (*Sesamum indicum*), groundnut, sunflower (*Helianthus annuus*), and linseed (*Linum usitatissimum*). The government is further planning to extend the TRFA program in the rice fallows of southern, north-eastern, and Himalayan states up to 2022 with new norms of assistance (MoA&FW 2019a). Under National Food Security Mission-Pulses (NFSM-Pulses), the government is also promoting areas under pulse cultivation for more production and increased SOC content in rice fallows. In this context, soils of India have a large potential for expansion under various upland pulses and oilseeds in different

TABLE 13.6
Potential Crops for Rice Fallows in Different States

Crop	States
Lentil	Assam, West Bengal, Bihar, Odisha, Eastern Uttar Pradesh, Chhattisgarh, and Jharkhand
Pea	Jharkhand, Chhattisgarh, Eastern Uttar Pradesh, and Northern Madhya Pradesh
Chickpea	Chhattisgarh, Bihar, and Jharkhand
Green gram	Odisha, Chhattisgarh, Jharkhand, Bihar, Andhra Pradesh, Tamil Nadu, and Karnataka
Black gram	Coastal Andhra Pradesh, Tamil Nadu, Karnataka, and Odisha
Grass pea (Lathyrus)	Tal area of Bihar, Chhattisgarh, and West Bengal
Cluster bean	Andhra Pradesh, Tamil Nadu, and Karnataka
Lablab bean (*Lablab purpureus*)	Andhra Pradesh, Tamil Nadu, and Karnataka
Mustard	Eastern Uttar Pradesh, Bihar, and Jharkhand
Sesame/linseed	Odisha, Chhattisgarh, West Bengal, Jharkhand
Groundnut	Char area of Bihar, Mahananda of Odisha, Brahmaputra Valley of Assam, and coastal Andhra Pradesh

agroecological conditions (Table 13.6). Although, the amount of SOC sequestered in these eco-regions depends upon the quantity of crop residues being returned back to the soil.

Farmers are encouraged to grow pulses by establishing cluster demonstrations, distributing quality seed, providing improved types of machineries/tools/technologies, irrigation tools, and other facilities. To support this scheme, other government plans like MGNREGA (Mahatma Gandhi National Rural Employment Guarantee Act), RKVY, and PMKSY (Pradhan Mantri Krishi Sinchai Yojana), are also contributing through developing water harvesting structures like farm ponds. Several institutional and organizational projects are ongoing in different regions of the country to strengthen pulses and oilseed cultivation in fallow soils. For example, the National Food Security Mission (NFSM) is promoting the cultivation of pulses (i.e., lentil, chickpea, and lathyrus) in rice fallow leaving positive impacts on building C stock in these regions (Kumar et al. 2019). The ICAR-Indian Institute of Pulses Research (ICAR-IIPR) is providing technological and scientific support for managing and expanding areas under pulses in these regions (Kumar et al. 2019). Likewise, the Central Research Institute for Dryland Agriculture (CRIDA) in collaboration with NARS launched a project in 2010 on "Enhancing Lentil Production for Food, Nutritional Security and Improved Rural Livelihood" for Bihar, Assam, Eastern Uttar Pradesh, and West Bengal. Similarly, another exclusive project (i.e., "Enhancing Grass Pea Production for Safe Human Food, Animal Feed, and Sustainable Rice–Based Production System in India")is ongoing in Bihar, West Bengal, Eastern Uttar Pradesh, and Chhattisgarh states with the support of NFSM in the International Center for Agricultural Research in the Dry Areas (ICARDA) (Kumar et al. 2019).

13.9 CARBON STATUS OF DRYLAND AGRICULTURE IN INDIA

As much as 175 M ha of land in India is prone to degradation by different processes (FAO 2017) that need to be checked urgently by adopting principles of conservation agriculture. In these areas, the average rainfall is lower than the potential evapotranspiration with less than 0.65 aridity index (Safriel et al. 2005). The lack of water availability and land degradation in these areas hamper crop productivity that negatively affect SOC and SOM stocks. Adoption of conservation agriculture, crop rotations, crop residue retention, and cover crops show a positive impact on soil organisms, weed seeds predation, biodiversity, and other ecosystem services (WBCSD 2018). Thus, practicing improved agricultural methods and soil management can lead to SOC sequestration and improvement of soil health in drylands.

13.9.1 Management of Degraded Lands

Soil erosion involves breakdown of aggregates, which exposes the once hidden SOM to the oxidizing microorganisms and release of CO_2 into the atmosphere. Reduction in plant biomass and its roots is the main factor, associated with the decline in SOC contents in degraded lands. Further, salt-affected soils of Haryana, Andhra Pradesh, and West Bengal may have SOC content of lower than 5 g kg^{-1} (Lal 2004). Hence, the higher the erosion losses the lower is the SOC contents, the poorer the soil health and the lower the agronomic yields. So, focused research and policies should be emphasized on integrated agricultural and livestock management. With implementation of various agronomic and land management practices, drylands have a huge potential for enhanced C budget in the soil so that these areas are also referred to as "bright spots" for C sequestration (UNEP 2011; FAO 2017). The target of achieving significant C sequestration in degraded dryland soil is feasible, which in turn will directly benefit society and the environment.

13.9.2 Government Action Plans for Dryland Agriculture

The GOI is promoting dryland agriculture through different schemes, programs, projects, and missions in one or the other part of the country. For dryland agriculture promotion schemes like MGNREGA and the National Agriculture Development Programme (NADP) promote activities for land development/management and construction of water harvesting structures. So various water harvesting structures like farm ponds, community ponds, percolation ponds, and check dams have been constructed at field level. This helps to uplift dryland farming, as almost 60% of farmers depend on dryland farming for livelihood. Several initiatives based on natural resource management (NRM) have been launched to promote agriculture growth in rain-fed areas. A judicious watershed management allows capturing the rainfall and runoff water at the field level, thus extending the availability of water in drylands. In this direction, schemes such as "Drought Prone Areas Programme" and the "Desert Development Programme" were initiated in 1994–95. Under both schemes, watershed development was targeted to control desertification by rejuvenating the natural resources in dryland areas. National Watershed Development Project for Rain-fed Areas (NWDPRA) was implemented at the state level in 1990–91 to intensify the efforts for agricultural improvement in rain-fed areas based on the concept of integrated watershed management and sustainable farming systems. In 2000–2001, it was further merged into the Macro Management of Agriculture (MMA) scheme. Currently, this is being implemented in 28 states and two Union Territories sponsored by the central government. For watershed development projects, the common guidelines issued by the National Rain-fed Area Authority (NRAA) are followed. Currently, all the watershed management schemes have been brought under the confinements of the PMKSY. But under PMKSY also, many sub-schemes have been structured that are handled by different agriculture, water resources, and rural development ministries, all with their management strategies and separate budgets. Under the NAPCC, the National Mission for Sustainable Agriculture (NMSA) is one of the initiatives to promote sustainable agriculture in India.

13.9.3 Policies for Promoting Trees/Shrubs Plantation for Soil Carbon Restoration

Trees absorb atmospheric CO_2 during photosynthesis and utilize it to form branches, stems, and roots. The introduction of trees causes a marked increase in SOC level as the C introduced from trees are more resistant to decomposition as compared to other herbaceous crop plants (Kumar et al. 2018). So, forest areas are capable to absorb CO_2 quickly from the air and store it as C sink for long periods (Figure 13.8).

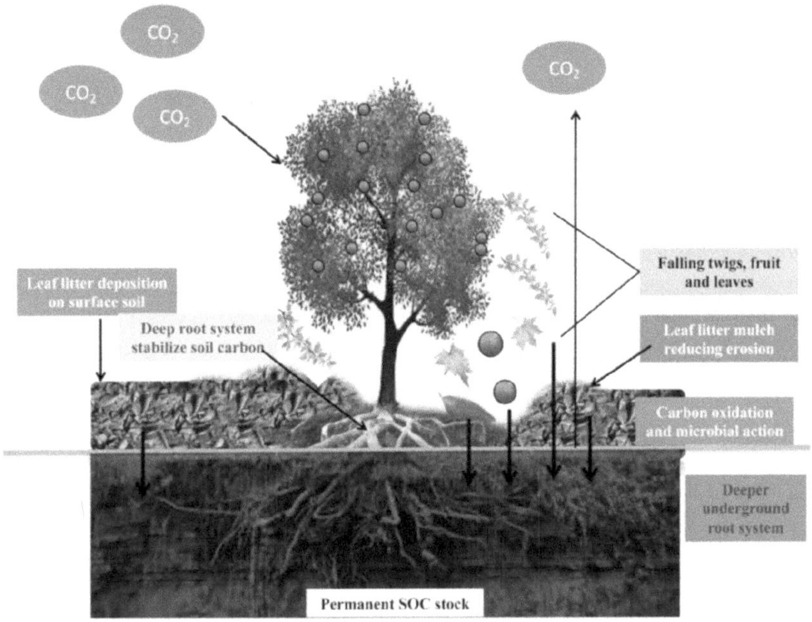

FIGURE 13.8 Trees enhance soil carbon sequestration.

Agriculture and forestry in India are significantly contributing to soil C sequestration. Thus, adopting different strategies and policies to enhance C sequestration can enrich the national C economy. So, in this direction, the government has taken various initiatives such as MGNREGA, National Mission of Green India, National REDD+ (Reducing Emissions from Deforestation and Forest Degradation) Programme, and different forests policies. In September 2005, the Indian Ministry of Rural Development proposed the livelihood security act named MGNREGA. The scheme was put forward for achieving India's climate change target under the Paris Climate Change Agreement in 2016. Under the scheme, the provision is to provide a minimum of 100 days of employment to at least a single member of an eligible family for any unskilled manual work assigned. Thus, to mitigate climate change, activities like afforestation and fruit orchards creation have been undertaken at block and village levels. Different interventions under MGNREGA have sequestered ~0.62 Pg of CO_2 equivalent in 2017–18 and it is predicted that by 2030 it will help to achieve sequestration of about 1.97 Pg CO_2 equivalent that is 8% of India's target (Hindustan Times 2018). The UNFCCC launched the REDD+ Programme globally in 2005 to utilize forest's potential in C sequestration to mitigate climate change. So, India has initiated its national REDD+ Strategy to address forest degradation and deforestation issues and enhance ecosystem C stock. Under the Paris agreement, India has established a target to capture 2.5–3.0 Pg of CO_2 by 2030 through afforestation and planting of trees (MoEF 2012). Therefore, the forest department is making intensive efforts in this direction to capture and store C in soil over the long term. Because forests play an important role in natural C sequestration, India has developed a robust policy and legal framework to preserve and maintain its forests. To strengthen the forest cover in the country, several enactments launched include Wildlife Protection Act (1972), Forest Conservation Act (1980), Environment Protection Act (1986), and the National Forest Policy (NFP) (1988). Under NFP, GOI plans to extend the area under forest cover by ~1% yr^{-1} and from 24% to 33% between 2005 and 2012. From the viewpoint of increased C sequestration, these policies were aimed at expanding and improving the plantation coverage and preventing deforestation of already existing forest area. During the ninth and tenth five-year plan, the government initiated the National Afforestation Programme (NAP) in integration with agricultural and rural development intending to increase 1.0 M ha of forest cover area every year. Together, all these forest policies have contributed significantly to the net CO_2 sink.

13.10 RESEARCH EVIDENCE ON CARBON DYNAMICS IN AGROFORESTRY MODELS

Agroforestry is a key model for managing the natural resources, increasing production of agriculture, and achieving the national vegetal cover for sustainable production and development. However, the degree of its performance and outputs depends upon the site conditions, the ecological amplitude of species, compatibility of trees, crops, and other components under the given set of environments. The sole production of either crop or tree both have some limitations in terms of soil, resource utilization, economics, environment, and ecological aspects. In this context, agroforestry has positive outputs which include efficient nutrient cycling, reduction in loss of soil and nutrients, improvement of the soil characteristics (physical, chemical, and biological), conservation of resources and achievement of food security and its sustainability (Jhariya et al. 2012).

Agroforestry has good potential to rehabilitate degraded lands, improve productivity and restore soil fertility. Agroforestry reportedly contributes up to 50% demand for fuelwood, more than 60% of small timber, nearly 75% of plywood, 60% of paper pulp raw material, and up to 10% of green fodder requirement besides the needs of the household (Nair et al. 2010). Presently, agroforestry in India is practiced on 25.3 M ha (approx. 8.20% of the country's area) (Dhyani 2018). Further, nearly 15% of the total cultivated area has a diverse form of agroforestry which comprises 11.2% irrigated and 16.5% of rain-fed areas. In arid and semi-arid regions, common tree species in the agroforestry system comprise *Acacia nilotica, A. tortilis, Ailanthus excelsa, Azadirachta indica, Albizia, Celtis australis, Dalbergia sissoo, Grewia optiva, Leucaena leucocephala, Morus alba, Prosopis cineraria, Ziziphus,* spp. etc. The higher plants and shrubs significantly enhance fodder supply in arid and semi-arid regions especially during the lean or dry periods when there is least or non-availability of green fodder. Therefore, silvo-pastoral system is one of the promising options for degraded lands (Dhyani 2018). For restoration of alkali soils through silvo-pastoral approaches, Dagar et al. (2001) indicated that tree species such as *A. nilotica, Prosopis juliflora* and *Tamarix articulata* have much potential. Dagar and colleagues also observed that incorporation of grasses like *Brachiaria mutica, Chloris gayana*, and *Leptochloa fusca* are also promising and productive. Nair et al. (2010) also reported that agroforestry systems have the C sink potential of 0.29–15.21 Mg ha^{-1}yr^{-1} aboveground and 30–300 Mgha^{-1}C up to 1m depth of soil. Nair and colleagues also observed that agroforestry has the potential of contributing toward 12% of the global terrestrial C stock. Therefore, it is essential to adopt agroforestry on underutilized areas that can be utilized efficiently through its adoption and development by linking to the World Bank initiatives in Community Development Carbon Fund and the Bio Carbon Fund toward resilient ecosystem (Nair et al. 2010).

Managing C through diverse forestry schemes is pivotal to enhancing the amount of C sequestered in vegetation and soil C pools. Agroforestry systems have a large potential of C sequestration, with a range of 25–96 Mg C ha^{-1}, depending upon the species and site interactions, biomass production, and ecological amplitude of the species (Dhyani 2018). Several researchers have confirmed that agroforestry is a promising system for increasing plant and soil C stocks as well as adapting and mitigating climate change. For example, Newaj and Dhyani (2008) reported that tropical agroforestry can sequester up to 12–228 Mg ha^{-1} C with an average of 95 Mg C ha^{-1}.

The SOC and C mitigation potential in various land-use systems (Figure 13.9) indicates that the higher values are under the alternate land-use practices than the farming and fallow land-use (Reddy 2002). The highest SOC was recorded in the order: agri-silviculture (19.9 Mg ha^{-1}) >silvi-pasture (17.5 Mg ha^{-1}) >agri-silvi-horticulture (17.0 Mg ha^{-1}) >*Leucaena leucocephala* (15.7 Mg ha^{-1}) >*Acacia albida* (15.2 Mg ha^{-1}) >*Eucalyptus camaldulensis* (13.2 Mg ha^{-1}) >*Tectona grandis* (12.5 Mg ha^{-1}) >*Dendrocalamus strictus* (11.7 Mg ha^{-1}) >*Azadirachta indica* (11.4 Mg ha^{-1}) > agricultural

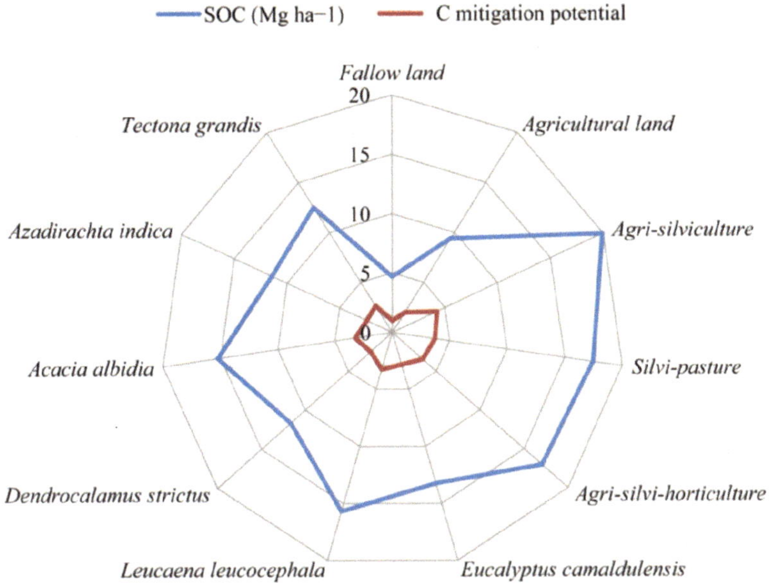

FIGURE 13.9 Carbon mitigation and soil organic carbon under various land-use in India (Data source: Reddy 2002).

system (9.4 Mg ha^{-1}). Further, C mitigation potential is reportedly higher in agri-silviculture (4.2 Mg ha^{-1}) than that under fallow and agricultural land (Reddy 2002).

13.10.1 Agroforestry Policies in India

The policy and legal instruments are the key pillars for the successful implementation of technology and the development of the country. India is the first nation across the world to have a separate policy in agroforestry (i.e., National Agroforestry Policy 2014). However, some earlier policies also addressed and covered the agroforestry development in India such as Green India Mission, National Policy for Farmers, National Bamboo Mission, National Agriculture Policy, and NPF. The new policy on agroforestry emphasizes and promotes agroforestry development and income to the farming community, reduces the risk of climate change, offers various diversified products and outputs, and creates avenues of farming community through employment generation and reduces the pressure on natural forest and resources. Further, these policies are initiated to address the main motto of covering 1/3rd area under vegetation cover for sustainability (Chavan et al. 2015).

In the Indian scenario, Chhattisgarh and Madhya Pradesh initiated the agroforestry policy to combat poverty and offer alternative economic gains to the farming community through various schemes, such as *Lok Vaniki* in Madhya Pradesh, to accomplish the forest in private farm or land. In other states, the government has taken initiatives regarding home gardens, bund planting, etc. These policies are framed to increase economic gains, enhance environmental sustainability, control soil erosion, reduce land and environmental degradation, and increase an efficient resource utilization and C sequestration, especially in the small and marginal farming communities. The recent agroforestry policy reflects positive impacts of agroforestry extension, developing and shifting the outlooks of farmers toward agroforestry. This new paradigm will contribute toward scaling up the agroforestry for C sequestration and climate mitigation in India (Chavan et al. 2015).

13.11 PROMOTION OF COMPOSTING/ VERMICOMPOSTING AT THE BLOCK LEVEL

Demonstrations related to composting must be conducted at each block level by the agricultural officers with the help of State Agricultural Universities, Krishi Vigyan Kendras (KVKs known as farmers' science center), Agricultural Research Institutes/Stations, and the State Agricultural Departments. Frequent visits, training, and financial assistance shall be provided to the farmers to make them aware of the practical benefits of organic farming and the resultant improvements in soil C status. The demonstration could involve the methods of preparation and use of compost, vermicompost, and other organic inputs. Besides, farmers must be trained by demonstrating micro-level trials utilizing waste decomposer, proper incorporation of crops residues and stubbles in soil, legume promotion in crop rotation, balancing the input-output of farm products, effective utilization of fallow period by growing leguminous green manuring, and other promotional activities (Kumar et al. 2020). In general, farmers hesitate to adopt a technology/methodology unless or until they see the practical significance of that particular technique. In that situation, these trials could better advertise the importance of soil health and organic farming and thus can be proved as an excellent trainer of the farmers at village levels. To make this program successful, it is necessary to take farmers into confidence, and for this, the honesty, behavior, and dedication of concerned Agricultural Officer to his/her responsibility are important. The Department of Rural Development and Blocks in Punjab is facilitating the land for the establishment of biomass-based units throughout the state (Kumar et al. 2015a).

13.12 POLICIES FOR SETTING UP COMPOSTING PLANTS AT LARGE SCALE

India annually generates about 5.0 Pg of crop residues, of which the cereal crops i.e., rice, wheat, maize, millets, etc. contribute 70% (3.62 Pg) of residue. Of this, rice and wheat contribute 34 and 22% of residue respectively and a large proportion of that is being brunt on the farm (The Economics Time 2019). Some of the crop residues produced are used as fodder and fuel and for other domestic purposes. Yet, about 1.40 Pg of crop residues are surplus (Bhuvaneshwari et al. 2019). In addition, about 3.20 Pg of cattle manure, 0.25 Pg poultry manure, and about 0.25 Pg of domestic wastes are being produced annually in the country. The municipal solid wastes also greatly contribute to cumulative organic wastes having an estimated annual production of 0.274 Pg including 0.045 Pg of wastes from fruits and vegetables. At the same time, about 7.60 Pg of potato (*Solanum tuberosum*) residues are produced annually throughout the country (MoA & FW 2020).

Thus, there is ample scope for the establishment of composting units at a large scale at the tehsil/ district level to promote their application and enhancing soil C stock. The country is generating a huge amount of agricultural wastes/raw organic materials including crop residues which can be effectively utilized after well decomposing in composting units for input into soil. According to the NPMCR. These agro-wastes including crop residues can be beneficially utilized again in farming as input after proper processing and composting. Therefore, it is crucial to create awareness among the farmers to establish vermicomposting units for recycling organic wastes rather than in-field burning of residues. The government has launched several schemes and programs to promote the addition of organic inputs to the soil. Some of the central sector schemes, by which financial assistance is provided to the farmers through the State government, are National Project on Organic Farming (NPOF) (2004), NFSM (2007–08), RKVY (August 2007), Mission for Integrated Development of Horticulture (MIDH) (2014–15), and NMSA (2014–15). Under NPOF, a 33% subsidy is provided on total establishment cost up to a maximum ceiling of ₹ 60 lakhs (~US$ 0.08 million) through National Bank for Agriculture and Rural Development (NABARD) for the installment of compost production unit from agricultural wastes (Indian Botanists 2014).

13.13 POLICIES TO EXPAND THE AREA UNDER ORGANIC FARMING

A high priority must be given to formulating and implementing novel policies for encouraging investment in research, technology transfer, farmers training, knowledge sharing, capacity building, and application of advanced technologies based on the region-specific available raw organic materials. The government is already working to promote balanced fertilization through soil test-based and judicious application of inorganic and organic manuring to increase soil C but to make the program successful still more efforts are needed in this direction. The GOI has also focused on the transformation of conventional farming to organic farming indifferent parts of the country on basis of site-specific ecological conditions. Further, attention was also given by the GOI to the concept of balanced fertilization including inorganic and organic fertilizers. But still, it had a challenge as small and marginal farmers who are doing farming on leased land do not have their own livestock and are not able to access organic manures like compost, farmyard manure, etc. For that GOI have to review the legal documents on organic manures to establish an appropriate legal structure on organic inputs and other issues such as taxes, credits, and new technologies in farming. It is important to frame the long-term policies on a priority basis to promote the production and application of organic inputs in crop production, livestock farming, city and food processing waste, and other raw materials such as peat, seaweed. The emphasis should be given to the balanced input use in areas where organic farming alone is not possible due to the unavailability of raw materials, such as in arid regions which are poor in soil fertility and prevalence of high temperatures can oxidize SOM content.

The GOI has started several programs to promote the usages of C-based input in agriculture and enhance organic production. Some important programs and schemes (Yadav 2017) of GOI include Paramparagat Krishi Vikas Yojana (PKVY), RKVY, Organic Value Chain Development in North Eastern Region Scheme (OVCDNERS), NPOF, National Programme for Organic Production (NPOP), NMSA, Soil Health Management Scheme (SHMS), and Soil Health Card (SHC). The PKVY is a cluster-based scheme under the NMSA mission to promote organic agriculture in 0.2 M ha areas covering 10,000 clusters (20 ha each). The government has sanctioned a total amount of ₹ 300 crores (~US $ 40.63millions) under this scheme during 2015–16. In this scheme, financial support of ₹ 50,000 (~US $ 677) ha^{-1} per farmer was allocated for three years to bring 20 M ha of land under organic cultivation during 2015–16 and 2017–18 (Yadav 2017). Organic farming and production of domestic organic inputs are being promoted through rational resources under this scheme. Besides, the scheme is focused to optimize the use of local natural resources for C-based input generation, to use indigenous techniques of crop management, and to sustain soil health. The scheme OVCDNERS was launched to promote organic farming in the north-eastern regions of the country. An amount of ₹ 115 crores (~US $ 15millions) was released to promote this scheme. The NPOP was launched by the GOI in 2001 to promote organic farming in 11 States (i.e., Rajasthan, Madhya Pradesh, Maharashtra, Gujarat, Tamil Nadu, Kerala, Karnataka, Uttaranchal, Mizoram, Sikkim, and Nagaland) (Yadav 2017). Under NMSA mission, the SHMS promotes integrated nutrient management (INM) for balanced plant nutrition focusing more the on use of locally available organic inputs such as compost, vermicompost, farmyard manure, green, manures, etc. to accelerate soil C level and soil health. Besides, Soil Testing Laboratories (STLs) are also established to recommend soil test-based fertilizer applications. As a part of the Swachh Bharat Mission, a policy on promoting the use of city compost was designed in 2016 by the Department of Fertilizers. Under this, financial assistance of ₹ 1500 (~ US $ 32) is given for the production and consumption of 1 Mg of city compost. In 2017–18, a total of 90.7 Pg of city compost has been marketed by the fertilizer companies. Besides, ICAR through the scheme of Network Project on Organic Farming is facilitating research in 20 centers covering 16 states to advance region-specific packages and practices of organic farming in crops and cropping systems. But the major concern is that till now no scheme and program is being promoted which has directly focused on restoring SOC stocks in India.

13.14 PRIVATE ORGANIZATIONS FOR INVESTMENT IN SOIL CARBON RESTORATION

The quantification and verification of C benefits under different management practices over time will motivate private investors to invest in soil C sequestration. Private investment in soil will be promoted up to an extent if the government clearly declares it to be new business opportunities. Private organizations/companies can promote the instruments and implements that enable farmers to adopt soil C sequestration practices such as CA implements and types of machinery. For that, an appropriate price on C is essential and that is an effective way to increase soil C sequestration. The public and private investments can also be attracted by documenting that soil C sequestration practices are directly linked with biodiversity, adaptation, mitigation, water management, and the economics of farming and rural communities. It can be accelerated by demonstrating C sequestration potential of best management practices through field trials, demonstration projects, regional and state trials, and farmer's training programs, and in the coming years, it may be one of the top priorities. The private companies together with the public sector may compensate farmers for SOC sequestration at least during the initial period of land use change toward new management practices. It is of great interest to see if existing payment programs and subsidies can be adjusted to prize the C accrual approaches.

13.15 RECOMMENDED POLICY AGENDA FOR SOIL CARBON RESTORATION

The amount of C sequestrated in soil depends on land-use change with same management approaches and on land-use decisions. Presently, several policies are promoting management decision to improve crop production and environmental quality, and attain enhanced SOC stocks. Regulatory standard policies (command-and-control guidelines) should be designed to encourage farmers for adopting best management practices or land-use decisions to increase soil C content and stocks. These regulatory policies should maintain standard rules and regulations to promote the adoption of C-intensive practices. Soil C sequestration can be considered indirect pecuniary support to the farmers in three important ways as follows:

Firstly, there should be a provision of direct incentive payment to the farmers or land managers for switching from conventional tillage to conservation agriculture. However, the C-intensive practices may be specific to a particular region of the country following the spatial heterogeneity of available resources. Therefore, the choice can be given to the farmers to choose the management practices among the recommended C-intensive practices or land-use decisions in accordance with the suitability of a particular approach in the field and other defined or undefined regions.

The farmers' incentive should be on per-hectare bases as the minimum amount per Mg of C sequestration that will encourage the farmers to achieve SOC storage. Appropriate policies can endorse SOC sequestration in parts of the country where the production losses can be counterbalanced by the incentives received from the sale of per hectare and per Mg of C sequestration. Besides, the sequestered C can be sold to the industries and companies interested in the reduction of GHGs emissions at the current market price. In Australia and Canada, for example, some government-subsidized, incentive-based compensation projects and trading related to soil C sequestration are being implemented (Paustian et al. 2019). However, implementation is lacking at a wider scale. In both per hectare and per Mg payment scheme, the following assumptions are used to evaluate the management decisions made by the farmers: 1) Each hectare of land is managed separately and that strategies followed in that field do not affect the nearby field, 2) Each farmer takes only one crop, and 3) Farmers have a fixed and sufficient capital to take crop using either no-till or conventional tillage. Although the spatial heterogeneity in the field, soil features, and their interference in assessing the change in soil C by adopting newer management practices or land use are some of the very issues that complicate the formulation of dynamic policies. Also, even after these policies being efficient in C sequestration, their execution and authentication could be challenging, as they require intensive knowledge in the understanding of connection between SOC sequestration and land quality, use and management as well as the guesstimate of SOC baseline. A policy can be efficient only if it endorses the espousal of best management practices to increase soil C sequestration by taking into consideration the overall economics of the program.

13.16 CONSIDERATIONS FOR POLICIES FRAMEWORK ON SOIL CARBON RESTORATION

- Advance policy agendas to stop the destruction of existing C-rich soils through deforestation, land degradation, and detrimental land-use change, etc.
- Document the efficacious C-intensive management approaches and demonstrate benefited farms.
- Identify or approach the progressive farmers who are already taking care of agroecology and C sequestration. Publish and popularize their success stories.
- Critically analyze data on existing subsidies that have a negative impact on soil C sequestration and communicate the outcomes in original and high-impact ways.
- Consider the multi-stakeholder approaches and accordingly advertise the governmental agenda that will improve soil C sequestration in addition to the employment creation, food security, environmental and social services.
- Reward the farmers and other stakeholders who adopt C-based practices by initiating an award program on a specific policy.
- Make the brief policy papers in simple and local language for their better understanding of the C sequestration program to the producers.
- Establish a protocol for tracing, quantification, and verification of historical change in soil C status under different best management practices over the period.
- Monitor and timely evaluate schemes for their workability, corruption in implementation.
- Document feedback from farmers concerning easiness, incentives, and economics of activities and accordingly find the gaps for further modification and improvement in existing policies.
- Conduct farmers' training programs at micro-level to create awareness on the financial support provided by the government to promote C inputs in farming by the expansion of recommended management practices.
- Follow timely discussions or open communication and sponsored crusade to land managers and monitoring officials to adopt soil C sequestration in the employment agenda, rural economics, and development.
- Establish a common platform for knowledge sharing among the farmers at the local, regional, and national levels through e-learning, seminars, conferences, symposiums, and other social platforms.
- Design governmental policy and C market in a way that includes small farmers, land tenure, land access and protect their interest.
- Organize a summit on agroecological approaches on soil C sequestration including farmers, rural communities, scientists, policymakers, and planners.
- Advance a communication expedition that exposes the subsidies on C-exhaustive practices and positively publicize the benefits of soil C, soil health, and their best management practices.
- Draft a declaration of rules and regulations to direct policies and programs on soil C sequestration.
- Form a strong virtual network among the farmers, land managers, and their associations to prepare an action plan and look for available funding to promote C-based management practices.

- Promote research and developmental activities on carbonaceous practices and land-use decisions across the agroecological regions and most soil types.
- Carry out case studies on farmers and land managers who actively engaged in best management practices for the soil and communicate their stories to planners and democrats.
- Build a local C market for the selling of C offset of farmers.
- Publish the soil C prospectus and invite the investors to invest in the field for C sequestration.
- Start work with developmental banks to reveal success stories on soil C sequestration to entice public and private investors.
- Prioritize work with community-based finance, and help the farmers and producers by providing small loans and grants for adopting best management practices in farming.

The following key points need to be considered to move forward on SOC sequestration:

- A strong and comprehensive political vision.
- An appropriate business proposal for land managers and farmers.
- A robust business proposal and track-record of triumph among private and public investors.
- Financial assistance and policy framework to make payment sturdier in programs and schemes of ecosystem services that include producers and attract more investments.
- Legal provisions to resolve controversies over land and C tenure.

The government or an organization may invest in the restoration of SOC stock only when the following elements are combined: 1) improved productivity, 2) risk management capacity, 3) improved marketing facilities, 4) financial assistance to C offsets, and 5) governmental subsidies and financial incentives. Therefore, considering the importance of soil C, this chapter is mainly focused on the effective plans and policies for soil management and C restoration in the agricultural soils that will eventually make the country more secure against climate change and vulnerability to soil degradation. It will also support the aims of "Sustainable Development Goals" for a better country and the planet in general.

ACKNOWLEDGEMENT

This chapter formulated has been through the financial support for faculty under the Institution of Eminence (IoE) scheme No. 6031, Banaras Hindu University, Varanasi (UP) – 221005, India.

REFERENCES

Bakker, R., E. Pedersen, G. va den Berg, R. Stewart, W. Lok, and J. Bouma. 2012. Impact of wind turbine sound on annoyance, self-reported sleep disturbance and psychological distress. *Science of the Total Environment* 425: 42–51.

Bhatt, R., S. S. Kukal, M. A. Busari, S. Arora, and M. Yadav. 2016. Sustainability issues on rice-wheat cropping system. *Inter Soil and Water Conservation Research* 4(1): 68–83. DOI: 10.1016/j.iswcr.2015.12.001.

Bhattacharrya, T., D. K. Pal, M. Velayutham, P. Chandran, and C. Mandal. 2000. Total carbon stock in Indian soils: Issues, priorities and management. In: *Land Resource Management for Food and Environment Security (ICLRM)*. New Delhi: Soil Conservation Society of India, 1–46.

Bhuvaneshwari, S., H. Hettiarachchi, and N. Meegoda Jay. 2019. Crop residue burning in India: Policy challenges and potential solutions. *International Journal of Environmental Research and Public Health* 16(5): 832. DOI: 10.3390/ijerph16050832.

Blazewicz, S. J., E. Schwartz, and M. K. Firestone. 2014. Growth and death of bacteria and fungi underlie rainfall-induced carbon dioxide pulses from seasonally dried soil. *Ecology* 95: 1162–1172.

Blume, E., M. Bischoff, J. M. Reichert, T. Moorman, A. Konopka, and R. F. Turco. 2002. Surface and subsurface microbial biomass, community structure and metabolic activity as a function of soil depth and season. *Applied Soil Ecology* 20(3): 171–181.

Bronick, C. J., and R. Lal. 2005. Soil structure and management: A review. *Geoderma* 124(1/2): 3–22.

Cayuela, M. L., L. Van Zwieten, B. P. Singh, S. Jeffery, A. Roig, and M. A. Sánchez-Monedero. 2014. Biochar's role in mitigating soil nitrous oxide emissions: A review and meta-analysis. *Agriculture Ecosystem & Environment* 191: 5–16.

Chavan, S. B., A. Keerthika, S. K. Dhyani, A. K. Handa, R. Newaj, and K. Rajarajan. 2015. National Agroforestry Policy in India: A low hanging fruit. *Current Science* 108: 1826–1834.

Chisti, Y. 2008. Biodiesel from microalgae beats bioethanol. *Trends in Biotechnology* 26(3): 126–131. DOI: 10.1016/j.tibtech.2007.12.002.

Chiu, W. Y., C. Y. Kao, C. H. Chen, T. C. Kuan, S. C. Ong, and C. S. Lin. 2008. Reduction of CO_2 by a high-density culture of *Chlorella* sp. in a semicontinuous photobioreactor. *Bioresource Technology* 99: 3389–3396. DOI: 10.1016/j. biortech.2007.08.013.

Chowdhury, Z. Z., K. Pal, W. A. Yehye, S. Sagadevan, S. T. Shah, G. A. Adebisi, E. Marliana, R. F. Rafique, and R. B. Johan. 2017. *Pyrolysis: A Sustainable Way to Generate Energy from Waste, Pyrolysis*, Mohamed Samer. Intechopen. DOI: 10.5772/intechopen.69036.

Conference Report. 2017. Summary Report of Conference Held at Chantilly, France on Sequestering Carbon in Soil: Addressing the Climate Threat from May 3–5, 2017. Available at file:///C:/Users/Sandeep%20Kumar/Desktop/Soil%20Carbon/Final-Report-Sequestering-Carbon-in-Soil.pdf.

Dagar, J. C., G. Singh, and N. T. Singh. 2001. Evaluation of forest and fruit trees used for rehabilitation of semiarid alkali/sodic soils in India. *Arid Soil Research and Rehabilitation* 15(2): 115–133.

Datta, A., M. A. Emmanuel, N. K. Ram, and S. Dhingra. 2020. *Crop Residue Management: Solution to Achieve Better Air Quality*. New Delhi: TERI.

Day, D., R. J. Evans, J. W. Lee, and D. Reicosky. 2005. Economical CO_2, SO_2, and NO_2 capture from fossil–fuel utilization with combined renewable hydrogen production and large-scale carbon sequestration. *Energy* 30(14): 2558–2579.

Dhyani, S. K. 2018. Agroforestry in Indian perspective. In: *Agroforestry Opportunities for Enhancing Resilience to Climate Change in Rainfed Areas*, eds. R. G. Rajeshwar, M. Prabhakar, G. Venkatesh, I. Srinivas, R. K. Sammi. Hyderabad, India: ICAR-Central Research Institute for Dryland Agriculture.

Doddabasawa, B. M., M. Chittapur, and M. Mahadeva. 2019. Comparison of carbon footprint of traditional agroforestry systems under rainfed and irrigated ecosystems. *Agroforestry Systems* 94: 465–475. DOI: 10.1007/s10457-019-00413-4.

Eagle, A. J., and L. P. Olander. 2012. Greenhouse gas mitigation with agricultural land management activities in the United States—A side-by-side comparison of biophysical potential. *Advances in Agronomy* 115: 79–179. DOI: 10.1016/B978-0-12-394276-0.00003-2.

FAO. 2017. *Soil Organic Carbon: The Hidden Potential*. Rome: Food and Agriculture Organization of the United Nations.

FAO. 2019. *Moving Forward on Food Loss and Waste Reduction*. Rome. Licence: CC BY-NC-SA 3.0 IGO. The State of Food and Agriculture.

Food and Agriculture Organization of the United Nations Statistics (FAOSTAT). 2020. http://www.fao.org/faostat/en/#data/GT.

Feeney, D. S., J. W. Crawford, T. Daniell, P. D. Hallett, N. Nunan, K. Ritz, M. Rivers, and I. M. Young. 2006. Three-dimensional microorganisation of the soil-root-microbe system. *Microbial Ecology* 52(1): 151–158.

Fierer, N., A. S. Allen, J. P. Schimel, and P. A. Holden. 2003. Controls on microbial CO_2 production: A comparison of surface and subsurface soil horizons. *Global Change Biology* 9(9): 1322–1332.

Gan, Y., C. Liang, Q. Chai, R. L. Lemke, C. A. Campbell, and R. P. Zentner. 2014. Improving farming practices reduces the carbon footprint of spring wheat production. *Nature Communications* 5: 5012–5012.

Giller, K. E., M. Corbeels, J. Nyamangara, B. Triomphe, F. Affholder, E. Scopel, and P. Tittonell. 2011. A research agenda to explore the role of conservation agriculture in African smallholder farming systems. *Field Crops Research* 124(3): 468–472.

GOI. 2019. *Report of the Committee Review of the Scheme. Promotion of Agricultural Mechanisation for In-Situ Management of Crop Residue in States of Punjab, Haryana, Uttar Pradesh and NCT of Delhi. Government of India*. New Delhi: Ministry of Agriculture and Farmers Welfare Krishi Bhawan, May 2019.

Haugen-Kozyra, K., and T. Goddard. 2009. Conservation agriculture protocols for greenhouse gas offsets in a working carbon markets. Paper presented at the IV World Congress on Conservation Agriculture, 3-7 February 2009. New Delhi, India.

Hindustan Times. 2018. Govt proposes 100-day work scheme to fight climate change. December 07, 2018. https://www.hindustantimes.com/india-news/govt-proposes-100-day-work-scheme-to-fight-climate-change/story-WpR89EJXrS3Y5NnDR8lpZP.html.

Hinsinger, P., A. G. Bengough, D. Vetterlein, and I. M. Young. 2009. Rhizosphere: Biophysics, biogeochemistry and ecological relevance. *Plant and Soil* 321(1–2): 117–152.

Ho, S. H., W. M. Chen, and J. S. Chang. 2010. Scenedesmus obliquus CNW-N as a potential candidate for CO2 mitigation and biodiesel production. *Bioresource Technology* 101(22): 8725–8730. DOI: 10.1016/j.biortech.2010.06.112.

Hulatt, C. J., and D. N. Thomas. 2011. Productivity, carbon dioxide uptake and net energy return of microalgal bubble column photobioreactors. *Bioresource Technology* 102(10): 5775–5787.

Hungate, B. A., R. L. Mau, E. Schwartz, J. G. Caporaso, P. Dijkstra, N. van Gestel, B. J. Koch, C. M. Liu, T. A. McHugh, J. C. Marks, E. M. Morrissey, and L. B. Price. 2015. Quantitative microbial ecology through stable isotope probing. *Applied and Environmental Microbiology* 81(21): 7570–7581.

IIPR. 2016. *Annual Report on Promotion of Pulses in NEH Region, 2015-16.* Kanpur, Uttar Pradesh 208 024 (India): ICAR-Indian Institute of Pulses Research, 56 pp.

Indian Botanists. 2014. *Avail Benefits from Government Schemes for Organic Farming.* February 21, 2014. http://www.indianbotanists.com/2014/02/avail-benefits-from-government-schemes.html.

Jangir, C. K., D. Panghaal, S. Kumar, R. S. Meena, and Prince. 2017. Enriching soil carbon stock through mitigating soil erosion. In: *Agricultural, Allied Sciences & Biotechnology for Sustainability of Agriculture, Nutrition & Food Security*, eds. Ratnesh Kumar Rao, P. K. Sharma, M. Raghuraman, J. K. Singh. Mahima Publications, ISBN: 978-81-926935-8-3, 415–419.

Jansson, J. 2011. Towards "Tera-Terra": Terabase sequencing of Terrestrial Metagenomes. *ASM MIcrobe Magazine.*

Jhariya, M. K., V. K. Nayak, and K. N. Yadaw. 2012. Rainwater harvesting—An approach towards water management. *International Research Journal Lab to Land* 4(13): 46–48.

Kallenbach, C. M., S. D. Frey, and A. S. Grandy. 2016. Direct evidence for microbial-derived soil organic matter formation and its ecophysiological controls. *Nature Communications* 7:13630.

Kassam, A., T. Friedrich, F. Shaxson, H. Bartz, I. Mello, J. Kienzle, and J. Pretty. 2014. The spread of Conservation Agriculture: Policy and institutional support for adoption and uptake. *Field Actions Science Reports* 7. http://journals.openedition.org/factsreports/3720.

Kell, D. B. 2011. Breeding crop plants with deep roots: Their role in sustainable carbon, nutrient and water sequestration. *Annals of Botany* 108(3): 407–418.

Kimble, J. M., R. Lal, and R. F. Follett. 2002. *Agricultural Practices and Policies for Carbon Sequestration in Soil.* Boca Raton; London; New York; Washington, DC: Lewis Publishers. A CRC Press Company.

Koch, B. J., A. T. McHugh, M. Hayer, E. Schwartz, S. J. Blazewicz, P. Dijkstra, N. Gestel, J. C. Marks, R. L. Mau, E. M. Morrissey, J. Pett-Ridge, and B. A. Hungate. 2018. Estimating taxon-specific population dynamics in diverse microbial communities. *Ecosphere.*9(1):e02090. doi:10.1002/ecs2.2090.

Kumar, K., C. N. Dasgupta, B. Nayak, P. Lindblad, and D. Das. 2011. Development of suitable photobioreactors for CO_2 sequestration addressing global warming using green algae and cyanobacteria. *Bioresource Technology* 102(8): 4945–4953. DOI: 10.1016/j.biortech.2011.01.054.

Kumar, P., S. Kumar, and L. Joshi. 2015a. Policies for restricting the agriculture residue burning in Punjab. In: *Socioeconomic and Environmental Implications of Agricultural Residue Burning*, eds. P. Kumar, S. Kumar, and Laxmi Joshoi. New Delhi: Springer, Briefs in Environmental Science. Springer, Singapore. DOI: 10.1007/978-81-322-2014-5_6.

Kumar, P., S. Kumar, and L. Joshi. 2015b. The extent and management of crop stubble. In: *Socioeconomic and Environmental Implications of Agricultural Residue Burning. Springer Briefs in Environmental Science*, eds. P. Kumar, S. Kumar, and Laxmi Joshoi. New Delhi: Springer Briefs in Environmental Science, Springer, Singapore. DOI: 10.1007/978-81-322-2014-5_2.

Kumar, R., J. S. Mishra, P. K. Upadhyay, and H. Hans. 2019. Rice fallows in the eastern India: Problems and prospects. *Indian Journal of Agricultural Sciences* 89(4): 567–577.

Kumar, R., J. S. Mishra, S. Mondal, R. S. Meena, P. K. Sundaram, B. P. Bhatt, R. S. Pan, R. Lal, K. Saurabh, N. Chandra, S. K. Samal, H. Hans, and R. K. Raman. 2021. Designing an ecofriendly and carbon-cum-energy efficient production system for the diverse agroecosystem of South Asia. *Energy* 214: 118860. DOI: 10.1016/j.energy.2020.118860.

Kumar, S., R. S. Meena, and R. Lal. 2018. *Role of Legumes in Soil Carbon Sequestration. In Book: Legumes for Soil Health and Sustainable Management*, eds. R. S. Meena, A. Das, G.S. Yadav, R. Lal. Springer publication, Singapore, 109–138. DOI: 10.1007/978-981-13-0253-4_4.

Kumar, S., R. S. Meena, R. Datta, S. K. Verma, G. S. Yadav, G. Pradhan, A. Molaei, G. K. M. Mustafizur Rahman, and H. A. Mashuk. 2020. Legumes for carbon and nitrogen cycling: An organic approach. In *Carbon and Nitrogen Cycling in Soil*, eds. R. Datta, R. S. Meena, S. I. Pathan, and M. T. Ceccherini. Springer publication, Singapore, 337–375. DOI: 10.1007/978-981-13-7264-3_10.

Lal, R. 1998. *Soil Quality and Agricultural Sustainability*. Advances in Soil Science. Boca Raton, FL: CRC Press, 17–30.
Lal, R. 2002. Soil carbon dynamics in cropland and rangeland. *Environment and Pollution* 116(3): 353–362. DOI: 10.1016/S0269-7491(01)00211-1.
Lal, R. 2004a. Soil carbon sequestration in India. *Climatic Change* 65(3): 277–296.
Lal, R. 2004b. Soil carbon sequestration impacts on global climate change and food security. *Science* 304(5677): 1623–1627.
Lal, R. 2010a. Soil quality and ethics: The human dimension. In: *Food Security and Soil Quality*. Advances in Soil Science, eds. R. Lal, and B. A. Stewart, 301–308. Boca Raton: Taylor & Francis (CRC Press).
Lal, R. 2010b. Managing soils and ecosystems for mitigating anthropogenic carbon emissions and advancing global food security. *BioScience* 60(9): 708–721.
Lal, R. 2011. Sequestering carbon in soils of agroecosystems. *Food Policy* 36: 533–539.
Lal, R., J. Kimble, R. F. Follet, and C. V. Cole. 1998. *The Potential for U.S. Cropland to Sequester Carbon and Mitigate Greenhouse Effect*. Chelsea, MI: Ann Arbor Press.
Lambers, H., C. Mougel, B. Jaillard, and P. Hinsinger. 2009. Plant–microbe–soil interactions in the rhizosphere: An evolutionary perspective. *Plant and Soil* 321(1–2): 83–115.
Loveland, P., and J. Webb. 2003. Is there a critical level of organic matter in the agricultural soils of temperate regions: A review. *Soil and Tillage Research* 70(1): 1–18.
Maedal, K., M. Owadai, N. K. Kimura, and I. Karubd. 1995. Pergamon CO2 fixation from the flue gas on coal-fired thermal power plant by microalgae to screen microalgae which are suitable for direct CO2 fixation, microalgae. *Energy Conversion and Management* 36: 717–720.
Maheswarappa, H. P., V. Srinivasan, and R. Lal. 2011. Carbon footprint and sustainability of agricultural production systems in India. *Journal of Crop Improvement* 25(4): 303–322.
Meena, R. S., S. Kumar, R. Datta, R. Lal, V. Vijaykumar, M. Brtnicky, M. P. Sharma, G. S. Yadav, M. K. Jhariya, C. K. Jangir, S. I. Pathan, T. Dokulilova, V. Pecina, and T. D. Marfo. 2020. Impact of agrochemicals on soil microbiota and management: A review. *Land MDPI* 9(2): 34. DOI: 10.3390/land9020034.
Meena, R. S., V. Kumar, G. S. Yadav, and T. Mitran. 2018. Response and interaction of *Bradyrhizobium japonicum* and Arbuscular mycorrhizal fungi in the soybean rhizosphere: A review. *Plant Growth Regulation* 84(2): 207–223.
Meena, R. S., and R. Lal. 2018. *Legumes for Soil Health and Sustainable Management*. Springer, Singapore. ISBN 978-981-13-0253-4 (eBook), ISBN: 978-981-13-0252-7(Hardcover), eds. R. S. Meena, A. Das, G .S. Yadav, R. Lal, Springer publication 541. DOI: 10.1007/978-981-13-0253-4_10.
Meena, R. S., R. Lal, and G. S. Yadav. 2020a. Long term impacts of topsoil depth and amendments on soil physical and hydrological properties of an Alfisol in Central Ohio, USA. *Geoderma* 363: 1141164.
Meena, R. S., R. Lal, and G. S. Yadav. 2020b. Long-term impact of topsoil depth and amendments on carbon and nitrogen budgets in the surface layer of an alfisol in Central Ohio. *CATENA* 2020194: 104752.
Meena, R. S., S. Kumar, and G. S. Yadav. 2019. *Soil Carbon Sequestration in Crop Production. In Book: Nutrient Dynamics for Sustainable Crop Production*, eds. R. S. Meena. Springer publication, Singapore, 1–39. DOI: 10.1007/978-981-13-8660-2_1.
Miglierina, A. M., J. O. Iglesias, M. R. Landriscini, J. A. Galantini, and R. A. Rosell. 2000. The effects of crop rotation and fertilization on wheat productivity in the Pampean semiarid region of Argentina. *Soil and Tillage Research* 53: 129–135.
Ministry of Power. 2020. *Bureau of Energy Efficiency*. Ministry of Power, Government of India. https://beeindia.gov.in/content/climate-change.
MoA&FW. 2018. *Pulses Revolution from Food to Nutritional Security. Crops Division*. New Delhi 110 001: Ministry of Agriculture & Farmers Welfare, Department of Agriculture, Cooperation & Farmers Welfare, and Krishi Bhavan.
MoA&FW. 2019. *Report of the Committee Report of the Committee Report of the Committee Review of the Scheme*. Promotion of agricultural mechanisation for in-situ management of crop residue in states of Punjab. Uttar Pradesh and NCT of Delhi, Haryana. Ministry of Agriculture and Farmers Welfare. Department of Agriculture, Cooperation & Farmers Welfare, Krishi Bhawan, New Delhi.
MoA&FW. 2019a. National Conference on Agriculture for Rabi Campaign, 2018-19. Ministry of Agriculture & Farmers Welfare, Department of Agriculture, Cooperation & Farmers Welfare, Krishi Bhavan, New Delhi—110 001. http://farmer.gov.in/imagedefault/conf/agenda_notes.pdf
MoA&FW. 2020. *Operational Guidelines for Capital Investment Subsidy Scheme for Vegetable and Fruit Market Waste Compost, and Biofertilizers—Biopesticides Production Units*. Under National Project on Organic Farming, New Delhi 110 001: Department of Agriculture and Cooperation, Ministry of Agriculture, Government of India.

MoEF. 2012. *India's Second National Communication to the United Nations Framework Convention on Climate Change*. Ministry of Environment and Forest, Government of India. http://unfccc.int/resource/docs/natc/indnc2.pdf

National Academy of Agricultural Sciences (NAAS). 2016. Policy paper 64 and Singh NP Et Al. (2016). *Indian Journal of Genetics and Plant Breeding* 76(4), 388–398.

Nair, P. K. R., V. D. Nair, K. B. Mohan, and J. M. Showalter. 2010. Carbon sequestration in agroforestry systems. *Advances in Agronomy* 108: 237–307.

Negoro, M., N. Shioji, K. Miyamoto, and Y. Miura. 1991. Growth of microalgae in high CO_2 gas and effects of SOx and NOx. *Applied Biochemistry and Biotechnology* 28/29: 877–886.

Newaj, R., and S. K. Dhyani. 2008. Agroforestry for carbon sequestration: Scope and present status. *Indian Journal of Agroforestry* 10: 1–9.

Nielsen, U. N., E. Ayres, D. H. Wall, and R. D. Bardgett. 2011. Soil biodiversity and carbon cycling: A review and synthesis of studies examining diversity-function relationships. *European Journal of Soil Science* 62(1): 105–116.

National. Policy for Management of Crop Residue (NPMCR). 2014. http://agricoop.nic.in/sites/default/files/NPMCR_1.pdf (accessed on 12 June 2020).

OECD/IEA/NEA/ITF. 2015. *Aligning Policies for a Low-Carbon Economy*. Paris: OECD Publishing. DOI: 10.1787/9789264233294-en.

Olivier, J. G. J., K. M. Schure, and J. A. H. W. Peters. 2017. *Trends in Global CO_2 and Total Greenhouse Gas Emissions: 2017 Report*. The Hague: PBL Netherlands Environmental Assessment Agency. https://www.pbl.nl/sites/default/files/downloads/pbl-2017-trends-in-global-co2-and-total-greenhouse-gas-emissons-2017- report_2674_0.pdf.

Pal, D. K., S. P. Wani, and K. L. Sahrawat. 2015. Carbon sequestration in Indian soils: Present status and the potential. *Proceedings of the National Academy of Sciences, India* 85: 337–358. DOI: 10.1007/s40011-014-0351-6

Pandher, L. K., R. K. Gupta, and S. S. Kukal. 2020. Soil organic carbon, its fractions and soil organic carbon stocks under different land use systems in Typic Ustrochrepts of northwest India. *Tropical Ecology* 61(2): 258–266.

PAU. 2020. *The Package of Practices for the Crops of Punjab Kharif 2020. Half Yearly.*

Paustian, K., S. Collier, J. Baldock, R. Burgess, J. Creque, M. DeLonge, J. Dungait, et al. 2019. Quantifying carbon for agricultural soil management: from the current status toward a global soil information system. *Carbon Management* 10(6): 567–587. DOI: 10.1080/17583004.2019.1633231.

Pinheiro, É. F. M., D. V. B. de Campos, B. F. de Carvalho, L. H. C. dos Anjos, and M. G. Pereira. 2015. Tillage systems effects on soil carbon stock and physical fractions of soil organic matter. *Agricultural Systems* 132: 35–39. DOI: 10.1016/j.agsy.2014.08.008.

Pratt, K., and D. Moran. 2010. Evaluating the cost-effectiveness of global biochar mitigation potential. *Biomass and Bioenergy* 34(8): 1149–1158.

Rahaman, M. S. A., L. H. Cheng, X. H. Xu, L. Zhang, and H. L. Chen. 2011. A review of carbon dioxide capture and utilization by membrane integrated microalgal cultivation processes. *Renewable and Sustainable Energy Reviews* 15(8): 4002–4012. DOI: 10.1016/j.rser.2011.07.031.

Ramachandra, T. V., and Shwetmala. 2012. Decentralised carbon footprint analysis for opting climate change mitigation strategies in India. *Renewable and Sustainable Energy Reviews* 16(8): 5820–5833.

Rasmussen, P. E., S. L. Albrecht, and R. W. Smiley. 1998. Soil C and N changes under tillage and cropping systems in semi-arid Pacific Northwest agriculture. *Soil and Tillage Research* 47(3–4): 197–205.

Razzak, S. A., M. M. Hossain, R. A. Lucky, A. S. Bassi, and H. De Lasa. 2013. Integrated CO_2 capture, wastewater treatment and biofuel production by microalgae culturing—A review. *Renewable and Sustainable Energy Reviews* 27: 622–653. DOI: 10.1016/j.rser.2013.05.063.

Reddy, S. N. 2002. *Evaluation of Different Land Use Systems in Carbon Sequestration*. MSc thesis submitted to ANGRAU.

Roller, B. R. K., S. F. Stoddard, and T. M. Schmidt. 2016. Exploiting rRNA operon copy number to investigate bacterial reproductive strategies. *Nature Microbiology* 1(11): 16160–16160.

Safriel, U., Z. Adeel, D. Niemeijer, J. Puigdefabregas, R. White, R. Lal, M. Winslow, J. Ziedler, S. Prince, E. Archner, and C. King. 2005. Dryland systems. In: *Ecosystems Human Well-Being. Findings of the Conditions Trends Working Group of the Millennium Ecosystem Assessment* 1, eds. R. Hassan, R. J. Scholes, N. Ash, 623–662. Washington, DC, USA: Island Press.

Sahoo, U. K., S. L. Singh, A. Gogoi, A. Kenye, and S. S. Sahoo. 2019. Active and passive soil organic carbon pools as affected by different land use types in Mizoram, Northeast India. *PLOS ONE* 14(7): e0219969. DOI: 10.1371/journal.pone.0219969.

Sainju, U. M., J. D. Jabro, and T. Caesar-TonThat. 2010. Tillage, cropping sequence, and nitrogen fertilization effects on dryland soil carbon dioxide emission and carbon content. *Journal of Environment Quality* 39(3): 935–945. DOI: 10.2134/jeq2009.0223.

Sakai, N., Y. Sakamoto, N. Kishimoto, M. Chihara, and I. Karube. 1995. Chlorella strains from hot springs tolerant to high temperature and high CO_2. *Energy Conversion and Management* 36(6–9): 693–696.

Sánchez Mirón, A., M. C. C. García, A. C. Gómez, F. G. Camacho, E. M. Grima, and Y. Chisti. 2003. Shear stress tolerance and biochemical characterization of *Phaeodactylum tricornutum* in quasi steady-state continuous culture in outdoor photobioreactors. *Biochemical Engineering Journal* 16: 287–297.

Saravanan, A. P., T. Mathimani, G. Deviram, K. Rajendran, and A. Ugazhendhi. 2018. Biofuel policy in India: A review of policy barriers in sustainable marketing of biofuel. *Journal of Cleaner Production* 193: 734–747.

Scialabba, N., and M. M. Lindenlauf. 2010. Organic agriculture and climate change. *Renewable Agriculture and Food Systems* 25(2): 158–169.

Singh, B. K., R. D. Bardgett, P. Smith, and D. S. Reay. 2010. Microorganisms and climate change: Terrestrial feedbacks and mitigation options. *Nature Reviews in Microbiology* 8(11): 779–790.

Singh, J., N. Singhal, S. Singhal, M. Sharma, S. Agarwal, and S. Arora. 2018. Environmental implications of rice and wheat stubble burning in north-western states of India. In: *Advances in Health and Environment Safety*, eds.N. A. Siddiqui, S. M. Tauseef, and K. Kamal Bansal. Singapore: Springer, 47–55. DOI: 10.1007/978-981-10-7122-5?nosfx=y.

Smith, W. N., R. L. Desjardins, and B. Grant. 2001. Estimated changes in soil carbon associated with agricultural practices in Canada. *Canadian Journal of Soil Science* 81(2): 221–227.

Sohi, S. P., E. Krull, E. Lopez-Capel, and R. Bol. 2010. A review of biochar and Its use and function in soil. *Advances in Agronomy*, 105: 47–82. DOI: 10.1016/s0065-2113(10)05002-9.

Srinivasa Rao, Ch., G. R. Chary, B. Venkateswarlu, et al. 2009. *Carbon Sequestration Strategies in Rainfed Production Systems in India*. CRIDA, 102.

Srinivasarao, Ch. 2011. Nutrient management strategies in rainfed agriculture: Constraints and opportunities. *Indian Journal of Fertilizers* 7(4): 12–25.

Srinivasarao, Ch., A. N. Deshpande, B. Venkateswarlu, R. Lal, A. K. Singh, S. Kundu, K. P. R. Vittal, P. K. Mishra, J. V. N. S. Prasad, U. K. Mandal, and K. L. Sharma. 2012. Grain yield and carbon sequestration potential of post monsoon sorghum cultivation in vertisols in the semi arid tropics of central India. *Geoderma* 175–176: 90–97.

Srinivasarao, Ch., B. Vankateswarlu, R. Lal, A. K. Singh, and K. Sumanta. 2013. Sustainable management of soils of dryland ecosystems for enhancing agronomic productivity and sequestering carbon. *Advances in Agronomy* 121: 253–325.

Sung, K. D., J. S. Lee, C. S. Shin, S. C. Park, and M. J. Choi. 1999. CO_2 fixation by *Chlorella* sp. KR-1 and its cultural characteristics. *Bioresource Technology* 68(3): 269–273. DOI: DOI: 10.1016/S0960-8524(98)00152-7.

Tang, D., W. Han, P. Li, X. Miao, and J. Zhong. 2011. CO_2 biofixation and fatty acid composition of *Scenedesmus obliquus* and *Chlorella pyrenoidosa* in response to different CO_2 levels. *Bioresource Technology* 102(3): 3071–3076.

The Economic Times. 2019. Niti Aayog working on policy to end crop residue burning. February 26, 2019. https://economictimes.indiatimes.com/news/economy/agriculture/niti-aayog-working-on-policy-to-end-crop-residue-burning/articleshow/68161554.cms?from=mdr.

Trivedi, P., I. C. Anderson, and B. K. Singh. 2013. Microbial modulators of soil carbon storage: Integrating genomic and metabolic knowledge for global prediction. *Trends in Microbiology* 21(12): 641–651.

Ugwu, C. U., H. Aoyagi, and H. Uchiyama. 2008. Photobioreactors for mass cultivation of algae. *Bioresource Technology* 99(10): 4021–4028.

UNEP. 2011. *Towards a Green Economy: Pathways to Sustainable Development and Poverty Eradication*. Nairobi: United Nations Environment Programme.

Uroz, S., P. Ioannidis, J. Lengelle, A. Cébron, E. Morin, M. Buée, and F. Martin. 2013. Functional assays and metagenomic analyses reveals differences between the microbial communities inhabiting the soil horizons of a Norway spruce plantation. *PLOS ONE* 8(2): e55929. DOI: 10.1371/journal.pone.0055929.

Van, D. H., S. H. Vervaeren, and N. Boon. 2012. Flue gas compounds and microalgae: (bio-) chemical interactions leading to biotechnological opportunities. *Biotechnology Advances* 30(6): 1405–1424.

Vermeulen, S., D. Bossio, J. Lehmann, P. Luu, K. Paustian, C. Webb, F. Augé, I. Bacudo, T. Baedeker, T. Havemann, C. Jones, K. King, M. Reddy, I. Sunga, M. Von Unger, and M. Warnken. 2019. A global agenda for collective action on soil carbon. *Nature Sustainability* 2(1): 2–4. DOI: 10.1038/s41893-018-0212-z.

WBCSD. 2018. *The Business Case for Investing in Soil Health*. World Business Council for Sustainable Development. https://docs.wbcsd.org/2018/12/The_Business_Case_for_Investing_in_Soil_Health.pdf.

Wilson, G. W. T., C. W. Rice, M. C. Rillig, A. Springer, and D. C. Hartnett. 2009. Soil aggregation and carbon sequestration are tightly correlated with the abundance of arbuscular mycorrhizal fungi: Results from long-term field experiments. *Ecology Letters* 12(5): 452–461. DOI: 10.1111/j.1461-0248.2009.01303.x.

Yadav, G., A. Karemore, S. K. Dash, and R. Sen. 2015. Performance evaluation of a green process for microalgal CO_2 sequestration in closed photobioreactor using flue gas generated in-situ. *Bioresource Technology* 191: 399–406.

Yadav, G., and R. Sen. 2017. Microalgal green refinery concept for biosequestration of carbon dioxide vis-à-vis wastewater remediation and bioenergy production. *Recent Technological Advances in Climate Research* 17: 188–206.

Yadav, G., K. D. Sukanta, and R. Sen. 2019. A biorefinery for valorization of industrial waste-water and flue gas by microalgae for waste mitigation, carbon-dioxide sequestration and algal biomass production. *Science of the Total Environment* 688: 129–135.

Yadav, G., R. Kumar, and R. Sen. 2007. Fermentation techniques in bioenergy production. In: *Marine Bioenergy: Marine Bioenergy: Trends and Developments*, eds. S. K. Kim, and C. G. Lee, 111–134, eBook ISBN 9780429076282, DOI: 10.1201/b18494-10.

Yadav, M. 2017. Towards a healthier nation: Organic farming and government policies in India. *International Journal of Advanced Research and Development* 2(5): 153–159.

14 No-Till Farming in the Maghreb Region
Enhancing Agricultural Productivity and Sequestrating Carbon in Soils

Rachid Mrabet, Rachid Moussadek, Mina Devkota, and Rattan Lal

CONTENTS

14.1 INTRODUCTION

In the Maghreb region, productivity of rainfed and irrigated crops has been threatened due to social and technological constraints, water scarcity, low and variable rainfall, reduced soil fertility and quality, and improper agricultural management. Consequently, complex links between food insecurity and environmental, climatic, social, and economic factors are increasingly being manifested. Hence, it is important for sustainable intensification of agriculture production systems to feed the growing population in the region, and it is now at the top of the agenda of policy makers. Sustainable intensification involves increasing the productivity of agriculture while minimizing any negative economic, social, or environmental consequences (The Montpellier Panel, 2013). The critical component of this pathway is the utilization of the existing land to produce greater yields, better nutrition, and higher net incomes for farmers while reducing over reliance on pesticides and fertilizers and lowering emissions of harmful greenhouse gases or GHGs (Godfray et al. 2010). In this regard, like in other regions of the world, the Maghreb is implementing obligations related to most conventions, protocols, and principles to which it is committed including the 2015 Paris Agreement and the Sustainable Development Goals (SDGs) of the United Nations. Hence, countries in the Maghreb

DOI: 10.1201/9781003102762-14

region have adopted the concept of sustainable development, which has also created opportunities for improvement and protection of soil and other natural resources.

Countries of the dryland regions (e.g. Australia, Argentina, USA, and South Africa) are increasingly adopting no-till (NT) systems on diverse soils and cropping systems. They have also developed advanced technologies to support farmers' use of NT systems (Kassam et al. 2020). However, countries in the Maghreb region are still lagging in adoption of these systems on a large scale and may benefit by learning from the experiences of countries in the dryland regions regarding systems to reverse the downward spiral of soil and environmental degradation.

Therefore, regional and technical cooperative projects based on NT systems have been widely initiated under the auspices of some international organizations (FERT, 2018). In these projects, NT farming is based on three interlinked principles: (1) continuous no or minimum mechanical soil disturbance, (2) crop species diversification through crop sequences, associations, and rotations, and (3) year-round ground cover and crop residue management. The overall strategy is to regenerate soil functions, increase productivity, and enhance the farmer's resilience and economy. These cooperative projects have been based upon research and studies conducted since the 1980s by national research centers (Mrabet et al. 2012). However, the major challenge that still remains to be addressed is mainstreaming the benefits of NT into development planning and extension activities. The objective of this chapter is to collate and synthetize the available information on research in relation to improvements in achieving sustainability and mitigating anthropogenic climate change—through adoption of NT systems over existing conventional agriculture systems (e.g. crop productivity, soil quality, erosion control, and carbon emissions and sequestration). This chapter builds upon some previous review articles which have already deliberated the differential effects of NT versus conventional tillage (CT) on soils, crops, environment and the farmer's economy (Mrabet, 2011; Boulal et al. 2014; El Gharras et al. 2017); it also includes the expertise of different researchers through their publications and projects. The principal finding of this synthesis is that the buildup of soil organic matter (SOM) content by using NT systems offers significant potential to contribute to the reversal of soil degradation and the attendant increased crop and soil productivity and mitigation of climate change by sequestering atmospheric carbon dioxide (CO_2) in the soil. The terms NT and conservation agriculture (CA) are used interchangeably in this chapter.

14.2 AGRICULTURE, SOIL, AND CLIMATE

The Maghreb region is in the northern part of Africa and comprises Algeria, Libya, Mauritania, Morocco, and Tunisia. The region covers a total land area of approximately 6 million km^2 with a total harvested area of ~17.7 million hectare (M ha) which represents less than 3% of the total land area of the region (Table 14.1). The data in Table 14.1 show that availability of land is limited and more than 90% of production area is under rainfed systems. The agriculture sector is critical to the Maghreb economy and represents 12.4% of its gross domestic product (GDP) (Table 14.1). The population of these five countries is 100 million.

Most of the region is located within arid and semi-arid climates characterized by high temperature and low precipitation with pronounced rainfall variability and extreme events (drought, heat). The mean average rainfall does not exceed 158 mm with Morocco receiving up to 346 mm per year and Libya, the driest, getting 56 mm/year. Climate change has and will impact agricultural productivity through increases in mean temperatures, alterations in rainfall patterns, expansion in frequent occurrences of weather extremes (including temperature, precipitation, drought, heat), and altered patterns of pest pressure. In addition, drought and rainfall variability are expected to be more spatially heterogeneous. Such alterations may exacerbate the already existing water scarcity (Bazza et al. 2018). The region is recognized as a "hot spot" for climate change and is predicted to reduce rainfall by 10–30% and increase temperatures by 1.2–1.8°C by 2025 (Driouch et al. 2013).

The Maghreb region is characterized by diversity of agroecological systems and a complex mixture of mountainous, desert, Saharan, pastoral, agro-pastoral, rainfed, and irrigated farming

TABLE 14.1
Population, Land and Agricultural Area, Agricultural Share in GDP, Mean Annual Rainfall, Rainfed Area (% of Total Agricultural Area), and Fertilizer Use by Maghreb Country (Compiled from FAO, 2020)

Country	Population (million)	Land Area (M km^2)	Rainfall (mm)	Agricultural Share in GDP (%)	Area Harvested Crops (M ha)	Agricultural Area Per Capita (ha)	Land Area Equipped for Irrigation (,000 ha)	Rainfed Land as Percentage of Total Agricultural Area	Fertilizer Use (kg/ha)
Algeria	42.2	2.38	89	12	5.1	0.12	1,360	98.6	22.3
Libya	6.7	1,76	56	3	0.8	0.12	400	97.0	11.7
Mauritania	4.4	1.03	92	20	0.4	0.29	45	90.0	-
Morocco	36.03	0.71	346	16	7.9	0.22	1,530	95.2	71.1
Tunisia	11.6	0.16	207	11	3.5	0.30	476	96.0	59.3
Total or average	100.93	6.05	158	12.4	17.7	0.21	3,811	95.36	41.1

systems. Hence, the type and degree of degradation and desertification vary from one country to another within the region. However, in all landscapes, soil erosion (by water and wind) and land degradation processes are being exacerbated because of unsustainable management by farmers and pastoralists (Ryan et al. 2006; Ryan, 2011). Yet, the latter are the stewards of the land resources (Ryan et al. 2006; Lagacherie et al. 2017). Thus, their actions are adversely affecting food availability, distribution, and overall food insecurity. The annual costs of soil degradation were estimated at about 1.2% of GDP in Algeria, 0.4% in Morocco, and 0.5% in Tunisia (World Bank, 2019).

The Maghreb region is a hot spot in terms of import, which corresponded to 54% of food and feed needs in 2015 (Le Mouël et al. 2015). Yet, the magnitude of food import will be aggravated by climate change, limited resources, population growth, and change in dietary preferences. At the same time, farmers, both of small or large land holdings across the region, have been and are in the front line of all environmental and social pressures and are disrupting impacts of extreme events driven by climate change.

Agricultural production in this region is highly variable due to unpredictable rainfall and periodic droughts. Cereals (wheat and barley) are dominant crops and yields are in the range of 1–2 Mg/ha and are characterized by large yield gaps mainly driven by intensive soil tillage, low rates of fertilizer application, and cereal monocropping (Devkota and Yigezu, 2020). These extractive practices degrade soil quality and create negative water, nutrient, and carbon budgets in the soil and the agricultural ecosystems. Yet, the low rate of fertilizer input (Table 14.1) is aggravating the nutrient deficits.

The Maghreb region has highly diverse soils, and predominant soils of the region include Entisols, Aridisols, and Inceptisols. These soils are relatively coarse textured, low in soil organic matter (SOM) content, inherent soil fertility, structural stability, and plant available water-holding capacity (Lal, 2013). Consequently, these soils have low resilience, are easily degraded and are prone to drought. Water poverty is widespread in this region, and most countries are already below the water "poverty line" and suffer from associated environmental degradation, pollution, and social problems (Bazza et al. 2018; World Bank, 2019).

14.3 NO-TILL FARMING: ADOPTION AND SPREAD

In the Maghreb region, NT systems are practiced in major crops (i.e. cereals, food legumes, forages, oil seed crops). National NT research-development initiatives were implemented in the region at different times and with diverse visions. Thus, available literature and studies on the adoption and adaptation of CA are limited in North Africa. Despite large development efforts across the Maghreb region, there is no comprehensive assessment of the adoption of the NT system, however, Kassam et al. (2020) reported that the rhythm of adoption of CA is still low and not increasing significantly with time. In 2008/09, arable land area under CA was estimated at 4000, 6000, and 100 ha in Morocco, Tunisia, and Algeria, respectively. In 2015/16, the land area under CA area was 10,500, 12,000, and 5,600 for the same countries, respectively (Kassam et al. 2020) (Figure 14.1). Furthermore, the adoption of NT in those three countries is also fragmented throughout the country. NT systems are yet neither experimented nor adopted in Libya and Mauritania.

14.3.1 Algeria

NT systems were first experimented in 2002 on the high plains of Setif province with cereals and legumes crops (Bouzerzour et al. 2006; Rouabhi et al. 2016) and later transferred to the Constantine, Oum El Bouaghi, Guelma, and Annaba regions. Belaid (2015) reported that in 5 years, area under NT systems reached 7,000 ha, and it was mostly adopted by large farmers (>100 ha) while mobilizing 20 direct seeders. Much of this interest was a result of the availability of NT seeders, and the technical and financial support provided by extension services of agriculture. Since then, government subsidy programs have been implemented at different levels to accelerate the adoption process.

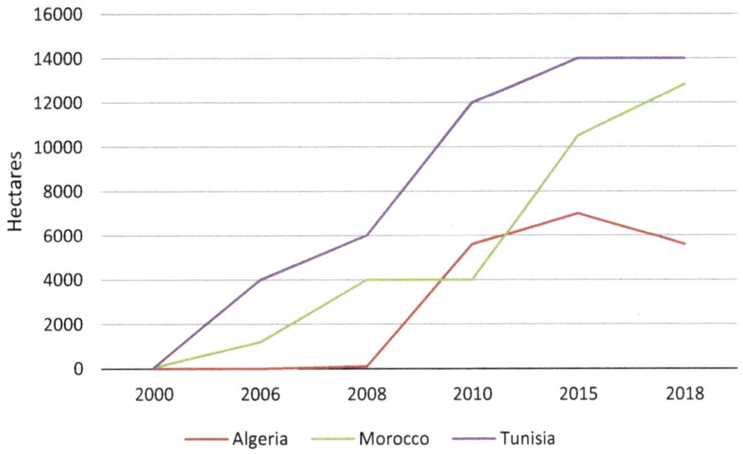

FIGURE 14.1 Temporal trends in the adoption of no-till farming in the Maghreb region (hectare).

However, adoption of NT has faced socioeconomic and technical constraints (Rouabhi et al. 2016, 2019) because of the mindset of farmers and policy makers, competition of crop residues for other uses, weed infestation, lack of availability of suitable and versatile direct seeders, and difficulty in changing paradigms and behaviors of stakeholders. The development of a local low-cost no-till seeder called "Boudour" with a collaborative effort from ICARDA and government support, adoption of the NT system is likely to increase in the country in near future (Idoudi et al. 2020).

14.3.2 Morocco

Research on water conservation started during the early 1980s in semi-arid Morocco. The goal was to encourage farmers to adopt best options for managing soils and crops under erratic rainfall. The first adoption studies were initiated in 1998 with support from FAO, CIRAD, and government's extension services and development agencies. In 2011, the area under NT farming was <5,000 ha and adopted primarily by large farms that had access to the imported NT seed drills.

El Gharras et al. (2017) reported that NT use by farmers spread over an area of 9,000 ha. NT adoption increased to 12,826 ha by 2018, of which 8,033 ha were managed by 800 smallholder farmers (<10 ha) and 4,793 ha by 120 large farmers (>10 ha) (Mrabet, 2018). In 2019, the OCP group and Mohammed VI Polytechnic University (Benguerir) launched the program of technical support from INRA, the El Moutmir initiative, for a large-scale adoption of NT systems in Morocco to reach 2,000 farmers and a land area of 10,000 ha. This program has mobilized 35 NT seeders and 600 innovation platforms in 18 provinces. Thus, land area under NT in Morocco increased to around 25,000 ha and used more than 105 NT seeders (Table 14.2).

A study by World Bank (2014), in the Central Morocco region, indicated that the area under the NT system could be increased to 60,000 ha by 2025. Bonzanigo et al. (2016) estimated that 40% of farmers in the Central Morocco region could adopt NT systems. Based on research findings and accumulated experience, the department of agriculture proposed a road map to promote and implement NT systems in 1.5 M ha during the next decade in cereal-based systems under the new agricultural strategy called Green Generation (2020–2030). This road map proposes NT development to ensure sustainability through an array of parallel and supporting interventions (i.e. investment in education and capacity development, rural community and farmer organizations empowerment and mobilization, awareness creation among stakeholders, legal and institutional piloting frameworks, research and extension, technology, machinery and seeds, marketing, and enterprise diversification and public–private partnership strengthening).

TABLE 14.2
Area Under No-Tillage (NT) and Number of NT Seeders Used in Different Regions of Morocco (AMAC, 2021)

	Area		Seeders	
Region	**Hectares**	**Percent**	**Number**	**Percent**
Fes-Meknes	9000	36.6	36	34.3
Casablanca-Settat	7300	29.7	24	22.9
Rabat-Sale-Kenitra	4670	19	28	26.7
Beni Mellal-Khenifra	2100	8.5	6	5.7
Marakech-Safi	1144	4.7	7	6.7
Tanger-Tetouan-El Hoceima	315	1.3	3	2.9
Draa-Tafilelt	48	0.2	1	1
Total	24577	100	105	100

14.3.3 Tunisia

In Tunisia, NT systems were first introduced during the period from 1970–80 through import of seed drills from the USA (Cheikh M'hamed et al. 2018). Beginning with 1999, land area under NT increased rapidly during the first years of adoption using seed drills from Brazil with support from Grain Tunisian Company (Cotugrain) (27 ha in 1999 and 12,000 ha in 2010) (Richard, 2005; FERT, 2016). Cheikh M'Hamed et al. (2018) estimated the land area under NT at ~14,000 ha and adopted by more than 200 farmers and using around 107 NT seeders (which represents 0.93% of total cereal production area). The study by Dhehibi et al. (2018) reported that most farmers were aware of the benefits of NT, yet only a minority had adopted CA practices, likely due to the countervailing strength of perceived risks. When faced with limited resources due to project termination, a proportion of CA adopters shift to reduced tillage or move completely back to traditional tillage systems. Bahri et al. (2019) identified 260,000 ha as priority areas for NT adoption, which represents one-third of the total area under cereal production in Tunisia.

14.4 DIVERSIFYING CROP ROTATION AND NARROWING THE PRODUCTIVITY GAPS

Most of the agricultural land in the Maghreb region is currently producing below its capacity. Crop yields are either stagnating, or the yield growth rates are lower than required to meet the growing food demand. In addition, perception by farmers, with fewer resource endowments and prone to high risks in adopting NT/CA systems and input use, are among the major drivers of large yield gaps.

For centuries, farmers in the region were obliged to simultaneously manage soils of low fertility, water scarcity, and crop failures by use of traditional soil, water, and crop management practices (Debaeke and Aboudrare, 2004). Presently, farmers are experiencing a decline in production while also losing harmony with natural resources (i.e. soils). In other words, the benefit–cost ratio under traditional farming systems is too low to encourage subsistence and smallholder farmers (<5 ha) to use improved seeds, fertilizers, and pesticides. Lack of adoption of modern innovation is also attributed to relatively high farm gate price, the low market price of food crops and risk of low agronomic efficiency of applied input due to variable rainfall amounts and distribution. Thus, there is a strong need to redesign such unsustainable farming systems to achieve high profitability and efficiency by implementing innovative incentives and policy measures. Accordingly, NT systems have been proposed and experimented for over 4 decades to meet this challenge and reverse the declining trend in the Maghreb region.

During the last four decades, several NT research and on-farm trials have been implemented in three countries (Algeria, Morocco, Tunisia) under contrasting climates, soils, and farming systems (Mrabet, 2011; Boulal et al. 2014: El Gharras et al. 2017; FERT, 2018). The available literature on the impact of NT systems on crop productivity shows that higher and stable yields are among the principal benefits to farmers in the Maghreb region (Mrabet et al. 2012; Kassam et al. 2012; Boulal et al. 2014). High and stable yields under NT systems have been more pronounced during years in which the accumulated rainfall is below normal. Overall, crop productivity under NT systems increases over time with a range of 0.2–146% compared to that under the CT systems (Tables 14.3–14.5). A report by FERT (2018) showed an average crop yield increase under NT ranging from 8% to 24% compared to CT in farmers' fields.

Such improvements in productive potential of soils under NT as compared to CT are due to amelioration in soil resilience, functions, and productivity, as is discussed in the sections below. In fact, NT systems increase water conservation and crop water use efficiency (Jemai et al. 2013; Mrabet, 2002a). Thus, large-scale adoption of these systems is needed in the Maghreb region to reduce adverse impacts of drought and enhance crop production (Mrabet, 2011a; Boulal et al. 2014). Such results of improvements have been confirmed by Sun et al. (2020) through use of a meta-analysis based on 1,970 paired studies (NT vs CT). The data of this meta-analysis showed that increases in crop productivity by adoption of NT are greater in dry regions than in humid regions.

In Morocco, earlier research on soil water conservation that started in 1983 indicated superiority of NT grain production compared to that of CT systems (Bouzza, 1990; Kacemi, 1992; Mrabet, 1997). It was also observed that soil conditions under NT systems favor more vigorous and healthier plants that are resilient to a range of biotic or edaphic stresses (Mrabet, 1997). Under continuous wheat or wheat–fallow, mean yields with NT, over a 10-year period at Sidi El-Aydi station (clay soil, 358 mm) and a 19-year period at Jemma Shaim station (clay soil, 270mm), were higher or equal to those under reduced tillage using sweep, with both being significantly better than that under CT (Mrabet, 2011a).

Using a line-source irrigation system to create varying soil moisture regimes, Mrabet (1997) observed that the NT system improved seedling establishment and growth of corn and wheat. Corn

TABLE 14.3
Grain Yield of Annual Crops (Mg/ha) Under No-Tillage (NT) and Conventional Tillage (CT) Systems in Different Regions of Morocco (Mrabet et al. 2012)

Region and Average Rainfall (mm)	Soil Type	Rotation	NT	CT	Average Change (%)	Duration (Years)
Abda (270)	Vertisol	Wheat–fallow	3.10	2.40	29	19
	Vertisol	Continuous Wheat (CW)	1.60	1.60	0	19
Chaouia (358)	Mollisol	CW	2.47	2.36	5	4
	Vertisol	Wheat–fallow	3.70	2.60	42	10
	Vertisol	CW	1.90	1.40	35	10
	Mollisol	Different rotations	2.21	1.90	16	9
	Vertisol	Wheat-chickpea	1.87	0.76	146	3
	Rendzina	Wheat-chickpea	2.53	1.47	72	9
Zaers (410)	Vertisol	Wheat-lentils	1.99	1.41	41	4
	Entisol		2.99	2.72	10	4
	Alfisol		2.71	2.49	9	4
Saiss (438)	Vertisol	Different rotations	2.55	2.49	2	4
	Alfisol		2.72	2.74	0	4
Gharb (570)	Vertisol	CW	2.80	2.26	24	3

TABLE 14.4
Wheat Grain Yield (Mg/ha) Under No-Tillage (NT) and Conventional Tillage (CT) Systems in Different Regions of Algeria

Site/Climate Regime	Soil Type	Rotation	NT	CT	Average Change (%)	Duration Years	References
Oued Smar /Mitidja (580 mm)	Clay loam	Wheat-berseem clover	2.78	2.78	0	5	Abdalaoui et al. 2010
Setif (458 mm)	steppic brown soil	Continuous wheat	3.02	2.54	19	2	Chennafi et al. 2011
Setif/ (458 mm)		28 farms with varying rotations	2.64	2.15	23	3	Rouabhi et al. 2019

TABLE 14.5
Wheat Grain Yield (Mg/ha) Under No-Tillage (NT) and Conventional Tillage (CT) Systems in Tunisia

Site	Rotation	NT	CT	Average Change (%)	Duration Years	References
Kodiat/ Jendouba (535 mm)	Wheat-barley	2.80	2.71	3	5	Mouelhi et al. (2016)
	Wheat-oat	2.73	2.61	5	5	
	Wheat-fababean	3.76	3.43	10	5	
Bourabia/(470 mm)	Wheat-Vetch	2.54	2.49	0.2	2	Souissi et al. (2020)
El Kef (450 mm)	Wheat-fababean Wheat-barley-fababean	2.69	2.57	5	3	Chaieb et al. (2020)

Note: Values in brackets indicate average annual rainfall (mm).

yields under NT increased by 8–22% compared to that under CT systems depending on the soil moisture regimes.

Based on a short-term experiment In the Setif region, Chennafi et al. (2011) observed that NT systems with appropriate level (60%) of crop residues maintained at soil surface improved yield components and obtained a higher grain yield of durum wheat compared to that under CT. Chennafi and colleagues also reported that the yield components improved with increased levels of crop residues. These trends in yield improvements were also confirmed by Jemai et al. (2013), Fellahi et al. (2013) and Mrabet (2002a). Chennafi et al. (2011) also concluded that concerns by farmers of decreased crop yield during the early years of conversion to NT practices were not supported with an adequate amount of crop residue cover. In the latter case, crop yields under NT systems did not decrease even during the early periods. Similarly, Benkherbache et al. (2012) reported improved physiological parameters (low stomata resistance and higher relative water content) of wheat, higher grain yields and better morphological features under NT than under CT even during the first year of conversion. These results of positive crop growth and yield are critical to addressing concerns of farmers for adoption of NT systems irrespective of climate conditions.

A meta-analysis conducted by Souissi et al. (2018) in semi-arid Tunisia revealed that crop yields were reduced by NT systems which were based primarily on elimination of mechanical tillage and seedbed preparation but without retention of crop residues on the surface as mulch. Souissi et al. (2018) found that crop yields under NT were similar to those under CT systems for most climatic regimes of Tunisia and concluded that using all three NT principles (direct seeding, crop residue retention, and crop rotation) is essential to increasing agronomic production. A system-based approach is essential to a successful adoption of NT (Lal, 2015), especially by small land holders and resource-poor farmers of Africa (Lal, 2016). A minimum of 60% of residue cover under NT is needed to improve yield of durum wheat while conserving water in soil profile compared to that under CT systems and that NT systems with bare soil surface are detrimental to wheat growth and development (Fellahi et al. 2013).

Based on a 10-year study comparing NT and CT, Angar et al. (2011) observed that wheat grain yield gradually increased under NT in semi-arid and sub-humid areas of Tunisia. In the sub-humid regions, yield of durum wheat under NT was on average higher by 800 kg/ha compared to that under CT. In the lower semi-arid zone, the difference was 350 kg/ha, and reaching up to 700 kg/ha in the higher semi-arid zone.

It has also been widely reported that several crops can be successfully grown in sequence under NT systems in the contexts of diverse climate of the Maghreb region. Among such crops are oil seed (sunflower and sesame), food legumes (chickpea, bean, lentil, fababean and soybean), forages (vetch), and most cereals (corn, bread and durum wheat, barley, sorghum, and oat) (Aboudrare et al. 2006; Ben-Hammouda et al. 2007; Mrabet, 2011a; M'Hedbi et al. 2004; Mekhlouf et al. 2011; Bouguendouz, 2010). In the Setif region (Algeria), the results by Bouguendouz (2010) indicated that NT improved barley grain yield by 31% and reduced the cost by 28.3%, compared to the CT technique. For improved integration of livestock in cereal systems, new species of forage crops (vetch, triticale, etc.), and crop mixture (triticale-vetch, oats-vetch, pea/triticale) were included in crop rotations and were identified as a highly suitable options for the marginal wheat systems (Chiekh M'Hamed et al. 2016; Abidi et al. 2021). Abidi et al. (2021) observed that grazing of sheep on vetch or on vetch and wheat stubbles produced greater carcass yield than those grazing on wheat stubbles alone. These data show that farmers in North Africa are relying mostly on wheat straw, and stubbles are neither adequately feeding their sheep nor improving the quality of their soils (Mrabet et al. 2012; Moujahed et al. 2015).

Mrabet (2011b) concluded that wheat–fallow rotation is an important option in dry years in some areas of Morocco. However, food legumes (lentils) or barley-based rotations are recommended in better environments for using NT systems. In addition, the author recommended that NT with partial residue cover could be adopted for the mixed crop-livestock systems in semi-arid areas for the purpose of guaranteeing the yield of both grains and feed. In the Setif research station (Algeria), Mekhlouf et al. (2011) reported that lentils in rotation with wheat improved both grain yield of wheat and yield components under NT and CT systems compared to cereal monoculture systems (wheat-oat).

In a study carried in the semi-arid conditions of the El Kef (Tunisia), Rezgui et al. (2014) found that the yield of grains under NT conditions increased by 20% in durum wheat and nearly tripled in fababeans in years of rainfall deficit. Mouelhi et al. (2016) reported higher durum wheat yields over 5 years under NT compared to that under CT and observed the highest yields under wheat-fababean compared to wheat-barley and wheat-oat rotations for both NT and CT systems.

It is widely known that growing disease-resistant and crop varieties adapted to drought and heat stress is the basic requirement for higher growth, quality, and yields of crops. Yet, farmers in the Maghreb region are not using certified seeds widely (Devkota and Yigezu 2020). A low adoption rate of improved varieties may reduce benefits from NT systems and keep the level of production insufficient to satisfy food requirements of the growing population.

A simulation study by Bahri et al. (2019) using the APSIM model showed that the sustainable production of durum wheat under conditions of climate change in Tunisia is possible through the adoption of NT practices mainly with maintenance of crop residues in both sub-humid and semi-arid areas. Bahri and colleagues found that NT is more effective than CT (under semi-arid and sub-humid conditions) in enhancing wheat yield by 15% and water use efficiency by 18% and 13% for both regions, respectively.

One of the main conclusions from the study of tillage systems and rotation by Mrabet (2011b) on stability analysis is that crop yields under the NT system were less influenced by adverse growing conditions than that under the CT system, particularly under low rainfall. These results indicate that improved soil quality under NT systems enhanced stability of wheat yield by reducing the impact of adverse growing conditions.

Reduced yields under NT compared to that under CT systems were mostly explained by partial adoption of NT principles and the prevalence of bare soil surface (Bahri et al. 2018; Souissi et al. 2018) or by the necessity of transition phase for soil recovery and yield improvement vis-à-vis rainfall variability (Pittelkow et al. 2015).

14.5 CARBON MANAGEMENT AND SEQUESTRATION

Soil organic carbon (SOC) is the most reported soil attribute from tillage experiments since it is the keystone soil quality indicator, being inextricably linked to other soil properties. SOC is also a crucial contributor to food production, mitigation, and adaptation to climate change and the achievement of SDGs (Lal et al. 2018; FAO and ITPS, 2020). It is widely used to assess the soil recovery and rehabilitation when shifting to a new form of agriculture or production system (i.e. NT systems) (Mrabet et al. 2003). Thus, it is important to realize that enhancing the SOC pool and maintaining it above the threshold of 1.1% in the surface layer is essential to advancing food security in soils of the tropics and sub-tropics (Aune and Lal, 1998).

The data in Figure 14.2 show that soils of the Maghreb region are low in SOC stock with an average of 14.6 Mg/ha and minimum, maximum, and standard deviation of 1 Mg/ha, 251Mg/ha, and 7.3 Mg/ha, respectively. In general, soils of Morocco have relatively higher SOC stock in the Maghreb region (Figure 14.2). It has been estimated that soils of West Asia and North Africa (WANA) may have lost 6–12 Pg C because of historic land use and management (Lal, 2002). However, soils of WANA can sequester 200–400 Tg C/ yr or 20% of the technical potential of global dry land ecosystems. The cumulative SOC sink capacity may be 3–7 Pg C or 0.2–0.4 Pg C/ yr through remedial interventions to control or reverse desertification (Lal, 2002).

Sequestration of SOC is an important climate mitigation option with numerous co-benefits of enhancing agricultural production, improving water resources, and strengthening biodiversity. It is also one of a few strategies that could be applied on a large scale and at low cost (Paustian et al. 2016). Thus, increasing net soil carbon storage by even a few percent represents a significant carbon

FIGURE 14.2 Soil organic carbon map of the Maghreb region (Mg/ha) in 0–30 cm depth (Adapted from FAO and ITPS, 2020).

sink potential. The global mitigation potential for increasing SOC stocks in mineral soils is estimated to be in the range of 1.3–5.1 Gt CO_2 eq/year (Smith et al. 2020).

International organizations and research institutions are promoting NT management to mitigate greenhouse gas (GHG) emissions through SOC storage. Worldwide, several researchers have reported positive effects of NT systems on SOC content, with sustained sequestration depending on climate, crop productivity, and amount of residues returned to soils (Jai et al. 2019; Mirzabaev et al. 2019; Mbow et al. 2019). Sun et al. (2020) reported that in dry lands, NT systems have a large potential for SOC sequestration, and it can also enhance system resilience to climate-related stress.

Based on a meta-analysis, González-Sánchez et al. (2019) reported that the potential of SOC sequestration by NT systems in diverse regions in Africa can be 143 Tg of C/ yr, or 524 Tg of CO_2 / yr. Furthermore, González-Sánchez and colleagues estimated that the rate of SOC sequestration in the Mediterranean region of Africa with NT systems in annual and woody crops can be 0.44 and 1.29 Mg/ha/yr, respectively.

Based on an earlier study in semi-arid Morocco, Mrabet et al. (2001a) observed that SOC content increased by 13.6% under NT systems compared to 3.3% under the CT system over an 11-year period, which corresponded to a SOC sequestration rate of 0.5 Mg/ha/yr. Increases in SOC content were also related to higher aggregation, more plant nutrient pools, reduced pH and enriched quality of SOM. After 4 and 11 years of experimentation, the experimental data showed that the soil surface under NT sequestered more particulate C than that under CT systems (Bessam and Mrabet, 2003; Saber and Mrabet, 2002). This SOM fraction also contributes to nutrient cycling and availability. Above all, humic acids and humin contents were significantly higher under NT compared to those under CT, but fulvic acid concentrations were significantly lower under NT than CT in a Vertisol (Moussadek et al. 2014). Increases of the humic acids than those of fulvic acid are particularly important in the stabilization of structural aggregates (Moussadek et al. 2011a). The relative decrease in fulvic acid under NT compared to CT may be attributed to the humification process which is favored by residue retention under NT that led to an increase of the most stable fractions (humic acids and humin).

The data from a study of comparing three soils in Central Morocco by Moussadek et al. (2014) showed that SOC stock in surface horizon (0–30 cm) after 5 years differed significantly among tillage systems and soil types. The SOC content was significantly higher under NT compared to CT (10% more in Vertisol and 8% more in Cambisol), but no significant difference was observed in the Luvisol. Average SOC stock within the 0–30 cm depth was 29.4 and 27.4 Mg /ha under NT and CT, respectively. The highest SOC stock (31.9 Mg/ ha) was observed in Vertisols under NT management.

Mrabet et al. (2004) reported that SOC content under NT increased by 0.036% and 0.015% of every Mg/ ha of residue added (0–6 Mg / ha) at 0–5 cm and 5–10 cm depths, respectively. Mrabet and colleagues also reported that SOC stock at 0–2.5 cm depth increased by 5.62 and 7.21 Mg/ ha under NT after 4 and 11 years of implementing NT, respectively.

In northern Tunisia, Brahim et al. (2009) observed that conversion to an NT system can sequester SOC at the rate of up to 0.9 Mg C/ha/yr. Brahim and colleagues found that SOC stock after 7 years of NT systems with wheat-corn rotation was 18.1 Mg /ha in the top 10 cm layer compared to 11.7 Mg/ha under CT, an increase of 35% upon conversion from CT to NT. Brahim et al. (2009) noted that crop residues were modest at the soil surface. However, in the same region, Ben Moussa-Mechraoui et al. (2010) observed a slight but not significant increase in SOC content after 4 years of NT as compared to CT systems.

In Algeria, and especially south of Setif, Bouzrara et al. (2011) reported a significant difference in SOC content between NT (2.97%), reduced tillage (1.98%) and CT (2.0%) at the surface 0–10 cm layer. Based on a study on a sub-humid site in central Algeria, Abdellaoui et al. (2011) reported an improvement in soil properties in the 0–30 cm layer under NT compared to that under CT with regard to an increase in water-holding capacity and soil porosity and decrease in bulk density. Increase in SOM content was reflected in a higher aggregation and better resistance of NT soil to erosion and rapid wetting compared to that under CT and reduced tillage practices.

Based on data from a study in a semi-arid region of north-western Tunisia, Jemai et al. (2012) reported that SOC content was significantly increased by NT after 3 and 7 years compared to that under a CT system involving moldboard plowing at 0–10 and 0–20 cm depth, respectively. However, a greater increase in SOC content was observed in the uppermost 10 cm soil layer in the field under NT for 7 years.

In general, there is a trend toward a stratification of SOC in the surface soil under NT, without any change in the deeper layers (Mrabet (2002b) in Morocco; Brahim et al. 2009 and Jamai et al. (2012) in Tunisia; Bouzrara et al. (2011) in Algeria). Besam and Mrabet (2003) reported that SOC stock in (0–25 mm) layer increased from 5.6 Mg/ha to 7.2 Mg/ ha under NT, after 4 and 11 years, respectively. For the same layer, however, SOC stock did not change under CT over the same periods. Bessam and Mrabet (2003) reported that the increase in SOC stock under NT was 3.5 and 3.4 Mg/ha more than that under CT in the 0–200 mm layer, after 4 and 11 years, respectively. Use of NT systems also affected SOM content of aggregates in all classes of a Calcixeroll soil (Mrabet et al. 2003).

Angar et al. (2011) reported an increase in SOC content of 0–40 cm depth after 10 years of NT and CT in three contrasting bioclimatic regions of Tunisia. Angar and colleagues did not observe any decline in SOC in sub-soil layers under NT in comparison to that under CT over the entire duration of the experiment.

The efficiency of an NT system for SOC sequestration can be enhanced by using a combination of crop residue retention with high intensity wheat rotation and reduction of summer fallowing. Low SOM levels in Morocco (i.e. <1–2%) can be increased under some cropping systems by using NT technology. For example, use of forage legumes in rotation with cereals can increase SOM content (Mrabet et al. 2012; Siddique et al. 2012). Sequestration of SOC also depends on quantity and quality of residues returned to the soil through diversification of crop rotation (Mrabet et al. 2003; Mrabet, 2008). Belmekki et al. (2014) reported that SOM content under NT increased from 3.08% to 3.36% (increase of 10%) by increasing crop residue level from 50% to 100%.

Upon conversion to NT systems, changing from continuous monocropping to diversified crop rotations involving leguminous crops (lentils, vetch, fababean, and chickpea) can increase the net SOC sequestration. Crops with deep root systems (i.e. forage crops, safflower, rapeseed, etc.) can also sequester SOC in the deeper soil layers, which is not readily mineralized.

The data in Table 14.6 show a general trend that NT systems conserve and enhance SOC levels, especially in the upper 0–10 cm layer for most sites and soils in the Maghreb region. However, the large range in the SOC sequestration rates indicates that the magnitude of change is likely to be highly site-specific and also depends on the duration of NT systems. Based on a simulation study, Bahri et al. (2019) reported an increase of SOC accumulation (0.13 Mg/ha/yr and 0.18 Mg/ha/yr) under NT systems for two contrasting climatic regions of Tunisia.

14.6 IMPROVING SOIL QUALITY AND FUNCTIONS

Soil quality is an integrated function of biological, chemical, and physical attributes. Synthesis of data of available results from several experiments on NT systems conducted in the Maghreb region indicates improvement of soil quality attributes. Such soil quality improvements are associated with increase in SOM content and reduced risks of soil erosion over time. Such regeneration and reversal of degradation processes are critically important to ensuring fertile and productive soils (Lal et al. 2018; Lal, 2016).

Adoption of an NT system based on three principles can contribute to improvement of soil quality. Retention of crop residues is important to enhancing SOM and nutrients contents while changing the cycling and transformation pathways which impact other properties (i.e. biological, biochemical, physical, chemical, and hydrological). A continuous addition of crop residues over time restores soil quality, increases its resilience, and improves its fertility (Ranaivoson et al. 2017). Minimum disturbance of soil stabilizes such dynamic transformations. Crop diversification is critical to the NT systems through reduction in biotic and abiotic stresses, providing energy source of

TABLE 14.6
Soil Organic Carbon as Affected by No-Tillage (NT) and Conventional Tillage (CT) Systems in Selected Sites in Maghreb Countries

Country	Site/Soil Type	Horizon mm	Duration Years	NT %	CT %	Annual Rainfall mm	References
Morocco	Calcixeroll/Sidi El Aydi	0–25	11	3.98	1.64	358	Mrabet et al. (2001a)
	(Chaouia)/	0–50	-	3.36	2.52	358	Belemekki et al. (2013)
	Calcixeroll/Gdana (Chaouia)	0–50	6	3.20	2.68	350	Belemekki et al. (2014)
	Vertisol/Jemma Shaim	0–25	32	1.91	1.30	270	Laghrour et al.
	Vertisol/Merchouch	0–100	11	2.22	1.71	410	(2019)Moussadek
	Vertisol/Merchouch	0–70	6	2.05	1.47	410	(2012)
	Chaouia/Vertisol	0–200	1–6	2.63	2.25	350	Aboutayeb et al. (2020)
Algeria	Calcisol/Sersour	0–100	-	2.97	2.00	300	Bouzrara et al.
	Limestone/Setif	0–100	4	2.89	2.44	374	(2011)Mekhlouf et al. (2013)
	Oued Smar/Clay Loam (Central Algeria, Sub-humid)	0–100	5	1.99	1.79	671	Abdellaoui et al. (2011)
Tunisia	Loamy clay/Beja	0–100	3	1.70	1.20	426	Chibani et al. (2018)
	Sandy loam/Kef	0–100	9	1.55	0.95	334	
	North	0–100	5	1.64	0.95	-	Bahri et al. (2017)
	Sub-humid/Mateur	0–100	10	2.31	1.93	-	Angar et al. (2011)
	Semi-arid/Krib	0–100	10	2.17	1.5	-	
	Semi-arid/Borj Elifa	0–100	10	2.31	1.93	-	

microbial communities for transforming organic matter and restoring soil quality (i.e. aggregation, nutrient pools, biological diversity). All these components must be managed judiciously by integrating livestock in low-input farming systems (Lal, 2020).

NT systems can also increase uptake, conservation, and use of available soil water by reducing evaporation and runoff, enhancing water distribution in the soil profile, and improving plant available water-holding capacity of soils. Such benefits in water balance are achieved and even enhanced with maintenance of crop residue mulch at the soil surface. (Mrabet, 1997) and Moussadek (2012) reported higher soil water potentials under NT compared to CT and hence lower soil water evaporation due to elimination of soil disturbance and maintenance of crop residues at the soil surface. The direct impact of such improvements in soil water properties is the increased crop yields and quality even in droughty years of below normal rainfall (Mrabet, 2008). The advantage of NT in terms of higher yield and water use efficiency originates from increased transpiration and reduced evaporation because of the effect of residue mulch on the surface (Chennafi et al. 2011; Mrabet, 2002a).

Several studies have shown increases in either aggregation index or mean weight diameter (MWD) under NT compared to CT and that aggregation stability increased with increased residue cover, mainly in the topsoil layers (Mrabet et al. 2004, 2012; Bouzrara et al. 2011; Rezgui et al. 2014; Belmekki et al. 2014; Chibani et al. 2018). Mrabet et al. (2004) also reported that even without residues, soil under NT management had higher levels of aggregation than that under CT. Mrabet (2002b) and Mrabet et al. (2012) observed higher MWD and aggregation index at the surface of a self-mulching swelling clay soil under NT than that under CT.

In comparison to CT, Zergui et al. (2014) reported that use of an NT systems increased structural stability by 75% and improved soil moisture content by 19%. Similar results were reported by several researchers (Chibani et al. 2018; Laghrour et al. 2015, 2016, 2019; Bouzrara et al. 2011). An increase in soil aggregation also indicates that SOC is conserved and protected and allows SOM to function as a reservoir of plant nutrients for crops and energy source for soil biota.

Mrabet (2002b) reported that use of NT systems transformed soil surface with higher aggregation over the wheat growing season and helped the crop to withstand drought and risks of soil erosion, leading to an increase in crop productivity. Mrabet (2008) and Belmekki et al. (2014) also stated that NT systems enhance SOC accumulation within micro-aggregates which in return form macro-aggregates. This shift of SOC within micro-aggregates is important to a long-term carbon storage in soil. Such marked differences in SOM, particulate organic matter and aggregate stability between NT and CT can be attributed to enhanced biological activity (including the activities of fungus and fauna).

Bouzrara et al. (2011) reported a recovery of equilibrium in soil after only 4 years of continuous NT systems, with increase in earthworm activity despite compaction of the topsoil under NT compared with that under CT practices. Similar findings were also reported by Errouissi et al. (2011) who observed enriched fauna and earthworms under NT compared to CT in some soils of northern Tunisia. It is also widely recognized that crop residues generate a more complex biological system and stable microclimatic conditions relative to soil moisture and temperature regimes, making a better habitat for soil fauna (Kennedy et al. 2004). Such changes in richness, abundance, and diversity of fauna and earthworms have strong and positive consequences to soil microbial, biochemical, physical, chemical, and hydrological properties which translate into better ecosystem services and higher productivity. NT systems are also associated with an improvement in the diversity of fungal populations in soils (Hamim et al. 2012).

It is also widely recognized that an increase in SOM content with adoption of NT systems is associated with an increase in soil nitrogen content (Mrabet et al. 2012; Chibani et al. 2018). Mrabet et al. (2001a) and Ben Moussa-Machraoui et al. (2010) reported that conversion to NT significantly improved soil content of K_2O, P_2O_5, and N in semi-arid regions of Morocco (after 11 years) and in northwest Tunisia (after 4 years). However, while comparing soils under NT and CT systems under on-farm conditions, Aboutayeb et al. (2020) observed that adopting NT systems for one to 6 years did not cause a significant difference in 0–20 cm layer for total and NO_3^- nitrogen contents, P, K^+, Na^+, Ca^{2+}, CEC, pH, and EC. Aboutayeb and colleagues attributed these trends to low levels of crop residues returned to soils under NT and continuous monocropping of wheat. In other words, NT systems can improve soil quality and generate other benefits only when used in conjunction with appropriate crop and soil management practices based on the three basic principles.

The absence of soil mixing in NT systems leads to the stratification of immobile nutrients in the surface layer of the soil profile (Mrabet et al. 2012). Such a stratification can be both an advantage and constraint to crop production in dry areas of the Maghreb region. Stratification of nutrients is important to improving fertilizer management, because concentration of nutrients in the root zone is important to crop vigor and growth and increase in availability of plant nutrients may alleviate any deficiency or misapplication. However, in case of prolonged drought, confinement of roots to the surface layer may adversely affect crop growth aggravating the effects of water deficit.

Therefore, amelioration of soil fertility may require a different strategy of fertilizer management under NT compared to that under CT systems (Mrabet, 2008). For example, Mrabet et al. (2001b) observed that wheat produced under an NT system may receive more nutrients from decomposition of SOM and acidification in the seed zone. Over time, however, wheat-based NT systems may require lesser amounts of fertilizer nutrients due to greater efficiency of nutrient cycling relative to that under CT systems. Crop rotation also differentially affects N content under different tillage systems. Mrabet et al. (2001a) reported that continuous monocropping of wheat and triennial rotation (wheat-forage-fallow) led to a higher total N under NT than that under CT systems. Souissi et al. (2020) reported that even though water use efficiency was not affected by tillage systems, nitrogen use efficiency was better under NT compared to that under CT systems. Some other studies have indicated that under NT systems, complex rotations based on food legume enhance the availability of N, and hence, optimize the fertilizer management (Vanlauwe et al. 2014; Boulakia et al. 2020).

Review of available literature also indicates that soil bulk density may increase under NT systems (Mekhlouf et al. 2013; Mrabet et al. 2012; Jemai et al. 2012; Abdellaoui et al. 2011), with some differences among studies because of differences in soil type, duration of the study, and time of sampling. The data from several studies involving the impact of NT compared to CT on soil bulk density show that high bulk density under NT may be due to the consolidation of soil over time because of minimum mechanical disturbance. Moussadek (2012) and Mekhlouf et al. (2013) also reported high soil compaction levels under NT compared to CT systems. However, the impact is not stable over time as it changes during the growing season and over the long term because of the progressive increase in SOM content. In semi-arid Morocco, Lahlou et al. (2005) reported that change in soil bulk density after 8 years of NT systems was accompanied by a concomitant modification in pore volume and increase in pore size distribution as compared to those in soil under a CT system. Such improvements in pore size distribution within the seed zone may also be the cause of a better water infiltration process and high level of stored water in a soil under the NT system. Consequently, the overall soil structure in the topsoil is satisfactory for crop growth under NT and does not limit crop growth and yields.

In summary, re-carbonization of soil under NT systems is essential to improving soil processes that impact agronomic productivity and environment quality. Restoration and enhancements of soil biological processes and soil-water properties increase resilience of farming systems against climate change and other stresses (Moussadek and Mrabet, 2017).

14.7 SOIL EROSION CONTROL AND PREVENTION

Tillage-led erosion is among the major threats to soil security in the Maghreb region, which is known for a high risk of soil erosion and losses due to soil mismanagement, intensive tillage, residue removal by over-grazing and loss of vegetation cover (e.g. 2,000–4,000 Mg km^{-2}. yr^{-1} in Algeria; Demmak, 1982). In Tunisia, 3 million ha of land is threatened by erosion, and 50% is seriously affected (Bahri et al. 2016). In Algeria, the total area threatened by water erosion is estimated at 10 million ha (Benguerai, 2011). Soils of the Maghreb region are highly prone to runoff and erosion because of low SOM content and aggregation leading to a decrease water infiltration and hydrodynamic properties during rainfall events (especially after long and hot summers) (Mrabet, 2008; Ryan et al. 2006). Consequently, severe soil erosion and deposition are key processes influencing the distribution and amount of carbon storage in soils and watersheds (Lagacherie et al. 2017).

Ogle et al. (2019) reported that the positive impact of the NT system in minimizing risks of soil erosion on cropland is globally recognized, and its adoption is essential to prevent or reverse the overall process of soil degradation. Similar observations are also observed in the Maghreb region (Moussadek et al. 2011b). Hence, conversion to NT systems for management and control of soil erosion can also reduce losses of SOC in water- or wind-transported sediments and contribute to reducing emissions and sequestration of carbon (Moussadek and Mrabet, 2017).

A rainfall simulator was used on a Vertisol by Moussadek et al. (2011a) to compare water runoff and soil loss in CT and NT systems with crop residues removed (NT0) and with 50% of crop residues returned to the soil surface (NT50). The data in Figure 14.3, depicting the effects of two successive rainfall simulations of 36 and 60 mm h^{-1}, indicate that runoff rates were 3–4 times lower under NT50 compared to NT0 and CT. Moussadek et al. (2011a) also reported a reduction of 50% in soil losses with NT50 compared to that under CT.

Infiltration of rainwater is improved in soil under the NT system, which also increases plant available water capacity in both heavy-textured soils (Bouzza, 1990) and in structurally weak soils through prevention of crust formation because the presence of crop residue mulch. Hence, for most soils, water infiltration properties are improved under the NT system. Increases in the rate of water infiltration under NT are due to better aggregate stability and bioporosity. For a Vertic Mollisol in

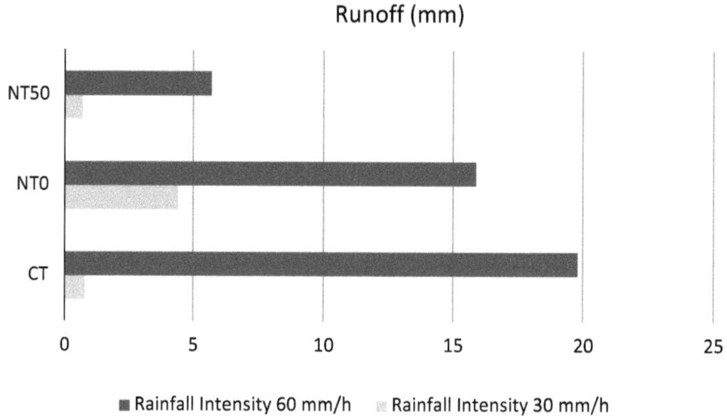

FIGURE 14.3 Runoff as affected by tillage and NT residue management under two rainfall intensities in a clay soil (Central Morocco) (NT50=No tillage with 50% residue; NT0= bare, no tillage without residue; CT= Conventional Tillage) (Moussadek et al. 2011a).

the Chaouia region, water infiltration rates under well-managed NT are higher over extended periods than under CT systems (Mrabet, 2008). These results agree with those reported by Dimanche (1997) for the Saiss region on a clayey soil. In Tunisia, Angar et al. (2011) reported that the rainwater infiltration rate under NT increased from 41 L/h to 73 L/h in sub-humid areas, from 32 L/h to 40 L/h in lower semi-arid, and from 19 L/h to 43 L/h in higher semi-arid areas, during the course of the experiment. In contrast, water infiltration rates under CT were lower approaching 20 L/h and stayed constant in soil of the semi-arid site and increased slightly for that in the more humid site (18–29 L/h) at the end of the experiment).

Synthesis of the data from several studies indicates that continuous use of mechanical tillage (both inversion and mixing) leads to development of structurally unstable soils, and conversion to an NT system does not lead to recovery from degradation of soil porosity and infiltration capacity over a short duration. For example, Abdellaoui et al. (2011) and Mekhlouf et al. (2013) reported lower water infiltration rates under NT compared to those under CT even though SOM contents increased and soil bulk density values were lower under NT in the top 10 cm layer. Moussadek (2012) studied water dynamic in soils under contrasting NT and CT systems, and reported that hydraulic conductivity and sorptivity were much higher under CT compared to those in soil under NT. Such differences were attributed to higher compaction and more bulk density in soil under NT which did not recover from severe historical degradation by conventional tillage and residue removal through grazing and balling.

On the basis of studies on soil in two contrasting regions of Tunisia, Bahri et al. (2019) reported that NT systems are more effective in enhancing soil resilience and minimizing risks of water erosion (1.7 Mg/ha/yr and 4.6 Mg/ha/yr of soil loss). In another study, Bahri et al. (2016) reported that after 5 years of NT systems, higher aggregate stability reduced risks of soil erosion compared to that under CT systems. In fact MWD for soil under NT was 16% higher than under CT. Mrabet et al. (2003) also concluded that the development of a good structure in the surface layer improves water infiltration, movement and distribution, and leading to positive effects on evaporation and erosion control.

14.8 REDUCE DIRECT AND INDIRECT CO_2 EMISSIONS

Adoption of NT systems have also been reported to reducing greenhouse gas (GHG) emission and thus mitigating climate change (Moussadek et al. 2011b; Sanz-Cobeña et al. 2017). Mrabet et al.

(2001a) reported an accumulated SOC stock corresponding to an increase at the rate of 1.4 Mg CO_2e ha^{-1} yr^{-1} In comparison, Jemai et al. (2012) reported the accumulation of SOC under NT in the range of 1.47–2.79 Mg CO_2e ha^{-1} yr^{-1}.

The data presented in Figure 14.4 show a drastic and continuous reduction in CO_2 emissions under the NT system compared to that under the CT system in the semi-arid areas of Morocco (Moussadek et al. 2011b). The data also showed that the three tillage practices (Chisel, Stubble plow, and Disk plow) result in rapid oxidation to CO_2 and loss to the atmosphere. In addition, retaining residues at the soil surface and avoidance of tillage in NT systems reduces the mineralization process of SOM and increases carbon accumulation as compared to that under residue incorporation under CT systems which facilitate decomposition and oxidation.

In the semi-arid regions of Morocco, El Gharras et al. (2004) found that NT systems reduce the number of field operations and days that are needed to plant crops compared to those needed for CT systems, which in turn, reduce fuel usage and associated GHG emissions. With NT, reduced use of tractors and powered farm equipment results in lower emissions of GHGs. Indeed, the NT system is about eight times more efficient in fuel consumption than CT and eliminating tillage can be more energy efficient than eliminating herbicide use. Studies by Bourarach (1989) and Chekli (1991) showed that machinery and fuel costs tend to be lower in NT than CT systems, with CT systems requiring 3.2 times more fuel. Labbaci et al. (2015) also reported reduced fuel use under NT by 22–33 liters/ha compared to CT systems. Labbaci and colleagues also reported that NT farmers save both time and money from cultivation of wheat compared to that for CT based systems. Bahri et al. (2018) also reported that NT systems reduced energy consumption by 16% compared to CT systems.

A study conducted in northern Algeria documented that fuel consumption was higher in a CT system with 90 L/ha of gasoline compared with 44 L/ha in NT (Fenni, 2013). With a similar pattern of labor duration and allocation, soil preparation operations consumed more fuel in CT at 49 L/ha (54.34%) while harvesting, baling, and freight operations were the most consumers of fuel in NT, with 34 L/ha (69.4%) (Rouabhi et al. 2019). Based on a study in the northeastern region of Algeria, Fenni et al. (2015) found that direct seeding of wheat under NT can reduce the use of fuel by 50–70%, and of machinery use by 60% relative to that by CT. Zaghouane et al. (2006) reported that the working time in direct seeding in NT systems is ten times less important than that in CT systems. Differential time use in farm operations among tillage systems leads to an appreciable gain in time, less wear and tear on the machinery, and, consequently, fuel economy. Such improvements in energy efficiency under NT systems are also reducing GHG emissions from the agricultural sector.

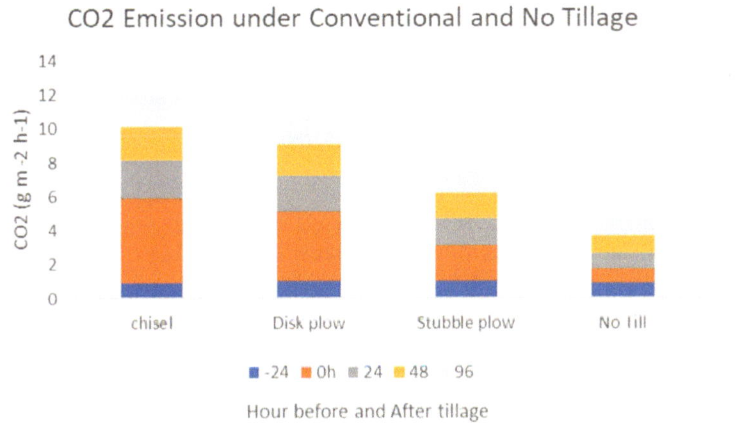

FIGURE 14.4 Soil CO_2 flux with tillage systems in Central Morocco (Moussadek et al. 2011b).

14.9 LAND SUITABILITY FOR NO TILLAGE IN MAGHREB: CASE STUDY FROM MOROCCO

NT systems were introduced in Morocco by INRA in 1980 by conducting field experiments to document NT climate-smart properties (economic benefits, reduced soil erosion, improved soil quality, and high and stable crop yields (Mrabet et al. 2012)). Moroccan farmers have yet to adopt these systems on a large scale. It is therefore important to develop a suitability study of CA/NT systems in the country and the region. INRA and ICARDA launched a pilot study in Central Morocco (the largest cereal production basin in the country) using a "target" area and "match" location to identify which areas in Morocco could be managed with CA. Land suitability approach, based on biophysical factors, including soil properties, climate, and land use, is used to compare and analyze the potential adaptation of the CA system.

In general, soils with fine soil texture, good internal drainage, and low salinity are better adapted for NT practice in cereal-based systems. Similarly, soils with fine texture and poor internal drainage are considered moderately suitable lands for adoption of the NT system (Busari et al. 2015). Finally, soils unsuitable for NT are saline and poorly drained with compaction problems (plow pan). The application of reduced tillage is more often recommended for shallow soils and those with a high gravel content. A similar soil guide for use of NT/CA systems was proposed by Lal (1985, 1987) for soils of the tropics and can be tailored to Mediterranean soils. McGarry et al. (2000) also stated that NT systems are particularly attractive for use in clayey soils in order to minimize compaction and promote natural processes which enhance formation of soil structure through shrink-swell cycles that increase soil water storage and improve beneficial soil fauna relative to CT systems. According to the suitability map analysis, 63% of total arable lands in Central Morocco (or 3.8 million ha) are highly to moderately suitable for the adoption of NT (Figure 14.5; Moussadek et al. 2016). Conversion to an NT system on these lands could benefit nearly 10 million inhabitants and contribute to significant C sequestration potential as well as mitigation of GHG emissions.

Additional research is needed to facilitate greater adoption of CA in the Maghreb region. Specific researchable priorities include: the number of farmers and households in a region, machinery available, etc., and other biophysical and socioeconomic factors which could make the NT technology an excellent tool in facilitating greater adoption in the Maghreb region.

14.10 CONCLUSIONS

The Maghreb, an agriculture-dependent region and climate change hot spot, is facing a range of challenges, such as soil mining and degradation, water scarcity, low crop productivity and large yield gaps, soil structural problems, and recurrent droughts and high temperatures. These challenges could increase demand for food production. The problem of food and nutritional insecurity will be aggravated by changes in dietary preferences and diversity of food types. These and other emerging issues necessitate significant transformations in policy, technology, resources, and the issues of human dimensions which lead to an increase in crop yields while lowering the environmental impact of agriculture and food systems. Faced with significant deficits in agricultural production, research institutes and international organizations have promoted NT/CA in the Maghreb region for more than four decades, since 1980. Such research and development projects have produced high quality and intensive but fragmented results on a wide range of indicators of sustainability. Among such indicators, this chapter has analyzed crop performances, soil carbon storage and sequestration, erosion, and sediment losses as well as direct and indirect emissions of $C0_2$ and other GHGs. Results collated and synthesized support the following conclusions:

1. NT systems can play a double role in mitigating climate change through carbon sequestration and CO_2 emission abatement and in enhancing agriculture resilience to droughts through better water conservation and use efficiency and functioning of soils.

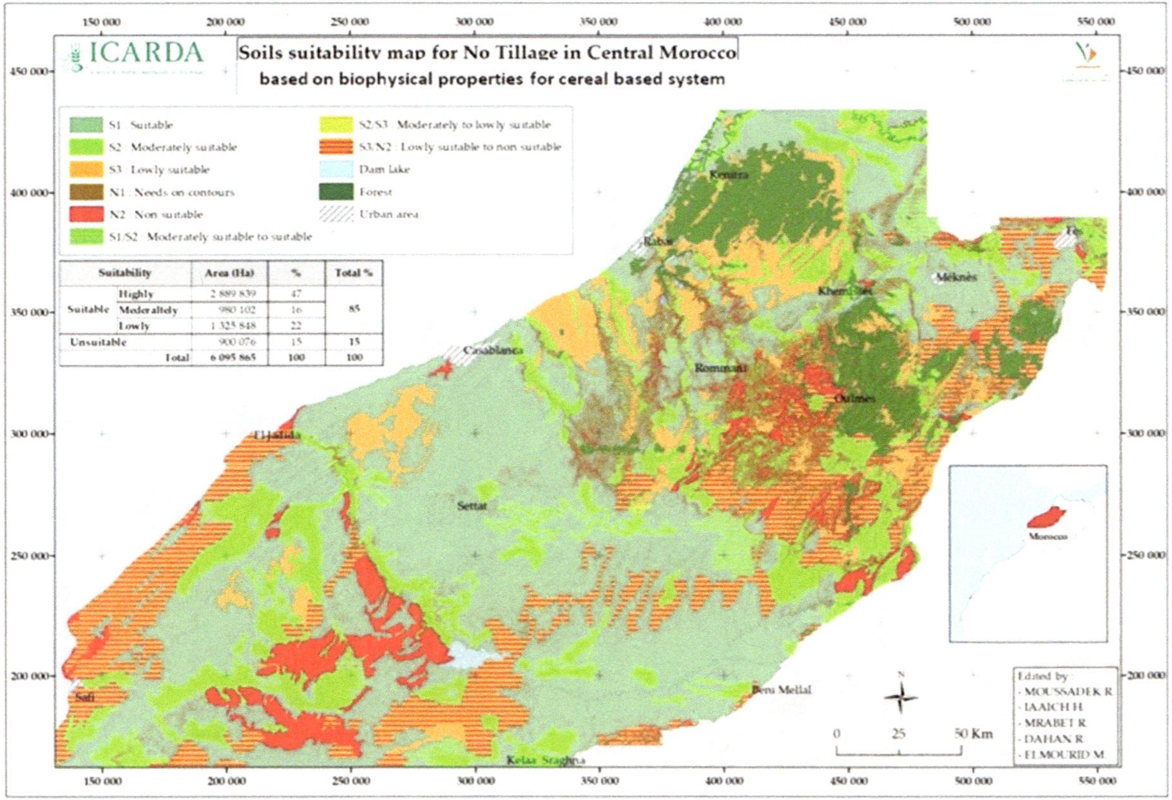

FIGURE 14.5 Soil suitability map for No tillage in Central Morocco (Moussadek et al. 2016).

2. Economic and environmental benefits of NT are confirmed and unequivocal, with regard to agronomic, biophysical, and climate mitigation.
3. NT technologies can make significant contributions to sequestering carbon in soils and, as negative emission technologies, help to meet GHG mitigation targets.
4. Intensive soil tillage, straw exportation, and excessive grazing are major causes of degradation of soil quality and decline in yield of crops in the Maghreb region.
5. Using NT systems based on three principles (elimination of tillage, retention of crop residues and straw as mulch on soils, and integration of crop-livestock through crop diversification and crop residue management) can restore soil quality and recycle plant nutrients.
6. NT research done in the Maghreb region since the early 1980s has shown that crop yields and SOM contents can simultaneously be improved with NT systems.
7. Some co-benefits from carbon sequestration under NT systems include greater fertility of soils, less runoff and erosion of fragile soils, and increased water storage in the root zone. Such benefits lead to restoration and rehabilitation of degraded soils and increase in plant biodiversity.
8. NT systems also reduce fuel consumption and reliance on machinery and fertilizers over time, which helps the agriculture sector to become carbon neutral.
9. Promoting widespread adoption of NT systems in the Maghreb region as a priority at all levels will certainly deliver climate change mitigation through large scale soil carbon sequestration.
10. Adoption of NT systems has a large potential in the Maghreb region. Widespread adoption is constrained by the fact that while NT research is carried out by national institutions, adoption efforts rely mainly on funding from international organizations that support specific time-bound projects. Uptake of NT systems by farmers during the project depends on availability of resources which in post-project period cannot support small-holders and resource-poor farmers.
11. Soils with fine texture, good internal drainage, and low salinity are more adapted for NT practice in cereal-based systems.
12. A high priority to promote adoption of CA in the Maghreb region is an urgent need for improving farmer's enabling environment and to instigate concrete coordinated efforts by all stakeholders including policy decision makers to constitute a functional legal framework and associated mechanisms for enhancing and enforcing CA at the farm gate and for promoting NT in carbon sequestration and climate change adaptation projects.

ACKNOWLEDGMENTS

The authors would like to thank ICARDA for providing great support to the development and spread of CA in the Maghreb region as well as all no-tillage research teams at the national research and education institutions in each country who provided results used to prepare this paper.

REFERENCES

Abdellaoui, Z., H. Teskrat, A. Belhadj et O. Zaghouane. 2011. Étude comparative de l'effet du travail conventionnel, semis direct et travail minimum sur le comportement d'une culture de blé dur dans la zone subhumide. *Options Méditerranéennes : Série A* 96:71–87.

Abidi, S., S. Benyoussef and H. Ben Salem. 2021. Foraging behaviour, digestion and growth performance of sheep grazing on dried vetch pasture cropped under conservation agriculture. *Journal of Animal Physiology and Animal Nutrition (Berl)* 105:51–58. doi:10.1111/JPN.13456.

Aboudrare, A., P. Debaeke, A. Bouaziz and H. Chekli. 2006. Effects of soil tillage and fallow management on soil water storage and sunflower production in a semi-arid Mediterranean climate. *Agricultural Water Management* 83(3):183–196.

Aboutayeb, R., B. El Yousfi and O. El Gharras. 2020. Impact of No-Till on physicochemical properties of vertisols in Chaouia region of Morocco. *Eurasian Journal of Soil Science* 9(2):119–125.

AMAC 2021. *Rapport sur la dissémination du semis direct au Maroc. Atelier pour la présentation de la feuille de route semis direct sur 1,5 Mha-2020–2030.* INRA, Rabat.

Angar, H., H. Ben Haj Salah et M. Ben Hammouda. 2011. Semis direct et semis conventionnel en Tunisie : Les résultats agronomiques de 10 ans de comparaison. *Options Mediterranéennes: Série, A.* 96:53–59.

Aune, J. and R. Lal. 1998. Agricultural productivity in the tropics and critical limits of properties of oxisols, ultisols and alfisols. *Tropical Agriculture (Trinidad)* 74:96–103.

Bahri, H., M. Annabi, R. Chibani, H. Cheick M'Hamed and T. Hermessi 2016 Can conservation agriculture reduce the impact of soil erosion in northern Tunisia ? *Geophysical Research Abstracts* 18, EGU2016-12234, 2016 EGU General Assembly.

Bahri, H., M. Annabi, A. Saoueb, et al. 2017. Can conservation agriculture sequester soil carbon in the long term in northern Tunisia? Recent advances in environmental science from the Euro-Mediterranean and surrounding regions, Euro-mediterr. Conference Environment Integrability 22–25 November, Sousse, Tunisia, pp. 69–71.

Bahri, H., M. Annabi, H. Cheikh M'hamed, R. Chibani, M. Chtourou, N. Riahi et L. Ben Becher. 2018. Evaluation de l'impact agro-environnemental et de la durabilité de l'adoption du semis direct au Nord de la Tunisie. *Annales de l'INRAT* 91:98–111.

Bahri, H., M. Annabi, H. Cheikh M'hamed and A. Frija. 2019. Assessing the long-term impact of conservation agriculture on wheat-based systems in Tunisia using APSIM simulations under a climate change context. *Science of the Total Environment* 692:1223–1233.

Bazza, M., M. Key and C. Knuston. 2018. Drought characteristics and management in North Africa and the Near East. *FAO Water Reports* 45. http://www.fao.org/3/CA0034EN/ca0034en.pdf.

Belaid, D. 2015. Algérie: Le semis direct, une revolution technique en agriculture. http://agriculturealgerie.blogspot.com/2015/07/algerie-le-semis-direct-une-revolution.html?m=1.

Belmekki, M., R. Mrabet, R. Moussadek, O. Ibenhalima, M. Boughlala, M. El Gharous and B. Bencharki. 2013. Effects of tillage and cropping systems on the structural stability and soil organic matter in semi-arid areas of Morocco. *International Journal of Innovation and Applied Studies* 4:322–333.

Belmekki, M., M. El Gharous, O. EL Gharras, M. Boughlala, O. Iben Halima and B. Bencharki. 2014. Tillage effects on basic properties of a calcareous soil under Moroccan semi-arid climate. *International Journal of Advanced Research in Engineering and Technology* 5(3):130–146.

Benguerai, A. 2011. *Evolution du phénomene de désertification dans le Sud oranais (Algérie).* PhD Thesis. Abou Bekr Belkaïd de Tlemcen University.

Ben-Hammouda, M., K. M'Hedhbi, M. Kammassi and H. Ghouili 2007. Direct drilling: An agro-environmental approach to prevent land degradation and sustain production. Proceedings: International Workshop on Conservation Agriculture for Sustainable Land Management to Improve the Livelihood of People in Dry Area. Damascus-Syria, pp. 37–48.

Benkherbache, N., R. Benniou, M. Merat, T. Mekhalfia and M. Makhlouf. 2012. From minimum tillage to no tillage, reaction of Waha, Variety of Durum Wheat in Algerian semi-arid region. *The Online Journal of Science and Technology* 2(3):17–22.

Ben Moussa-Machraoui, S.B., F. Errouissi, M. Ben-Hammouda and S. Nouira. 2010. Comparative effects of conventional and no-tillage management on some soil properties under Mediterranean semi-arid conditions in northwestern Tunisia. *Soil and Tillage Research* 106(2):247–253.

Bessam, F. and R. Mrabet. 2003. Long-term changes in soil organic matter under conventional and no-tillage systems in semiarid Morocco. *Soil Use and Management* 19(2):139–143.

Bonzanigo, L., C. Giupponi and R. Moussadek. 2016. Conditions for the adoption of conservation agriculture in Central Morocco: An approach based on Bayesian network modeling. *Italian Journal of Agronomy* 11(1s):665. doi:10.4081/ija.2016.665.

Boulakia, S., F. Tivet, O. Husson and L. Séguy. 2020. Nutrient management practices and benefits in Conservation Agriculture systems. In: A. Kassam (ed.), *Advances in Conservation Agriculture. Volume 2: Practice and Benefits*. Burleigh Dodds Science Publishing, Cambridge, pp. 1–29.

Boulal, H., M. El Mourid, H. Ketata and A. Nefzaoui. 2014. Conservation agriculture in North Africa. In: R.A. Jat, K.L. Sahrawat and A.H. Kassam (eds.). *Conservation Agriculture: Global Prospects and Challenges*. CABI, Wallingford, UK, pp. 293–310.

Bouguendouz, A. 2010. Effet de trois itinéraires techniques sur l'élaboration du rendement de l'orge (Hordeum vulgare L.) sous conditions semi-arides des hautes plaines Sétifiennes. Actes des IVièmes RMSD, Setif, 3-5 mai 2010. Recherches Agronomiques, INRAA. N° Spécial:161–164.

Bourarach, E.H. 1989. *Mécanisation du travail du sol en céréaliculture pluviale : Performances techniques et aspects économiques dans une région semi-aride au Maroc.* PhD Thesis, IAV Hassan II. Rabat, Maroc.

Bouzza, A. 1990. *Water Conservation in Wheat Rotations under Several Management and Tillage Systems in Semiarid Areas.* Ph.D. Dissertation. University of Nebraska, Lincoln.

Bouzerzour, H., S. Mahnane and M. Makhlouf. 2006. Une association pour une agriculture de conservation sur les hautes plaines orientales semi-arides d'Algérie. *Options méditerranéennes: Série A* 69:107–111.

Bouzrara, S., M.E.H. Ould Ferroukh and A. Bouguendouz. 2011. Influence du semis direct et des techniques culturales simplifiées sur les propriétés d'un sol de la ferme pilote Sersour (Sétif). *Options Méditerranéennes : Série A* 96:123–129.

Brahim, N., T. Gallali and M. Bernoux. 2009. Effects of agronomic practices on the soil carbon storage potential in Northern Tunisia. *Asian Journal of Agricultural Research* 3(3):55–66.

Busari, M.A., S.S. Kukal, A. Kaur, R. Bhatt and A.A. Dulazi. 2015. Conservation tillage impact on soil, crop and the environment. *International Soil and Water Conservation Research* 3(2):119–129.

Chaieb, N., M. Rezguia, S. Ayed, et al. 2020. Effects of tillage and crop rotation on yield and quality parameters of durum wheat in Tunisia. *Journal of Animal and Plant Sciences* 44(2):7654–7676.

Cheikh M'hamed, H., H. Bahri and M. Annabi. 2018. Conservation agriculture in Tunisia: historical, current status and future perspectives for rapid adoption by smallholder farmers. Second African Congress on Conservation Agriculture. 9–12 October 2018. Johannesburg, South Africa, pp. 57–60.

Cheikh M'hamed, H., M. Annabi, S. Ben Youssef et H. Bahri. 2016. L'agriculture de conservation est un système de production permettant d'améliorer l'efficience de l'utilisation de l'eau et de la fertilité du sol. *Annales de l'INRAT* 89:68–71.

Chekli, H. 1991. *Eléments du choix des séquences d'installation de la culture du blé dans la région de Meknès. Modification des états structuraux et aspects énergétiques.* PhD Thesis. IAV Hassan II. Rabat, Morocco.

Chennafi, H., A. Hannachi, O. Touahria, et al. 2011. Tillage and residue management effect on durum wheat [Triticum turgidum (L.) Thell. ssp. turgidum conv. durum (Desf.) MacKey] growth and yield under semi-arid climate. *Advances in Environmental Biology* 5(10):3231–3240.

Chibani, R., H. Bahri, M. Annabi and H. Cheikh M'hamed. 2018. L'agriculture de conservation comme alternative pour améliorer la résistance des sols à l'érosion hydrique dans le Nord de la Tunisie. *Annales de l'INRAT* 91:86–97.

Debaeke, P. and A. Aboudrare. 2004. Adaptation of crop management to water-limited environments. *European Journal of Agronomy* 21(4):433–446.

Demmak, A. 1982. *Contribution à l'étude de l'érosion et des transports solides en Algérie septentrionale.*

Devkota, M. and Y.A. Yigezu. 2020. Explaining yield and gross margin gaps for sustainable intensification of the wheat-based systems in a Mediterranean climate. *Agricultural Systems* 185. doi:10.1016/j.agsy.2020.102946.

Dhehibi, B., C. Zucca, A. Frija, N. Shinan and S.N. Kassam. 2018. Biophysical and Econometric analysis of adoption of soil and water conservation techniques in the semiarid region of Sidi Bouzid (Central Tunisia). *New Medit* 2:15–28.

Dimanche, P.H. 1997. *Impacts des différents itinéraires techniques de travail du sol sur la dégradation des sols argileux dans la région de Meknès, Maroc.* Ph.D. Thesis. Louvain-La-Neuve, Belgium.

Driouch, F., S. Rached and T. El Hairech 2013. Climate variability and change in North African countries. In: Sivakumar, et al. (Eds.), *Climate Change and Food Security in West Asia and North Africa.* Springer, pp.161–172.

El Gharras, O., M. El Mourid and H. Boulal. 2017. Conservation Agriculture in North Africa: Experiences, achievements and challenges. In: A. Kassam, S. Mkomwa and T. Friedrich (Eds.), *Conservation Agriculture for Africa: Building Resilient Farming Systems in a Changing Climate.* CAB International, Wallingford, UK, pp. 127–138.

El Gharras, O., A. Ait Lhaj et M. Idrissi. 2004. Développement d'un semoir non-labour industriel. In: *Actes des 2èmes rencontres méditerranéennes sur le semis direct, Tabarka, Tunisie.*

Errouissi, F., S. Ben Moussa-Machraoui, M. Ben-Hammouda and S. Nouiraab. 2011. Soil invertebrates in durum wheat (Triticum durum L.) cropping system under Mediterranean semi arid conditions: A comparison between conventional and no-tillage management. *Soil and Tillage Research* 112(2):122–132.

FAO 2020. *World Food and Agriculture—Statistical Pocketbook 2020.* Rome.

FAO and ITPS. 2020. *Global Soil Organic Carbon Map V1.5: Technical Report.* Rome, FAO. doi:10.4060/ca7597en

Fellahi, Z., A. Hannachi, H. Chennafi, M. Makhlouf and H. Bouzerzour. 2013. Effets des résidus et du travail du sol sur la production de la biomasse et le rendement du blé dur (Triticum durum Desf. variété MBB)

en lien avec l'utilisation de l'eau dans les conditions semi-arides des Hautes Plaines Sétifiennes. *Revue de l'Agriculture* 06:03 –11.

Fenni, M., K. Nadjem and A. Mekhlouf 2015. Carbon sequestration by direct seeding of wheat in Setif high plains (North East of Algeria). In I. Dincer, C. Coplan, O. Kizilkan and M. Ezan (eds.), *Progress in Clean Energy*. Vol. 1. Cham: Springer. doi:10.1007/978-3-319-16709-1_45.

Fenni, M. 2013. Socio-economic and environmental benefits of direct seeding of wheat in Sétif high plains (North East of Algeria). WEI International Academic Conference, Istanbul, Turkey. West East Institute (WEI), Westchester, PA, USA, p. 47.

FERT, 2016. *Le semis direct en Tunisie: Situation actuelle et perspectives. FERT-APAD-TAE-INGC*. 66p.

FERT 2018. L'agriculture de conservation au Maghreb: Les agriculteurs font évoluer leurs pratiques. https://www.agricord.org/sites/default/files/2018_med_capi_ac_maghreb_fr.pdf (accessed December 15, 2020).

Godfray, H.C., J.R. Beddington, I.R. Crute, et al. 2010. Food security: The challenge of feeding 9 billion people. *Science* 327(5967):812–818. doi:10.1126/science.1185383.

González-Sánchez, E.J., O. Veroz-González, G. Caneway, et al. 2019. Meta-analysis on carbon sequestration through conservation agriculture in Africa. *Soil and Tillage Research* 190:22–30.

Hamim, A., A. Essahat, S. El Othmani, et al. 2012. Etude de l'effet de la rotation et du travail du sol sur la densité spécifique des champignons mycorhiziens arbusculaires. International Congress: Microbial Biotechnology for Development (MICROBIOD 2), Marrakech, Morocco.

Idoudi, Z., N. Louahdi, M. Devkota, et al. 2020. *Public-Private Partnership for Enhanced Conservation Agriculture Practices: The Case of Boudour Zero-Till Seeder in Algeria*. International Center for Agricultural Research in the Dry Areas (ICARDA).

Jai, G., E. Shevliakova, P. Artaxo, et al. 2019. Land–climate interactions. In: P.R. Shukla, J. Skea, E. Calvo Buendia, et al. (eds.), *Climate Change and Land: An IPCC Special Report on Climate Change, Desertification, Land Degradation, Sustainable Land Management, Food Security, and Greenhouse Gas Fluxes in Terrestrial Ecosystems*. In press.

Jemai, I., N. Ben Aissa, S. Ben Guirat, T. Gallali and T. Gallali. 2012. On-farm assessment of tillage impact on the vertical distribution of soil organic carbon and structural soil properties in a semiarid region in Tunisia. *Journal of Environmental Management* 113:488–494.

Jemai, I., N. Ben Aissa, S. Ben Guirat, M. Ben-Hammouda and G. Gallali. 2013. Impact of three and seven years of no-tillage on the soil water storage, in the plant root zone, under a dry subhumid Tunisian climate. *Soil and Tillage Research* 126:26–33.

Kacemi, M. 1992. *Water Conservation, Crop Rotations, and Tillage Systems in Semiarid Morocco*. Ph.D. Diss. Colorado State University, Fort Collins.

Kassam, A., R. Derpsh and T. Friedrich 2020. Development of conservation agriculture systems globally. In: A. Kassam (ed.), *Advances in Conservation Agriculture. Volume 1: Systems and Science*. Burleigh Dodds Science Publishing, Cambridge, UK, pp. 1–56.

Kassam, A.H., T. Friedrich, R. Derpsch, et al. 2012. Conservation Agriculture in the dry Mediterranean climate. *Field Crops Research* 132:7–17.

Kennedy, A.C., T.L. Stubbs and W.F. Schillinger. 2004. Soil and crop management effects on soil microbiology. In: F. Magdoff and R.R. Weil (eds.), *Soil Organic Matter in Sustainable Agriculture. Advances in Agroecology*. CRC Press, pp. 295–326.

Labbaci, T., P. Dugué, H. Kemoun et D. Rollin. 2015. Innovation et action collective : le semis direct des cultures pluviales au Moyen Sébou (Maroc). *Cahiers Agricultures* 24(2):76–83.

Lagacherie, P., J. Álvaro-Fuentes, M. Annabi, et al. 2017. Managing Mediterranean soil resources under global changes: Expected trends and mitigation strategies. *Regional Environmental Change* 18:66–675. doi:10.1007/s10113-017-1239-9.

Laghrour, M., R. Moussadek, M. Mekkaoui, et al. 2015. Impact of no-tillage on physical proprieties of a clay soil in Central Morocco. *Journal of Material and Environmental Science* 6:391–396.

Laghrour, M., R. Moussadek, R. Mrabet, et al. 2016. Long and midterm effect of conservation agriculture on soil properties in dry areas of Morocco. *Applied and Environmental Soil Science*, 9 pages:Article ID 6345765. doi:10.1155/2016/6345765.

Laghrour, M., R. Moussadek, R. Mrabet et M. Mekkou. 2019. Effet à moyen et à long terme du semis direct sur la matière organique, la stabilité structurale et la compaction des sols argileux au Maroc. *Revue Marocaine des Sciences Agronomiques et Vétérinaires* 7(2):356–362.

Lahlou, S., M. Ouadia, O. Malam Issa, O.Y. Le Bissonnais et R. Mrabet. 2005. Modification de la porosité du sol sous les techniques culturales de conservation en zone semi-aride Marocaine. *Etude et gestion des sols* 12(1):69–76.

Lal, R. 1985. A soil suitability guide for different tillage systems in the tropics. *Soil and Tillage Research* 5(2):179–196.

Lal, R. 1987. Managing soils of sub-Saharan Africa. *Science* 236(4805):1069–1076.

Lal, R. 2002. Carbon sequestration in dryland ecosystems of West Asia and North Africa. *Land Degradation and Development* 13(1):45–59.

Lal, R. 2013. Climate change and soil quality in the WANA region. In: Sivakumar et al. (Eds.), *Climate Change and Food Security in West Asia and North Africa*. Springer, pp.55–74.

Lal, R. 2015. A system approach to conservation agriculture. *Journal of Soil and Water Conservation* 70(4):82A–88A.

Lal, R. 2016. Potential and challenges of conservation agriculture in sequestration of atmospheric CO2 for enhancing climate resilience and improving productivity of small landholder farms. *CAB Reviews*. 11009. doi:10.1079/PAVSNNR2016.

Lal, R. 2020. Integrating animal husbandry with crops and trees. *Frontiers in Sustainable Food Systems* 4:113. doi: 10.3389/fsufs.2020.00113

Lal, R., R. Horn and T. Kosaki (Eds.). 2018. *Soil and the Sustainable Development Goals*. Catena-Schweizerbart, Stuttgart, 196pp.

Le Mouël, C., A. Forslund, P. Marty, et al. 2015. *Le système agricole et alimentaire de la région Afrique du Nord – Moyen-Orient à l'horizon 2050: Projections de tendance et analyse de sensibilité*. Rapport final de l'étude réalisée pour le compte de Pluriagri, Paris et Rennes: INRA-DEPE & INRA-SAE2, 133 p.

Mbow, C., C. Rosenzweig, L.G. Barioni, et al. 2019. Food security. In: P.R. Shukla, et al. (eds.), *Climate Change and Land: an IPCC Special Report on Climate Change, Desertification, Land Degradation, Sustainable Land Management, Food Security, and Greenhouse Gas Fluxes in Terrestrial Ecosystems*. Intergovernmental Panel on Climate Change (IPCC), Geneva, Switzerland.

McGarry, D., B.J. Bridge and B.J. Radford. 2000. Contrasting soil physical properties after zero and traditional tillage of an alluvial soil in the semi-arid subtropics. *Soil and Tillage Research* 53(2):105–115.

Mekhlouf, A., N. Rouag, R. Boukhadra, S. Chenni, M. Fenni and M. Makhlouf. 2013. Influence of cropping systems on soil properties in semi-arid conditions of Setif, Algeria. *Journal of Agricultural Science and Technology A* 3:653–658.

Mekhlouf, A., M. Makhlouf, A. Achiri, A. Ait Ouali et S. Kourougli. 2011. Etude comparative de l'effet des systèmes de travail du sol et des précédents culturaux sur le sol et le comportement du blé tendre en conditions semi arides. *Agriculture* 2:52–64.

M'hedbi, K., S. Chouen and J.F. Richard. 2004. Historique du semis direct en Tunisie. Amorce du passage de l'agriculture conventionnelle ‡ l'agriculture de conservation. Dans : Actes des Deuxièmes Rencontres Méditerranéennes du Semis Direct. *Tabarka (Tunisie)*, 19-22 janvier 2004:32–37.

Mirzabaev, A., J. Wu, J. Evans, et al. 2019. Desertification. In: P.R. Shukla, J. Skea, E. Calvo Buendia, et al. (eds.), *Climate Change and Land: An IPCC Special Report on Climate Change, Desertification, Land Degradation, Sustainable Land Management, Food Security, and Greenhouse Gas Fluxes in Terrestrial Ecosystems*. In press.

Mouelhi, B., S. Slim, S. Arfaoui, A. Boussalmi and F. Ben Jeddi. 2016. Effet du mode de semis et de la rotation culturale sur les paramètres de croissance et les composantes de rendement du blé dur (Triticum durum Desf.) variété « Karim ». *Journal of New Sciences, Agriculture and Biotechnology* 28(11):1638–1648.

Moujahed, N., S. Abidi, S.B.E.N. Youssef, C. Darej, M. Chakroun and H. Ben Salem. 2015. Effect of stocking rate on biomass variation and lamb performances for barley stubble in Tunisian semiarid region and under conservation agriculture conditions. *African Journal of Agricultural Research* 10(50):4584–4590.

Moussa-Machraoui, S., F. Ben, Errouissi, M. Ben-Hammouda and S. Nouira. 2010. Comparative effects of conventional and no-tillage management on some soil properties under Mediterranean semi-arid conditions in northwestern Tunisia. *Soil and Tillage Research* 106(2):247–253.

Moussadek, R. 2012. *Impacts de l'agriculture de conservation sur les propriétés et la productivité des vertisols du Maroc Central*. PhD Diss. Ghent University.

Moussadek, R., R. Mrabet, R. Dahan, A. Zouahri, M. El Mourid and E.V. Ranst. 2014. Tillage system affects soil organic carbon storage and quality in Central Morocco. *Applied and Environmental Soil Science*, 8 pages:Article ID 654796. doi:10.1155/2014/654796.

Moussadek, R., H. Iaaich, R. Mrabet, R. Dahan and M. El Mourid. 2016. *Land Suitability to Conservation Agriculture in Central Morocco*. Morocco Collaborative Grant Program report. INRA-ICARDA. 157p.

Moussadek, R., R. Mrabet, P. Zante, et al. 2011a. Impact of tillage and residue management on the soil properties and water erosion of a Mediterranean Vertisol. *Canadian Journal of Soil Science* 91(4):627–635.

Moussadek, R., R. Mrabet, R. Dahan, et al. 2011b. Effect of tillage practices on the soil carbon dioxide flux during fall and spring seasons in a Mediterranean vertisol. *Journal of Soil Science and Environmental Management* 2(11):362–369.

Moussadek, R. and R. Mrabet 2017. Carbon management and sequestration in dryland soils of Morocco: Nexus approach. Global Symposium on Soil Organic Carbon, 2017. Food and Agriculture Organization of the United Nations, Rome, Italy, pp. 497–500.

Mrabet, R. 2018. Information and data collections for preparation of synthesis report of the general status of adoption of conservation agriculture (CA) in Africa Towards 2ACCA: Case of Morocco. 3p. Second African Congress on Conservation Agriculture. 9–12 October 2018. Johannesburg, South Africa.

Mrabet, R., R. Moussadek, A. Fadlaoui and E. Van Ranst. 2012. Conservation agriculture in dry areas of Morocco. *Field Crops Research* 132:84–94.

Mrabet, R. 2011a. No-tillage agriculture in West Asia & North Africa. In P.G. Tow, I.M. Cooper, I. Partridge and C.J. Birch (eds.), *Rainfed Farming Systems*. Springer, Dordrecht Netherlands. ISBN: 978-1-4020-9131-5, pp. 1015–1042.

Mrabet, R. 2011b. Effects of residue management and cropping systems on wheat yield stability in a semiarid Mediterranean clay soil. *American Journal of Plant Sciences* 2(2):202–216.

Mrabet, R. 2008. *No-Tillage Systems for Sustainable Dryland Agriculture in Morocco*. Institut National de la Recherche Agronomique (INRA). 153pp.

Mrabet, R., S. Lahlou, Y. Le Bissonnais and O. Duval. 2004. Estimation de la stabilité structurale des sols semi-arides marocains: Influence des techniques culturales simplifiées. *Bull. du Réseau Erosion* 23:405–415.

Mrabet, R., A. El-Brahli, F. Bessam and I. Anibat. 2003. No-tillage technology: Research review of impacts on soil quality and wheat production in semiarid Morocco. *Options Méditerranéennes* 60:133–138.

Mrabet, R. 2002a. Wheat yield and water use efficiency under contrasting residue and tillage management systems in a semiarid area of Morocco. *Experimental Agriculture* 38(2):237–248.

Mrabet, R. 2002b. Stratification of soil aggregation and organic matter under conservation tillage systems in Africa. *Soil and Tillage Research* 66(2):119–128.

Mrabet, R., N. Saber, A. El-Brahli, S. Lahlou and F. Bessam. 2001a. Total, particulate organic matter and structural stability of a Calcixeroll soil under different wheat rotations and tillage systems in a semiarid area of Morocco. *Soil and Tillage Research* 57(4):225–235.

Mrabet, R., K. Ibno-Namr, F. Bessam and N. Saber. 2001b. Soil chemical quality changes and implications for fertilizer management after 11 years of no-tillage wheat production systems in semiarid Morocco. *Land Degradation and Development* 12(6):505–517.

Mrabet, R. 1997. *Crop Residue Management and Tillage Systems for Water Conservation in a Semiarid Area of Morocco*. PhD Diss. Colorado State Univ., Fort Collins, CO. USA.

Ogle, S.M., C. Alsaker, J. Baldock, et al. 2019. Climate and soil characteristics determine where no-till management can store carbon in soils and mitigate greenhouse gas emissions. *Scientific Reports* 9(1):11665. doi:10.1038/s41598-019-47861-7.

Paustian, K., J. Lehmann, S. Ogle, D. Reay, G.P. Robertson and P. Smith. 2016. Climate-smart soils. *Nature* 532(7597):49–57. doi:10.1038/nature17174.

Pittelkow, C.M., B.A. Linquist, M.E. Lundy, et al. 2015. When does no-till yield more? A global meta-analysis. *Field Crops Research* 183:156–168.

Rezgui, M., M. Mechri and A. Gharbi. 2014. Effet du travail du sol sur les propriétés physiques du sol et sur le rendement de la fèverole et du blé dur cultivés sous les conditions semi-arides du Kef. *Annales de l'INRAT* 87:1–10.

Richard, F. 2005. Le Semis Direct en Tunisie. La Gazette des SCV au Cirad.

Ranaivoson, L., K. Naudin, A. Ripoche, F. Affholder, L. Rabeharisoa and M. Corbeels. 2017. Agroecological functions of crop residues under conservation agriculture. A review. *Agronomy for Sustainable Development* 237(4):26.

Rouabhi, A., A. Laouar, A. Mekhlouk and B. Dhehibi. 2019. Socioeconomic assessment of no-till in wheat cropping system: A case study in Algeria. *New Medit* 18(1):53–64.

Rouabhi, A., B. Dhehibi, A. Laouar, M. Houmoura and F. Sebaoune. 2016. Adoption perspectives of direct seeding in the high plains of Sétif-Algeria. *Journal of Agriculture and Environmental Sciences* 5:53–64.

Ryan, J., E. de Pauw, H. Gomez and R. Mrabet 2006. Drylands of the Mediterranean zone: Biophysical resources and cropping systems. In: G.A. Peterson, et al. (Eds), *Dryland Agriculture* (2nd edition), American Society of Agronomy Monograph N° 23, pp.577–624. LCCCN: 2006922050

Ryan, J. 2011. Rainfed farming systems in the West Asia & North Africa (WANA) region. In: P.G. Tow, I.M. Cooper, I.M., I. Partridge, I. and C.J. Birch (eds.), *Rainfed Farming Systems*. Springer, Dordrecht, Netherlands, ISBN: 978-1-4020-9131-5, pp.65–393.

Saber, N. and R. Mrabet. 2002. Impact of no-tillage and crop sequence on selected soil quality attributes of a vertic calcixeroll soil in Morocco. *Agronomie* 22(5):451–459.

Sanz-Cobeña, A., L. Lassaletta, E. Aguilera, et al. 2017. Strategies for greenhouse gas emissions mitigation in Mediterranean agriculture: A review. *Agriculture, Ecosystems and Environment* 238:5–24.

Siddique, K.H.M., C. Johansen, N.C. Turner, et al. 2012. Innovations in agronomy for food legumes. A review. *Agronomy for Sustainable Development* 32(1):45–64.

Smith, P., K. Calvin, J. Nkem, et al. 2020. Which practices co-deliver food security, climate change mitigation and adaptation, and combat land-degradation and desertification? *Global Change Biology* 26(3):1532–1575. doi:10.1111/gcb.14878.

Souissi, A., H. Bahri, H. Cheikh Mhamed, et al. 2020. Effect of tillage, previous crop and N fertilization on agronomic and economic performances of durum wheat (*Triticum durum* Desf.) under rainfed semi-arid environment. *Agronomy*, August 2020. 10:1161. doi:10.3390/agronomy10081161.

Souissi, A., H. Bahri, H. Cheikh M'hamed and M. Annabi. 2018. Une méta-analyse sur les effets de la fertilisation azotée sur le rendement et l'efficience de l'utilisation de l'azote chez le blé dur en Tunisie. *Annales de l'INRAT* 91:62–85.

Sun, W., J.G. Canadell, L. Yu, et al. 2020. Climate drives global soil carbon sequestration and crop yield changes under conservation agriculture. *Global Change Biology* 26(6):3325–3335. doi:10.1111/gcb.15001.

The Montpellier Panel. 2013. Sustainable Intensification: A New Paradigm for African Agriculture. London Agric. Impact, 36. https://workspace.imperial.ac.uk/africanagriculturaldevelopment/Public/Montpellier Panel Report 2013 - Sustainable Intensification - A New Paradigm for African Agriculture.pdf.

Vanlauwe, B., J. Wendt, K.E. Giller, B. Gerard and C. Nolte. 2014. A fourth principle is required to define Conservation Agriculture in sub-Saharan Africa: The appropriate use of fertilizer to enhance crop productivity. *Field Crops Research* 155:10–13.

World Bank 2014. *Réalisation de l'étude sur le semis direct dans le cadre du projet d'intégration du changement climatique dans la mise en œuvre du plan Maroc Vert (PICCPMV). Phase 2 : Feuille de route pour le développement et promotion du semis direct au niveau des exploitations agricoles.* Document n° : 943-N711-14b. The World Bank, Washington, DC, USA.

World Bank 2019. *The World Bank. Sustainable Land Management and Restoration in the Middle East and North Africa Region-Issues, Challenges, and Recommendations.* Washington, DC.

Zaghouane, O., Z. Abdellaoui et D. Houassine. 2006. Quelles perspectives pour l'agriculture de conservation dans les zones céréalières en conditions algériennes ? *Options Méditerranéennes : Série A* 69:183–187.

15 No-Till Farming for Managing Soil Organic Matter in Semiarid, Temperate Regions

Synergies, Tradeoffs, and Knowledge Gaps

Qiuping Peng and David R. Huggins

CONTENTS

DOI: 10.1201/9781003102762-15

ABBREVIATION LIST

AEC: agroecological class
ASP: aspirational
BAU: business as usual
C: carbon
CA: conservation agriculture
CAF: Cook Agronomy Farm
CMI: carbon management index
CRI: carbon rejuvenate index
CT: conventional tillage
DIN: dissolved inorganic nitrogen
DOC: dissolved organic carbon
DOM: dissolved organic matter
DON: dissolved organic nitrogen
DRP: dissolved reactive phosphorus
EC: electrical conductivity
ES: ecosystem service
EXP: export of carbon in harvested biomass
FAO: food and agriculture organization of the United Nations
GHG: greenhouse gas
GPP: gross primary production
GWP: global warming potential
IMP: import via seeding in C budget
IPCC: Intergovernmental Panel on Climate Change
iPNW: inland Pacific Northwest
ITSP: Intergovernmental Technical Panel on Soils
K: potassium
LTAR: Long-Term Agroecosystem Research
MBC: microbial biomass carbon
MRT: mean residence time
N: nitrogen
NBI: nitrogen balance index
NEE: net ecosystem exchange C flux.
NP: native prairie
NPP: net primary production
NT: no-tillage
P: phosphorus
PA: precision agriculture
PM: particulate matter
POC: particulate organic carbon
POXC: permanganate oxidizable carbon
Ra: autotrophic respiration

Reco: ecosystem respiration
Rh: heterotrophic respiration
SC: static chamber
SIC: soil inorganic carbon
SL: shallow lysimeters
SOC: soil organic carbon
SOM: soil organic matter

15.1 INTRODUCTION

15.1.1 Semiarid, Temperate Regions of the World

Semiarid regions cover 17.7% (about 2,625 Mha) of the earth's land area with annual precipitation ranging from 200 mm to 700 mm (Kašanin-Grubin et al. 2018) and aridity indices (a ratio of total annual precipitation to potential evapotranspiration) between 0.20 and 0.50 (Lal 2004a). Based on the Köppen-Geiger climate classification (Beck et al. 2018), semiarid temperate regions are mainly located in eastern and western Europe, northeast China, the Great Plains region of North America, semiarid pampas in South America, southern Africa, and the Mediterranean areas of Europe, North America, and Australia (Garcia-Franco et al. 2018; Beck et al. 2018) and supported about 14.4% of the world's population in the year 2000 (Hassan et al. 2005).

Global maps of cropland distribution (Thenkabail et al. 2012, https://www.croplands.org/app/map?lat=36.31513&lng=-271.23046875000006&zoom=3) and the Köppen-Geiger climate classification (Beck et al. 2018) indicate that semiarid, temperate regions of Europe, Australia, and New Zealand primarily produce wheat (*Triticum aestivum L.*) and barley (*Hordeum vulgare*) along with other mixed crops of corn (*Zea mays L.*), rice (*Oryza sativa*), and soybean (*Glycine max*); while in northeastern China, mixed crops dominate. In the semiarid regions of North America, corn and soybean are the main crops in the southern Great Plains and are primarily replaced by wheat and barley in the northern Great Plains of the US and Canada (Thenkabail et al. 2012).

Many arid and semiarid regions have become drier as observed from 1860 to 2000 (McMichael et al. 2004) and are projected to be hotter and drier in terms of monthly mean air temperature and precipitation, particularly in the coastal areas of countries with Mediterranean climates (Beck et al. 2018; Ragab and Prudhomme 2002). An exception is in the North American semiarid regions where historical climate data (1948–2008) indicates that newly formed semiarid regions have developed as arid regions became wetter (Huang et al. 2016). Here, precipitation increases have continued as projected, coupled with increasing temperature (Beck et al. 2018).

Conservation agriculture (CA) has been developed to address soil erosion and effective water use by minimizing soil disturbance, maintaining soil surface cover, and diversifying cropping systems (Kassam et al. 2015). Globally, CA is practiced on about 180 M ha of cropland—about 12.5% of cropland in 2016—with the largest extents of adoption in South and North America (Kassam et al. 2019). In recent decades, CA has been adopted over approximately 35.6 M ha in the US, with half of the area in continuous no-tillage (NT), and accounting for 10–12% of US cropland (Kassam et al. 2015).

15.1.2 Objectives

The over-arching question to be addressed in this chapter is whether NT agriculture can meet the broad demands of sustainable intensification (Kleinman et al. 2018) and provide multiple, diverse ecosystem services (ES). Inevitably, as no agricultural system is perfect, there will be synergies, tradeoffs, and knowledge gaps associated with various NT performance outcomes. Importantly, NT addresses soil erosion and significantly curtails the foremost soil degradation process of the world (FAO 2015). Adoption of NT, however, also changes multiple soil properties, impacts many ongoing

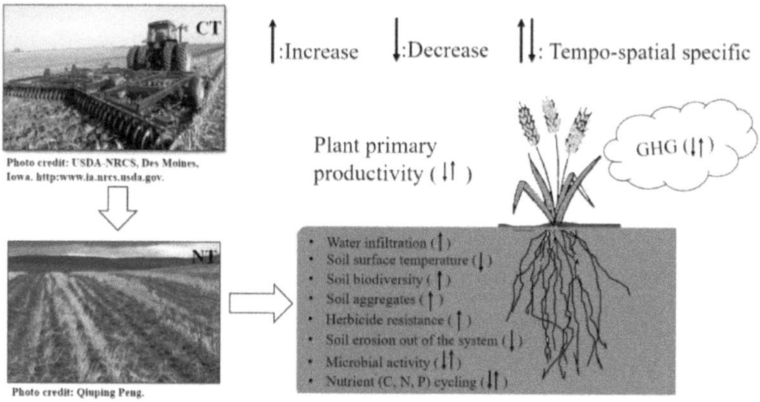

FIGURE 15.1 Response of soil physical, chemical, and biological properties, plant productivity, and greenhouse gas emission to no-tillage (NT) after conversion from conventional tillage (CT).

processes in the soil-plant-air continuum (Figure 15.1), and has many socioeconomic and biophysical challenges (Huggins and Reganold 2008). Complex feedback loops affect typical ESs outcomes, including provisioning, regulating, supporting and cultural services under NT (FAO, 2015).

What are the broader, collective effects of NT agriculture considering sustainable intensification goals? Numerous review papers (meta-analyses) have globally or regionally assessed NT impacts on specific soil properties (e.g. soil organic carbon (SOC), microbial diversity, phosphorus) (Luo et al. 2010; Daryanto et al. 2017; Sun et al. 2020; Ogle et al. 2019; Wang and Zou 2020) and crop yield (Pittelkow et al. 2015; Sun et al. 2020). These studies provide important insights toward evaluating NT benefits, tradeoffs, and knowledge gaps. Specific objectives of this chapter are: (1) to further the synthesis of previous findings regarding NT performance in semiarid and temperate regions; and (2) to illustrate the value of long-term, site-specific research that pursues a broader evaluation of NT performance outcomes. This chapter is based on the data obtained from the R.J. Cook Agronomy Farm (CAF) (46.78°N, 117.09°W), one of 18 USDA Long-Term Agroecosystem Research (LTAR) sites. As a case study, it is limited with respect to fully assessing multiple, diverse ESs, but it is still compelling by way of illustrating the diverse, unique, site-specific nature of soil, water, air, and crop outcomes under NT in the semiarid region of the inland Pacific Northwest (Figure 15.2). Furthermore, we seek to illustrate the value of long-term, more transdisciplinary research in furthering a more comprehensive understanding of the complexity underlying the answer to a seemingly simple question: can NT meet the expectations of sustainable intensification?

15.2 GENERAL THREATS TO PRODUCTIVITY, SOIL, WATER, AIR, AND ASSOCIATED KNOWLEDGE GAPS IN SEMIARID AGRICULTURE

15.2.1 Crop Yield under No-Tillage Systems

Pittelkow and coworkers (2015) reported that crop yield response to NT is primarily influenced by crop category, followed by the aridity index, residue management, NT duration, and N fertilizer rate. Here, NT reduced yield by 5.1% on average across 50 crops; however, NT performed best under rainfed conditions in dry climates producing equal or higher yield than conventional tillage (CT). Likewise, Sun et al. (2020) also found that arid regions—where the humidity index <40—can benefit the most from NT with increased crop yield and enhanced SOC sequestration relative to local CT. Semiarid to humid regions, however, are likely to increase SOC under NT with no effect on crop yield, while cold humid and tropical humid regions experience yield losses without positive

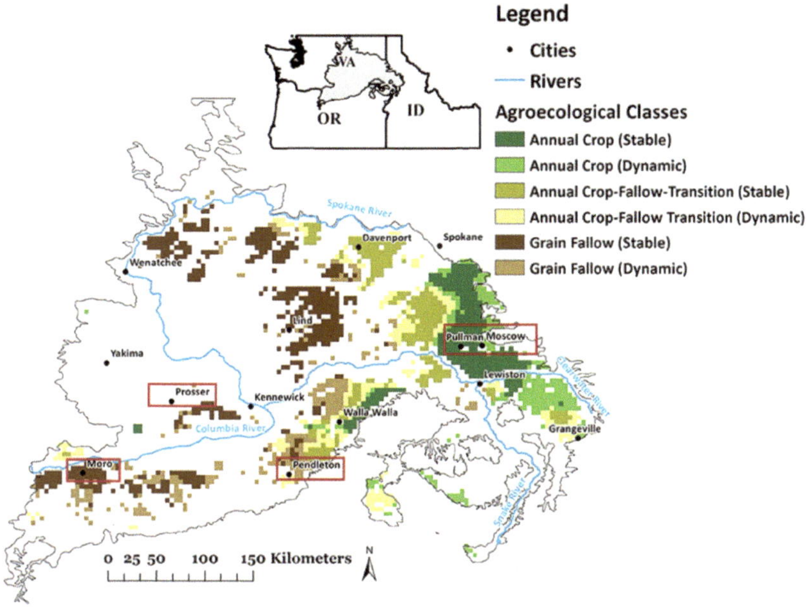

FIGURE 15.2 Agroecological classes (AECs) of years 2007–2014 at 4-km resolution. The sites mentioned in the case study session are highlighted with red boxes (Mapsource: Kaur et al. 2017; reused with permission).

effects on SOC under NT. These insightful findings summarized from two meta-analyses using thousands of paired observations globally, also acknowledged the importance of regional climate, site-specific soil conditions and management influences on both yield and SOC sequestration under NT (Pittelkow et al. 2015; Sun et al. 2020).

15.2.2 Soil Organic Carbon

Most of the world's soil resources are in only fair, poor, or very poor condition and will likely continue to degrade unless concerted actions are taken (FAO 2015). The most serious causes of degradation that threaten global soil function are soil erosion, SOC decline, and nutrient imbalance (nutrient excess or deficiency) (FAO 2015), and these are often interrelated. Other degradation processes of concern are soil acidification, soil contamination, waterlogging, soil compaction, soil sealing, salinization, and loss of soil biodiversity (FAO 2015; FAO and ITPS 2020; IPCC 2020). Soil degradation constitutes a quiet global crisis, driven by human action, where depletion of natural capital diminishes the capacity to generate ESs and agroecosystem resilience. Ultimately, soil degradation is symptomatic and represents a failure of adaptive capacity, the ability to manage resilience in the face of system stresses, uncertainties, and unprecedented change. From an NT farming perspective, soil erosion, SOC decline, nutrient imbalance, soil acidification, soil contamination, and loss of soil biodiversity are arguably most relevant.

Croplands in semiarid, temperate regions were primarily derived from conversion of native prairie over the past 300 years (Clay et al. 2014). It is well known that cultivation of native systems leads to SOC loss, with most losses observed in the topsoil (Cochran et al. 2007; Deng et al. 2016; Guo and Gifford 2002; Purakayastha et al. 2008; Wang et al. 2008; Wiesmeier et al. 2014; Yan et al. 2012). Guo and Gifford (2002) compiled 74 worldwide investigations of land-use change impacts on SOC stocks and reported average declines of 42% for native forest to cropland conversions and 59% SOC reductions following conversion of pastureland to cropland. The greatest SOC loss (78%)

was found in areas with an annual precipitation of 400–500 mm (semiarid), while the magnitude of cumulative SOC loss peaked (85%) 30–50 years following conversion from pasture to crop land (Guo and Gifford 2002). In terms of SOC loss rate, 0.53–0.84 Mg C ha^{-1} yr^{-1} were reported after conversion from native prairie to semiarid agricultural land in the inland Pacific Northwest (iPNW) (Brown and Huggins 2012), close to the global average of 0.89 Mg C ha^{-1} yr^{-1} reported for converting grassland to farmland (Deng et al. 2016). The rapid decline of SOC is associated with: (1) less carbon input from crop biomass compared to native vegetation (Huggins et al. 1998); and (2) tillage disturbance that accelerates soil organic matter decomposition and promotes the physical export of SOC from cultivated fields by erosion processes (Post and Kwon 2000).

No-tillage can contribute to agricultural sustainability by controlling soil erosion while, and in many cases, benefiting SOC sequestration and crop yield (Lal et al. 1999; West and Post 2002). The accrual of SOC in NT has been attributed to decreases in decomposition related to reduced disturbance and lower temperatures (Janzen et al. 1998), physical protection from increased formation of soil aggregates (Beare et al. 1994; Bossuyt et al. 2002; Sithole et al. 2019), C input increases from greater crop productivity (Ogle et al. 2012), and decreased physical transport of soil C from the field as tillage is reduced and erosion curtailed (Schuller et al. 2007; Williams et al. 2009; Huggins et al. 2014). Recent studies, however, have found inconsistent changes in SOC stocks at different soil profile depths following conversion from tillage-based systems to NT (Ogle, Swan, and Paustian 2012; Luo, Wang, and Sun 2010; Baker et al. 2007; Kuhn et al. 2016; Sun et al. 2020). For instance, Luo et al. (2010) concluded, from a 69-paired experiment meta-analysis, that adoption of NT compared to conventional tillage (CT) resulted in a significant increase in SOC stocks in the surface at 0–10 cm, a decline in SOC from 20 cm to 40 cm, no significant change in SOC for soil below 40 cm, and overall SOC stocks in most cases either stable or with insignificant increases. Differences of SOC between NT and CT were also reported to decrease with soil depth (Sun et al. 2020) and to have largely occurred around 10 years of NT (Kuhn et al. 2016). Discrepancies in SOC stock changes due to a tillage regime can be attributed to different sampling depths, SOC concentration and bulk density measurement procedures, stock calculations via equal sampling depth or equivalent soil mass (Wendt and Hauser, 2013), duration of NT adoption (Luo et al. 2010; Kuhn et al. 2016), climate condition, and soil type (Sun et al. 2020, Ogle et al. 2019). Overall, the lack of subsoil SOC data compared to topsoil data is a serious knowledge gap for understanding and assessing SOC sequestration potential of agricultural systems (Lorenz and Lal 2005; Chirinda et al. 2014; Angst et al. 2018). In addition, from a SOC sequestration perspective, the fate of SOC that is exported from fields via erosion processes compared to SOC remaining in place is largely unknown, although degradation of soil via surface water by erosion and carbon-enriched sediment clearly does not meet broader sustainable agricultural goals.

Increases in SOC under NT primarily occur in "young" fractions (Janzen et al. 1998) and are subject to loss with a return to greater soil disturbance (VandenBygaart and Kay 2004). For instance, VandenBygaart and Kay (2004) found that about 3 Mg C ha^{-1} was lost within 18 months after moldboard plowing a sandy loam soil, which accounted for two-thirds of the SOC gain after 22 years of NT. The increase of "young" and labile carbon after NT management is also supported by studies investigating various fractions of SOC. For example, Awale et al. (2013) found that, compared to CT, NT had significantly greater SOC concentrations, coarse particulate organic carbon (POC), permanganate oxidizable carbon (POXC), and 30-day cumulative mineralizable carbon in a silty clay soil in North Dakota. Kennedy and Schillinger (2006) found that NT contained 5.5–9.8% more soil aggregates >1000 μm, 30% greater SOC at toeslope positions, and more fungal biomarkers in NT than CT in the semiarid Palouse region of the iPNW. Purakayastha et al. (2008) also found that NT increased the SOC concentration, POC, and microbial biomass carbon (MBC) compared to CT in the Palouse region. Regardless of methods used for fractionating SOC, carbon in larger-sized aggregates, POC, MBC, POXC, and total mineralized carbon all assess relatively labile SOC constituents. Further research is required on the persistence of these labile SOC constituents under modified soil disturbance regimes, climate change, increased biomass harvest, and other factors from a total soil profile perspective.

15.2.3 Soil Erosion

Semiarid regions susceptible to water and wind erosion are 213.2 M ha and 150.2 M ha on a global basis, and 24.4 M ha and 27.3 M ha, respectively, are distributed in North America (Reich et al. 1999). Global maps of predicted vulnerability to wind and water erosion indicate that semiarid temperate zones of the Great Plains region of North America have no less than moderate levels of erosion risk, while water erosion risks in Mediterranean areas are much higher than in the Great Plains region (Reich et al. 1999). Soil erosion processes related to heterogeneous topography and tillage have resulted in substantial soil loss (Busacca et al. 1993; Reganold et al. 1987; Van Oost et al. 2005) and SOC redistribution. Globally, erosion processes in cropland transport SOC ranging from 0.47 Pg C yr^{-1} to 0.61 Pg C yr^{-1} (Van Oost et al. 2007). Soil erosion from tillage operations (tillage erosion) can be the same order of magnitude or higher than erosion rates from water. Lindstrom et al. (1992) reported that moldboard plowing transported an estimated 30 Mg ha^{-1} yr^{-1} of soil from upland convex slope positions and exposed lighter colored, lower SOC subsoils on many landscapes and is similar to reported water erosion soil loss rates of 31 Mg ha^{-1} yr^{-1} (Busacca et al. 1993) and 17.0±18.9 Mg ha^{-1} yr^{-1} (Li et al. 2018). While tillage erosion from primary and secondary tillage operations have been documented (Van Muysen and Govers 2002; Van Muysen et al. 2006; Li et al. 2007), soil displacement and erosion caused by high disturbance NT drill operations in conservation management systems remains largely unquantified (Zhang et al. 2009).

In semiarid regions of the iPNW, where precipitation is less than 300 mm yr^{-1}, annual fallow is a common practice used to recharge available soil water and stabilize winter wheat yield (Schillinger and Bolton 1993; Pannkuk et al. 1997). Airborne particulate matter (PM), however, derived from wind-driven erosion of the unprotected soil surface during fallow periods poses public health and safety issues and degrades soil productivity (Saxton et al. 2001; Sharratt and Schillinger 2005). An NT cropping system would help reduce PM10 and PM2.5 (inhalable particles with an aerodynamic diameter ≤10 µm and 2.5 µm, respectively) by minimizing soil disturbance and leaving the soil surface covered by crop residues. Among nine cropland covers, soil and PM10 loss in the fallow/idle cropland is at least 1.2 and at most 20 times greater compared to cropland having surface residue cover (Pi et al. 2020a). Additionally, Pi et al. (2020b) compared the effect of the standing residue of major commodity crops on wind erosion in the iPNW and found that in 110, 20, and 16 standing residue elements m^{-1} of row for winter wheat, spring canola (*Brassica napus*), and chickpea (*Cicer arietinum*), soil loss was reduced by 73.3, 53.4, and 60.9%, respectively, compared with a bare soil surface without residue.

15.2.4 Water Quality Management and "Leaky" Agricultural Systems

Sediment accumulation in waterbodies from surface runoff and soil erosion, is agriculture's primary pollutant (Dabney et al. 2001). Sediment transport by global rivers as a consequence of soil erosion has been estimated at 2.3±0.6 billion metric tons (Gt) per year (Syvitski et al. 2005). The fate of soil and SOC that is redistributed within fields or exported to surface waters under NT as well as other soil disturbance practices are largely unknown and remain a significant knowledge gap. Approximately 0.4–0.6 Gt of carbon was deposited annually into aquatic ecosystems and buried in depressional sites across the world (Lal 2004b). The eroded sediments are also enriched with essential crop nutrients (e.g. N, P, K, and S), thereby negatively impacting the economy and the environment (Bashagaluke et al. 2018). For example, a recent field study reported losses of N, P, and potassium (K) due to erosion processes of 34, 12, and 13 kg ha^{-1}, respectively, with a monetary value over three seasons ranging from 31 US$ ha^{-1} to 67 US$ ha^{-1} (Bashagaluke et al. 2018).

NT can significantly reduce sediment and associated nutrient transport and delivery to aquatic systems by reducing surface runoff (Williams et al. 2009; DeLaune and Sij 2012). Increased water infiltration under NT, however, can result in greater transport of water and soluble agrichemicals to groundwater and/or surface waters via subsurface artificial drains (Bakhsh et al. 2002; Bjorneberg et al. 1996; Jenkins et al. 2009). In addition, decomposition of crop residues and animal manures that remain at or near the soil surface under NT can be sources of nutrient pollution via surface runoff to adjacent waters (Kleinman et al. 2009). In fact, in 1974 about 17.32 M ha farmland in the US has been drained (Jaynes and James 2007) to decrease the potential for surface waterlogging related with agricultural traffic compaction as well as for improving the timeliness of field operations to maintain productivity. After converting to NT, subsurface drainage can also have unintended consequences, such as accelerating decomposition of soil organic matter (SOM) and translocation of dissolved carbon (inorganic and organic) and other soluble soil constituents with increased water infiltration (Bellmore et al. 2015; Manninen et al. 2018).

Reliance of NT on agrichemicals (Huggins and Reganold 2008) for augmenting crop nutrition and managing weed, insect, and soil-borne disease can have unintended consequences on water quality. In semiarid agricultural regions of Argentina, where more than 91% of lands are under NT, Mas et al. (2020) found 20 out of 30 herbicides and four secondary metabolites that had a detection frequency of more than 10% out of the 353 water samples collected from cisterns, wells, and dams whereas, for some compounds (e.g. atrazine, hydroxyatrazine), the detection frequency was up to 99%. Here, 73% of the samples exceeded the limit of 0.5 $\mu g L^{-1}$ established by the European Union for the sum of organic contaminants, though almost all measured values were below the US Environmental Protection Agency (EPA) limits. While shifting to NT may not necessarily increase the amount of herbicide used compared to CT (Friedrich 2005), associated changes in a hydrologic regime and other system properties justify the need for further research on agrichemical persistence and its fate, in watersheds and the food system, for ecological and human health.

The development of irrigated agriculture in arid and semiarid regions poses a threat to environmental and agricultural sustainability due to induced waterlogging and soil salinization (Singh 2015). More than one-third of the world's irrigated land is estimated to have soil waterlogging, compaction, and/or associated salinization issues (Heuperman et al. 2002; Singh 2015). The combined negative influence on root development, crop growth, and yield from waterlogging and soil salinization (Singh 2015; Saqib et al. 2004; Datta and De Jong 2002; de San Celedonio et al. 2016) creates a challenge for future crop options and management in semiarid temperate regions. Salts and NO_3-N resulting from anthropogenic sources have degraded the groundwater quality of the Ogallala aquifer in the Great Plains (Mehta et al. 2000; Chaudhuri and Ale 2014). Furthermore, depletion of aquifers and conversion from irrigated to dryland farming are current challenges in semiarid regions. For example, depletion of the Ogallala aquifer in the Texas high plains decreased irrigated cropland from 697,410 ha in 1979 to 576, 870 ha in 1989 and resulted in conversions to dryland cropping systems (Terrell et al. 2002). Consequently, management practices that improve soil health and increase water use efficiency, such as the incorporation of reduced tillage or NT, while transitioning to dryland agricultural systems are encouraged as the aquifer supply continues to decrease (Cano et al. 2018).

Soil salination is prevalent in arid and semiarid regions and saline soils are estimated at 397 M ha of global land area (Koohafkan and Stewart 2008). Application of water containing salts, weathering of minerals in soils, organic matter decline, and inefficient water use by crops are major factors promoting the accumulation of salts in the soil (Miller et al. 1981; Jordán et al. 2004). For instance, "saline seep" in the North American Great Plains in the early 1960s, frequently occurred in dryland, crop-fallow, farming systems following conversion of native prairie as unconsumed precipitation led to greater soluble-salt concentrations in root zones and lower landscape positions (Miller et al. 1981; Halvorson 1988). Solutions include improving soil drainage, and increasing consumptive water use by eliminating fallow and intensifying the cropping system (Halvorson 1988).

15.2.5 Soil Acidification and Soil Inorganic Carbon

Globally, acid soil (soil pH <5.5) is widespread and occurs on approximately 30% of the total ice-free land area, varying from 3,777 M ha to 3,950 M ha (Sumner and Noble 2003). The main areas of acid soil are in humid northern temperate and tropic zones with high precipitation, whereas alkaline soils are commonly found in arid climates (FAO/UNESCO 2007, https://data.apps.fao.org/map/catalog/srv/eng/catalog.search#/metadata/3f9be529-5672-4286-86c7-a63c7da39853; Slessarev et al. 2016). The degree of soil acidity varies depending on factors such as parent material, climate, and SOM. In particular, water balance impacts soil pH values with an abrupt transition of soil pH from alkaline to acid occurring when mean annual precipitation exceeds mean annual potential evapotranspiration and augments leaching losses (Slessarev et al. 2016). In agricultural lands, anthropogenic activities have enhanced soil acidification through intensive removal of aboveground biomass, application of synthetic N fertilizers, and increasing deposition of atmospheric N and S (Pennock et al. 2015). Furthermore, many agricultural systems have less annual evapotranspiration and are far "leakier" than the prior native vegetation which promotes leaching of soluble bases (Ca^{2+}, Mg^{2+}, K^{+}) and nitrate-N (Wilcke and Lilienfein 2005). The collective consequence is increased concern for accelerated soil acidification in semiarid regions (McFarland and Huggins 2015; Jones et al. 2019).

Related to soil acidification of semiarid regions is the fate of soil inorganic carbon (SIC), particularly in estimations of soil carbon (both organic and inorganic) sequestration. Globally, soils are estimated to contain 950 Gt of inorganic carbon in the upper one meter with significant amounts distributed in arid and semiarid regions (Lal 2004b; Duniway et al. 2010). In addition to SIC losses driven by soil acidification from synthetic N fertilizer use, soil water dynamics are crucial to changes in SIC stocks, influencing carbonate leaching and transport of solutes (e.g. protons, nitrates, etc.) that facilitate soil carbonate weathering and losses of SIC to the atmosphere as CO_2. For example, approximately 51% of the surface soil of cultivated land in China has experienced SIC loss, with the most significant losses in dryland farmed, irrigated, and paddy soils (Wu et al. 2009).

No-tillage and residue retention (Dao 1993; Fuentes et al. 2004; TerAvest et al. 2015), often coupled with increased earthworm populations (Willoughby et al. 1997; Willoughby and Kladivko 2002), can increase water filtration compared to CT. Increased water infiltration and preferential flow under NT could adversely impact the soil's pH and SIC dynamics in topsoil and subsoil. For instance, declines in soil pH in deep soil layers (100-200 cm) were reported to be significantly greater than upper soil layers (0–80 cm) in a long-term (12 years) N fertilizer application dryland system under tillage, as the SOC and SIC in calcareous soil profiles respond to soil acidification quite heterogeneously (Jin et al. 2018). Besides the downward movement of proton reaction with SIC resulting in SIC loss, the downward movement of Ca^{2+} and Mg^{2+}, can also lead to an increase of carbonate formation during dry periods. Soil profile pH in semiarid regions is therefore a complex balance, often governed by factors influencing carbonate weathering or formation. Overall, agricultural activities, such as NT, that increase infiltration and the potential leaching and loss of soluble ions (e.g. nitrate) will lead to continued long-term soil acidification and associated contributions to the atmospheric CO_2 of semiarid regions. Addressing soil acidification that has moved beyond the tillage zone and deeper into the soil profile can also be difficult to resolve. Consequently, developing preventive strategies by coupling NT with agricultural intensification that increases overall consumptive water use (e.g. cover crops, shifts to perennial crops) should be a major research priority for semiarid regions. Worth noting, liming of soils to correct soil acidity is a common, effective practice; however, it will result in CO_2 emissions as $CaCO_3$ reacts with protons (Majumder et al. 2018). But the total emission of greenhouse gases (GHGs) in terms of CO_2-equivalent may not necessarily increase due to a reduction in N_2O emissions associated with pH improvement (Majumder et al. 2018). Consequently, NT systems in semiarid regions that are "leaky" and reliant on synthetic N fertilizer warrant further long-term research to assess acidification, impacts on stocks of SIC and SOC in subsoil, liming practices, and associated GHG emissions.

15.2.6 Soil Biodiversity

Increases in soil fauna under NT compared to CT have often been reported, especially the density of earthworms, micro-anthropods, and microbial biomass (Adl et al. 2006; Arai et al. 2018; Edwards et al. 1993; House and Parmelee 1985; Johnson-Maynardet al. 2007; Kladivko et al. 1997). For example, yearly increases in soil microbial biomass, nematode abundance, mite diversity, abundance of protozoa (e.g. *Gymnamoebae and Testacea*) were reported to occur under NT, while the abundance of Collembola peaked before 4–5 years (Adl et al. 2006). Sapkota et al. (2012) found a 71% increase of microbial biomass, and a higher abundance and diversity of micro-arthropods, after 15 years of NT compared to CT. Additionally, changes in soil pH induced by NT were found to be important drivers for microbial composition throughout the soil profile (Lewis et al. 2018; Schlatter et al. 2020), with unknown effects on processes, such as C and N cycling.

Shallow dwelling endogeic species of earthworms (e.g. *Aporrectodea trapezoids)* and deep-dwelling anecic species (e.g. *Lumbricus terrestris*) were both observed to increase under NT (Johnson-Maynard et al. 2007, Kladivko et al. 1997). Earthworm and insect feeding habits aid the decomposition of plant materials (Bohlen et al. 1997), while burrowing behavior increases soil aggregate formation and stability and affects C concentration in the aggregates (Fonte et al. 2007), creates localized drilosphere effects on soil C and nutrient concentrations (Shuster et al. 2001; Don et al. 2008; Brown et al. 2000; Andriuzzi et al. 2013), and increases preferential paths for crop roots and the movement of water, nutrients, and pollutants. While it is largely agreed that earthworm activity can transform soil function under NT, quantifying field-scale impacts on crop production, soil formation, and function has remained elusive.

Besides the diversity change in the belowground food web, aboveground biodiversity of pollinators for crops and fruit trees can be a benefit of more diversified conservation farming systems. Pollinators perform a crucial ecological function that supports a significant fraction of agriculture production, as 87 of the leading food crops and 35% of global food production benefit from this ES (Klein et al. 2007). For instance, incorporation of oilseed rape (*Brassica napus*) and field beans (e.g. *Vicia Faba*) via crop rotation or intercropping, can attract flower visitors, such as bumble bees, honey bees, butterflies, hoverflies, and other insects, potentially offsetting the continued decline in abundance and diversity of crop pollinators, both wild and managed (Garratt et al. 2014; Kovács-Hostyánszki et al. 2017; Ollerton 2017). Conservation farming systems with diverse crop rotations can also suppress soil-borne pathogens over time (Peters et al. 2003; Schlatter et al. 2017). Increasing the biodiversity of conservation farming systems to realize multiple ESs is a yet-to-be-realized and quantified potential with many knowledge gaps.

15.2.7 Herbicide Resistance

Glyphosate-resistant crop use has been credited with the wide adoption of conservation tillage in the US, however, herbicide-resistant weeds have become a major agronomic problem (Price et al. 2011; Dentzman and Burke 2021). Globally, the number of unique glyphosate-resistant weeds are led by Australia, Argentina, and the US with 21, 16, and 17, respectively (Heap, I. The International Herbicide-Resistant Weed Database, http://weedscience.org/Pages/GeoChart.aspx). Concern regarding herbicide resistance under NT has driven some farmers to return to tillage-based mechanical weed management, regardless of the benefits NT provides for soil health, water conservation, and effects to mitigate climate change. For example, to address herbicide resistance, periodic moldboard plowing (once every 8–10 years) was introduced in dryland NT systems of semiarid Australia and reduced annual ryegrass seedlings by over 95% compared with other tillage management without harming crop yield (Douglas and Peltzer 2004). As a recent meta-analysis reported, occasional tillage did not affect crop yields, pH, and microbial activity, but did improve erosion resistance and weed control while alleviating soil compaction; but it reduced SOC (Peixoto et al. 2020). Therefore, management of the soil disturbance regime through strategic use of tillage is a complex situation with numerous potential outcomes.

15.2.8 Greenhouse Gas Emissions

Agriculture is a major anthropogenic source of CH_4 and N_2O GHG emissions and is responsible for one-quarter of the world's GHG emissions (Forster et al. 2007; Robertson et al. 2000, https://ourworldindata.org/emissions-by-sector#). In semiarid, dryland agricultural soils, CO_2 and N_2O contribute to GHG emissions, while CH_4 can be a net sink when no livestock are included (Yeboah et al. 2016). Soil biophysical changes (e.g. soil moisture, temperature, microbial activities) under NT (Figure 15.1), combined with the legacy effect of historical managements, can contribute to field spatial and temporal variability and result in disputability and large uncertainty about GHG emissions, C sink, or the source (Lal 2003; Doetterl et al. 2016). The static chamber (SC) method has been widely used for decades to monitor the temporal pattern of GHG emissions, along with the co-located measurements of soil properties and environmental variables, to identify the biophysical drivers of GHG production (Duchemin et al. 1999; Regina and Alakukku 2010; Yu et al. 2013; Collier et al. 2014; Kostyanovsky et al. 2019), but the spatial coverage is quite limited. Long-term monitoring of multiple GHG emissions concurrently covering a larger spatial extent at field scale, however, are only possible with the advanced technology of flux tower-based micrometeorological approaches to address the high degree of spatial and temporal variability in GHG emissions, or "hot spots" and "hot moments" (McClain et al. 2003; Yu et al. 2013; Leon et al. 2014; Waldo et al. 2019). Comparisons of multiple GHG emissions concurrently monitored by eddy covariance (EC) flux tower (Figure 15.3, http://www.cafltar.org/data/) between CT and NT cropping systems at larger spatial scale and longer temporal period incorporating multiple crops seasons are currently quite scarce (Chi et al. 2016; Waldo et al. 2019); therefore, more interdisciplinary research is required to contribute to more accurate carbon and nitrogen monitoring.

FIGURE 15.3 Different systems used to monitor greenhouse gas emission and crop growth at the Cook Agronomy Farm (CAF) Notes: (a) Continuous automated chamber system in CAF-CT in 2015; (b) Eddy covariance flux tower setup in business as usual (BAU) system in 2017; (c) Phonecam picture captured in real time for crop growth monitoring on 2 Apr 2021; (d) Eddy covariance flux tower locations and associated footprint coverage in CAF-CT and CAF-NT for June 2014. The background Google maps during the footprint evaluation period do not represent the real land cover situations (vegetated or bare). (Figures are used with permission).

15.3 CASE STUDY OF COOK AGRONOMY FARM (CAF) AND OTHER SITES IN THE IPNW: TRADEOFFS ABOUT THE SOIL-PLANT-AIR CONTINUUM AND WATER QUALITY UNDER NT

15.3.1 Study Site of CAF

The dryland farming region of the iPNW encompasses 2.2 M ha of eastern Washington, northern Idaho, and northeast Oregon (US) (Figure 15.2). Farmers in the iPNW annually produce over 3.54 Tg (teragram) of wheat that exceeds $500 million in value (reacchpna.org). Imposing projected future climate scenarios (representative concentration pathway, RCP 8.5) onto current cropping systems indicated that a large spatial shift from stable to dynamic agroecological classes (AECs) would occur in iPNW areas (Kaur et al. 2017). Furthermore, greater use of annual fallow—with more soil erosion hazard and less cropping system diversity and intensification that counter agricultural sustainability goals—would be expected and have stimulated long-term research on diversified conservation farming systems (Kaur et al. 2017). In 2012, the USDA established a national Long-Term Agroecosystem Research (LTAR) network to explore sustainable agroecosystem intensification, and the R.J. Cook Agronomy Farm (CAF, 37 ha) (cafltar.org) was one of ten sites selected to initiate the network (Spiegal et al. 2018). Field-scale cropping system research at the CAF was established in 1998 to assess and advance NT farming systems' ability to provide diverse ecosystem services. In 2017, an adjacent field under long-term reduced tillage (since 1998) was paired with the original continuous NT field to establish paired catchments and compare management outcomes of business as usual (BAU, 15 ha), and aspirational practice (ASP). The BAU field is managed using traditional three-year crop rotation (winter wheat-spring wheat-chick pea) under reduced tillage (fall chisel plow following winter wheat, field cultivation in spring prior to sowing spring wheat; otherwise, NT planted for all other crops), with uniform application of agrichemicals. The ASP is continuous NT using precision N management and a more diversified, but still wheat-based, three-year crop rotation that includes winter wheat, spring wheat, chickpea (Figure 15.4a, c), and spring canola (Figure 15.4b, d). A non-aligned, systematic, randomized grid of 369 geo referenced sample locations was established in 1998 on the long-term NT site (Figure 15.5). A low-disturbance, double-disk NT drill (Great Plains®) was used for direct-seeding all crops through the 2008 harvest, afterward a high disturbance hoe-type NT drill (Horsch® with Anderson® openers) was used through 2020. In 2016, a non-aligned, systematic, randomized grid of 151 geo-referenced sample locations was established in the BAU field with associated small catchment (Figure 15.5). Reported here are representative research on crop productivity, N use efficiency (NUE), GHG flux, water quality, soil carbon storage, and soil acidification in CAF and other sites in iPNW (Figure 15.2). More details about the data collection are available in the original papers.

15.3.2 Crop Yield and Nitrogen Balance Index

In 2019, winter wheat yield averaged 6.5 Mg ha^{-1} for the NT ASP and was 34% greater compared to BAU, which averaged 4.8 Mg ha^{-1} (Figure 15.5a). Considerable yield variability typically occurs, here ranging from just under 0.3 Mg ha^{-1} to over 10 Mg ha^{-1} across both BAU and ASP, with low yielding areas associated with eroded upland soils. The N balance index (NBI), calculated as the ratio of N harvested in the grain divided by N applied in the fertilizer, was 33% greater in ASP where it averaged 0.65 compared to 0.49 for BAU (Figure 15.5b). An NBI greater than 0.6 indicates that N uptake efficiency was greater than 50%, which was assumed in an overall NUE goal (Huggins et al. 2010). The greater yield and NBI in ASP compared to BAU indicates the ASP system has the potential to increase crop yield as in other semiarid regions (Pittelkow et al. 2015; Sun et al. 2020) and to reduce negative effects on the environment with respect to N management.

FIGURE 15.4 Current popular cropping rotation systems in iPNW Notes: (a) Chickpea seedlings in spring wheat residue within no-tillage (NT), (b) Canola seedlings in winter wheat residue in NT, (c) Chickpea in bloom (bottom part of the image), (d) Canola in bloom. (All images are credited to Qiuping Peng).

15.3.3 Soil Organic Matter and pH

In 2017, SOM concentration sampled from the surface 0–10 cm at 30 of the geo-referenced points in each catchment ranged from 16 g kg^{-1} to 35 g kg^{-1} in BAU and from 25 g kg^{-1} to 48 g kg^{-1} for ASP (Figure 15.6a). Target SOM from a yield perspective is around 34 g kg^{-1} (2% SOC concentration) (Oldfield et al. 2019). Differences in SOM concentration can be attributed to reduced soil erosion, greater plant biomass C return from increased productivity (Figure 15.6a), and surface stratification effects of the NT ASP compared to BAU. An earlier study based on this field found that the estimated mean soil erosion rate using the Revised Universal Soil Loss Equation (RUSLE) was ten times greater for a CT scenario (5.6–10 Mg ha^{-1} yr^{-1}) than NT (0.5–1Mg ha^{-1} yr^{-1}), using a similar crop rotation (Huggins et al. 2014). The spatial pattern of SOM concentrations is evidence of historical soil erosion impacts that have physically transported SOM to lower field locations and to adjacent surface waters while reducing soil depth and yield potential. Low SOM areas are mainly located on or near summit positions where even under NT ASP, SOM recovery to target levels has not been fully achieved after 19 years of continuous NT (Figure 15.6a).

Acidification of surface (0–10 cm) soil was more prevalent in ASP than BAU (Figure 15.6b). Soil pH (1:1 with water) ranged from 4.5 to 7.5 for the two fields, with more than 30% of ASP areas less than 5.0. The critical soil pH threshold for wheat is 5.2 (Mahler 2002; Ghimire et al. 2017); however, soil pH thresholds were established from a sample depth of 0–30 cm and are likely not applicable to low soil pH that is stratified at the surface. The overall higher soil pH for the BAU than ASP is due to the dilution of acidification effects from mixing of surficial soil with underlying soil with greater pH as a consequence of tillage operations. Further evidence of high subsoil pH is expressed by soil pH greater than 6.5 found in historically eroded field locations of ASP where subsoil with $CaCO_3$ is exposed. The use of synthetic N fertilizers, a major driver of soil acidification in croplands, has significantly contributed to the decline of soil pH from near neutral under native

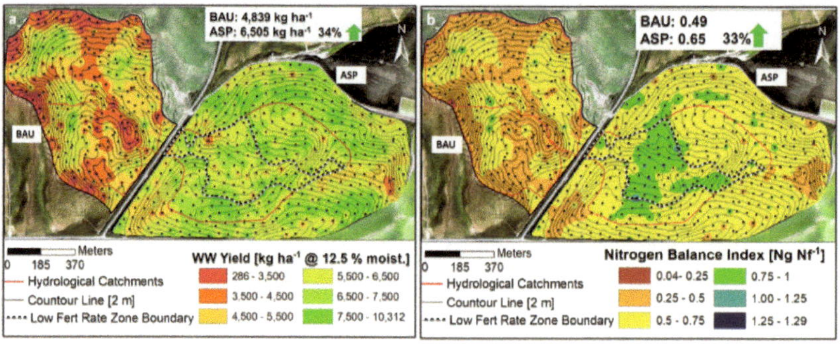

FIGURE 15.5 Spatial pattern of winter wheat (ww) yield in 2019 (a) and nitrogen balance index (b) in BAU field and ASP field. BAU: business as usual, ASP: aspirational practice (Source: Huggins, unpublished).

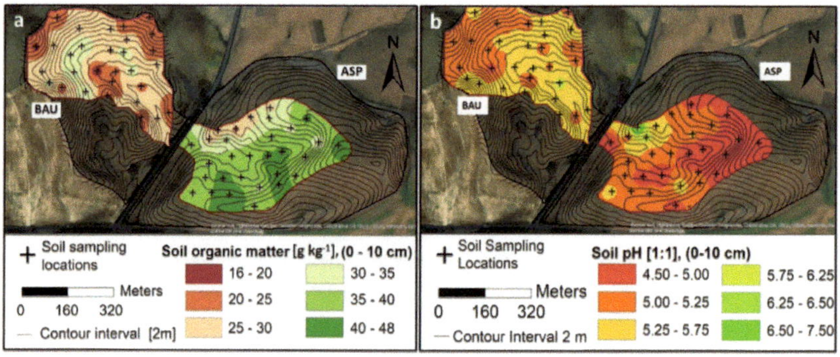

FIGURE 15.6 Soil organic matter concentration (a) and soil pH (b) spatial distribution across the catchment in both business as usual (BAU) and aspirational practice (ASP) systems (Source: Huggins unpublished).

conditions to less than 6.0 in many locations of the annually cropped AEC (Mahler 2002, http://pnwsteep.wsu.edu/directseed/conf2k2/dscmahler.htm). Continuous NT promotes soil pH acidification near the depth of N fertilizer placement (0–10 cm) where pH can be less than 4.5 and Al toxicity can occur (McFarland and Huggins 2015). Soil pH, however, usually increases with depth even within the surface 30 cm, and crop growth may not be seriously impacted until soil acidification occurs at deeper depths. Average soil pH for CAF-NT showed that in the surface 30-cm soil, pH tended to continuously decline over 17 years of continuous NT (Table 15.1). In contrast, significant increases of pH occurred at 60 cm, 120 cm, and 150 cm, though the increase was not significant at a 90-cm depth (Table 15.1). The pH increase in the subsoil is likely due to leaching of bases from the topsoil (data not shown). Amelioration of acidified soil with lime applications has been slow by area farmers, as lime is expensive, and applications have not consistently resulted in crop yield responses. Determining appropriate lime application strategies that consider the high spatial variability of soil pH, both horizontally and vertically within a field, is a challenging area of current research.

15.3.4 Soil Organic Carbon Sequestration in Quantity and Quality

The complexity of abiotic and biotic drivers that determine the persistence of SOM, varying from millennia to days or seconds, limits our ability to predict soil responses to changes related to agricultural management and climate (Schmidt et al. 2011). The spatial-temporal change of SOC stocks (0–10 cm) under continuous NT at CAF is strong evidence of the complication/inconsistency of

TABLE 15.1
Average Soil pH Change over Time at Different Depths at the Cook Agronomy Farm (Data source: Huggins, unpublished)

	Soil pH			
Depth (cm)	**1998**	**2008**	**2015**	***p*-value**
0–10	5.27	5.16	5.07	0.28
20–30	6.27	6.16	6.05	0.26
20–30	6.27	6.15	6.05	0.26
60	6.37	6.59	6.76	0.06
90	6.68	6.89	7.05	0.11
120	6.70	7.00	7.23	0.03
150	6.74	7.15	7.47	0.005

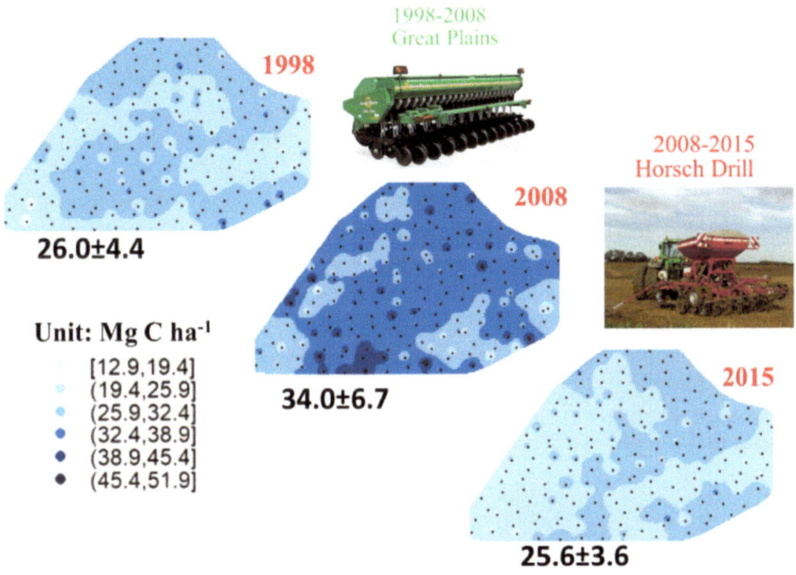

FIGURE 15.7 Changes in spatial distribution and stocks of soil organic carbon (SOC) (0–10 cm) for the Cook Agronomy Farm (CAF) under continuous no-tillage (NT) starting in 1998. Notes: Year 1998 (baseline, top left); 2008 (middle); and 2015 (bottom right); different NT drills used; The numbers below the maps are the SOC stocks mean ± standard deviation for the field.(Sources: Huggins and Peng, unpublished).

evaluating SOC sequestration without taking site- and time-specific soil conditions and management into account (Figure 15.7). During the first 10 years of continuous NT that used a relatively low-disturbance NT drill with double-disk openers, more than 80% of areas across the field experienced SOC stock increases (0–10 cm), and the average stock changed from 26.0±4.4 Mg C ha^{-1} in 1998 to 34.0±6.7 Mg C ha^{-1} in 2008. Surface stocks of SOM declined, however, during the subsequent seven years to 25.6±3.6 when sampled in 2015 and were attributed to the switch to a high disturbance drill with hoe-type openers, negating surface increases of SOC stocks after 17 years (Figure 15.7). The relatively labile nature of SOC that is influenced by management, with the exception of physical SOC transport via erosion, stimulated interest to investigate SOC fractions that are sensitive to changes in management as characterized by mean residence time (MRT) derived from long-term laboratory incubation.

This method contrasts with many assessments of SOC changes after land-use conversion that have focused on total SOC concentration (Deng et al. 2016; Wang et al. 2008; Yan et al. 2012) or SOC fractionation using physical, chemical, or combined analyses (Bossuyt et al. 2002; Lavallee et al. 2019; Purakayastha et al. 2008; Wang et al. 2009; Wiesmeier et al. 2014). Given that the lability of SOC fractionated by physical and chemical methods or combined are questionable, especially when fractionated parts fail to be directly linked with active and inert pools used in many C turnover models (Poeplau et al. 2018), using biological-based methods to fraction SOC can provide insight to soil functions and direct linkage to process-oriented models. Only a few studies, however, have investigated the turnover rates of SOC from the biological function perspective (Cochran et al. 2007; Toosi et al. 2012; Crow and Sierra 2018). To quantify long-term tillage effects on SOC pools with different MRT, soil samples from the surface 20 cm under CT, NT, and native prairie (NP) for three soil series (Naff, Palouse, and Thatuna) were collected and incubated under controlled temperature (21°C) and at field capacity (26% gravimetric water content) for at least 180 days (see Purakayastha et al. 2008, for details on sample collection and site description). Unlike common methods that use non-fixed turnover rates in first-order decay models to simulate soil respiration rate over time, three optimized and fixed turnover rates were derived from an incubation study with more than 500 soil samples from the CAF (see Peng 2020). The specific MRT converted from the optimized turnover rate for different labile SOC pools was MRT= 5 days, 60 days, and 2 years, and the corresponding labile pool concentration with specific MRT was labeled as C1, C2, and C3, while the remaining C subtracted C1, C2, and C3 from the total SOC was labeled as C4. The labile SOC pools (C1, C2, and C3) were combined with the more recalcitrant SOC pool (C4) to develop a SOC rejuvenate index (CRI) that was modified from a C management index (CMI), originally proposed by Blair et al. (1995), to evaluate the capacity of management systems to promote soil quality. The most important argument supporting the newly proposed CRI is to emphasize biologically defined active C, in contrast to the physical or chemical methods used to calculate CMI. The CRI is a SOC pool fraction-weighted overall indicator, as calculated by Eqs. 15.1–4, that directly evaluate the managed site as compared to a reference site regarding SOC of labile and stable pools. Here, the reference site is NP and the CRI value is 1. A CRI in a cultivated site that approaches or exceeds 1 has labile and stable fractions that are similar to NP.

$$\text{Lability of C}\left(\text{L}\right) = \frac{\text{C1} + \text{C2} + \text{C3}}{\text{SOC}} \tag{15.1}$$

$$\text{Lability Index}\left(\text{LI}\right) = \frac{\text{C1} + \text{C2} + \text{C3 in sample soil}}{\text{C1} + \text{C2} + \text{C3 in references soil}} \tag{15.2}$$

$$\text{Stability Index}\left(\text{SI}\right) = \frac{\text{C4 in sample soil}}{\text{C4 in reference soil}} \tag{15.3}$$

$$\text{SOC Rejuvenate Index}\left(\text{CRI}\right) = \text{LI*L} + \text{SI*}\left(1 - \text{L}\right) \tag{15.4}$$

where C1, C2, and C3 are the incubation-derived labile pool concentrations with corresponding MRT= 5 days, 60 days, and 2 years, respectively. C4 is the remaining slow C (C4=SOC-C1-C2-C3).

Comparing CT, NT, and NP showed that MRT of 5 days, 60 days, and 2 years defined labile SOC fractions that were sensitive to management (Table 15.2). Here, NP had the largest SOC pools with MRT = 5 days (C1) for all soils sampled, while NT often had larger labile SOC pools with MRT =

TABLE 15.2
Soil Organic Carbon (SOC) Fractions (C1, C2, C3, and C4) at Different Depths from Different Soil Series under Various Management Systems (Source: Peng, 2020; reproduced with permission)

Soil Series	Landscape Position	Management	Depth (cm)	C1	C2 (mg kg⁻¹)	C3	C4	Lability of C	Lability Index	Stability Index	SOC Rejuvenate Index
Palouse	Summit	NP	0–10	337	356	3141	44186	0.08	1.00	1.00	1.00
			10–20	153	72	1380	28165	0.05	1.00	1.00	1.00
		CT	0–10	223	92	2192	9568	0.21	0.65	0.22	0.31
			10–20	102	0	780	7168	0.11	0.55	0.25	0.29
		NT	0–10	187	587	3284	15875	0.21	1.06	0.36	0.51
			10–20	89	363	1440	12397	0.14	1.18	0.44	0.54
Naff	Sideslope	NP	0–10	274	395	2004	22133	0.11	1.00	1.00	1.00
			10–20	135	9	1237	13759	0.09	1.00	1.00	1.00
		CT	0–10	208	17	1482	4402	0.28	0.64	0.20	0.32
			10–20	102	5	647	2916	0.21	0.55	0.21	0.28
		NT	0–10	193	543	3508	18788	0.20	1.59	0.85	1.01
			10–20	100	425	1130	13396	0.11	1.20	0.97	1.00
Thatuna	Footslope	NP	0–10	438	528	4987	52032	0.10	1.00	1.00	1.00
			10–20	159	54	1965	32312	0.06	1.00	1.00	1.00
		CT	0–10	70	442	1361	22558	0.08	0.31	0.43	0.42
			10–20	32	277	1898	19433	0.10	1.01	0.60	0.64
		NT	0–10	209	525	3404	19682	0.18	0.69	0.38	0.44
			10–20	109	371	1642	15991	0.12	0.97	0.49	0.56

Notes: NP: native prairie; CT: conventional tillage; NT: no-tillage; C1: labile SOC pool with mean residence time (MRT) = 5 days; C2: labile SOC pool with MRT = 60 days; C3: labile SOC pool with MRT = 2 years; C4: remaining more recalcitrant SOC.

60 days and 2 years than CT and NP for Naff and Palouse soils. Stable carbon (MRT >2 years) was greatest in NP, despite various SOC levels, and was the dominant fraction of SOC regardless of land management. Overall, the LI, SI, and CRI were sensitive indicators of soil C quality that occurred under different land-use and tillage management. Using LI, SI, and CRI as indicators, NT resulted in greater LI, SI, and CRI than CT and greater LI but less SI and CRI than NP for Palouse and Naff soils. For Thatuna soil, however, LI, SI, and CRI were all greater under NP, while NT and CP were lower and comparable (Table 15.2). These data illustrate that soil C restoration occurred differently for SOC pools of different MRT and was soil-specific and management-sensitive.

15.3.5 Greenhouse Gas Emission and the Carbon Budget

High spatial and temporal variabilities and complex interactions between biophysical factors and nutrient availability make GHG emission measurement at the field scale challenging (Waldo et al. 2019; Kostyanovsky et al. 2018). Newly developed, robust, and portable setups for automated concurrent measurement of both CO_2 and N_2O emissions have made continuous GHG monitoring possible (Figure 15.3a) (Kostyanovsky et al. 2018). Coupled with the advancement of tower-based methods (Figure 15.3b), this has enabled monitoring of GHG emissions under NT and CT in short- and long-term periods associated with natural or artificial rainfall events, fertilizer applications, and tillage regimes. The following illustrates key findings regarding CO_2 and N_2O emission across dryland farming systems of the iPNW.

15.3.5.1 Short-Term GHG Emissions under NT and CT

Kostyanovsky et al. (2019) investigated N_2O and CO_2 emissions following a short-term disturbance (precipitation and fertilizer application event) in agricultural systems using automated chamber-based methods. Four study sites representing three AECs and with different treatments of NT and CT management are summarized here (Figure 15.2 and Table 15.3). Three micro-plot treatments were applied to both CT and NT plots and monitored for N_2O and CO_2 emissions over the subsequent 22 hours. Treatments applied to NT and CT were (a) control (no water or N additions, DRY), (b) simulated precipitation to achieve approximately 80% water-filled pore space (H_2O), (c) simulated precipitation plus fertilizer additions of 150 kg N ha^{-1} as NH_4NO_3 (H_2O+N). Monitoring N_2O and CO_2 emissions over three days showed the diurnal flux dynamics and indicated that the largest flux occurred within the first day, decreasing over time (Figure 15.8).

Since the first 22 hours had the highest emissions for both N_2O and CO_2 and appeared adequate for capturing treatment differences, the cumulative flux over 22 hours following applied treatments was summarized to show short-time GHG emissions under NT and CT management (Table 15.3). Strong effects of tillage, water, and N treatments on N_2O emissions occurred over the 22-hour period (Kostyanovsky et al. 2019). Water and N treatments in CT plots produced cumulative N_2O fluxes greater than the corresponding plots with 2, 10, and 12 years of NT by 142%, 36%, and 202%, respectively. The site with 31 years of NT and CT comparison had the lowest N_2O flux among all sites but had over 4 times higher N_2O fluxes in NT than CT. Across four sites and three treatments, NT had higher CO_2 flux than CT, except for two treatments at the Prosser site and one treatment at the Moro site where NT had slightly lower CO_2 flux than CT. Additionally, following water and N fertilizer addition, NT tended to produce more N_2O than CT when fallow was incorporated after 31 years of NT in the cropping system; but without fallow, CT produced higher levels of N_2O than NT (data not shown). Stepwise multiple regression analysis indicated that N_2O flux was positively related to soil N availability, while CO_2 flux was positively related to labile SOC. Hot moments of N_2O emissions likely occur during other seasons in addition to summer and warrant further research to adequately compare effects of NT versus CT.

TABLE 15.3
Cumulative Emissions of N_2O and CO_2 During the First Day (0–22 h) of Measurements across the Sites with Different Crop Rotation and Tillage Management Systems (Modified from: Kostyanovsky et al. 2019; reproduced with permission)

			Treatment					
			DRY	H_2O	H_2O+N	DRY	H_2O	H_2O+N
Site	Cropping + Tillage Management	Year Established (Duration of NT)	N_2O-N (g ha^{-1})			CO_2-C (kg ha^{-1})		
Moro, OR	ww-fallow NT[a]	2003 (10 years NT)	0.36	15.49	24.04	6.49	57.31	61.39
Pendleton, OR	ww-fallow NT	1982 (31 years NT)	1.07	1.40	7.01	16.76	51.62	49.92
Genesee, ID	ww-sb-sl NT	2000 (12 years NT)	0.27	5.57	12.71	11.70	71.66	52.23
Prosser, WA[b]	ww-SwCn-potato NT	2011 (2 years NT)	2.34	5.46	9.29	43.74	38.96	37.21
Moro, OR	ww-fallow CT		0.42	25.61	32.73	8.18	56.70	45.50
Pendleton, OR	ww-fallow CT		0.19	0.14	1.15	12.84	22.19	23.98
Genesee, ID	ww-sb-sl CT		0.93	19.98	38.35	9.50	36.69	37.00
Prosser, WA	ww-SwCn-potato CT		0.85	8.66	22.54	36.25	40.82	45.18

[a] ww: winter wheat; sb: spring barley; sl: spring legume; SwCn: sweet corn; CT: conventional tillage; NT: no-tillage.
[b] Prosser site is an irrigated system.

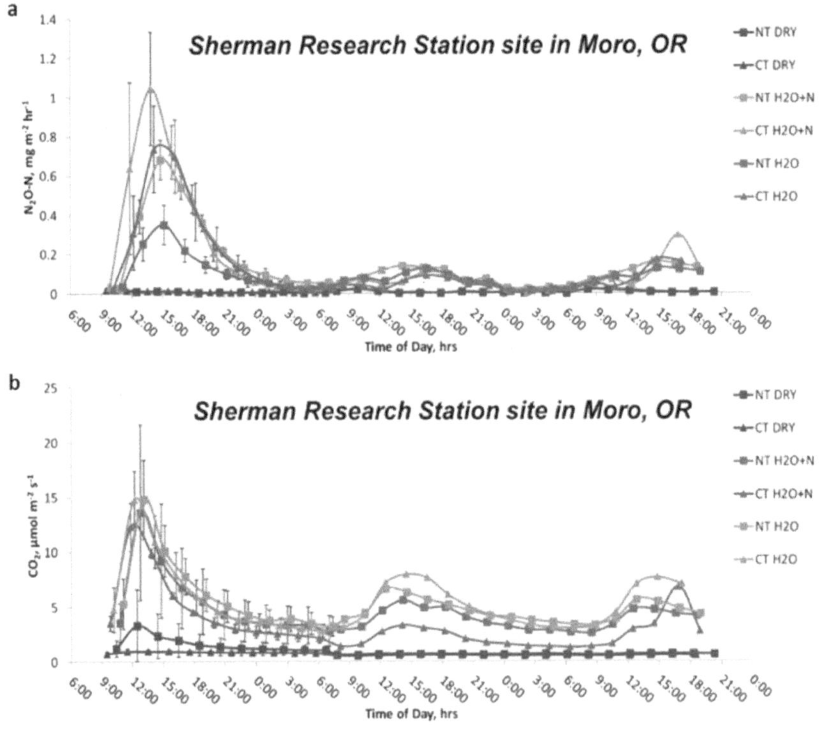

FIGURE 15.8 Emissions of N_2O (a) and CO_2 (b) during the first three days of measurements with and without additions of 150 kg N ha^{-1} and water (H_2O) to no-tillage (NT) and conventional tillage (CT) management systems at Moro, OR site (Source: Kostyanovsky et al. 2019; reused with permission).

15.3.5.2 Annual Field Monitoring of GHGs and Carbon Budgets

Eddy covariance (EC) towers were installed at CAF-NT and CAF-CT in 2011 (after 13 years of NT and CT) to monitor CO_2 flux (Figure 15.3d). In addition, during 2015, EC towers were equipped to simultaneously monitor N_2O and CO_2 with a larger spatial heterogeneity coverage ~2500 m^2 and compared to co-located automated static chambers covering about 360 m^2. Complete descriptions of the automated chamber system, the eddy covariance systems, and auxiliary measurements are in Kostyanovsky et al. (2018), Chi et al. (2016, 2017), and Waldo et al. (2019).

Ecosystem respiration (Reco), the efflux of CO_2 from ecosystem to the atmosphere, includes soil respiration and aboveground plant respiration (Barba et al. 2018). Soil respiration is defined as CO_2 produced by the biological activity of plant roots, microbes, and soil fauna and is the second largest C flux in terrestrial ecosystems (Raich and Schlesinger 1992; Reichstein and Beer 2008; Schlesinger and Andrews 2000). The annual global CO_2 flux from soil is estimated to average 68±4 Pg C yr^{-1}, and range from 60 g C m^{-2} yr^{-1} to 1260 g C m^{-2} yr^{-1} across different biomes (Raich and Schlesinger 1992). At the CAF, Reco was used as a proxy for GHG CO_2 emissions. Over four successive cropping years (2012–2016) in CAF-NT and CAF-CT, Reco for CAF-NT was significantly ($p<0.05$) lower than CAF-CT (Figure 15.9 and Table 15.4). The amount of annual Reco under NT varied with crop species, but, on average, NT reduced Reco by 182 g C m^{-2} yr^{-1} compared to CT. Notably higher Reco peaks were observed in CT compared to NT during the off-growing season mainly after rainfall events (Figure 15.9b). As crop species were identical and local meteorologic conditions were similar between CAF-NT and CAF-CT, farming practices of tillage, but also slight differences in applied fertilizer contributed to management differences (Table 15.3).

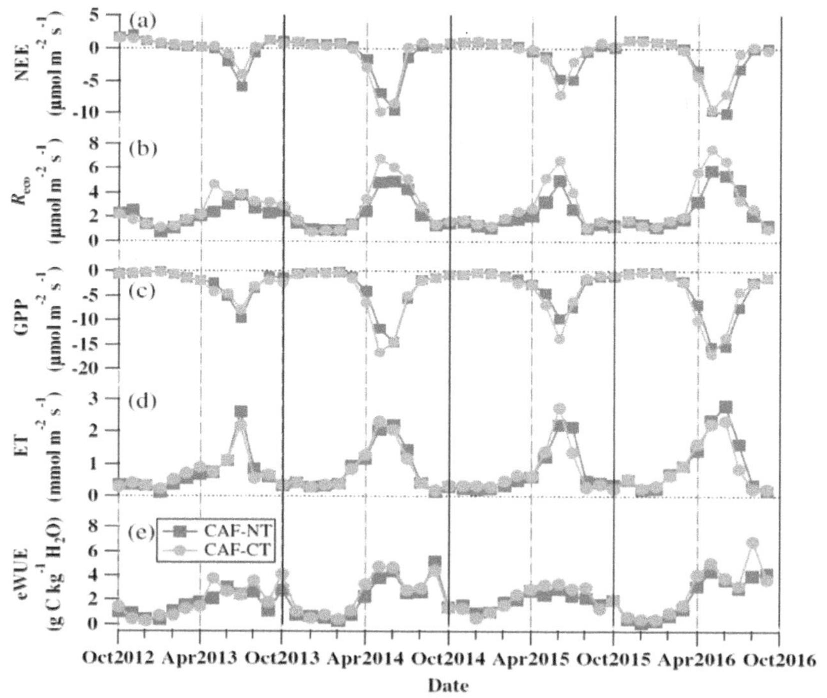

FIGURE 15.9 Monthly mean net ecosystem exchange of CO_2 (NEE), total ecosystem respiration (Reco), gross primary productivity (GPP), evapotranspiration (ET) and monthly ecosystem water use efficiency (eWUE) at CAF-NT and CAF-CT sites from October 2012–September 2016 (Source: Chi et al. 2017; reused with permission).

In terms of C balance, it is simplified and calculated following the equations below (Waldo et al. 2016):

$$\text{NECB} = -\text{NEE} + \text{IMP} - \text{EXP} \tag{15.5}$$

$$\text{NEE} = \text{GPP} - \text{Reco} \tag{15.6}$$

$$\text{Reco} = \text{Ra} + \text{Rh} \tag{15.7}$$

Where NECB is the net ecosystem C balance, NEE is net ecosystem exchange, IMP is import of C via seeding, fertilizer, and amendments and EXP is export of C in harvested biomass. In Chi et al. (2017), IMP is minimal and not considered in the balance equation. The GPP is gross primary productivity, Reco is total ecosystem respiration, Ra and Rh is autotrophic and heterotrophic respiration, respectively. All variables in daily scale are in unit of g C m^{-2} d^{-1}, the sum of it is in unit of g C m^{-2}.

Positive NECB indicates C accumulations in the system as a net C sink, and when NECB is negative, the system is a net C source. Either system serving as a C sink or source seems quite variable by year, crop type, and site: for example, CT at CAF was a larger carbon source in the year 2013 but a smaller sink in the years 2015 and 2016 compared to NT, while in 2014 CT was a slightly larger sink than NT (Table 15.4).

The annual cumulative N_2O-N flux monitored with automated static chambers from 1 Oct 2014 through 30 Sep 2015 was 3.9±0.22 kg ha^{-1} for CAF-NT and 5.6±0.34 kg ha^{-1} for CAF-CT (Table 15.4). The EC flux method showed similar N_2O-N emissions between CAF-NT and CAF-CT from 1 Apr

TABLE 15.4

Mean Annual Air Temperature (Tair), Precipitation, Length of Main Growing Seasons (MGS), Annual Net Ecosystem Exchange of CO_2 (NEE), Gross Primary Productivity (GPP), Total Ecosystem Respiration (Reco), Export of Harvest C (EXP), Net Ecosystem Carbon Balance (NECB), Flux-Gradient Based Eddy Covariance Measured N_2O-N Flux and Static Chamber (SC), Measured N_2O-N Flux across Sites in Different Years (Adapted from: Chi et al. (2017) and Waldo et al. (2019); reused with purchased permission)

Year	Site (crop)	Tair (°C)	Precipitation (mm)	Length of MGS (days)	NEE	GPP	Reco	EXP	NECB	Monitor Time	EC Flux	SC Flux
					CO_2-C (g m^{-2} yr^{-1})					N_2O-N (kg ha^{-1})		
2012	CAF-NT (ww)	8.9	496	134	−527	−1407	890	287	230			
2013	CAF-NT (sg)	9.2	539	121	−20*	−851	831*	68	−48			
	CAF-CT (sg)	9.1	539	121	117*	−848	965*	86	−203			
2014	CAF-NT (ww)	9	455	127	−450	−1343	893*	232	218			
	CAF-CT (ww)	8.7	455	127	−521	−1613	1092*	269	252			
2015	CAF-NT (sc)	10.8	467	110	−210	−980	770*	67	143	1 Apr–30 Sep	2.4 ± 0.7	1.9 ± 0.14
	CAF-CT (sc)	10.5	467	110	−162	−1170	1008*	77	85		2.1 ± 0.2	4.8 ± 0.33
	CAF-NT (sc)									1 Oct 2014–30 Sep 2015	N/A	3.9 ± 0.22
	CAF-CT (sc)										N/A	5.6 ± 0.34
2016	CAF-NT (ww)	10	527	136	−671	−1657	986*	256	415			
	CAF-CT (ww)	9.8	527	136	−533	−1677	1144*	283	251			

Notes: f: fallow; ww: winter wheat; sg: spring garbanzo; sc: spring canola; CAF: Cook Agronomy Farm; NT: no-tillage; CT: conventional tillage; MGS: main growing season length; Tair: annual mean of air temperature; NEE: annual net ecosystem exchange of carbon; GPP: gross primary productivity; Reco: total ecosystem respiration; EXP: export of harvest carbon; NECB: net ecosystem carbon balance. EC flux: eddy covariance flux; SC flux: static chamber flux; N/A: data not available.

*Significant difference between NT and CT ($p < 0.05$).

through 30 Sep 2015 (main growing season) but differed appreciably from the static chamber measurements for the same time period (Table 15.4, Figure 15.10), although similar temporal patterns were observed (Figure 15.10). In contrast, major emissions occurred following spring N fertilization with elevated daily fluxes prolonged for almost two months from 5 Apr until 15 Jun (Figure 15.10). Daily maximum N_2O-N emissions ranged from 54±26 g N_2O-N ha^{-1} d^{-1} to 202±65 g N_2O-N ha^{-1} d^{-1} in CAF-NT, and from 94±48 g N_2O-N ha^{-1} d^{-1} to 136±67 g N_2O-N ha^{-1} d^{-1} in CAF-CT by EC and SC methods. Applied N Fertilizer, rainfall, and temperature increases stimulated N_2O emissions. During the late fall and winter (1 Oct 2014–1 Apr 2015), N_2O-N emissions were still measurable in both sites, and CAF-NT had larger emissions by the SC method compared to CT (Figure 15.10b, d).

The N_2O-N emission can be converted into CO_2-equivalent using the global warming potential (GWP) factors of 265 as in Waldo et al. (2019). The annual N_2O-N emissions measured by the SC method during the crop year of 1 Oct 2014–30 Sep 2015 was equivalent to 44.3 and 63.6 g CO_2-C

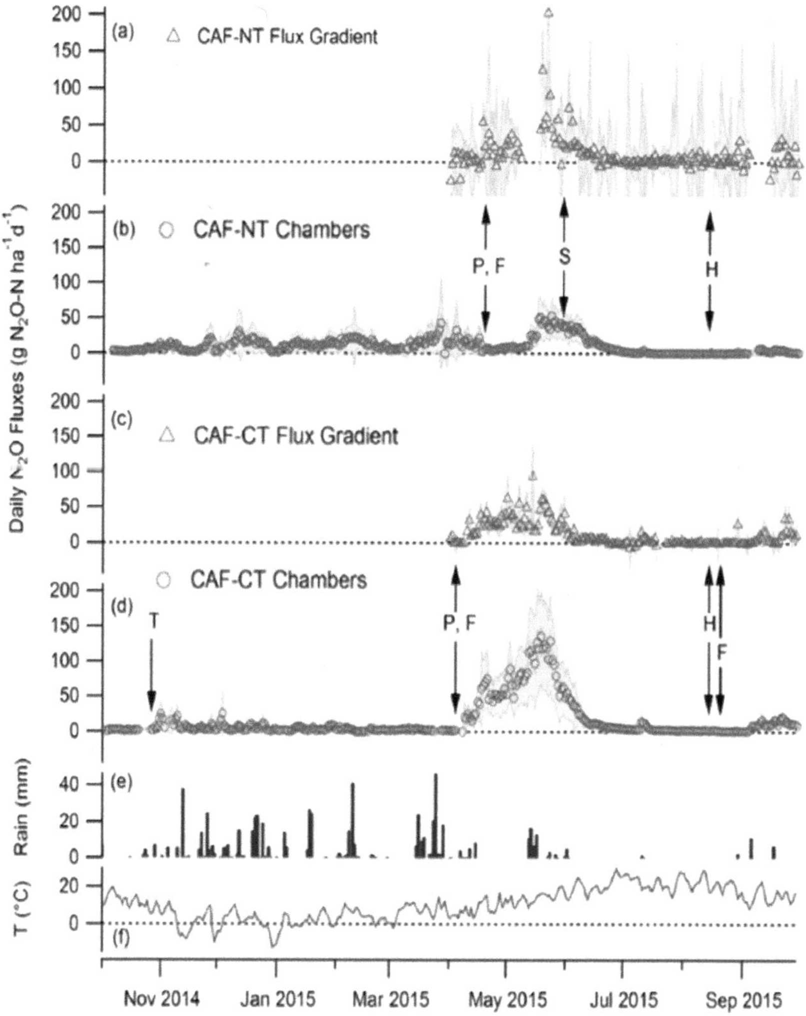

FIGURE 15.10 Time series of N_2O fluxes Notes: (a) CAF-NT FG (Flux gradient); (b) CAF-NT chambers; (c) CAF-CT FG; (d) CAF-CT chambers; (e) Precipitation; (f) temperature for the 2015 crop year; Management activities are indicated by T= Tillage, P = planting, F= fertilization, S= herbicide spray application, and H= harvest (Source: Waldo et al. 2019; reused with permission).

m^{-2} yr^{-1} for CAF-NT and CAF-CT respectively, compared to the corresponding NECB of 173 and 85 g CO_2-C m^{-2} yr^{-1} during the same time period. Consequently, the C sequestered in both systems were offset by 26% in CAF-NT and 75% in CAF-CT. The uncertainty in gap-filling of the EC flux data, however, could result in uncertainty of the offset percentage of C sequestration by N_2O emissions. The soil C sequestration rate in continuous NT cropping systems in iPNW regions has been reported as 20 g C m^{-2} yr^{-1} (Brown and Huggins 2012) and 16–23 g C m^{-2} yr^{-1}(Stöckle et al. 2012). The equivalent C emission rate (as CO_2 equivalent) from N_2O-N flux would more than offset the soil C sequestration rate, as SOC approaches a new steady state after 15–20 years of continuous NT (Pittelkow et al. 2015; Ogle et al. 2012; Waldo et al. 2019; Ogle et al. 2019).

Overall, the long- and short-term influence of CT and NT on GHG emissions showed high spatial and temporal variability. Future research should further characterize field-scale drivers of "hot" fluxes of GHGs in space and time over the course of longer-term crop rotations and associated management practices. More diversified crop rotations and precision N management that increases N use efficiency will likely influence GHG emissions. Process-level understanding of GHG flux would also benefit from linking broader field monitoring of C and N cycling and flows to modeling efforts to improve overall accuracy of C and N budgets (Figure 15.11).

15.3.6 Water Quality

The dryland farming region of the iPNW has a Mediterranean, semiarid climate, with wet, relatively cold winters and hot, dry summers, with an average of 59% of annual precipitation received in the winter season from November through March (Western Regional Climate Center, 2008; Singh et al.

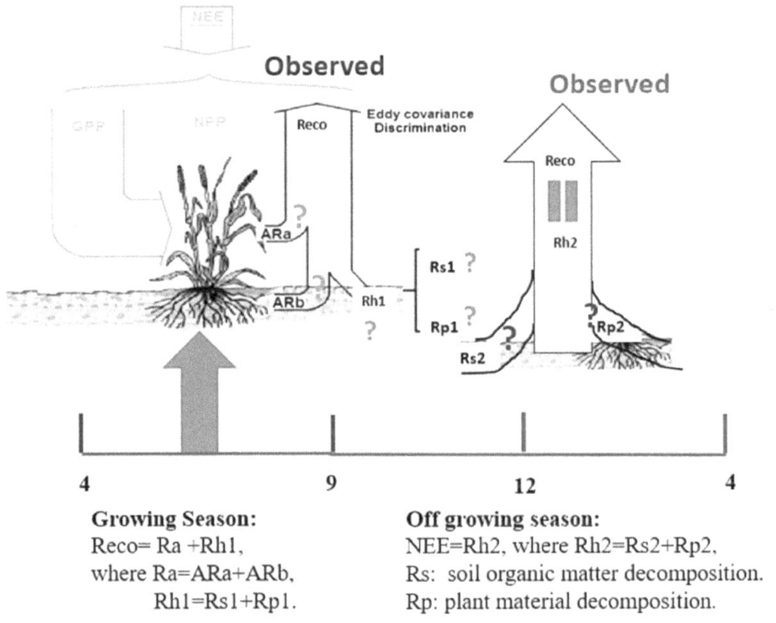

FIGURE 15.11 Diagram illustrating components contributing to annual ecosystem respiration. Abbreviations: NEE: net ecosystem exchange C flux; GPP: gross primary production; NPP: net primary production; Reco: ecosystem respiration; ARa: aboveground autotrophic respiration; ARb: belowground autotrophic respiration; Rh: heterotrophic respiration; Rs: respiration from soil organic matter decomposition; Rp: respiration from plant material decomposition. Numbers "1" and "2" after the Rp and Rs represent "growing season" and "off-growing season," respectively. Numbers 4, 9, 12 under the line are months. "?" are flux from each component that needs to be monitored (Sources: Peng, 2020; Picture was modified from Suleau et al. 2004); reused with permission).

2009). Frequent freeze-thaw cycles (on annual average, around 100), combined with rain events, can quickly saturate soil and exacerbate water runoff and erosion with maximum soil loss rates of 225–450 Mg ha^{-1} yr^{-1} (USDA 1978; McCool and Roe 2005). Artificial subsurface drains are often used to improve drainage along intermittent waterways to address early spring waterlogging, primarily in the annual cropping region (Figure 15.2). Monitoring water flows from outlets of subsurface artificial drains can provide a spatial integration of catchment-scale processes useful for understanding nutrient and chemical delivery from agricultural management systems (Zehe and Flühler 2001; Keller et al. 2008; Bakhsh et al. 2002; Jenkins et al. 2009).

Water samples derived from a 12 ha artificially drained catchment in the CAF-NT (Figure 15.6), where complexities of soil and terrain interact with pronounced hydrologic seasonality, have been monitored from Sep 2000 through 2021. Water dissolved organic carbon (DOC) and different species of N and P have been monitored periodically, and details of sampling methods are described by Keller et al. (2008), Kelley et al. (2017), and Ortega-Pieck et al. (2020). The following highlights the potential environmental risks faced by dryland NT farming under a Mediterranean climate.

15.3.7 Dissolved Organic Matter and Nitrogen in Artificial Drainage under NT

Dissolved organic carbon (DOC), dissolved organic nitrogen (DON, calculated as the difference between total dissolved nitrogen and total dissolved inorganic nitrogen or DIN (NO_3^- + NH_4^+)), C:N ratio and electrical conductivity (EC), dissolved organic matter (DOM), and spectroscopic characteristics (FI and β/α) were monitored over four years (2011-2014) in both soil and water from artificial drainage (Bellmore et al. 2015) (Figure 15.12). Here, FI is an indicator of the relative contributions of plant or terrestrially derived DOM (FI~1.2) versus microbially derived DOM (FI~1.8) in aquatic ecosystems, and β/α is a DOM freshness index where β represents recently produced DOM while α is more decomposed organic compounds.

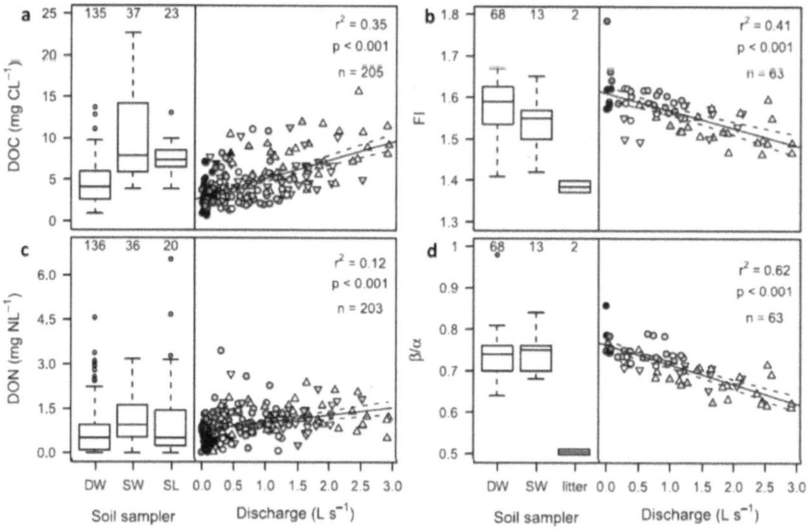

FIGURE 15.12 Dissolved organic carbon (DOC) (a), dissolved organic nitrogen (DON) (c), dissolved organic matter (DOM) spectroscopic characteristics (FI and β/α) (b and d) in soil water (boxplots) and drain discharge (scatterplots) Abbreviations: DW: deep wells; SW: shallow wells; SL: shallow lysimeters during 2012–2013 water year. Notes: Numbers above boxplots are sample sizes. Drain discharge samples are identified by flow conditions: base flow (open circle), rising limb (triangle), or falling limb (inverted triangle) and season: fall (black), winter (white), spring (light gray), and summer (dark gray). Dotted lines are 95% confidence intervals around regression models.(Source: Bellmore et al. 2015; reused with permission).

Briefly, shallow wells (SW) were installed to sample shallow soil water, and deep wells (DW) were installed to sample permanent groundwater where a soil argillic horizon occurred. In locations with no argillic horizon, shallow lysimeters (SL) were installed at 0.5 m below the soil surface, and a deep well (DW) was installed and screened from the bedrock to approximately a 1-m depth.

Concentrations of DOC and DON tended to be higher in surface soil water samples than in deep samples (Figures 15.12a, c boxplots). Additionally, DOC and DON concentrations from the subsurface drainage were weakly correlated with discharge (Figures 15.12a, c), which had substantial seasonal variability. Drainage flow ranged from 0 L $s^{-1}$to 3 L s^{-1}, and high-flow events contributed to 48% and 25% of total annual flow in 2012 and 2013, respectively (Bellmore et al. 2015). Estimated annual DOC-C and DON-N fluxes varied from 1.0 kg C ha^{-1} to 8.6 kg C ha^{-1} and from 0.23 kg N ha^{-1} to1.8 kg N ha^{-1} from 2001 to 2013 (excluding 2002, 2008, and 2010), respectively (Bellmore et al. 2015). Annual flow-weighted DOC and DON both were positively correlated with annual runoff (Figure 15.13). The relationship between DOC, DON concentrations and discharge, and annual runoff (Figures 15.12 and 15.13), strongly suggested that exported DOM from this drainage catchment is highly regulated by hydrologic dynamics.

Regarding DOM quality, the FI and β/α both showed strong negative relationships with discharge (Figures 15.12b, d), indicating that contributions of recently produced microbial products to the DOM pool declines as discharge increases. Comparison of DOM concentrations and characteristics in soil water samples reflects deep sources from soil during low flow (fall and summer) and shallow sources during high flow (spring and winter) (Bellmore et al. 2015). The latter phenomenon could be due to macropore connectivity created from vertical earthworm and/or root channels that remain relatively undisturbed in NT and potentially open to the surface, which could result in recent organic additions and topsoil DOM being rapidly transported to artificial subsurface drains (Johnson-Maynard et al. 2007; Bellmore et al. 2015).

15.3.8 Inorganic Nitrogen (NO_3^-) in the Subdrainage Water under NT

Discharge in the NT CAF catchment (approximately 12 ha) showed seasonal patterns, with peak flows occurring in March and low flows in the summer from June until October (Keller et al. 2008).

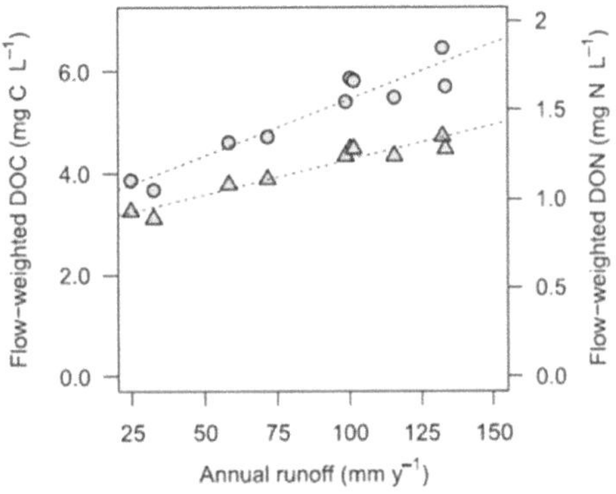

FIGURE 15.13 Relationship between flow-weight dissolved organic carbon [DOC] (circle) and dissolved organic nitrogen [DON] (triangles) and annual runoff from 2001 to 2013 (2002, 2008, and 2010 removed because of biased sampling events and fewer observations). For DOC, r^2=0.89, p <0.001, for DON, r^2=0.92, p<0.001 (Source: Bellmore et al. 2015; reused with permission). Figure 15.14 Daily precipitation (a), measured artificial drain (AD) discharge and estimated AD discharge (b), and NO_3^- concentration (c) in the AD discharge from October 2000 to September 2013. The NO_3^- concentration (d), $\delta^{15}N\text{-}NO_3^-$ (e) and $\delta^{18}O\text{-}NO_3^-$ (f) in the drainage water over April 2005–October 2010 are displayed in more detail. The shade areas are representing the high-discharge season (Jan–May) (Adapted from: Kelley et al. 2017 and Kelley et al. 2013; reused with permission).

Estimated discharge values ranged from <10 $m^3\ d^{-1}$ to a high of ~350 $m^3\ d^{-1}$ over water years of 2001–2013 (Figure 15.14b). The NO_3^- concentration in drainage waters showed a similar seasonal pattern to water discharge, with high concentrations of up to 32 mg L^{-1} occurring simultaneously with high-discharge rates and a baseline of 3–4 mg L^{-1} during low flows of summer and fall (Keller et al. 2008). The mean annual N loss as NO_3^- in drainage water was 15 kg N ha^{-1}, whereas the average annual loss of N as DON over the same period was 1.3 kg N ha^{-1} (Bellmore et al. 2015; Kelley et al. 2017). The proportion of NO_3^--N exported via drainage water accounted for 3–24% of applied fertilizer N from 2001 to 2013 (omitting the years 2002 and 2008) with a mean of 15%. The mean proportion of applied N lost via artificial drainage increased, however, to 21% with a range of 16–24% from 2010 to 2013, though the applied N rate decreased from 115 kg N $ha^{-1}\ yr^{-1}$ to 77 kg N $ha^{-1}\ yr^{-1}$ for the last four years (Kelley et al. 2017).

The observed "lag" between increased NO_3^--N loss flux and reduced N fertilizer application over the last four years, as discussed by Kelley et al. (2017), is primarily attributed to the difference in N sources. Those NO_3-N sources include the largest N reservoir of SON via the mineralization and nitrification process and legacy of accumulated and transformed fertilizer N that has not cycled through the SON pool. The rationale that nitrification of NH_4^+-based fertilizer was a major contributor to N export in subsurface drainage was mainly supported by the measured variation of $\delta^{15}N$-NO_3^-, $\delta^{18}O$-NO_3^- and NO_3^- flux in drainage water over 2006–2010 (Figures 15.14d, e, f). The $\delta^{15}N$-NO_3^- varied during the high and low-discharge seasons from −0.5‰ to +9‰ with a volume-weighted mean of 2.1‰, whereas $\delta^{18}O$-NO_3^- did not show clear seasonal fluctuation, averaging −1.3‰ throughout the study. The range of stable isotopes ($\delta^{15}N$-NO_3^- and $\delta^{18}O$-NO_3^-) in NO_3^- from various sources reported in literature, combined with the temporal pattern of the $\delta^{15}N$-NO_3^- and $\delta^{18}O$-NO_3^- and NO_3^--N flux (Figures 15.14 e, f), suggests that sources of NO_3^- loss during the high-discharge season is mainly from nitrification of NH_4^+ fertilizer, while mineralized SON is the primary source during the low-discharge season (Kelley et al. 2017).

15.3.9 Dissolved Reactive Phosphorus (DRP) in Subsurface Artificial Drainage Water under NT

Dissolved reactive phosphorus (DRP, orthophosphates) in artificial subsurface drainage from CAF catchment under NT was monitored over three water years from 2015 to 2018 (Ortega-Pieck et al. 2020). The concentration of DRP varied seasonally with low values (<0.05 mg L^{-1}) during seasonally low drainage discharge but larger than 0.1 mg L^{-1} during winter and early spring when drainage discharge was greater (Figure 15.15). High DRP concentration in drainage waters generally occurred in Jan but declined over time and did not simultaneously peak with drainage discharge as did DOC and nitrate (Figures 15.14 and 15.15). The estimated average DRP load discharged via the artificial drain line over three years was 0.11 kg $ha^{-1}\ yr^{-1}$. The DRP daily flux ranged from 0 mg L^{-1} to 0.21 mg L^{-1} and mostly remained above 0.05 mg L^{-1} (Figure 15.15). The acceptable total P level ranges from 4 µg L^{-1} to 35 µg L^{-1} for the lakes at Columbia Basin Ecoregions under the Clean Water Act (EPA, https://www.epa.gov/wqs-tech/state-specific-water-quality-standards-effective-under-clean-water-act-cwa#tb2). As a fraction of total P, excessive exported DRP during the monitoring period at CAF-NT (predominantly DRP concentration >0.05 mg L^{-1}) into surface waters, consequently, would impair water quality and lead to potential eutrophication. A complimentary laboratory leaching study with intact soil cores (0–0.1 m, 0.2–0.3m, 0.5–0.6 m) from CAF showed that all soil cores have high P sorption potential regardless of landscape positions (toeslope versus summit), but much of the DRP in the drainage is derived from P hotpots located in toeslope positions, where the total P, water extractable P and Mehlich-3 P are significantly higher than summit landscape positions (Ortega-Pieck et al. 2020).

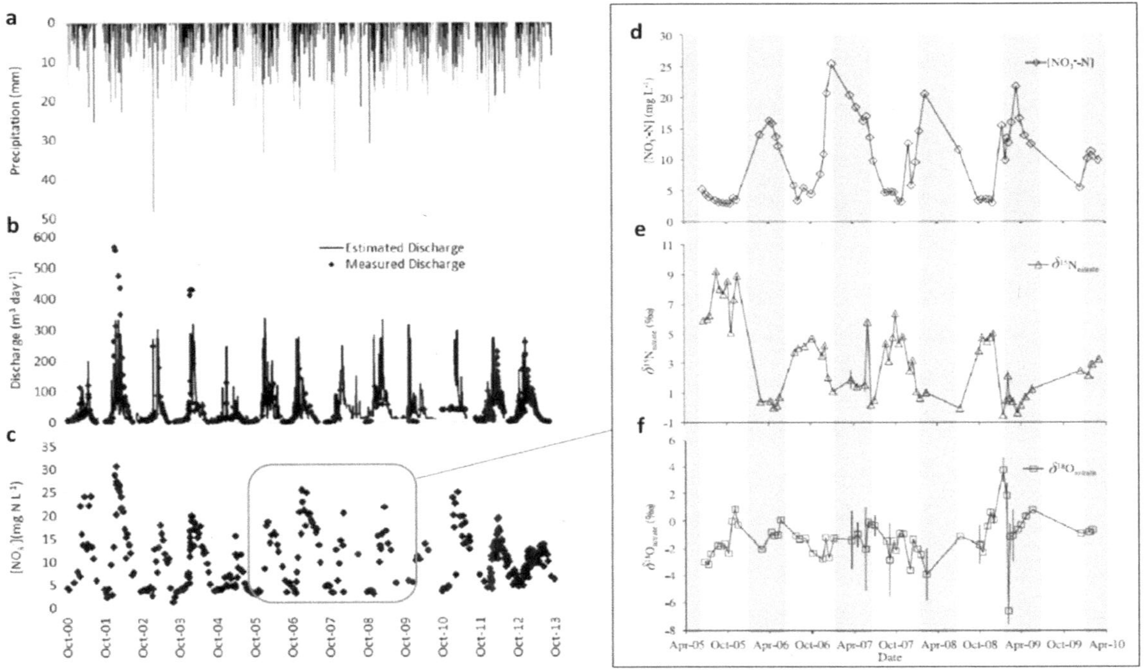

FIGURE 15.14 Daily precipitation (a), measured artificial drain (AD) discharge and estimated AD discharge (b), and NO_3^- concentration (c) in the AD discharge from October 2000 to September 2013. The NO_3^- concentration (d), $\delta^{15}N\text{-}NO_3^-$ (e) and $\delta^{18}O\text{-}NO_3^-$ (f) in the drainage water over April 2005–October 2010 are displayed in more detail. The shade areas are representing the high-discharge season (Jan–May) (Adapted from: Kelley et al. 2017 and Kelley et al. 2013; reused with permission).

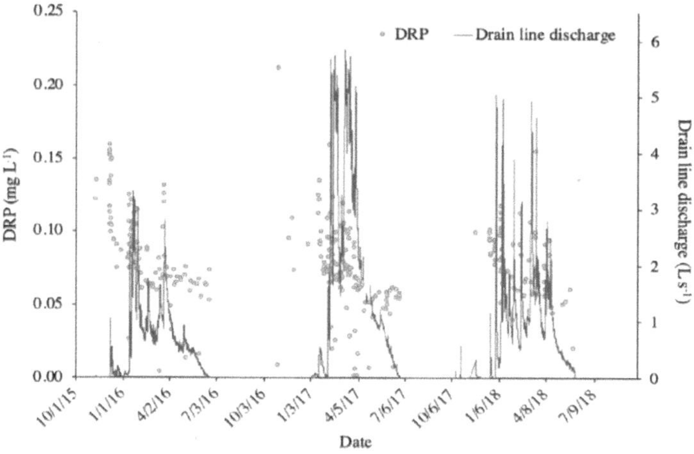

FIGURE 15.15 Dissolved reactive phosphorus (DRP) concentration (circle) and drain line discharge (solid line) at the Cook Agronomy Farm measured from October 2015 through October 2018 (Source: Ortega-Pieck et al. 2020; reused with permission).

15.3.10 Comparison of Drainage Water between NT and CT

The estimated DOC, DON, NO_3^- flux and dissolved reactive phosphorus (DRP) loads in the tile drainage water from CAF during specific measured water years varied from 1.0 kg C ha^{-1} yr^{-1} to 8.6 kg C ha^{-1} yr^{-1}, from 0.23 kg N ha^{-1} yr^{-1} to 1.8 kg N ha^{-1} yr^{-1}, from 4 kg N ha^{-1} yr^{-1} to 22 kg N ha^{-1} yr^{-1}, and from 0.09 kg P ha^{-1} yr^{-1} to 0.15 kg P ha^{-1} yr^{-1}, respectively. Those corresponding values are lower than the reported average leaching rate of DOC (15.4 kg C ha^{-1} yr^{-1}), DON (1.5–4.5 kg N ha^{-1} yr^{-1}), NO_3^- (5.9–35.3 kg N ha^{-1} yr^{-1}) across different crops under the NT system on a well-drained sandy loam soil in the humid temperate climate of southwestern Michigan (Hussain et al. 2020) but higher for DRP (0.077 kg P ha^{-1} y^{-1}) under a corn-soybean cropping NT system in clay loam soils in southwestern Ontario (Zhang et al. 2017). Compared to the N_2O-N emission rate of 0.72–0.98 kg N ha^{-1} yr^{-1} from CAF, which accounted for 0.71–1.1% of applied fertilizer N (Waldo et al. 2019), leached inorganic N is one of the major fertilizer N loss pathways with most of nutrient loss occurring during the high-flow seasons of late winter and early spring when crop coverage is absent or limited. Those findings strongly suggest that management practices aiming to improve use efficiency under NT are required to decrease nutrient loss and water contamination risks, such as the use of perennial crops with deep root systems (Keller et al. 2008), winter-hardy crops (winter wheat, canola, pea, barley, etc.) (Huggins et al.1989), or cover crops (Magdoff and Van Es 2000) with small biomass without sacrificing too much moisture for following spring cash crop, and increased NUE with precision agriculture (Kelley et al. 2013).

15.4 POTENTIAL INNOVATION PRACTICES TO MAXIMIZE THE BENEFITS OF NT AGRICULTURAL SYSTEMS

Challenges faced by NT systems in semiarid regions include lack of cropping system diversification and intensification. In turn, these challenges contribute to nutrient losses to water and air, soil acidification, and herbicide resistance. A few considerations are proposed here for further investigation.

15.4.1 Cropping System Diversification and Intensification

The continued leaching of residual inorganic nitrogen (primarily NO_3^-) into deeper soil horizons below root zones and into surface and ground waters can contribute to soluble base and agrichemical losses and soil acidification, promoted by greater water infiltration that occurs under NT. Greater cropping system intensification and diversification such as intercropping characterized by different crop's root morphology has been demonstrated to be effective in mitigating NO_3^- leaching in greenhouse simulation studies (Ding et al. 2021). In addition, system diversification and intensification through integration of crops and livestock and use of cover, forage, and perennial crops has been demonstrated to effectively improve winter wheat yield, lower N fertilizer rates, and improve SOC compared to a conventional NT cropping system (Wachter et al. 2019). Here, soil profile (0–150 cm) SOC stocks and net profit return were favored by a reduced tillage organic system that integrated livestock and perennial alfalfa in the Palouse region of eastern Washington (Wachter et al. 2019). As an integral and purposeful component of agroecosystems, incorporating perennial plants is a promising tool to improve ES. Scientific studies on spatial configuration, amount and type of perennial vegetation at watershed scales, however, are lacking, and further efforts are required to support perennialization practices and evaluation of associated ESs (Asbjornsen et al. 2014). Future efforts are recommended to address this significant knowledge gap and explore different perennialization strategies as well as quantifying the benefits of perennialization to ecosystem services as suggested in Figure 15.16.

15.4.2 NT with Precision Agriculture (PA) Techniques

Precision agriculture (PA), is conceptualized as a system that tailors management to site-specific, field-scale variability to improve efficiency and outcomes and to potentially advance sustainable agriculture via application of industrial technologies (Stafford 2000; Zhang 2002). Field-observed patterns of soil erosion, areas contributing to water quality issues, and variability in crop productivity provide a basis for a conservation-oriented PA system that reduces nutrient loss and can promote profitability, efficiency, and flexibility (Kitchen et al. 2005; Lerch et al. 2005; Tomer et al. 2013). Following a four-year plot-scale study in a semiarid region in the northwestern Indo-Gangetic Plains, significantly higher average biomass productivity, net economic returns, and biomass water use efficiency were observed—in NT and site-specific nutrient management with PA—more than CT and farmers' fertilizer practice (Parihar et al. 2017). Field-scale grain production and profitability comparisons between a long-term PA system and conventional management found that PA can reduce temporal yield variation though spatial yield was not altered (Yost et al. 2017). The PA

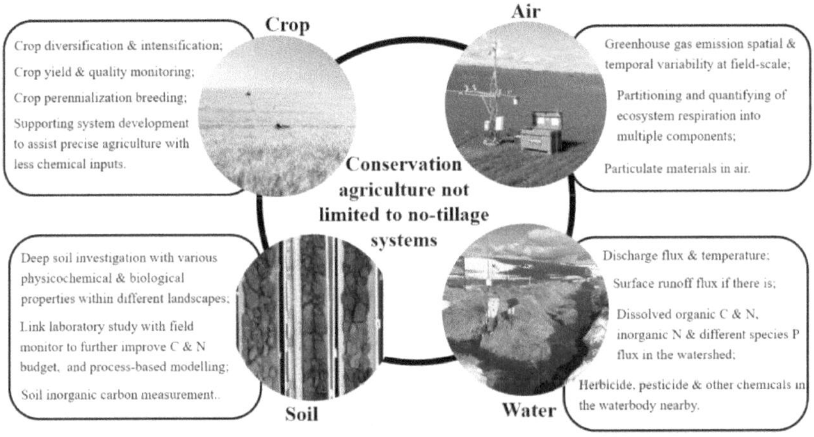

FIGURE 15.16 Conceptual diagram of recommendations for comprehensive evaluations and further research on conservation agriculture (Sources: Images of crop and water, credited to Qiuping Peng; image of air, credited to Ian Leslie; images of soil, credited to Caven Halsey).

system sustained profits in the majority of the field without subsidies for enhanced environmental protection, but profit variations were dependent on crop and evaluation periods and associated grain prices (Yost et al. 2019). Fertilizer placement to reduce surface broadcast of nutrients and apply fertilizer in deep bands is another potential strategy to mitigate GHG emissions without adversely affecting yield (Rychel et al. 2020). Besides PA-based use of fertilizers and pesticides to efficiently use agrichemicals and potentially reduce input costs, the application of PA strategies for seeding, irrigation, and tillage also holds the potential to optimize resource use efficiencies and profitability. Future research efforts that quantify multiple ES and economic tradeoffs and synergies at various temporal and spatial scale are needed (Lal 2015; Spiegal et al. 2018).

15.4.3 Lime Application and Occasionally Tillage

Soil acidification, largely driven by continued use of synthetic N fertilizers and associated soil pH stratification under NT is a continuing soil degradation concern (Tian and Niu 2015; Pennock et al. 2015). Application of lime (Blevins et al. 1978; Brown et al. 2008; Flower and Crabtree 2011), gypsum (Caires 2011), biochar (Teutscherova et al. 2017; Chintala et al. 2014) and lime mixed with other organic compounds (Tao et al. 2019) can effectively raise soil pH; it also can influence crop yield (Blevins 1978; Brown et al. 2008; Bezdicek et al. 2003; Caires et al. 2011), emissions of N_2O and CO_2 (Obia et al. 2015; Sheng and Zhu 2018; Wu et al. 2018), and SOC sequestration (Inagaki et al. 2017; Lehmann 2006). These effects, however, are not consistent and vary with materials, amounts, crop species, climate, soil depth, and other site-specific factors. Application of materials aimed at ameliorating soil acidity, interact with soil solutions and minerals, alter microbial activity and nutrient cycling, and can contribute to the uncertainty of yield, GHG emission, and SOC status.

Occasional tillage in an NT system has been demonstrated to effectively redistribute the accumulated total and labile C in the surface (0–2.5 cm) into deeper layers (5–10 cm), without significant loss on an equivalent soil mass basis compared to long-term NT (Quincke et al. 2007). However, an on-farm study has revealed that occasional tillage in a long-term NT can lead to significantly higher soil bulk density, somewhat higher soil penetration resistance, lower water-stable aggregate and equilibrium infiltrate rate, and lower grain yield compared to NT (Stavi et al. 2011). The most recent paper reviewed about occasional tillage under NT systems suggested that occasional tillage has limited effects on soil properties (pH, SOC and P, aggregation, water infiltration, soil MBC), GHG fluxes, and yield and weed control that lasts less than 2 years; and the benefits depend on tillage type, timing, depth, and frequency (once every 5–10 years of NT) (Blanco-Canqui and Wortmann 2020).

Solid research data on lime application and occasional tillage, influences on ESs, and approaches to improve the effectiveness of lime materials are lacking. Further work targeting different climate zones and management systems is needed to provide insights to address uncertainties associated with quantifying ES benefits.

15.5 CONCLUSION

Conservation agriculture, in particular NT management, has been adopted worldwide to overcome the concerns of the growing constraint of water, land degradation associated with erosion, and climate change. More and more evidence has shown that the effects of no-tillage on crop yield, SOC sequestration, GHG emissions, and water quality is dependent on site-specific pedology-climate conditions. In this book chapter, the synthesized previous studies and a case study from CAF from a cereal-based long-term research site located in the iPNW were used to demonstrate and discuss the benefits and potential tradeoffs of NT systems in dryland temperate regions. The case study of CAF showed that continuous NT increased SOC at the surface as well as crop yields while reducing N_2O emissions compared to reduced tillage. Surficial SOC, however, was reduced by high-disturbance NT drills and substantial N and P losses occurred through artificial subsurface drains and does give an alarm to consider the surficial soil disturbance created under NT and to focus on more effective fertilizer management.

In a broader perspective, agroecosystem management impacts system sustainability as a result of influencing often site-specific physico-chemical and biological drivers, which are key factors in regulating nutrient cycling, crop productivity, GHGs emission, water quality, and adaptation to climate change in the soil-plant-air continuum. Therefore, fully understanding the synergies and tradeoffs of NT management, in the context of socioeconomic drivers and environmental impacts, is crucial for long-term land management policy implementation and cropping system design at regional scales to optimize resource and adaptive capacities. A flexible framework is recommended with evaluation metrics for NT (Figure 15.16), to comprehensively assess NT impacts on ecosystem services and dynamic process response to NT in the soil-plant-air continuum. This evaluation is required to further develop policies promoting sustainable intensification to restore/rejuvenate the vulnerable semiarid ecosystem and increase its resilience to future uncertainties including weather extremes and a changing climate. Furthermore, some promising strategies contributing to sustainability of dryland NT agroecosystem proposed in Section 4 need further innovation and investigation. Summaries and case study examples here are relevant and applicable to other cropping systems in the world to promote more transdisciplinary efforts at field-scales.

ACKNOWLEDGEMENT

This research was a contribution from the Long-Term Agroecosystem Research (LTAR) network. LTAR is supported by the United States Department of Agriculture.

BIBLIOGRAPHY

Adl, Sina M., David C. Coleman, and Frederick Read. 2006. "Slow Recovery of Soil Biodiversity in Sandy Loam Soils of Georgia after 25 Years of No-Tillage Management." *Agriculture, Ecosystems and Environment* 114(2–4): 323–34.

Andriuzzi, Walter S., Thomas Bolger, and Olaf Schmidt. 2013. "The Drilosphere Concept: Fine-Scale Incorporation of Surface Residue-Derived N and C around Natural Lumbricus Terrestris Burrows. " *Soil Biology and Biochemistry* 64: 136–38.

Angst, Gerrit, Johanna Messinger, Maria Greiner, Werner Häusler, Dietrich Hertel, Kristina Kirfel, Ingrid Kögel-Knabner, Christoph Leuschner, Janet Rethemeyer, and Carsten W. Mueller. 2018. "Soil Organic Carbon Stocks in Topsoil and Subsoil Controlled by Parent Material, Carbon Input in the Rhizosphere, and Microbial-Derived Compounds." *Soil Biology and Biochemistry* 122: 19–30.

Arai, Miwa, Toshiko Miura, Hiroshi Tsuzura, Yukio Minamiya, and Nobuhiro Kaneko. 2018. "Two-Year Responses of Earthworm Abundance, Soil Aggregates, and Soil Carbon to No-Tillage and Fertilization." *Geoderma* 332: 135–41.

Awale, Rakesh, Amitava Chatterjee, and David Franzen. 2013. "Tillage and N-Fertilizer Influences on Selected Organic Carbon Fractions in a North Dakota Silty Clay Soil." *Soil and Tillage Research* 134: 213–22.

Asbjornsen, Heidi, V. Hernandez-Santana, M. Liebman, J. Bayala, J. Chen, M. Helmers, C. K. Ong, and A. Lisa Schulte. 2014. "Targeting Perennial Vegetation in Agricultural Landscapes for Enhancing Ecosystem Services." *Renewable Agriculture and Food Systems* 29(2): 101–25.

Baker, John M., Tyson E. Ochsner, Rodney T. Venterea, and Timothy J. Griffis. 2007. "Tillage and Soil Carbon Sequestration—What Do We Really Know?." *Agriculture, Ecosystems and Environment* 118(1–4): 1–5.

Bakhsh, Allah, Rameshwar S. Kanwar, Theodore B. Bailey, Cynthia A. Cambardella, D. L. Karlen, and Thomas S. Colvin. 2002. "Cropping System Effects on NO3–N Loss with Subsurface Drainage Water." *Transactions of the ASAE. American Society of Agricultural Engineers* 45(6): 1789.

Barba, Josep, Alejandro Cueva, Michael Bahn, Greg A. Barron-Gafford, Benjamin Bond-Lamberty, Paul J. Hanson, Aline Jaimes, et al. 2018. "Comparing Ecosystem and Soil Respiration: Review and Key Challenges of Tower-Based and Soil Measurements." *Agricultural and Forest Meteorology* 249: 434–43.

Bashagaluke, Janvier Bigabwa, Vincent Logah, Andrews Opoku, Joseph Sarkodie-Addo, and Charles Quansah. 2018. "Soil Nutrient Loss through Erosion: Impact of Different Cropping Systems and Soil Amendments in Ghana." *Plos One* 13(12): e0208250.

Beare, M. H., P. F. Hendrix, M. L. Cabrera, and D. C. Coleman. 1994. "Aggregate-Protected and Unprotected Organic Matter Pools in Conventional-And No-Tillage Soils." *Soil Science Society of America Journal* 58(3): 787–95.

Beck, Hylke E., Niklaus E. Zimmermann, Tim R. McVicar, Noemi Vergopolan, Alexis Berg, and Eric F. Wood. 2018. "Present and Future Köppen-Geiger Climate Classification Maps at 1-Km Resolution." *Scientific Data* 5: 180214.

Bellmore, Rebecca A., John A. Harrison, Joseph A. Needoba, Erin S. Brooks, and C. Kent Keller. 2015. "Hydrologic Control of Dissolved Organic Matter Concentration and Quality in a Semiarid Artificially Drained Agricultural Catchment." *Water Resources Research* 51(10): 8146–64.

Bezdicek, D. F., T. Beaver, and D. Granatstein. 2003. "Subsoil Ridge Tillage and Lime Effects on Soil Microbial Activity, Soil PH, Erosion, and Wheat and Pea Yield in the Pacific Northwest, USA." *Soil and Tillage Research* 74(1): 55–63.

Bjorneberg, David L., Ramesh S. Kanwar, and Stewart W. Melvin. 1996a. "Seasonal Changes in Flow and Nitrate-N Loss from Subsurface Drains." *Transactions of the ASAE. American Society of Agricultural Engineers* 39(3): 961–67.

Blanco-Canqui, Humberto, and Charles S. Wortmann. 2020. "Does Occasional Tillage Undo the Ecosystem Services Gained with No-Till? A Review." *Soil and Tillage Research* 198: 104534.

Blair, Graeme J., Rod D. B. Lefroy, and Leanne Lisle. 1995. "Soil Carbon Fractions Based on Their Degree of Oxidation, and the Development of a Carbon Management Index for Agricultural Systems." *Australian Journal of Agricultural Research* 46(7): 1459–66.

Blevins, R. L., L. W. Murdock, and G. W. Thomas. 1978. "Effect of Lime Application on No-Tillage and Conventionally Tilled Corn 1." *Agronomy Journal* 70(2): 322–26.

Bohlen, Patrick J., Robert W. Parmelee, David A. McCartney, and Clive A. Edwards. 1997. "Earthworm Effects on Carbon and Nitrogen Dynamics of Surface Litter in Corn Agroecosystems." *Ecological Applications* 7(4): 1341–49.

Bossuyt, Heleen, Johan Six, and Paul F. Hendrix. 2002. "Aggregate-Protected Carbon in No-Tillage and Conventional Tillage Agroecosystems Using Carbon-14 Labeled Plant Residue." *Soil Science Society of America Journal* 66(6): 1965–73.

Brown, George G., Isabelle Barois, and Patrick Lavelle. 2000. "Regulation of Soil Organic Matter Dynamics and Microbial Activity in the Drilosphere and the Role of Interactions with Other Edaphic Functional Domains." *European Journal of Soil Biology* 36(3–4): 177–98.

Brown, Tabitha, and D. R. Huggins. 2012. "Soil Carbon Sequestration in the Dryland Cropping Region of the Pacific Northwest." *Journal of Soil and Water Conservation* 67(5): 406–15.

Brown, Tabitha T., Richard T. Koenig, David R. Huggins, James B. Harsh, and Richard E. Rossi. 2008. "Lime Effects on Soil Acidity, Crop Yield, and Aluminum Chemistry in Direct-Seeded Cropping Systems." *Soil Science Society of America Journal* 72(3): 634–40.

Busacca, A. J., C. A. Cook, and D. J. Mulla. 1993. "Comparing Landscape-Scale Estimation of Soil Erosion in the Palouse Using Cs-137 and RUSLE." *Journal of Soil and Water Conservation* 48(4): 361–67.

Caires, E. F., H. A. W. Joris, and S. Churka. 2011. "Long-Term Effects of Lime and Gypsum Additions on No-Till Corn and Soybean Yield and Soil Chemical Properties in Southern Brazil." *Soil Use and Management* 27(1): 45–53.

Cano, Amanda, Agustín Núñez, Veronica Acosta-Martinez, Meagan Schipanski, Rajan Ghimire, Charles Rice, and Charles West. 2018. "Current Knowledge and Future Research Directions to Link Soil Health and Water Conservation in the Ogallala Aquifer Region." *Geoderma* 328: 109–18.

Chaudhuri, Sriroop, and Srinivasulu Ale. 2014. "Long Term (1960–2010) Trends in Groundwater Contamination and Salinization in the Ogallala Aquifer in Texas." *Journal of Hydrology* 513: 376–90.

Chi, Jinshu, Sarah Waldo, Shelley Pressley, Patrick O'Keeffe, David Huggins, Claudio Stöckle, William L. Pan, Erin Brooks, and Brian Lamb. 2016. "Assessing Carbon and Water Dynamics of No-Till and Conventional Tillage Cropping Systems in the Inland Pacific Northwest US Using the Eddy Covariance Method." *Agricultural and Forest Meteorology* 218: 37–49.

Chi, Jinshu, Sarah Waldo, Shelley N. Pressley, Eric S. Russell, Patrick T. O'Keeffe, William L. Pan, David R. Huggins, Claudio O. Stöckle, Erin S. Brooks, and Brian K. Lamb. 2017. "Effects of Climatic Conditions and Management Practices on Agricultural Carbon and Water Budgets in the Inland Pacific Northwest USA." *Journal of Geophysical Research: Biogeosciences* 122(12): 3142–60.

Chintala, Rajesh, Javier Mollinedo, Thomas E. Schumacher, Douglas D. Malo, and James L. Julson. 2014. "Effect of Biochar on Chemical Properties of Acidic Soil." *Archives of Agronomy and Soil Science* 60(3): 393–404.

Chirinda, Ngonidzashe, Lars Elsgaard, Ingrid K. Thomsen, Goswin Heckrath, and Jørgen E. Olesen. 2014. "Carbon Dynamics in Topsoil and Subsoil along a Cultivated Toposequence." *CATENA* 120: 20–8.

Clay, David E., Sharon A. Clay, Kurtis D. Reitsma, Barry H. Dunn, Alexander J. Smart, Gregg G. Carlson, David Horvath, and James J. Stone. 2014. "Does the Conversion of Grasslands to Row Crop Production in Semi-Arid Areas Threaten Global Food Supplies?" *Global Food Security* 3(1): 22–30.

Cochran, R. L., H. P. Collins, Ann Kennedy, and D. F. Bezdicek. 2007. "Soil Carbon Pools and Fluxes after Land Conversion in a Semiarid Shrub-Steppe Ecosystem." *Biology and Fertility of Soils* 43(4): 479–89.

Collier, Sarah M., Matthew D. Ruark, Lawrence G. Oates, William E. Jokela, and Curtis J. Dell. 2014. "Measurement of Greenhouse Gas Flux from Agricultural Soils Using Static Chambers." *Journal of Visualized Experiments: JoVE* 90(90): e52110.

Crow, Susan E., and Carlos A. Sierra. 2018. "Dynamic, Intermediate Soil Carbon Pools May Drive Future Responsiveness to Environmental Change." *Journal of Environmental Quality* 47(4): 607–16.

Dabney, S. M., J. A. Delgado, and D. W. Reeves. 2001. "Using Winter Cover Crops to Improve Soil and Water Quality." *Communications in Soil Science and Plant Analysis* 32(7–8): 1221–50.

Dao, Thanh H. 1993. "Tillage and Winter Wheat Residue Management Effects on Water Infiltration and Storage." *Soil Science Society of America Journal* 57(6): 1586–95.

Daryanto, Stefani, Lixin Wang, and Pierre André Jacinthe. 2017. "Meta-Analysis of Phosphorus Loss from No-Till Soils." *Journal of Environmental Quality* 46(5): 1028–37.

Datta, K. K., and C. De Jong. 2002. "Adverse Effect of Waterlogging and Soil Salinity on Crop and Land Productivity in Northwest Region of Haryana, India." *Agricultural Water Management* 57(3): 223–38.

DeLaune, P. B., and J. W. Sij. 2012. "Impact of Tillage on Runoff in Long Term No-Till Wheat Systems." *Soil and Tillage Research* 124: 32–5.

de San Celedonio, Romina Paola, Leonor Gabriela Abeledo, J. M. Brihet, and Daniel Julio Miralles. 2016. "Waterlogging Affects Leaf and Tillering Dynamics in Wheat and Barley." *Journal of Agronomy and Crop Science* 202(5): 409–20.

Deng, Lei, Guang-yu Zhu, Zhuang-sheng Tang, and Zhou-ping Shangguan. 2016. "Global Patterns of the Effects of Land-Use Changes on Soil Carbon Stocks." *Global Ecology and Conservation* 5: 127–38.

Dentzman, Katherine, and Ian Cristofer Burke. 2021. "Herbicide Resistance, Tillage, and Community Management in the Pacific Northwest." *Sustainability* 13(4): 1937.

Ding, Yu, Xing Huang, Yong Li, Haiyang Liu, Qichun Zhang, Xingmei Liu, Jianming Xu, and Hongjie Di. 2021. "Nitrate leaching losses mitigated with intercropping of deep-rooted and shallow-rooted plants." *Journal of Soils and Sediments* 21(1): 364–75.

Doetterl, Sebastian, Asmeret Asefaw Berhe, Elisabet Nadeu, Zhengang Wang, Michael Sommer, and Peter Fiener. 2016. "Erosion, Deposition and Soil Carbon: A Review of Process-Level Controls, Experimental Tools and Models to Address C Cycling in Dynamic Landscapes." *Earth-Science Reviews* 154: 102–22.

Don, Axel, Bert Steinberg, Ingo Schöning, Karin Pritsch, Monika Joschko, Gerd Gleixner, and Ernst-Detlef Schulze. 2008. "Organic Carbon Sequestration in Earthworm Burrows." *Soil Biology and Biochemistry* 40(7): 1803–12.

Douglas, Alexandra, and Sally C. Peltzer. 2004. "Managing Herbicide Resistant Annual Ryegrass (Lolium rigidum Gaud.) in No-Till Systems in Western Australia Using Occasional Inversion Ploughing." In: Proceedings of the 14th Australian Weeds Conference, 300–3.

Duchemin, E., M. Lucotte, and R. Canuel. 1999. "Comparison of Static Chamber and Thin Boundary Layer Equation Methods for Measuring Greenhouse Gas Emissions from Large Water Bodies §." *Environmental Science and Technology* 33(2): 350–57.

Duniway, Michael Cohrs, Jeffrey E. Herrick, and H. Curtis Monger. 2010. "Spatial and Temporal Variability of Plant-Available Water in Calcium Carbonate-Cemented Soils and Consequences for Arid Ecosystem Resilience." *Oecologia* 163(1): 215–26.

Edwards, William M., M. J. Shipitalo, L. B. Owens, and W. A. Dick. 1993. "Factors Affecting Preferential Flow of Water and Atrazine through Earthworm Burrows under Continuous No-Till Corn." *Journal of Environmental Quality* 22(3): 453–57.

FAO (Food and Agriculture Organization). 2015. "Status of the World's Soil Resources." http://www.fao.org/3/i5199e/I5199E.pdf.

FAO and ITPS. 2020. *State of Knowledge of Soil Biodiversity: Status, Challenges and Potentialities, Summary for Policy Makers*. http://www.fao.org/3/cb1929en/CB1929EN.pdf.

FAO/UNESCO (Food and Agriculture Organization/ The United Nations Educational, Scientific and Cultural Organization). 2007. Soil Map of the World. https://data.apps.fao.org/map/catalog/srv/eng/catalog.search#/metadata/3f9be529-5672-4286-86c7-a63c7da39853.

Flower, K. C., and W. L. Crabtree. 2011. "Soil PH Change after Surface Application of Lime Related to the Levels of Soil Disturbance Caused by No-Tillage Seeding Machinery." *Field Crops Research* 121(1): 75–87.

Fonte, Steven J., Angela Y. Y. Kong, Chris van Kessel, Paul F. Hendrix, and Johan Six. 2007. "Influence of Earthworm Activity on Aggregate-Associated Carbon and Nitrogen Dynamics Differs with Agroecosystem Management." *Soil Biology and Biochemistry* 39(5): 1014–22.

Forster, Piers, Venkatachalam Ramaswamy, Paulo Artaxo, Terje Berntsen, Richard Betts, David W. Fahey, James Haywood, Judith Lean, David C. Lowe, and Gunnar Myhre. 2007. "Changes in Atmospheric Constituents and in Radiative Forcing. Chapter 2." In: *Climate Change 2007. The Physical Science Basis.*

Friedrich, Theodor. 2005. "Does No-Till Farming Require More Herbicides?" *Outlooks on Pest Management* 16(4): 188.

Fuentes, Juan P., Markus Flury, and David F. Bezdicek. 2004. "Hydraulic Properties in a Silt Loam Soil under Natural Prairie, Conventional Till, and No-Till." *Soil Science Society of America Journal* 68(5): 1679–88.

Garcia-Franco, Noelia, Eleanor Hobley, Rico Hübner, and Martin Wiesmeier. 2018. "Chapter 23-Climate-Smart Soil Management in Semiarid Regions." In: *Soil Management and Climate Change*, eds. María Ángeles Muñoz and Raúl Zornoza. Academic Press, 349–368.

Garratt, Michael P. D., Duncan J. Coston, C. L. Truslove, M. G. Lappage, C. Polce, R. Dean, J. C. Biesmeijer, and Simon G. Potts. 2014. "The Identity of Crop Pollinators Helps Target Conservation for Improved Ecosystem Services." *Biological Conservation* 169(100): 128–35.

Ghimire, Rajan, Stephen Machado, and Prakriti Bista. 2017. "Soil PH, Soil Organic Matter, and Crop Yields in Winter Wheat–Summer Fallow Systems." *Agronomy Journal* 109(2): 706–17.

Guo, Lanbin B., and R. M. Gifford. 2002. "Soil Carbon Stocks and Land Use Change: A Meta Analysis." *Global Change Biology* 8(4): 345–60.

Hassan, Rashid, Robert Scholes, and Neville Ash. 2005. *Ecosystems and Human Well-Being: Current State and Trends*. Washington: DC(USA) Island Press.

Heap, I. The International Herbicide-Resistant Weed Database. Online. Tuesday, January 19, 2021. http://weedscience.org/Pages/GeoChart.aspx.

Heuperman, Albertus Frederik, Arjun S. Kapoor, and Harry W. Denecke. 2002. *Biodrainage: Principles, Experiences and Applications*. Food & Agriculture Org. No. 6.

Halvorson, A. D. 1988. "Role of Cropping Systems in Environmental Quality: Saline Seep Control." *Cropping Strategies for Efficient Use of Water and Nitrogen* 51: 179–91.

House, Garfield J., and Robert W. Parmelee. 1985. "Comparison of Soil Arthropods and Earthworms from Conventional and No-Tillage Agroecosystems." *Soil and Tillage Research* 5(4): 351–60.

Huang, Jianping, Mingxia Ji, Yongkun Xie, Shanshan Wang, Yongli He, and Jinjiang Ran. 2016. "Global Semi-Arid Climate Change over Last 60 Years." *Climate Dynamics* 46(3–4): 1131–50.

Huggins, D., W. Pan, and J. Smith. 2010. "Yield, Protein and Nitrogen Use Efficiency of Spring Wheat: Evaluating Field-Scale Performance Chapter 17." 2010: 2010–001.

Huggins, D. R., G. A. Buyanovsky, G. H. Wagner, J. R. Brown, R. G. Darmody, T. R. Peck, G. W. Lesoing, M. B. Vanotti, and L. G. Bundy. 1998. "Soil Organic C in the Tallgrass Prairie-Derived Region of the Corn Belt: Effects of Long-Term Crop Management." *Soil and Tillage Research* 47(3–4): 219–34.

Huggins, D. R., W. L. Pan, and J. L. Smith. 1989. Improving Yield, Percent Protein, and N Use Efficiency of No-Till Hard Red Spring Wheat through Crop Rotation and Fall N Fertilization. In Proceedings of the 40th Annual Far West Regional Fertilizer Conference. Coeur d'Alene, Idaho.

Huggins, David R., Chad E. Kruger, Kathleen M. Painter, and David P. Uberuaga. 2014. "Site-Specific Trade-Offs of Harvesting Cereal Residues as Biofuel Feedstocks in Dryland Annual Cropping Systems of the Pacific Northwest, USA." *Bioenergy Research* 7(2): 598–608.

Huggins, David R., and John P. Reganold. 2008. "No-Till: The Quiet Revolution." *Scientific American* 299(1): 70–7.

Hussain, Mir Zaman, G. Philip Robertson, Bruno Basso, and Stephen K. Hamilton. 2020. "Leaching Losses of Dissolved Organic Carbon and Nitrogen from Agricultural Soils in the Upper US Midwest." *Science of the Total Environment* 734: 139379. https://doi.org/10.1016/j.scitotenv.2020.139379.

Inagaki, Thiago Massao, João Carlos de Moraes Sá, Eduardo Fávero Caires, and Daniel Ruiz Potma Gonçalves. 2017. "Why Does Carbon Increase in Highly Weathered Soil under No-Till upon Lime and Gypsum Use?" *Science of the Total Environment* 599: 523–32.

IPCC. 2020. A Special Report on Climate Change and Land, Summary for Policymakers. https://www.ipcc.ch/srccl/ and https://www.ipcc.ch/site/assets/uploads/sites/4/2020/02/SPM_Updated-Jan20.pdf.

Janzen, H. H., C. A. Campbell, R. C. Izaurralde, B. H. Ellert, N. Juma, W. B. McGill, and R. P. Zentner. 1998. "Management Effects on Soil C Storage on the Canadian Prairies." *Soil and Tillage Research* 47(3–4): 181–95.

Jaynes, D. B., and D. E. James. 2007. *The Extent of Farm Drainage in the United States*. US Department of Agriculture.

Jenkins, Michael B., Dinku M. Endale, Harry H. Schomberg, Peter G. Hartel, and Miguel L. Cabrera. 2009. "17β-Estradiol and Testosterone in Drainage and Runoff from Poultry Litter Applications to Tilled and No-Till Crop Land under Irrigation." *Journal of Environmental Management* 90(8): 2659–64.

Jin, Shaofei, Xiaohong Tian, and Hesong Wang. 2018. "Hierarchical Responses of Soil Organic and Inorganic Carbon Dynamics to Soil Acidification in a Dryland Agroecosystem, China." *Journal of Arid Land* 10(5): 726–36.

Johnson-Maynard, J. L., K. J. Umiker, and S. O. Guy. 2007. "Earthworm Dynamics and Soil Physical Properties in the First Three Years of No-Till Management." *Soil and Tillage Research* 94(2): 338–45.

Jones, Clain, Rick Engel, and Kathrin Olson-Rutz. 2019. "Soil Acidification in the Semiarid Regions of North America's Great Plains." *Crops & Soils* 52(2): 28–56.

Jordán, M. M., J. Navarro-Pedreno, E. García-Sánchez, J. Mateu, and P. Juan. 2004. "Spatial Dynamics of Soil Salinity under Arid and Semi-Arid Conditions: Geological and Environmental Implications." *Environmental Geology* 45(4): 448–56.

Kaur, Harsimran, David R. Huggins, Richard A. Rupp, John T. Abatzoglou, Claudio O. Stöckle, and John P. Reganold. 2017. "Agroecological Class Stability Decreases in Response to Climate Change Projections for the Pacific Northwest, USA." *Frontiers in Ecology and Evolution* 5: 74.

Kašanin-Grubin, Milica, Francesca Vergari, Francesco Troiani, and Marta Della Seta. 2018. "Chapter 3-The Role of Lithology: Parent Material Controls on Badland Development."In: *Badlands Dynamics in a Context of Global Change*, eds. Estela Nadal-Romero, Juan F. Martínez-Murillo and Nikolaus J. Kuhn. Elsevier, 61–109.

Kassam, Amir, Theordor Friedrich, R. Derpsch, and J. Kienzle. 2015. "Overview of the Worldwide Spread of Conservation Agriculture." Field Actions Science Reports. *The Journal of Field Actions* 8.

Kassam, A., T. Friedrich, and R. Derpsch. 2019. "Global Spread of Conservation Agriculture." *International Journal of Environmental Studies* 76(1): 29–51.

Keller, C. Kent, Caroline N. Butcher, Jeffrey L. Smith, and Richelle M. Allen-King. 2008. "Nitrate in Tile Drainage of the Semiarid Palouse Basin." *Journal of Environmental Quality* 37(2): 353–61.

Kelley, Christopher J., C. Kent Keller, Erin S. Brooks, Jeffrey L. Smith, Cailin Huyck Orr, and R. D. Evans. 2017. "Water and Nitrogen Movement through a Semiarid Dryland Agricultural Catchment: Seasonal and Decadal Trends." *Hydrological Processes* 31(10): 1889–99.

Kelley, Christopher J., C. Kent Keller, R. D. Evans, C. H. Orr, Jeffrey L. Smith, and Benjamin A. Harlow. 2013. "Nitrate–Nitrogen and Oxygen Isotope Ratios for Identification of Nitrate Sources and Dominant Nitrogen Cycle Processes in a Tile-Drained Dryland Agricultural Field." *Soil Biology and Biochemistry* 57: 731–38.

Kennedy, Ann C., and William F. Schillinger. 2006. "Soil Quality and Water Intake in Traditional-Till vs. No-Till Paired Farms in Washington's Palouse Region." *Soil Science Society of America Journal* 70(3): 940–9.

Kitchen, N. R., K. A. Sudduth, D. B. Myers, R. E. Massey, E. J. Sadler, R. N. Lerch, J. W. Hummel, and H. L. Palm. 2005. "Development of a Conservation-Oriented Precision Agriculture System: Crop Production Assessment and Plan Implementation." *Journal of Soil and Water Conservation* 60(6): 421–30.

Kladivko, Eileen J., Neela M. Akhouri, and Glenn Weesies. 1997. "Earthworm Populations and Species Distributions under No-Till and Conventional Tillage in Indiana and Illinois." *Soil Biology and Biochemistry* 29(3–4): 613–15.

Klein, Alexandra-Maria, Bernard E. Vaissiere, James H. Cane, Ingolf Steffan-Dewenter, Saul A. Cunningham, Claire Kremen, and Teja Tscharntke. 2007. "Importance of Pollinators in Changing Landscapes for World Crops." *Proceedings of the Royal Society B: Biological Sciences* 274(1608): 303–13.

Kleinman, P. J. A., Sheri Spiegal, J. R. Rigby, S. C. Goslee, J. M. Baker, B. T. Bestelmeyer, R. K. Boughton, et al. 2018. "Advancing the Sustainability of US Agriculture through Long-Term Research." *Journal of Environmental Quality* 47(6): 1412–25.

Kleinman, Peter J. A., Andrew N. Sharpley, Lou S. Saporito, Anthony R. Buda, and Ray B. Bryant. 2009. "Application of Manure to No-Till Soils: Phosphorus Losses by Sub-Surface and Surface Pathways." *Nutrient Cycling in Agroecosystems* 84(3): 215–27.

Koohafkan, Parviz, and Bobby Alton Stewart. 2008. *Water and Cereals in Drylands*. London: Routledge. DOI: https://doi.org/10.4324/9781849773744.

Kostyanovsky, K. I., D. R. Huggins, C. O. Stockle, S. Waldo, and B. Lamb. 2018. "Developing a Flow through Chamber System for Automated Measurements of Soil N_2O and CO_2 Emissions." *Measurement* 113: 172–80.

Kostyanovsky, Kirill I., David R. Huggins, Claudio O. Stockle, Jason G. Morrow, and Isaac J. Madsen. 2019. "Emissions of N_2O and CO_2 Following Short-Term Water and n Fertilization Events in Wheat-Based Cropping Systems." *Frontiers in Ecology and Evolution* 7: 63.

Kovács-Hostyánszki, Anikó, Anahí Espíndola, Adam J. Vanbergen, Josef Settele, Claire Kremen, and Lynn V. Dicks. 2017. "Ecological Intensification to Mitigate Impacts of Conventional Intensive Land Use on Pollinators and Pollination." *Ecology Letters* 20(5): 673–89.

Kuhn, Nikolaus J., Yaxian Hu, Lena Bloemertz, Jin He, Hongwen Li, and Philip Greenwood. 2016. "Conservation Tillage and Sustainable Intensification of Agriculture: Regional vs. Global Benefit Analysis." *Agriculture, Ecosystems and Environment* 216: 155–65.

Lal, R., R. F. Follett, J. Kimble, and C. V. Cole. 1999. "Managing US Cropland to Sequester Carbon in Soil." *Journal of Soil and Water Conservation* 54(1): 374–81.

Lal, Rattan. 2003. "Soil Erosion and the Global Carbon Budget." *Environment International* 29(4): 437–50.

Lal, Rattan. 2004a. "Carbon Sequestration in Dryland Ecosystems." *Environmental Management* 33(4): 528–44.

Lal, Ratan. 2004b. "Soil Carbon Sequestration Impacts on Global Climate Change and Food Security." *Science* 304(5677): 1623–27.

Lal, Ratan. 2015. "16 Challenges and Opportunities in Precision Agriculture." *Soil-Specific Farming: Precision Agriculture* 22: 391.

Lavallee, J. M., R. T. Conant, M. L. Haddix, R. F. Follett, M. I. Bird, and E. A. Paul. 2019. "Selective Preservation of Pyrogenic Carbon across Soil Organic Matter Fractions and Its Influence on Calculations of Carbon Mean Residence Times." *Geoderma* 354: 113866.

Lehmann, Johannes, John Gaunt, and Marco Rondon. 2006. "Bio-Char Sequestration in Terrestrial Ecosystems–A Review." *Mitigation and Adaptation Strategies for Global Change* 11(2): 403–27.

Leon, Elievf, Rodrigo Vargas, Stephen Bullock, Eulogio Lopez, Alan Rodrigo Panosso, and Newton La Scala Jr. 2014. "Hot Spots, Hot Moments, and Spatio-temporal Controls on Soil CO2 Efflux in a Water-Limited Ecosystem." *Soil Biology and Biochemistry* 77: 12–21.

Lerch, R. N., N. R. Kitchen, R. J. Kremer, W. W. Donald, E. E. Alberts, E. J. Sadler, K. A. Sudduth, D. B. Myers, and Fessehaie Ghidey. 2005. "Development of a Conservation-Oriented Precision Agriculture System: Water and Soil Quality Assessment." *Journal of Soil and Water Conservation* 60(6): 411–21.

Lewis, Ricky W., Victoria P. Barth, Todd Coffey, Carol McFarland, David R. Huggins, and Tarah S. Sullivan. 2018. "Altered Bacterial Communities in Long-Term No-Till Soils Associated with Stratification of Soluble Aluminum and Soil PH." *Soil Systems* 2(1): 7.

Li, Sheng, David A. Lobb, and Michael J. Lindstrom. 2007. "Tillage Translocation and Tillage Erosion in Cereal-Based Production in Manitoba, Canada." *Soil and Tillage Research* 94(1): 164–82.

Li, Xia, Gregory W. McCarty, Douglas L. Karlen, and Cynthia A. Cambardella. 2018. "Topographic Metric Predictions of Soil Redistribution and Organic Carbon in Iowa Cropland Fields." *CATENA* 160: 222–32.

Lindstrom, Michael J., W. W. Nelson, and T. E. Schumacher. 1992. "Quantifying Tillage Erosion Rates Due to Moldboard Plowing." *Soil and Tillage Research* 24(3): 243–55.

Lorenz, Klaus, and Rattan Lal. 2005. "The Depth Distribution of Soil Organic Carbon in Relation to Land Use and Management and the Potential of Carbon Sequestration in Subsoil Horizons." *Advances in Agronomy* 88: 35–66.

Luo, Zhongkui, Enli Wang, and Osbert J. Sun. 2010. "Can No-Tillage Stimulate Carbon Sequestration in Agricultural Soils? A Meta-Analysis of Paired Experiments." *Agriculture, Ecosystems and Environment* 139(1–2): 224–31.

Magdoff, Fred, and Harold Van Es. 2000. "Building Soils for Better Crops." *Sustainable Agriculture Network Beltsville.*

Mahler, R. L. 2002. "Impacts and Management of Soil Acidity under Direct-Seed Systems: Status and Effects on Crop Production." In: *Spokane Ag Expo and Pacific Northwest Farm Forum*. Spokane, Washington.

Majumder, Shyam Prasad, Anantha Krishna Chaitanya, Ashim Datta, Dhaneshwar Padhan, Shrikant Badole, and Biswapati Mandal. 2018. "Chapter 11-Dynamics of Carbon and Nitrogen in Agricultural Soils: Role of Organic and Inorganic Sources."In: *Soil Management and Climate Change*, eds. María Ángeles Muñoz and Raúl Zornoza. Academic Press, 151–69.

Manninen, Noora, Helena Soinne, Riitta Lemola, Laura Hoikkala, and Eila Turtola. 2018. "Effects of Agricultural Land Use on Dissolved Organic Carbon and Nitrogen in Surface Runoff and Subsurface Drainage." *Science of the Total Environment* 618: 1519–28.

Mas, Laura I., Virginia C. Aparicio, Eduardo De Gerónimo, and José L. Costa. 2020. "Pesticides in Water Sources Used for Human Consumption in the Semiarid Region of Argentina." *SN Applied Sciences* 2(4): 1–18.

McClain, Michael E., Elizabeth W. Boyer, C. Lisa Dent, Sarah E. Gergel, Nancy B. Grimm, Peter M. Groffman, Stephen C. Hart, et al. 2003. "Biogeochemical Hot Spots and Hot Moments at the Interface of Terrestrial and Aquatic Ecosystems." *Ecosystems* 6(4): 301–12.

McCool, D. K., and R. D. Roe. 2005. "Long-Term Erosion Trends on Cropland in the Pacific Northwest." In: *2003, 2004, 2005 Pacific Northwest Region Papers*, 1. American Society of Agricultural and Biological Engineers.

McFarland, Carol, and David R. Huggins. 2015. "Acidification in the Inland Pacific Northwest." *Crops & Soils* 48(2): 4–12.

McMichael, Anthony J., Diarmid Campbell-Lendrum, Sari Kovats, Sally Edwards, Paul Wilkinson, Theresa Wilson, Robert Nicholls, Simon Hales, Frank Tanser, David Le Sueur, Michale Schlesinger and Natasha Andronova. 2004. "Chapter 20- Global Climate Change". In: *Comparative Quantification of Health Risks : Global and Regional Burden of Disease Attributable to Selected Major Risk Factors,* eds. E. Majid, A. D. Lopez, A. A. Rodgers, and C. J. L. Murray. Geneva, Switzerland: World Health Organisation, 1543–650.

Mehta, Sunil, Alan E. Fryar, and Jay L. Banner. 2000. "Controls on the Regional-Scale Salinization of the Ogallala Aquifer, Southern High Plains, Texas, USA." *Applied Geochemistry* 15(6): 849–64.

Miller, M. R., P. L. Brown, J. J. Donovan, R. N. Bergatino, J. L. Sonderegger, and F. A. Schmidt. 1981. "Saline Seep Development and Control in the North American Great Plains-Hydrogeological Aspects." *Agricultural Water Management* 4(1–3): 115–41.

Obia, Alfred, Gerard Cornelissen, Jan Mulder, and Peter Dörsch. 2015. "Effect of Soil PH Increase by Biochar on NO, N_2O and N_2 Production during Denitrification in Acid Soils." *PLOS ONE* 10(9): e0138781.

Ogle, Stephen M., Cody Alsaker, Jeff Baldock, Martial Bernoux, F. Jay Breidt, Brian McConkey, Kristiina Regina, and Gabriel G. Vazquez-Amabile. 2019. "Climate and Soil Characteristics Determine Where No-Till Management Can Store Carbon in Soils and Mitigate Greenhouse Gas Emissions." *Scientific Reports* 9(1): 1–8.

Ogle, Stephen M., Amy Swan, and Keith Paustian. 2012. "No-Till Management Impacts on Crop Productivity, Carbon Input and Soil Carbon Sequestration." *Agriculture, Ecosystems and Environment* 149: 37–49.

Oldfield, Emily E., Mark A. Bradford, and Stephen A. Wood. 2019. "Global Meta-Analysis of the Relationship between Soil Organic Matter and Crop Yields." *Soil* 5(1): 15–32.

Ollerton, Jeff. 2017. "Pollinator Diversity: Distribution, Ecological Function, and Conservation." *Annual Review of Ecology, Evolution, and Systematics* 48(1): 353–76.

Ortega-Pieck, Aline, Jessica Norby, Erin S. Brooks, Daniel Strawn, Alex R. Crump, and David R. Huggins. 2020. "Sources and Subsurface Transport of Dissolved Reactive Phosphorus in a Semiarid, No-Till Catchment with Complex Topography." *Journal of Environmental Quality* 49(5): 1286–97.

Pannkuk, Chris D., Robert I. Papendick, and Keith E. Saxton. 1997. "Fallow Management Effects on Soil Water Storage and Wheat Yields in the Pacific Northwest." *Agronomy Journal* 89(3): 386–91.

Parihar, Chitar Mal, S. L. Jat, A. K. Singh, K. Majumdar, M. L. Jat, Y. S. Saharawat, S. Pradhan, and B. R. Kuri. 2017. "Bio-Energy, Water-Use Efficiency and Economics of Maize-Wheat-Mungbean System under Precision-Conservation Agriculture in Semi-Arid Agroecosystem." *Energy* 119: 245–56.

Peixoto, Devison Souza, Lucas de Castro Moreira da Silva, Laura Beatriz Batista de Melo, Raphael Passaglia Azevedo, Brunno Cassiano Lemos Araújo, Teotônio Soares de Carvalho, Silvino Guimarães Moreira, Nilton Curi, and Bruno Montoani Silva. 2020. "Occasional Tillage in No-Tillage Systems: A Global Meta-Analysis." *Science of the Total Environment* 745: 140887.

Peng, Qiuping. 2020. *Labile Organic Carbon in Soil Profiles after Long-Term No-Tillage: Laboratory to Field-Scale and Comparison with Native Prairie*. PhD Diss. Washington State University.

Pennock, Dan, N. McKenzie, and L. Montanarella. 2015. "Status of the World's Soil Resources." *Technical Summary*. FAO, Rome, Italy.

Peters, R. D., A. V. Sturz, M. R. Carter, and J. B. Sanderson. 2003. "Developing Disease-Suppressive Soils through Crop Rotation and Tillage Management Practices." *Soil and Tillage Research* 72(2): 181–92.

Pi, Huawei, David R. Huggins, John T. Abatzoglou, and Brenton Sharratt. 2020a. "Modeling Soil Wind Erosion from Agroecological Classes of the Pacific Northwest in Response to Current Climate." *Journal of Geophysical Research: Atmospheres* 125(2): e2019JD031104.

Pi, Huawei, Nicholas P. Webb, David R. Huggins, and Brenton Sharratt. 2020b. "Critical Standing Crop Residue Amounts for Wind Erosion Control in the Inland Pacific Northwest, USA." *CATENA* 195: 104742.

Pittelkow, Cameron M., Bruce A. Linquist, Mark E. Lundy, Xinqiang Liang, Kees Jan van Groenigen, Juhwan Lee, Natasja van Gestel, Johan Six, Rodney T. Venterea, and Chris van Kessel. 2015. "When Does No-till Yield More? A Global Meta-Analysis." *Field Crops Research* 183: 156–68.

Poeplau, Christopher, Axel Don, Johan Six, Michael Kaiser, Dinesh Benbi, Claire Chenu, M. Francesca Cotrufo, et al. 2018. "Isolating Organic Carbon Fractions with Varying Turnover Rates in Temperate Agricultural Soils–A Comprehensive Method Comparison." *Soil Biology and Biochemistry* 125: 10–26.

Post, Wilfred M., and Kyung C. Kwon. 2000. "Soil Carbon Sequestration and Land-Use Change: Processes and Potential." *Global Change Biology* 6(3): 317–27.

Price, Andrew J., K. S. Balkcom, S. A. Culpepper, J. A. Kelton, R. L. Nichols, and H. Schomberg. 2011. "Glyphosate-Resistant Palmer Amaranth: A Threat to Conservation Tillage." *Journal of Soil and Water Conservation* 66(4): 265–75.

Purakayastha, T. J., D. R. Huggins, and J. L. Smith. 2008. "Carbon Sequestration in Native Prairie, Perennial Grass, No-Till, and Cultivated Palouse Silt Loam." *Soil Science Society of America Journal* 72(2): 534–40.

Quincke, J. A., C. S. Wortmann, M. Mamo, T. Franti, and R. A. Drijber. 2007. "Occasional Tillage of No-Till Systems: Carbon Dioxide Flux and Changes in Total and Labile Soil Organic Carbon." *Agronomy Journal* 99(4): 1158–68.

Ragab, Ragab, and Christel Prudhomme. 2002. "SW—Soil and Water: Climate Change and Water Resources Management in Arid and Semi-Arid Regions: Prospective and Challenges for the 21st Century." *Biosystems Engineering* 81(1): 3–34.

Raich, James W., and William H. Schlesinger. 1992. "The Global Carbon Dioxide Flux in Soil Respiration and Its Relationship to Vegetation and Climate." *Tellus: Series B* 44(2): 81–99.

Reganold, John P., Lloyd F. Elliott, and Yvonne L. Unger. 1987. "Long-Term Effects of Organic and Conventional Farming on Soil Erosion." *Nature* 330(6146): 370.

Regina, Kristiina, and Laura Alakukku. 2010. "Greenhouse Gas Fluxes in Varying Soils Types under Conventional and No-Tillage Practices." *Soil and Tillage Research* 109(2): 144–52.

Reich, Paul, Hari Eswaran, and Fred Beinroth. 1999. "Global Dimensions of Vulnerability to Wind and Water Erosion." In: *Sustaining the Global Farm*, eds. D. E. Stott, R. H. Mohtar, and G. C. Steinhardt, Selected Papers from the 10th International Soil Conservation Organization Meeting, Purdue University and USDA-ARS National Soil Erosion Research Laboratory, 838–46.

Reichstein, Markus, and Christian Beer. 2008. "Soil Respiration across Scales: The Importance of a Model–Data Integration Framework for Data Interpretation." *Journal of Plant Nutrition and Soil Science* 171(3): 344–54.

Robertson, G. Philip, Eldor A. Paul, and Richard R. Harwood. 2000. "Greenhouse Gases in Intensive Agriculture: Contributions of Individual Gases to the Radiative Forcing of the Atmosphere." *Science* 289(5486): 1922–25.

Rychel, Katrin, Katharina HE Meurer, Gunnar Börjesson, Monika Strömgren, Gizachew Tarekegn Getahun, Holger Kirchmann, and Thomas Kätterer. 2020. "Deep N Fertilizer Placement Mitigated N 2 O Emissions in a Swedish Field Trial with Cereals." *Nutrient Cycling in Agroecosystems*: 1–16.

Sapkota, Tek Bahadur, Marco Mazzoncini, Paolo Bàrberi, Daniele Antichi, and Nicola Silvestri. 2012. "Fifteen Years of No till Increase Soil Organic Matter, Microbial Biomass and Arthropod Diversity in Cover Crop-Based Arable Cropping Systems." *Agronomy for Sustainable Development* 32(4): 853–63.

Saqib, Muhammad, Javaid Akhtar, and Riaz Hussain Qureshi. 2004. "Pot Study on Wheat Growth in Saline and Waterlogged Compacted Soil: II. Root Growth and Leaf Ionic Relations." *Soil and Tillage Research* 77(2): 179–87.

Saxton, Keith, David Chandler, and William Schillinger. 2001. "Wind Erosion and Air Quality Research in the Northwest US Columbia Plateau: Organization and Progress." In: The 10th International Soil Conservation Organization Meeting, 766–70.

Schillinger, William F., and Floyd E. Bolton. 1993. "Fallow Water Storage in Tilled vs. Untilled Soils in the Pacific Northwest." *Journal of Production Agriculture* 6(2): 267–69.

Schlatter, Daniel C., Kendall Kahl, Bryan Carlson, David R. Huggins, and Timothy Paulitz. 2020. "Soil Acidification Modifies Soil Depth-Microbiome Relationships in a No-Till Wheat Cropping System." *Soil Biology and Biochemistry* 149: 107939.

Schlatter, Daniel, Linda Kinkel, Linda Thomashow, David Weller, and Timothy Paulitz. 2017. "Disease Suppressive Soils: New Insights from the Soil Microbiome." *Phytopathology* 107(11): 1284–97.

Schlesinger, William H., and Jeffrey A. Andrews. 2000. "Soil Respiration and the Global Carbon Cycle." *Biogeochemistry* 48(1): 7–20.

Schmidt, Michael W. I., Margaret S. Torn, Samuel Abiven, Thorsten Dittmar, Georg Guggenberger, Ivan A. Janssens, Markus Kleber, et al. 2011. "Persistence of Soil Organic Matter as an Ecosystem Property." *Nature* 478(7367): 49.

Schuller, Paulina, Desmond E. Walling, Alejandra Sepúlveda, Alejandra Castillo, and Inés Pino. 2007. "Changes in Soil Erosion Associated with the Shift from Conventional Tillage to a No-Tillage System, Documented Using 137Cs Measurements." *Soil and Tillage Research* 94(1): 183–92.

Sharratt, B., and W. Schillinger. 2005. "The Columbia Plateau Wind Erosion/Air Quality Project: Predicting and Controlling Windblown Dust." In: Atmospheric Sciences and Air Quality Conference, American Meteorological Society. http://Ams. Confex. Com/Ams/Pdfpapers/89381. Pdf [Accessed 30 March 2009]. Citeseer.

Sheng, Yaqi, and Lizhong Zhu. 2018. "Biochar Alters Microbial Community and Carbon Sequestration Potential across Different Soil PH." *Science of the Total Environment* 622: 1391–99.

Shuster, W. De, S. Subler, and E. L. McCoy. 2001. "Deep-Burrowing Earthworm Additions Changed the Distribution of Soil Organic Carbon in a Chisel-Tilled Soil." *Soil Biology and Biochemistry* 33(7–8): 983–96.

Singh, Ajay. 2015. "Soil Salinization and Waterlogging: A Threat to Environment and Agricultural Sustainability." *Ecological Indicators* 57: 128–30.

Singh, Prabhakar, Joan Q. Wu, Donald K. McCool, Shuhui Dun, Chun-Hsu Lin, and John R. Morse. 2009. "Winter Hydrologic and Erosion Processes in the US Palouse Region: Field Experimentation and WEPP Simulation." *Vadose Zone Journal* 8(2): 426–36.

Sithole, Nkanyiso J., Lembe S. Magwaza, and Guy R. Thibaud. 2019. "Long-Term Impact of No-Till Conservation Agriculture and N-Fertilizer on Soil Aggregate Stability, Infiltration and Distribution of C in Different Size Fractions." *Soil and Tillage Research* 190: 147–56.

Slessarev, E. W., Y. Lin, N. L. Bingham, J. E. Johnson, Y. Dai, J. P. Schimel, and O. A. Chadwick. 2016. "Water Balance Creates a Threshold in Soil PH at the Global Scale." *Nature* 540(7634): 567–69.

Spiegal, Sheri, Brandon T. Bestelmeyer, David W. Archer, David J. Augustine, Elizabeth H. Boughton, Raoul Keith Boughton, Michel A. Cavigelli, et al. 2018. "Evaluating Strategies for Sustainable Intensification of US Agriculture through the Long-Term Agroecosystem Research Network." *Environmental Research Letters* 13(3), 034031.

Stafford, John V. 2000. "Implementing Precision Agriculture in the 21st Century." *Journal of Agricultural Engineering Research* 76(3): 267–75.

Stavi, I., R. Lal, and L. B. Owens. 2011. "On-Farm Effects of No-Till versus Occasional Tillage on Soil Quality and Crop Yields in Eastern Ohio." *Agronomy for Sustainable Development* 31(3): 475–82.

Stöckle, C., Stewart Higgins, Armen Kemanian, Roger Nelson, David Huggins, Jordi Marcos, and Harold Collins. 2012. "Carbon Storage and Nitrous Oxide Emissions of Cropping Systems in Eastern Washington: A Simulation Study." *Journal of Soil and Water Conservation* 67(5): 365–77.

Suleau, Marie, Christine Moureaux, Delphine Dufranne, Pauline Buysse, Bernard Bodson, Jean-Pierre Destain, Bernard Heinesch, Alain Debacq, and Marc Aubinet. 2011. "Respiration of Three Belgian Crops: Partitioning of Total Ecosystem Respiration in Its Heterotrophic, above-and below-Ground Autotrophic Components." *Agricultural and Forest Meteorology* 151(5): 633–43.

Sumner, Malcolm E., and Andrew D. Noble. 2003. "Soil Acidification: The World Story." In:*Handbook of Soil Acidity*, ed. Z. Rengel, 1–28. New York: Marcel Dekker.

Sun, Wenjuan, Josep G. Canadell, Yu Lijun, Yu Lingfei, Zhang Wen, Pete Smith, Tony Fischer, and Huang Yao. 2020. "Climate Drives Global Soil Carbon Sequestration and Crop Yield Changes under Conservation Agriculture." *Global Change Biology* 26(6): 3325–35.

Syvitski, James P. M., Charles J. Vörösmarty, Albert J. Kettner, and Pamela Green. 2005. "Impact of Humans on the Flux of Terrestrial Sediment to the Global Coastal Ocean." *Science* 308(5720): 376–80.

Tao, Haiying, William L. Pan, Paul Carter, and Ku Wang. 2019. "Addition of Lignin to Lime Materials for Expedited PH Increase and Improved Vertical Mobility of Lime in No-Till Soils." *Soil Use and Management* 35(2): 314–22.

TerAvest, Dan, Lynne Carpenter-Boggs, Christian Thierfelder, and John P. Reganold. 2015. "Crop Production and Soil Water Management in Conservation Agriculture, No-Till, and Conventional Tillage Systems in Malawi." *Agriculture, Ecosystems and Environment* 212: 285–96.

Terrell, Bonnie L., Phillip N. Johnson, and Eduardo Segarra. 2002. "Ogallala Aquifer Depletion: Economic Impact on the Texas High Plains." *Water Policy* 4(1): 33–46.

Teutscherova, Nikola, Eduardo Vazquez, Alberto Masaguer, Mariela Navas, Kate M. Scow, Radomir Schmidt, and Marta Benito. 2017. "Comparison of Lime-And Biochar-Mediated PH Changes in Nitrification and Ammonia Oxidizers in Degraded Acid Soil." *Biology and Fertility of Soils* 53(7): 811–21.

Thenkabail, Prasad S., Jerry W. Knox, Mutlu Ozdogan, Murali Krishna Gumma, Russell G. Congalton, Z. T. Wu, Cristina Milesi, et al. 2012. “Assessing Future Risks to Agricultural Productivity, Water Resources and Food Security: How Can Remote Sensing Help?” *PE&RS, Photogrammetric Engineering & Remote Sensing* 78(8): 773–82.

Tian, Dashuan, and Shuli Niu. 2015. “A Global Analysis of Soil Acidification Caused by Nitrogen Addition.” *Environmental Research Letters* 10(2), 024019.

Tomer, Mark D., Sarah A. Porter, David E. James, Kathleen M. B. Boomer, Jill A. Kostel, and Eileen McLellan. 2013. “Combining Precision Conservation Technologies into a Flexible Framework to Facilitate Agricultural Watershed Planning.” *Journal of Soil and Water Conservation* 68(5): 113A–20A.

Toosi, Ehsan R., Peter W. Clinton, Michael H. Beare, and David A. Norton. 2012. “Biodegradation of Soluble Organic Matter as Affected by Land-Use and Soil Depth.” *Soil Science Society of America Journal* 76(5): 1667–77.

USDA. 1978. *Palouse Cooperative River Basin Study.* Available at pnwsteep.wsu.edu/resourcelinks/Palouse_Basin_Study.pdf (verified 20 Feb. 2009). Washington State Univ., Pullman.

Van Muysen, W., and Gerard Govers. 2002. “Soil Displacement and Tillage Erosion during Secondary Tillage Operations: The Case of Rotary Harrow and Seeding Equipment.” *Soil and Tillage Research* 65(2): 185–91.

Van Muysen, W., Kristof Van Oost, and Gerard Govers. 2006. “Soil Translocation Resulting from Multiple Passes of Tillage under Normal Field Operating Conditions.” *Soil and Tillage Research* 87(2): 218–30.

Van Oost, Kristof, Gerard Govers, Timothy A. Quine, Goswin Heckrath, Jorgen E. Olesen, Steven De Gryze, and Roel Merckx. 2005. “Landscape-scale Modeling of Carbon Cycling under the Impact of Soil Redistribution: The Role of Tillage Erosion.” *Global Biogeochemical Cycles*, 19: GB4014. doi:10.1029/2005GB002471.

Van Oost, Kristof, T. A. Quine, Gerard Govers, Steven De Gryze, J. Six, J. W. Harden, J. C. Ritchie, et al. 2007. “The Impact of Agricultural Soil Erosion on the Global Carbon Cycle.” *Science* 318(5850): 626–29.

VandenBygaart, A. J., and B. D. Kay. 2004. “Persistence of Soil Organic Carbon after Plowing a Long-Term No-Till Field in Southern Ontario, Canada.” *Soil Science Society of America Journal* 68(4): 1394–402.

Wachter, Jonathan M., Kathleen M. Painter, Lynne A. Carpenter-Boggs, David R. Huggins, and John P. Reganold. 2019. “Productivity, Economic Performance, and Soil Quality of Conventional, Mixed, and Organic Dryland Farming Systems in Eastern Washington State.” *Agriculture, Ecosystems and Environment* 286, 106665.

Waldo, Sarah, Jinshu Chi, Shelley N. Pressley, Patrick O’Keeffe, William L. Pan, Erin S. Brooks, David R. Huggins, Claudio O. Stöckle, and Brian K. Lamb. 2016. “Assessing Carbon Dynamics at High and Low Rainfall Agricultural Sites in the Inland Pacific Northwest US Using the Eddy Covariance Method.” *Agricultural and Forest Meteorology* 218: 25–36.

Waldo, Sarah, Eric S. Russell, Kirill Kostyanovsky, Shelley N. Pressley, Patrick T. O’Keeffe, David R. Huggins, Claudio O. Stöckle, William L. Pan, and Brian K. Lamb. 2019. “N_2O Emissions from Two Agroecosystems: High Spatial Variability and Long Pulses Observed Using Static Chambers and the Flux-Gradient Technique.” *Journal of Geophysical Research: Biogeosciences* 124(7): 1887–904.

Wang, Jinyang, and Jianwen Zou. 2020. “No-Till Increases Soil Denitrification via Its Positive Effects on the Activity and Abundance of the Denitrifying Community.” *Soil Biology and Biochemistry*, 142: 107706.

Wang, Zhi-Ping, Xing-Guo Han, and Ling-Hao Li. 2008. “Effects of Grassland Conversion to Croplands on Soil Organic Carbon in the Temperate Inner Mongolia.” *Journal of Environmental Management* 86(3): 529–34.

Wang, Yu Jia Xiao-Ling, Xiao-Gang Li, Rui-Jun Long, Qifu Ma, Feng-Min Li, and Ya-Jie Song. 2009. “Effects of Land Use on Soil Total and Light Fraction Organic, and Microbial Biomass C and N in a Semi-Arid Ecosystem of Northwest China.” *Geoderma* 153(1–2): 285–90.

Wang, Zhi-Ping, Xing-Guo Han, and Ling-Hao Li. 2008. “Effects of Grassland Conversion to Croplands on Soil Organic Carbon in the Temperate Inner Mongolia.” *Journal of Environmental Management* 86(3): 529–34.

Wendt, J. W., and S. Hauser. 2013. ”An Equivalent Soil Mass Procedure for Monitoring Soil Organic Carbon in Multiple Soil Layers.” *European Journal of Soil Science* 64(1): 58–65.

West, Tristram O., and Wilfred M. Post. 2002. “Soil Organic Carbon Sequestration Rates by Tillage and Crop Rotation.” *Soil Science Society of America Journal* 66(6): 1930–46.

Western Regional Climate Center. 2020. *Western U.S. Climate Historical Summaries.* Available at https://wrcc.dri.edu/Climsum.html [Accessed 28 August 2020].

Wiesmeier, Martin, Peter Schad, Margit von Lützow, Christopher Poeplau, Peter Spörlein, Uwe Geuß, Edzard Hangen, Arthur Reischl, Bernd Schilling, and Ingrid Kögel-Knabner. 2014. "Quantification of Functional Soil Organic Carbon Pools for Major Soil Units and Land Uses in Southeast Germany (Bavaria)." *Agriculture, Ecosystems and Environment* 185: 208–20.

Wilcke, Wolfgang, and Juliane Lilienfein. 2005. "Nutrient Leaching in Oxisols under Native and Managed Vegetation in Brazil." *Soil Science Society of America Journal* 69(4): 1152–61.

Williams, John D., Hero T. Gollany, Mark C. Siemens, Stewart B. Wuest, and Daniel S. Long. 2009. "Comparison of Runoff, Soil Erosion, and Winter Wheat Yields from No-Till and Inversion Tillage Production Systems in Northeastern Oregon." *Journal of Soil and Water Conservation* 64(1): 43–52.

Willoughby, G. L., and E. J. Kladivko. 2002. "Water Infiltration Rates Following Reintroduction of Lumbricus terrestris into No-Till Fields." *Journal of Soil and Water Conservation* 57(2): 82–8.

Willoughby, Gregory L., Eileen J. Kladivko, and M. Reza Savabi. 1997. "Seasonal Variations in Infiltration Rate under No-Till and Conventional (Disk) Tillage Systems as Affected by Lumbricus terrestris Activity." *Soil Biology and Biochemistry* 29(3–4): 481–84.

Wu, Di, Mehmet Senbayram, Huadong Zang, Ferhat Ugurlar, Salih Aydemir, Nicolas Brüggemann, Yakov Kuzyakov, Roland Bol, and Evgenia Blagodatskaya. 2018. "Effect of Biochar Origin and Soil PH on Greenhouse Gas Emissions from Sandy and Clay Soils." *Applied Soil Ecology* 129: 121–27.

Wu, Haibin, Zhengtang Guo, Qiong Gao, and Changhui Peng. 2009. "Distribution of Soil Inorganic Carbon Storage and Its Changes Due to Agricultural Land Use Activity in China." *Agriculture, Ecosystems and Environment* 129(4): 413–21.

Yan, J. I. A. O., X. U. Zhu, Zhao JiaoHong, and Yang WenZhu. 2012. "Changes in Soil Carbon Stocks and Related Soil Properties along a 50-Year Grassland-to-Cropland Conversion Chronosequence in an Agro-Pastoral Ecotone of Inner Mongolia, China." *Journal of Arid Land* 4(4): 420–30.

Yeboah, Stephen, Renzhi Zhang, Liqun Cai, Min Song, LingLing Li, Junhong Xie, Zhuzhu Luo, Jun Wu, and Jun Zhang. 2016. "Greenhouse Gas Emissions in a Spring Wheat–Field Pea Sequence under Different Tillage Practices in Semi-Arid Northwest China." *Nutrient Cycling in Agroecosystems* 106(1): 77–91.

Yost, M. A., N. R. Kitchen, K. A. Sudduth, R. E. Massey, E. J. Sadler, S. T. Drummond, and M. R. Volkmann. 2019. "A Long-Term Precision Agriculture System Sustains Grain Profitability." *Precision Agriculture* 20(6): 1177–98.

Yost, M. A., N. R. Kitchen, K. A. Sudduth, E. J. Sadler, S. T. Drummond, and M. R. Volkmann. 2017. "Long-Term Impact of a Precision Agriculture System on Grain Crop Production." *Precision Agriculture* 18(5): 823–42.

Yu, Kewei, April Hiscox, and Ronald D. DeLaune. 2013. "Greenhouse Gas Emission by Static Chamber and Eddy Flux Methods." *Methods in Biogeochemistry of Wetlands* 10: 427–37.

Zehe, Erwin, and Hannes Flühler. 2001. "Preferential Transport of Isoproturon at a Plot Scale and a Field Scale Tile-Drained Site." *Journal of Hydrology* 247(1–2): 100–15.

Zhang, J. H., Z. A. Su, and X. J. Nie. 2009. "An Investigation of Soil Translocation and Erosion by Conservation Hoeing Tillage on Steep Lands Using a Magnetic Tracer." *Soil and Tillage Research* 105(2): 177–83.

Zhang, Naiqian, Maohua Wang, and Ning Wang. 2002. "Precision Agriculture—A Worldwide Overview." *Computers and Electronics in Agriculture* 36(2–3): 113–32.

Zhang, T. Q., C. S. Tan, Y. T. Wang, B. L. Ma, and T. Welacky. 2017. "Soil Phosphorus Loss in Tile Drainage Water from Long-Term Conventional-And Non-Tillage Soils of Ontario with and without Compost Addition." *Science of the Total Environment* 580: 9–16.

Index

D

O

P

Q

R

T

U